Proceedings of the 1997 Noise and Vibration Conference
Volume 2

P-309

All SAE papers, standards, and selected books are abstracted and indexed in the Global Mobility Database.

Published by:
Society of Automotive Engineers, Inc.
400 Commonwealth Drive
Warrendale, PA 15096-0001
USA
Phone: (412) 776-4841
Fax: (412) 776-5760
May 1997

Permission to photocopy for internal or personal use, or the internal or personal use of specific clients, is granted by SAE for libraries and other users registered with the Copyright Clearance Center (CCC), provided that the base fee of $7.00 per article is paid directly to CCC, 222 Rosewood Drive, Danvers, MA 01923. Special requests should be addressed to the SAE Publications Group. 0-7680-0012-2/97$7.00.

Any part of this publication authored solely by one or more U.S. Government employees in the course of their employment is considered to be in the public domain, and is not subject to this copyright.

No part of this publication may be reproduced in any form, in an electronic retrieval system or otherwise, without the prior written permission of the publisher.

ISBN 0-7680-0012-2
SAE/P-97/309
Library of Congress Catalog Card Number: 97-66360
Copyright 1997 Society of Automotive Engineers, Inc.

Positions and opinions advanced in this paper are those of the author(s) and not necessarily those of SAE. The author is solely responsible for the content of the paper. A process is available by which discussions will be printed with the paper if it is published in SAE Transactions. For permission to publish this paper in full or in part, contact the SAE Publications Group.

Persons wishing to submit papers to be considered for presentation or publication through SAE should send the manuscript or a 300 word abstract of a proposed manuscript to: Secretary, Engineering Meetings Board, SAE.

Printed in USA

GENERAL COMMITTEE
1997 NOISE & VIBRATION CONFERENCE

Bernard J. Challen, General Chairman
Ralph K. Hillquist, Chairman Emeritus
Waldemar Semrau, Vice Chairman
Kalyan S. Bagga
Robert J. Bernhard
Francis H. K. Chen
Michael L. Dinsmore
John Feng
Robert F. Hand
Jen Y. Her
Timothy Hirabayashi
Deane B. Jaeger
Sunil K. Jha
Raymond A. Kach
Yury Kalish
Jay H. Kim
John Mathey
Thomas L. Mitchell
Mark J. Moeller
Ivan E. Morse, Jr.
James M. Nieters
Robert E. Powell
Pranab Saha
Ernst Schonthal
John F. Schultze
Richard F. Schumacher
James R. Schwaegler
Ahmet Selamet
William J. Sobkow
Steven R. Sorenson
James K. Thompson
Terry S. Trasatti
Paul E. Waters
Silvia I. Zorea

TABLE OF CONTENTS

Volume 1

971870   **Reducing Compression Brake Noise** 1
Thomas E. Reinhart and Thomas J. Wahl
Cummins Engine Co.

971871   **Multivariate Analysis of Engine Noise and Exhaust Emissions from an Ethanol Fuelled Diesel Engine** 9
Grover Zurita Villarroel, Esbjörn Pettersson, and Anders Ågren
Luleå University of Technology

971872   **An Analytical/Empirical Approach to Sound Quality Evaluation for Exhaust Systems** 19
Patricia Hetherington and William Hill
Walker Manufacturing Division of Tenneco Automotive

971873   **Wave Propagation in Catalytic Converters: A Preliminary Investigation** 27
A. Selamet and V. Easwaran
Ohio State Univ.
J. M. Novak
Ford Motor Co.

971874   **Combustion Induced Powertrain NVH - A Time-Frequency Analysis** 33
T. N. Patro
Ford Motor Co.

971875   **A General Procedure for the Analysis of Gas Pulsations in Thin Shell Type Gas Cavities with Special Attention to Compressor Manifolds** 45
Peter C.-C. Lai
Copeland Corp.
Werner Soedel
Purdue Univ.

971876   **Air Cleaner Shell Noise Reduction with Finite Element Shape Optimization** 59
John A. White Jr. and Jack C. Webb
Delphi Automotive Systems

971877   **Acoustic Performance Analysis of an Air Handling System** 65
Morris Y. Hsi
Ford Motor Co.
Fred Périé
Mecalog

971878   **Elastic Porous Materials for Sound Absorption and Transmission Control** 77
J. Stuart Bolton
Purdue Univ.
Yeon June Kang
Seoul National Univ.

971879	**Quarter Lambda Absorbers Rejuvenated as Multi-Frequency In-Plane Arrays** 93

Roberto Frosio and Robert H. van Ligten
Rieter Automotive Management AG

971881	**Effects of Sound Absorption on Speech Intelligibility in an Automotive Environment** 101

Anthony P. Visintainer
Chrysler Corp.
Jeff A. VanBuskirk
Rieter Automotive North America Inc.

971882	**New Low-Odor Phenolic-Resin Molded Insulators: Product and Process Technology** 109

A. Aggarwal, H. Khan, S. Tessendorf, and T. Lamb
Rieter Automotive North America

971883	**Alternate Binders for Molded Acoustical Insulators** 115

William VanBemmel and Hameed Khan
Rieter Automotive North America, Inc.

971884	**Reclaimed Fiber Acoustical Composites - Addressing Today's Recycling Challenges** 119

Tony M. Lamb, Steve Tessendorf, and Hameed Khan
Rieter Automotive North America

970148	**Importance of Durability on Acoustical Performance for Carpet Underlay** 123

Kevin A. Buck and Randy A. Wellman
ICI Polyurethanes

971885	**Maximize Sound Transmission Loss Through Double Wall System with Optimized Coupling** 127

Jerry Tillery, Marek L. Szary, and Maciej Noras
Southern Illinois Univ.

971886	**Active Vibro-Acoustic Control in Automotive Vehicles** 131

Rahmat A. Shoureshi
Colorado School of Mines
James L. Vance and Lanre Ogundipe
Cooper Tire and Rubber Co.
Klaus Schaaf
Volkswagen AG
Günter Eberhard and Hans-Jürgen Karkosch
ContiTech Formteile GmbH

971887	**Active Boom Noise Control** 137

Reza Kashani
The University of Dayton
David Naastad
Independent Consultant

971888	**Piezoceramic Based Actuator/Sensor Arrays for Active Noise Control of Machinery** 147

Robert N. Jacques
Active Control eXperts (ACX), Inc.

971890   **Influence of Confined Vibrations on Sensor and Actuator Optimization** 155
Yin-Tsan Shih, David J. Tarnowski, and Daryoush Allaei
QRDC, Inc.

971891   **Active Control of Wind Noise Using Robust Feedback Control** 165
Craig M. Heatwole, Matthew A. Franchek, and Robert J. Bernhard
Purdue Univ.

971892   **New Vibration System Using Magneto-Spring** 175
Etsunori Fujita, Yumi Ogura, and Yutaka Sakamoto
Delta Tooling Co., Ltd.
Shigeo Honda
Hiroshima Univ.

971893   **Chassis Dynamometer Simulation of Road Noise** 187
Sung-Ping Cheng, David Griffiths, and P. Perry Gu
Ford Motor Co.
Cuiping Li
Troy Design, Inc.

971894   **Case Studies Involving the Identification of Problematic Impulsive Effects on Vibration Signals** 193
D. Storer and G. Guerci
Centro Ricerche FIAT
C. Morari and V. Tanfoglio
TRW Automotive
X. Lauwerys and K. Reybrouck
MONROE Europe

971895   **Verification of a Miniature Reverberation Room for Sound Absorption Measurements Using Corner Microphone Technique** 205
Richard A. Kolano and Jeff A. Kleckner
Kolana and Saha Engineers, Inc.

971896   **Development of a New Sound Transmission Test for Automotive Sealant Materials** 209
J. Stuart Bolton and Richard J. Yun
Purdue Univ.
Joseph Pope
Pope Engineering Co.
David Apfel
L and L Products, Inc.

971897   **An Overview of the European Research Project DIANA** 217
W. Hendricx and D. De Vis
LMS International N.V.
H. Ghesquière and J. Meillier
Renault Research Centre
G. Toniato and D. Storer
Fiat Research Center
M. Niedbal and M. Hermanski
Technical University Bielefeld
B. Jones and S. Roberts
Motor Industry Association

971898   **Collins & Aikman Technical Centre** **227**
     A. E. D. Hughes
       Collins & Aikman
     Bill Gastmeier
       Howe Gastmeier Chapnik Ltd.

971899   **NVH Research Facilities at The Ohio State University: Existing Facilities and Envisioned Enhancements** **233**
     Donald R. Houser and Rajendra Singh
       Ohio State Univ.

971900   **Design, Construction and Application of a World Class Vehicle Acoustic Test Facility** **241**
     F. K. Brandl
       AVL LIST GmbH

971901   **Modelling the Dissipative Effect of Seal Air Hole Spacing and Size on Door Closing Effort** **249**
     Yuksel Gur and Kenneth N. Morman
       Ford Motor Co.

971902   **Analysis of Door and Glass Run Seal Systems for Aspiration** **255**
     Yuksel Gur and Kenneth N. Morman
       Ford Motor Co.
     Niranjan Singh
       Automated Analysis Corp.

971903   **Sound Transmission Through Primary Bulb Rubber Sealing Systems** **271**
     L. Mongeau and R. Danforth*
       Purdue Univ.

971904   **Investigation of Acoustic Leakage of Vehicle Dash Pass-Through Components** **279**
     Kent K.H. Fung and Jason J. Zhu
       Collins & Aikman

971905   **Acoustic Absorption in Vehicles and the Measurement of Short Reverberation Times** **287**
     Gordon L. Ebbitt and Najam A. Rauf
       Lear Corp.

971906   **A Predictive Model for the Interior Pressure Oscillations from Flow Over Vehicle Openings** **295**
     L. Mongeau, D. Brown,* and H. Kook
       Purdue Univ.
     S. Zorea
       Ford Motor Co.

971907   **Determination of Vehicle Interior Sound Power Contribution Using Sound Intensity Measurement** **305**
     Shaobo Yang, Jian Pan, Prakash Sabnis, and Ashok Khubchandani
       Ford Motor Co.

971908   **Acoustical Drain Plugs in Body Cavity Sealer "Baffles"** **311**
     Pranab Saha
       Kolano and Saha Engineers, Inc.
     Dave Jones
       Sika Corp.

971909   **What Really Affect Customer Perception? - A Window Regulator Sound Quality Example** 317

Lijian Zhang and Alicia Vértiz
Delphi Interior and Lighting Systems

971910   **Measures to Quantify the Sharpness of Vehicle Closure Sounds** 321

Anthony J. Champagne and Scott A. Amman
Ford Motor Co.

971911   **Automated Production Noise Testing of Power Steering Pumps** 325

Daniel Bleitz and Christian Fernholz
Ford Motor Co.
Stephen Ivkovich
IMPACT Engineering, Inc.

971912   **Statistical Analysis of Vehicle High Mileage NVH Performance** 335

E. Y. Kuo and L. Liaw
Ford Motor Co.

971913   **Experimental Body Panel Contribution Analysis for Road Induced Interior Noise of a Passenger Car** 351

Wim Hendricx
LMS Engineering Services
Y. B. Choi, S. W. Ha, and H. K. Lee
Hyundai Motor Co.

971914   **Computer Simulation of In-Vehicle Boom Noise** 357

Wayne Stokes, John Bretl, and Alun Crewe
Structural Dynamics Research Corp.
Woo Sun Park, Jae Young Lee, and Myung Sik Lee
Daewoo Motor Co.

971915   **Vehicle Interior Noise and Vibration Reduction Using Experimental Structural Dynamics Modification** 365

Yong-Hwa Park and Youn-sik Park
KAIST

971916   **The Booming Noise Reduction of a Vehicle Using Effectiveness Analysis** 373

Jae-Eung Oh, Jun H. Cho, and Hae S. Lee
Hanyang Univ.
Tae U. Kim
DAEWOO Motor Co.

971917   **Analysis of Flap Side-Edge Flowfield for Identification and Modeling of Possible Noise Sources** 379

Mehdi R. Khorrami and Bart A. Singer
High Technology Corp.
M. A. Takallu
Lockheed/Martin Engineering and Sciences Co.

971918   **The French High Speed Train (TGV) Pantograph's Aeroacoustics** 387

A. Fages and C. Bertrand
GEC ALSTHOM Transport

971919 Noise Radiation from Axial Flow Fans 399
Shigong Su and Sean F. Wu
Wayne State Univ.
Hemant Shah
Ford Motor Co.

971920 Experimental Investigaton of Radio Antenna Wind Noise 409
Farokh Kavarana and Susan Stroope*
Defiance Testing & Engineering Services
Park-hoon Young
Hyundai America Technical Center Inc.

971921 The Wind Noise Modeller 417
Gary S. Strumolo
Ford Motor Co.

971922 Experimental Assessment of Wind Noise Contributors to
Interior Noise 427
Jen Y. Her, Mingder Lian, and James J. Lee
Ford Motor Co.
Jim Moore
Noise & Vibration Consultant

971923 Pressure Fluctuations in a Flow-Excited Door Gap Cavity
Model 439
L. Mongeau, J. Bezemek,* and R. Danforth#
Purdue Univ.

971924 An Alternative Approach to Robust Design: A Vehicle Door
Sealing System Example 447
R. S. Thomas and G. J. Ehlert
Ford Motor Co.

971925 Current Technologies in the Acoustical Rear Parcel Shelf 455
A. Aggarwal and H. Khan
Rieter Automotive North America, Inc.

971926 Assessing Headliner and Roof Assembly Acoustics 463
Richard E. Wentzel
Ford Motor Co.
Edward R. Green
Roush Anatrol

971927 Correlation of Various Test Methodologies with Vehicle Seat
Acoustical Performance 475
John C. Gagliardi
Lear Corp.

971928 Experimental Development of a Unique Door Cavity Sound
Absorber 485
Steve Balinski
Saturn Corp.
Del Thompson
3M Corp.

971930 Damping Efficiency of Ribbed Panels with Different Damping
Materials 495
Yang Qian, Anuj Aggarwal, and Hameed Khan
Rieter Automotive North America, Inc.

971931   **Acoustical Study of Cavity Fillers for Vehicle Applications** 503
Anthony P. Visintainer
  Chrysler Corp.
Pranab Saha
  Kolano and Saha Engineers, Inc.

971932   **In-Line Application of Solid NVH Material** 511
Maurice Lande
  L and L Products Inc.

Volume 2

971933   **Robust Design of Elastic Mounting Systems** 517
Thomas Vietor, Rolf Deges, Norbert Hampl, and Karl-Heinz Bürger
  Ford Werke AG

971934   **Improvements in the Optimization of Dynamic Vibration Absorbers** 525
James L. Swayze
  Ford Motor Co.
Eugene R. Rivin
  Wayne State Univ.

971936   **Study of Nonlinear Hydraulic Engine Mounts Focusing on Decoupler Modeling and Design** 535
Thomas J. Royston
  University of Illinois at Chicago
Rajendra Singh
  Ohio State Univ.

971937   **Optimization of Vibration Isolators for Marine Diesel Engine Applications** 547
Stanley R. Walker
  Barry Controls, A Unit of Applied Power Inc.
John G. Foscolos, Jr.
  Cummins Marine, A Division of Cummins Engine Corp.

971938   **The Evolution of the 1997 Chevrolet Corvette Powertrain Mounting System** 553
Brad Saxman, Brian Deutschel, Keith Weishuhn, and Kenneth Brown
  General Motors Corp.

971940   **Dynamic Simulation of Engine-Mount Systems** 561
Chung-Ha Suh and Clifford G. Smith
  University of Colorado-Boulder

971942   **Automotive Powerplant Isolation Strategies** 573
R. Matthew Brach
  Ford Motor Co.

971943   **Vibration Analysis of Metal/Polymer/Metal Laminates - Approximate Versus Viscoelastic Methods** 579
Lezza A. Mignery and Edward J. Vydra
  Material Sciences Corp.

971944   **Hybrid Substructuring for Vibro-Acoustical Optimisation: Application to Suspension - Car Body Interaction** 591
         K. Wyckaert and M. Brughmans
            LMS International NV
         C. Zhang and R. Dupont
            Renault S.A.

971945   **Acoustic Analysis of Vehicle Ribbed Floor** 599
         Kevin Zhang, Jihe Yang, Lung Wu, and Walt Mazur
            Ford Motor Co.
         Xiandi Zeng and Xiaoye Gu
            Automated Analysis Corp.

971946   **A Comparison Between Passive Vibration Control by Confinement and Current Passive Techniques** 607
         Daryoush Allaei, Yin-Tsan Shih, and David J. Tarnowski
            QRDC, Inc.

971947   **An Indirect Boundary Element Technique for Exterior Periodic Acoustic Analysis** 615
         S. T. Raveendra, B. K. Gardner, and R. Stark
            Automated Analysis Corp.

971948   **Modal Content of Heavy-Duty Diesel Engine Block Vibration** 621
         Deanna M. Winton and David R. Dowling
            University of Michigan

971949   **H∞ Control Design of Experimental State-Space Modeling for Vehicle Vibration Suppression** 631
         Zhongyang Guo, Itsuro Kajiwara, and Akio Nagamatsu
            Tokyo Institute of Technology
         Tsutomu Sonehara
            Isuzu Advanced Engineering Center, Ltd.

971950   **Acoustic Modeling and Optimization of Seat for Boom Noise** 639
         Yang Qian and Jeff VanBuskirk
            Rieter Automotive North America, Inc.

971951   **Experimental Verification of Design Charts for Acoustic Absorbers** 649
         Mardi C. Hastings and Richard D. Godfrey
            Ohio State Univ.

971953   **Panel Contribution Study: Results, Correlation and Optimal Bead Pattern for Powertrain Noise Reduction** 653
         Anbarasu Nachimuthu and Karen M. Carnago
            Chrysler Corp.
         Farshid Haste
            Automated Analysis Corp.

971954   **Prediction of Radiated Noise from Engine Components Using the BEM and the Rayleigh Integral** 659
         A. F. Seybert
            University of Kentucky
         D. A. Hamilton and P. A. Hayes
            Cummins Engine Co.

971955 **Interior Noise Prediction Process for Heavy Equipment Cabs** 665
Andrew F. Seybert, Tiemin Hu, David W. Herrin, and Robert S. Ballinger
University of Kentucky

971956 **Noise Source Identification in a Highly Reverberant Enclosure by Inverse Frequency Response Function Method: Numerical Feasibility Study** 671
Drew A. Crafton, Jingdong Ding, and Jay H. Kim
University of Cincinnati

971957 **Estimation of a Structure's Inertia Properties Using a Six-Axis Load Cell** 677
Mark Stebbins, Jason Blough, Stuart Shelley, and David Brown
University of Cincinnati

971958 **Practical Aspects of Perturbed Boundry Condition (PBC) Finite Element Model Updating Techniques** 685
Michael Yang, Randall Allemang, and David Brown
University of Cincinnati

971959 **Prediction of Powerplant Vibration Using FRF Data of FE Model** 695
Yukitaka Takahashi, Toshibumi Suzuki, and Masayoshi Tsukahara
Honda Research and Development Co., Ltd.

971960 **Heavy Duty Diesel Engine Noise Reduction Using Torsional Dampers on Fuel Pump Shafts** 703
Neil Hutton
Holset Engineering Co., Ltd.

971961 **Attenuation of Engine Torsional Vibrations Using Tuned Pendulum Absorbers** 713
Steven W. Shaw, Vishal Garg, and Chang-Po Chao
Michigan State Univ.

971962 **Influence of Tensioner Friction on Accessory Drive Dynamics** 723
M. J. Leamy, N. C. Perkins, and J. R. Barber
University of Michigan
R. J. Meckstroth
Ford Motor Co.

971963 **Development of an Isolated Timing Chain Guide System Utilizing Indirect Force Measurement Techniques** 731
Wayne Nowicki
Roush Anatrol
Eric Sheffer
Ford Motor Co.

971964 **Dynamic Analysis of Layshaft Gears in Automotive Transmission** 739
Teik C. Lim and Donald R. Houser
Ohio State Univ.

971965 **Numerical Methods to Calculate Gear Transmission Noise** 751
W. Hellinger, H. Ch. Raffel, and G. Ph. Rainer
AVL List GmbH

971966   **Gear Noise Reduction of an Automatic Transmission Through Finite Element Dynamic Simulation** 761
     Brian Campbell, Wayne Stokes, and Glen Steyer
       Structural Dynamics Research Corp.
     Mark Clapper, R. Krishnaswami, and Nancy Gagnon
       Ford Motor Co.

971967   **Influence of the Valve and Accessory Gear Train on the Crankshaft Three-Dimensional Vibrations in High Speed Engines** 773
     Hideo Okamura
       Sophia Univ.
     Kenichi Yamashita
       Mitsubishi Motors Ltd.

971968   **Overview of the Experimental Approach to Statistical Energy Analysis** 783
     Benjamin Cimerman
       Vibro-Acoustic Sciences Inc.
     Tej Bharj
       Ford Motor Co.
     Gerard Borello
       InterAC

971969   **Methods to Estimate the Confidence Level of the Experimentally Derived Statistical Energy Analysis Model: Application to Vehicles** 789
     L. Hermans, K. De Langhe, and L. Demeestere
       LMS International NV

971970   **Statistical Energy Analysis of Airborne and Structure-Borne Automobile Interior Noise** 801
     Alan V. Parrett and John K. Hicks
       General Motors Corp.
     Thomas E. Burton
       Vibro-Acoustic Sciences, Inc.
     Luc Hermans
       LMS International

971972   **Statistical Energy Analysis for Road Noise Simulation** 811
     Mark J. Moeller and Jian Pan
       Ford Motor Co.

971973   **SEA Modeling and Testing for Airborne Transmission Through Vehicle Sound Package** 819
     Robert E. Powell
       Ford Motor Co.
     Jason Zhu
       Collins & Aikman Corp.
     Jerome E. Manning
       Cambridge Collaborative, Inc.

971974   **High Frequency NVH Analysis of Full Size Pickups Using "SEAM"** 831
     Martin Botz and Bijan Khatib-Shahidi
       Ford Motor Co.

971975 **Statistical Energy Analysis of Noise and Vibration from an Automotive Engine** 835
Cliff Kaminsky
Vibro-Acoustic Sciences, Inc.
Robert Unglenieks
Roush-Anatrol, Inc.

971976 **A Standardized Scale for the Assessment of Car Interior Sound Quality** 843
Rudolf Bisping, Sönke Giehl, and Martin Vogt
S.A.S. Systems

971977 **Engine Sound Quality in Sub-Compact Economy Vehicles: A Comparative Case Study** 849
Brian Chapnik and Brian Howe
HGC Engineering Ltd.

971978 **Binaural "Hybrid" Model for Simulation of Engine and Wind Noise in the Interior of Vehicles** 861
K. Genuit and N. Xiang
HEAD acoustics GmbH

971979 **A New Tool for the Vibration Engineer** 867
R. C. Meier Jr., N. C. Otto, W. J. Pielemeier, and V. Jeyabalan
Ford Motor Co.

971980 **Commercial Van Diesel Idle Sound Quality** 873
Anthony J. Champagne and Nae-Ming Shiau
Ford Motor Co.

971981 **Pitch Matching for Impulsive Sounds** 879
Richard J. Fridrich
General Motors Corp.

971982 **Linearity of Powertrain Acceleration Sound** 887
Norman Otto and John Feng
Ford Motor Co.
Robert Cheng
University of Michigan
Eric Wisniewski
Edsel Ford High School

971983 **The Effect of Powertrain Sound on Perceived Vehicle Performance** 891
Michael A. Blommer, Scott A. Amman, and Norman C. Otto
Ford Motor Co.

971984 **The Investigation of a Towed Trailer Test for Passenger Tire Coast-By Noise Measurement** 897
James K. Thompson
Automated Analysis Corp.
Thomas A. Williams
Hankook Tire

971985 **Prediction of Vehicle Radiated Noise** 905
Luigi Pilo
FIAT Auto
Francesco Gamba and Bernard J. Challen
Cornaglia Research Center

971986   **Time Dependent Correlation Analysis of Truck Pass-by-Noise Signals** **915**
Herman Van der Auweraer, Luc Hermans, and Dirk Otte
LMS International NV
Manfred Klopotek
Scania Trucks

971987   **A Doppler Correction Procedure for Exterior Pass-By Noise Testing** **923**
Renaat Vancauter
LMS International NV

971988   **Two-Microphone Measurements of the Acoustical Properties of SAE and ISO Passby Surfaces in the Presence of Wind and Temperature Gradients** **931**
Troy J. Hartwig and J. Stuart Bolton
Purdue Univ.

971989   **Methods of Passby Noise Predictionin a Semi-Anechoic Chamber** **947**
S. -H. Park and Y. -H. Kim
KAIST
B. -S. Ko
DaeWoo Motors Co. Ltd.

971990   **An Assessment of the Tire Noise Generation and Sound Propagation Characteristics of an ISO 10844 Road Surface** **955**
Paul R. Donavan
General Motors

971991   **Temperature Dependency of Pass-By Tire Road Noise** **965**
Satoshi Konishi, Toshiaki Fujino, Naotaka Tomita, and Toshio Ozaki
Bridgestone Corp.

971992   **An Integrated Numerical Tool for Engine Noise and Vibration Simulation** **971**
B. Loibnegger and G. Ph. Rainer
AVL List GmbH
L. Bernard, D. Micelli, and G. Turino
Fiat Research Centre

971993   **Valvetrain Unbalance and Its Effects on Powertrain NVH** **981**
Joseph L. Stout
Ford Motor Co.

971994   **The Effect of Cranktrain Design on Powertrain NVH** **989**
J. Querengaesser and J. Meyer
Ford-Werke AG
E. Schaefer and J. Wolschendorf
FEV Motorentechnik

971995   **Experiments and Analysis of Crankshaft Three-Dimensional Vibrations and Bending Stresses in a V-Type Ten-Cylinder Engine: Influence of Crankshaft Gyroscopic Motions** **999**
Jouji Kimura and Kazuhiro Shiono
Isuzu Motors Ltd.
Hideo Okamura and Kiyoshi Sogabe
Sophia Univ.

971996 **Experiments and Analyses of the Three-Dimensional Vibrations of the Crankshaft and Torsional Damper in a Four-Cylinder In-Line High Speed Engine** 1009
　　Takeo Naganuma
　　　NOK-MEGULASTIK Co., Ltd.
　　Hideo Okamura and Kiyoshi Sogabe
　　　Sophia Univ.

971998 **A New Generation of Condenser Measurement Microphones** 1021
　　Gunnar Rasmussen
　　　G.R.A.S. Sound and Vibration
　　Ernst Schøntal
　　　PCB Piezotronics

971999 **Accelerometer Calibration** 1027
　　Ernst Schonthal and David M. Lally
　　　PCB Piezotronics, Inc.

Volume 3

972000 **A New Approach for the On-Road Data Acquisition and Analysis System** 1031
　　John Mathey, David Tao, Nancy Chen, Mark Maskill, and Ken Horste
　　　Ford Motor Co.

972001 **Implementation of a Third Generation Sound Power Test for Production Earthmoving Machinery** 1037
　　Mark B. Sutherland
　　　Caterpillar, Inc.

972002 **A Simple QC Test for Knock Sensors** 1043
　　S. Gade, S. Møllebjerg Matzen, and H Herlufsen
　　　Brüel and Kjær

972003 **Wavelet Transform Analysis of Measurements of Engine Combustion Noise** 1049
　　Grover Zurita Villarroel and Anders Ågren
　　　Luleå University of Technology

972004 **Transient Engine Vibration Analysis by Using Directional Wiger Distribution** 1057
　　C. -W. Lee and Y.-S. Han
　　　KAIST

972005 **Time Scale Re-Sampling to Improve Transient Event Averaging** 1063
　　Jason R. Blough, Susan M. Dumbacher, and David L. Brown
　　　University of Cincinnati

972006 **The Time Variant Discrete Fourier Transform as an Order Tracking Method** 1073
　　Jason R. Blough and David L. Brown
　　　University of Cincinnati
　　Håvard Vold
　　　Vold Solutions

972007   **Theoretical Foundations for High Performance Order Tracking with the Vold-Kalman Tracking Filter** 1083
     Håvard Vold and Michael Mains
       Vold Solutions, Inc.
     Jason Blough
       University of Cincinnati

972008   **Development of a Comparison Index and a Database for Sea Model Results** 1089
     Charles Birdsong and Clark Radcliffe
       Michigan State Univ.

972009   **Structure-borne Noise Prediction Using an Energy Finite Element Method** 1095
     Fernando Bitsie and Robert Bernhard
       Purdue Univ.

972010   **Statistical Energy Methods for Mid-Frequency Vibration Transmission Analysis** 1103
     Sungbae Choi, Christophe Pierre, and Matthew P. Castanier
       University of Michigan

972011   **The Effects of Linear Microphone Array Changes on Computed Sound Exposure Level Footprints** 1109
     Arnold W. Mueller
       NASA, Langley Research Center
     Mark R. Wilson
       Lockheed Martin Space Mission Systems and Serivces

972012   **Utilization of a Chassis Dynamometer for Development of Exterior Noise Control Systems** 1127
     John J. Todd
       Collins and Aikman
     Richard F. Schumacher
       General Motors Corp.

972013   **Tire Noise Reduction Treatment for a Passenger Car Used as a Tow Vehicle for Pass-by Noise Testing** 1135
     John R. Harris
       Continental-General Tire, Inc.
     Thomas A. Williams
       Hankook Tire

972014   **Pass-by Noise Modelling with Boundary Elements** 1143
     J. M. Auger and M. A. Hamdi
       STRACO S.A.
     G. Amadasi and E. Gilimondi
       SCS Controlli e sistemi

972015   **Sophistication of Noise Measurement Regulations for Powered Vehicles in the EU** 1147
     Heinrich Steven
       FIGE GmbH

972016   **Evaluation of Neural Networks as a Technique for Correlating Vehicle Noise with Subjective Response** 1163
David Fish
MIRA

972017   **Power Window Sound Quality - A Case Study** 1171
John N. Penfold
Ford Motor Co.

972018   **The Creation of a Car Interior Noise Quality Index for the Evaluation of Rattle Phenomena** 1177
Gernot Weisch and Wolfgang Stücklschwaiger
AVL LIST GmbH
Alvaro Alves de Mendonca, Nilton T. S. Monteiro, and Luis Alves dos Santos
FORD do Brazil

972020   **Spectrogram Analysis of Accelerometer-Based Spark Knock Detection Waveforms** 1183
David Scholl, Terry Barash, Stephen Russ, and William Stockhausen
Ford Motor Co.

972021   **A QC System for Testing Engine Knock and Whine on Motorcycle Engine Assemblies** 1191
Alexander Bozmoski
Harley-Davidson Motor Co., NVH Engineering
Roger Upton
Bruel and Kjær

972022   **Simulation of Radiated Noise from a Transmission Side Cover** 1197
Glen Steyer and Brian Campbell
Structural Dynamics Research Corp.

972023   **The Use of Pre-Test Analysis Procedures for FE Model/Test Correlation of a Transmission Side Cover** 1203
Mark Donley and Wayne Stokes
Structural Dynamics Research Corp.

972024   **Transmission Side Cover Design Optimization for NVH Part 1: Shell Curvature Studies** 1211
Glen Steyer, Chih-Hung Chung, and Brian Brassow
Structural Dynamics Research Corp.

972025   **Transmission Side Cover Design Optimization for NVH Part 2: Geometric Optimization Studies** 1217
Chih-Hung Chung, Glen Steyer, and Brian Brassow
Structural Dynamics Research Corp.

972027   **Brake Judder Analysis: Case Studies** 1225
Mohamed Khalid Abdelhamid
Bosch Braking Systems

972028   **Mode Shape of a Squealing Drum Brake** 1231
Johan Hultén
Chalmers University of Technology
John Flint and Thomas Nellemose
A/S Roulunds Fabriker

972030 **Signal Processing for Shift Feel Simulation on the Ford Vehicle Vibration Simulator** 1247
W.J. Pielemeier, V. Jeyabalan, and N.C. Otto
Ford Motor Co.

972031 **Measurement of the Rotational Vibrations of RWD Output Shafts and Characterization of the Resulting Effect on Passenger Perceived Noise** 1261
Jeffery S. Williams and Brian K. Wilson
Structural Dynamics Research Corp.
David T. Hanner
General Motors Proving Grounds

972032 **Steady State Reverberation Time Measurement** 1269
Steve Sorenson
Lear Corp.

972033 **Vold-Kalman Order Tracking: New Methods for Vehicle Sound Quality and Drive-Train NVH Applications** 1275
Håvard Vold
Vold Solutions, Inc.
James Deel
Structural Dynamics Research Corp.

972034 **Application of Noise Path Analysis Technique to Transient Excitation** 1283
P. Perry Gu and Joe Juan
Ford Motor Co.

972035 **Measurement of Transient Vibrational Power Flow in a Car Door Panel Using Intensity Technique** 1289
Naoya Kojima and Hai Zhou
Yamaguchi Univ.
Seiji Fujibayasi and Toru Hirayu
Yamaguchi Univ.

972036 **Noise Source Identification in Thermal Systems Using Transient Spectral Analysis** 1297
Keith A. Temple
Lennox Industries, Inc.
Scott Hommema, James D. Jones, and Victor W. Goldschmidt
Purdue Univ.

972037 **Determining Sound Power for Automotive Applications** 1301
Robert Hickling
Sonometrics Inc.
Peng Lee, Alexei Goumilevski, and Wei Wei
Univerity of Mississippi

972038 **Developing a Test Procedure for Compression Brake Noise** 1315
Thomas J. Wahl and Thomas E. Reinhart
Cummins Engine Co.

972039 **Development of an Engine System Model for Predicting Structural Vibration and Radiated Noise of the Running Engine** 1327
Shung H. Sung, Donald J. Nefske, and Francis H. K. Chen
General Motors Research and Development Center
Michael P. Fannin
General Motors Powertrain Group

972040   **Development Stages for Reducing Noise Emissions of the New OM 904 LA Commerical Vehicle Diesel Engine** 1333
Frank W. Leipold and Ralph A. Zima
Mercedes-Benz AG

972041   **Predictive Design Support in the Achievement of Refined Power for the Jaguar XK8** 1341
S. H. Richardson
Jaguar Cars
D. H. Riding
Perkins Technology Consultancy

972042   **Meeting Future Demands for Quieter Commercial Powertrain Systems** 1351
Christian V. Beidl and Alfred Rust
AVL LIST GmbH

972043   **A Numerical Approach for Piston Secondary Motion Analysis and its Application to the Piston Related Noise** 1361
Teruo Nakada
ISUZU Advanced Engineering Center, Ltd.
Atsushi Yamamoto and Takeshi Abe
ISUZU Motors Ltd.

972044   **Noise and Vibration Technology for the Perkins V6 HSDI Demonstration Engine** 1371
Robert Southall and Malcolm Trimm
Perkins Technology Ltd.

972045   **Sound Power Approximation for Rectangular Ribbed Plates Subject to Harmonic Excitation** 1381
Jack C. Webb
Delphi Energy and Engine Management Systems

972046   **Modeling of Airborne Tire Noise Transmission into Car Interior by Using the Vibro-acoustic Reciprocity and the Boundary Element Method** 1389
Jeong-Guon Ih and Bong-Ki Kim
Korea Advanced Institute of Science and Technology
Gi-Jeon Kim
Kumho Tire Co.

972047   **Application of Nearfield Acoustical Holography to Tire/Pavement Interation Noise Emissions** 1397
Richard J. Ruhala and Courtney B. Burroughs
Pennsylvania State Univ.

972048   **The Laboratory Simulation of Tyre Noise** 1407
Steven Jorro and Andrew Tambini
MIRA, Motor Industry Research Association

972049   **A Model Study of How Tire Construction and Materials Affect Vibration-Radiated Noise** 1415
Paul Bremner
Vibro-Acoustic Sciences Inc.
John Huff and J. Stuart Bolton
Purdue Univ.

972050   **Transient Tyre Noise Measurements Using Time Domain Acoustical Holography** 1423
Ernst-Ulrich Saemann
Continental A.G.
Jørgen Hald
Brüel and Kjær A/S

972052   **Vehicle Powertrain Noise Diagnosis Using Acoustic Holography Techniques** 1431
T. N. Patro
Ford Motor Co.

972053   **The Use of Nearfield Acoustical Holography (NAH) and Partial Field Decomposition to Identify and Quantify the Sources of Exterior Noise Radiated from a Vehicle** 1449
Hiroshi Takata, Takuo Nishi, and Weikang Jiang
Isuzu Motors Ltd.
J. Stuart Bolton
Purdue Univ.

972054   **Modeling and Analysis of Automotive Transmission Rattle** 1457
Yu (Michael) Wang
University of Maryland

972055   **Joint Performance of Injection Molded Thermoplastic Bosses Containing Post Consumer Recyclate: Possible Squeak and Rattle Implications** 1463
Martin A. Trapp, Rick Rozmus, and Al Dapoz
Ford Motor Co.

972056   **Frictional Behavior of Automotive Interior Polymeric Material Pairs** 1479
Norm Eiss
Virginia Tech
Edward Lee
Cryovac, Division of W. R. Grace
Martin Trapp
Ford Motor Co.

972057   **A CAE Methodology for Reducing Rattle in Structural Components** 1497
Shang-Rou Hsieh, Victor J. Borowski, and Jen Yuan Her
Ford Motor Co.
Steven W. Shaw
Michigan State Univ.

972058   **Identification and Elimination of Steering Systems Squawk Noise** 1503
Giacomo Sciortino and Jahanshah Bamdad-Soofi
Delphi Automotive Systems

972059   **Quantitative Prediction of Rattle in Impacting System** 1509
Jen Her and Shang-Rou Hsieh
Ford Motor Co.
Wei Li and Alan Haddow
Michigan State Univ.

972060   **Intermittent Modal Vibration and Squeal Sounds Found in Electric Motor-Operated Seat Adjusters** 1517
      D. J. Pickering and T. L. Rachel
        ITT Automotive

972061   **"Next Generation" Means for Detecting Squeaks and Rattles in Instrument Panels** 1527
      William M. Rusen, Edward L. Peterson, and Richard E. McCormick
        MB Dynamics, Inc.
      Richard Byrd
        Ford Motor Co.

972062   **Impulsive Sound Analysis of an Automotive Engine Using a Two-Stage ALE** 1533
      Sang-Kwon Lee and Paul Robert White
        ISVR, University of Southampton

972063   **Engineering Vehicle Sound Quality** 1545
      David C. Quinn and Ruediger Von Hofe
        AVL List GmbH

972064   **Layered Fibrous Treatments for a Sound Absorption and Sound Transmission** 1553
      Heng-Yi Lai, Srinivas Katragadda, and J. Stuart Bolton
        Purdue Univ.
      Jonathan H. Alexander
        3M Occupational Health and Environmental Safety Division

972065   **Application of a Laser Vibrometer for Automotive Aeroacoustic Analysis** 1561
      M. M. Marchi, A. Petniunas, and D. E. Everstine
        Ford Motor Co.

951375   **Application of Noise Control and Heat Insulation Materials and Devices in the Automotive Industry** 1571
      Timothy Hirabayashi
        General Motors Corp.
      David McCaa
        CertainTeed Corp.
      Robert Rebandt
        Ford Motor Co.
      Phillip Rusch
        Harley-Davidson Motor Co.
      Pranab Saha
        Kolano and Saha Engineers, Inc.
      SAE Acoustical Materials Committee
      SAE Thermal Materials Committee

971933

Robust Design of Elastic Mounting Systems

Thomas Vietor, Rolf Deges, Norbert Hampl, and Karl-Heinz Bürger
Ford Werke AG

Copyright 1997 Society of Automotive Engineers, Inc.

ABSTRACT

NVH (Noise, Vibration, Harshness) is one of the main attributes in a passenger car. One of the key systems responsible for the overall NVH behaviour is the powertrain mounting system. The optimal layout of this system leads to the definition of a stochastic optimization problem. In this paper the background of the complexity of the powertrain mounting system is highlighted. In a first approach the optimization problem is solved with Design of Experiments (DOE) methods.

INTRODUCTION

The development of a passenger car is a multidisciplinary task. The vehicle has to fulfill demands out of different attributes like vehicle dynamics, driveability, acoustics, thermal and heat management, safety, crash, economics. This paper focuses on acoustics and general Noise, Vibration and Harshness (NVH) of the vehicle. This area is currently one of the main attributes responsible for overall performance and customer perception of a passenger car. Very often demands out of this area are conflicting with other areas. A main problem is the variability of mechanical measures describing the NVH performance of a car. To overcome this, the extension of the conventional deterministic optimization problem to a stochastic optimization problem is necessary. The powertrain mounting system, which is the physical link between powertrain and vehicle, is one of the most critical systems with respect to NVH. This paper concentrates on aspects for the optimal layout of the powertrain mounting system. Because of the complexity of this single task in a first approach, a sensitivity calculation of the system is performed with Design of Experiments (DOE) methods.

Vehicle system concepts (e.g. body structure, front- and rear suspension, powertrain mounting systems, etc.), which are selected in an early program phase, have significant influence on NVH performance of the vehicle. It is almost impossible to solve NVH concerns resulting from selection of poor concepts in a later program phase. A good understanding of powertrain dynamics and its interaction within the vehicle is needed to design and realize mounting systems that allow the company to reach its NVH leadership goals. The selection of the powertrain mounting concept and the design of powertrain mounting system is a highly complex task which requires involvement of several areas. In the following, demands for powertrain mounting systems, theoretical background, an overview on the design process as well as some practical details of the development process will be given.

DEMANDS FOR POWERTRAIN MOUNTING SYSTEMS

The powertrain mounts are the main links between powertrain and body structure. The main functional demands for the powertrain mounting system are:

- Support of the powertrain under all load cases, e.g.
  - gravity weight
  - drive torque
  - tip in / back out (torque change)
  - switch on / switch off
  - acceleration, cornering
  - road impacts
- Isolation of vibration due to
  - Engine excitation [1], e.g. idle, acceleration, cruise, overrun
  - Driving maneuvers, e.g. torque change, switch on / switch off
  - Road and wheel excitation

The selection of the powertrain mounting system is heavily restricted by program assumptions as well as corporate demands:

- Powertrain concept or concepts
  - Front-wheel-drive, rear-wheel-drive and/or all-wheel-drive
  - N-S or E-W powertrain installation
  - Engine architecture, e.g. I4, V6, I3, I5, etc.
  - Gas- or Diesel Engine
  - Manual or automatic transmission
- Available package space (rock angle)
- Crash requirements
- Cost requirements / Investments
- Feasibility and manufacturing demands

- Durability
- Company cross-carline strategy: reduced com-plexity and unique parts

All demands for powertrain mount systems as given above have to be taken into account for selection of the mounting concepts. Many of the demands are contradictory.

EFFECTS AND VEHICLE SYSTEMS INFLUENCED BY THE POWERTRAIN MOUNT SYSTEM

A series of phenomena within the vehicle are mainly influenced by the powertrain mounting system. The main NVH effects are given in Tab. 1

Powertrain mounts have effects over a wide frequency range, excitation amplitudes vary from several millimeters in the low frequency range to micrometer in the high frequency range. As already stated above, the selection of the engine mount system has severe influences on several other vehicle systems.

SYSTEMS ENGINEERING APPROACH

Customer wants and customer satisfaction are the key points during development process of a new vehicle [2]. Extensive market research activities including benchmarking, customer drives and translation from customer wording to objective measurables (quality function deployment QFD) result in target values for vehicle attribute performance. With respect to powertrain NVH, these total vehicle targets are including interior noise level at specified driving conditions or idle vibration of the seat track. All vehicle development work e.g. vehicle system selection, vehicle system optimization and component design is based on these target values, which become program objectives during the development process after confirmation of vehicle system selection.

NVH Effect	Frequency Range
Drive noise and vibration	20 - 500 Hz
Idle shake, vibration and boom	5 - 50 Hz
Road induced shake	10 - 15 Hz
Take-off judder	10 - 30 Hz
Drive-away harshness	20 - 100 Hz
Switch on/off vibration	5 - 20 Hz
Tip in / back out	3 - 10 Hz
Steering column shake	25 - 40 Hz
Powertrain boom and harshness	50 - 500 Hz

Table 1. NVH Effects Influenced by Powertrain Mounting System.

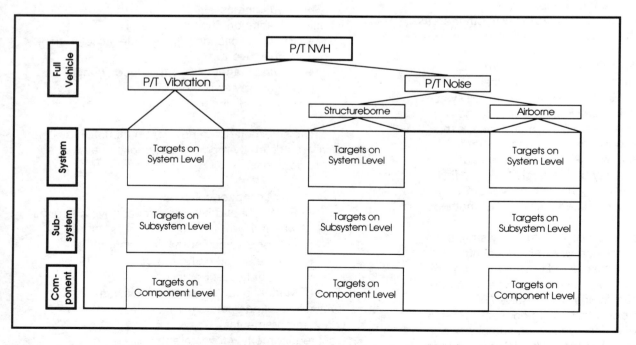

Figure 1. Target Setting for System and Components - Systems Engineering (P/T = Powertrain).

Target cascading, which is mainly supported and performed by CAE techniques, is the process of deriving system and subsystem targets based on total vehicle targets. Some aspects of target cascading for powertrain NVH, e.g. source identification, source ranking and transfer path analysis can be seen from Fig. 1. The final outcome of these exercises are targets for excitation levels and transfer characteristics as well as targets and design specifications for components.

Target cascading is performed for all vehicle attributes. If system and component targets derived for different aspects are contradictory, trade-offs have to be performed in an early program state to ensure high overall customer satisfaction.

Specifications for each vehicle system can be derived only by taking into account total vehicle performance. One main objective during NVH development work is to avoid coincident resonances within the vehicle. It is well known that best isolation of vibration can be achieved if excitation frequencies are well above resonance frequencies. With respect to powertrain mounting system characteristics, rigid body modes of the powertrain should be as low as possible, but above 1/2 engine order frequency and should be well separated from overall vehicle system resonances. Fig. 2 shows characteristic resonance and excitation frequencies of the main vehicle systems. There is only a small frequency window available for rigid powertrain modes.

For competitive vehicle NVH performance, it is not sufficient to simply design a powertrain mounting system which has just all rigid powertrain modes within the frequency band marked in Fig. 2. Experience shows that mode shape characteristics and frequencies of modes play an important role for the quality of a powertrain mounting system.

POWERTRAIN MOUNT SYSTEM DESIGN PROCESS

Several disciplines have to be utilized during the powertrain mount system selection and design process. This process is heavily driven by CAE analyses [3], as there is no hardware available in this early development stage. The powertrain mount system design process is illustrated in Fig. 3.

- Powertrains are defined in program assumptions, full vehicle performance requirements, e. g. NVH

targets, are derived from the benchmarking and target setting process.

- An initial proposal for the powertrain mount positions as well as powertrain mount stiffnesses for idle load is made based on body, powertrain and chassis design, available package as well as experience from former vehicle programs (bookshelf knowledge).

- These proposals have to be confirmed by CAE analyses. As in this early stage of the development program, no detailed body information is available, first CAE calculations are performed on a linear, 'grounded' powertrain CAE model. The first set of design variables is optimized to fulfill modal demands.

- Linear mount characteristics are not sufficient because of package constraints (e.g. clearance for maximum rock angle). Therefore, progressive mounts characteristics have to be designed. Non-linear, progressive mount stiffnesses with idle stiffnesses as confirmed above are selected, which should ensure the following criteria:

  ◆ smooth transition from linear to progressive characteristics range,
  ◆ enter non-linear characteristics range in low gears only,
  ◆ mount stiffness for drive as low as possible,
  ◆ use of maximum feasible deflection at maximum engine load only,
  ◆ feasibility to achieve desired characteristic.

CAE checks of maximum rock angle and powertrain mount forces of non-linear mounts can be performed with non-linear models only, which are mainly grounded powertrain models.

- If all targets and requirements are met, a 'total vehicle system CAE model' has to be used to check, if full vehicle NVH targets for all phenomena influenced by powertrain mounts are met as well as to optimize mount characteristics as well as body and chassis performance.

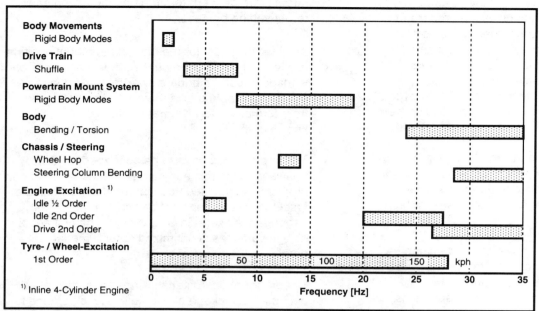

Figure 2. Typical Excitation- and Resonance Frequencies of a Vehicle.

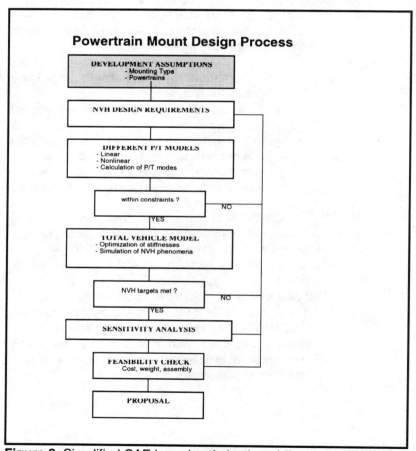

Figure 3. Simplified CAE based optimization of Engine Mount Systems.

If no detailed body information is available at the first total vehicle calculations, CAE runs can be performed using surrogate body models or simplified body CAE models. It is necessary to use full vehicle CAE models from this step on as rigid body modes of grounded powertrains can significantly differ from modes of the same powertrain with identical mounts in a vehicle. Therefore, a post-optimization of the above developed mount characteristics is necessary. The following CAE tools are widely used to support powertrain mount design process:

ABAQUS: Standard Finite Element Program for linear and nonlinear calculation.

ADAMS: Standard tool for Vehicle Dynamic applications. For NVH the limited upper frequency limit constraints the use rigidly. Currently best validated tool for preload and modal calculation for grounded powertrain models. Covers geometric and material non-linearities. Very convenient use of measured progression curves is possible.

NASTRAN: Standard Finite Element Program originally created for linear analysis. Most CAE engineers are well experienced in this tool. Currently only tool capable to calculate with frequency dependent stiffnesses.

MOTRAN: Set of programs developed by FORD, which are used for NVH analysis of vehicles. It allows combination of modal models with simple elements, e.g. linear springs.

HYBRID TECHNIQUES: Techniques, which allow combination of calculated (CAE) and measured (test) data.

Idle as well as drive vibrations can be calculated with help of standard dynamic FE packages. One sample result of such a calculation of steering wheel idle vibrations is given in Fig. 4. It shows vibrations for the same load for two different sets of engine mounts.

Estimates for total vehicle NVH performance above 80 Hz are not currently available with standard FE tools. A method to verify system performance in a higher frequency range is to use hybrid techniques for so-called 'Noise Path Analysis' tools.

- If all NVH targets are met, a sensitivity analysis follows, which identifies main influence factors and enables definition of feasible tolerances or requires a complete redesign to achieve a more robust powertrain mount system. After a feasibility check, the final proposal will be realized as hardware and will be tested in prototypes or demonstration vehicles.
- The described development process leads to a fast possibility to find an optimal layout of the engine mount system from NVH point of view. During the design process, it is possible and often necessary to jump back to an arbitrary earlier step. In each step and as soon as available, consideration of requirements from other areas like Vehicle Dynamics, Driveability, component areas and the supplier is necessary. This rises the need for parallel optimization for several attributes. It is impossible to solve this multi-criteria optimization problem purely by CAE. So a well defined process must be established with participation of all related disciplines including the manufacturer directly from the beginning. By this way only, necessary trade-off decisions are possible in an early stage.

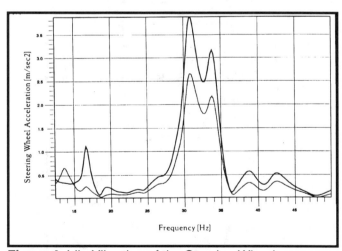

Figure 4. Idle Vibration of the Steering Wheel.

TORQUE ROLL AXIS SYSTEM

Since introduction of E-W installed powertrains in front-wheel driven vehicles, a series of mounting systems has been developed. Design of the first systems has been mainly dominated by package reasons whereas in the last years, the general trend has moved towards to so-called 'Torque Roll' axis (TRA) systems. These are powertrain mounting systems mainly based on inertia characteristics of the powertrain.

The 'Torque Roll axis' (TRA) is the theoretical axis on which a free powertrain rotates if subjected to torque fluctuations about the crankshaft. The orientation of this torque roll axis is defined only by inertia properties of the powertrain, the center of gravity is located on this axis (Fig. 5). The roll axis, which is the axis of rotation of the mounted powertrain, if it is subjected to torque fluctuations, is not identical with the torque roll axis. The roll axis is dependent on powertrain inertia, mount stiffness and excitation frequency, whereas the (static) rock axis depends only on the mount stiffness. The idea of so-called TRA mounting systems is to minimize the difference between roll axis and TRA axis in order to achieve minimum dynamic mount forces in idle, which are mainly caused by combustion pressure torque fluctuations.

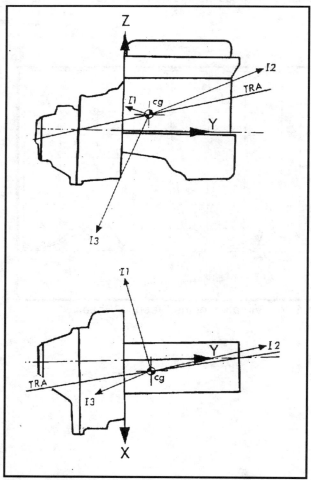

Figure 5. Torque Roll Axis and Principal Axis.

SCATTERING VARIABLES

It is necessary for customer satisfaction to design and manufacture a robust vehicle. A great scattering of different parameters describing the overall NVH performance of the vehicle is definitely not tolerable. Identified main parameters responsible for the scattering of the interior noise and vibrations at certain driver contact points in the vehicle are:

- Engine mount idle stiffnesses. Here a tolerance level of $\pm x$ % is realistic for prototype parts and $\pm y$ % for production parts. x and y are depending on the kind of mount (hydraulic or rubber).
- Engine mount dynamic stiffnesses. Here $\pm 2*y$ % seems to be a realistic value.
- Maximum preload values due to weight of the powertrain and/or engine torque. A deviation of $\pm$ (5...10) % seems to be a realistic lower tolerance bound. This scattering is critical because of large deviations of the dynamic engine mount stiffnesses at various preloads.
- Load variation in idle with AC, PS and electrical consumers on/off.
- Mounting positions.

Fig. 4 shows calculated accelerations at the steering wheel in z-direction for two different engine mount idle stiffness sets. The original set with nominal stiffnesses shows accelerations at a lower level than the second set. Stiffness data of the second set are varied within the production tolerances. The calculations are performed for a total vehicle CAE model which includes a detailed modal model of the flexible body.

FORMULATION OF AN OPTIMIZATION PROBLEM

For the optimal layout of the engine mount system an intuitive approach is unsatisfying. So the formulation of an optimization problem is necessary as well as the use of an optimization procedure for the solution of the problem.

DEFINITION OF A STOCHASTIC OPTIMIZATION PROBLEM [5] - A simplified definition of the continuous, stochastic optimization problem is given in the following.

$$\text{"Min"} \, \mathbf{f}_A(\mathbf{Z})$$
$$\mathbf{x} \in D \quad (8.1)$$

with

$$f_{A_i}(\mathbf{Z}) = k_1 \, E\!\left(f_i(\mathbf{Z})\right) + k_2 \, V\!\left(f_i(\mathbf{Z})\right),$$
$$\mathbf{Z}^T = (\mathbf{X}^T, \mathbf{P}^T) \quad (8.2)$$

$$D = \left\{ \mathbf{x} \in R^n \mid h_i = 0 \; \forall \, i = 1,\ldots,m_{st}; \right.$$

$$P_{fk} = P\big[g_k(\mathbf{Z}) < 0\big] < P_{k_{max}} \; \forall \, k = 1,\ldots,n_g, \qquad (8.3)$$

$$\left. x_{kl} \leq x_k \leq x_{ku} \; \forall \, k = 1,\ldots,n \right\}$$

the vector of inequality constraints $\mathbf{g}^T = (g_1, g_2,\ldots,g_{n_g})$
and

f_{A_i} augmented objective function, here interior noise or vibrations,

g_k inequality constraints as a function of stochastic variables, here frequency conditions,

h_i equality constraints $h_i = h_i(\mu)$, here the system equations,

$\mathbf{Z}$ vector of the stochastic variables,

$\mathbf{X}$ vector of the stochastic design variables, here mount stiffnesses or coordinates,

$\mathbf{P}$ vector of the stochastic parameters,

$E(f(\mathbf{Z}))$ expected value of the objective function,

$V(f(\mathbf{Z}))$ variance of the objective function,

k_1, k_2 weighting factors,

D feasible design space,

P_{fk} failure probability of the k-th inequality constraint,

$P_{k_{max}}$ feasible value of the failure probability,

n_g number of stochastic inequality constraints,

x_{kl}, x_{ku} lower and upper bounds of the design variables, respectively.

Here $\mathbf{X} \equiv \mathbf{P}$ is assumed.

STOCHASTIC OPTIMIZATION PROCEDURES
- The optimization problem defined in equation (8.1) can be solved by means of different procedures. By integrating a stochastic optimization procedure into the optimization procedure SAPOP [4], a complete and extensive optimization environment is available that additionally allows to use further optimization strategies in combination with stochastic optimization [5]. Thus, the stochastic optimization problem in (8.1) is transformed into a quasi-deterministic optimization problem with reliability constraints. Here, fulfillment of constraints is determined by stating probabilities, which is considered during optimization. On the other hand, the deviation of the objective function owing to the stochastic distribution of the design variables and constraints is neglected.

PRACTICAL SOLUTION - For the solution of the mentioned optimization problem it is necessary to have a structural model of the total vehicle. This includes:

- detailed Finite Element (FE) body model,
- concept FE chassis model,
- detailed FE powertrain model for interior noise calculations,
- cavity FE or Boundary Element model (only for interior noise calculations).

The CPU time for one structural analysis is of the magnitude of several hours on a supercomputer. So it is obvious that a stochastic optimization is only possible:

- with the help of simplified models where sufficient,
- with the use of model reduction techniques like superelements or modal models,
- the limitation to a small number of variables.

ROBUST DESIGN WITH DOE METHODS

With CAE methods it is possible in a very early stage of the program to investigate the robustness of the total vehicle and different subsystems, e.g the engine mount system. One of the methods is the Design of Experiments (DOE) [6, 7] methodology. In combination with Response Surface Methodologies (RSM), which can be used with CAE models and tests very effectively, also an optimization is possible. Two different approaches are:

- Design of a robust engine mount system. This is only possible in the concept phase of the program.

- Investigate the robustness of a given system, identification of sensitive parameters and formulation of feasible tolerances for the mechanical quantities of the engine mount system like stiffnesses and the manufacturing process.

The first approach is for sure the best but often not possible. The second one is applicable in a running development with a given concept but can only improve a given system. DOE allows to quantify the influence of varying dynamic stiffness of each powertrain mount for each direction on both powertrain forces as well as vehicle NVH performance. Changing stiffness of one mount in one direction only can influence forces of other mounts drastically.

Fig. 6 shows adjusted response curves as a result of a typical DOE study. The structural model contains a full vehicle detailed model but with a modal representation of the body. This reduces the structural analysis time by a factor of about 20. In a first step no interactions between the variables is assumed. The meaning of the variables is as follows for a system with three engine mounts.

- $x_1,\ldots,x_n$ dynamical stiffnesses of the engine mounts,
- $y_1,\ldots,y_m$ coordinates of the engine mount locations.

So in total m+n independent variables are used. The adjusted response curve is a measure for the vibrations at certain points in the vehicle like seat track or steering wheel. For the vehicle development the target is to minimize this value. From Fig. 6 it is obvious that the influence of the variables x_2, x_3, x_6 is dominating the other variables. In a further step a new DOE is necessary including interactions but limited to the main identified variables. With this result the RSM is used to find an optimal design and to solve the stochastic optimization problem defined in equations (8.1-8.3).

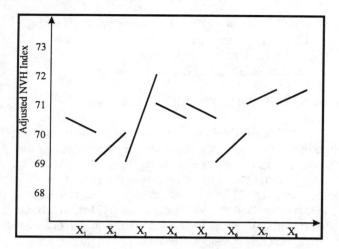

Figure 6. Adjusted Response Curves for 8 variables.

CONCLUSION

The engine mount system is one of the key systems for the overall NVH behavior of the vehicle. This paper presents in a short form the mechanical background of this system and explains the stochastic character of the optimization problem. Because of the complex structural model only a simplified approach for the solution is applicable. With the proposed procedure the identification of the main sensitive parameters is possible. In the later development the optimization of these parameters is necessary. Currently the investigation of all parameters is state of the art. With the given methods in future a drastical reduction of the effort is possible.

REFERENCES

[1] Mass, H., Klier, H.: Kräfte, Momente und deren Ausgleich in der Verbrennungskraftmaschine, Springer-Verlag, Wien, 1981.
[2] Eichhorn, U., Sauerwein, D., Schmitz, T., de Vlugt, A., Teubner, H.-J.: Kundenorientierte Entwicklung am neuen Ford Fiesta, ATZ 97 (1995) 9, S. 522 -531.
[3] Bürger, K.-H.: Analysen des Schwingungsverhaltens des Gesamtfahrzeuges mit Konzeptmodellen, in 'Computergestützte Berechnungsverfahren in der Fahrzeugdynamik', VDI, 1991.
[4] Eschenauer, H.A.; Koski, J.; Osyczka, A.: Multicriteria Design Optimization. Springer, Berlin, 1990.
[5] Vietor, T.: Stochastic Optimization in Mechanical Engineering. In: Marti, K. (ed.): Special Edition of Operations Research "Structural Reliability and Stochastic Structural Optimization". To be published in 1997.
[6] Bandemer, H.; Bellmann, A.: Statistische Versuchsplanung. Teubner, Leipzig, 1994.
[7] N.N.: RS/DISCOUVER Reference Manual. BBN-Softw. Prod., Cambridge, 1992.

971934

Improvements in the Optimization of Dynamic Vibration Absorbers

James L. Swayze
Ford Motor Co.

Eugene R. Rivin
Wayne State Univ.

Copyright 1997 Society of Automotive Engineers, Inc.

ABSTRACT

Improved formulas are developed for the optimum tuning of damped vibration absorbers whose motion may or may not be in-line with the motion axis of the main mass. J. Ormondroyd and J. P. Den Hartog (1928), studied a class of dynamic vibration absorbers whose motion is in-line with that of the main mass and reported the existence of optimum tuning and damping values for the absorber mass along with supporting numerical results. An asymptotic solution for the optimum damping of J. Ormondroyd's and J. P. Den Hartog's absorber was presented by J. E. Brock (1946); however, his solution breaks down as the mass ratio increases. In this paper, an exact solution for the optimum damping of J. Ormondroyd's and J.P. Den Hartog's absorber is given along with an analysis of vibration absorbers whose motion is not in-line with the motion axis of the main mass. This work was made possible by the use of a symbolic algebra program which was not available to the earlier investigators.

1. INTRODUCTION

The push to make lighter vehicles has given rise to some unique Noise, Vibration and Harshness (NVH) problems and solutions. Fundamentally, the cure for vibration is to remove the source. In a vehicle, this is almost impossible to do. Another method is resonance avoidance which is, essentially, a modal alignment strategy for a vehicle or a mechanical system. In this method, either the modes of the vehicle, the system and the subsystem are never aligned with each other or they are positioned in such a way that none of them will ever be excited by the excitation source. Active control can also be used to reduce unwanted vibrations, however, it is too costly for most automotive applications. Direct application of damping, either viscous or hysteretic, if a resonance condition exists, is another highly effective method to reduce vibrations in order to improve NVH. Application of damping is a passive means of controlling vibration. Another passive method, which is the topic of

this paper, is the application of dynamic vibration absorbers.

Dynamic absorbers can fill two roles. First they can be used to reduce linear or torsional periodic motions including: vibration of accessory drive (FEAD) components and crank shafts of internal combustion engines, steering column/wheel shake, ship stabilization, and etc.. Secondly, dynamic absorbers can be used to reduce the force transmitted by a machine or a system to its foundation (e.g., exhaust systems, shock towers, rotating machinery, and even suspension systems on vehicles [1,2,3]).

Dynamic vibration absorbers work by reducing the vibration of the main mass (or system) by means of creating dynamic action opposing the disturbing force. If designed and applied at the frequency where the vibrations are of greatest concern, an absorber will exert a force equal and opposite to the disturbing force on the main mass, thus canceling the vibration of the main mass.

Known publications deal with dynamic vibration absorbers whose motion axes are in-line with the motion axis of the main mass. One of the earliest applications of a dynamic vibration absorber was by Frahm in 1911 [3]. Ormondroyd and Den Hartog [3] were the first to numerically discover the existence of optimum tuning and optimum damping for absorbers whose motion axes are in-line with the motion of the main mass. Brock [4,5] was the first to find an analytical solution for the optimum damping and tuning of Ormondroyd's and Den Hartog's dynamic absorber; however, for the case of optimum tuning, Brock's optimum damping solution breaks down when the absorber mass to main mass ratio exceeds 0.2. In 1968, Snowdon [6] expanded on Brock's solution and obtained a solution which was good for all absorber mass ratios.

This paper extends the previous work by considering an absorber whose motion axis is not in-line with that of the main mass. Only application of dynamic absorbers to lumped mass systems is considered; however, the obtained results can be used for analysis of more complicated systems.

2. ABSORBER MODEL

In the simplest form, vibrations of a system can be modeled using the mass, spring, and damper system shown in Figure 1. From the free-body diagram the equation of motion is

$$\ddot{x}_1 + 2\omega_1 \zeta_1 \dot{x}_1 + \omega_1^2 x_1 = f(t)/M \qquad (2.1)$$

where $\omega_1 = \sqrt{K/M}$ and $\zeta_1 = C/(2M\omega_1)$ is the damping ratio. The solution to equation (2.1) contains two parts: a homogeneous solution ($f(t) = 0$) and a forced solution ($f(t) \neq 0$). The solution of the homogeneous equation is initial condition dependent and decays exponentially with time for $\zeta_1 \neq 0$. It is sometimes referred to as the transient solution. On the other hand, the forced solution (often referred to as the steady state or particular solution) does not vanish with time. In this work, the forced solution will be our primary interest, and in particular we shall concentrate on the response to harmonic excitation of the form $f(t) = Fe^{i\omega t}$. [7]

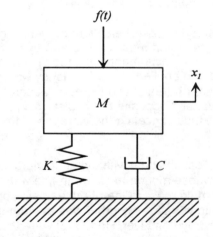

Figure 1 - *Single-degree-of-freedom vibrating system.*

Letting the steady state solution take the form

$$x_1(t) = X_1(i\omega) e^{i\omega t} \qquad (2.2)$$

inserting this and its appropriate derivatives into equation (2.1), dividing through by $e^{i\omega t}$ and solving for $X_1(i\omega)$ gives

$$X_1(i\omega) = \frac{\dfrac{F}{\omega_1^2 M}}{1 - (\omega/\omega_1)^2 + i2\zeta_1 \omega/\omega_1}. \qquad (2.3)$$

The frequency response is then given by

$$H_1(i\omega) = \frac{X_1(i\omega)}{\dfrac{F}{\omega_1^2 M}} \qquad (2.4)$$

$$= \frac{1}{1 - (\omega/\omega_1)^2 + i2\zeta_1 \omega/\omega_1}$$

The magnitude of the frequency response or magnification factor is given by

$$|H_1(i\omega)| = \frac{1}{\left(\left(1 - (\omega/\omega_1)^2\right)^2 + \left(2\zeta_1 \omega/\omega_1\right)^2 \right)^{1/2}}. \qquad (2.5)$$

The forced response then becomes

$$x_1(t) = \frac{F}{\omega_1^2 M} |H_1(i\omega)| e^{i(\omega t + \phi)} \qquad (2.6)$$

where

$$\phi = \tan^{-1} \frac{\text{Im}[H_1(i\omega)]}{\text{Re}[H_1(i\omega)]}$$

$$= \tan^{-1} \frac{-2\zeta_1 \omega/\omega_1}{1 - (\omega/\omega_1)^2} \qquad (2.7)$$

is the phase angle. Plots of the magnification factor versus the frequency ratio $\Omega = \omega/\omega_1$ are shown in Figure 2 for various damping ratios. When the frequency ratio Ω is such that

$$\Omega_{res} = \frac{\omega}{\omega_1} = \left(1 - 2\zeta_1^2\right)^{1/2}, \qquad (2.8)$$

a peak or maxima occurs in the magnification plot. Ω_{res} is often referred to as the damped resonance frequency. The amplitude at the peak is a function of the damping ratio. Increases in the damping ratio decrease the peak amplitude. When $\zeta_1 \to 0$, the abscissa of the peak approaches $\Omega = 1$ and its amplitude approaches infinity. In such a case, the system is said to approach an undamped resonance condition characterized by violent vibrations. When the excitation must operate at frequencies near a resonance, a vibration absorber tuned to the resonance of the main mass frequently provides an effective means of reducing vibration amplitudes to acceptable limits.

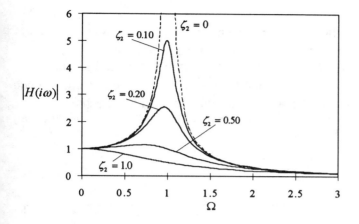

Figure 2 - *Plot of the magnification factor for the dynamic system shown in Figure 1 versus the forcing frequency for various damping ratios.*

Adding a dynamic vibration absorber (auxiliary mass-spring-damper system *m-k-c*) to the main mass of the single degree of freedom system in Figure 1 creates a new system having two degrees of freedom as shown in Figure 3. The absorber's effect is to dramatically reduce the motion of the main mass at or near the frequency at which the absorber is tuned. Ideally, the absorber's motion axis should be parallel to or in-line with the motion axis of the main mass as in Figure 3. However, sometimes it is not possible to mount an absorber in such an ideal configuration. In these situations, what happens to the effectiveness of the absorber when its motion axis is not in-line or parallel to the motion axis of the main mass?

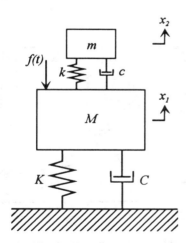

Figure 3 - *Schematic representation of a damped vibration absorber whose motion-axis is in-line with the motion-axis of the main mass M.*

Consider the model of an absorber in Figure 4 whose motion axis is not in-line or parallel to the motion axis of the main mass. The main mass M is constrained to move in the x_1 direction. Its motion is resisted by a linear spring K and a viscous damper C. Attached to the main mass is a damped vibration absorber of mass m constrained to move in the x_2 direction at an angle θ from the x_1 axis. The absorber is attached to the main mass M by a damping element c and spring k arranged in parallel as shown in Figure 4.

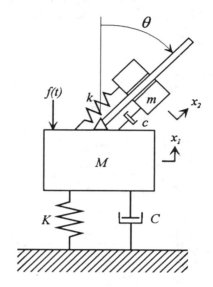

Figure 4 - *Damped vibration absorber whose motion-axis is inclined to the motion-axis of the main mass M.*

2.1 EQUATIONS OF MOTION - If the damping element in the diagram of Figure 4 is considered to be a viscous damper, the equations of motion are given by

$$\begin{aligned}(1+\mu\sin^2\theta)\ddot{x}_1 &+ 2\omega_1\zeta_1\dot{x}_1 \\ &+ 2\mu\omega_2\zeta_2\cos\theta(\cos\theta\,\dot{x}_1 - \dot{x}_2) \\ &+ \mu\omega_2^2\cos\theta(\cos\theta\,x_1 - x_2) \\ &+ \omega_1^2 x_1 = f(t)/M\end{aligned} \quad (2.9)$$

$$\begin{aligned}\mu\ddot{x}_2 &+ 2\mu\omega_2\zeta_2(\dot{x}_2 - \dot{x}_1\cos\theta) \\ &+ \mu\omega_2^2(x_2 - x_1\cos\theta) = 0\end{aligned} \quad (2.10)$$

where

$$\begin{aligned}\mu &= m/M \\ \omega_1 &= \sqrt{K/M} \\ \omega_2 &= \sqrt{k/m} \\ \zeta_1 &= C/(2M\omega_1) \\ \zeta_2 &= c/(2m\omega_2) \\ f(t) &= Fe^{i\omega t}\end{aligned} \quad (2.11)$$

Assuming the solution to x_1 and x_2 to be of the form

$$x_1(t) = X_1(i\omega)e^{i\omega t} \qquad (2.12)$$

$$x_2(t) = X_2(i\omega)e^{i\omega t}, \qquad (2.13)$$

inserting Eq. (2.12) and (2.13) and their appropriate derivatives into Eqs. (2.10) and (2.11), and solving for $X_1(i\omega)$ and $X_2(i\omega)$ leads to the following response functions:

$$\frac{X_1}{\frac{F}{\omega_1^2 M}} = \frac{1}{\left\{\begin{array}{l}1+\mu\omega_t^2\cos^2\theta \\ -(1+\mu\sin^2\theta)\Omega^2 \\ +i2\Omega(\zeta_1+\mu\omega_t\zeta_2\cos^2\theta) \\ -\mu\dfrac{\left[\omega_t^2\cos\theta+i2\zeta_2\omega_t\cos\theta\Omega\right]^2}{\left[\omega_t^2-\Omega^2+i2\omega_t\zeta_2\Omega\right]}\end{array}\right\}} \qquad (2.14)$$

$$\frac{X_2}{\frac{F}{\omega_1^2 M}} = \frac{\left[\omega_t^2+i2\zeta_2\omega_t\Omega\right]}{\left[\omega_t^2-\Omega^2+i2\zeta_2\omega_t\Omega\right]}\frac{X_1\cos\theta}{\frac{F}{\omega_1^2 M}} \qquad (2.15)$$

where

$$\omega_t = \frac{\omega_2}{\omega_1}$$

$$\Omega = \frac{\omega}{\omega_1}.$$

As mentioned above, the absorber's task is to reduce the motion of the main mass at the frequency at which the absorber has been tuned. However, unless the original system is at or near resonance, the addition of a dynamic absorber may not make much sense since the vibration amplitudes away from resonance are usually inconsequential. For that reason, we shall confine our study to that for which $\omega_t \approx 1$.

For $\zeta_1 = 0$, it is interesting to follow what happens when the damping of the absorber is increased. Figures 5 and 6 show plots of the magnitudes $\left|\dfrac{X_1}{F/\omega_1^2 M}\right|$ and $\left|\dfrac{X_2}{F/\omega_1^2 M}\right|$ versus Ω for values of $\zeta_2 = 0$ to $\zeta_2 \to \infty$ for $\omega_t = 1$, $\theta = 0°$ and a fixed value of the mass ratio. Figure 7 shows the effect of θ on the magnitude of $\left|\dfrac{X_1}{F/\omega_1^2 M}\right|$.

For $\zeta_2 = 0$ and $\theta = 0°$, the system has two undamped resonances at

$$\Omega^2 = \left(1+\frac{\mu}{2}\right)\pm\sqrt{\mu+\frac{\mu^2}{4}} \qquad (2.16)$$

When $\zeta_2 \to \infty$, the two masses become clamped and the result is a system described by a one degree of freedom system with a mass $M + m$. Somewhere between zero and infinity, ζ_2 reaches an optimum value at which point the resonant amplitudes reach a minimum. The optimum value for this case will now be discussed.

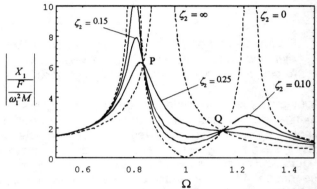

Figure 5 - Response magnitude of X_1 versus Ω for $\omega_t = 1$, $\mu = 0.2$, $\theta = 0°$ and $\zeta_1 = 0$ for various values of ζ_2. Points P and Q represent fixed points whose locations and magnitudes are invariant with damping ζ_2.

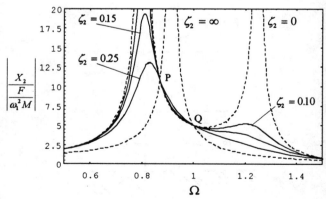

Figure 6 - Response magnitude of X_2 versus Ω for $\omega_t = 1$, $\mu = 0.2$, $\theta = 0°$ and $\zeta_1 = 0$.

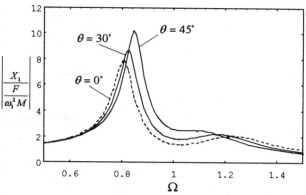

Figure 7 - Response magnitude of X_1 versus Ω showing the effect of θ for $\omega_t = 1$, $\mu = 0.2$, $\zeta_1 = 0$ and $\zeta_2 = 0.15$.

2.1.1 Determination of Optimum Absorber Tuning and Damping

Ormondroyd and Hartog [3], in 1928, were the first to report the existence of an optimum tuning and damping for a damped vibration absorber whose absorber mass motion axis was in-line with that of the main mass (Figure 3 or Figure 4 when $\theta = 0°$). Their analysis was performed using numerical techniques. In 1946, Brock [4] reported an approximate analytical solution for the tuning and damping of Ormodroyd's and Hartog's absorber when the damping of the main mass is zero, i.e. $\zeta_1 = 0$. In practice, there is always be some damping associated with the main mass. However, Ormondroyd and Hartog [3] have proven that the introduction of small damping in the main mass system has a negligible effect.

To derive the optimum tuning and damping of the absorber system in Figure 4 when $\zeta_1 = 0$, it is important to note in Figure 5 that all curves intersect at points P and Q (so called fixed points) independent of the damping ratio ζ_2. If their locations are calculated, the problem is solved, for the most favorable response curve is one which passes with a horizontal tangent through the maximum of the two points P or Q. The lowest obtainable resonant amplitudes occurs at the resonant frequency of that point.

Setting $\zeta_1 = 0$ in Eq. (2.14), the response magnitude of the main mass can be written as

$$\left| \frac{X_1}{F/\omega_1^2 M} \right| = \sqrt{\frac{A^2 + B^2 \zeta_2^2}{C^2 + D^2 \zeta_2^2}} \qquad (2.17)$$

where

$$A = \omega_t^2 - \Omega^2$$
$$B = 2\omega_t \Omega$$
$$C = (1 + \mu \sin^2 \theta)\Omega^4 - (1 + \omega_t^2 + \mu \omega_t^2)\Omega^2 + \omega_t^2 \qquad (2.18)$$
$$D = -2\omega_t(1 + \mu)\Omega^3 + 2\omega_t \Omega$$

Expression (2.17) becomes independent of ζ_2 if $A^2/C^2 = B^2/D^2$ giving

$$\left(\frac{\omega_t^2 - \Omega^2}{(1 + \mu \sin^2 \theta)\Omega^4 - (1 + \omega_t^2 + \mu \omega_t^2)\Omega^2 + \omega_t^2} \right)^2$$
$$= \left(\frac{1}{1 - (1 + \mu)\Omega^2} \right)^2 \qquad (2.19)$$

The squares on each sign of Eq. (2.19) can be removed if $\pm$ is added on the right-hand side of the equal sign. With the plus sign, Eq.(2.19) can be written as

$$(\omega_t^2 - \Omega^2)(1 - (1 + \mu)\Omega^2) =$$
$$(1 + \mu \sin^2 \theta)\Omega^4 - (1 + (1 + \mu)\omega_t^2)\Omega^2 + \omega_t^2 \qquad (2.20)$$

or

$$(1 + \mu)\Omega^4 = (1 + \mu \sin^2 \theta)\Omega^4 \qquad (2.21)$$

which yields the trivial solution $\Omega^4 = 0$. In other words, at $\Omega = 0$ the frequency response function amplitude given by Eq. (2.14) is 1, independent of ζ_2, simply because the motion of the main mass is so slow that there is no chance for a damping force to build up.

With the minus sign, after some algebra, Eq. (2.19) can be written as

$$\Omega^4 - \frac{1 + (1 + \mu)\omega_t^2}{\left[1 + \mu \frac{(1 + \sin^2 \theta)}{2}\right]} \Omega^2$$
$$+ \frac{\omega_t^2}{\left[1 + \mu \frac{(1 + \sin^2 \theta)}{2}\right]} = 0 \qquad (2.22)$$

which is quadratic in terms of Ω^2. Solving for the roots of Eq. (2.22) gives

$$\Omega_{1,2}^2 = \frac{1+(1+\mu)\omega_t^2}{2+\mu(1+\sin^2\theta)}$$
$$\pm \frac{\sqrt{(1+(1+\mu)\omega_t^2)^2 - (4+2\mu(1+\sin^2\theta))\omega_t^2}}{2+\mu(1+\sin^2\theta)} \quad (2.23)$$

Substituting these roots into the magnitude of the response given by Eq. (2.14) and equating the two expressions to determine the optimum tuning is very tedious. However, since the fixed points P and Q are independent of ζ_2, ζ_2 can be chosen to simplify the response function. The simplest form of expression (2.14) occurs when $\zeta_2 \to \infty$ giving

$$\frac{X_1}{F/\omega_1^2 M} = \left| \frac{1}{1-(1+\mu)\Omega^2} \right| e^{-i\phi(\Omega_{1,2})} \quad (2.24)$$

Substituting Ω_1 and Ω_2 into this equation gives

$$\frac{1}{1-(1+\mu)\Omega_1^2} = -\frac{1}{1-(1+\mu)\Omega_2^2} \quad (2.25)$$

where the negative comes from the fact that $e^{-i\phi(\Omega_1)} = 1$ and $e^{-i\phi(\Omega_2)} = -1$. After some algebra, Eq. (2.25) becomes

$$\Omega_1^2 + \Omega_1^2 = \frac{2}{1+\mu} \quad (2.26)$$

Recalling the fact that the negative coefficient of the middle term in a quadratic equation is equal to the sum of its roots, from Eq. (2.22), $\Omega_1^2 + \Omega_2^2$ can also be written as

$$\Omega_1^2 + \Omega_2^2 = \frac{1+(1+\mu)\omega_t^2}{1+\mu\frac{(1+\sin^2\theta)}{2}} \quad (2.27)$$

Substitution of (2.27) into Eq. (2.26) gives the optimum tuning

$$(\omega_t^2)_{opt} = \frac{\sqrt{1+\mu\sin^2\theta}}{1+\mu} \quad (2.28)$$

A plot of ω_t versus $1/\mu$ for $\theta = 0°$, $30°$ and $45°$ is shown in Figure 8. The location of the fixed points is given by

$$\Omega_{1,2}^2 = \frac{1 \pm \sqrt{\frac{\mu(2\cos^2\theta + \mu(1-\sin^4\theta))}{\left(2+\frac{3}{2}\mu-\frac{\cos 2\theta}{2}\mu\right)^2}}}{1+\mu} \quad (2.29)$$

The magnitude of the frequency response at the optimum tuning is given by

$$G_{1,2} = \left| \frac{X_1}{\frac{F}{\omega_1^2 M}} \right| = \sqrt{\frac{4+\mu(3-\cos 2\theta)}{2\mu\cos^2\theta}} \quad (2.30)$$

Now the determination of the optimum damping involves the general frequency response equation given by Eq (2.17). Solving Eq. (2.17) for ζ_2 gives

$$\zeta_2^2 = \frac{A^2 - G^2 C^2}{G^2 D^2 - B^2} \quad (2.31)$$

Substituting locations of the fixed points given by Eq (2.23), the optimum tuning given by Eq. (2.28), the expression for the magnitude of the frequency response at the fixed points $G_{1,2}$ and the expressions for A, B, C, and D into Eq. (2.31) leads to an indeterminate expression $0/0$ because the locations of the fixed points are independent of the damping. If, however, a slightly perturbed frequency for the fixed points is used in the form of $\Omega_{1,2}^* = \Omega_{1,2} + \varepsilon$, a finite ratio between higher-order terms in ε can be obtained for the optimum damping as $\varepsilon \to 0$. Alternatively, l'Hopital's Rule could be applied to Eq. (2.31) but involves more work to obtain the same end results.

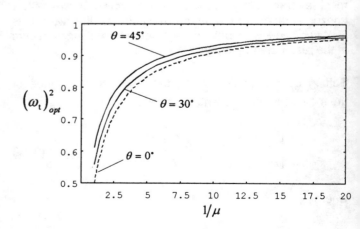

Figure 8 - Optimum tuning, $(\omega_t)^2_{opt}$, from Eq. (2.28) versus $1/\mu$ for various θ.

Substitution of $\Omega_{1,2}^* = \Omega_{1,2} + \varepsilon$ into Eq. (2.31) transforms Eq. (2.31) into

$$\zeta_2^2 = \frac{a_0 + a_1\varepsilon + a_2\varepsilon^2 + \ldots + a_8\varepsilon^8}{b_0 + b_1\varepsilon + b_2\varepsilon^2 + \ldots + b_6\varepsilon^6} \quad (2.32)$$

Since Eq. (2.32) must assume an indeterminate form $0/0$ when $\varepsilon = 0$, then $a_0 = b_0 = 0$, and limit of (2.32) as $\varepsilon \to 0$ is given by

$$\zeta_2^2 = \frac{a_1}{b_1} \quad (2.33)$$

where

$$\begin{aligned}
a_1 =\ & 4\omega_t^2\Omega_{1,2} - 4\Omega_{1,2}^3 - 4\omega_t^2\Omega_{1,2}G_{1,2}^2 \\
& - 4\omega_t^4\Omega_{1,2}G_{1,2}^2 - 4\mu\omega_t^4\Omega_{1,2}G_{1,2}^2 \\
& + 4\Omega_{1,2}^3G_{1,2}^2 + 16\omega_t^2\Omega_{1,2}^3G_{1,2}^2 \\
& + 12\mu\omega_t^2\Omega_{1,2}^3G_{1,2}^2 + 4\omega_t^4\Omega_{1,2}^3G_{1,2}^2 \\
& + 8\mu\omega_t^4\Omega_{1,2}^3G_{1,2}^2 + 4\mu^2\omega_t^4\Omega_{1,2}^3G_{1,2}^2 \\
& - 12\Omega_{1,2}^5G_{1,2}^2 - 6\mu\Omega_{1,2}^5G_{1,2}^2 \\
& - 12\omega_t^2\Omega_{1,2}^5G_{1,2}^2 - 18\mu\omega_t^2\Omega_{1,2}^5G_{1,2}^2 \\
& - 6\mu^2\omega_t^2\Omega_{1,2}^5G_{1,2}^2 + 8\Omega_{1,2}^7G_{1,2}^2 \\
& + 8\mu\Omega_{1,2}^7G_{1,2}^2 + 3\mu^2\Omega_{1,2}^7G_{1,2}^2 \\
& - 4\mu\omega_t^2\Omega_{1,2}^3G_{1,2}^2\cos(2\theta) \\
& + 6\mu\Omega_{1,2}^5G_{1,2}^2\cos(2\theta) \\
& + 6\mu\omega_t^2\Omega_{1,2}^5G_{1,2}^2\cos(2\theta) \\
& + 6\mu^2\omega_t^2\Omega_{1,2}^5G_{1,2}^2\cos(2\theta) \\
& - 8\mu\Omega_{1,2}^7G_{1,2}^2\cos(2\theta) \\
& - 4\mu^2\Omega_{1,2}^7G_{1,2}^2\cos(2\theta) \\
& + \mu^2\Omega_{1,2}^7G_{1,2}^2\cos(4\theta)
\end{aligned} \quad (2.34)$$

$$\begin{aligned}
b_1 =\ & 8\omega_t^2\Omega_{1,2} - 8\omega_t^2\Omega_{1,2}G_{1,2}^2 + 32\omega_t^2\Omega_{1,2}^3G_{1,2}^2 \\
& + 32\mu\omega_t^2\Omega_{1,2}^3G_{1,2}^2 - 24\omega_t^2\Omega_{1,2}^5G_{1,2}^2 \\
& - 48\mu\omega_t^2\Omega_{1,2}^5G_{1,2}^2 - 24\mu^2\omega_t^2\Omega_{1,2}^5G_{1,2}^2
\end{aligned} \quad (2.35)$$

Substituting $\Omega_{1,2}$ and $G_{1,2}$ given by equations (2.29) and (2.30), gives two expressions for ζ_2^2 which are too long to be of any use here. However, a useful average between the two results given by Eq. (2.33), i.e. $\dfrac{(\zeta_2^2)_1 + (\zeta_2^2)_2}{2}$, results in a more useful form

$$\zeta_2^2 = \frac{3\mu\cos^4\theta(-2 - \mu + \mu\cos 2\theta)}{\begin{array}{l}2(1+\mu)(\mu - 8\cos^2\theta \\ \quad - 6\mu\cos^2\theta + 2\mu\cos 2\theta \\ \quad + 2\mu\cos^2\theta\cos 2\theta \\ \quad + \mu\cos^2 2\theta)\end{array}} \quad (2.36)$$

A plot of the optimum damping given by Eq. (2.36) for optimum tuning given by Eq. (2.28) for $\theta = 0°$, $30°$ and $45°$ is given in Figure 9. Generally, the optimum damping value increases with increasing μ and decreases with increasing θ.

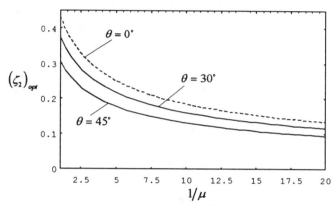

Figure 9 - Optimum damping, $(\zeta_2)_{opt}$, from Eq. (2.36) versus $1/\mu$ for various θ.

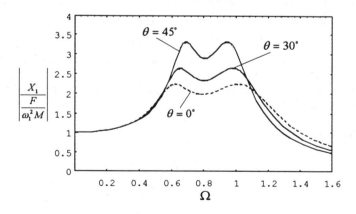

Figure 10 - Optimized response magnitude of X_1 versus Ω for $\mu = 0.5$, $\zeta_1 = 0$, $\omega_t = (\omega_t)_{opt}$ and $\zeta_2 = (\zeta_2)_{opt}$ for various values of θ.

Using the optimum tuning given by Eq. (2.28) and the optimum damping above, the response function for the main mass may now be determined from Eq. (2.14). A plot of the optimum response for the case when $\mu = 0.5$ and for $\theta = 0°$, $30°$ and $45°$ is shown in Figure 10. In general, response levels near the tuning frequency increase with increasing θ. At frequencies much higher than the tuning frequency, the response decreases with increasing θ.

2.1.2 Determination of Optimum Damping for Constant Tuning
Another case that may be of interest is the so called case of constant tuning in which the undamped natural frequency of the absorber is the same as the natural frequency of the main mass, i.e. $\omega_t = 1$. Little recommendation is given for this case, however, in some circumstances such restrictions may be imposed. By an analysis similar to that above, the following results for $\Omega_{1,2}$ and $G_{1,2}$ are obtained:

$$\Omega_{1,2}^2 = \frac{2+\mu \pm \sqrt{\mu(\mu+2\cos^2\theta)}}{2+\frac{\mu}{2}(3-\cos 2\theta)} \quad (2.37)$$

$$G_{1,2} = \left| \frac{2+\frac{\mu}{2}(3-\cos 2\theta)}{-\left(\mu^2+\frac{\mu}{2}(3+\cos 2\theta)\right) \pm (1+\mu)\sqrt{\mu^2+\mu(1+\cos 2\theta)}} \right| \quad (2.38)$$

It can be seen that the ordinate of P given by G_1 is always greater than that of Q given by G_2. Therefore, it is desirable for the slope to reach zero at point P for an optimal absorber design. Setting $\omega_t = 1$ and substituting Ω_1 and G_1 given by Eqs. (2.37) and (2.38), into (2.34) and (2.35) respectively, yields a complicated expression for the optimum damping of a constant tuned absorber. A good approximation is given by the Taylor series expansion:

$$\begin{aligned}\zeta_2^2 &= 0.08677 + 0.4343(\mu-0.2) \\ &- 0.05635(\mu-0.2)^2 - 0.09315(\mu-0.2)^3 \\ &+ \big[0.1092 + 1.029(\mu-0.2) \\ &\quad + 2.099(\mu-0.2)^2 \\ &\quad - 1.008(\mu-0.2)^3\big]\theta^2\end{aligned} \quad (2.39)$$

which is accurate to within 15% of the exact solution to ζ_2 for $0 < \mu \leq 0.5$ and $0° < \theta \leq 30°$. Figure 11 shows the effect of μ and θ on the optimum damping solution. As in the case of optimum tuning, the optimum damping value increases with increasing μ and decreases with increasing θ. Also, with higher μ the optimum damping ratio becomes more sensitive to θ.

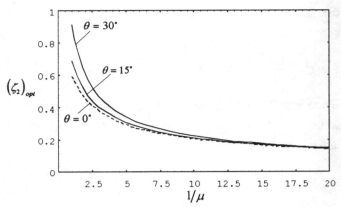

Figure 11 - *Plot of the optimum damping for the case of $\omega_t = 1$ versus $1/\mu$ for various angles of θ shown.*

3. CONCLUSION

Optimizing the performance of dynamic absorbers involves considering the mass ratio, μ, the tuning frequency, ω_t, the damping ratio of the absorber mass, ζ_2, and the angle θ between the motion-axes of the main mass and absorber mass. The mass ratio μ controls the bandwidth of absorber effectiveness. A higher mass ratio leads to a broader bandwidth of absorber effectiveness. Tuning has an effect on the location and magnitude of resonant response peaks. For $\omega_t = 1$, the response peak at the lower resonance frequency was always greater than the response peak at the higher resonance frequency. As ω_t is increased from $\omega_t = 1$, the response peak at the lower resonance frequency increases in amplitude and frequency location, and the response peak at the high frequency decreases in amplitude but increases its frequency location. Conversely, as ω_t is decreased from $\omega_t = 1$, the response peak at the lower resonance frequency decreases both in amplitude and location, and the response peak at the high frequency increases in amplitude and location. Existence of an optimum absorber tuning was proven. Optimum absorber tuning was shown to be dependent on μ and θ. Somewhere between zero and infinity, the damping was shown to reach an optimum value at which point the resonant amplitudes reach a minimum. The optimum damping value was shown to be dependent on μ, θ and the tuning ω_t. In general, the response levels near the tuning frequency are increasing with increasing θ as shown in Figure 10. At frequencies much higher than the tuning frequency, the response decreases with increasing θ.

As a final remark, it is recommended that the optimum absorber tuning given by Eq. (2.28) and the optimum absorber damping given by Eq. (2.36) be used with as large as possible absorber mass ratio, μ, and as small as possible angle θ for optimum absorber benefit, especially when the excitation is broadband in nature.

REFERENCES

[1] J. P. Den Hartog 1985 *Mechanical Vibrations*. New York: Dover Publications, Inc..

[2] K. Winkler 1988 *Society of Automotive Engineers* **No. 8800076**. Guidelines for Optimizing Vibration Mass Dampers.

[3] J. Ormondroyd and J.P Den Hartog 1928 *Transactions ASME* **50**, 9-22. The Theory of the Dynamic Vibration Absorber.

[4] J. E. Brock 1946 *Transactions ASME* **A284**. A note on the Damped Vibration Absorber.

[5] J.E. Brock 1949 *Journal of Applied Mechanics* **16**, 86-92. Theory of the Damped Dynamic Vibration Absorber for Inertial Disturbances.

[6] J. C. Snowdon 1968 *Vibration and Shock in Damped Mechanical Systems*. New York: John Wiley \& Sons, Inc..

[7] L. Meirovitch 1986 *Elements of Vibrational Analysis*. New York: McGraw-Hill.

971936

Study of Nonlinear Hydraulic Engine Mounts Focusing on Decoupler Modeling and Design

Thomas J. Royston
University of Illinois at Chicago

Rajendra Singh
Ohio State Univ.

Copyright 1997 Society of Automotive Engineers, Inc.

ABSTRACT

Decoupler nonlinearities of the automotive hydraulic engine mount affect its isolation performance and the transmission of structure-borne noise. The kinematic gap nonlinearity of the decoupler is examined in considerable detail in the context of the quarter car model. It is shown that while modeling it with a "softened" nonlinear expression may only moderately affect predicted system behavior at the excitation frequency, it can significantly alter it at higher harmonics, changing the predicted level of structure-borne noise transmission. Studies of multi-harmonic motion and vibratory power transmission under sinusoidal and composite excitation conditions confirm that, in fact, use of a decoupler with a "softened" nonlinearity improves performance.

INTRODUCTION

MOTIVATION AND BACKGROUND - Trends in automobile system design toward lighter and more flexible support frames and increasing engine operating speeds combined with increased market emphasis on passenger comfort have focused attention on the impact of vibratory loads and structure-borne sound being transmitted through the engine mount to the chassis. Fundamental tradeoffs exist in the design of mounts. First, they must be relatively stiff and provide significant damping to control large amplitude engine motion caused by rough road conditions, abrupt vehicle acceleration/deceleration and cornering. Such events excite low frequency transient engine vibration (less than 30 Hz) and, in particular, can excite the engine mounting resonance, typically around 10 Hz. Second, for small amplitude excitation over a larger frequency range (up to several hundred Hertz), caused by imbalance forces associated with engine operation, a compliant and lightly damped mount is desirable to minimize the level of structure-borne noise traveling to the passenger compartment. To satisfy these conflicting design criteria,

simple elastomeric (rubber) mounts which possess relatively linear stiffness and damping behavior are being replaced by more complex hydraulic mounts which exhibit amplitude- and frequency-dependent nonlinear behavior. The typical hydraulic mount utilizes a decoupler which is, essentially, an amplitude-dependent mechanism that switches the mount between high stiffness and damping behavior and low stiffness and damping behavior. High stiffness and damping are achieved via a fluid inertia track which acts as a tuned absorber at the fundamental engine mounting resonance. The inertia track is short-circuited by the decoupler during periods of low amplitude motion.

For continued improvement in design more realistic theoretical models are necessary to understand the dynamics of hydraulic engine mount operation. While the mount is clearly nonlinear, most prior investigations have employed linearized models. See Singh et al. [1] and Colgate et al. [2] for excellent reviews of the literature. While such models lead to an easily solvable set of differential equations, they lack robustness in describing the engine mounting system's performance over a realistic range of operating conditions. A situation of particular interest is the engine mount's ability to provide isolation from structure-borne noise at frequencies in the audio range in the presence of large amplitude motion at sub-audio frequencies; i.e. control and isolation simultaneously. A linear model formulation inherently assumes that system behavior at different frequencies is completely independent. However, for the hydraulic engine mount, large amplitude motion at low frequencies will continuously engage and disengage the inertia track which would otherwise not be engaged if the low amplitude, higher frequency motion were the only motion present. In addition to the issue of decoupler modeling accuracy, investigators have also questioned the appropriateness of its design, particularly based on its poor performance under composite excitation conditions [2-3]. Researchers have argued intuitively that a softened decoupler nonlinearity may be more appropriate.

535

Robustly modeling the dynamics of the hydraulic engine mount requires a nonlinear description, not only for the decoupler, but also for fluid flow through the inertia track and hysteretic behavior of the mount rubber [4]. However, recent studies by Colgate et al. [2] and Kim and Singh [5-6] have specifically focused on the strong nonlinearities associated with the decoupler. While attempts at linearization based on energy and squeeze film principles have had some success, more accurate system description over a wide range of operating conditions, including low frequency large amplitude motion, still requires a nonlinear formulation. Nonlinear studies have employed direct time numerical integration for the solution of the resulting differential equations. This is a time-consuming process that typically does not lead to much physical insight into the nature of the system behavior. Additionally, if one also wishes to understand the coupled interaction of the engine mount with realistic multi-degree-of-freedom (MDOF) models for the remainder of the vehicle, one is left with an ever increasingly difficult task of simultaneously integrating numerous coupled differential equations.

How the performance of the hydraulic engine mount is assessed is also important. The most common index has been its dynamic stiffness [1-4,6]. Some articles, in the context of a simplified vehicle model, have also used accelerance, chassis acceleration, force transmissibility and other motion and/or force-related frequency response criteria [4-7]. A limitation of these types of descriptors is that they assume the system is linear, or at least that system response at frequencies other than the excitation frequencies is negligible. It is argued that the "low-pass filter principle" justifies this approach. If the only strong nonlinearity is very localized, being related to the decoupler, and there is sufficient damping in the system, then other frequency components of its response, assumed to be at higher harmonics, will be attenuated by the low-pass filtering effect of inertial systems. This assumes that the decoupler operation does not lead to significant subharmonic behavior and that the supporting structure is nonresonant at higher frequencies. Fundamental studies of vibration isolation in linear systems have shown that support structure dynamics can play a crucial role in isolator or mount performance [8-10]. Such studies have also shown that the best performance descriptor for an isolator is vibratory power transmission which is based on both force and motion. Assessing vibratory power transmission through a nonlinear isolator, like the engine mount, has added complexity since transmission occurs at multiple harmonics of the excitation frequency [11].

In a recent article by the authors [7], an alternative computational strategy based on the Galerkin method was proposed for the efficient analysis of complex mechanical systems with local nonlinearities. The automotive engine mounting system may be considered as such a system where the mount represents a local nonlinearity. While other strong nonlinearities associated with engine and chassis dynamics may exist, for the purpose of assessing engine mount dynamics, these components may be approximated as multi-degree-of-freedom linear subsystems. In this article, a decoupler-equipped hydraulic engine mounting system is analyzed using this Galerkin-based computational method. Issues of decoupler modeling and design and mount performance under sinusoidal and composite excitation conditions in the context of the vehicle system are investigated.

OBJECTIVES - The objectives of this paper are to: (1) apply the efficient Enhanced Galerkin Method [7] to the decoupler-equipped hydraulic engine mounting system, (2) consider alternative modeling formulations and decoupler designs, and (3) assess mount performance via calculation of total (multi-harmonic) vibratory power transmission for harmonic and composite (dual harmonic) excitation conditions. Only vertical motion is considered; however, methodologies outlined here could easily be extended to the multiple axes motion case.

ENGINE MOUNTING SYSTEM EQUATIONS

Consider the automotive hydraulic engine mount system of Figure 1. This system and the theoretical model described below are based on several papers co-authored by Kim and Singh [4-6] covering theoretical and experimental studies of hydraulic mounts employing a decoupler and inertia track between two fluid chambers. Kim and Singh's model was only validated for excitation frequencies below 50 Hz. A detail cut-away of the mount is shown in Figure 2. For a periodic excitation at frequency ω, system response is assumed to be periodic with super-harmonic content up to the N_p^{th} order and sub-harmonic content up to the N_b^{th} order. Equations are written as a function of the nondimensional time variable $\tau = \omega t / N_b$ or in terms of their frequency response at $\omega_n' = n\omega/N_b$, $n = 1, ..., N_b N_p$. Governing equations are defined around static equilibria; hence, the static (gravitational) force is not present in the following formulation. Stiffness and damping elements, k_r and b_r respectively, account for the rubber portion of the mount. They are moderately nonlinear and are typically given as frequency-dependent parameters. Nominal values are provided in Table 1. Since their amplitude dependence is minimal, the force associated with the mount rubber at a particular response frequency ω_n' can be expressed as follows where $j = \sqrt{-1}$:

$$F_r(\omega_n') = \left[j\frac{\omega_n'}{N_b} b_r(\omega_n') + k_r(\omega_n') \right] \left(y_e(\omega_n') - y_s(\omega_n') \right). \quad (1)$$

The force from the fluid components acting on the engine and chassis is given by the following expression where A_p refers to the mount's equivalent fluid piston area, p_1 denotes its upper chamber fluid pressure, and $\overline{p}$ is the static equilibrium pressure in the fluid chambers:

$$F_f(\tau) = A_p \left[p_1(\tau) - \overline{p} \right]. \quad (2)$$

(a)

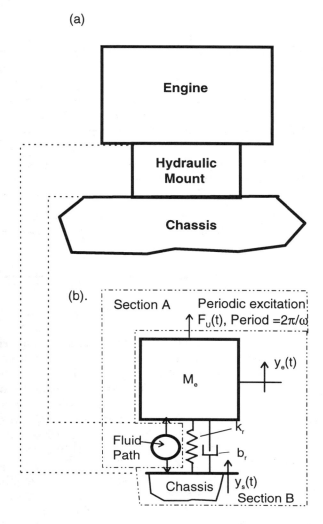

Figure 1. Engine mount system.

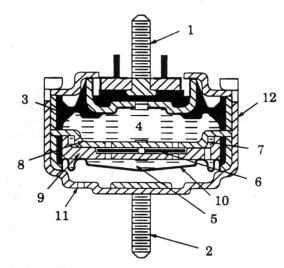

Figure 2. Engine mount components. (1) & (2) mounting studs, (3) rubber which supports engine weight, (4) upper and (5) lower chambers filled with glycol fluid mixture, (6) decoupler, (7) inertia track, (8) upper and (9) lower plates which define decoupler gap, (10) lower chamber thin rubber bellow, (11) air breather, and (12) canister. See Kim and Singh [6] for a more complete description.

Table 1. Parameter Values for Engine Mounting System

$A_1 = A_2 = 0.2726$ cm^2	$m_s = 270$ kg
$A_d = 2.3 \times 10^{-3}$ m^2	$m_e = 122.7$ kg
$A_p = 5.027 \times 10^{-3}$ m^2	$\bar{p} = 116.4$ kPa
$b_r = 1000$ N-s/m	$p_{atm} = 101.232$ kPa
$b_s = 1400$ N-s/m	$\rho_g = 1.06 \times 10^{-3}$ kg/cm^3
$C_{de} = 0.65$	$\bar{V}_1 = 0.715$ cm^3
$k_r = 2.7 \times 10^5$ N/m	$\bar{V}_2 = 28.251$ cm^3
$k_s = 2 \times 10^4$ N/m	$\bar{V}_{air} = 4$ cm^3
$\ell_1 = 16.1$ cm, $\ell_2 = 5$ cm	$\omega_s = \sqrt{k_s/m_s}$
$\eta_1 = 1.9 \times 10^{-3}$ kPa-s^2/cm^3	$\varsigma_s = b_s/2\sqrt{k_s m_s}$
$\eta_2 = 3.4 \times 10^{-3}$ kPa-s^2/cm^3	

In the production-grade hydraulic mount studied in the references [4-6], the inertia track consisted of two fluid paths, one of which is referred to as the leakage path. First order nonlinear differential equations were used to model the relationship between the pressure differential between the lower (2) and upper (1) fluid chambers and the resulting flow q_i through each path, $i = 1, 2$:

$$\frac{N_b}{\omega I_i}[p_2(\tau) - p_1(\tau)] \\ -\frac{\eta_i N_b}{I_i \omega} q_i(\tau)^2 \text{sign}[q_i(\tau)] - \dot{q}_i(\tau) = 0 \quad (3)$$

Here, $\cdot$ denotes $d/d\tau$, $I_i = \rho_g \ell_i / A_i$ is the effective fluid inertia and η_i is an experimentally measured fluid resistance parameter. Also, ρ_g refers to the mount fluid density, ℓ_i denotes the ith inertia track length, and A_i is i^{th} inertia track cross sectional area. Flow through the decoupler orifice q_d is given by a similar nonlinear first order differential equation,

$$\frac{N_b}{\omega I_d}[p_2(\tau) - p_1(\tau)] \\ -\frac{\eta_d(\tau) N_b}{\omega I_d} q_d(\tau)^2 \text{sign}[q_d(\tau)] - \dot{q}_d(\tau) = 0 \quad (4)$$

where the expression for η_d is based upon turbulent flow,

$$\eta_d(\tau) = \left(\frac{1}{C_{de} A_{de}(\tau)}\right)^2 \frac{\rho_g}{2}, \quad (5)$$

with C_{de} denoting the discharge coefficient and $A_{de}(\tau)$ denoting the effective decoupler area. A kinematic model of the decoupler behavior is as follows. In the decoupled state, $A_{de}(\tau) = A_d$, and in the coupled state, $A_{de}(\tau) = 0$, i.e. $q_d(\tau) = 0$. The total volume flow through the decoupler orifice is denoted as v_d. Thus, we have $\dot{v}_d(\tau) = q_d(\tau)$. The decoupler free volume gap is given by $V_{gap} = A_d\Delta_d$ where Δ_d is the decoupler path length. At static equilibrium, the decoupler plate floats in the center. Hence, for $|v_d| \leq V_{gap}/2$ the decoupler plate does not block flow and $A_{de}(\tau) = A_d$. Under cyclic loading, starting from the equilibrium position, a positive pressure differential $[p_2(\tau) - p_1(\tau)] > 0$ will result in an increase in $v_d(\tau)$. When $v_d(\tau) = V_{gap}/2$, $A_{de}(\tau) = 0$ and hence v_d will not exceed $V_{gap}/2$. When the direction of flow reverses and $[p_1(\tau) - p_2(\tau)] < 0$, the decoupler plate becomes unseated and again $A_{de}(\tau) = A_d$ until $v_d(\tau) = -V_{gap}/2$ at which time $A_{de}(\tau) = 0$. As the flow reverses again, the process repeats itself. The effective orifice area can be expressed logically as follows: $A_{de}(\tau) = A_d$ if $|v_d| \leq V_{gap}/2$ or $v_d[p_1 - p_2] > 0$. Otherwise, $A_{de}(\tau) = 0$. The total flow between the two fluid chambers is given by the following first order linear differential equation,

$$\frac{N_b}{\omega}(q_1(\tau) + q_2(\tau) + q_d(\tau)) - \dot{v}(\tau) = 0, \tag{6}$$

where v represents increments in the upper and lower chamber volumes from the $p_1 = p_2 = \overline{p}$ condition. The remaining equations relating pressure and volume in this lumped parameter fluid model are given below [4]:

$$p_2(\tau) = 5.26\text{x}10^{-3}V_2(\tau)^{2.5} - 8.9\text{x}10^{-8}V_2(\tau)^6$$
$$+1.41\text{x}10^{-8}V_2(\tau)^{6.5} + p_{atm},$$

$$p_1(\tau) = \begin{cases} -6.4V_1(\tau) + 29.2V_1(\tau)^{7/6} + p_{atm} & V_1(\tau) \geq 0, \\ p_{atm}\overline{V}_{air}/(\overline{V}_{air} + |V_1(\tau)|) & V_1(\tau) < 0, \end{cases}$$

$$V_1(\tau) = \overline{V}_1 + v(\tau) - A_p[y_e(\tau) - y_s(\tau)], \quad V_2(\tau) = \overline{V}_2 - v(\tau).$$

$$\text{(7a-d)}$$

Here, V_1 and V_2 denote the hydraulic engine mount upper and lower fluid chamber volumes, respectively, p_{atm} denotes atmospheric pressure, V_{air} is the air volume trapped in the upper fluid chamber, and y_e and y_s denote the engine and chassis vertical motions, respectively, at their connection points to the mount. An overhead bar on a variable indicates the static equilibrium condition.

For the engine and chassis, regardless of the number of degrees of freedom used to model them, the relationship between the force and motion at the interface to the fluid components of the mount can always be expressed as a transfer function in the frequency domain. The mount rubber dynamics can also be expressed this way. Consider harmonic motion of frequency ω_n'. Then, we will have harmonic displacement and force response at these connection points of the following form where ~ denotes a complex-valued amplitude:

$$y_e(\tau) = \tilde{y}_e e^{j\omega_n'\tau}, \quad y_s(\tau) = \tilde{y}_s e^{j\omega_n'\tau},$$
$$F_e(\tau) = \tilde{F}_e e^{j\omega_n'\tau}, \quad F_s(\tau) = \tilde{F}_s e^{j\omega_n'\tau} \tag{8a-d}$$

Consequently, a transfer function in the frequency domain between displacement and force at the connection points to Section (A) of Figure 1b can be written as such:

$$\tilde{T}(\omega_n') = \begin{bmatrix} \dfrac{\tilde{y}_e}{\tilde{F}_e} & \dfrac{\tilde{y}_e}{\tilde{F}_s} \\ \dfrac{\tilde{y}_s}{\tilde{F}_e} & \dfrac{\tilde{y}_e}{\tilde{F}_s} \end{bmatrix}. \tag{9}$$

For the simplest case, where the engine mass is assumed to be rigid and the chassis is modeled as a single-degree-of-freedom (SDOF) linear system, we have the following expression where m_s, b_s, and k_s denote the mass, linear viscous damping coefficient and stiffness coefficient of the chassis:

$$\tilde{T}(\omega_n') =$$
$$\begin{bmatrix} k_r(\omega_n') - \omega_n'^2 m_e + j\omega_n' b_r(\omega_n') & -k_r(\omega_n') - j\omega_n' b_r(\omega_n') \\ -k_r(\omega_n') - j\omega_n' b_r(\omega_n') & k_r(\omega_n') + k_s - \omega_n'^2 m_s + j\omega'[b_r(\omega_n') + b_s] \end{bmatrix}$$

$$\text{(10)}$$

MODELING AND COMPUTATIONAL STRATEGIES

MODELING ISSUES - If the single degree-of-freedom chassis model is used, it is fairly straightforward to analyze the mounting system response using direct time numerical integration techniques. The system is composed of four first-order and two second-order differential equations. Of course, the mount rubber nonlinearity, which is defined in the frequency domain must be approximated; but, this is one of the weaker system nonlinearities. For more complex chassis models, where many degrees of freedom are considered, numerical integration will become very complex and inefficient. Additionally, it would be beneficial if experimentally obtained mobility or impedance data for the chassis at the engine mount point could be directly used in simulation studies. In a previous article, an efficient nonlinear solution method for complex systems with local nonlinearities was developed [7]. The method is based on the Galerkin procedure and employs order reduction to reduce the number of degrees of freedom to be solved using an iterative strategy, and continuation to aid in parametric studies. The advantage of the method is that complex linear chassis models of many degrees of freedom, either based on theory or experimental transfer function data are easily incorporated with minimal additional cost in solution time.

Previously, this strategy was applied to a simplified engine mount without a decoupler and only one inertia track [7]. Several difficulties are encountered when trying to solve the decoupler-equipped, multi-path inertia track equations using the Galerkin procedure. Explicit, differentiable analytical expressions are needed. An approximation for the logic-based decoupler equations is proposed based on physical reasoning. If there is any compliance in the decoupler orifice plate or the mechanical stops one should approximate its force vs. position relationship as a spring with a backlash or deadspace regime and stiffness k_d. In general, piecewise discontinuity in the Galerkin code is tolerable. However, the degree of nonlinearity using this expression is very high and consequently, many frequency components will need to be assumed in the periodic solution. A further approximation using a polynomial stiffness expression is more easily handled. Several alternative stiffness formulations are shown in Figure 3. Equations are as follows where γ refers to the order of the polynomial.

Spring with backlash model

$$F_d = \begin{cases} k_d(2v_d/A_d\Delta_d - 1) & 2v_d/A_d\Delta_d > 1 \\ 0 & |2v_d/A_d\Delta_d| \leq 1 \\ k_d(2v_d/A_d\Delta_d + 1) & 2v_d/A_d\Delta_d < -1 \end{cases} \quad (11)$$

Polynomial Spring model

$$F_d = k_d(2v_d/A_d\Delta_d)^\gamma \quad (12)$$

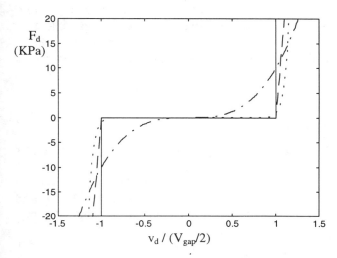

Figure 3. Decoupler stiffness modeling options. Key: ——— kinematic decoupler model, — — — piecewise linear stiffness decoupler model, k_d = 200 KPa, - - - - polynomial stiffness decoupler model with γ = 21, k_d = 1 KPa, ——·—— polynomial stiffness decoupler model with γ = 3, k_d = 10 KPa.

Urabe and Reiter developed the Galerkin procedure for either first or second-order differential equations. The hydraulic mounting system model, excluding base dynamics, consists of four first-order differential equations for the fluid processes and one second-order differential equation for the mount rubber - engine mass degree of freedom. While, both first- and second-order differential equations can be simultaneously handled using the procedure, some modifications are proposed for the sake of computational efficiency. Additionally, direct calculation of v_d is needed to implement Equations (11) and (12).

Two first-order differential equations representing the volume of fluid flow through a single inertia track were cascaded by Royston and Singh [7] to obtain one second-order differential equation. This reduces the complexity of the Galerkin implementation since the order of the problem (in terms of the sizes of arrays that must be handled) is the same for each additional first or second order equation. In the present case, the situation is more complicated since three paths exist between two fluid chambers (two inertia tracks and one decoupler path). Representing the total fluid volume flow through each path with a second order differential equation would result in an unrestrained degree of freedom. While the "spring" force of the decoupler acts to restore the total flow through it to zero, no such force is applied on either inertia path. Consequently, total volume flow through either inertia path is unrestrained, which allows a nonzero mean circulating fluid flow. While this condition has little consequence in the physical system dynamics, it is a source of numerical instability in the Galerkin method. To remedy this difficulty, it is proposed to approximate the two in-parallel inertia paths (i = 1, 2) with one equivalent path (e) using the following dynamic relationships:

$$\eta_e = \frac{\eta_1 \eta_2}{\left(\sqrt{\eta_1} + \sqrt{\eta_2}\right)^2}, \quad I_e = \rho_g \ell_e / A_e,$$

$$\ell_e = \ell_1 \ell_2 / (\ell_1 + \ell_2), \quad A_e = A_1 = A_2. \quad (13a\text{-}d)$$

This further reduces the equations for the fluid processes to only two degrees of freedom with nonlinearities defined in the time domain which are given as follows:

$$\frac{N_b^2}{\omega^2 I_e}[p_2(\tau) - p_1(\tau)] - \frac{\eta_e}{I_e}\dot{v}_e(\tau)^2 \text{sign}[\dot{v}_e(\tau)] - \ddot{v}_e(\tau) = 0,$$

$$\frac{N_b^2}{\omega^2 I_d}[p_2(\tau) - p_1(\tau)] - \frac{\eta_d}{I_d}\dot{v}_d(\tau)^2 \text{sign}[\dot{v}_d(\tau)]$$

$$-\frac{N_b^2}{\omega^2 I_d}F_d(\tau) - \ddot{v}_d(\tau) = 0 \quad (14a\text{-}b)$$

COMPUTATIONAL METHOD - With the above modifications, application of the Galerkin strategy is briefly reviewed. As indicated in Figure 1b, the system can be divided into two sections, (A) and (B). Section (A), comprising the fluid components of the mount, contains the N_v = 2 second order differential equations (14a-b) with nonlinearities defined in the time domain. These two expressions are the determining equations, $\mathbf{d}^v(\tau)$, for the Galerkin method. Section (B), the engine, chassis and rubber components of the system, consists of N_w linear differential equations as well as equations with nonlinearities defined in the frequency domain. These N_w equations will be dependent on the displacement vector $\mathbf{w}$ or state variables in Section (B). This section can be analyzed completely in the frequency domain using linear algebraic methods. The connection between Section (B) and Section (A), as indicated above, is described by the vector $\mathbf{y}(\tau) \equiv [y_e(\tau) \; y_s(\tau)]^T$ which is a linear mapping of $\mathbf{w}(\tau)$. Here, superscript T denotes the transpose. Likewise, the force vector at this connection is described by an N_y = 2 dimensional vector $\mathbf{F}_y(\tau) \equiv [F_e(\tau) \; F_s(\tau)]^T$. Hence, for Section (B), a frequency dependent transfer function may be defined as follows:

$$\tilde{\mathbf{T}}_y(\omega') \equiv \frac{\tilde{\mathbf{y}}}{\tilde{\mathbf{F}}_y}(\omega'). \quad (15)$$

Now, the $N_b N_p^{th}$ approximate solution to the problem will have the following form:

$$\mathbf{v}(\tau) = \mathbf{a}_0^v + \sum_{n=1}^{N_b N_p} \mathbf{a}_{2n-1}^v \sin(n\tau) + \mathbf{a}_{2n}^v \cos(n\tau),$$

$$\mathbf{v} \equiv \left[v_e, v_d \right]^T, \; \mathbf{a}_n^v \equiv \left[a_n^{v_e}, a_n^{v_d} \right]^T,$$

$$\mathbf{y}(\tau) = \mathbf{a}_0^y + \sum_{n=1}^{N_b N_p} \mathbf{a}_{2n-1}^y \sin(n\tau) + \mathbf{a}_{2n}^y \cos(n\tau),$$

$$\mathbf{y} \equiv \left[y_e, y_s \right]^T, \; \mathbf{a}_n^y \equiv \left[a_n^{y_e}, a_n^{y_s} \right]^T. \qquad \text{(16a-b)}$$

By substituting expressions (16a-b) into equations (14a-b) and numerically calculating the Fourier coefficients of $\mathbf{d}^v(\tau)$, we obtain the following $(4N_b N_p + 1) \times (N_v)$ nonlinear algebraic determining equations for finding the values of the coefficients $\alpha^v \equiv \left[\mathbf{a}_0^v \; \mathbf{a}_1^v \; \cdots \; \mathbf{a}_{4N_b N_p}^v \right]$ and $\alpha^y \equiv \left[\mathbf{a}_0^y \; \mathbf{a}_1^y \; \cdots \; \mathbf{a}_{4N_b N_p}^y \right]$:

$$\mathbf{D}_i^v(\alpha) \equiv \mathcal{F}_i \left[\mathbf{d}^v(\tau) \right] = 0, \qquad i = 0, \ldots, 4N_b N_p,$$

where

$$\alpha \equiv \left[\alpha^v \;\; \alpha^y \right]^T, \; \mathcal{F}_0[d(\tau)] \equiv \frac{1}{2N_f} \sum_{n_f=1}^{2N_f} d\left(\tau_{n_f} \right),$$

$$\mathcal{F}_{2n-1}[d(\tau)] \equiv \frac{1}{N_f} \sum_{n_f=1}^{2N_f} d\left(\tau_{n_f} \right) \sin\left(n\tau_{n_f} \right),$$

$$\mathcal{F}_{2n}[d(\tau)] \equiv \frac{1}{N_f} \sum_{n_f=1}^{2N_f} d\left(\tau_{n_f} \right) \cos\left(n\tau_{n_f} \right), \; n = 1, \ldots, 2N_b N_p,$$

and

$$\tau_{n_f} = \frac{2n_f - 1}{2N_f} \pi \text{ with } N_f \geq 2N_b N_p, \; n_f = 1, \ldots, N_f.. \qquad \text{(17)}$$

The remaining $(4N_b N_p + 1) \times (N_y)$ determining equations which are needed take the following form. Here, *Re* and *Im* refer to the real and imaginary parts, respectively:

$$D_0^{y_r^{mn}}(\alpha) \equiv \sum_{z=1}^{N_y} \left\{ Re\left[\tilde{T}_y^{r,z}(0) \right] a_0^{F_z} \right\} - a_0^{y_r} = 0,$$

$$D_{2n-1}^{y_r}(\alpha) \equiv \sum_{z=1}^{N_y} \left\{ \begin{array}{l} Re\left[\tilde{T}_y^{r,z}\left(\frac{n\omega}{N_b} \right) \right] a_{2n-1}^{F_z} \\[4pt] + Im\left[\tilde{T}_y^{r,z}\left(\frac{n\omega}{N_b} \right) \right] a_{2n}^{F_z} \end{array} \right\} - a_{2n-1}^{y_r} = 0,$$

$$D_{2n}^{y_r}(\alpha) \equiv \sum_{z=1}^{N_y} \left\{ \begin{array}{l} Re\left[\tilde{T}_y^{r,z}\left(\frac{n\omega}{N_b} \right) \right] a_{2n}^{F_z} \\[4pt] - Im\left[\tilde{T}_y^{r,z}\left(\frac{n\omega}{N_b} \right) \right] a_{2n-1}^{F_z} \end{array} \right\} - a_{2n}^{y_r} = 0,$$

where $r = 1, \ldots, N_y$, $n = 1, \ldots, 2N_b N_p$ and
$$a_i^{F_z} \equiv \mathcal{F}_i \left[F_z(\tau) \right], \; i = 0, \ldots, 4N_b N_p. \qquad \text{(18a-c)}$$

Consequently, using this order reduction technique, the number of coupled nonlinear algebraic equations to be iteratively solved remains fixed at $N=(N_y+N_v) \times (4N_b N_p + 1)$ regardless of the number of equations N_w describing motion in Section (B). Once the nonlinear solution is

obtained, the response of any variable in the Section (B) is quickly found by simple linear algebraic calculations.

The Galerkin method employs an iterative method to solve the coupled nonlinear algebraic equations, minimizing the sum of the squares of the determining equations in the frequency domain (a least squares approach). Further details of the method can be found in the references [7,12].

POWER FLOW COMPUTATION - Assuming a periodic response with fundamental frequency ω/N_b, the spectral content of vibratory power flow throughout the system may be calculated from the Galerkin procedure results. For a given dynamic displacement variable $z(\tau)$, there is an associated constraint force variable $F_z(\tau)$, both of which can be expressed in the following series forms:

$$z(\tau) = a_0^z + \sum_{n=1}^{N_b N_p} a_{2n-1}^z \sin(n\tau) + a_{2n}^z \cos(n\tau),$$

$$F_z(\tau) = a_0^{F_z} + \sum_{n=1}^{N_b N_p} a_{2n-1}^{F_z} \sin(n\tau) + a_{2n}^{F_z} \cos(n\tau). \qquad \text{(19a-b)}$$

The associated vibratory power flow can then be formulated from the inner product of the force and velocity by summing respective harmonic contributions:

$$P(\omega) = \sum_{n=1}^{N_b N_p} \frac{n\omega}{2N_b} \left[-a_{2n-1}^{F_z} a_{2n}^z + a_{2n}^{F_z} a_{2n-1}^z \right]. \qquad \text{(20)}$$

For example, vibratory power flow through the mount into the automotive chassis will be given by Equations (19-20) where:

$$F_z(\tau) = F_r(\tau) - F_f(\tau), \; a_i^{F_z} \equiv \mathcal{F}_i \left[F_z(\tau) \right], \text{ and } a_i^z = a_i^{y_s}. \qquad \text{(21)}$$

Vibratory power flow through the rubber component or the fluid component of the mount may be considered separately by using $F_z(\tau) = F_r(\tau)$ or $F_z(\tau) = -F_f(\tau)$, respectively.

RESULTS AND DISCUSSION

MODELING AND DESIGN ISSUES - In the previous section, several modifications to the engine mount model were proposed to make the problem tractable using the enhanced Galerkin method. The impact of these changes is assessed via comparison of different modified cases for harmonic excitation using the SDOF base model. System parameter values used in this study are provided in Table 1. In Figure 4, base acceleration $\ddot{y}_s$ at the fundamental harmonic of the excitation force $F_u(t) = 100 \sin(\omega t)$ N for $3 < \omega/2\pi < 20$ Hz is shown for four different decoupler models and for the rubber mount alone which excludes fluid elements. Here, note that the hydraulic mount acts like a tuned absorber at the engine mounting resonance, attenuating the otherwise large peak near $\omega/2\pi = 10$ Hz.

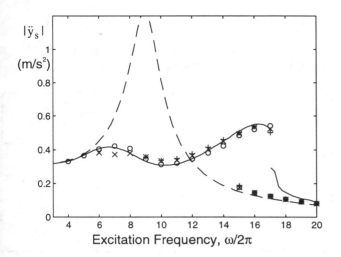

Figure 4. Mounting system frequency response. Chassis vertical acceleration at the excitation frequency, $\ddot{y}_s(\omega)$, for $\Delta_d = 0.7$ mm and $F_u(t) = 100\sin(\omega t)$ N. Key: — — — rubber mount, o o o kinematic decoupler model (numerical integration), x x x piecewise-linear stiffness decoupler model (numerical integration), + + + polynomial stiffness decoupler model with $\gamma = 21$ (numerical integration), ——— polynomial stiffness decoupler model with $\gamma = 3$ (Galerkin solution & numerical integration).

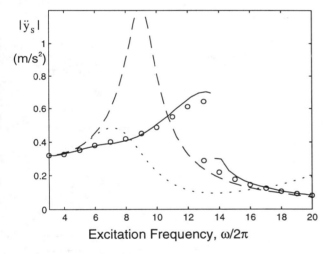

Figure 5. Mounting system frequency response. Chassis vertical acceleration at the excitation frequency, $\ddot{y}_s(\omega)$, for $\Delta_d = 1.4$ mm and $F_u(t) = 100\sin(\omega t)$ N. Key: — — — rubber mount, o o o kinematic decoupler model (numerical integration), ——— polynomial stiffness decoupler model with $\gamma = 3$ (Galerkin solution & numerical integration), - - - - inertia mount with no decoupler ($\Delta_d = 0$ mm) (Galerkin Solution & numerical integration).

The decoupler models agree fairly well over this frequency range. All four models, even the kinematic one based exactly on Kim and Singh's equations [4-6], indicate a hardening stiffness associated with the decoupler. For the models with stronger nonlinearity, a jump phenomena is also evident. In other words, there are frequency regimes where multiple stable solutions exist. For numerical integration, converging to these solutions is sometimes only possible by starting at a frequency outside of the multi-solution regime and slowly incrementing the frequency and using the steady state solution from the previous frequency as an initial condition for the incremented frequency. The weakest nonlinear decoupler model did not predict as strong of a jump phenomena, but did have a region of numerical instability where a solution, either stable or unstable, could not be found with the Galerkin method. Hence, there is a break in the solution curve.

As expected, the response bears similarity to that shown in Figure 9a of Kim and Singh [6] for the case of $\Delta_d = 0.7$ mm. However, Kim and Singh's simulations and their experimental studies do not indicate the presence of jump behavior, even though their graphical results lead one to suspect that such behavior may have existed. For their experimental studies either jump phenomena did exist but were not measured or possibly, the actual mount decoupler is more suitably modeled as a softer stiffness nonlinearity.

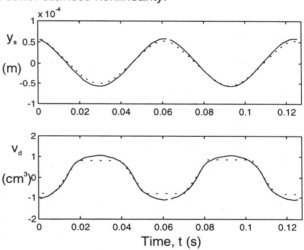

Figure 6. Mounting system time response. Chassis motion, $y_s(t)$, and the total volume flow through the decoupler orifice, $v_d(t)$, for $\Delta_d = 0.7$ mm and $F_u(t) = 100\sin(15.86 \times 2\pi t)$ N (large amplitude solution). Key: - - - - kinematic decoupler model (numerical integration), ——— polynomial-stiffness decoupler model with $\gamma = 3$ (Galerkin solution & numerical integration).

In Figure 5, further confirmation of the robustness of the modified decoupler model and of the jump phenomena are provided by repeating the test conditions of Figure 4 with a decoupler length $\Delta_d = 1.4$ mm. This also compares well with Figure 9a of Kim and Singh [6]. In this figure, the response of the hydraulic mount without a decoupler (an inertia track mount) is also shown. Here, the usefulness of the decoupler is evident as it attenuates higher system response levels as the frequency increases. The qualitatively good agreement for this case with Kim and Singh [6] indicates that the equivalent inertia track path is a reasonable approximation for the dual path configuration of the actual mount.

In Figure 6, selected time plots of the decoupler total volume displacement v_d and the chassis acceleration $\ddot{y}_s$ are provided to illustrate two points. First, the low-pass filter effect is evident as high frequency motion of the chassis has been significantly curtailed. Second, it is clear that "softening" the decoupler model also lowers the high frequency excitation of the system. In Figure 7, this same trend is observed where higher harmonic response to harmonic excitation is shown for the "kinematic" decoupler model and the "softened" decoupler model. This figure also raises questions with respect to how hydraulic mount performance is assessed. Clearly, it seems that ignoring the system response at frequencies other than the excitation frequency may be inappropriate. This issue is addressed in the next section.

decoupler model does predict less vibration transmission at higher frequencies.

Table 2. Vibratory Power Flow for Different Mount Models. Given SDOF Chassis

Power Quantity* (Watts)	rubber mount (linear system)	inertia track without decoupler	with inertia track and decoupler, $\Delta_d = 0.7$ mm	
			kinematic model	poly-stiffness model, $\gamma = 3$
Input primary harmonic only	0.398	0.244	0.341	0.347
Into Chassis primary harmonic	8.6×10^{-2}	3.55×10^{-2}	4.01×10^{-2}	3.89×10^{-2}
(% of TOTAL)	(100)	(99.9)	(98.5)	(99.5)
super-harmonics	0	2.65×10^{-5}	5.93×10^{-4}	2.08×10^{-4}
(% of TOTAL)	(0)	(0.1)	(1.5)	(0.5)
TOTAL	8.6×10^{-2}	3.55×10^{-2}	4.07×10^{-2}	3.91×10^{-2}
(% of Power input)	(21.6)	(14.5)	(11.9)	(11.3)

* Averaged over $3 < \omega/2\pi < 20$ Hz.

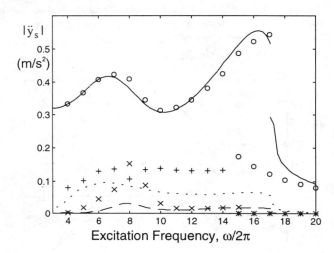

Figure 7. Mounting system frequency response. Chassis acceleration at the excitation frequency and its first two harmonics, $\ddot{y}_s$, for $\Delta_d = 0.7$ mm and $F_u(t) = 100\sin(\omega t)$ N. Key: Kinematic decoupler model; o o o 1st, x x x 2nd, + + + 3rd harmonics (numerical integration), polynomial-stiffness decoupler model with $\gamma = 3$; ——— 1st, —— —— 2nd, - - - - 3rd harmonics (Galerkin solution & numerical integration).

VIBRATORY POWER TRANSMISSION - Using equations (19-21), several vibratory power transmission variables are calculated and graphed in Figures 8 and 9 for different decoupler configurations and models. Power related quantities are also provided in Table 2. Directions of positive vibratory power transmission in Figure 10 show that the fluid component of the engine mount absorbs vibratory energy at the excitation frequency coming from both the engine and the chassis. However, it also acts as a source of vibratory energy at higher frequencies which is then transmitted and dissipated throughout the rest of the system. Vibratory power results support the conclusion based on motion descriptor analysis that the "softened" nonlinear

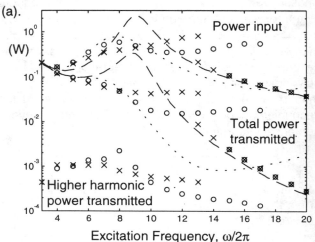

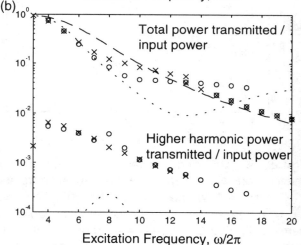

Figure 8. Mounting system frequency response. Vibratory power transmission for different decoupler configurations using the kinematic decoupler model (numerical integration). a. Power input, total and higher harmonic power transmission into chassis. b. Ratio of total and higher harmonic power transmission to power input. Here, $F_u(t) = 100\sin(\omega t)$ N. Key: —— —— rubber mount, - - - - inertia mount with no decoupler ($\Delta_d = 0$ mm), o o o o $\Delta_d = 0.7$ mm, x x x x $\Delta_d = 1.4$ mm.

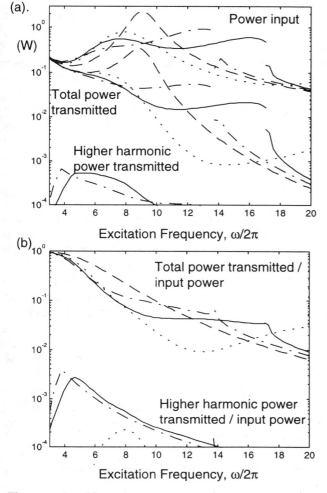

Figure 9. Mounting system frequency response. Vibratory power transmission for different decoupler configurations using the polynomial stiffness decoupler model with $\gamma = 3$ (Galerkin solution). a. Power input, total and higher harmonic power transmission into chassis. b. Ratio of total and higher harmonic power transmission to power input. Here, $F_u(t) = 100\sin(\omega t)$ N. Key: — — — rubber mount, - - - - inertia mount with no decoupler ($\Delta_d = 0$ mm), ———— $\Delta_d = 0.7$ mm, — · — $\Delta_d = 1.4$ mm.

Results also show that, on a quantitative basis, the contribution to the total vibratory energy transmission of the higher harmonic components is negligible with respect to the primary harmonic, supporting performance assessment methods based on the "low pass filter" assumption. This fact must be qualified by two remarks, however. First, the frequency of transmission in addition to its level is of interest. Results reported here indicate that, even for primary engine excitations below 20 Hz, structure-borne noise is transmitted through the mount in the audible frequency range. Even if the relative energy is low, its perceived level in terms of radiated sound in the passenger compartment may still be significant. Second, in this example case, the chassis model is a simplistic, SDOF model, with a 40 dB/decade attenuation in mobility above its resonance frequency near 1.36 Hz. Other base models with less frequency attenuation or high frequency resonant conditions may lead to different comparative conclusions. This is discussed in detail in the context of the source - nonlinear path - resonant receiver problem posed in reference [13].

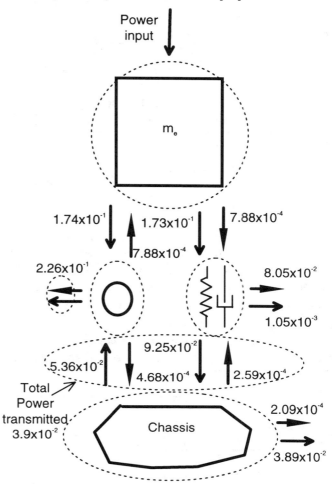

Figure 10. Mounting system schematic showing directions of positive vibratory power flow for the SDOF chassis model and the polynomial-stiffness decoupler model with $\gamma = 3$. Key: ⟶ primary harmonic, ⟶ higher harmonics.

COMPOSITE EXCITATION - In prior articles, such as Colgate et al. [2], the lower and higher excitation frequencies in composite studies, ω_l and ω_u, respectively, were selected to not be commensurate to maintain independence of phase. In this study, due to the constraints of the Galerkin Method, the two excitation frequencies are commensurate. Thus, there is a phase dependence, which the effects of are not addressed in this article. Use of a multi-base-frequency Galerkin formulation [14-15] to avoid phase dependence is left for future studies. The composite excitation is of the following form where $a^{U_1} = 100$ N and $a^{U_{10}} = 100$ N:

$$F_u(t) = a^{U_1}\sin(\omega t) + a^{U_{10}}\sin(10\omega t) \quad 3 < \omega/2\pi < 20 \text{ Hz.} \tag{22}$$

This is comparable to studies of composite displacement inputs to the mount. The higher frequency component produces an amplitude of relative motion across the mount that is on the order of one tenth of that of the lower frequency component.

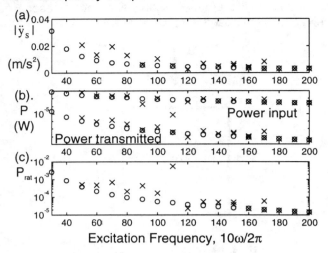

Figure 11. Mounting system frequency response to composite excitation. Vibratory power transmission into the SDOF chassis model (Kinematic decoupler model, numerical integration). Here, $F_u(t) = a^{U_1}\sin(\omega t) + 100\sin(10\omega t)$ N and $\Delta_d = 0.7$ mm. Key: o o o $a^{U_1} = 0$, x x x $a^{U_1} = 100$. a. Chassis vertical acceleration, $\ddot{y}_s$, at higher excitation frequency. b. Power transmission at higher excitation frequency. c. Power transmission ratio at higher excitation frequency.

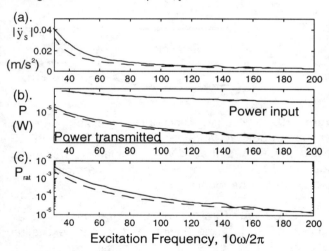

Figure 12. Mounting system frequency response to composite excitation. Vibratory power transmission into the SDOF chassis model (Polynomial stiffness decoupler model with $\gamma = 3$, Galerkin solution). Here, $F_u(t) = a^{U_1}\sin(\omega t) + 100\sin(10\omega t)$ N and $\Delta_d = 0.7$ mm. Key: o o o $a^{U_1} = 0$, x x x $a^{U_1} = 100$. a. Chassis vertical acceleration, $\ddot{y}_s$, at higher excitation frequency. b. Power transmission at higher excitation frequency. c. Power transmission ratio at higher excitation frequency.

In Figures 11 and 12 quantities related to vibratory power transmission at the higher excitation frequency, with and without the lower excitation frequency present, are shown for the SDOF chassis model and either the kinematic or polynomial decoupler models, respectively. From the graphical results, it is clear that in both cases the presence of the low frequency excitation component does alter system dynamics at the higher excitation frequency. The mount's ability to restrict structure-borne acoustic power transmission is degraded under the composite excitation condition. For the kinematic decoupler case, this performance degradation is more severe at some excitation frequencies.

CONCLUSION

This study has made a number of contributions to the modeling and design of decoupler-equipped hydraulic engine mounts. Several theoretical models were formulated focusing on the highly nonlinear decoupler dynamics. With the aid of a Galerkin-based computational method, mount performance was assessed based on total (multi-harmonic) vibratory power transmission under harmonic and composite (dual) harmonic excitation conditions. Several key findings are reported in this article, including the following:

• While modeling the decoupler with a softened nonlinear expression only moderately alters the predicted system behavior at the fundamental harmonic, it significantly alters it at higher harmonics.

• A jump phenomenon has been predicted, which is associated with the stiffness hardening effect of the decoupler.

• Use of the modified decoupler model has raised the issues of (a) whether or not a more softened nonlinearity actually produces better system response and (b) whether or not the actual hydraulic mount is better described by a more softened nonlinearity.

• With respect to mount performance, studies of multi-harmonic motion and power transmission show that it is negligible on a quantitative basis relative to the fundamental harmonic. Structure-borne noise, nonetheless is generated in the mount from sub-audio excitations due to nonlinearities and, even if its relative energy is low, its perceived level in terms of radiated sound in the passenger compartment may be significant.

• Studies of vibratory power transmission under composite excitation support prior observations in the literature that decoupler performance is degraded and that a softer decoupler nonlinearity may result in improved performance.

Clearly, further study is needed. Several future research issues have been identified. Experimental studies are necessary to determine the actual nature of the decoupler, including whether or not jump phenomena may actually occur in practice. Decouplers which exhibit weaker nonlinear behavior, such as rubber membranes, should also be tested for superior performance under composite excitation conditions. Experimental mobility data taken from actual automotive chassis's should be applied using the enhanced Galerkin method. For example, a conceptual study of the affect of multi-degree of freedom support structure dynamics on the performance of nonlinear mounting systems is offered in reference [13]. There, it is shown that a conventional SDOF base model predicts superior performance relative to an MDOF base model since the mobility is less for the SDOF case. For an MDOF base model, more significant levels of structure-borne noise are transmitted at higher harmonics of the excitation frequency, especially when these harmonics coincide with base resonant frequencies. Also, vibratory power transmission under composite excitation conditions with noncomensurate excitation frequencies should be investigated. Finally, alternatives to using a simple force source to represent engine imbalance should be considered in simulation studies.

REFERENCES

1   R. Singh, G. Kim, and P.V. Ravindra 1992. *Journal of Sound and Vibration* **158**, 219-43. Linear Analysis of Automotive Hydro-Mechanical Mount with Emphasis on Decoupler Characteristics.

2   J. E. Colgate, C.-T. Chang, Y.-C. Chiou, W. K. Liu and L. M. Keer 1995. *Journal of Sound and Vibration* **184**, 503-28. Modeling of a Hydraulic Engine Mount Focusing on Response to Sinusoidal and Composite Excitations.

3   T. Ushijima, K Takano and H. Kojima 1988. *Society of Automotive Engineers,* Technical Paper No. 880073. High Performance Hydraulic Mount for Improving Vehicle Noise and Vibration.

4   G. Kim and R. Singh 1993. *Transactions of the American Society of Mechanical Engineers, Journal of Dynamic Systems, Measurement, and Control* **115**, 482-7. Nonlinear Analysis of Automotive Hydraulic Engine Mount.

5   G. Kim and R. Singh 1992. *Proceedings of the third ASME Symposium on Transportation Systems,* Anaheim, CA **DSC-44**, 165-80. Resonance, Isolation and Shock Control Characteristics of Automotive Nonlinear Hydraulic Engine Mounts.

6   G. Kim and R. Singh 1995. *Journal of Sound and Vibration* **179**, 427-53. A Study of Passive and Adaptive Hydraulic Engine Mount Systems With Emphasis on Nonlinear Characteristics.

7   T. J. Royston and R. Singh, 1996. *Journal of Sound and Vibration* **194**, 243-263. Periodic Response of Mechanical Systems with Local Nonlinearities Using an Enhanced Galerkin Technique.

8   H. G. D. Goyder and R. G. White 1980. *Journal of Sound and Vibration* **68**, 59-75. Vibrational Power Flow From Machines into Built-Up Structures, Part I: Introduction and Approximate Analyses of Beam and Plate-Like Foundations.

9   H. G. D. Goyder and R. G. White 1980. *Journal of Sound and Vibration* **68**, 97-117. Vibrational Power Flow From Machines into Built-Up Structures, Part III: Power Flow Through Isolation Systems.

10   R.J. Pinnington and R.J. White 1981. *Journal of Sound and Vibration* **75**, 179-197. Power Flow Through Machine Isolators to Resonant and Non-Resonant Beams.

11   T. J. Royston and R. Singh 1996. *Journal of Sound and Vibration* **194**, 295-316. Optimization of Passive and Active Nonlinear Vibration Mounting Systems Based on Vibratory Power Transmission.

12   M. Urabe and A. Reiter 1966. *Journal of Mathematical Analysis and Applications* **14**, 107-40. Numerical Computation of Nonlinear Forced Oscillations by Galerkin's Procedure.

13   T. J. Royston and R. Singh, 1997. (in press) *Journal of the Acoustical Society of America.* Vibratory Power Flow Through a Nonlinear Path into A Resonant Receiver.

14   A. Ushida and L. O. Chua 1984. *IEEE Transactions on Circuits and Systems* **CAS-31**, 766-779. Frequency Domain Analysis of Nonlinear Circuits Driven by Multi-Tone Signals.

15   Lee, M.-r., C. Padmanabhan and R. Singh 1994. *Transactions of the American Society of Mechanical Engineers, Journal of Dynamic Systems, Measurement, and Control.* Dynamic Analysis of a Brushless D.C. Motor Using a Modified Harmonic Balance Method.

971937

Optimization of Vibration Isolators for Marine Diesel Engine Applications

Stanley R. Walker
Barry Controls, A Unit of Applied Power Inc.

John G. Foscolos, Jr.
Cummins Marine, A Division of Cummins Engine Corp.

Copyright 1997 Society of Automotive Engineers, Inc.

ABSTRACT

The objective of this paper is to examine the properties of vibration isolators that are used in marine diesel engine applications, develop a design that optimizes these properties to provide the best possible isolation performance, and perform testing in actual marine installations to verify the performance increase.

The properties that will be examined include mount stiffness in three orthogonal and three rotational axes, system vibrational mode shapes and deflections, fatigue life vs. isolation tradeoffs, resilient material optimization, and manufacturability.

INTRODUCTION

The paper will present a computer analysis to predict and optimize the free body six degree of freedom response of a typical marine diesel engine mounted on resilient isolators. When the optimum stiffnesses in each axis have been determined, it will examine the mount configurations required to provide these stiffnesses.

Finally, the paper will present the results of vibration testing of the optimized mounts in marine engine installations and compare measured performance with both computer predictions and the performance of other existing marine vibration isolators.

MAIN SECTION

To design an effective engine vibration isolator, we must first review the factors that determine the performance of any isolation system. The simplest representation of an isolation system is that of a single degree of freedom spring-mass system as shown in Figure 1. The isolator (spring) functions by accepting the motion of the mass, at the disturbing frequency

generated by the mass, and attempts to respond to it at the natural frequency of the spring-mass system.

If the spring-mass frequency is sufficiently lower than the disturbing frequency, as shown in Figure 2, then the force transmitted through the spring will be reduced. Keeping the isolator frequency less than 1/3 of the disturbing frequency will achieve almost 90% isolation, which is usually sufficient in most engine isolation applications.

The two main sources of the disturbing vibration produced by a marine internal combustion engine are the torsional dynamic pulses produced during cylinder firing and the unbalanced forces caused by reciprocating pistons or rotating crankshaft and rod masses. With a 4 cycle engine, the firing frequency can be calculated as:

f_d = RPM x number of cylinders /2
(1 power stroke per 2 RPM)

(**Equation 1**. 4 cycle engine firing frequency)

Unbalanced forces are usually negligible in engines with more than 6 cylinders, but in engines of 4 cylinders or less, they can be a significant source of vibration, and must be addressed in isolation system design.

Also essential to the correct design of the isolators is the location of the principle axis of inertia. Figure 3 gives the translational and rotational modes of an engine; x, y, and z are the principle axes of inertia, so rotation about these axes exhibits the least moment of inertia. The engine will tend to rotate about these axes whenever it is disturbed, since it takes the course of least resistance and additional energy would be required to move in any other axes

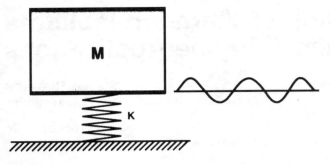

(**Figure 1**. Single degree of freedom system with base excitation)

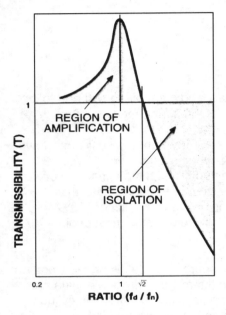

(**Figure 2**. Transmissibility curve for an isolated system. fd = forcing frequency, fn = natural frequency)

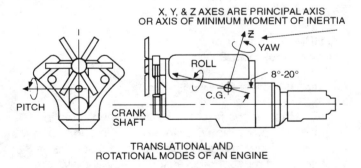

(**Figure 3**. Translational and rotational modes of a typical engine)

Effective engine isolation depends on the relationship of the isolators to the principle axis of inertia, <u>not</u> the crankshaft centerline. The design of the isolators should provide minimum restraint in rotation about the principle inertia axis. Cylinder firing disturbances (firing frequency) make the engine vibrate about the principle roll axis of inertia (x axis in Figure 3). This is the axis of least inertia, and torque impulses are most disturbing; therefore the softest mode of vibration (stiffness) of the isolation system should be in the rotational (torsional) mode.

If a mount is to isolate vibration, it must be able to deflect. And the more it moves or deflects, generally, the better the isolation. But if the support structure in the boat lacks sufficient stiffness, the relative motion across the isolator may be inadequate. This would result in reduced isolation efficiency provided by the mounting system. A general rule is that the support structure stiffness should be at least ten times the isolator stiffness. This will guarantee that 90% or more of the total system deflection occurs in the isolator, and excessive deflections and possible fatigue of the structure will be eliminated.

The isolators in a marine propulsion system are subjected to a number of dynamic and static loading conditions. First, they must support the static weights of the engine, transmission, and auxiliary equipment. Second, they must react the dynamic forces caused by wave slap, cornering loads, and docking impacts. Third, unlike in vehicular installations, the isolators must often react the full thrust load of the drive propellers acting along the prop shaft. And finally, they must isolate all the vibration produced by the engine, as discussed in the previous paragraphs. As we shall see, optimizing the isolator's performance involves a tradeoff in characteristics to produce the best overall performance in any application.

The ideal marine isolation system should be as soft as practical in the roll axis to provide the maximum degree of isolation of engine vibration, but still stiff enough to react the torque produced by the engine at maximum throttle without excessive windup, or excessive lateral motion which would cause problems when stuffing boxes are used.

Practically speaking, this usually results in a rotational natural frequency that must be less than 20 Hz for effective isolation at engine idle, with 10 Hz being an ideal lower limit for maximum isolation without excessive roll axis deflections. Along the axis of prop thrust, the isolator should be as stiff as possible, since there is little engine vibration generated in this direction and limitation of fore-aft engine motion caused by thrust is the main concern. This usually results in a translational minimum frequency above 10 hz, but should be reviewed for each application based on the maximum thrust generated by the particular engine. As a rule of thumb, maximum thrust (in pounds) is usually around 15 times the horsepower rating of the engine.

In existing practice, the most common type of isolator used to mount 4, 6, and 8 cylinder gas or diesel engines in marine applications uses an elastomeric element to provide the desired stiffnesses for acceptable isolation of all engine vibrational modes.

Due to the nature of elastomers, an element loaded in shear will generally be 1/5 the stiffness, or less, of the same element loaded in compression. Because of this effect, it is desirable to design elastomeric isolators with their compression axis in line with the thrust direction of the propeller, to react the high expected loads with minimal deflection, and the shear axes in line with the direction of engine roll, to provide a low stiffness and good isolation of the torque pulsations produced by the primary firing frequency of the engine An example of this type of isolator is shown in figure 4.

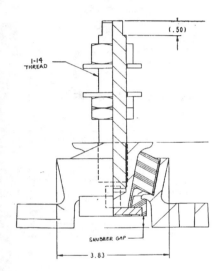

(**Figure 4**. Elastomer isolator optimized for use in marine engine applications)

When a typical 4 cylinder diesel engine, as shown in fig.5, is mounted on this type of isolator with appropriate elastomer stiffness, the system natural frequencies can be determined by the use of a six degree of freedom analysis computer program. These programs are generally available as part of many commercially available dynamic analysis software programs. The results using SHVIB, a proprietary analysis program developed by Barry Controls, are as follows: (See Appendix 1 for a typical SHVIB run, showing both input and output formats)

MODE	X trans.	Y trans.	Z trans.
f_n (HZ)	10.601	5.969	8.368
f_n (RPM)	636	358	502

MODE	X rot. (roll)	Y rot. (pitch)	Z rot. (yaw)
f_n (HZ)	14.818	25.241	23.100
f_n (RPM)	889	1514	1386

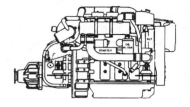 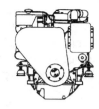

(**Figure 5**. Typical marine diesel engine)

Since the firing frequency of a 4 cycle, 4 cylinder engine is 4 x RPM/2, (from Equation 1) then at an idle speed of 700 RPM, the primary disturbing frequency would be 1,400 RPM. The equation for calculating the transmissibility of an isolation system (neglecting damping) is:

$$T = [\ 1\ /\ [1-(f_d/f_n)^2\ ]^2\ ]^{.5}$$

(**Equation 2**. Transmissibility of isolation system)

Therefore this system will provide 20% isolation of the engine firing frequency (1400 RPM) at 700 RPM idle, and 90% isolation by 2700 RPM, while maximizing the thrust axis stiffness to minimize fore-aft deflections, and providing sufficiently high stiffnesses in all other directions to minimize engine motion under transient dynamic inputs.. Also, since the isolators are designed to be much softer in the side to side direction, they are more likely to be closer to the recommended 10% of the stiffness of the support structure, an important consideration in today's trend toward lighter weight boat construction.

Once the desired mount stiffnesses have been determined by analysis, the proper size mount can be determined by examining the elastomer size, shape, and durometer required to meet the stiffness, deflection, and fatigue requirements of the installation. To meet the stiffnesses determined by the analysis, the best indicator of required mount size is the isolator shear pad dimensions. The equation for the stiffness of an isolator in shear is:

$$K_{shear} = (G_s\ A)\ /\ T$$

(**Equation 3**. Stiffness of isolator in shear)

where: G_s = elastomer shear modulus (usually between 50 and 250), A = elastomer pad area and T = elastomer pad thickness. When this equation is combined with the requirement that the maximum compressive stress in the elastomer pad due to thrust loading be kept below 500 psi. (35.15 kg/cm^2), the pad dimensions can be iteratively determined to meet both requirements, and the relative mount size can be determined.

Generally, at this point the last check that must be made to finalize the mount design is to determine the fatigue life of the elastomer pad under the expected life cycle deflections in the marine application. The fatigue life of an elastomer pad subjected to periodic strains can be roughly predicted by the following power law equation: $N = (K/E)^b$

(**Equation 4**. Fatigue life of elastomer pad subjected to periodic strains)

where: N = number of cycles to failure, K = constant (depending on elastomer material, usually around 1,000), E = % shear strain on each cycle and b = constant, depending upon elastomer material (usually between 4 and 5). Once the predicted fatigue life is compared to the required life in the application and the mount size increased, if necessary, the design can be finalized and prototypes can be made and tested to insure the performance is as predicted.

It is at this stage that optimization for manufacturability must be performed. Care should be taken that the design incorporates an elastomer configuration that can be efficiently molded and assembled into support housings. Support castings and forgings are preferred over weldments, to reduce costs and maximize corrosion protection. Ideally, the mount assembly should be done by automated equipment to minimize labor. Testing has been conducted on optimized isolators in various pleasure boat applications between 32 and 39 feet of length, mounting Cummins Marine diesel engines with horsepowers in the 150-420 horsepower range. Typical vibration results are shown in figures 6, 7 & 8.

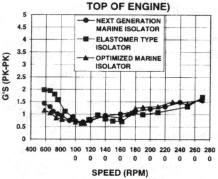

(**Figure 7**. A comparison of engine vibration using the three types of vibration isolators. At 600-1000 RPM it shows improved engine vibration with the optimized marine isolator compared to the elastomer style)

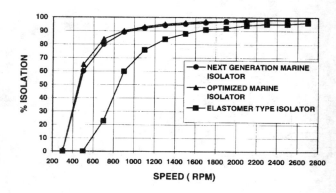

(**Figure 8**. A comparison of three types of vibration isolators. The percent isolation was calculated using computer software for determining the proper engine mounting systems. The results were based on the different X,Y and Z stiffnesses of each isolator along with the proper location of the mounting systems)

CONCLUSION

Since real time fatigue testing can take many years, accelerated laboratory fatigue testing has been conducted to verify the predicted fatigue life of the optimized isolators. A pair of isolators was subjected to 5 million cycles of sinusoidal fatigue loading to a peak force of 4,500 pounds (2,045 kilograms) at a frequency of 10 Hz. This input represented the cumulative fatigue damage expected in 10 years of actual marine use. At the conclusion of the test, there was no change in the mount's stiffness or frequency, and only slight surface abrasion was visible on the mount. The results to date have all shown that when properly analyzed and designed, isolation mounts for marine applications can be produced that optimize both vibration isolation and fatigue life, in a cost effective and compact space envelope.

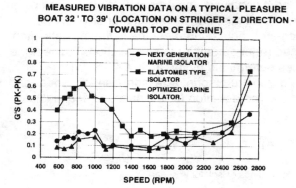

(**Figure 6**. A comparison of 3 types of vibration isolators on a typical 6 cylinder engine. At 600-1400 RPM it clearly shows the large difference in engine vibration between the optimized marine mount and the elastomer type mount that is transmitted into the hull)

APPENDIX 1. SHVIB COMPUTER ANALYSIS

PROGRAM OMEGA: NATURAL FREQUENCIES/MODE SHAPE

DATE: 18 JUL 1996   **TIME**: 03:43 PM

TITLE : CUMMINS MARINE 6CTA8.3 WITH 27391-4 SEAMOUNTS (65 DURO)

WEIGHT (LBS) = 1820.000

	X	Y	Z
CG LOCATION (IN) =	23.330	13.000	8.680
MOMENT OF INERTIA (LBS-IN-SEC**2) =	270.000	660.000	570.000

EULER ANGLES OF MASS (DEG)

	PHI	THETA	PSI
	0.000	0.000	0.000

OUTPUT IN MASS COORDINATES: ISOLATORS

ISOL NO.	TYPE NO.	EULER ANG. NO.	COORDINATES (IN)		
			X	Y	Z
1	1	1	0.0000	0.0000	0.0000
2	1	1	42.5200	0.0000	0.0000
3	1	1	42.5200	26.0000	0.0000
4	1	1	0.0000	26.0000	0.0000

ISOLATOR STIFFNESS TYPES

TYPE	STIFFNESS (LBS/IN)		
	X	Y	Z
1	7250.000	2250.000	2250.000

EULER ANGLES

NUMBER	PHI	THEA	PSI (Degrees)
1	0.00000	0.00000	0.00000

NATURAL FREQUENCIES

MODE	1 (Fore-Aft)	2 (Lateral)	3 (Vertical)	4 (Roll)	5 (Pitch)	(Yaw)	6
FREQ(HZ)	8.796	5.544	6.883	14.931	17.913	20.032	
(RPM)	527.730	332.662	412.989	895.846	1074.763	1201.911	

MODE SHAPE X,Y,Z (IN)   ROT X,ROT Y,ROT Z (RADIANS)

X	1.000000	0.000000	-0.121205	0.000000	-1.000000	0.000000
Y	0.000000	1.000000	0.000000	1.000000	0.000000	-0.393719
Z	0.200098	0.000000	1.000000	0.000000	0.044830	0.000000
ROT X	0.000000	-0.041620	0.000000	0.421160	0.000000	-0.092613
ROT Y	0.058001	0.000000	-0.009717	0.000000	0.122066	0.000000
ROT Z	0.000000	0.001431	0.000000	0.021733	0.000000	1.000000

APPENDIX 2. GUIDELINES FOR MARINE ENGINE ISOLATION USING
SIX DEGREE OF FREEDOM ANALYSIS

1. Modes 1 & 2 should be approximately 10 Hertz of less for good translational isolation. Frequencies lower than 10 Hertz will give better isolation, but because of the greater associated motions, a larger isolator is required to avoid fatigue problems.

2. Mode 4 (roll frequency) should be below 15 Hertz for good isolation of the primary cylinder firing torque frequency. (See #4 below) A roll frequency lower than 10 Hertz will give excellent isolation, but care must be taken to react the maximum engine torque windup deflections in the isolator (usually with auxiliary snubber).

3. Compare Mode 5 (pitch) to Mode 1 (vertical translation). The pitch frequency should be at least 1.5 times higher than the translation frequency to avoid mode coupling and excessive engine deflections.

4. Examine the Mode Shapes of 4, 5 & 6. Ideally, there will be little coupling between the modes (i.e. the mode shape motion will predominate in one direction with relatively little associated motion in the other 2 translational directions). If excessive coupling occurs, then even with a low roll frequency, the engine firing frequency may excite a pitch or yaw frequency that could transmit excessive vibration. If the system is highly coupled, then modes 4, 5 and 6 should all be below 15 Hertz.

NATURAL FREQUENCIES

MODE	1 (Fore-Aft)	2 (Lateral)	3 (Vertical)	4 (Roll)	5 (Pitch)	5. (Yaw)	6
FREQ(HZ)	8.796	5.544	6.883	14.931	17.913	20.032	
(RPM)	527.730	332.662	412.989	895.846	1074.763	1201.911	

MODE SHAPE X,Y,Z (IN)   ROT X,ROT Y,ROT Z (RADIANS)

X	1.000000	0.000000	-0.121205	0.000000	-1.000000	0.000000
Y	0.000000	1.000000	0.000000	1.000000	0.000000	-0.393719
Z	0.200098	0.000000	1.000000	0.000000	0.044830	0.000000
ROT X	0.000000	-0.041620	0.000000	0.421160	0.000000	-0.092613
ROT Y	0.058001	0.000000	-0.009717	0.000000	0.122066	0.000000
ROT Z	0.000000	0.001431	0.000000	0.021733	0.000000	1.000000

971938

The Evolution of the 1997 Chevrolet Corvette Powertrain Mounting System

Brad Saxman, Brian Deutschel, Keith Weishuhn, and Kenneth Brown
General Motors Corp.

Copyright 1997 Society of Automotive Engineers, Inc.

ABSTRACT

Benchmarking of competitive vehicles and lessons learned from development of the 4th generation (C4) Chevrolet Corvette provided the starting point for the design of the 1997 5th generation (C5) Chevrolet Corvette powertrain mounting system. Synthesis methods were used to evaluate initial powertrain mounting systems, while development and analysis methods were utilized to refine the selected system in successive generations of prototype vehicles. The 1997 Chevrolet Corvette powertrain mounting design is seen to be a marked improvement over the previous generation.

INTRODUCTION

The 1997 Chevrolet Corvette's unique body structure, powertrain configuration, and all new aluminum engine presented many challenges and opportunities to develop an improved powertrain mounting system. The Corvette team used extensive Voice of the Customer research to identify "well built" as a key product characteristic. Noise, vibration and harshness of the vehicle are critical factors in the customers impression of "well built". The powertrain's mounting system and structural properties greatly influence the noise, vibration and harshness of the vehicle. This paper will describe the evolution of the Corvette powertrain mounting system from the initial concepts to the development of the final production powertrain mounting configuration.

BENCHMARKING

The C5 Corvette Team measured several C4 vehicles and benchmarked numerous competitive vehicles. This assessment, along with Voice of the Customer input was combined to identify key noise and vibration issues that were used to optimize the powertrain mounting system These issues included:

Powertrain Isolation
- Driveline Imbalance
- Axle Whine
- Idle Shake
Ride Character
- Rough Road Shake
- Impact Isolation

(ref: definition section)

Design requirements were established in an effort to meet or exceed customer expectations in these areas.

POWERTRAIN MOUNTING

The fourth generation Corvette (C4) had inadequate isolation between the powertrain and body due to the very stiff bushings at the rear of the powertrain that attached the axle beam to the body. Body and powertrain attachment mobility's at these locations were also inadequate. The C4 vehicle layout (Figure 1) required the axle beam bushings to serve two functions: isolate driveline energy from the body and provide high lateral stiffness between the suspension and body. These opposing requirements limited their ability to provide adequate powertrain isolation. This configuration, coupled with relatively low vehicle bending and torsion modes, created noise and vibration issues which the Corvette team struggled with since 1984.

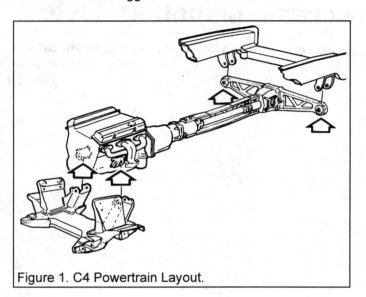

Figure 1. C4 Powertrain Layout.

The layout of the 1997 Chevrolet Corvette (Figure 2) driveline and suspension does not require the powertrain mounts to serve this dual function. Handling loads are managed by the suspension system, leaving powertrain and road inputs to be managed by the powertrain mounting system.

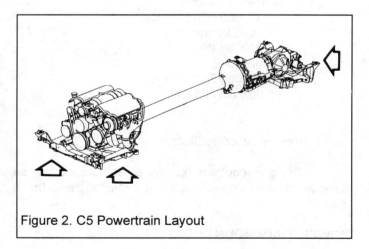

Figure 2. C5 Powertrain Layout

POWERTRAIN STRUCTURE SPECIFICATIONS

Early in the program the transmission was moved to the rear of the vehicle, just forward of the differential. This was done to provide better vehicle weight distribution, additional occupant footwell space, and provide packaging space needed to meet vehicle structure requirements. This powertrain configuration has low structural modes due to it's distributed lumped masses at the front and rear of the vehicle(Figure 2).

The vehicle mode map(Figure 3) was used to set the initial targets for the primary modes of the powertrain structure. The objective was to minimize interaction with vehicle structural modes and powertrain excitation at idle.

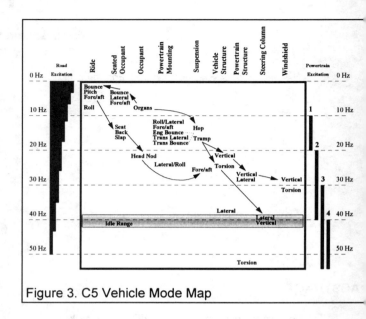

Figure 3. C5 Vehicle Mode Map

To minimize the interaction between the powertrain structure and engine combustion forces at idle, the lowest flexural modes of the powertrain were targeted below engine fourth order excitation. The targeted first lateral and vertical bending frequencies were 26 & 28 Hz respectively. The first torsion mode of the powertrain was targeted for 60 Hz which placed it above fourth order excitation.

Normal modes analysis of this powertrain configuration, with a packageable beam section connecting the engine and transmission, indicated that the frequency of the lowest structural mode would be below 30 Hz. Analysis also indicated nodal lines at the engine and transmission/differential C.G. locations.

The C5 body structure[1] was also a completely new design, developed simultaneously with the powertrain system. The body had a 1st bending target of 21 Hz with nodal lines at the front and rear shock towers, and a 1st torsion target of 23 Hz with nodal lines through the tunnel and at the front edge of the seats. Decoupling the powertrain structure from combustion forces at idle left little separation (approximately 3 Hz) between the primary bending modes of the powertrain and body structure. This small separation made placement of the powertrain mounts at the powertrain structural bending nodal lines critical to the overall performance of the mounting system.

The front nodal lines between the structure and powertrain lined up very well and set the fore/aft location of the engine mounts. In the rear, there was also good agreement between the body and powertrain nodal lines. However, due to design constraints with the carryover transmission, a location forward or rearward of the transmission case was required. The front location was preferred due to its closer proximity to the true nodal position.

POWERTRAIN RIGID BODY MODES SPECIFICATION

Using the vehicle system model, powertrain mount rates were optimized to minimize the vehicles response to road and driveline excitation. The model indicated that the powertrain's front and rear lumped masses could be tuned independently to improve road shake and impact response.

Placement of the rigid body roll mode was dictated by the need to decouple engine excitation (1, 1.5, and 2nd order) from exciting the powertrain roll mode at idle. It was also required that the primary powertrain rigid body modes: engine bounce, transmission bounce, and engine roll be greater than 90 % decoupled.

POWERTRAIN MOUNT SPECIFICATION

Mount Rate Specification- To adequately isolate the engine and yet control the system during road and powertrain inputs, targets were established for stiffness and damping under various mount displacements. High dynamic stiffness was required for rough road powertrain control while low dynamic stiffness was desired for powertrain isolation. In addition a maximum dynamic stiffness requirement of 2450 N/mm was established to address high frequency isolation in the 30 to 800 Hz range. It was evident that these specifications could only be met with the use of hydraulic mounts.

Attachment Structure Specification - With the maximum mount stiffness for high frequency isolation defined, both body attachment stiffness and powertrain attachment stiffness targets could be defined. The target was to maintain at least a 20 dB mismatch between the mount mobility and attachment structure input mobility levels. This corresponded to an attachment stiffness of ten times the mount stiffness from 30 to 800 Hz.

Mount Location- The ability to obtain an input mobility of ten times the maximum mount rate at the mount attachments depends upon the location of the mounts. Mounts need to be located where short, stout powertrain brackets and body structure could be implemented that would achieve the mobility targets. Mounts must also be located near the powertrain structure primary bending nodal lines to minimize excitation of structural modes.

POWERTRAIN MOUNTING SELECTION PROCESS

Table 1 defines the powertrain mounting systems considered for the 1997 Chevrolet Corvette. These mounting systems were evaluated against the vehicle and powertrain mounting specifications previously discussed. Noise and vibration performance of each system was evaluated using finite element lumped parameter models. Systems 1,2,3, and 4 were eliminated due to their multiple shortcomings of additional crossmembers and mounts. The vehicle program had neither the mass, money, or packaging

Table 1: Powertrain Mounting Proposals

System	Number of Engine Mounts	Number of Engine Sub-frames	Number of Transmission Mounts	Number of Transmission Sub-Frames	Layers of Transmission Isolation	Powertrain Structure
1	3	2	3	2	1	None
2	4	1	3	2	1	None
3	3	2	4	1	1	None
4	4	1	4	1	1	None
5	2	1	2	1	1	C-Section
6	2	1	2-Low Front	1	1	Torque Tube
7	2	1	2-Low Rear	1	1	Torque Tube
8	2	1	2-High	0	1	Torque Tube
9	2	1	1-Low Front	1	1	Torque Tube
10	2	1	1-Low Rear	1	1	Torque Tube
11	2	1	2-High	0	2	Torque Tube
12	2	1	2-Low Front	1	2	Torque Tube
13	2	1	2-Low Rear	1	2	Torque Tube

room for these additional components.

The powertrain systems remaining utilized either a C-beam or torque tube design. Both had nearly identical packaging requirements for mounts and sub-frames. However, the systems differed in how they transferred powertrain loads to the body.

The C-beam powertrain structure reacts drive torque through the powertrain mounts. The reaction loads are low due to the fore/aft span between the transmission mount and engine mounts. However, because of the low torsional stiffness of the C-beam, engine torque was also reacted through the powertrain mounts. These engine torque loads resulted in substantial mount forces which would excite the vehicle structure.

The torque tube powertrain structure also reacts drive torque through the powertrain mounts, similar to the C-beam structure. The engine torque, however, is managed within the powertrain structure and not reacted through the powertrain mounts. Managing engine torque within the powertrain structure also aided in the handling of the vehicle. The body structure is not being twisted under hard acceleration, making left and right hand turns symmetric.

For the above reasons the C-beam concept, system 5, was dropped, leaving only proposals with a torque tube powertrain structure to be considered. System 8 was dropped due to is low assembly rating.

Detailed finite element system modeling of both rough road and powertrain excitation confirmed that the front transmission mount location was optimal. It was more important for the mount to be near the powertrain bending node than the body's bending mode. Analysis also indicated that a single transmission mount was desirable because it minimized the transfer of torsional powertrain vibration to the body. Thus, system 9 was identified as the best powertrain mounting system.

At this stage in the program, adequate high frequency isolation performance at the rear of the driveline was an open issue. The optimized rate with the single conventional elastomeric mount at the transaxle was high. This high rate was of concern for isolation of driveline imbalance and axle whine. The system model could not address these concerns, so the double isolation system (11) was retained as a viable option to be evaluated in the mule vehicles.

Packaging Compromises to the Powertrain Mounting System - While analysis showed that a low front transmission mount location was desired from a noise and vibration performance perspective the rear suspension ride spring, in a last minute change, occupied the same space. After many packaging studies, the conflict between the mount and rear suspension spring could not be resolved. Therefore, all mounting systems utilizing the front lower position for the transmission mount were abandoned(6,9,12).

The four remaining systems (7,10,11 and 13) each having the transmission mount at a less than optimal location, were analyzed using the vehicle system model. Mount rates were optimized to minimize the vehicle response for the following conditions:

- Idle Shake
- Rough Road Shake Control
- Powertrain Imbalance
- Powertrain Isolation

VEHICLE DEVELOPMENT PROCESS

With the three major sub-systems, body, powertrain and chassis, now having a mainstream design, it was time to build hardware to confirm the analytical results that had lead the design to this point. The first in the hardware succession were two mule vehicles. These vehicles were followed by successive phases of prototype vehicles, each more refined and complete in content.

With hardware for evaluation existing, subjective and experimental test methods were utilized to select the final mounting system from the four remaining mounting proposals (7,10,11,13). Utilizing the mule vehicles, each mounting system was evaluated with the optimized mounting rates for the key noise and vibration characteristics.

MULE VEHICLES

The mule vehicles, though originally built to confirm the structural integrity of the body, were also used to evaluate the remaining powertrain mounting systems. The major powertrain issues to be resolved on the mule vehicles included:

- Single versus double isolation of the transaxle
- Three powertrain mounts versus four mounts

The need for a double isolation system at the transaxle was cause for considerable debate among the members of the Corvette team. The primary purpose of the double isolated system was elimination of axle whine on the C5 Corvette, a known weakness of the C4 Corvette. Due to the level of mule vehicle content, no roof, evaluations of double and single isolation systems were inconclusive. This fact, coupled with the Corvette's mass, cost and packaging constraints eliminated the double isolated four mount systems (11,13) from further consideration.

With the double isolation issue decided, the debate over three versus four mount systems took center stage. The major concerns with the three mount system were powertrain motion control, axle whine and the potential for significant idle lumpiness due to it's low powertrain rigid body roll mode. The obvious advantages were the mass, cost and packaging improvements the three mount system provided.

To aid in the decision process a risk assessment chart was developed comparing the three and four mount systems (Table 2). This chart enabled the team to evaluate the advantages and disadvantages of each system and define the relative importance given to each requirement.

Table 2: Risk Assessment

Requirement	Importance	3 Mount	4 Mount
Idle Isolation			
Instability	6	-	+
4th Order	5	+	+
Axle Whine	8	-	-
Road Response			
Impact	8	Fore/Aft + Vert/Lat -	Fore/Aft - Vert/Lat +
Steady State	8	Fore/Aft + Vert/Lat -	Fore/Aft - Vert/Lat +
Powertrain	8	Fore/Aft + Vert/Lat -	Fore/Aft - Vert/Lat +
Cost	6	+	-
Mass	9	+	-

Upon evaluation of the risk assessment chart, the powertrain integration team decided that the concerns with the three mount system could be managed, when compared against the cost and mass penalties. Therefore, the three mount system (10) was recommended as the mainstream design.

The next task was to determine the sensitivity of the vehicle to driveline imbalance. This testing identified the automatic transmission vehicles to have a high level of sensitivity to driveline imbalance. The sensitivity of the automatic vehicles was a dramatic departure from the C4 vehicles, in which manual transmission vehicles were inherently worse for balance due to the effect of the dual mass flywheel.

The location of the torque converter in the rear of the C5 provided two potentially significant sources of powertrain imbalance: engine and torque converter. The torque converter imbalance produced significant response levels due to the excitation of 2nd order vertical and lateral bending modes of the powertrain. A commitment from the Powertrain Division to meet the engine balance specification shifted the focus from engine balance to torque converter balance. Internal changes to the torque converter, in an effort to improve balance, became a priority.

The mule vehicles were also used to assess the potential to achieve mount attachment input mobility targets for the remaining powertrain mounting systems. This parameter was tracked continuously through each vehicle iteration to insure that the mobility targets were achievable.

Idle shake was also evaluated but little was learned due to non-representative powertrain hardware and control system. Road response data on the mule vehicles showed that they were close to target PSD levels.

PROTO I VEHICLES

The Proto I vehicles were the first vehicles representing the production intent architecture and driveline hardware. They included front and rear cast aluminum sub-frame, and powertrain mounts at the mainstream locations. The powertrain mounting consisted of three mounts; two hydraulic engine mounts and a conventional elastomer mount located rearward of the transaxle. Assessment of the Proto I vehicles generated the following list of concerns:

- Inadequate mobility at mount attachments on the front & rear cradle
- Mount optimization(Rate, Damping, Travel)
- Inadequate rough road shake control
- Objectionable powertrain periods

The Proto I body side powertrain mount attachment mobility levels were significantly below the target. Proposals for improvement were analyzed and hardware was built to improve the cradle mobility.

High priority was also placed on optimization of the mounts from a rate, damping and travel perspective, to insure the mainstream mounting system would achieve our goals. Mount displacement data was used to optimize mount tuning. Mount displacements were evaluated under numerous vehicle test conditions with and without internal mount stops.

During the process of measuring mount displacements and numerous subjective vehicle rides, objectionable rough road shake was observed. A series of operating deformed shapes and additional displacement measurements showed rough road memory shake to be tied to transaxle bounce. Evaluation of several styles of mounts indicated that the target level of control and isolation could only be achieved with a hydraulic mount. Incorporation of a decoupler provided significant improvement to powertrain isolation and course road noise. This configuration became mainstream: two hydraulic engine mounts and one hydraulic transaxle mount.

The Proto I vehicles also provided the first chance to assess axle whine performance with axles from the chosen supplier. A correlation study between differential vibration and interior noise was initiated. Transmissibility across the transaxle mount was assessed and determined to be substantially below target levels. Lower rate mounts were evaluated to improve transmissibility.

PROTO II & III VEHICLES

The arrival of Proto II and Proto III vehicles allowed the continuation of the refinement process. The isolation characteristics of the powertrain were again assessed including mount location mobility, axle whine, road response and interior noise performance.

Assessment of the Proto II vehicles generated the following list of concerns:

- Objectionable buzz in vehicle just above idle
- Inadequate powertrain isolation
- Inadequate rough road shake control

The objectionable buzz was determined to be the 1st torsion mode of the powertrain coupling with an acoustic cavity mode at approximately 60 Hz. A reduction in the torque tube wall thickness decoupled the powertrain torsion mode from the cavity mode. This torque tube change combined with a tuned absorber mounted on the axle, eliminated the noise complaint.

The Proto II level vehicles incorporated updated sub-frame designs which provided adequate input mobility levels. Mount rates were further refined to optimize powertrain isolation and rough road shake. These changes produced transmissibility levels across the powertrain mounts which met or exceeded target levels.

The addition of the transaxle hydraulic mount required an interface bracket between the mount and the differential. The bracket design was enhanced until the 20 dB mobility separation requirement was achieved.

PROTO III VALIDATION

These vehicles were used to confirm the production intent design. Modal analysis, mobility testing, idle shake and axle whine testing was conducted. The final mount tuning was also completed.

Figures 4 through 7 highlight the success of the 1997 Chevrolet Corvette powertrain mounting system in achieving targeted levels for rough road shake, idle shake, attachment mobility and axle whine.

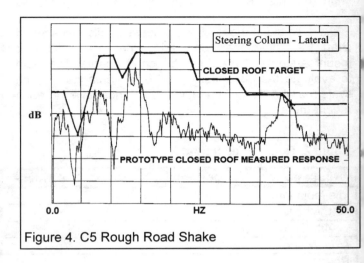

Figure 4. C5 Rough Road Shake

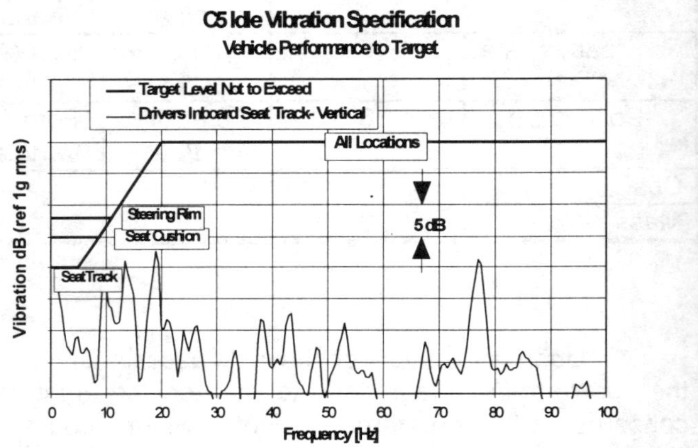

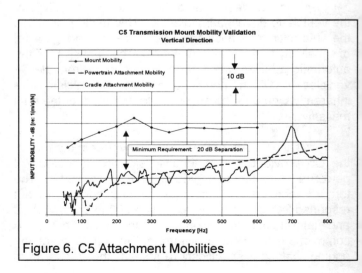

Figure 5. C5 Idle Shake

Figure 6. C5 Attachment Mobilities

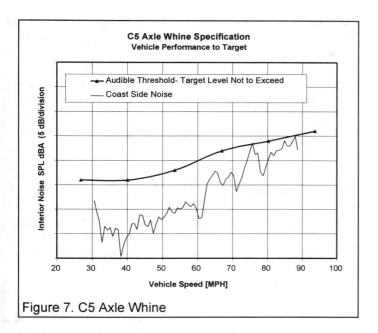

Figure 7. C5 Axle Whine

CONCLUSIONS

The performance of the 1997 Chevrolet Corvette demonstrates a substantial improvement in road shake control and powertrain isolation compared to the C4. This was accomplished by continuously tracking performance against customer based vehicle and subsystem specifications throughout the development of the vehicle. The final result being a 1997 Corvette that not only surpasses the C4, but also exceeds customer expectations.

DEFINITIONS

Driveline Imbalance - High priority was placed on reducing the vehicle response to imbalance of the driveline rotating components. Inherently, the general layout of the driveline results in bending and torsional resonant frequencies which will couple with powertrain and driveline imbalance forces. For this reason stringent imbalance limitations were specified for all powertrain and driveline components.

To lower vehicle sensitivity, powertrain mounts were initially located at the driveline bending nodal locations to minimize energy transfer into the body. Mount stiffness requirements at engine first order frequencies were used to closely monitor isolation performance.

Axle Whine - Axle whine is a tonal noise generated by the meshing energy of the differential hypoid gear set, and is predominantly a structureborne noise source into the vehicle. The requirement for the C5 program was to have no audible axle whine in the vehicle. Thus the target level for axle whine inside the vehicle is the audible threshold.

Masking of tones with broadband noise using critical bandwidth criteria was used to generate the audible threshold. Specifications for source energy, mount isolation, and vehicle acoustic sensitivity at the mounts were rolled down from the target level. The frequency range of interest for axle whine is 300-800 Hz, so high priority was placed on meeting mount isolation and body acoustic sensitivity targets in this range.

Idle Shake - Vehicles with a pleasing engine idle response require good isolation of engine firing and inertial energy to reduce idle buzz in the seat and steering wheel. They also require good control of low frequency random inputs generated by combustion variation to minimize "lumpiness". This primarily occurs when the engine is not loaded at idle, with the worst case being manual transmission vehicles.

Many vehicle functional areas must be balanced to provide a pleasing response during engine idle. These include full vehicle, steering column and driveline dynamic properties. Powertrain mount locations, orientation, rate, damping are also important. The powertrain combustion process also plays a primary roll in minimizing a "lumpy" idle.

Vehicle level targets were specified for idle performance which attempted to capture both idle buzz at firing frequencies and idle "lumpiness" at lower frequencies. The profile of the target generated from benchmarking several vehicles is shown in Figure 5.

Rough Road Shake - Targets for vibration performance at the occupant interfaces were established by subjective and objective benchmarking of competitive vehicles. This produced target response levels at the steering column and seat track using acceleration power spectral density (PSD) curves. This target setting process indicated reductions up to 15 dB would be required at some locations to achieve the program goals. This level of improvement required the integration of the critical vehicle subsystems, including body, powertrain, and chassis, to meet targeted levels. No single subsystem could be optimized to achieve the target. Therefore, rough road shake performance ranked high on the overall attribute list for all candidate mounting configurations in consideration during the initial design stages of the program.

All powertrain mounting systems were evaluated against these PSD target curves using the vehicle system model to predict results. These specifications were also used to set the response limits for analytical simulations and mount rate optimization.

Mobility - The driveline isolation goal for C5 was transformed to input mobility requirements. Mobility is a measure of dynamic stiffness in terms of velocity per unit force. The requirement is to maintain at least 20 dB of separation between the mobility level of the mounts and the mobility level of the attaching body side and powertrain side structure. Refer to Figure 6. The 20 dB mobility separation requirement is specified for a frequency range from 30 to 800 Hz. This is the primary range of structureborne energy through the powertrain mounts into the vehicle. Using this technique to achieve an isolation target allows independent component specifications to be allocated for mounts, body structure, and powertrain side mounting brackets. The transmissibility (ε) across powertrain mounts with a targeted 20 dB mobility separation is -16 dB using the equation below:

$$\varepsilon = \frac{\left| M_{body} + M_{powertraain} \right|}{\left| M_{body} + M_{powertrain} + M_{isolator} \right|} \qquad [2]$$

where: M_{body} = Body Attachment Input Mobility
$M_{powertrain}$ = Powertrain Attachment Input Mobility
$M_{isolator}$ = Mount Dynamic Mobility

ACRONYMS, ABBREVIATIONS

C5: Fifth generation Chevrolet Corvette beginning with the 1997 model year.

C4: Fourth generation Chevrolet Corvette model years 1984 through 1996

Hz: Hertz, cycles per second

FEA: Finite Element Analysis

CAE: Computer Aided Engineering

PSD: Power Spectral Density

dB: Decibel

ACKNOWLEDGMENTS

As discussed throughout this paper, the development of the powertrain mounting system was a multi-disciplinary process requiring the work of many of our colleagues. Our team consisted of many disciplines that worked together with the single goal of making the 1997 Chevrolet Corvette the best ever.

The vehicle platform engineering team recognized the importance of powertrain mounting on both powertrain isolation and road isolation. The mounting system was always kept a high priority throughout the vehicle program. The vehicle engineering team made use of both analytical and experimental techniques as an integral part of the design process which greatly contributed to the evolution of the powertrain mounting system.

Structural analysts provided stiffness and mobility analysis that enabled us to study many alternatives at the component level. Many ideas came from the analysis work of Mary Dona, Laurie Decker and Paul Juras.

Experimental testing was provided by Kyle Tucker and George Foltz.

We would also like to thank both Steve Wolfe and Dennis Schwerzler for sharing their years of powertrain mounting experience and expertise with us.

REFERENCE

1.   S. D. Longo, E. D. Moss and B. W. Deutschel, "The 1997 Chevrolet Corvette Structure Architecture Synthesis", SAE Paper 970089,1997

2.   Beranek, L.L.,"Noise and Vibration Control", Chapter 13, pg. 414-417

971940
Dynamic Simulation of Engine-Mount Systems

Chung-Ha Suh and Clifford G. Smith
University of Colorado-Boulder

Copyright 1997 Society of Automotive Engineers, Inc.

ABSTRACT

This paper presents a simplified method to determine the vibrational amplitude produced by a four-cylinder engine when supported by a viscoelastic foundation. The pistons, connecting rods, and crankshaft are modeled as rigid bodies connected to the foundation by standard industrial rubber mounts which are modeled with spring and damping elements whose location, orientation, and stiffness can be easily configured for specific design analysis, or optimized to reduce vibration and noise in a design. The paper is presented in a general algorithmic form, with a detailed numerical example emphasizing the three parts of the analysis: the mechanism kinematics, the shaking force and torque determination on the engine block, and the three-dimensional vibrational analysis.

INTRODUCTION

A major emphasis in the design of automotive engines, as with other high-speed mechanisms, is limiting the unbalanced loads transmitted from the mechanism to its supports, which can lead to irreparable damage and/or harsh environmental conditions. The recent trend in automotive design of decreasing the mass of the engine components to achieve greater fuel economy only exacerbates the situation. Using viscoelastic mounts to isolate the mechanism from its foundation is a solution that has found wide acceptance in an abundance of applications in the automotive, aerospace and computer industries, and in nature in biomechanical systems.

This paper presents the use of the displacement matrix method [1][2] in determining the vibrational amplitude caused by unbalanced dynamic forces being transmitted from a four-cylinder engine supported by a viscoelastic foundation. The viscoelastic mounts are modeled by massless springs and dampers with three degrees of translational resistance, and the pistons, connecting rods, crankshaft, and engine block are modeled as rigid bodies. The three parts of the analysis -- the kinematic analysis of the engine mechanism, the shaking force and torque determination on the engine block, and the three-dimensional vibrational analysis are described in detail, and numerical results are presented for a specific case.

BASIC ENGINE SIMULATION PROBLEM DEFINITION

The following is a typical problem one could face in analyzing the vibrational amplitude of an engine:

For the given data for an inline four-cylinder two-cycle engine (see Figs. 1 and 2), find the displacement, velocity, and acceleration of all components of the engine mechanism, find the components of the shaking force and torque acting on the engine block, and finally apply the shaking force and torque to the engine block to determine the vibrational amplitude.

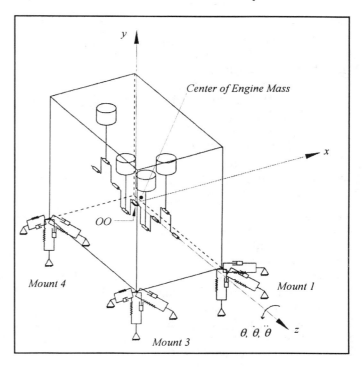

Figure 1. Four-cylinder engine-mount system.

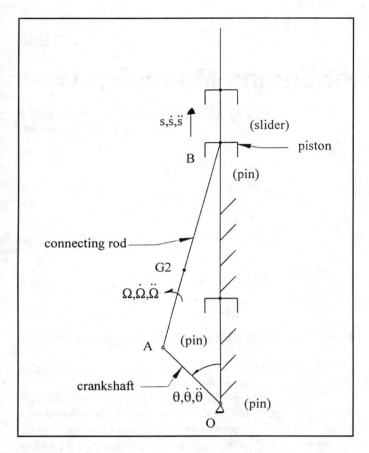

Figure 2. One cylinder of engine modeled as slider-crank mechanism.

Given Data:
- Starting angular velocity of the crankshaft is 100.0 rpm
- Engine firing order is 1-3-4-2
  Initial crankshaft angle for
    Cylinder 1: $\theta_1 = 360°$
    Cylinder 2: $\theta_2 = \theta_2 - 540°$
    Cylinder 3: $\theta_3 = \theta_1 - 180°$
    Cylinder 4: $\theta_4 = \theta_1 - 360°$
- The resisting torque on the crankshaft MR is given by
  $MR = CA + (CB + CC*\dot\theta) * \dot\theta$

where CA = 11.0526819, CB = 0.0072757, CC = 0.0000609, and $\dot\theta$ is the angular velocity of the crankshaft in rad/sec.
Simulate the engine from startup at time t = 0.0 sec to 4.0 sec in increments of 0.01 sec.

Input Data of Combustion
CR = Compression ratio = 9.5
AR = Admission ratio (AR = 1.0 for Otto Cycle) = 1.0
KC = Polytropic Exponent for compression = 1.32
KE = Polytropic Exponent for Expansion = 1.32

Input Data of Engine Specification (in SI units)
R = crank radius = 0.041 m
L = connecting rod length = 0.131 m
A = distance from crank pin to mass center of connecting rod = 0.028069 m
C = distance from crank axis to mass center of crank = 0.0 m
JC = moment of inertia crank assembly about its c.g. (center of gravity) = 0.003585326 kg*m²
JG = moment of inertia of connecting rod about its c.g. = 0.0016035 kg*m²

m1 = mass of crank assembly = 1.94264 kg
m2 = mass of connecting rod = 0.518841 kg
m3 = mass of piston (including piston pin) = 0.3260 kg
AP = area of piston = 0.0045 m²
DZ = distance from front of block to no. 1 piston center = 0.0555 m
DP = distance between piston centers (bore pitch) = 0.082 m

Input Data of Different Pressures at Different Angular Velocities
(Pressure units are in Pascals)
Table values:
PM1 = 2939076
PEX1 = 103986   at RPM(1) = 1000 rpm
PIN1 = 79461

PM2 = 4275198
PEX2 = 182956.5   at RPM(2) = 3000 rpm
PIN2 = 52974

PM3 = 5363127
PEX3 = 180013.5   at RPM(3) = 4000 rpm
PIN3 = 50031

PM4 = 5424930
PEX4 = 187371   at RPM(4) = 5500 rpm
PIN4 = 39240

PM5 = 1139235.3
PEX5 = 98100   at RPM(5) = 6800 rpm
PIN5 = 98100

Where: PM = Maximum Pressure, PEX = Exhaust Pressure, PIN = Intake Pressure,
PA = Atmospheric Pressure = 101352.9297

The pressure data for a particular angular velocity between the tabulated values is given by:

For $\dot\theta < RPM(1)$, $P = P(1)$
For $RPM(i) < \dot\theta < RPM(i+1)$

$$P = P(i) + [\dot\theta - RPM(i)]\frac{P(i+1) - P(i)}{RPM(i+1) - RPM(i)}, \quad i=1,2,3,4$$

For $\dot\theta > RPM(5)$, $P = P(5)$

where i is the index value of the tabulated angular velocities, P is PM, PEX or PIN, and the $\dot\theta$ is in units of rpm.

Additional Crankshaft Information
The crankshaft is modeled as being attached to the engine block at point O, the origin of the block local coordinate system, and the crankshaft axis coincides with the local z axis of the block (see Fig. 1). In this reference frame, the intersection point of the cylinder and crankshaft axes are:
Crank 1: (0,0,0.123), Crank 2: (0,0,0.041), Crank 3: (0,0,-0.041), Crank 4: (0,0,-0.123).

Engine Block Data
Assumption: The origin of the principal axes xyz is located at the center of mass, and the principal axes coincide with the local

coordinate system used for the block, as shown in Fig. 1.

M = mass of engine block = 280 kg

I_x, I_y, I_z = principal moment of inertia components of engine block = 8.60, 15.73, 14.41 kg*m^2

$P(P_x, P_y, P_z)$ = center of mass of engine block in unsprung (zero gravity) position = (0.01, 0.02, 0.02)

α, β, γ = orientation of block in terms of traditional Euler angles in unsprung position = 0, 0, 0

Engine Mount Data

As can be seen from Fig 1., there are four rubber mounts -- one each at the bottom corners of the engine block. The rubber mounts are modeled with springs and dampers attached from the foundation to the moving engine block, and each mount provides three degrees of translational resistance. Tables 1-4 give the mount specifications for the unsprung position of the engine block.

In the tables, SK is the spring constant (N/m), CD is the damping constant (N*s/m), and OL is the original length of the spring (m).

The kinematic analysis of the engine will now be described.

ENGINE KINEMATICS

As described in Paul [3], the distance d is the length of an idealized right circular cylinder of area AP which has a volume equal to that of the gas trapped between the cylinder head and the piston crown. The three values of d_1, d_2 and d_5, where the subscript indicates the state of the four-stroke cycle, are significant to the analysis of the problem.

$d_1 = (2R)/(CR-1)$ at top dead center (t.d.c.)
$d_2 = d_1 + 2R$ at bottom dead center (b.d.c.)
$d_5 = AR*d_1$ at combustion point

and in general

$d = d_1 + L + R - s$

where s is the piston's position from the crankshaft axis (see Fig. 2).

The kinematics will be analyzed using the reduction of system method described in Suh and Radcliffe [1]. The reduced torque for the engine assembly is

$$T^* = MP - MR \qquad (1)$$

where MP is the sum of turning moments of the gas force for each cylinder given by the equation

$$MP = -PF*\dot{s} \qquad (2)$$

where PF is the piston force (P-PA)AP and $\dot{s}$ is the piston velocity. MR is the resisting torque on the crankshaft and the appropriate equation was given in the problem definition.

The fundamental angle is defined for each crank as

$$\theta_f = \theta - 4\pi \cdot INT\left\{\frac{\theta}{4\pi}\right\}$$

and the pressure on each piston at a particular fundamental angle is found from the set of equations:

if $0.0 < \theta_F \le \pi$ then P = PIN

if $\pi < \theta_f < 2\pi$ then $P = PIN\left(\dfrac{d_2}{d}\right)^{KC}$

if $\theta_F = 2\pi$ then $P = \dfrac{\left[PIN(CR)^{KC} + PM\right]}{2}$

if $2\pi < \theta_F$ and $d \le d_5$ then P = PM

if $2\pi < \theta_F$ and $d > d_5$ then $P = PM\left(\dfrac{d_5}{d}\right)^{KE}$

if $\theta_F = 3\pi$ then $P = \dfrac{\left[PM\left(\dfrac{d_5}{d}\right)^{KE} + PEX\right]}{2}$

if $\theta_F > 3\pi$ then P = PEX
if $\theta_F = 0.0$ or $\theta = 4\pi$ P = PA

Reducing the engine assembly to a shaft rotating about its longitudinal axis with velocity $\dot{\phi}^*$ and moment of inertia I^* results in

$$\frac{1}{2}I^*(\dot{\phi}^*)^2 = \frac{1}{2}\sum_{i=1}^{4}\left[m_2\dot{G2}_i^2 + JG\dot{\Omega}_i^2 + m_3\dot{s}_i^2\right]$$
$$+ \frac{1}{2}JC*\dot{\theta}^2 \qquad (3)$$

where G2 is the mass center of the connecting rod. Now, since either the inertia I^* or the angular velocity of the reduced system may be set arbitrarily, it is convenient, in this case, to set the angular velocity $\dot{\phi}^*$ equal to the angular velocity of the crankshaft $\dot{\theta}$. Setting $\dot{\phi}^* = \dot{\theta}$ in Eq. (3), the reduced inertia I^* can be rewritten as

$$I^* = \sum_{i=1}^{4}\left[m_2\frac{\dot{G2}_i^2}{\dot{\theta}^2} + JG\frac{\dot{\Omega}_i^2}{\dot{\theta}^2} + m_3\frac{\dot{s}_i^2}{\dot{\theta}^2}\right] + JC \qquad (4)$$

The differential equation of motion is now

$$\ddot{\theta} = \frac{T^* - \dfrac{1}{2}\dfrac{dI}{d\theta}\dot{\theta}^2}{I^*} \qquad (5)$$

and $dI^*/d\theta$ is calculate numerically as

$$\frac{dI^*}{d\theta} = \frac{I^*(\theta + \Delta\theta) - I^*(\theta)}{\Delta\theta} \qquad (with\ \Delta\theta = 0.0001)$$

To numerically determine the terms in Eqs. (1), (2), (4) and (5), the kinematic displacement and velocity analysis must be performed. The values of $\dot{G2}$, $\dot{\Omega}$, and $\dot{s}$, and the coordinates of the displaced point G2 need to be found. The following analysis is valid for each cylinder when θ is the crank angle for the appropriate cylinder.

First, the new position of point A is found as (see Fig. 2)

AX = AX0*cosθ-AY0*sinθ
AY = AX0*sinθ+AY0*cosθ

where AX0 = 0.0, AY0 = R are the initial coordinates of the crankpin.

Table 1. Mount 1 specifications.					
Element	SK	CD	Block Point	Ground Point	OL
1	70000	2000	(0.1,-0.1,0.18)	(0.11,-0.1,0.18)	0.01
2	253667	5000	(0.1,-0.1,0.18)	(0.1,-0.11,0.18)	0.01
3	70000	2000	(0.1,-0.1,0.18)	(0.1,-0.1,0.19)	0.01

Table 2. Mount 2 specifications.					
Element	SK	CD	Block Point	Ground Point	OL
1	70000	2000	(0.1,-0.1,-0.18)	(0.11,-0.1,-0.18)	0.01
2	253667	5000	(0.1,-0.1,-0.18)	(0.1,-0.11,-0.18)	0.01
3	70000	2000	(0.1,-0.1,-0.18)	(0.1,-0.1,-0.19)	0.01

Table 3. Mount 3 specifications.					
Element	SK	CD	Block Point	Ground Point	OL
1	70000	2000	(-0.1,-0.1,0.18)	(-0.11,-0.1,0.18)	0.01
2	253667	5000	(-0.1,-0.1,0.18)	(-0.1,-0.11,0.18)	0.01
3	70000	2000	(-0.1,-0.1,0.18)	(-0.1,-0.1,0.19)	0.01

Table 4. Mount 4 specifications.					
Element	SK	CD	Block Point	Ground Point	OL
1	70000	2000	(-0.1,-0.1,-0.18)	(-0.11,-0.1,-0.18)	0.01
2	253667	5000	(-0.1,-0.1,-0.18)	(-0.1,-0.11,-0.18)	0.01
3	70000	2000	(-0.1,-0.1,-0.18)	(-0.1,-0.1,-0.19)	0.01

Then point B, in terms of angle Ω of the connecting rod (which is unknown), at the connecting rod side is

$$\begin{bmatrix} BX1 \\ BY1 \\ 1 \end{bmatrix} = \begin{bmatrix} \cos\Omega & -\sin\Omega & AX - AX0\cos\Omega + AY0\sin\Omega \\ \sin\Omega & \cos\Omega & AY - AY0\cos\Omega - AX0\sin\Omega \\ 0 & 0 & 1 \end{bmatrix} \begin{bmatrix} BX0 \\ BY0 \\ 1 \end{bmatrix}$$

where BX0 = 0.0, BY0 = R+L are the initial coordinates of the piston.

One the other hand, point B at the piston side is

$$BX1' = 0.0$$
$$BY1' = s \quad (unknown)$$

To solve for the two unknown values of s and Ω, the two equations BX1=BX1' and BY1=BY1' are solved simultaneously. After Ω and s are found, the displacement matrix of the connecting rod can be operated on point G2 to get the displaced coordinates

$$\begin{bmatrix} G2X \\ G2Y \\ 1 \end{bmatrix} = \begin{bmatrix} \cos\Omega & -\sin\Omega & AX - AY0\cos\Omega - AX0\sin\Omega \\ \sin\Omega & \cos\Omega & AY - AY0\cos\Omega - AX0\sin\Omega \\ 0 & 0 & 1 \end{bmatrix} \begin{bmatrix} G2X0 \\ G2Y0 \\ 1 \end{bmatrix}$$

The velocity matrix of the connecting rod now needs to be found in order to find the angular velocity of the connecting rod $\dot{\Omega}$ and the piston velocity $\dot{s}$. The velocity of point A is found as

$$\dot{AX} = -AY\dot{\theta}$$
$$\dot{AY} = AX\dot{\theta}$$

Then forming the velocity matrix of the connecting rod in terms of the unknown $\dot{\Omega}$ and operating on point B, the velocity of point B is obtained in terms of $\dot{\Omega}$ as

$$\begin{bmatrix} \dot{BX} \\ \dot{BY} \\ 0 \end{bmatrix} = \begin{bmatrix} 0 & -\dot{\Omega} & \dot{AX} + AY\dot{\Omega} \\ \dot{\Omega} & 0 & \dot{AY} - AX\dot{\Omega} \\ 0 & 0 & 0 \end{bmatrix} \begin{bmatrix} BX \\ BY \\ 1 \end{bmatrix}$$

One the other hand, point B's velocity at the piston side is

$$\dot{BX}' = 0.0$$
$$\dot{BY}' = \dot{s}$$

The two unknowns $\dot{\Omega}$ and $\dot{s}$ can be found by solving the two equations

$$\dot{BX} = \dot{BX}'$$
$$\dot{BY} = \dot{BY}'$$

After $\dot{\Omega}$ and $\dot{s}$ are found the values of $(\dot{G2X}, \dot{G2Y})$ are found by operating the velocity matrix of the connecting rod on point G2

$$\begin{bmatrix} \dot{G2X} \\ \dot{G2Y} \\ 0 \end{bmatrix} = \begin{bmatrix} 0 & -\dot{\Omega} & \dot{AX} + AY\dot{\Omega} \\ \dot{\Omega} & 0 & \dot{AY} - AX\dot{\Omega} \\ 0 & 0 & 0 \end{bmatrix} \begin{bmatrix} G2X \\ G2Y \\ 1 \end{bmatrix}$$

Now, Eq. (4) for the reduced inertia and Eq. (1) for the reduced torque are numerically determined and these values may be substituted into Eq. (5), the equation of motion for the reduced system or crankshaft. By numerically integrating this equation twice, the angular velocity $\dot{\theta}$ and angular displacement θ are found for the crankshaft at the next time point.

Knowing the current displacement θ, velocity $\dot{\theta}$, and acceleration $\ddot{\theta}$ of the crankshaft, the acceleration analysis of the mechanism may now be performed. The acceleration matrix of the connecting rod needs to be found in order to find the angular acceleration of the connecting rod $\ddot{\Omega}$, and the piston's acceleration $\ddot{s}$. The acceleration of point A is found as

$$\ddot{AX} = -AX\dot{\theta}^2 - AY\ddot{\theta}$$
$$\ddot{AY} = AX\ddot{\theta} - AY\dot{\theta}^2$$

Then forming the acceleration matrix of the connecting rod in terms of the unknown $\ddot{\Omega}$ and operating on point B, the acceleration of point B is obtained as

$$\begin{bmatrix} \ddot{BX} \\ \ddot{BY} \\ 0 \end{bmatrix} = \begin{bmatrix} -\dot{\Omega}^2 & -\ddot{\Omega} & \ddot{AX} + AX\dot{\Omega}^2 + AY\ddot{\Omega} \\ \ddot{\Omega} & -\dot{\Omega}^2 & \ddot{AY} + AY\dot{\Omega}^2 - AX\ddot{\Omega} \\ 0 & 0 & 0 \end{bmatrix} \begin{bmatrix} BX \\ BY \\ 1 \end{bmatrix}$$

One the other hand, point B's acceleration at the piston side is

$$\ddot{BX}' = 0.0$$
$$\ddot{BY}' = \ddot{s}$$

The two unknowns $\ddot{\Omega}$ and $\ddot{s}$ can be found by solving the two equations:

$$\ddot{BX} = \ddot{BX}'$$
$$\ddot{BY} = \ddot{BY}'$$

After $\ddot{\Omega}$ and $\ddot{s}$ are found the values of $(\ddot{G2X}, \ddot{G2Y})$ are found by operating the acceleration matrix of the connecting rod on point G2

$$\begin{bmatrix} \ddot{G2X} \\ \ddot{G2Y} \\ 0 \end{bmatrix} = \begin{bmatrix} -\dot{\Omega}^2 & -\ddot{\Omega} & \ddot{AX} + AX\dot{\Omega}^2 + AY\ddot{\Omega} \\ \ddot{\Omega} & -\dot{\Omega}^2 & \ddot{AY} + AY\dot{\Omega}^2 - AX\ddot{\Omega} \\ 0 & 0 & 0 \end{bmatrix} \begin{bmatrix} G2X \\ G2Y \\ 1 \end{bmatrix}$$

This completes the kinematic analysis of the engine mechanism.

SHAKING FORCE AND TORQUE DETERMINATION

Figure 3 shows the joint reaction forces and moments for a single cylinder of the engine. T_f is the friction torque on the crankshaft, and N is the normal force exerted on the piston by the cylinder wall. The net force components (Fx',Fy') exerted by the engine on the frame at point O and the torque Tz' on the frame (about the z axis of the engine block) are (see Fig. 4)

$$Fx' = N - F_{ox} \tag{6}$$
$$Fy' = PF - F_{oy} \tag{7}$$
$$Tz' = -N*s \tag{8}$$

where a prime superscript indicates the xyz local coordinate system of the engine block. Through straightforward application of Newton's equations of motion, these components may be

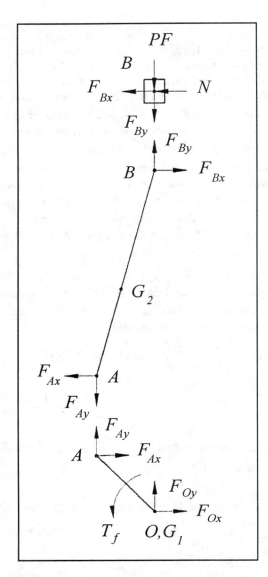

Figure 3. Joint reaction forces for one cylinder of the engine mechanism.

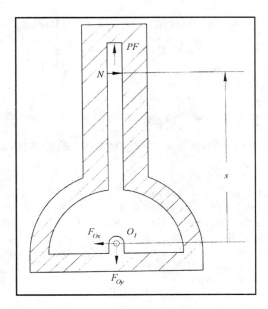

Figure 4. Shaking forces acting on the engine block.

determined. Now considering the net forces and moments of all four cylinders acting on the engine at point OO, the equations are (see Fig. 5)

$$FFx' = \sum_{i=1}^{4} Fx_i' \qquad FFy' = \sum_{i=1}^{4} Fy_i' \qquad (9,10)$$

$$TTx' = -\sum_{i=1}^{4} z_{Oi} Fy_i' \qquad TTy' = \sum_{i=1}^{4} z_{Oi} Fx_i' \qquad (11,12,13)$$

$$TTz' = \sum_{i=1}^{4} Tz_i'$$

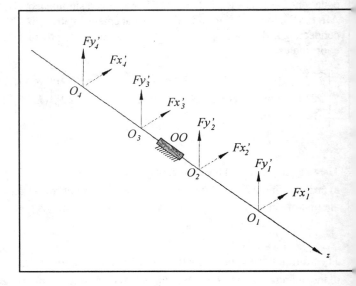

Figure 5. Components of the shaking forces.

THREE-DIMENSIONAL VIBRATIONAL ANALYSIS

When a mechanism is connected to a foundation by means of a rubber mount the degrees of freedom between the mechanism as a whole and the foundation is the maximum of six, or it is kinematically unconstrained, and correspondingly the number of differential equations involved is also the maximum of six. In the foregoing section, a general yet elementary approach to formulating these differential equations is presented.

Three-Dimensional One-Body Elasto-Dynamic System

Figure 6 illustrates the system discussed in this section. A rigid body A having mass M, principal axes xyz, and moments of inertia about the principal axes xyz of (I_x, I_y, I_z), is connected to a fixed base with massless springs and dampers. The base is fixed with respect to the global axes XYZ.

Assumptions and Developments

Assumptions:

(1) The origin of the principal axes (or moving axes) xyz is located at the center of mass of the rigid body A.
(2) The attachments of the spring-dampers to the rigid body and the base are ideal, frictionless ball and socket joints.
(3) The spring-dampers are massless.

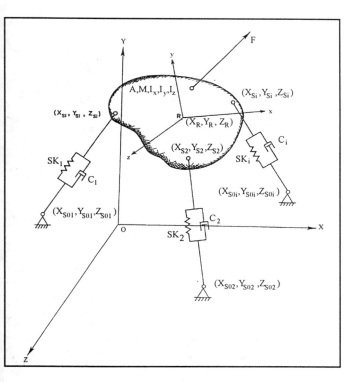

Figure 6. Notations used for three dimensional elasto-dynamic system.

Developments:
(1) Displacement matrix
Any displacement $[D]_{12}$ of a rigid body from position 1 to position 2 can be replicated by the following two displacements: (Refer to Fig. 7)
(i) Displacement $[D]_{10}$ such that the rigid body at position 1 $(G_1x_1y_1z_1)$ is displaced to the position 0 (OXYZ)
(ii) Displacement $[D]_{02}$ such that the rigid body at position 0 (OXYZ) is displaced to the position 2 $(G_2x_2y_2z_2)$ which is the present position.

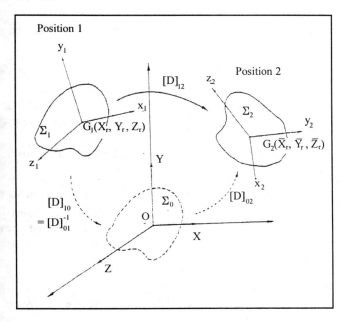

Figure 7. Displacement sequences.

Combining these two displacements results in the total displacement matrix

$$[D]_{12} = [D]_{02} [D]_{10} \quad (14)$$

or

$$[D]_{12} = [D]_{02} [D]_{01}^{-1} \quad (15)$$

Also,

$$[D]_{01} = [D_T]_{01} [D_R]_{01} \quad (16)$$

$$[D]_{02} = [D_T]_{02} [D_R]_{02} \quad (17)$$

where

$$[D_R]_{01} = \begin{bmatrix} & & & | & 0 \\ & [R_{\alpha_0 \beta_0 \gamma_0}] & & | & 0 \\ & & & | & 0 \\ -- & -- & -- & | & -- \\ 0 & 0 & 0 & | & 1 \end{bmatrix}$$

$$[D_R]_{02} = \begin{bmatrix} & & & | & 0 \\ & [R_{\alpha \beta \gamma}] & & | & 0 \\ & & & | & 0 \\ -- & -- & -- & | & -- \\ 0 & 0 & 0 & | & 1 \end{bmatrix}$$

Here, the (3x3) rotation matrix $[R_{\alpha\beta\gamma}]$ is an Euler angle matrix and includes the following three rotations:
The first is a rotation through an angle α about the Z axis, followed by a second rotation through an angle β about the displaced Y axis, and then a final rotation through an angle γ about the displaced X axis. The angles α, β, and γ are used as traditional Euler angles as defined in Greenwood [4]. The order of rotations lead to the following matrix equations

$$\begin{bmatrix} \bar{x}_1 \\ \bar{y}_1 \\ \bar{z}_1 \end{bmatrix} = \begin{bmatrix} \cos\alpha & \sin\alpha & 0 \\ -\sin\alpha & \cos\alpha & 0 \\ 0 & 0 & 1 \end{bmatrix} \begin{bmatrix} \bar{x}_0 \\ \bar{y}_0 \\ \bar{z}_0 \end{bmatrix} \quad (18)$$

$$\begin{bmatrix} \bar{x}_2 \\ \bar{y}_2 \\ \bar{z}_2 \end{bmatrix} = \begin{bmatrix} \cos\beta & 0 & -\sin\beta \\ 0 & 1 & 0 \\ \sin\beta & 0 & \cos\beta \end{bmatrix} \begin{bmatrix} \bar{x}_1 \\ \bar{y}_1 \\ \bar{z}_1 \end{bmatrix} \quad (19)$$

$$\begin{bmatrix} \overline{x_3} \\ \overline{y_3} \\ \overline{z_3} \end{bmatrix} = \begin{bmatrix} 1 & 0 & 0 \\ 0 & \cos\gamma & \sin\gamma \\ 0 & -\sin\gamma & \cos\gamma \end{bmatrix} \begin{bmatrix} \overline{x_2} \\ \overline{y_2} \\ \overline{z_2} \end{bmatrix} \qquad (20)$$

Subsequently,

$$\begin{bmatrix} \overline{x_3} \\ \overline{y_3} \\ \overline{z_3} \end{bmatrix} = \begin{bmatrix} \cos\alpha & -\sin\alpha & 0 \\ \sin\alpha & \cos\alpha & 0 \\ 0 & 0 & 1 \end{bmatrix} \begin{bmatrix} \cos\beta & 0 & \sin\beta \\ 0 & 1 & 0 \\ -\sin\beta & 0 & \cos\beta \end{bmatrix} \begin{bmatrix} 1 & 0 & 0 \\ 0 & \cos\gamma & -\sin\gamma \\ 0 & \sin\gamma & \cos\gamma \end{bmatrix} \begin{bmatrix} \overline{x_0} \\ \overline{y_0} \\ \overline{z_0} \end{bmatrix}$$

$$= \begin{bmatrix} \cos\alpha\cos\beta & -\sin\alpha\cos\gamma +\cos\alpha\sin\beta\sin\gamma \\ \sin\alpha\cos\beta & \cos\alpha\cos\gamma +\sin\alpha\sin\beta\sin\gamma \\ -\sin\beta & \cos\beta\sin\gamma \end{bmatrix} \qquad (21)$$

$$\begin{bmatrix} \sin\alpha\sin\gamma +\cos\alpha\sin\beta\cos\gamma \\ -\cos\alpha\sin\gamma +\sin\alpha\sin\beta\cos\gamma \\ \cos\beta\cos\gamma \end{bmatrix} \begin{bmatrix} \overline{x_0} \\ \overline{y_0} \\ \overline{z_0} \end{bmatrix}$$

The rotation matrix $[R_{\alpha\beta\gamma}]$ can now be defined from the equation

$$\begin{bmatrix} \overline{x_3} \\ \overline{y_3} \\ \overline{z_3} \end{bmatrix} = [R_{\alpha\beta\gamma}]^{-1} \begin{bmatrix} \overline{x_0} \\ \overline{y_0} \\ \overline{z_0} \end{bmatrix} \qquad (22)$$

$[D]_{01}$ and $[D]_{02}$ are now obtained from Eqs. (16), (17), (21) and (22) as

$$[D]_{01} = \begin{bmatrix} d_{11} & d_{12} & d_{13} & | & X_r \\ d_{21} & d_{22} & d_{23} & | & Y_r \\ d_{31} & d_{32} & d_{33} & | & Z_r \\ ---- & ---- & ---- & ---- & ---- \\ 0 & 0 & 0 & | & 1 \end{bmatrix} \qquad (23)$$

where

$d_{11} = \cos\alpha_0 \cos\beta_0$
$d_{12} = \sin\alpha_0 \cos\gamma_0 + \cos\alpha_0 \sin\beta_0 \sin\gamma_0$, etc. as shown in Eq. (21)

and

$$[D]_{02} = \begin{bmatrix} a_{11} & a_{12} & a_{13} & | & \overline{X_r} \\ a_{21} & a_{22} & a_{23} & | & \overline{Y_r} \\ a_{31} & a_{32} & a_{33} & | & \overline{Z_r} \\ ---- & ---- & ---- & ---- & ---- \\ 0 & 0 & 0 & | & 1 \end{bmatrix} \qquad (2\bullet$$

where

$a_{11} = \cos\alpha \cos\beta$

$a_{12} = \sin\alpha \cos\gamma + \cos\alpha \sin\beta \sin\gamma$, etc.
      as shown in Eq. (21)

VELOCITY MATRIX - Angular velocity $\overline{\omega}(\omega_x,\omega_y,\omega_z)$ defined about the xyz axes of the moving reference frame. Th relationship between the angular velocity $\overline{\omega}(\omega_x,\omega_y,\omega_z)$ and th Euler angular velocity components $(\dot\alpha,\dot\beta,\dot\gamma)$ is [5]

$$\begin{bmatrix} \dot\alpha \\ \dot\beta \\ \dot\gamma \end{bmatrix} = \begin{bmatrix} 0 & \dfrac{\sin\gamma}{\cos\beta} & \dfrac{\cos\gamma}{\cos\beta} \\ 0 & \cos\gamma & -\sin\gamma \\ 1 & \tan\beta\sin\gamma & \tan\beta\cos\gamma \end{bmatrix} \begin{bmatrix} \omega_x \\ \omega_y \\ \omega_z \end{bmatrix} \qquad (2\bullet$$

NOMENCLATURE - Capital letters are used for th expressions which refer to the fixed (ground) XYZ coordina system, while lower case letters refer to the moving xyz system X_{si},Y_{si},Z_{si} the given coordinates of the connecting points to th rigid body (point si) of the i-th spring-damper with respect to th XYZ frame

$\overline{X_{si}},\overline{Y_{si}},\overline{Z_{si}}$ — the calculated coordinates of the displace position of point si

$\dot{X_{si}},\dot{Y_{si}},\dot{Z_{si}}$ — the calculated X, Y and Z velocity components point si on the rigid body with respect to the XYZ frame

X_r,Y_r,Z_r — the given initial coordinates of the center of mas point R (origin of principal axes) on the moving rigid body wi respect to the XYZ frame

$\overline{X_r},\overline{Y_r},\overline{Z_r}$ — the calculated coordinates of the displaced positic of point R

$\dot{X_r},\dot{Y_r},\dot{Z_r}$ — the calculated X, Y and Z velocity components the center of mass of point R with respect to the XYZ frame

X_{soi},Y_{soi},Z_{soi} — the given coordinates of the connecting poir point soi, of the i-th spring-damper to the base with respect to th XYZ frame

Ol_i — the original length of the i-th spring
Pl_i — the new length of the i-th spring
Ux_{si},UY_{si},UZ_{si} — the direction cosines of the line from point to point soi

UX_{fi},UY_{fi},UZ_{fi} — the direction cosines of the line from point fi to point foi

FSX_i,FSY_i,FSZ_i — the components of the spring forces FS_i of the i-th spring

FDX_i,FDY_i,FDZ_i — the components of the damping force FD_i of the i-th damper

FX,FY,FZ — total force components with respect to the XYZ frame

TMX,TMY,TMZ — moments of all forces about the center of mass, referred to the XYZ frame

XT,YT,ZT — the components of moments of all forces about the center of mass with respect to the moving xyz frame

$\alpha_0,\beta_0,\gamma_0$ — initial Eulerian angles from the axes of the ground frame to the initial principal axes

α,β,γ — present Eulerian angles from the axes of the ground frame to the present principal axes

$\omega_{x0},\omega_{y0},\omega_{z0}$ — initial angular velocity about principal axes

$\omega_x,\omega_y,\omega_z$ — components of present angular velocity vector in the moving frame

$\omega_X,\omega_Y,\omega_Z$ — components of present angular velocity vector in the ground frame

$\dot{\alpha},\dot{\beta},\dot{\gamma}$ — components of present angular velocity vector of its own rotated axes

M — the mass of the rigid body

I_x,I_y,I_z — the moments of inertia around the principal axes x, y and z

k_i — the spring constant of the i-th spring-damper

d_i — the damping coefficient of the i-th spring-damper

$[D]_{02}$ — the displacement matrix of the rigid body from the "zero" position to the present position

$[D]_{01}$ — the displacement matrix of the rigid body from the "zero" position to the initial position

$[D]_{12}$ — the displacement matrix of the rigid body from the initial position to the present position

$[R]_{01}$ — the rotation part of $[D]_{01}$

$[R]_{02}$ — the rotation part of $[D]_{02}$

$[\dot{D}]$ — the velocity matrix of the rigid body in the present position

STEPS FOR FORMULATION AND COMPUTER PROGRAMMING - In order to form the Newton-Euler dynamic equations, the forces and moments acting on the rigid body must be summed. To find the force in a spring, the displaced position of its connecting point $(\overline{X}_{si},\overline{Y}_{si},\overline{Z}_{si})$ must be determined. The velocity of the connecting point $(\dot{\overline{X}}_{si},\dot{\overline{Y}}_{si},\dot{\overline{Z}}_{si})$ must also be determined to find the force exerted by the damper. This displaced point $(\overline{X}_{si},\overline{Y}_{si},\overline{Z}_{si})$ also determines the direction of the spring and damping force.

The original positions (X_{si},Y_{si},Z_{si}), (OO_X,OO_Y,OO_Z), (X_r,Y_r,Z_r), the initial Eulerian angles $(\alpha_0,\beta_0,\gamma_0)$ (referring to the ground system), the initial linear and angular velocities of the rigid body, $(\dot{X}_r,\dot{Y}_r,\dot{Z}_r)$ and $(\omega_{x0},\omega_{y0},\omega_{z0})$, and the information about the shaking force and moment of the engine are all given.

The dynamic analysis of the viscoelastic system is to then determine the unknowns $\overline{X}_r,\overline{Y}_r,\overline{Z}_r$, α, β, γ, $\dot{\overline{X}}_r$, $\dot{\overline{Y}}_r$, $\dot{\overline{Z}}_r$, ω_x,ω_y, and ω_z continuously at all times (present position at any time t).

The following is the step-by-step procedure showing the formulation of the system.

(Step 1) Form the displacement matrix $[D]_{12}$ of the body from the initial to the present position as given in Eq. (14).

The matrix contains the six unknowns: $\overline{X}_r$, $\overline{Y}_r$, $\overline{Z}_r$, α, β, and γ.

(Step 2) Form the velocity matrix $[\dot{D}]$ of the body in the present position.

When the angular velocity vector $\overline{\omega}$ is expressed in the moving and ground system, its components $(\omega_x,\omega_y,\omega_z)$ and $(\omega_X,\omega_Y,\omega_Z)$ are related to each other as

$$\begin{bmatrix}\omega_X\\\omega_Y\\\omega_Z\end{bmatrix} = [R]_{02}\begin{bmatrix}\omega_x\\\omega_y\\\omega_z\end{bmatrix}$$

where $[R]_{02}$ is the rotation part of $[D]_{02}$

$$[R]_{02} = \begin{bmatrix}a_{11} & a_{12} & a_{13}\\a_{21} & a_{22} & a_{23}\\a_{31} & a_{32} & a_{33}\end{bmatrix}$$

Then the velocity matrix is

$$[\dot{D}] = \left[\begin{array}{ccc|c}0 & -\omega_Z & \omega_Y & \overline{X}_r+\omega_Z\overline{Y}_r-\omega_Y\overline{Z}_r\\\omega_Z & 0 & -\omega_X & \overline{Y}_r-\omega_Z\overline{X}_r+\omega_X\overline{Z}_r\\-\omega_Y & \omega_X & 0 & \overline{Z}_r+\omega_Y\overline{X}_r-\omega_X\overline{Y}_r\\\hline 0 & 0 & 0 & 0\end{array}\right]$$

Now $[\dot{D}]$ is expressed in terms of another six unknown variables: $\dot{\overline{X}}_r$, $\dot{\overline{Y}}_r$, $\dot{\overline{Z}}_r$, $\omega_x,\omega_y,\omega_z$, as well as $\overline{X}_r$, $\overline{Y}_r$, and $\overline{Z}_r$.

(Step 3) Find coordinates of connecting and force-applying points in the present position.

$$\begin{bmatrix}\overline{X}_{si}\\\overline{Y}_{si}\\\overline{Z}_{si}\\1\end{bmatrix} = [D_{12}]\begin{bmatrix}X_{si}\\Y_{si}\\Z_{si}\\1\end{bmatrix},\quad \begin{bmatrix}\overline{OO}_X\\\overline{OO}_Y\\\overline{OO}_Z\\1\end{bmatrix} = [D]_{12}\begin{bmatrix}OO_X\\OO_Y\\OO_Z\\1\end{bmatrix}$$

(Step 4) Find velocities of connecting points in the present position

$$\begin{bmatrix}\dot{\overline{X}}_{si}\\\dot{\overline{Y}}_{si}\\\dot{\overline{Z}}_{si}\\0\end{bmatrix} = [\dot{D}]\begin{bmatrix}\overline{X}_{si}\\\overline{Y}_{si}\\\overline{Z}_{si}\\1\end{bmatrix}$$

(Step 5) Calculate the geometry in the present position.
1. Present length of the spring PL_i

$$PL_i = [(\overline{X}_{si} - X_{soi})^2 + (\overline{Y}_{si} - Y_{soi})^2 + (\overline{Z}_{si} - Z_{soi})^2]^{\frac{1}{2}}$$

2. Elongation of spring length ΔL_i

$$\Delta L_i = PL_i - OL_i$$

where OL_i is the original length and

$$OL_i = [(X_{si} - X_{soi})^2 + (Y_{si} - Y_{soi})^2 + (Z_{si} - Z_{soi})^2]^{\frac{1}{2}}$$

3. Direction cosines of the line from si to soi

$$UX_{si} = \frac{X_{soi} - \overline{X}_{si}}{PL_i}$$

$$UY_{si} = \frac{Y_{soi} - \overline{Y}_{si}}{PL_i}$$

$$UZ_{si} = \frac{Z_{soi} - \overline{Z}_{si}}{PL_i}$$

(Step 6) Calculate forces whose components are referred to the ground system.

1. Spring force and its components (FSX_i, FSY_i, FSZ_i)

$$FS_i = SK_i (\Delta L_i)$$

$$FSX_i = FS_i UX_{si}$$

$$FSY_i = FS_i UY_{si}$$

$$FSZ_i = FS_i UZ_{si}$$

2. Applied shaking force components in global coordinate system

$$\begin{bmatrix} FFX \\ FFY \\ FFZ \end{bmatrix} = [R_{12}] \begin{bmatrix} FFx' \\ FFy' \\ 0 \end{bmatrix}$$

3. Damping force and its components (FDX_i, FDY_i, FDZ_i)

Let ∇u_i be the component of the velocity of point Si along the line Si-Soi.

$$\nabla u_i = -\dot{\overline{X}}_{si} UX_{si} - \dot{\overline{Y}}_{si} UY_{si} - \dot{\overline{Z}}_{si} UZ_{si}$$

Then

$$FD_i = CD_i \nabla u_i$$

$$FDX_i = FD_i UX_{si}$$

$$FDY_i = FD_i UY_{si}$$

$$FDZ_i = FD_i UZ_{si}$$

4. Summation of all forces

$$FX = \sum_{i=1}^{NS} (FSX_i + FDX_i) + FFX$$

$$FY = \sum_{i=1}^{NS} (FSY_i + FDY_i) + FFY$$

$$FZ = \sum_{i=1}^{NS} (FSZ_i + FDZ_i) + FFZ$$

(Step 7) Moments of forces about mass center with respect to th ground system (TMX, TMY, TMZ)
1. Arm of spring-damping force:

$$DXS_i = \overline{X}_{si} - \overline{X}_1$$

$$DYS_i = \overline{Y}_{si} - \overline{Y}_1$$

$$DZS_i = \overline{Z}_{si} - \overline{Z}_1$$

2. Arm of shaking force application point:

$$DXF = \overline{OO}_X - \overline{X}_1$$

$$DYF = \overline{OO}_Y - \overline{Y}_1$$

$$DZF = \overline{OO}_Z - \overline{Z}_1$$

3. Components of moments of forces:

$$TMX = \sum_{i=1}^{NS} [(FSZ_i + FDZ_i)DYS_i - (FSY_i + FDY_i)DZS_i]$$

$$+ (FFZ * DYF - FFY * DZF)$$

$$TMY = \sum_{i=1}^{NS} [(FSX_i + FDX_i)DZS_i - (FSZ_i + FDZ_i)DXS_i]$$

$$+ (FFX*DZF - FFZ*DXF)$$

$$TMZ = \sum_{i=1}^{NS} [(FSY_i + FDY_i)DXS_i - (FSX_i + FDX_i)DYS_i]$$

$$+ (FFY*DXF - FFX*DYF)$$

(Step 8) Moments of forces about the mass center with respect to the moving system (XT, YT, ZT)

$$\begin{bmatrix} XT \\ YT \\ ZT \end{bmatrix} = [R]_{02}^{-1} \begin{bmatrix} TMX \\ TMY \\ TMZ \end{bmatrix}$$

where

$$[R]_{02}^{-1} = \begin{bmatrix} a_{11} & a_{21} & a_{31} \\ a_{12} & a_{22} & a_{32} \\ a_{13} & a_{23} & a_{33} \end{bmatrix}$$

(Step 9) Add shaking moment components

$$XTT = XT + TTx'$$

$$YTT = YT + TTy'$$

$$ZTT = ZT + TTz'$$

(Step 10) Set up the 12 first order differential equations

$$(1) \quad \frac{d\overline{X}_r}{dt} = \dot{\overline{X}}_r$$

$$(2) \quad \frac{d\overline{Y}_r}{dt} = \dot{\overline{Y}}_r$$

$$(3) \quad \frac{d\overline{Z}_r}{dt} = \dot{\overline{Z}}_r$$

Equation (25) results in

$$(4) \quad \frac{d\alpha}{dt} = \frac{\sin\gamma}{\cos\beta}\omega_y + \frac{\cos\gamma}{\cos\beta}\omega_z$$

$$(5) \quad \frac{d\beta}{dt} = \cos\gamma\,\omega_y - \sin\gamma\,\omega_z$$

$$(6) \quad \frac{d\gamma}{dt} = \omega_x + \tan\beta\sin\gamma\,\omega_y + \tan\beta\cos\gamma\,\omega_z$$

$$(7) \quad \frac{d\dot{\overline{X}}_r}{dt} = \frac{FX}{M}$$

$$(8) \quad \frac{d\dot{\overline{Y}}_r}{dt} = \frac{FY}{M}$$

$$(9) \quad \frac{d\dot{\overline{Z}}_r}{dt} = \frac{FZ}{M}$$

and finally from the Euler equations of motion

$$(10) \quad \frac{d\omega_x}{dt} = \frac{XTT + (I_y - I_z)\omega_y\omega_z}{I_x}$$

$$(11) \quad \frac{d\omega_y}{dt} = \frac{YTT + (I_z - I_x)\omega_z\omega_x}{I_y}$$

$$(12) \quad \frac{d\omega_z}{dt} = \frac{ZTT + (I_x - I_y)\omega_x\omega_y}{I_z}$$

Now that the equations are in first-order form a standard ordinary differential equation solver can be used to simulate the engine motion.
(Note: The static equilibrium position resulting from the spring-mass-damper system loaded by gravity can be found by solving the algebraic static equilibrium equations formulated in a similar manner for the position and orientation of the engine block.)

RESULTS

Selected numerical results for the engine mechanisms are given below. Figures 8, 9, and 10 plot the global X,Y,Z coordinates of the center of mass of the engine block as a function of time. The vibrational amplitude for this approximation method was found to be within 10% of the value as determined when checked by Rasna Corporation's MECHANICA-Motion software.

Static Position/Orientation of Engine Block
$X = 0.010136 \quad Y = 0.017236 \quad Z = 0.020017$
$\alpha = -0.145651° \quad \beta = 0.008871° \quad \gamma = 0.088023°$

Dynamic Analysis of Engine Block
time = 0.05 s

X = 0.01028   Y = 0.01728   Z = 0.02002

FFx' = 0 FFy' = -1541.66502 N

TTx' = 0   TTy' = 0   TTz' = -19.39545 N*m

time = 0.1 s

X = 0.01068   Y = 0.01722   Z = 0.02002

FFx' = 0 FFy' = 1124.33698 N

TTx' = 0   TTy' = 0   TTz' = -377.57809 N*m

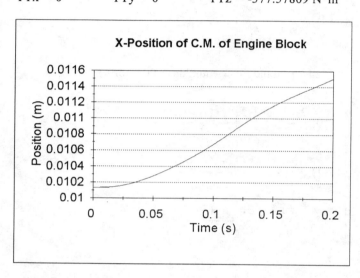

Figure 8. Position of x-coordinate of engine block as a function of time.

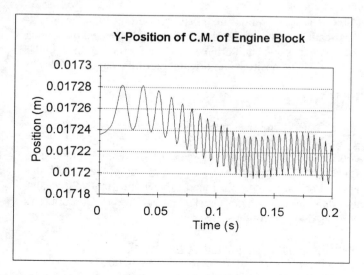

Figure 9. Position of y-coordinate of engine block as a function of time.

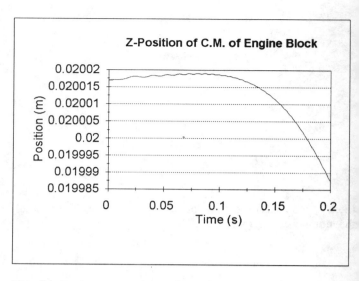

Figure 10. Position of z-coordinate of engine block as a function of time.

CONCLUSION

The displacement matrix method, available since 1966, has been demonstrated as an effective tool for simulating engine dynamics. It requires no commercial software but simply a non-linear algebraic equation solver and a numerical integrator (ODE solver). Since few dynamic simulations of engine or engine mount mechanisms have been published, particularly from the automobile industry due to strong international competition, the numerical example has been provided in detail and may be checked using any method.

REFERENCES

[1] Suh, C. H. and Radcliffe, C. W., Kinematics and Mechanism Design. John Wiley and Sons, New York, 1978.

[2] Kang and Suh, "Synthesis and Analysis of Spherical-Cylindrical (SC) Link in the McPherson Strut Suspension Mechanism," ASME Journal of Mechanical Design, Vol. 116, pp. 599-606, 1994.

[3] Paul, B., Kinematics and Dynamics of Planar Machinery Prentice-Hall, Inc., Englewood Cliffs, New Jersey, 1979.

[4] Greenwood, Donald T., Principles of Dynamics, 2nd Edition. Prentice-Hall, Inc, Englewood Cliffs, New Jersey, 1988.

[5] Suh, C. H., Computer Aided Design of Mechanism - Part B. Johnson Publishing Company, Boulder, Colorado, 1992.

971942

Automotive Powerplant Isolation Strategies

R. Matthew Brach
Ford Motor Co.

Copyright 1997 Society of Automotive Engineers, Inc.

ABSTRACT

Recently an increase in interest has occurred in automotive powerplant mounting. Evidence of this growth is the increase in the number of publications on the topic. The majority of this renewed interest has come from predicting and understanding the response of hydraulic engine mounts and the application of optimization techniques to the problem of powertrain vibration isolation, and occasionally to the combination of these two topics. However, it appears that these analytical techniques have been sufficiently developed and correlated to actual powertrain systems to have found widespread use by the automotive manufacturers. Subject to timing and packaging constraints, the more traditional mounting system design strategies are typically utilized. These strategies include natural frequency placement, torque axis mounting and elastic axis mounting. This paper presents a comprehensive review of these three strategies including a discussion of the assumptions associated with each method. In addition, the center of percussion mounting strategy, applicable to the isolation of transient inputs to the powertrain, is discussed in detail, including the technical basis for the theory.

INTRODUCTION

Since the inception of the automotive industry, isolation of the vibration of the engine from the rest of the vehicle has been desirable. Various isolation schemes have been devised to deal with this problem. Four strategies have been identified in the literature to address this problem. These strategies are: torque axis mounting, elastic axes mounting, natural frequency placement-optimization, and center of percussion mounting. A review of the literature indicates that the first two techniques are more widely used and investigated by vehicle manufacturers. All four techniques are described here in detail to establish a comprehensive background about engine mounting strategies that consider the motion of the engine as a rigid body only. Each description contains references that present the development and theory associated with each method. In each case, discussion

regarding the assumptions made in the development of the method is presented. Where appropriate, the limitations of a method are also discussed. Other powerplant isolation strategies exist that consider the flexural motion of the automotive engine and drivetrain. These methods (Bolton-Knight, 1971) are outside the scope of this paper.

AUTOMOTIVE POWERPLANT ISOLATION STRATEGIES

NATURAL FREQUENCY PLACEMENT-OPTIMIZATION - Optimization algorithms have been applied to the problem of isolation of engine vibrations. The typical objective of this analysis is to move the rigid body natural frequencies of the powerplant away from the frequencies of the input sources. This alters the amount of modal coupling by changing the mode shapes and presumably reducing the displacements of the response, thereby reducing the transmitted forces. Numerous papers have been written about this topic, including Johnson and Subhedar, (1979); Bernard and Starkey, (1983); Geck and Patton, (1984); Spiekerman, Radcliffe, and Goodman, (1985); Staat, (1986); Saitoh and Igarashi, (1989), and Bretl (1993).

In these studies the powerplant is typically modeled as a rigid body with six degrees of freedom mounted on resilient supports. These supports are commonly attached to ground and the system is referred to as a "grounded" system. The design parameters usually include the stiffness values of the mount, and the mount locations and orientations. Viscous damping is occasionally included in the system model. However, it is usually treated as a fixed design parameter. These methods depend largely on the ability to predict or determine the natural frequencies of the powerplant on its mounts. This task can be difficult, and the results of this isolation method are only effective near the natural frequency. As such, this technique is used primarily to address vibrational problems that are due to the engine idle frequency, or one of its associated orders, being close to a system rigid body natural frequency. Modifications of this approach have been implemented that determine the stiffness values of the mount by directly minimizing the forces transmitted through the mounts to the vehicle structure at idle (Oh, Lim and Lee, 1991; Swanson, Wu, and Ashrafiuon, 1993).

More recently, optimization methods have been used to analyze a rigid body affixed to resilient mounts attached to a flexible support (Ashrafiuon, 1993; Lee, Yim, and Kim, 1995). In another recent publication, optimization techniques were applied to a system model that takes into account the general rotational motion of the engine as a rigid body (Snyman, et. al., 1995).

TORQUE AXIS MOUNTING - The torque axis, also referred to as the torque roll axis, is defined as the resulting fixed axis of rotation of an unconstrained three dimensional rigid body (i.e. either free or supported elastically on very soft springs) when a torque is applied along an axis not coincident with any of the principal axes (Timpner, 1965; Fullerton, 1984). For the case of an automotive engine, the axis about which torque is applied is the crankshaft axis. This axis is rarely coincident with a principal axis of the engine. It is hypothesized (Timpner, 1966) that the disturbances transferred to the vehicle can be reduced by positioning the engine mounts such that the engine oscillates predominantly about this torque axis. The two references above each present a method for determining the location of the torque axis of an engine.

In order to better understand the concept of torque axis, consider the general equations of motion governing the rotational behavior of a rigid body in three dimensions acted on by the moment vector **M**. These equations are given by (Greenwood, 1988):

$$M_x - I_{xx}\dot{\omega}_x + I_{xy}(\dot{\omega}_y - \omega_x\omega_z) + I_{xz}(\dot{\omega}_z + \omega_x\omega_y)$$
$$+ (I_{zz} - I_{yy})\omega_y\omega_z + I_{yz}(\omega_y^2 - \omega_z^2)$$

$$M_y - I_{xy}(\dot{\omega}_x + \omega_y\omega_z) + I_{yy}\dot{\omega}_y + I_{yz}(\dot{\omega}_z + \omega_x\omega_y)$$
$$+ (I_{xx} - I_{zz})\omega_x\omega_z + I_{xz}(\omega_z^2 - \omega_x^2) \qquad (1)$$

$$M_z - I_{xz}(\dot{\omega}_x - \omega_y\omega_z) + I_{yz}(\dot{\omega}_y + \omega_x\omega_z) + I_{zz}\dot{\omega}_z$$
$$+ (I_{yy} - I_{xx})\omega_x\omega_y + I_{xy}(\omega_x^2 - \omega_y^2)$$

where I_{ij} are the elements of the rigid body inertia matrix, ω_x, ω_y, ω_z are the angular velocity components and $\dot{\omega}_x$, $\dot{\omega}_y$, $\dot{\omega}_z$ are the angular accelerations.

For the case of an automotive engine, it is common to assume the engine to be a rigid body on resilient supports that are attached to ground. It can be further assumed that the values of the moments, both the applied moments and the moments generated by the

reaction forces at the mounts, and the inertia properties of the engine are known or can be calculated. Then, for a given set of initial conditions, the vector of angular velocities, ω, can theoretically be solved for as a function of time by numerical integration of equations (1). In general, the magnitude and direction of the axis of rotation of the rigid body will also be a function of time. Thus, the torque axis as defined previously, may not exist. Closer investigation of these equations for a free rigid body system will lead to a better understanding of the rotational response of the powertrain on its mounts and whether a single axis of rotation exists for a rigid body.

While the existence of an axis meeting the definition of the torque axis may be debated, the application of the torque axis theory generally involves the premise of placing the rigid body response that is predominantly about the axis parallel to the crank-shaft, commonly referred to as the roll-response, at as low a frequency as possible. This separates it from the other modes of vibration. This mounting philosophy does not require the existence of an axis, but merely uses the axis as a means to describe the response at that frequency.

It is further noted that equations (1) are derived under the assumption that the inertia properties of the powerplant are independent of time. This is not the case for an automotive engine. The motion of the pistons, cranks, connecting rods, and the crankshaft make the inertia properties periodic functions of time, independent of the coordinate system. This further influences the time dependent nature of ω in an actual engine. However, this time dependence of the inertia properties is typically neglected in practice (Bachrach, 1995).

ELASTIC AXES - Elastic axes for an elastically supported rigid body system are those axes for which application of a force or torque, along or about a line produces only a corresponding translation or rotation on or about the same line. In the coordinate system defined by the elastic axes, the system response consists of decoupled translational and rotational modes. This decoupled state is frequently referred to as focalization.

The elastic axes of a rigid body on flexible supports can be determined using the flexibility matrix. Analytical modal decoupling of the flexibility matrix does not yield the elastic axes system because the eigenvectors do not typically span a physical space. Hence, the transformation to the elastic axes must be a physical coordinate transformation.

For a 2-dimensional system with three degrees of freedom, full decoupling can be accomplished since the six off-diagonal terms of the symmetric 3×3 flexibility matrix can be eliminated by two independent translations and one rotation. However, full decoupling of the symmetric 6×6 flexibility matrix for a 3-dimensional rigid body system with six degrees of freedom cannot be accomplished since 30 off-diagonal terms exist in the flexibility matrix (15 symmetric pairs) and only three independent coordinate translations and three independent coordinate rotations can be defined (Kim,

1991). This allows for only six symmetric off-diagonal pairs to be set to zero. Due to this limitation, effort has been spent in the area of partial decoupling of the modes of vibration (Derby, 1973; Ford, 1985, Xuefeng, 1991).

CENTER OF PERCUSSION MOUNTING - This isolation technique uses the mechanical phenomenon known as the center of percussion. One property of the center of percussion is that for a compound pendulum acted on by an impulse applied through the center of percussion and perpendicular to the line defined by the center of mass and the fixed point, no reactive impulse results at the fixed point. (A detailed presentation of the definition of the center of percussion and the properties associated with this point is given in Appendix A.)

The application to engine mounting follows directly. If the front and rear engine mounts are arranged such that their locations are reciprocal centers of percussion (see Appendix A), then an impulse to one mount from a road disturbance results in little or no reaction at the other mount (Wilson, 1959; Timpner, 1965; Bolton-Knight, 1971). This improves the overall isolation performance of the mounting system with respect to impulsive inputs since less vibrational input is imparted to the body of the vehicle. Although the application of this isolation scheme to actual mounting problems is typically not as simple as this description would indicate, placement of the engine mounts consistent with this theory will enhance their overall isolation effectiveness. This application of the center of percussion approach to powertrain mounting addresses the minimization of the reaction forces at the powertrain mounts to an externally applied impulsive load. This technique does not address the harmonic response of the system and is therefore supplemental to those methods that address the harmonic response of the powertrain.

Additionally, the system configuration used in the derivation of the theory is restricted to response in one plane. In the case of an automotive powertrain, the response of the system will be multi-planar even if the force applied by the front suspension is symmetric. This asymmetry of response is due to the lack of asymmetry of the mounting system and coupling. For this case, and the case of asymmetric loading of the front of the vehicle (only one wheel striking a pothole for example) the effectiveness of the technique as presented in this paper is not clear since the response will be multi-planar. No analysis has been done in this area in this paper or by others.

DISCUSSION

The complexity of the harmonic response of the automotive powerplant mounted on engine mounts in a vehicle cannot be understated. As mentioned in the previous sections, simplifications and assumptions are made in the development of the strategies currently used in industry to analyze this system. These simplifications and assumptions are made to facilitate the tractability of

the analysis and permit expectations about the response of the system to be more easily formulated. These assumptions include: the time invariant nature of the inertia properties of the powerplant, the body-side of the engine mounts affixed to a rigid structure (a grounded system), and that the response of the system exhibits the same characteristics that its planar counterpart does, such as in the case of Elastic Axis theory and Center of Percussion theory.

In addition to these assumptions, other assumptions are made that have not yet been mentioned. It is typically assumed in many of the analyses that the powerplant system includes the engine, the transmission and the mounts. Occasionally the subframe (if applicable) and frame/body of the vehicle is included in the analysis. The exhaust sub-system, the drive-shaft, and hose connections are almost never included in the system model. The inclusion of each of these sub-systems in the system model increases the number of degrees of freedom thereby increasing the complexity of the problem. It has been shown that the these additional connections can affect the response of the system (Spiekerman, Radcliffe, and Goodman, 1985).

Furthermore, it is typically assumed that the force-displacement characteristics of the engine mounts are linear. Although this assumption may be valid for elastomeric mounts for small displacements, hydraulic engine mounts are inherently nonlinear and their response characteristics have been investigated and found to exhibit quite complicated nonlinear behavior (Kim, Singh, and Ravindra, 1992; Kim and Singh, 1993). In addition, it has been shown analytically that the response of a three degree-of-freedom system comprised of a rigid body on mounts with linear force displacement characteristics can be nonlinear (Brach and Haddow, 1996).

Lastly, it is commonly assumed that the effects of temperature on the properties of the engine mounts, and hence the performance of the powertrain isolation system, are negligible.

CONCLUSION

All of these assumptions are important in the modelling of automotive powertrain vibration isolation strategies. However, what has not been established, or at least not published, is the relative importance or ramifications of these assumptions. Of equal importance is assessing that when these assumptions are made, whether the models based on these assumptions correlate with actual systems.

Evaluation of these assumptions must take place so that the validity of the models can be established. This evaluation typically takes place on the simplest system that still exhibits the fundamental response characteristics of interest. In the case of a powertrain system, this model would likely consist of an engine on mounts attached to ground. This six degree-of-freedom system, the response of which will still likely be quite complicated, can be

575

experimentally investigated effectively. Models of the engine-mount system have been established and correlation between the experimental response and the predicted analytical response can be done. With this simple system, the assumptions made in the development of the strategies discussed earlier that are currently used in the automotive industry can be evaluated.

With a fundamental understanding of the system response confirmed and a model that reliably predicts this response over a relevant range of parameters, parameter design studies can be performed to assess the performance of a mounting system against design requirements. Additional complexity can then be added to the system to broaden the range of parameters and circumstances over which the model is valid. This correlated model can then lead to more effective application to the optimization procedures already being applied to this system.

It should be noted that this correlated design process will not likely eliminate the fine tuning that inevitably occurs with the actual system. However, it can reduce the number of trials thereby reducing the time required to complete this fine tuning. This is important as it contributes to the overall reduction of the development time of the vehicle.

REFERENCES

Ashrafiuon, H., (1993). Design Optimization of Aircraft Engine-Mount Systems. *Journal of Vibrations and Acoustics*, v 115: 463-467.

Bachrach, Ben I., (1995). Personal communication at the Ford Motor Company, April 25, 1995.

Bernard, James E. and Starkey, John M., (1983). Engine Mount Optimization. SAE Paper 830257, Warrendale, PA.

Bolton-Knight, B. L., (1971). Engine Mounts: Analytical Methods to Reduce Noise and Vibration. *Instn. Mech. Engrs.*, C98/71.

Brach, R. Matthew and Haddow, A. G.(1996). The Nonlinear Response and Passive Vibration Isolation of Rigid Bodies. *J. of Machine Vibration*, v 5, n 15, 131-141.

Bretl, John, (1993). Optimization of Engine Mounting Systems to Minimize Vehicle Vibration. SAE Paper 931322, Warrendale, PA.

Derby, Thomas F., (1973). Decoupling the Three Translational Modes from the Three Rotational Modes of a Rigid Body Supported by Four Corner-located Isolators. *Shock and Vibration Bulletin*, Bulletin 43, Part 4: 91-108.

Ford, D. M., (1985). An Analysis and Application of a Decoupled Engine Mount System for Idle Isolation. SAE Paper 850976, SAE, Warrendale, PA.

Fullerton, Raymond R., editor, (1984). *Front Wheel Powertrain Mounts: Design Guide*. Ford Motor Company.

Geck, Paul E. and Patton, R. D., (1984). Front Wheel Engine Mount Optimization. SAE Paper 840736, Warrendale, PA.

Greenwood, D. T., (1988). *Principles of Dynamics*. Prentice-Hall, Englewood Cliffs, New Jersey.

Johnson, Stephen R. and Subhedar, Jay W., (1979). Computer Optimization of Engine Mount Systems. SAE Paper 790974, SAE, Warrendale, PA.

Kim, B. J., (1991). *Three Dimensional Vibration Isolation Using Elastic Axes*. M. S. Thesis, Michigan State University, East Lansing, MI.

Kim, G. and Singh, R., (1993). Nonlinear Analysis of Automotive Hydraulic Engine Mount. *Journal of Dynamic Systems, Measurements, and Control*, v 115: 482-487.

Lee, J. M., Yim, H. J., and Kim, J., (1995). Flexible Chassis Effects on Dynamic Response of Engine Mount Systems. SAE Paper 951094, SAE, Warrendale, PA.

Meriam, J. L., (1978). *Engineering Mechanics, Volume 2, Dynamics*. John Wiley & Sons, New York, New York.

Oh, T., Lim, J. and Lee, S. C., (1991). Engineering Practice in Optimal Design of Powertrain Mounting System for 2.0ℓ FF Engine. Proceedings for the Sixth International Pacific Conference on Automotive Engineering, Published by Korea Society of Automotive Engineers, Inc., 1638-3, Socho-dong, Seoul, South Korea: 607-616..

Saitoh, S. and Igarashi, M., (1989). Optimization Study of Engine Mounting Systems, 1989 ASME 12th Biennial Conference on Mechanical Vibration and Noise, Montreal, Quebec, Canada, September 17-21.

Singh, R., Kim, G. and Ravindra, P. V., (1992). Linear Analysis of Automotive Hydro-Mechanical Mount with Emphasis on Decoupler Characteristics. *Journal of Sound and Vibration*, v 158, n 2: 219-243.

Snyman, J. A., Heyns, P. S., and Vermeulen, P. J., (1995). Vibration Isolation of a Mounted Engine through Optimization. *Mech. Mach. Theory*, vol. 30, no. 1, pp 109-118.

Spiekerman, C. E., Radcliffe, C. J., and Goodman, E. D., (1985). Optimal Design and Simulation of Vibration Isolation Systems. *Journal of Mechanisms, Transmission and Automation in Design*, v 107: 271-276.

Staat, L. A., (1986). Optimal Design of Vibration Isolators Based on Frequency Response. M. S. Thesis, Michigan State University, East Lansing, MI.

Swanson, D. A., Wu, H. T., and Ashrafiuon, H., (1993). Optimization of Aircraft Engine Suspension Systems. *Journal of Aircraft*, v 30, n 6: 979-984.

Timpner, F. F., (1965). Design Considerations for Engine Mounting, (966B) SAE Paper 650093, Warrendale, PA.

Wilson, William Ker, (1959). *Vibration Engineering: A Practical Treatise on the Balancing of Engines, Mechanical Vibration, and Vibration Isolation*. Charles Griffin and Company Ltd., London.

Xuefeng, Jiang, (1991). A Study of Inclined Engine Mounting System Design Parameters, SAE Paper 912492, Warrendale, PA.

APPENDIX

The Center of Percussion[1/]

Consider a rigid body in a horizontal plane, fixed at one point, and moving under applied loads as shown in Figure A.1(a). This system is referred to as a compound pendulum (Greenwood, 1988). The free body diagram for this rigid body is shown in Figure A.1(b) which includes the reaction force at O and applied loads. The external force system can be replaced by the resultants $m\mathbf{a}$ and $I\alpha$. The vector quantity $m\mathbf{a}$ can be broken down into its components $r\omega^2$ and $r\alpha$ which act through the center of mass in the normal and tangential directions, respectively, as shown in Figure A.1(c). Another resultant force diagram can be obtained by moving the force $mr\alpha$ to point Q along the line OG (actually beyond point G) such that the resulting moment created about O equals $I_G\alpha$. $I_G\alpha$ can then be removed from the resultant force diagram as shown in Figure A.1(d). This condition can be written as:

$$I_G\alpha + mr^2\alpha = mr\alpha\ell \qquad (A.1)$$

[1/]The topic of the center of percussion is covered in many books on dynamics and vibrations. This information is presented here for completeness and also since no references were found that contained this information relative to powertrain mounting. The material presented here most closely follows that presented in Meriam (1978) and Greenwood (1988).

Replacing I_G by $k_G^2 m$ where k_G is the radius of gyration of the rigid body about G:

$$k_G^2 m\alpha + mr^2\alpha = mr\alpha\ell \qquad (A.2)$$

which results in

$$k_G^2 + r^2 = r\ell \qquad (A.3)$$

Solving this for ℓ:

$$\ell = \frac{k_G^2 + r^2}{r} = \frac{k_O^2}{r} \qquad (A.4)$$

where k_O is the radius of gyration of the rigid body about O. The point Q defined by this procedure is called the center of percussion of a body of fixed point O. Note that $\sum M_Q = 0$.

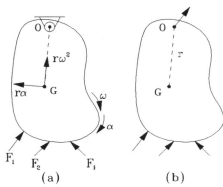

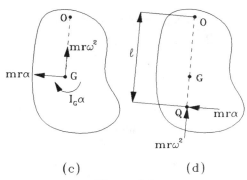

Figure A.1

The center of percussion has two interesting properties as shown by the following analysis. The first property involves the response of the rigid body to an impulsive load and the second shows a property about the natural frequency of a compound pendulum.

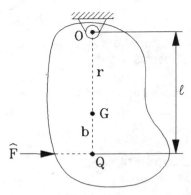

Figure A.2

Consider the same compound pendulum as shown in Figure A.1(a), initially at rest, impacted by an impulse $\hat{F} \perp OG$, as shown in Figure A.2. Since by definition, the impulse is equal to the change in momentum:

$$\hat{F} = p_2 - p_1 = m(v_{G_2} - v_{G_1}) \quad (A.5)$$

Since $v_{G_1} = 0$, let $v_{G_2} = v_G = \dfrac{\hat{F}}{m} = r\dot{\theta}$ where $\dot{\theta}$ is the angular velocity of the line OG. By definition, the angular impulse of $\hat{F}$ about G, $\hat{M}$, is:

$$\hat{M} = b\hat{F} \quad (A.6)$$

Also,

$$\hat{M} = H_2 - H_1 = I_G \dot{\theta} \quad (A.7)$$

Using the two previous results along with the fact that $H_1 = 0$ and $H_2 = H_G = I_G \dot{\theta}$, $\dot{\theta}$ can be solved for:

$$\dot{\theta} = \dfrac{b\hat{F}}{I_G} \quad (A.8)$$

Using this relationship and that $\dfrac{\hat{F}}{m} = r\dot{\theta}$ it can be shown that $\dfrac{I_G}{m} = k_G^2 = rb$. With $b = \ell - r$, the same relationship for the location of Q is obtained as before:

$$\ell = \dfrac{r^2 + k_G^2}{r} \quad (A.9)$$

This demonstrates that no impulsive reaction occurs at the point O for the applied impulse $\hat{F}$. Note that an impulsive reaction at O will occur if $\hat{F}$ is applied in any direction other than perpendicular to OC.

Now consider the natural frequency for small motion of the compound pendulum:

$$\omega_n = \sqrt{\dfrac{mgr}{I_O}} \quad (A.10)$$

Consider the natural frequencies of the compound pendulum for motion about points O and Q. Using the previous result, the following relationships are obtained:

$$(\omega_n)_O = \sqrt{\dfrac{mgr}{I_O}}$$
$$(\omega_n)_Q = \sqrt{\dfrac{mgb}{I_Q}} \quad (A.11)$$

Using the fact that $I_O = k_G^2 + r^2$, $I_Q = k_G^2 + b^2$, and that $k_G^2 = rb$, it can be shown that $(\omega_n)_O = (\omega_n)_Q$. This relationship illustrates the second property of the center of percussion which is that point O is the center of percussion for the body when Q is the center of oscillation and vice versa. Points O and Q are then referred to as reciprocal centers of percussion.

971943

Vibration Analysis of Metal/Polymer/Metal Laminates - Approximate Versus Viscoelastic Methods

Lezza A. Mignery and Edward J. Vydra
Material Sciences Corp.

Copyright 1997 Society of Automotive Engineers, Inc.

ABSTRACT

In this report, two finite element models are presented which predict the vibration characteristics of metal/polymer/metal laminates. The first model uses an approximate elastic solution, while the second model uses a viscoelastic solution. A finite element preprocessor was created to implement both models. With this preprocessor, four complex geometries and a simple plate are investigated. Predictions are made for natural frequencies, damping values, and frequency responses. In addition, the predictions for the plate and one of the geometries is compared to experimental results. It is shown that the two models predict natural frequencies well, but bound experimental damping values. The conservative estimate of damping is given by the viscoelastic model. It is further shown that if the geometry of the component resembles a beam, that both models agree. Based on these observations, recommendations are made to exclusively use the viscoelastic model in design analysis.

INTRODUCTION

Metal/polymer/metal (MPM) laminates have long been used to reduce vibrations, noise, and fatigue wear. In its symmetric form, the MPM laminate provides the most efficient way to introduce damping into a structure. The damping mechanism is the cyclic shear deformation of the adhesive layer [1,2]. The component will, then, benefit the most from the laminate if it resides in an area with high shear. This will cause the layers of the laminate to slide against each other, and thereby activate the damping mechanism. This sliding action is also beneficial for forming considerations. It allows the laminate to be easily stamped into complex shapes, in most cases utilizing the same tooling as the original metal component. However, if a straight substitution (laminate for metal) is made, strength and stiffness may suffer. Therefore, the laminate is best utilized in a shear dominated, low load situation.

While the laminated design can reduce vibrations efficiently, determining its dynamic characteristics can be quite challenging. The reduction in overall vibration level must be determined in order to design the part by vibration or noise characteristics. A vibration model of the laminated design is complicated by two major issues: the sliding action of the layers and the rate dependence of the core material. Concerning the first issue, the polymer's modulus is so much weaker than that of the metallic skins ($\geq 30:1$) that the layers are able to readily slide against each other. Unlike solid metal and fibrous composite laminates, this shearing action does not result in plane sections remaining plane. Therefore, each layer of the MPM laminate must be modeled separately to capture this sliding effect. Concerning the second issue, the polymeric core is viscoelastic and therefore frequency (as well as strain and temperature) dependent. The core modulus and damping values are different for every frequency seen in an application. The model must be able to capture this dependence in order to determine the proper behavior of the laminate.

A further concern in analyzing with any MPM laminate model is compatibility with three dimensional contoured designs. Several finite element realizations have been developed to study the vibration damping of these laminates [3,4]. Two of these models will be discussed and used in this report. The difficulty with these models is that they require a considerable amount of information to be used for three dimensional component design. This amount of detail can inhibit the use of the MPM laminate, even though it may be the best material choice to solve a resonant frequency problem. A finite element preprocessor is imperative to implement the models. A preprocessor was begun in a previous SAE paper [5]. A continuation of this design tool will be presented here.

579

This report will document the two finite element models [3,4] and show how they perform in component design. First the design process is discussed and how it can be complemented by finite element analysis. A description of the finite element models is then given. In one model, an elastic approximation is used for the core material, while the other model uses a viscoelastic characterization. Next, the finite element preprocessor will be presented. Both models and the preprocessor will then be demonstrated on four designs and a simple plate. In all cases, predictions will be made for natural frequencies, damping loss factors, and frequency responses. The plate will be investigated to determine how representative the models are of dynamic design testing. The predictions for one of the components will also be compared to design experiments.

It will be shown that the two models predict natural frequencies well, but bound experimental loss factor values. The lower bound, or conservative estimate, of damping is predicted by the viscoelastic model. It is further shown that only when the geometry resembles a simple beam structure do the two models agree. Recommendations are then made to use the viscoelastic model exclusively in component design.

THE DESIGN PROCESS

One of the first steps in designing a component for automotive noise and vibration is to select a material with desirable damping characteristics. A simple coupon sample, say a cantilever beam, is constructed from several materials, and the damping is compared between materials. In the case of an MPM laminate, the damping values are determined with an ASTM standard test (ASTM E-756) in which an electromagnetic force is applied to the free end of the beam. Several components are next constructed from those attractive materials. An impact modal test is conducted on the components at room temperature to determine a relative level of damping between designs. From these tests, those remaining attractive designs are further tested on a vehicle. The most desirable design is then chosen from the results of the vehicle test [6].

Finite element analysis can be used to compliment, or even reduce, the amount of testing required in the design process. The finite element analysis will predict results for the modal test, and thereby allow one to judge between designs in their true geometric form, without constructing a component. The analysis will also allow the designer to judge from the results at any temperature, a condition which is not always easily realized in an experimental test. Operating temperature testing can sometimes change the results of the modal test with components constructed from viscoelastic materials. The finite

element analysis may also be used to generate input for other design analyses, such as noise determination.

In this report, finite element models will be compared to the impact test. While the impact test can be complicated by the nonlinearity of highly damped structures, it provides a good indication of how the materials compare in the component geometry [1,2,5-8]. It is also a desirable test for design because it is quick and easy to perform.

An attempt was made to reduce the amount of nonlinearities in the impact tests here. In the case of the MPM laminate, the material can be sensitive to load level (or strain), temperature, and frequency. A variance between test results will be most sensitive to changes in temperature, rather than strain or frequency [1,2]. In the experimental tests reported here, a high temperature damping material (peak damping of approximately 80 C) is tested at room temperature (24 C). At this temperature, the material is stiff and should behave linearly. This will not show the maximum damping potential of the laminate, but will result in only a small amount of strain being developed in the viscoelastic layer. Coupled with this is the thickness of the viscoelastic layer. For the laminates under consideration, the adhesive layer is on the order of 0.05mm. This small thickness limits the amount of strain which can be developed in this layer. In addition, a low load level was applied in the tests, and was kept relatively constant between impacts. Therefore for all tests and analyses in this report, the results are considered linear with respect to load level.

FINITE ELEMENT MODELS

The goal of the finite element representation is to model enough of the geometry of the layered system to capture the sliding action of the laminate and enough of the frequency dependence of the core material to capture the viscoelasticity of the system. Using a conventional finite element program, the most ideal situation would involve modeling each layer of a structure by a number of three-dimensional solid elements. This would entail representing a laminated component by many rows of nodes through the thickness, as opposed to one row of through-thickness nodes for a sheet metal design. Each element would then be given material properties of the corresponding layers. For the metal skins, a constant modulus is given to every element. For the polymeric layer, however, a frequency dependent, complex modulus is needed (constant temperature assumed). The coupling of the detailed layers and the complex modulus results in a finite element problem that is practically unsolvable from a design viewpoint, due to a lack of computer memory and speed requirements.

More practical design models have been developed for the MPM laminate, two of which are shown in Figure 1 [3,4]. Each representation has its short comings in describing the deformation and material aspects of the problem, however, the result is a solvable problem in almost any workstation environment.

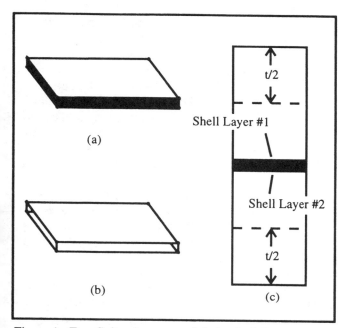

Figure 1. Two finite element models for the MPM laminate (a, b), as well as a through-the-thickness view of the element layer placement (c).

Both models represent the metal skins by two-dimensional offset shell elements, but differ in their representation of the polymeric core. Referring to the through-thickness view of the laminate in Figure 1c, at least two rows of nodes are necessary for the core (one at the top of the layer and one at the bottom) in order to capture the sliding action in the laminate. One could use basic shell elements for the skins, requiring two additional rows of nodes (the shell elements would reside at the midplane of the skins). Multi-point constraints (MPC's) are then needed to couple the elements together. However, using the offset shell elements allows for coincident nodes between the layers. The offset shell elements are placed directly on either side of the polymeric layer, requiring an offset designation of half the skin layer thickness (t/2). The offset shell elements are available in MSC-NASTRAN [9]. The offset shell elements eliminate the need for MPC's, resulting in a great memory savings for the MPM laminate problem (MPC's require a considerable amount of memory).

The models differ by their representation of the polymeric core. In Figure 1a, the core is modeled by three-dimensional solid elements, while in Figure 1b, the core is modeled by beam elements. The models also differ by their representation of the core modulus, a real elastic versus a complex viscoelastic representation. Accordingly, the finite element solution is different for both models. The elastic material uses natural frequency extraction and steady state dynamics solutions (real, modal solutions), while the rate dependence and the complex nature of the viscoelastic material requires a direct dynamics solution (direct integration, not a modal method). Both models are explained below in more detail.

APPROXIMATE ELASTIC MODEL - The approximate elastic model uses solid brick elements to describe the core of the laminate (Figure 1a). The three-dimensional elements require so much computer memory, that the modulus must be limited to an elastic, real value. Two approximations must then be made to account for the frequency dependence and damping of the core material. The first approximation is for the modulus. An iterative approach is required to capture the frequency dependence of the modulus. One determines the natural frequencies of the system by guessing what the natural frequency is and giving the finite element analysis the corresponding real modulus value. The natural frequency is then determined by the finite element program. Depending on the result, the modulus may need to be adjusted and the problem run again, until the frequency used for the modulus and the frequency obtained from the analysis agree (to within some percentage, say 10%). This iteration must be performed for every mode.

A further approximation is needed to determine the amount of damping available from the laminate. The loss factor is calculated per mode from a ratio of the strain energy in the core elements to the total strain energy, multiplied by the polymeric material's loss factor at that frequency. This popular treatment of the MPM system is known as the Modal Strain Energy Method (MSEM) [3,5-13].

If one additionally desires a frequency response curve, a modulus is chosen which is representative of the frequency range for the curve (a third approximation). Damping is applied per mode (global damping for the entire laminated structure) with the values from the MSEM calculations and a steady state dynamics solution (combination of natural modes) is done. In order to determine the vibration characteristics using this method, then, requires a considerable amount of calculations to be performed by the analyst, beyond the model description. In spite of the work requirement involved, the resulting analysis is computer time efficient. This can make the MSEM attractive for vibration design. However, it has been shown to be reliable only for relative comparisons [5].

VISCOELASTIC MODEL - The second model uses beam elements rather than solid elements (Figure

1b) for the polymeric layer. The beam elements require much less computer memory, so that a full complex viscoelastic solution may be used. In the material description portion of MSC-NASTRAN [9], a complex viscoelastic material model is available. One provides tables of the core modulus and loss factor values for frequency ranges covered in a solution. In this case, damping is accounted for via the core material properties. A direct frequency analysis (non-modal method) is then done to determine a frequency response curve. Natural frequencies are determined from the peaks of the response curve. Loss factors are determined using one of several methods available for modal analysis, say the 3dB method, or from Nyquist plots [2].

While no iteration is required to analyze the problem with this model, additional geometric description is required for the core as compared to the first model. For each beam, the area of the beam, the moment of inertia, and an orientation vector must be given. For a two-dimensional plate, these calculations are straight forward. For a three-dimensional geometry, with highly contoured curvature, these calculations are much more involved. A square cross-section is assumed for the beam as in Reference [4]. One must calculate the area of the connecting shell elements and factor it to the beams representing the core to determine the beam area and moment of inertia. The adhesive layer height corresponds to the beam length. The orientation vector requires determining the location of the beam with respect to the original coordinate system. These parameters can be determined easily in all designs if a preprocessor is used. This will be discussed below. Once these geometric values are determined, the solution of the viscoelastic problem is completely controlled by the finite element solver.

With the viscoelastic model, the frequency dependence of the core material is taken into account by the finite element solution. Viscoelastic damping is automatically included in the solution. Beyond the model description, the analysis results in relatively little analyst interaction to determine the vibration characteristics of a component. However, the computer execution time involved with the viscoelastic solution can be extensive, depending on the frequency resolution desired for the analysis. The automatic generation of a response curve makes the viscoelastic solution most attractive for noise design. It will be shown below that comparisons to experimental results also make it the analysis of choice for vibration design, in spite of the extensive computer execution time.

FINITE ELEMENT PREPROCESSOR

The finite element models of the MPM laminate contain offset shell elements, for the metallic skins, with different representations for the inner core. While these models capture the necessary ingredients for the description of the MPM design, their implementation in a three-dimensional component can prove to be as challenging as the analysis itself. In order to examine the four geometries as desired in this study, an extensive amount of drawing preparation is needed. According to the diagram in Figure 1, two rows of nodes, set one adhesive layer away from each other, 0.05mm, are needed to describe the geometry of any laminated component. In order to utilize this representation in design, one would first obtain two mathematical drawings of the part, also spaced one adhesive layer away from each other. One would then divide the drawings into equal finite element meshes, and then connect the two grid patterns with beam or solid elements. The material and element properties would then have to be given to each element. For the beam elements, this would require determining the area of each shell element the beam is connected to, as well as an orientation vector, for each beam.

This amount of detail could take a designer a great deal of time just to determine the parameters of the problem. If a design were to change, say a region was stiffened, the vibration characteristics of the MPM laminated component change as well [1,2]. Again an extensive amount of time would be necessary to determine the parameters for the new laminated design. This difficulty can make the laminate undesirable to use, even though it may be the best choice of material to alleviate a resonant frequency problem. However the development of a preprocessor can ease the use of the models.

The reports in References [5,10-13] detail the development of the preprocessor. The result is an approximate geometric model, which determines a MPM laminated system from any sheet metal design. The preprocessor duplicates the original sheet metal design one adhesive layer away from its initial location, determines parameters for all elements, generates material properties and solution control parameters. The preprocessor relies on standard geometric modeling techniques [14] to transform the mesh. A brief description is given below. Use of the preprocessor can be found in Reference [13].

A finite element model of the component is first made with two-dimensional shell elements. One then orients the normals on all the elements so that they point towards the inner core of the laminate. The geometric model contained in the preprocessor will then identify a new location for every node in the finite element mesh at the lower interface. For each node, the geometric model will translate all connecting elements down their normals to the new interface. A parametric cubic equation is used to describe each of the elements as surface patches in the preprocessor. After translation, the elements will be intersected and a new nodal point chosen based on this intersection. A demonstration of the transformation is shown in Figure 2. Figure 2 shows a finite element discretization of the shift cable bracket to be analyzed in this report. The insets in the figure show the details of the two layers of offset shell elements and the connecting beam elements (solid elements are defined between the layers for the MSEM analysis).

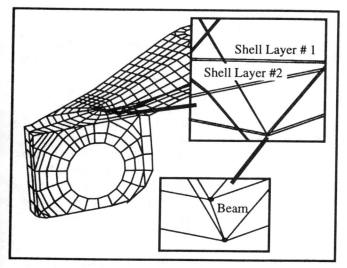

Figure 2. Offset shell/beam finite element model of a shift cable bracket. Insets show detail of the two layers of shell

Once the nodes are translated to the lower interface, the preprocessor connects the layers with either three-dimensional solid or beam elements. For the beam elements, the preprocessor also calculates the area of the connecting elements and an orientation vector for each beam. Beams are then given a square cross-section with this corresponding area (weighted per node in each element) and moment of inertia. The thickness of the core layer serves as the beam length. The area and orientation calculations are basic problems for the preprocessor to accomplish since it already has a mathematical description of every element, as well as the vector between the elements [14].

Material properties for the core are then determined by the preprocessor from another mathematical model [5]. The material model was determined from a cantilever beam test (ASTM E-756). For the MSEM, only one modulus is used per natural frequency extraction. For the viscoelastic method, a table is created which identifies the value of the modulus and loss factor for a corresponding frequency. Finally the preprocessor applies boundary conditions to the MPM construction and defines the solution parameters for the desired analysis.

By accomplishing these tasks, the preprocessor reduces the generation of a MPM laminated component to nothing more than the generation of a sheet metal design. This allows the design to be altered as often as its sheet metal counterpart, without a considerable amount of work to consider it as a laminate. Analysis with this preprocessor will enable the designer to anticipate the benefit of the design before manufacturing a component, thereby improving the design process. This preprocessor will be used in the next section to analyze several geometries with both finite element models. Without the preprocessor, examining four geometries would be considerably more involved, with time spent primarily in preparing the models for analysis.

GEOMETRIES

Four geometries (and a simple plate) will now be investigated to demonstrate the capabilities of the two MPM laminate models. The geometries are shown in Figure 3. Figure 3a shows a front engine cover, 3b a shift cable bracket, 3c a rear wheel house cover, and 3d a support-transmission planetary. Each part resides in an area dominated by shear loading and is well suited for a laminate design.

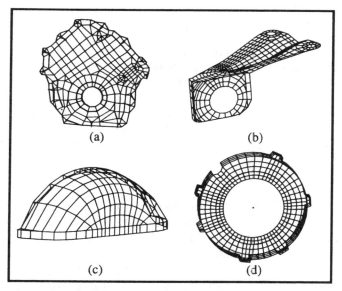

Figure 3. Geometries used for accuracy study: (a) front engine cover, (b) shift cable bracket, (c) rear wheel house.

The first geometry to be considered will be the front engine cover in Figure 3a. This part is currently manufactured from a steel laminate with an elevated temperature damping material [5,6]. The steel skins of the laminate are 0.762 mm thick, with core material 0.05 mm thick. The cover measures 349 mm from top to bottom. The operating temperature of the cover is 77 C. In the original analysis of this component [5,6], it was desired to improve the design's vibrational characteristics using a different laminate. An analysis was done using the MSEM which showed that the damping properties of this component could be improved if it was constructed from an aluminum rather than a steel laminate. Vehicle testing of the aluminum laminate has verified this result [6].

In the analysis to be presented here, only the steel laminate will be investigated. The skin and core thicknesses are as given above. Material properties for the steel and core can be found in Reference [5]. This component will be analyzed with both the MSEM and viscoelastic model. The results will be compared to a room temperature impact modal test with free and bolted boundary conditions.

The second geometry to be investigated is shown in Figure 3b. It is a new design of a shift cable bracket, which was desired to have reduced noise over a metal counterpart. The goal of the laminated design was to reduce vibration transmissibility at all frequencies, especially in the 3000-4000 Hz range. In addition, the laminated design was not supposed to greatly alter the static displacement due to a 222 N load applied at the hole. An asymmetric steel laminate with the same thickness as the original was chosen for the design. The asymmetric laminate was necessary so that the component could be easily stamped into its complex shape. All objectives were accomplished with the laminated design [15].

The asymmetric design will be used here to investigate the two methods (in Reference [15], only the MSEM solution was used). Skins for this laminate measure 1.0 and 2.0 mm. The core measures 0.05 mm. The maximum length along the top of the bracket is 146 mm. The bracket is cantilevered at the back bolt holes. The skin material is steel, with the core material a room temperature damper. Material properties can be found in Reference [5].

Figure 3c shows the third geometry to be investigated. This is a rear wheel house which was also desired to have reduced noise over its metal counterpart. A study was done to determine the amount of damping available from a MPM design [16]. This part will be analyzed here with symmetric steel skins, measuring 0.508 mm thick. The core material is the same room temperature damper used in the shift cable bracket. The geometry measures 842 mm across its back side. This geometry is clamped along its outer edge.

The final geometry to be investigated is the support-transmission planetary in Figure 3d. Again the laminated design was desired to reduce noise associated with the original steel planetary [17]. As with the bracket design, an asymmetric laminate was chosen because of stamping concerns. Steel skins were used, with thicknesses 2.4 mm and 0.762 mm. A core

thickness of 0.0508 mm was used in the analysis. The elevated temperature damping material of the front engine cover was used for the core. The operating temperature is 93 C. The planetary measures 168 mm across its widest point. The inner ring of the planetary is clamped in position to simulate its attachment to other structures.

All four geometries and a simple plate will be analyzed in the following section using the approximate and the viscoelastic method, as well as the preprocessor. The finite element meshes of Figure 3 were created using two dimensional shell elements. The finite element preprocessor was then used to create the necessary laminate designs and input files for the MSC-NASTRAN solutions. An extensive amount of model preparation time was saved due to the preprocessor. Without the preprocessor, a great deal of time would be needed to define the two skins of the laminate, as well as the interconnecting adhesive. With the preprocessor, the task reduces to little more than the generation of the sheet metal design. The next section of this report will investigate how the results of the analyses of these components compare to experimental values and to each other.

COMPARISON TO EXPERIMENTAL RESULTS

Two sets of experimental data will be investigated with the finite element models. The first set is for a room temperature impact modal test of a plate (300 x 125 mm) constructed of the same laminate as the front engine cover. The second set will be of the front engine cover itself. In each set, the geometry was struck with an impact hammer at several locations, and the acceleration was measured at several transfer location. The impact test was used to show how the finite element predictions compared to design experiments. The impact test is widely used to judge the relative damping of materials in component design and in numerous other studies already discussed (see the above section on the design process) [1,2,5-8]. While it can be plagued by nonlinearities due to the strain dependence of the viscoelastic material, it is believed that the conditions of the test (temperature well below the maximum damping value, low load level, and thin adhesive layer) will cause the experimental results to be linear.

In the testing reported on here, both the plate and engine cover geometries were tested while supported on a foam rubber mat. This simulates a free boundary condition. In addition, the front engine cover was tested while bolted to an engine test bed [6]. Each geometry was then analyzed using the MSEM and the viscoelastic method. Comparisons between the predicted and experimental frequency response, natural frequency, and loss factor of both geometries were then made.

PLATE RESULTS - Figure 4 shows the frequency response from one of the transfer points in the impact test of the plate sample. Also shown in the figure are predictions from the MSEM analysis and the viscoelastic method. It can be seen that both methods predict the frequency response well up to approximately 500 Hz.

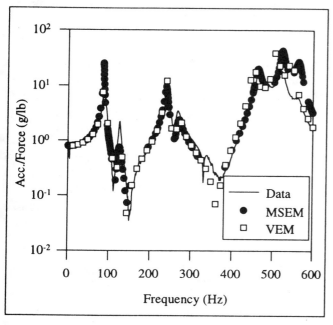

Figure 4. Frequency response predictions and data for the plate sample. MSEM indentifies results from the Modal Strain Energy Method. VEM identifies results from the viscoelastic method.

Near this frequency, discrepancies do appear between the models and the data. Three peaks are seen in both the data and the model predictions between 400 and 600 Hz, however the final data peak is not as distinguished as those predicted by the models. The third data peak also occurs at a slightly higher frequency than that predicted by either finite element model.

The discrepancy between this peak and the predictions could be a reflection of several factors. It may be caused by the coarseness of the models through the thickness of the plate. Each model contains only two rows of nodes through the thickness. Many more rows of nodes would be necessary to smoothly transition between the markedly different types of materials in the three layers of the laminate. This type of detail may be necessary to resolve the accelerations to agree with the data. The discrepancy may also be due to the impact test itself. As noted above, the impact test can lead to uncertainties because of the material's dependence on load amplitude. Perhaps a more sophisticated dynamics test, where the load is more controlled would improve the correlation between the data and predictions. Yet, for most of the frequency range, the models predict the data response curve quite well.

The results from this test are further examined in Figure 5. Figure 5 shows the natural frequency and loss factor from the plate modal test and the two solution methods. Due to the uncertainties with the third peak discussed above, it is not included in this figure. Referring now to Figure 5, natural frequencies for the impact test and the viscoelastic method were determined from the peaks of the corresponding response curves. Loss factors for the test and viscoelastic method were calculated using the 3dB method [2]. The natural frequency for the MSEM is a direct output of the finite element solution. The loss factor for this method is determined with the strain energy calculation discussed in the model review in a previous section.

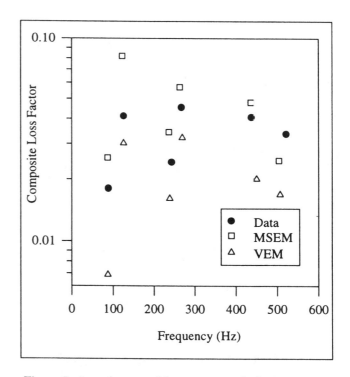

Figure 5. Loss factor and frequency predictions as compared to data for the plate sample.

Using Figure 5, the natural frequencies may be compared with the horizontal axis. The loss factor values may be compared with the vertical axis. It can be seen from this figure that both the MSEM and viscoelastic methods predict the natural frequency quite well, with varying amounts of agreement to test data for the loss factor. The MSEM analysis is seen to over predict the loss factor for all but one mode, while the viscoelastic method under predicts the experimental value.

It can be concluded that the two models predict natural frequencies well, but bound the experimental loss factors. It can be further concluded that using the viscoelastic method will result in a conservative estimate of the amount of damping available with this laminate since it gives a lower bound for the loss factor data. This would indicate that the viscoelastic method is more reliable for vibration analysis, giving a conservative estimate of the amount of available damping and a good estimate of the natural frequency.

FRONT ENGINE COVER RESULTS - A similar comparison will now be made with the front engine cover. For this geometry, two tests were conducted. The first test to be discussed is the free impact modal test and how it compares to the viscoelastic method. Figure 6 shows the response curve from the data and the viscoelastic method.

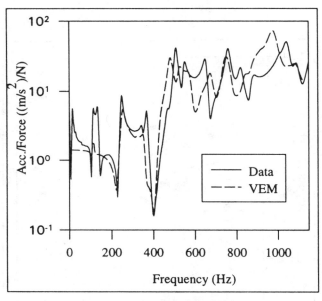

Figure 6. Finite element predictions of frequency response using the viscoelastic method. Data is shown for the front engine cover with free boundary conditions.

Note that only the viscoelastic method will be compared to the data for this test condition. This is due to the difficulty in generating results using the MSEM. As can be seen from the figure, a number of modes occur with the free boundary conditions. For every mode, the MSEM requires one to iterate for the natural frequency and then use a further calculation for the loss factor. When a number of modes are present, it can be difficult to distinguish between modes with this method because the modulus is varied constantly from iteration to iteration. Also, the frequency response curve is a secondary calculation from the MSEM. One must first determine all natural frequency and damping values, even if they are not needed for the particular response curve. It also requires assuming one constant modulus for the entire frequency span. The frequency response curve is a direct output of the viscoelastic method. For these reasons, the frequency response of the data will only be compared to the viscoelastic method. An examination of the MSEM with the data will be given for a bolted boundary condition below. For the bolted boundary condition, much fewer mode shapes are encountered in the response curves, making it a more ideal problem for the MSEM.

Continuing with the model evaluation, Figure 6 shows a comparison between response curves generated from the viscoelastic method and the experimental results with free boundaries. The predicted curve shows approximately the same curve as the experimental data.

There are some low level data peaks (< 200 Hz) that are not realized in the predicted curve. It also appears that the predictions are slightly shifted in natural frequency above 400 Hz. However, the shape of the response curve is very similar between the data and the viscoelastic method.

The discrepancy between the curves could be related to the similar points made above for the plate sample. It could be a reflection of the model lacking enough detail through the thickness. It could also be a reflection of the model coarseness in the plane of the component (not through the thickness, see Figure 3). It could also be a function of the beam descriptions, in that their length (adhesive layer thickness) to height (determined by the connecting shell areas) ratio is unacceptable for a beam model. Further, the data may be marred by nonlinearities because of the strain dependence of the material. The data may be improved by using a more controlled force input, such as an electromagnetic shaker. In spite of these discrepancies, the viscoelastic method appears to well represent this complex data curve.

A second data comparison was available from the original design testing of the component. During the design process, the cover was bolted to an engine chuck and impact tested. In this test, only three modes were present below 750 Hz. This makes the data analyzed easily by either the MSEM (hampered above due to iteration with multiple modes) or viscoelastic method.

Paralleling the free boundary condition test evaluation of Figure 6, a frequency response curve comparison is made in Reference [11] between the viscoelastic method and the data (the MSEM is not compared due to the difficulty in generating a frequency response with this method as discussed above).

Comparisons with this new boundary condition are similar to those for the free boundary test. The response curve shapes agree quite well between the data and the viscoelastic method, with an accurate prediction of the first mode's natural frequency. Yet, the second mode's natural frequency is under predicted by the viscoelastic method, giving once again a shifted curve.

This discrepancy is no doubt related to the difficulties with the model and test as stated above, coupled with the boundary condition difference between the predictions and the test. In the test, the cover was bolted in position, while in the prediction, the cover was pinned at the bolt holes.

As with the plate samples, the loss factor and natural frequencies are now calculated. Only values for the cover bolted to the engine chuck will be examined. As stated above, the complexity of the response curve with free boundary conditions makes this problem difficult for comparison using the MSEM. The results for the bolted cover test and predictions are shown in Figure 7. Like Figure 5, comparisons using the horizontal axis show the natural frequency predictions, while loss factor values can be compared on the vertical axis.

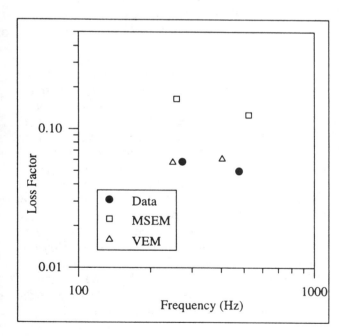

Figure 7. Loss factor and frequency predictions as compared to data for the front engine cover.

From this figure, it can be seen that the first natural frequency is predicted well by either method. The second natural frequency, however, is not predicted well. The MSEM over predicts the natural frequency, while the viscoelastic method under predicts the experimental value.

Referring first to the viscoelastic method, it is believed this discrepancy is primarily caused by the boundary conditions used in the analysis as discussed above. The experimental test had the cover bolted to an engine chuck, while in the analysis, this was simulated by pinning the model at the bolt holes. The bolting of the cover to the chuck will result in a much stiffer system than simulated in the predictions. This would cause the natural frequency of the test to occur at a higher value.

Referring now to the MSEM prediction of the second mode, it is seen that the natural frequency is over predicted by this method. This is believed to be caused by the iterative technique of the method. As discussed above, the iterative technique requires one to adjust the modulus of the polymeric material to match the natural frequency of the mode. Two modes were predicted at approximately the same natural frequency as the test data using this method. The value shown in the figure corresponded the best to the reported data natural frequency. The iterative calculations involved with this method and the complications in obtaining a frequency response curve are serious drawbacks to using the MSEM approximation.

Varying amounts of correspondence to data are seen from the vertical axis for the loss factor. The loss factor is greatly over predicted, by almost three magnitudes, when the MSEM analysis is used. Calculations with the viscoelastic method show excellent agreement with the experimental results. This result shows that as with the plate sample, the models bound the data loss factor values. These results also show that the viscoelastic method gives the more realistic result. With the accompanying plate data, and the iterative nature of the MSEM analysis discussed above, it is suggested that the viscoelastic method should be used in all design situations.

Indeed, the MSEM analysis has been criticized for its inability to predict test results [18]. This is not surprising since the strain energy method used to calculate the damping values is a great approximation and that the method does not use the complex modulus of the polymeric core. The original goal of the MSEM was to give a useful, though not necessarily exact, means to determine damping values, without the extensive solution time associated with the full viscoelastic analysis [3]. Early researchers into the constrained layer damping (MPM laminate) problem concluded that the modal loss factor for this system was related to a strain energy ratio at resonance. This inspired the development of the MSEM in which the loss factor is approximated from energy distributions of undamped mode shapes.

Other methods have been suggested to improve the short comings of the MSEM, in which some of the viscoelasticity is included in the analysis. A study in Reference [18] finds that if the undamped part of the system is sufficiently separated from the damped part, then the MSEM is accurate. Otherwise, the method does not guarantee a good prediction of the actual situation. Other researchers have indicated the viscoelasticity alters the vibration amplitudes and frequencies enough so that a full viscoelastic solution is always needed [19]. In Reference [3], where the MSEM is introduced, verification of the MSEM was demonstrated for several cantilever beam problems.

However, the MSEM does result in a much faster computer execution time as compared to the viscoelastic method. It also can predict relative comparisons between laminates quite well [5]. As mentioned above, this method was used to make a new design recommendation for the front engine cover laminate. Yet, the viscoelastic method is a more rigorous model, and requires much less interaction by the analyst to determine the vibration characteristics. Recall the MSEM requires a loss factor calculation, which is assumed for the entire component, plus an additional steady state dynamics solution, beyond the initial iterative normal modes solution. The viscoelastic method merely uses a direct frequency response solution, with the true complex modulus of the material. Comparisons between geometries with the two methods will be examined next.

COMPARISON OF OTHER GEOMETRIES

The MSEM analysis and viscoelastic method are now applied to the remaining geometries of Figure 3. The goal of the analysis in this section is to determine if the geometric shape of the sample will influence the correspondence of the two methods. The damping of the MPM laminate can vary greatly with geometry [1,2]. Two modes are investigated per geometry. The correspondence of the analysis methods is examined with the loss factor and the natural frequency.

The comparison between the methods is shown in Figure 8. Figure 8 shows two results for each of the geometries. As with the previous graphs of this type, loss factors are compared on the vertical axis and natural frequencies on the horizontal axis. From this figure it can be seen that the natural frequencies correspond quite well between models.

It can also be seen from this figure that loss factor predictions agree only in the case of the shift cable bracket (note the high frequency calculation is made to reflect the predictions necessary for the design consideration. The design required improvement in the 3000-4000 Hz range). Otherwise, as with the plate and front engine cover results, the MSEM analysis over predicts the loss factor from the viscoelastic analysis.

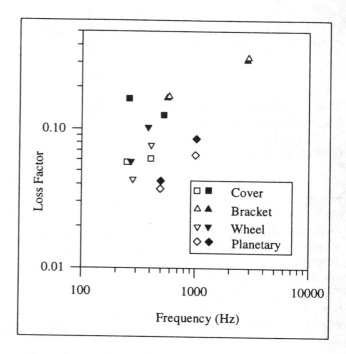

Figure 8. Loss factor and frequency comparisons between the two solution methods. Closed symbols identify the MSEM results, open symbols the VEM.

Based on these results, it appears that the MSEM will agree with the viscoelastic solution more for the bracket geometry than the other three-dimensional geometries. In the previous section, it was determined that material properties could effect the results from the MSEM, and again this may be the major contributor to the disagreement between the methods. Yet, Figure 8 suggests there is an additional factor influencing the MSEM, the geometry of the part. Therefore, the lack of rate dependent material properties may not account for the total difference between the methods. The MSEM determines the loss factor of the system by multiplying the core loss factor by the ratio of the strain energy in the core to total strain energy. It is believed that the strain energy ratio calculation is additionally responsible for the difference between the two methods.

The strain energy ratio used in the MSEM calculations has long been the basis for damping estimates for simple systems [1,3]. In the case of the shift cable bracket, the geometry could be easily approximated by a cantilever beam with a mass at the end. For such simple systems, a steady-state or near steady-state situation occurs in which the response of the system is controlled by a balance between the energy input and the energy dissipation. Excitation has a spatial distribution that matches that of a freely propagating wave down the beam. In such systems, the response is controlled by damping, and therefore the loss factor. A significant portion of the energy is attributed to the damped portion of the component. The strain energy ratio is then appropriate for the simple bracket system. In the more complex geometries, there are more pronounced contributions to the strain energy dissipation other than just that in the core elements. The systems have much more complicated mode shapes, and therefore more complicated strain energy distributions, so that the ratio calculation in not as reliable.

The original presentation of the MSEM was verified with observations of sandwich beam behavior [3]. It is therefore reasonable that geometries shaped and constrained similarly to beams, such as the bracket, will give good results as compared to the viscoelastic method. For the more complex geometries, a beam approximation is not valid, resulting in a poor comparison between methods. Coupled with the comparison to the experimental data, and the difficulty of the iterative technique discussed above, it is concluded that the MSEM analysis is advisable only for simple systems. In all other cases, the viscoelastic model will give a conservative estimate of the available damping and natural frequency and thereby predict more realistic results.

CONCLUSION

In this report, the reliability of two finite element methods has been investigated. The methods both predict the vibration characteristics of a metal/polymer/metal laminate. The first method uses an approximate elastic solution, known as the Modal Strain Energy Method (MSEM), to compute damped responses. The second method uses a full viscoelastic solution to describe the system. It was desired to show the capability of the two methods with a simple plate and several contoured three-dimensional geometries. Essential for this study was the development of a finite element preprocessor. The preprocessor made the application of these models possible in the contoured designs.

Comparisons to experimental values with the plate and a front engine cover demonstrated that the models predict the natural frequencies well, while bounding the experimentally determined damping values. The approximate method over predicts the amount of damping available from both geometries. Comparisons to experimental values with the viscoelastic solution show that it gives either a conservative estimate of the damping available, or agreement with experimental values. It is concluded that the viscoelastic method is more representative of the actual behavior. The lack of viscoelastic material properties and the iterative technique involved in the MSEM analysis are responsible for the method's inaccuracy.

A study was then done to determine if the geometry of the model could affect the predictions. Four contoured three-dimensional geometries were investigated: a front engine cover, a shift cable bracket, a rear wheel house cover, and a support-transmission planetary. It was shown that only in the case of the shift cable bracket do the results from the MSEM agree with the full viscoelastic solution. This is due to the geometry of the bracket itself. One of the underlying assumptions in the MSEM is that the damping is proportional to a strain energy ratio. The ratio has long been used for damping estimates of simple systems. Because the bracket can be easily approximated by a simple system, the MSEM agrees well with the viscoelastic solution.

In conclusion, two models are available for vibration analysis of MPM components. The reliability of the approximate method is dependent upon the amount of viscoelasticity in and the geometry of the component. It has been shown to be applicable only for simple geometries. In addition, the method requires considerable interaction from the analyst to determine the damping characteristics. The alternative viscoelastic method requires no interaction from the analyst and is reliable for all geometries, however, it can result in an extensive analysis execution time. In spite of this, the results shown here suggest it should always be used in design analysis of MPM laminates. Because of the detailed geometric description needed with this model, the finite element preprocessor developed as part of this study is imperative to using the viscoelastic model in component design.

REFERENCES

1. Drake, M.L., *Vibration Damping Short Course*, University of Dayton Research Institute, (1993).

2. Nashif, A.D., Jones, D.I.G., Henderson, J.P., *Vibration Damping*, John Wiley & Sons, New York (1985).

3. Johnson, C.D., Klenholz, D.A., and Rogers, L.C., "Finite Element Prediction of Damping in Beams with Constrained Viscoelastic Layers", *Shock Vib. Bull*, 51,71-80 (1981).

4. Lu, Y.P., Killian, J.W., and Everstine, G.C., A Finite Element Modeling Approximations for Damping Material Used in Constrained Damped Structures, J. Sound Vibr., 97, 352-354, (1984).
5. Mignery, L.A., "Reduced Vibration Design," Society of Automotive Engineers Technical Paper Series No. 951242, (1995).

6. Zsirai, F., General Motors Powertrain Engineering, personal communication, (1994).

7. Borchert, T., "Dynamical Behavior of the Disc Brake Pad," Society of Automotive Engineers Technical Paper Series No. 912656, (1991).

8. Rinsdorf, A., and Schiffner, K., "Practical Evaluation and FEM-Modelling of a Squealing Disc Brake," Society of Automotive Engineers Technical Paper Series No. 933071, (1993).

9. MSC/NASTRAN User's manual, Version 68, (1993).

10. Mignery, L.A., "Vibration Analysis of Metal/Polymer/Metal Components," Proceedings of the Design Engineering Technical Conference, DE-Vol. 84-3, Volume 3 - Part C, American Society of Mechanical Engineers, (1995).

11. Mignery, L.A., "Designing with Metal/Polymer/Metal Composites", proceedings to International Body Engineering Conference, Detroit, MI, (1996).

12. Material Sciences Corporation, "DAMP - A Finite Element Preprocessor for MPM Laminates I. Reference Guide," unpublished data (1996).

13. Material Sciences Corporation, "DAMP - A Finite Element Preprocessor for MPM Laminates I. User's Manual," unpublished data (1996).

14. Mortenson, M.E., *Geometric Modeling*, John Wiley & Sons, New York (1985).

15. Pre Finish Metals (Material Sciences Corporation, subsidiary), "Dynamic and Static Characteristics of a Laminated Shift Cable Bracket," unpublished data (1995).

16. Pre Finish Metals (Material Sciences Corporation, subsidiary), "Damping Design of a Rear Wheel House Inner Panel," unpublished data (1995).

17. Material Sciences Corporation, "Vibration Analysis of a Support-Transmission Planetary," unpublished data (1996).

18. McDaniel, J.G., and Ginsberg, J.H., "Fundamental Tests of Two Modal Strain Energy Methods," Technical Brief to Journal of Vibration and Acoustics, American Society of Mechanical Engineers, Vol. 118, No. 2 (1996).

19. Yi, S., Ahmad, M.F., and Hilton, H.H., "Dynamic Responses of Plates with Viscoelastic Free Layer Damping Treatment," Journal of Vibration and Acoustics, American Society of Mechanical Engineers, Vol. 118, No. 3 (1996).

971944

Hybrid Substructuring for Vibro-Acoustical Optimisation: Application to Suspension - Car Body Interaction

K. Wyckaert and M. Brughmans
LMS International NV

C. Zhang and R. Dupont
Renault S.A.

Copyright 1997 Society of Automotive Engineers, Inc.

ABSTRACT

For the prediction of the vibro-acoustical vehicle behaviour up to higher frequency ranges, modal approaches are not very applicable. Hybrid frequency response function based substructuring methods are therefore proposed, in which the high modal density components are represented by experimental data, and in which the lower density components are represented by finite element models. The frequency response function synthesis of the lower density component can be based on modal synthesis. In this paper, an application of coupling a rear twist beam suspension with a car body is discussed. In this case the vibro-acoustical behaviour of the car body is the high density component, the low density component is the suspension finite element model. Aspects of accuracy, related to truncation, influence of rotational degrees of freedom, symmetry of the experimental matrix, and prestraining of the suspension springs are discussed. The relationship between transfer path analysis and FRF based substructuring produces an understanding of the contribution of component behaviour to acoustical interior response.

INTRODUCTION

A topic of high interest in the automotive NVH is the ability to predict the interior acoustics of a car in early design stages. At the component level, the description of the vibratory behaviour can be approached by finite element models. The car body vibro-acoustic behaviour however is not easily represented by a numerical model, this is mainly due to the difficulty to model the fully trimmed car body structural behaviour. Therefore hybrid approaches can be very useful, where one combines test data where appropriate with numerical data. Typically the car body vibro-acoustics can be represented with test data;

subcomponents can be represented by numerical models. The calculation of the interaction between subcomponent and car body, allows evaluation of design alternatives at the subcomponent level on the acoustical interior pressure response.

In this paper, rear suspension- car body interaction is discussed. The rear suspension is represented by a finite element model, the car body by a set of measured frequency response functions (FRFs). The objective is to synthesise vibro-acoustical frequency response functions between 50 and 250 Hz between input at the wheel center and the acoustical response inside the car.

A constrained finite element model of the rear suspension has been supplied. This model is characterised by the fact that the suspension springs are staticly preloaded with the car body weight. The preloaded finite element model has been correlated with a test based modal analysis on the suspension component in preloaded constrained conditions. This resulted in an improved finite element model of the suspension component. A database of experimentally obtained frequency response functions on the fully trimmed car body (without suspension) has been acquired.

1. THEORETICAL BACKGROUND

The scope of the application is to predict the global frequency response function matrix between forces that are operational at the wheel centers of the rear suspension, and the interior pressure response in the car. In a first step the prediction implies the calculation of the interaction of the suspension component with the car body at the attachment points, and the calculation of the force transmissibility between the wheelcenters and the attachment points. This interaction is described by what is called the coupling kernel in the following equation that describes the relationship between the forces at the wheelcenter and the forces active at the connection of the car body and the suspension. Figure 1 shows the schematic of this approach.

$$\{f_s\} = \left([H_{ss}^A] + [H_{ss}^B] + [K]^{-1}\right)^{-1}[H_{si}^A]\{f_i\} \quad (1)$$

The variables in this equation are the following:

$\{f_s\}$ reaction forces in the connection points between suspension and car body

$[H_{ss}^A]$ compliance FRF matrix of the suspension in free-free conditions at the connection points

$[H_{ss}^B]$ compliance FRF matrix of the car body in free-free conditions at the connection points

$[K]$ stiffness matrix, containing the stiffness characteristics of the isolation elements

$[H_{si}^A]$ compliance FRF matrix, describing transfer in free-free suspension from wheelcenter to interface points

$\{f_i\}$ input forces at wheelcenter

In order to then predict the vibro-acoustical FRFs between input forces at the wheelcenters one must combine the reaction forces in the connection between car body and suspension with the vibro-acoustical FRFs,

$$\{p\} = [H_{ps}^B]\{f_s\} \quad (2)$$

with

$\{p\}$ pressure response in interior cavity

$[H_{ps}^B]$ vibro-acoustical FRFs on car body between connection points and interior pressure (without suspension)

When the car body compliance is much lower than the component compliance, equation (1) can be simplified as:

$$\{f_s\} = \left([H_{ss}^A] + [K]^{-1}\right)^{-1}[H_{si}^A]\{f_i\} \quad (3)$$

This equation reflects the constrained reaction forces in the interface due to an input force at the points i in the suspension component.

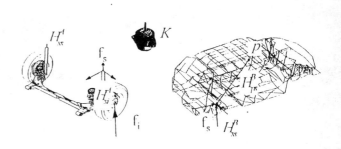

Figure 1. Schematic representation of coupling procedure

2. FINITE ELEMENT MODEL VALIDATION

In a first phase of the project, the finite element (FE) model of the suspension was built, and validated with measured test data on the prototype suspension component. In our FE model, the main component was built directly from the CAD geometry and others were modeled by using simplified elements like beams and springs. The MSC/Nastran model contains 2344 elements and 1562 nodes. The elements are of type BEAM (MSC/Nastran CBAR), SPRING (CELAS), TRIANGLE (CTRIA3), QUAD (CQUAD4), MASS (CONM2).

To simulate realistic boundary conditions, the constrained suspension model is preloaded with the vehicle load. In order to avoid the non-linearities caused by the shock absorbers, the latter have not been included in the FE model for the correlation and updating procedures with test data. The preload is particularly important to model the dynamical coil behaviour correctly. The model also contains the representation of the flexible connectors (bushings).

The test data for correlation were taken on the suspension in similar constrained conditions. Preload masses were used to apply representative vehicle load.

The model is schematically represented in figure 2. The top of the suspension springs is clamped to the ground. The rear axle connections (see B and D in the figure 2) to the ground are modelled by MSC/Nastran springs, representing the bushings. The connection between the main transversal beam (see EF on the figure 2) and the longitudinal beams are modelled by six springs (three in translation, three in rotation).

Three target modes for correlation were selected: 47 Hz (motion of transversal beam; 42 Hz computed), 81 Hz (bending mode; 80 Hz computed) and 132 Hz (rotation around the axis of the transversal beam; 138 Hz computed). The updating parameters were 36 in total:

- the stiffness of the connection between transversal and longitudinal beams in E and F

- the stiffness of the rear axle bushings at B and D
- the stiffness of the connections between the coils and the transversal beam.

In addition to this, the statical stiffness of the suspension coils was adjusted manually.

The updating criteria were the modes of the suspension (frequencies and modeshapes) which is directly related to the road noise to be treated. The quality of the FRF correlation after updating is a good illustration of the FE model validation. Convergence was obtained in four iterations. The interesting fact is that the frequency of the two first modes increases during the updating, while the third frequency decreases.

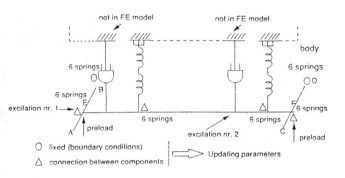

Figure 2. Schematic representation of the suspension FE model

3. DESCRIPTION OF DATA IN COUPLING PROCEDURE

For the calculation of the coupled vibro-acoustical FRFs between wheel center of the suspension and the acoustical response in the car, experimental FRF data were obtained on the trimmed car body without suspension, and the FRFs representing the suspension component were calculated.

The experimental data contained the following:

- vibratory frequency response functions among all connection points between car body and suspension. This makes a matrix of size 18 x 18 (6 connections, 3 degrees of freedom each).
- vibro-acoustical frequency response function matrix between excitation at all connection points and acoustical response at 3 positions inside the cavity. This makes a matrix of 18x3.

The constrained updated and validated FE model of the suspension was adapted in order to make free the suspension connection degrees of freedom. Due to the preconstrained condition of the suspension a special procedure was adopted, in which a static preloaded analysis was followed by a dynamical analysis in which the boundary conditions at the connection points were released. Two approaches for the free-free dynamical analysis were followed:

- Modal Frequency Response (SOL 111), after which the FRFs were synthesised by modal superposition (a 1 % overall damping was assumed)
- Direct Frequency Response (SOL 108), in which the uniform structural damping coefficient was set to 0.02 (a frequency step of 1 Hz was used)

The first methodology implies possible truncation errors on the FRFs, which can be shown to have in some cases a serious effect on the coupled results. Therefore care must be taken to include "enough" free-free modes. A more consistent approach is based on the inclusion of residual effects (reference [1]). In the actual case, the truncation can be completely avoided, by using the direct inversion approach. Due to the rather limited model size this was a feasible approach. A typical comparison between the two sets of FRFs (direct vs. modal) is shown in figure 3.

Figure 4 shows the MAC matrix between the preloaded and non-preloaded normal modes of the constrained FE model of the suspension. One can see a dominant diagonal, but also scatter. The scatter is mainly related to typical spring modes which behave very differently when being preloaded or not. This shows the importance of the preload condition on the dynamics of the (free-free) suspension.

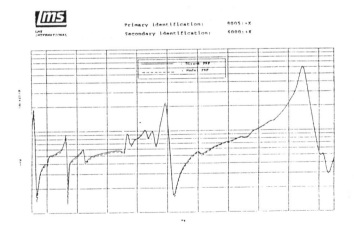

Figure 3. Suspension free-free FRF, obtained by modal synthesis, and compared to direct inversion

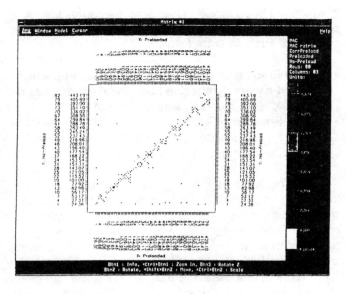

Figure 4. MAC matrix between preloaded and non-preloaded modes of the suspension FE model

4. EVALUATION OF COUPLING PROCESS

4.1. INFLUENCE OF DAMPING IN FE MODEL AND OF SYMMETRY OF EXPERIMENTAL MATRIX -

Considering the equations (1) and (2), one can see that the damping model that is assumed in the FE model is influencing the synthesised FRF matrices $[H]_{ss}^A$ and $[H]_{si}^A$. On the other hand, the symmetry of the experimental FRF matrix $[H]_{ss}^B$ can also be important for the coupling results.

For the evaluation of the effect of the damping model, an amount of 0.2, 1, and 2 % damping for all (FE) suspension modes is assumed, after which the vibro-acoustical FRFs between wheel center input and acoustical response is calculated. Figure 5 shows the three superposed results. It is very clear that the damping model very heavily influences the level of the vibro-acoustical predictions. Tentatively correct FE damping models are therefore prerogative.

Due to measurement errors (alignment of input, cross-sensitivity of sensors, ...) the experimental matrix $[H]_{ss}^B$ is often not completely reciprocal. Therefore, the effect of different matrix symmetrisation strategies is calculated on the vibro-acoustical FRFs. The three following schemes were implemented: one is based on the original experimental matrix, one is based on a copy of the lower diagonal matrix part to the upper diagonal part, and one is based on a copy of the upper diagonal matrix part to the lower part. Figure 6 shows the effects of the different symmetrisation strategies on the coupled calculation results. Possible non-reciprocity of the experimental matrix influences the levels of the results.

4.2. INVERSION OF COUPLING KERNEL -

For the inversion of the "kernel" matrix in equation (1), singular value decomposition techniques are used. This allows decomposition of the matrix in a diagonal singular value matrix, pre- and postmultiplied with the singular vectors, and allows to easily assess and manipulate the condition, by omission of certain singular values.

$$[H]_{\ker nel} = \{U\}[\Sigma]\{V\}^T$$

In this case the size of the matrix is 18x18 (6 connection points in 3 DOFs). This is reflected in figure 7, which shows a number of singular values of this matrix as function of frequency. Even though the ratio between the largest and smallest singular value is high, the smallest singular values still contain system information, and therefore must not be neglected in the inversion process. This is confirmed in figure 8 which shows the coupling results, taking into account certain thresholds for singular value elimination. Clearly underestimates and unrealistic dips in the coupled frequency response functions are obtained.

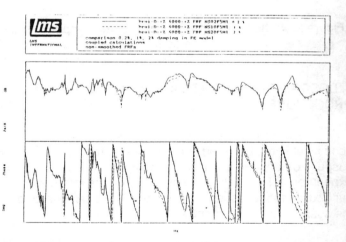

Figure 5. Effect of FE damping on suspension (0.2, 1, 2%) on coupled vibro-acoustical FRF between wheelcenter and acoustical interior response

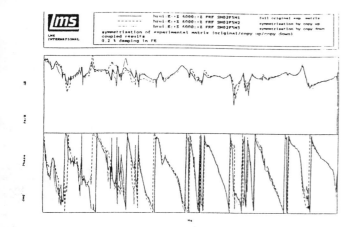

Figure 6. Effect of symmetrisation of experimental FRF matrix on coupled vibro-acoustical FRF between wheelcenter and acoustical interior response

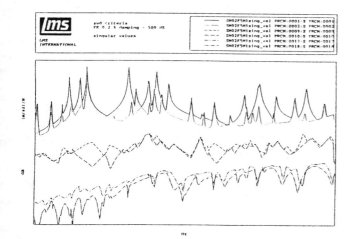

Figure 7. Singular values of the coupling kernel over the frequency range (2 largest, 2 intermediate, 2 smallest)

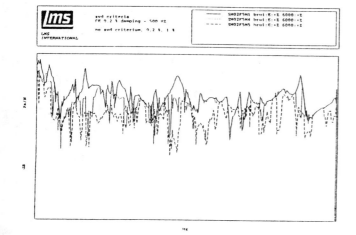

Figure 8. Effect of singular value threshold strategies on coupled vibro-acoustical FRF between wheelcenter and acoustical interior response

4.3. INFLUENCE OF ROTATIONAL VS. TRANSLATIONAL DEGREES OF FREEDOM

In order to evaluate the influence of rotational degrees of freedom, the force transmissibilities between wheel center input and the connections were calculated for the following three situations. First, the 3 translational degrees of freedom at the suspension connection points are coupled with the (measured) car body translational FRFs. The rotational degrees of freedom remain in this case non-constrained. In the second situation, the free-free suspension FE model is constrained (grounded) in 3 translational degrees of freedom in each of the connection points. Third, the free-free suspension model is constrained in 6 degrees of freedom (including the rotational degrees of freedom). This allows evaluation of the influence of the car body on the coupling results; and the influence of restraining the rotational degrees of freedom. Figure 9 shows the results. One can see, when comparing the first two situations, that the force transmissibility is affected by the car body, mainly in level. The effect of the rotational degrees of freedom is more limited, even though frequency deviations can be found. At around 100 Hz, for the rear axle connection point, frequency shifts up to 7 % are introduced. Of course the influence of rotational DOFs heavily depends on the actual rotational stiffnesses of the flexible bushings.

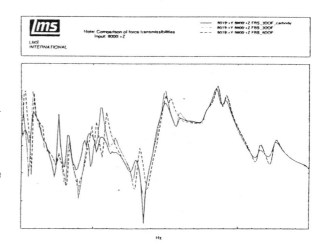

Figure 9. Comparison of force transmissibilities at the rear axle connection due to input at wheelcenter in z direction for the coupling to the car body (only translational DOFs), for the coupling to the ground (only translational DOFs), and for the coupling to the ground (translational and rotational DOFs).

4.4. TRUNCATION EFFECTS

The effect of truncation was analysed by comparing coupling results, based upon two sets of FE FRFs, synthesised from a limited number of modes: only taking free-free modes up to 300 Hz, vs. taking free-free modes up to 500 Hz. The frequency band for the results is considered between 50 and 250 Hz. In figure 10, truncation effects show due to neglected residual flexibility when only considering modes in a limited frequency band. Clearly resonances shift down when taking into account more modes. Truncation effects must therefore certainly be accounted for.

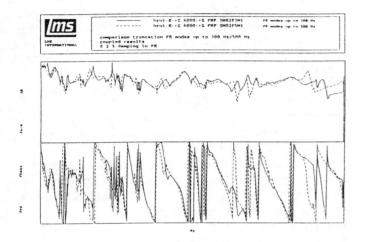

Figure 10. Comparison of coupled vibro-acoustical FRFs due to input at wheelcenter in z direction for modal synthesis of free-free FRFs in the FE suspension model up to 300 Hz and up to 500 Hz

4.5. DIRECT INVERSION AND VALIDATION RESULTS - In order to completely avoid truncation effects, direct inversion techniques have been used. Additionally, validation FRFs between wheel center input and acoustical response were measured on an assembled suspension car body.

Figure 11 compares the results of the validation measurements with the results of the coupling calculations based on FE FRFs obtained by direct inversion. Also the constrained force transmissibility approach (in which the car body is assumed to have zero compliance) is put into the comparisons. In general, the frequency content of the coupling results and the validation results is quite similar. Deviations in level can be seen, which can be explained by a number of reasons: the FE/test model correlation of the suspension was performed up to 130 Hz, the test conditions for validation measurements, the non-linearity of bushings and shock absorbers, the symmetry of the experimental matrix,...

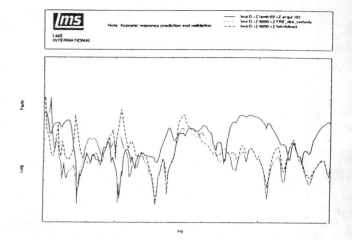

Figure 11. Comparison of coupled vibro-acoustical FRF due to input at wheelcenter in z direction: validation (full line), coupled results with car body (dotted), constrained results (in 6 DOFs) (dashed)

5. INTERPRETATION OF RESULTS

One can interpret the results in a transfer path analysis context. Based upon the substructuring approach, the interface forces in the connection points due to a unit input force at the wheel center are calculated. These forces are combined with the vibro-acoustical FRFs of the car body in order to obtain the contribution from each of the transfer paths in the suspension (see equation (2)).

5.1. FORCE TRANSMISSIBILITIES AT INTERFACE - Figure 12 shows the force transmissibilities in all connection points between car body and suspension. Two force transmissibilities are always dominant: the rear axle connection points. This is certainly true in the frequency bands 70-90 Hz, 130-200 Hz. Around 60 Hz, spring connections are important. In the frequency area around 100 Hz, the spring connections become almost equally dominant. In the frequency ranges over 230 Hz, the shock absorber connection force transmissibilities are getting more important, mainly in the x and y directions.

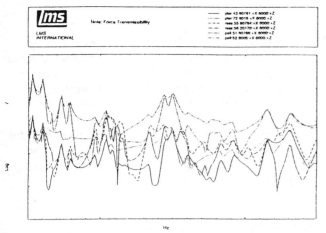

Figure 12. Force transmissibilities at connection points in x direction, due to unit force input at wheel center in z direction

For interpretation of these force transmissibilities, the connection is made to finite element modes of the suspension. Table 5.1 gives an overview of the more important modes. The dominancy of the force transmissibilities at the rear axle connection at 140-150 Hz can be explained by the modal behaviour of the rear axle. Also one can clearly see that above 230 Hz, the force transmissibilities at the shock absorber connections become important due to the modes of the shock absorber.

52.9 Hz	stabilisator mode, spring mode, global movement rear axle
72.6 Hz	spring mode
78.5 Hz	stabilisator mode
96.6 Hz	spring mode
97.3 Hz	spring mode, rear axle
103.9 Hz	rear axle mode, stabilisator mode
144.0 Hz	rear axle mode
149.5 Hz	rear axle mode
159.6 Hz	stabilisator, rear axle mode
214.9 Hz	rear axle and stabilisator
230.4 Hz	shock absorber mode
246.9 Hz	shock absorber mode

Table 5.1 Selection of more important modes of finite element suspension model.

5.2. TRANSFER PATH ANALYSIS

Combining the force transmissibilities with the vibro-acoustical FRFs, one can calculate the partial contributions from each of the paths (per unit input force at the wheelcenter) to the acoustical response in the interior. By summing all the path contributions, one can obtain a summed interior pressure in the car.

Figures 13 shows the contribution analysis for a microphone in the interior cavity. The figure shows the total summation of all the individual pressure contributions from the different connection points or transfer paths, together with the contributions (summed in 3 directions) from the globally most dominant paths, which are the right and left rear axle connections. Additionally in this figure the summed contribution of all the remaining paths (shock absorber and springs) is given. In most frequency ranges, the rear axle connections are indeed most important for the acoustical pressure response.

Focusing on the frequency problem around 150 Hz, the rear axle connections are dominant. To understand the problem in terms of a source/transfer approach, the interacting effects in the force transmissibilities and the vibro-acoustical FRFs must be looked at. Figure 14 shows the vibro-acoustical FRFs to a target pressure point in the cavity due to input at the rear axle connections. It is clear that in the 150 Hz problem area, peaks in the force transmissibilities coincide with peaks in the vibro-acoustical FRFs.

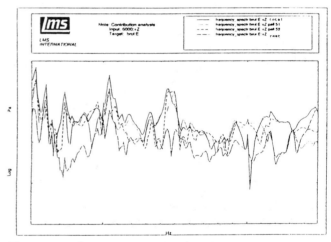

Figure 13. Contribution analysis in point in interior cavity for unit input force at wheel center: total sum of all paths, contribution of left and right rear axle connections, summed contribution of remaining paths

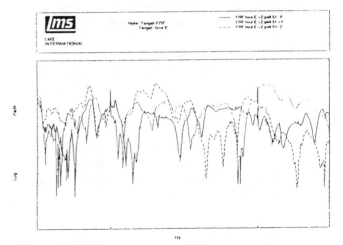

Figure 14. Vibro-acoustical FRF between the rear axle point and the acoustical response in the interior

5.3. MODIFICATION PREDICTION - To evaluate solutions for the problem around 150 Hz, modification predictions on the suspension were performed. A 20% increase in shell thickness of the rear axle was implemented and a prediction of the modified force transmissibilities was performed. An acoustical prediction due to unit input force at the wheelcenter is performed based on this modification.

Figure 15 gives the effect on the synthesised vibro-acoustical response as a result of the structural modification. Clearly by implementing this type of modification, one could resolve the specific acoustical problem.

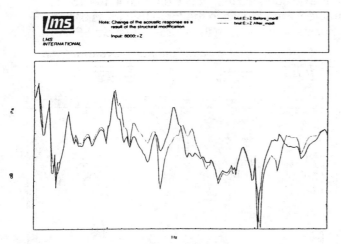

Figure 15. Acoustical response in the cavity before and after modification due to unit input force at wheelcenter

CONCLUSION

The FRF based hybrid substructuring approach is a powerful tool, in combination with modal interpretation as well as with transfer path analysis, to obtain information about possible interaction effects between subcomponents and car body. It is shown in this section that such an approach can be used to provide a strategy of structural modification in order to solve possible acoustical problems.

REFERENCES

[1] K. Wyckaert, K.Q. Xu, P. Mas, The Virtues of Static and Dynamic Compensations for FRF based Substructuring, 15th International Modal Analysis Conference, Orlando, 1997

[2] D. Otte, J. Leuridan, H. Grangier, R. Aquilina, Coupling of Structures using Measured FRFs by Means of SVD Based Data Reduction Techniques, Proceedings 9th International Modal Analysis Conference, Florida, 1990

[3] K. Wyckaert, G. Mc Avoy, P. Mas, Flexible Substructuring Based on Mixed Finite Element and Experimental Models: A Step Ahead of Transfer Path Analysis", Proceedings 14th International Modal Analysis Conference, Michigan, 1996

[4] Queckenberg, Hybrid Test-Analytical Modeling for Structure borne Noise Prediction, Proceedings of the 21st International Seminar on Modal Analysis (ISMA), Leuven, 1996

ACKNOWLEDGMENTS

With many thanks to Tom Martens and K. Q. Xu for their involvement in the project.

971945

Acoustic Analysis of Vehicle Ribbed Floor

Kevin Zhang, Jihe Yang, Lung Wu, and Walt Mazur
Ford Motor Co.

Xiandi Zeng and Xiaoye Gu
Automated Analysis Corp.

Copyright 1997 Society of Automotive Engineers, Inc.

ABSTRACT

Ribbed floor panels have been widely applied in vehicle body structures to reduce interior noise. The conventional approach to evaluate ribbed floor panel designs is to compare natural frequencies and local stiffness. However, this approach may not result in the desired outcome of the reduction in radiated noise.

Designing a "quiet" floor panel requires minimizing the total radiated noise resulting from vibration of the floor panel. In this study, the objective of ribbed floor panel design is to reduce the total radiated sound power by optimizing the rib patterns. A parametric study was conducted first to understand the effects of rib design parameters such as rib height, width, orientation, and density. Next, a finite element model of a simplified body structure with ribbed floor panel was built and analyzed. The structural vibration profile was generated using MSC/Nastran, and integrated with the acoustic boundary element model. Then, a boundary element analysis was performed to predict the total radiated sound power of the floor panel. With this approach, different ribbed floor panel designs were evaluated without hardware being built. The optimized ribbed floor panel design showed a significant reduction of the total sound power in the desired frequency range.

1. INTRODUCTION

Vehicle interior noise, vibration and harshness (NVH) has become increasingly important in recent years. A major factor in interior NVH is the vibrational and acoustic behavior of the vehicle body panels. In particular, body floor panels draw attention in NVH designs since floor panels are usually the major contributor of interior noise, which directly input loads from suspension, powertrain, exhaust system, and air-borne noise to the passenger compartment. In order to have a "quiet" vehicle interior with lighter weight, many automakers have been implementing floor pan designs with some type of ribbing or beading patterns [1]. The analytic approaches are preferred to experimental methods for evaluating ribbed floor designs because of very costly floor panel fabrication equipment. The conventional approach to evaluate ribbed floor panel designs has been to focus on stiffness and natural frequency. The outcome may not result in a reduction in total radiated noise.

A vibrating structural surface radiates sound. It is well known that the radiated acoustic power W is directly related to time and space averaged mean-square surface velocity V^2 and radiation efficiency σ [2], i.e.,

$$W = \rho c A \sigma V^2 \qquad (1)$$

where ρ is the air density, c is the speed of sound, and A is the area of the vibrating surface.

Apparently, total radiated sound power can be reduced by reducing the averaged surface vibration velocity and/or reducing noise radiation efficiency. The radiation efficiency is a non-dimensional quantity which relates the averaged surface velocity to the radiated noise. Its value depends on a vibration pattern and geometry of surface. Generally, it increases with frequency as the vibration pattern becomes more complex until it reaches a constant. Radiation efficiency increases also with stiffness. Usually, the higher the local stiffness, the higher the radiation efficiency. Thus, total radiated sound power is a better criterion than natural frequency or local stiffness for evaluating a rib pattern design.

This paper suggests that the criterion of ribbed floor panel design should be reduction of total radiated sound power. The goal of reducing the interior noise can be achieved by modifying the floor panel structure.

This paper presents an analytical approach to assess vehicle ribbed floor designs. The approach is of a vibro-acoustic method to calculate total radiated sound power and guide floor panel rib designs. A structural Finite Element (FE) model of a floor panel assembly is used to predict the vibration velocity of the floor panel under certain excitations. An acoustic Boundary Element(BE) model of the floor panel is used to calculate the total radiated sound power. The total radiated sound power from the floor panel was calculated and compared with different ribbed floor panel patterns.

2. Parametric Study for Rib Structure

A rib pattern includes several important parameters: the geometry of the cross section, the length, the density, and the orientation. Sometimes, the thickness is also a parameter. However, in this study, the thickness of the floor panel determines the thickness of the ribs. Figure 1 shows those parameters when ribs are built on a flat floor; $B0$ and $B1$ are the base and top width of the rib, H the height, L the length, and D can be used to express the density. The orientation of a rib is the angle between the rib and longitudinal axis of the vehicle. Based on the results of Zhang and *et al* [3], the presence of the rib increases the natural frequency of the ribbed structure. Changing any of the parameters has effects on the natural frequency. The geometry of cross section for individual rib is one of the important parameters. As shown in [3], increasing the height, or decreasing the width increases the frequency. However, the height and the width of a rib can not be arbitrarily changed because of the limitation of the manufacturability. After carefully considering all the factors, the cross section has been chosen as: $B0$=40mm, $B1$=18mm, H=10mm. This cross section will be used for all the rib patterns. The other parameters such as length, density and orientation will be changed

with the locations of the floor panel and with different designs.

3. The Calculation of the Total Radiated Sound Power

The vehicle interior noise is directly generated by the vibration of the body panels through various paths due to road, wind and powertrain input. Usually, the more the power (energy) radiated by those panels, the higher the interior noise level. To evaluate the rib pattern design, the total radiated sound power from the floor panel is calculated. It is a two-step procedure. First, a finite element model is used to calculate the vibration velocity of the floor panel under certain excitations. Then a boundary element model is used to calculate the total sound power radiated from the floor panel.

3.1. Structure Velocity Calculation

To simplify the problem and extend the analysis frequency range, a floor panel assembly model is used in this study. The floor assembly model includes entire floor panels, floor cross members, rockers, dash panels, rear wheel houses, under body rails and back panels. Figure 2 shows the FE model of the floor panel assembly.

Using this simplified model, the structure vibration velocity of the floor panel can be calculated up to 500 Hz. Figure 3 is the FE model of the baseline floor panel design. The FE models for different ribbed floor designs are built and put into the assembly to replace the baseline floor model.

The equation of motion for this floor panel assembly can be written as

$$[M]\{\ddot{x}\}+[C]\{\dot{x}\}+[K]\{x\}=\{f(t)\} \quad (2)$$

Where $[M]$, $[C]$, and $[K]$ is the mass, damping, and stiffness matrix; $\{x\}$ is the displacement vector and $\{f(t)\}$ is the force vector.

The excitations are applied to simulate tire/wheel inputs. The unit sinusoidal forces are applied at the two rear shock towers and the two ends of the front under body rails which are close to the front shock towers. They are labeled as *A*, *B*, *C*, and *D* in Fig. 2. The phase angles between the excitation forces are specified as following. The phase angle of the force at *A* is zero, 90 degrees at *B*, 180 degrees at *C*, and 270 degrees at *D*, The analysis frequency range is from 20 (Hz) to 500 (Hz).

MSC/Nastran was used to perform modal frequency response analysis. The damping ratio for each mode was specified as 3 percent in the frequency range of interest. The structure vibration velocity of the floor panels was output from Nastran, and inputted into the acoustic boundary element model as the boundary conditions for radiated sound power calculations.

3.2. Calculation of Total Radiated Sound Power

Equation (1) can not be directly used to calculate the sound power radiated from a complex structure such as a floor panel. The reason is that the radiation efficiency is not known. A vibro-acoustic approach has to be used to calculate the radiated sound power.

It is well known that the linear wave equation can be written as [4]:

$$\nabla^2 p = \frac{1}{c^2}\frac{\partial^2 p}{\partial t^2} \qquad (3)$$

where ∇^2 is the Laplacian operator, p is the acoustic pressure, and c is the speed of sound. By assuming p is time harmonic, i.e.,

$$p = Pe^{j\omega t}, \qquad (4)$$

equation (3) becomes Helmholtz equation

$$\nabla^2 P + k^2 P = 0 \qquad (5)$$

where k is the acoustic wave number,
$$k = \omega/c. \qquad (6)$$

Once the acoustic pressure in Eq. (5) is obtained for an acoustic field, the power from an source can be obtained. Equation (5) can be solved by numerical techniques such as finite element method or boundary element method. The boundary element method is selected in this study since its simplicity, accuracy, and capability for solving radiation problems.

An acoustical BE model was built for the floor panel which is shown in Fig. 4. This acoustic model was modified from the structure model of the floor panels. Since the acoustic wave length is larger than that of structure wave for same frequency, the acoustic model mesh is not as fine as the structure models.

COMET/Acoustics was used in this study. The indirect method was chosen to perform acoustic analysis.

To calculate the total radiated sound power, a set of data recovery points have to be defined which function the same as the microphones in an experiment . A spherical data recovery mesh was created which enclosed the floor panel. This mesh was used to calculate the total radiated sound from the floor panel.

Figure 5 shows the total radiated sound power of the floor panel as a function of the frequency. The solid line is for the baseline floor panel. As it can be seen, there are several peaks from 100 Hz to 300 Hz. It was desired to reduce these peaks by designing some rib patterns on the floor.

4. Evaluation of Different Ribbed Floor Panel Design

Based on the vehicle on-road test data, the interior noise needs to be reduced in the frequency range between 100 Hz and 300 Hz. The floor panel sound intensity plots of the baseline design were carefully reviewed to identify noise "hot" spots. The "hot " spots help us to identify the rib locations and orientations, and guide the rib pattern designs. Based on the information of radiated sound power, sound intensity and structure vibration velocity, several new designs of rib patterns have been made.

For each design, the analysis procedure was performed and the total radiated sound power was calculated. Then, the designs were compared with the baseline in terms of the total radiated sound power, the vibration velocity of the floor panel, and the acoustic intensity. It is not surprising that total radiated sound power is increased in one of the ribbed floor designs. The prediction of a ribbed floor panel design has been confirmed by an experiment. The analysis and experiment show it is not true that a stiffer floor pan is always better in terms of acoustic performance. The key is to find an optimal design to minimize structure vibration velocity and radiation efficiency at the same time.

After several design iterations, a good design has been reached. Figure 6 shows its rib patterns in a highlighted mode. The total radiated sound power for this ribbed floor panel is the dashed line in Fig. 5. It can be seen that the peaks on the baseline floor in the frequency range of 100 Hz to 300 Hz have been significantly reduced.

5. Conclusions and Discussions

In this paper, an analytic approach to evaluate ribbed floor panel designs is presented. In this approach, the objective of ribbed floor panel designs is to minimize the total radiated sound power. To calculate the total radiated sound power, the finite element method and boundary element method are used for vibro-acoustic analysis. MSC/Nastran is used to predict the structure vibration velocity of the floor panel by employing a FE model of the floor panel assembly. COMET/Acoustics is used to calculate the total radiated sound power by use of a BE model of the floor panel. The information of total radiated sound power, sound intensity, and structure vibration velocity is used to guide the designs. This approach has been successfully applied to a vehicle program for evaluating several floor rib pattern designs. And a promising ribbed floor panel design has been achieved.

6. References

[1] Oka, Y., H. Ono, and N. Hirako, "Panel Vibration Control for Booming Noise Reduction," In *Proceedings of the 1991 Noise and Vibration Conference*, SAE P-244, 1991, pp. 427-433.

[2] Fahy, Frank, *Sound and Structural Vibration: Radiation, Transmission and Response*, Academic Press, Lodon, 1993.

[3] Zhang, Y., K. Phaneuf, and William Hill, "Dynamic Analysis of Rib Patterns," In *Proceedings of the 1991 Noise and Vibration Conference*, SAE P-244, 1991, pp. 221-228.

[4] Kinsler, L. E., A. R. Frey, A. B. Coppens, and J. V. Sunders, *Fundamental of Acoustics*, 3rd Edition, John Wiley and Sons, New York, 1982.

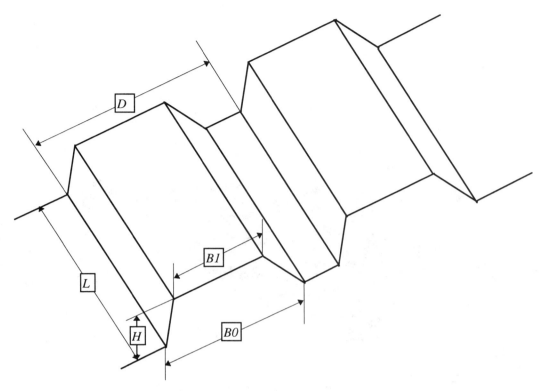

Figure 1. Definitions of rib parameters

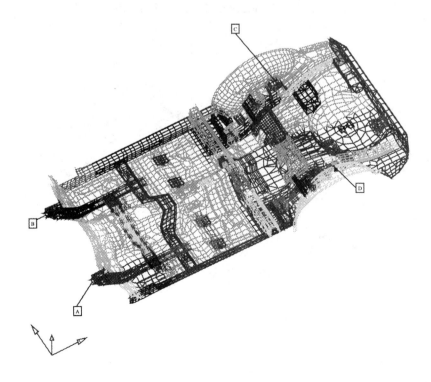

Figure 2. FE model of the floor panel assembly

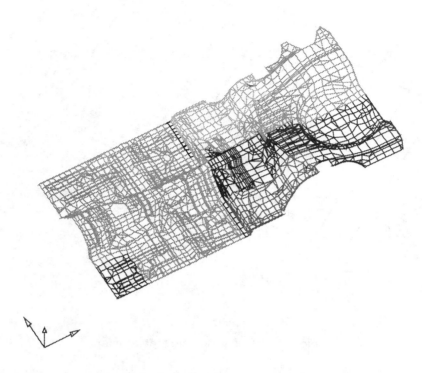

Figure 3. FE Model of the baseline floor panel

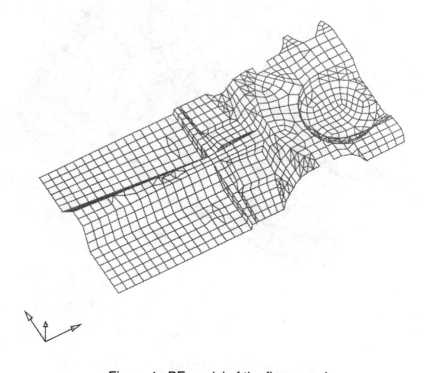

Figure 4. BE model of the floor panel

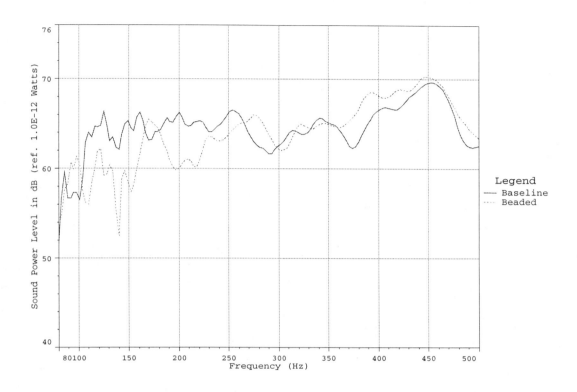

Figure 5. Total radiated sound power of baseline floor and ribbed floor

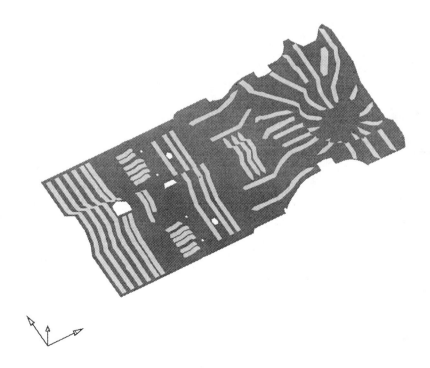

Figure 6. The rib pattern of the good ribbed floor panel design

971946

A Comparison Between Passive Vibration Control by Confinement and Current Passive Techniques

Daryoush Allaei, Yin-Tsan Shih, and David J. Tarnowski
QRDC, Inc.

Copyright 1997 Society of Automotive Engineers, Inc.

ABSTRACT

In this paper, an innovative passive vibration suppression method, namely Vibration Control by Confinement (VCC), is presented. The basic concept of VCC is mainly related to the phenomena of "Anderson localization". The present study is focused on the effectiveness of VCC by applying the method to the vibration suppression of a system comprised of a component resting on a beam-type supporting structure. The effectiveness of the three most common passive vibration suppression methods, namely absorption, isolation, and added damping, are then compared to that of VCC. Based on the results of this study it is shown that VCC is effective over a wider frequency range than those of common practices, and its performance is resonance independent. Furthermore, it is demonstrated that VCC is an effective and viable approach that not only has significantly better performance but also does not have as many limitations as current techniques.

INTRODUCTION

Vibration related issues have always been one of the major concerns in machine and structural design. In general excess vibrations are the result of motions, such as the of rotating and reciprocating parts, and dynamic loads, such as those due to impacts and fluid interactions, internal and/or external to a system. Generated vibrations are then propagated throughout machinery or structures and are often amplified due to inherent imperfections and defects in structures. Even though vibration related issues can be dealt with or at least minimized during the design stage, by design modifications, and/or retrofitting, often they are addressed after the system has been in use for some time. There are two main reasons for this: 1) many vibration related problems show themselves after the system has been in use for some time, and 2) wear and tear, lack of proper maintenance, and change in the operation conditions and/or environment of the system usually increase the vibration levels. Therefore, there has always been a need for more effective approaches with lower cost-to-performance ratio to lower the damaging and discomforting vibrations in machinery and structures. The purpose of this article is to introduce such a method.

Three of the most practiced passive vibration suppression methods can be grouped in three main areas: energy isolation, energy absorption, and energy dissipation through damping material and/or devices. Single and multiple degree-of-freedom systems have been developed to absorb vibrational energy. A variety of isolators have been designed to prevent the transmission of vibrational energy to or from structures and machinery. In other words, vibrational energy is trapped to one side of an isolator. When damping is added to a structure, energy is dissipated in the form of noise and/or heat. These conventional techniques can be implemented in their passive, active, hybrid, and semi-active form depending on the application, flexibility requirements, desired performance, and allowable cost.

There are many advantages for using the above passive methods. Vibration isolators, absorbers, and added damping elements are well understood and have relatively simple mathematical models to be incorporated in the design stage and used by designers and engineers. They are easy to manufacture and low cost to apply. However, they have a few important performance disadvantages. Isolators and absorbers are usually tuned to one or few selected resonant frequencies and therefore, they are most effective within a narrow band around the selected resonant frequencies. Their performance degrades away from the designed frequency ranges. In certain cases, they may even amplify undesired vibrations. As it was previously mentioned, the primary role of added damping in a structure is to take out more energy at a faster rate. Thus, their performance depends on how well and how much energy is delivered to the damping mechanism by the structure. Because structural vibrations are maximum at resonance, damping treatment methods are most effective at and near the resonant frequencies. Weight penalty is a concern when absorbers or added damping elements are used to reduce vibrations. Furthermore, most damping materials have a limited temperature range and perform better at higher frequencies. Therefore, more effective vibration suppression schemes with broader frequency ranges are desired.

In recent years it has been shown [1-4, 9-16] that vibration suppression/control is possible by confinement. The Vibration Control by Confinement (VCC) method is an innovative and effective vibration control approach explored by the authors since early 1990. The VCC approach is based on the theory that the flow of vibrational energy can be controlled and/or trapped within a specified region of a structure and thereby suppress the vibration levels over its remaining parts. Confinement can be implemented in the modal domain and/or physical domain. When vibrational energy confinement is implemented in the modal domain, the phenomenon is known as "Anderson Localization" [5]. In the physical domain, confinement can be induced by applied forces under certain controlled conditions [11-15]. It should be noted that localized modes may not necessary result in the confinement of the total vibration response of a structure. The VCC approach presented in this paper is based on the general form of confinement which can be induced in either modal or physical domain.

The VCC technique is applicable to those structures whose mission, safety, comfort, and/or performance critical components must be sustained at much higher degree of vibration reduction when compared with its non-critical parts. The focus of this work is to demonstrate the effectiveness of the VCC method when compared with the first three common practices described above. To achieve this goal, all four methods are applied to the same structure and their performance is rated based on an identical vibration reduction criterion.

The remaining sections of this paper are organized as follows. In sections two and three, the system under consideration and the application of the four vibration suppression methods are described. Our numerical results and discussion are presented in section four. Based on these results, conclusions and recommendations are made in section five.

PROBLEM DEFINTION

The structure under consideration is shown in Figure 1. It is composed of a critical component modeled as a single DOF system resting on a beam-type supporting structure. The boundary conditions of the supporting beam are assumed to be moment free (i.e., pinned) at both ends. The beam is also supported by a elastic brace. The critical component is placed on the left side and harmonic disturbances are applied to the right side of the supporting beam. The interface between the component and the supporting beam is modeled as a viscoelastic element. It is assumed that a single frequency harmonic disturbance is applied to supporting beam as shown in the Figure 1. Detailed system specifications are listed in Table 1. Note that the resonant frequency of the critical component (about 22 Hz) is designed to be lower than the first resonance (about 36 Hz) of the beam. The frequency range of interest in this study is 10 to 70 Hz, and therefore the first two modes of the beam have significant impact on the total response of the structure.

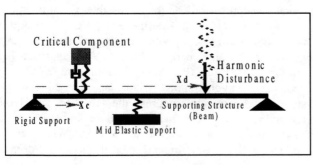

Figure 1 Schematic of the structure

Table 1 System Specifications

Supporting Structure (S-S Beam)	Material, E [psi], ρ [lb/in³]	Steel, 30e6, 0.283
	Dimensions [in]	L=100, W=8, T=1
	Mass [Slug]	7
	Material Damping [%]	2
	Elastic Support Location [in]	48
	Elastic Support Stiffness [lb/in]	1.0e10
	Natural Frequencies [Hz]	36.3, 58.0, 144.0
Critical Component	Location, Xc [in]	36
	Mass [Slug]	0.07
	Stiffness of its Mount [lb/in]	111.6
	Damping Factor [%]	5
	Natural Frequency [Hz]	22
Disturbance	Type	Sinusoidal
	Location, Xd [in]	86
	Magnitude [lb]	1000
	Frequency Range [Hz]	10 - 70
Vibration Suppression Requirements		Minimize Vibration Level at the Critical Component

The vibration suppression requirement is to maintain the vibration level of the component at its minimum. To achieve this objective, three vibration reduction approaches (isolation, absorption, and add damping) are simulated and the results are compared with the Passive Vibration Control by Confinement (PVCC) technique. In the following sections, the specifications for each of the conventional approaches and PVCC are discussed.

VIBRATION REDUCTION APPROACHES

VIBRATION ISOLATOR - Vibration isolation refers to the design of a resilient interface between the component and its supporting structure so as to attenuate unwanted disturbances exchanged between them. Vibration isolators can be passive and/or active. The passive isolators consist of resilient interface that can have two basic members: a spring and a damper. These two element are usually put in parallel. The spring is the resilient member and the damper is a energy dissipator. Metal springs, cork, pneumatic springs, and elastomer (rubber) springs are few examples of passive isolators available in the market.

Since the focus of this work is not the design of a isolator, the authors use the recommended isolator parameters given in reference [6]. The basic idea is to select the spring and damping parameters so that the transmissibility, which is defined as the ratio of the input force to the force transmitted to the isolated mass, is as low as possible at the system resonance. Since the component is subjected to base excitation, high damping levels may be disadvantageous frequencies above the resonant frequency. Therefore, one should compromise between the reduction of vibration at and above the resonant frequency. In the present work, we design the dashpot to achieve a reasonable balance between the transmissibility at and above the resonant frequency of the component.

At the resonant frequency, the transmissibility, T, of a isolator is given by Eq. 1.

$$T = \frac{transmitted\ force}{applied\ force} \equiv \frac{\sqrt{1+(2\xi)^2}}{2\xi} \qquad (1)$$

Based on a value of 20% for the damping factor, the transmissibility of the isolator is about 2.693 which can be considered an effective isolator. Based on the information in Table 1, since the component has its own damping factor of 5% at the interface, the isolator only adds 15% to interface damping. The detailed specifications of the isolator used in this study are shown in Table 2.

Table 2 Isolator Specifications

	Material	Rubber
Isolator was placed between the component and beam	Added Damping [%]	15
	Added Mass	Neglected
	Added Stiffness	Neglected
	Total Damping [%]	20
	Transmissibility	2.60

VIBRATION ABSORBER - In this section, a brief discussion on passive absorbers is presented because the focus of this work is not on the design or development of new absorbers. For more detailed information, the reader is referred to a recent survey [7] on passive, adaptive, active, and semi-active tuned vibration absorbers.

A vibration absorber or neutralizer is basically an added SDOF tuned to absorb the excess vibrational energy. They are most effective at the resonance of the machinery or component whose excess energy needs to be absorbed. Based on standard design practices [6-7], the natural frequency of the attached SDOF absorber is tuned to be the same as the excitation frequency while having an effective mass much smaller than that of the component. Since a range of frequency (i.e., 10 to 70 Hz) is of interest in this study, the natural frequency of the absorber is tuned to that of the component. Furthermore, the mass of the absorber is designed to be 1/12 of the component mass. The damping factor of the absorber is designed [6] to be 18.5%. Detailed specifications of the absorber are shown in Table 3.

Table 3 Absorber Specifications

	Material	Rubber
Absorber was placed on the component	Added Damping [%]	18.5
	Total Damping [%]	23.50
	Added Mass [Slug]	5.83e-03
	Added Stiffness [lb/in]	9.30
	Frequency [Hz]	22.0

LAYER DAMPING - The application of layer damping has now become one of the commonly practiced means of suppressing unwanted vibrations. Among the four major categories of passive damping treatment methods, viscoelastic materials, viscous devices, magnetic devices, and passive piezoelectric, about 85 percent of the passive damping treatments in commercial applications are based on viscoelastic materials [8] used in layered form on the surface of structures. The latter damping treatment is used in this comparison study.

One of the common damping methods with a relatively large weight penalty is the unconstrained or free-layer damping treatment. In this case, a high storage modulus with a high-loss-factor material is applied to the surface of the structure to absorb and dissipate sufficient amount of structural strain energy. Constrained layer treatments are surface treatments where the damping material is sandwiched between the base structure and a relatively stiff constraining layer. As the structure deforms, the constraining layer generates shear force in the damping material. This type of damping treatment is most commonly used to damp bending modes. Constrained layer damping techniques result in relatively smaller weight penalty than their corresponding unconstrained damping to achieve the same performance. In both of this cases, damping is incorporated into the model in terms of modal damping factor.

In the present comparison study, damping is added to the supporting beam structure by increasing the modal damping factor by 15%. In other words, the total structural damping, the sum of the material and layer damping, will be 17% (referred to as LD17%) as it is shown in Table 4. The weight penalty of the damping layer is neglected in this work.

Table 4 Layer Damping (LD17%) Specifications

	Material	Thin Film
Layer damping was added to the supporting beam structure	Added Damping [%]	15
	Total Damping [%]	17
	Added Mass	Neglected
	Added Stiffness	Neglected

VIBRATION CONTROL BY CONFINEMENT - Mode localization implies the confinement of the modal response to a limited segment of the structure, as opposed to a conventional modal response which extends throughout the structure. Vibration confinement implies that the total vibration response of the structure is confined to a limited region. The latter can be caused by either inducing mode localization or applying external forces whose magnitude may depend on displacement field and its derivatives. More details on the basic concept of mode localization and an extensive review of the state-of-art of the Vibration Control by Confinement (VCC) can be found in reference [3-4, 9-11].

The occurrence of mode localization and confinement of vibration energy in structures is now accepted because its presence and effects have been demonstrated in a variety of engineering structures [1-5, 9-15]. Until early 1990, however, it was believed that mode localization occurs only in nearly periodic/cyclic lattices and structures [5, 9-10]. Furthermore, the cause of the occurrence of the phenomenon in such structures was reported [9] to be due to unavoidable small material tolerances, defects caused during manufacturing processes, and/or assembly variations that lead to small differences between the individual components of nearly periodic structures. Thus, the phenomenon was often associated with its random behavior, uncontrollable features, skewing of the prediction models, and catastrophic failures that it can cause. In early 1990, the research community began questioning the validity of these reports. Now, it is theorized [1-4, 11-15] that the phenomenon is more general than previously thought. First, it is not limited to nearly periodic structures, rather mode localization can occur in any structure should the design parameters fall in certain ranges. Second, confinement of vibrations can be passively and/or actively induced in a controlled manner in both modal and physical domains resulting in the confinement of the global vibration response of structures. However, as it was previously thought, the occurrence of the phenomenon can be random due to the randomness in the manufacturing and/or design parameters. Third, on one hand, displacement and stress concentrations due to localized vibrations can be potentially harmful to the structure, on the other hand, confinement of vibrations can be advantageous if used to enhance the performance of the passive/active vibration/noise control systems. The focus of the present work is to demonstrate that passive VCC (PVCC) may be much more effective in suppressing vibrations when compared with conventional passive methods such isolators, absorbers, and added damping. Furthermore, due to its high performance and reduced cost, an optimum combination of PVCC with other passive vibration suppression methods has a great potential to be one of the most attractive passive approaches.

There are several ways [1-2, 11, 13-14] to induce PVCC. Variations in material and/or structural parameters can cause the occurrence of mode localization whose severity depends on the its sensitivity to the variation of such design parameters. In terms of material properties, variation in elastic constants and mass density can be selected as the design parameters to either avoid or induce mode localization. In terms of structural parameters, boundary conditions and coupling constants are among the parameters that can cause mode localization. A collection of methods to induce confinement (and mode localization) in structures has been reported in references [1-4, 9-15]. In the present work, one of the simplest means of inducing modal confinement is selected because the focus of the study is to demonstrate the effectiveness of PVCC rather than how to implement confinement.

Confined modes are induced by adding a torsional stiffener at the same location as the elastic support which is present in all cases (confined and unconfined). In order to induce an appropriate degree of confinement in the structure shown in Figure 1, the dependency of the confinement severity on the stiffness of the torsional spring is first evaluated. Figure 2 shows the degree of confinement for structures with different vibration characteristics and torsional stiffness. The degree of confinement is defined as the ratio of the maximum deflections determined at the left and right spans when the response is dominated by the second beam mode which is a critical mode for the transmission of energy to the component. The dimensionsless torsional stiffness is defined as the ratio of the stiffness of the elastic support to that of the beam as defined below.

$$\overline{K}_t = \frac{K_t}{(EI/L)} \qquad (2)$$

Where E is the elastic constant, I is the area moment of inertia of the cross section, and L is the effective length of the beam. The case of zero torsional stiffness, shown in Figure 2, represents the structure alone and the structure with an isolator or layer damping. The structure with an absorber is shown as the second bar in the figure because it adds some effective mass and stiffness to the structure, and therefore, affects the degree of confinement. The main point is that the above four cases have a noticeable confinement due to the asymmetry caused by the presence of the component and the off center position of the elastic support. The degree of confinement, however, can be effectively controlled by adjusting the torsional stiffness of the elastic support as shown in the Figure 2. In this work, a strong confinement is defined when the magnitude ratio is larger than 10 which does not occur unless the dimensionless stiffness is larger than 500. To induce a moderate confinement, a dimensionless stiffness of 100 (referred to as VCC100) is used in the remaining sections of this paper.

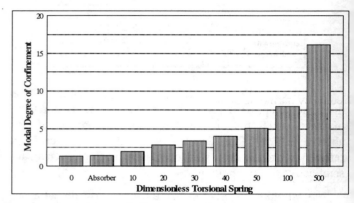

Figure 2 Severity of Confinement

VCC100 specifications are listed in Table 5. As it was expected, the natural frequencies of the beam undergo a relatively small shift due to the additional torsional stiffness. However, the beam still has two resonances within the frequency range of interest. Finally, as it was previously discussed, the addition of the grounded torsional spring, which controls the coupling between the spans of the beam, is only one of the methods to induce confinement. As long as the modes are confined as shown here, the results presented in this paper do not depend on the method used to induce confinement.

Table 5 VCC100 Specifications

Confinement was induced to trap vibrational energy away from the component	Confinement Device	Torsional Spring
	Added Damping	None
	Added Mass	None
	Dimensionless Torsional Stiffness	100
	Location [in]	48
	Beam Frequencies [Hz]	52.0, 61.2, 169

NUMERCIAL RESULTS

The numerical model used in this study is based on standard finite element method. The model consist of a beam with 50 elements, two DOF per node, and with moment free boundary conditions. The elastic support is incorporated in the model as a lumped stiffness element. The component was included in the model as an additional DOF connected to the beam via a spring and a dashpot as shown in Figure 1. Having entered the material properties and geometric characteristics of the structure, free vibration characteristics of the structure were first generated. Next, the forced vibration analysis of the structure was conducted in the modal domain and then transferred to physical coordinates described in the time domain. The total sampled time was chosen to be equal to the 2% settling time for the component as shown by Eq. (2).

$$\text{Sampled time} = 2\% \text{ Settling Time} = \frac{4}{\omega \xi} \quad (2)$$

In Eq. (2), the undamped natural frequency and damping factor of the component are represented by ω and ξ, respectively. The component was designed to have 5% damping factor and a natural frequency of 22 Hz (see Table 1) which gives a sampling time of about 0.5787 seconds. Within this sampled time, the maximum displacement, velocity, and acceleration of the component and its interface with the beam were determined. Finally percent reduction and dB reductions were calculated for comparison purposes. Due to page limitation, only displacement and one case of velocity comparison will be made in this paper.

The disturbance (see Table 1) was chosen to be a single frequency harmonic applied at the location shown in Figure 1. In order to have a comprehensive comparison between PVCC and the other three passive vibration suppression methods, the excitation frequency was set at 11, 22, 36.3, 44, 52, 58, 61.2, and 70 Hz resulting in eight data set. Note that 22 Hz is the resonant frequency of the component while the pair of 36.3 and 58 Hz and pair of 52 and 61.2 Hz are the lowest two resonant frequencies of the supporting beam with unconfined and confined modes, respectively. These selected excitation frequencies represent the worst possible cases for the structural vibrations. Also, the three conventional passive vibration reduction methods are known to be most effective at these frequencies (i.e., resonances). Two of the selected excitation frequencies, 52 and 61.2 Hz, represent the worst cases for the proposed PVCC approach (or VCC100 for this study). The authors believe that this collection of excitation frequencies provides a satisfactory set for a meaningful comparison.

As the first set of the numerical results, the lowest four modes whose natural frequencies are within the range of interest (i.e., below 70 Hz) are presented in Figures 3a to 3c. Figure 3a shows the mode shapes of the structure without the application of any vibration suppression method. This structure will be used as the baseline structure. Note that the first mode, with a frequency of 21.82 Hz, is the component dominated mode. Furthermore, these mode shapes remain the same when an isolator or layer damping is added to the structure. The mode shapes of the structure with the added absorber are shown in Figure 3b. In this case, the first mode with a frequency of 18.95 and the second mode with a frequency of 25.25 Hz are the component and absorber dominated modes, respectively. Finally the mode shapes of the structure with confined beam are shown in Figure 3c. Again, the first mode, with a frequency of 21.9 Hz, is the component dominated mode. Furthermore, it should be noted that half of the beam modes are confined to the left and the other half to the right. The latter is an advantageous for PVCC since the disturbance is applied to the opposite side where the component is located. The advantage is due to the fact that the modes that are confined to the segment, where the load is applied, do not excite the component while the modes that are confined to the component region do not participate strongly in the total vibration response of the beam. Thus, such confinement configuration makes an effective vibration reduction scheme. For each application, confinement configuration can be optimized. The latter issue is not within the scope of this paper.

It should be pointed out that based on the degree of confinement and the mode shapes shown in Figures 2 and 3c, respectively, the added torsional spring does not uncouple the two beam spans. However, it provides a weak coupling which is responsible for the moderate confinement in the beam structure. Dimensionless torsional spring constants larger than 1000 can result in uncoupled beam spans.

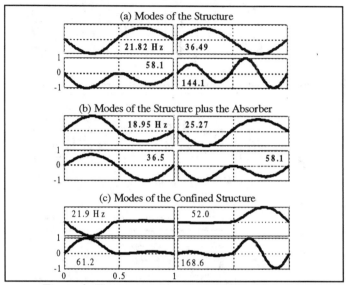

Figure 3 The lowest four modes of the structure

Figures 4 and 5 show sample data sets representing displacements in time domain when the excitation frequency is tuned to 22 Hz. Figure 4 displays the component displacement and Figure 5 displays the displacement at the interface between the component and the supporting beam. As it can be observed, in the case of the confined structure (i.e., VCC100), the component displacements are almost one order of magnitude smaller than the other cases. The latter is the result of significantly low displacement at the interface point as shown in Figure 5. It should be noted that the layer damping is not effective at the resonate frequency of the component (see Figure 4).

Figures 6 and 7 show the maximum displacement at the component and interface, respectively. The maximum displacements are extracted from data sets such as those shown in Figures 4-5. It is clearly shown that the PVCC (or VCC100, in this case) is far more effective in suppressing vibrations at both the component and interface. The isolator and absorber are most effective in reducing the vibrations at the component near its resonant frequency (22 Hz). However, as it was expected, they have practically no impact on the vibrations measured at the interface. On the other hand, the added layer damping (or LD17%, in this case) is most effective near the beam resonant frequencies (36.3 and 58 Hz). Only the PVCC approach offers high level performance

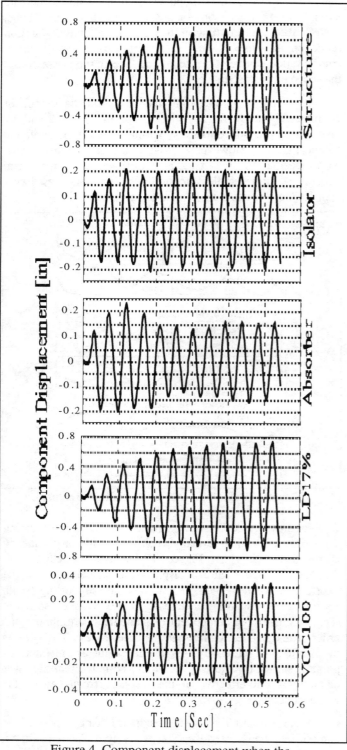

Figure 4 Component displacement when the disturbance frequency is at 22 Hz

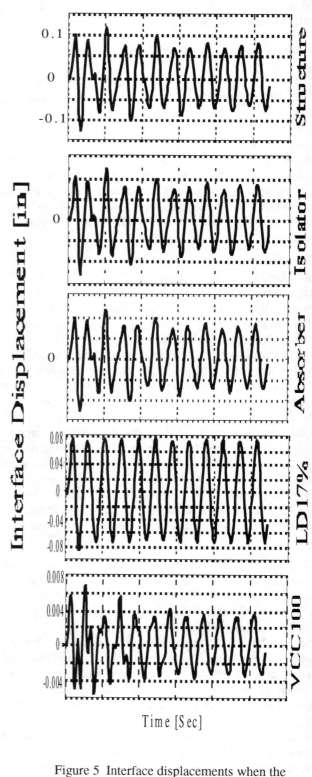

Figure 5 Interface displacements when the disturbance frequency is 22 Hz

in reducing vibrations at the component and the interface throughout the frequency range of interest (10 to 70 Hz). Even at its worst possible case (when the excitation frequency is near 61.2 Hz), PVCC performs better than the other three passive methods to meet the specified vibration requirement (i.e., minimum vibrations at the component).

The superiority of the PVCC with respect to the conventional passive methods is more clearly shown in Figures 8 and 9. The percent reductions of the maximum displacements from the baseline displacements (structure by itself) are displayed in these figures. When the excitation frequency is about 36.3 Hz (first beam resonance), the absorber increases (shown as negative reduction in the figure) the component vibration level. The latter is an inherent characteristic of absorbers.

Another potential benefit of applying PVCC to suppress vibration levels is to reduce the noise level that usually radiates from the supporting structures and or component. In other words, by reducing the velocity levels at the supporting beam structure, one expects to reduce the acoustic pressure levels in the near field. Figure 10 shows the dB reduction of the maximum velocity determined at the interface between the component and the beam. Application of PVCC reduces the velocity level by 4 to 37 dB. The added damping approach results in a velocity reduction of about 4 to 18 dB. VCC100 shows a lower performance than LD17% in the frequency range of 52 to 61.2 Hz in which fall two of the resonances of the confined beam (52 and 61.2 Hz) and one of the unconfined beam (58 Hz). The VCC100 has its lowest performance level at 52 and 61.2 Hz while LD17% has its highest performance at the frequency of 58 Hz. Note that the isolator and absorber have no effect on reducing the vibrations of the beam.

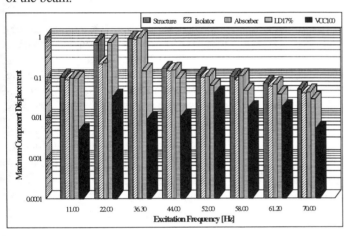

Figure 6 Maximum component displacement at selected disturbance frequencies

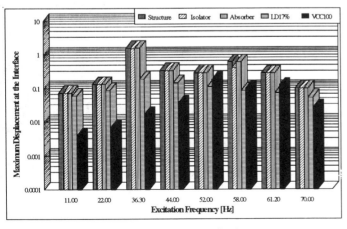

Figure 7 Maximum interface displacement at selected disturbance frequencies

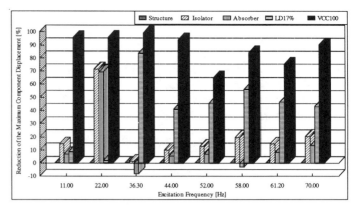

Figure 8 Reduction of the maximum component displacement at selected disturbance frequencies

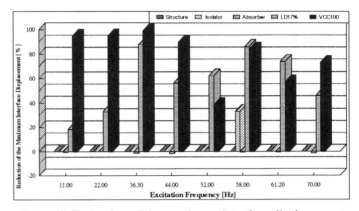

Figure 9 Reduction of the maximum interface displacement at selected disturbance frequencies

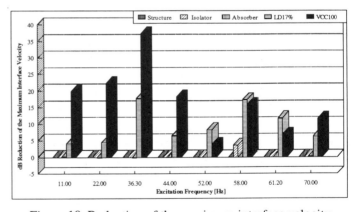

Figure 10 Reduction of the maximum interface velocity at selected disturbance frequencies

SUMMARY AND CONCLUSIONS

The purpose of this paper was to demonstrate that the PVCC (Passive Vibration Control by Confinement) approach is much more effective than currently practiced passive vibration suppression methods. To achieve this objective, the performance of PVCC was compared with three of the most commonly applied passive systems, namely isolators, absorbers, and layer damping. A beam-type structure carrying a critical component and resting on elastic supports was used as the baseline in this comparison study. To obtain numerical results for the worst possible cases, a frequency range was selected so that at least two of the beam resonances and the component resonant frequency are excited. Finally, the maximum displacement, velocity, and acceleration were determined during the same sampled time for each case. The discussion presented in this paper was based on the displacement and velocity comparisons.

Based on the results presented in this paper, it was clearly shown that PVCC had superiority over the conventional isolators, absorbers, and layer damping techniques. It was discussed that PVCC first controls the flow of energy and isolates the sensitive parts of the system while the other three methods focus on suppressing or dissipating the energy. It was demonstrated that PVCC was effective throughout the frequency range of interest while the other methods were most effective near the resonance of the component or the supporting beam structure. Furthermore, it was argued that application of PVCC to structures could result in reduction of radiated noise in near field.

To realize full benefits of the PVCC approach, a more elaborated comparison involving several application problems are being carried out by the authors. Furthermore, a comparison study between the PVCC approach, active version of VCC (AVCC), and current active vibration control methods is in progress at QRDC. The combination of the PVCC and current passive methods has the potential to result in revolutionary effective vibration suppression methods with affordable costs. The authors believe that PVCC is a viable methods and will find its way to the main stream commercial applications in the near future.

REFERENCE

[1] Allaei, D., "Application of Localized Vibration and Smart Materials in Controlling the Dynamic Response of Structures, Part III, " delivered to DARPA, Contract Number DAAH01-94-C-R001, 1994.

[2] Allaei, D., "Application of Localized Vibration and Smart Materials in Controlling the Dynamic Response of Structures, Part IV, delivered to DARPA, Contract Number DAAH01-94-C-R001, 1995.

[3] Allaei, D., "Vibration Control by Confinement: An Overview", 4th International Congress on Sound and Vibration, St. Petersburg, Russia, June 1996.

[4] Allaei, D., "Passive and Active Vibration Control by Confinement: An Overview", Society of Engineering Science, 33rd Annual Technical Meeting, October 1996.

[5] Anderson, P.W., "Absence of Diffusion in Certain Random Lattices." Physical Review, Vol. 109, March 1958.

[6] Hartog, D., "Mechanical Vibrations", 2nd edition, McGraw-Hill Book Company, 1940.

[7] Sun, J.Q., Jolly, M.R., and Norris, M.A., "Passive, Adaptive and Active Tuned Vibration Absorbers - A Survey", Transactions of the ASME Special 50th Anniversary Design Issue, Vol. 117(B), June 1995, p. 171-176.

[8] Johnson, C.D., "Design of Passive Damping Systems", Transactions of the ASME Special 50th Anniversary Design Issue, Vol. 117(B), June 1995, p. 171-176.

[9] Hodge, C.H., "Confinement of Vibration by Structural Irregularities," Journal of sound and Vibration, Vol. 28, No. 3, 1982, p. 411-424.

[10] Levine-West, M.B. and Salama, M.A., "Mode Localization Experiments on a Ribbed Antenna," AIAA J., 31(10), 1993.

[11] Tarnowski, D. J., Allaei, D., Landin, D., Chen, C., and Liu, S. "An innovative noise and vibration suppression technique based on the combination of vibration control by confinement and layer damping approaches," presented at the 3rd Joint Meeting of the Acoustical Societies of America and Japan, Hawaii, 1996.

[12] Yigit, A. and Choura, S., "Vibration Confinement in Flexible Structures via Alternation of Mode Shapes by Using Feedback," J. of S. & Vib. 179(4), 1995.

[13] Shelley, F.J. and Clark, W.W., "Active Mode Localization in Distributed Parameter Systems with Consideration of Limited Actuator Placement," 15th ASME Biennial Conf. On Vibration & Noise, 1995.

[14] Allaei, D., "Performance Comparison Between Vibration Control by Confinement and Conventional Techniques," to be presented Symposium on Active and Hybrid Structural Control, ASME 16th Biennial Conf. on Mech. Vib. & Noise, 1997.

[15] Shih, Y.T., Tarnowski, D.J., and Allaei, D., "Influence of Confined Vibration on Sensor and Actuator Optimization," SAE paper 97NV48, 1997.

[16] Clark, W.W., Sun, F., and Tarnowski, D.J., "On the Use of Vibration Confinement to Enhance Conventional Vibration Control," submitted to J. of Vib. and Control, 1997.

971947

An Indirect Boundary Element Technique for Exterior Periodic Acoustic Analysis

S. T. Raveendra, B. K. Gardner, and R. Stark
Automated Analysis Corp.

Copyright 1997 Society of Automotive Engineers, Inc.

ABSTRACT

The boundary element solution procedure for exterior periodic acoustic problems fails at frequencies associated with the eigenfrequencies of the corresponding interior problems. A new technique is developed to overcome this problem in the indirect boundary element method by expanding the integral equations through the application of multivalued impedance boundary conditions. The effectiveness of this newly developed UNequal Impedance technique for QUalitative Evaluation of acoustic response (UNIQUE) is demonstrated by applying this procedure for the solution a series of exterior acoustic problems.

Introduction

The use of boundary element method (BEM) for the solution of acoustic problems for a wide range of situations is well documented [1], [2], [3], [4], [5], [6], [7], [8], [9] [10], [11], [12], [13]. Two distinct, but formally equivalent, boundary element techniques have been applied in practice: a direct method that involves the modelling of actual physical variables and an indirect method that is expressed in terms of non-physical single and double layer potentials [10], [11], [14], [15], [16]. The use of the indirect method is not as widespread as the direct method in most areas of engineering applications since the indirect method models non-physical variables such as surface potentials. However, the indirect method is well suited for the analysis of acoustic problems since the single and double layer surface potentials are related to physical variables, pressure and velocity. Further, the indirect formulation allows the simultaneous modelling of acoustic domain on both sides of thin vibrating structures trivially.

Boundary element techniques for exterior periodic acoustic problems based on both the direct and indirect methods fail at frequencies associated with the eigenfrequencies of the corresponding interior problems. These failures do not occur at the same frequencies in direct and indirect methods. As an example, for a problem with Neumann boundary conditions, the direct method fails at eigenfrequencies of the associated interior Dirichlet problem, whereas the indirect method fails at eigenfrequencies associated with the interior Neumann problem. Yet, the elimination of this undesirable phenomena is essential in both methods. Techniques to overcome non-uniqueness in the direct version of the boundary element have been actively researched for many years and as a result various procedures have been developed to alleviate this deficiency in the direct method. One widely used method is the CHIEF method [4] in which an overdetermined system is obtained by combining the Helmholtz boundary integral equation collocated at surface nodes with integral equations at interior points that do not fall on the nodes of the eigenfunction of the interior Dirichlet problem. In an alternative method [5], the Helmholtz boundary integral equations and their normal derivatives are linearly combined through a complex coupling parameter to ensure uniqueness. Both of these methods have been improved subsequently [17], [18], [19], [20], [21], [22].

The availability of procedures to overcome the irregular frequencies in the indirect method is rather minimal. The only method that has been utilized with limited success is based on the application of a real non-zero admittance at additionally defined surface within the interior acoustic space [23]. The present work is aimed at developing an alternative procedure to overcome the undesirable behavior at irregular frequencies in the solution of exterior problems using the indirect boundary element method. This is achieved through the development of integral equations that permit the concurrent application of unequal impedance values at the structural-acoustic interface [25]. The resulting integral equations are solved by using a variational approach [24] [11] [25]. The effectiveness of this new technique, which we call UNIQUE (UNequal Impedance technique for QUalitative Evaluation of acoustic response), is demonstrated by applying this procedure for the the solution of problems with known analytical results as well for the solution of a realistic acoustic problem with complex geometry.

Governing Equations

The boundary element method is a numerical solution procedure based on the solution of integral equations derived through the application of the divergence theorem to the inner product of the governing differential equation and the Green's function. The governing differential equation for periodic acoustic problems is the well-known Helmholtz wave equation [26],

$$\nabla^2 p + k^2 p = 0 \qquad in \quad \Omega, \qquad (1)$$

where $k = \omega/c$ is the wave number, p is the acoustic pressure, ω is the circular frequency, c is the speed of sound and Ω is a bounded acoustic domain.

The acoustic velocity vector v_i is related to the acoustic pressure through the expression,

$$v_i = -\frac{1}{j\rho\omega} \frac{\partial p}{\partial x_i}, \qquad (2)$$

where ρ is the density of the acoustic medium and j is the complex constant $\sqrt{-1}$. The normal acoustic velocity v is then obtained as $v = v_i n_i$ where n_i is the normal vector at the surface point.

The indirect boundary element equation may be derived by applying the direct Helmholtz/Kirchoff integral equation to an actual problem domain as well as to an auxiliary exterior domain [14], [10]. The resulting indirect integral equation for acoustic pressure at a *field* point x within he domain can be expressed in terms of single and double layer potentials as

$$p_x = -\int_{\Gamma_Y} \left[G_{xY}\sigma_Y - \frac{\partial G_{xY}}{\partial n_Y}\mu_Y \right] d\Gamma_Y \qquad (3)$$

where Γ is the surface of the acoustic domain, n is the outward normal unit vector to the acoustic domain at a surface point (*source* point) Y and G is the Green's function. The Green's function for the three-dimensional free-space problem, expressed in terms of the wave number and the distance between the source and field point locations r is

$$G_{XY} = \frac{1}{4\pi r}e^{-jkr}. \qquad (4)$$

Further, the single layer potential σ and the double layer potential μ are related to the acoustic velocity and pressure jumps across the surface Γ as

$$\sigma = \frac{\partial p^+}{\partial n} - \frac{\partial p^-}{\partial n} = -j\rho\omega \left[v^+ - v^- \right], \qquad (5)$$

and

$$\mu = p^+ - p^-, \qquad (6)$$

where superscript $'+'$ indicates values on the positive side, which is defined by the outward normal vector at the surface of the domain Ω, and superscript $'-'$ indicates values on the opposite side.

The integral equation for normal acoustic velocity can be derived from the pressure integral equation by using the relationship given by equation (2) as

$$\frac{\partial p_x}{\partial x_i} = -j\rho\omega v_{xi} =$$
$$-\int_{\Gamma_Y} \left[\frac{\partial G_{xY}}{\partial x_i}\sigma_Y - \frac{\partial^2 G_{xY}}{\partial x_i \partial n_Y}\mu_Y \right] d\Gamma_Y. \qquad (7)$$

Boundary Conditions

The integral equations derived in the previous section are solved subject to the applied boundary conditions. The boundary condition for the Neumann problem can be expressed as

$$v^+ = v^- = \bar{v}, \qquad X \epsilon \Gamma_v. \qquad (8)$$

The boundary condition for the mixed problem can be expressed as

$$\begin{aligned} p^+ &= z^+ (v^+ - v_s), \\ p^- &= z^- (v^- - v_s), \qquad X \epsilon \Gamma_z. \end{aligned} \qquad (9)$$

In the above equations (8) and (9), $\bar{v}$ is the known value of acoustic velocity, v_s is the applied structural velocity and z^+, z^- are the specified impedance values on the positive and negative sides of the surface, respectively. The impedance conditions, given by equation (9), can be expressed in terms of specific admittance β using the relationship given by equation (2) as

$$\frac{\partial p^+}{\partial n} + j\,k\,\beta^+\,p^+ + j\rho\omega v_s = 0$$
$$\frac{\partial p^-}{\partial n} + j\,k\,\beta^-\,p^- + j\rho\omega v_s = 0 \qquad (10)$$

where

$$\beta^+ = -\frac{\rho c}{z^+},$$
$$\beta^- = -\frac{\rho c}{z^-}.$$

To substitute the boundary conditions into the integral equations, these conditions must be expressed in terms of surface potentials. Using the definition of these potentials, it is easily deduced that applied velocity corresponds to $\sigma = 0$. However, the impedance boundary condition is not reducible to a single relationship for the unequal impedance case. When $\beta^+ = \beta^- = \bar{\beta}$, we can easily show that the impedance condition is expressible as

$$\sigma + jk\bar{\beta}\mu = 0. \qquad (11)$$

Since $\bar{\beta}$ is specified, we have a single determinate relationship between the single and double layer potentials and this relationship can be easily incorporated into the boundary integral equations.

For the unequal impedance situation that is essential for the elimination of undesirable behavior at irregular frequencies, the impedance boundary conditions cannot be reduced to a single equations in terms of surface potentials. Instead, we can arrive at a pair of relationship from equation (10), using the definitions of single and double layer potentials given by equations (5) and (6) as

$$\frac{\sigma}{jk} = \beta^- p^- - \beta^+ p^+, \quad (12)$$

and

$$jk\beta^+\beta^-\mu = \beta^+ \frac{\partial p^-}{\partial n} - \beta^- \frac{\partial p^+}{\partial n} - j\rho\omega v_s(\beta^+ - \beta^-). \quad (13)$$

Surface Integral Equations

The required boundary integrals equations can be derived by substituting the boundary conditions into the integral equations. At a boundary point $X \in \Gamma_v$, equation 7 takes the form,

$$-j\rho\omega\bar{v}_X = \frac{\partial p_X}{\partial n_X} = FP \int_{\Gamma_v} \mu_Y \frac{\partial^2 G_{XY}}{\partial n_X \partial n_Y} d\Gamma_v$$

$$+ \int_{\Gamma_z} \left(\mu_Y \frac{\partial^2 G_{XY}}{\partial n_X \partial n_Y} - \sigma_Y \frac{\partial G_{XY}}{\partial n_X} \right) d\Gamma_z. \quad (14)$$

At the impedance surface, instead of one boundary condition equation, we have two and therefore, we need two set of equations to complete the formulation. Substituting the surface integral equations corresponding to equations (3) and (7) into the boundary condition relationship equations (12) and (13), the following set of equations is obtained:

$$\frac{j\sigma_X}{k} = \frac{1}{2}\mu_X(\beta_X^+ + \beta_X^-) + \Delta\beta \int_{\Gamma_v} \mu_Y \frac{\partial G_{XY}}{\partial n_Y} d\Gamma_v$$

$$+ \Delta\beta\, PV \int_{\Gamma_z} \mu_Y \frac{\partial G_{XY}}{\partial n_Y} d\Gamma_z - \Delta\beta \int_{\Gamma_z} \sigma_Y G_{XY} d\Gamma_z, \quad (15)$$

and

$$-jk\beta_X^+\beta_X^-\mu_X = \frac{1}{2}\sigma_X(\beta_X^+ + \beta_X^-)$$

$$- \Delta\beta \int_{\Gamma_v} \mu_Y \frac{\partial^2 G_{XY}}{\partial n_X \partial n_Y} d\Gamma_v - \Delta\beta\, FP \int_{\Gamma_z} \mu_Y \frac{\partial^2 G_{XY}}{\partial n_X \partial n_Y} d\Gamma_z$$

$$+ \Delta\beta\, PV \int_{\Gamma_z} \sigma_Y \frac{\partial G_{XY}}{\partial n_X} d\Gamma_z - j\rho\omega\Delta\beta v_s, \quad (16)$$

where $\Delta\beta = \beta^+ - \beta^-$. In equations (14), (15) and (16), PV and FP indicate integrals that must be interpreted in the sense of principal value and finite part integrals, respectively.

The simultaneous solution of equations (14), (15) and (16) constitutes the solution of the prescribed boundary value problem. In the present development these equations are solved through the application of a variational principle [25].

Numerical Examples

The derived integral equations and the numerical implementation procedure are validated by solving the problem of a vibrating plate at the midplane of a duct as shown in Figure 1.

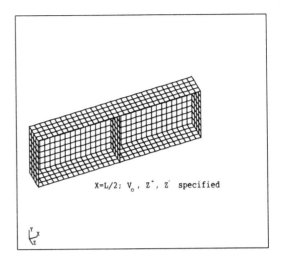

Figure 1: BEM Mesh for Duct

The plate was assumed to vibrate with a velocity of V_o. An impedance value of $0.1 * \rho c$ was specified at the left side of the plate and the impedance value at the right side was varied from a small value that depicts a pressure release situation to a large value that forces the acoustic velocity to be equal to the structural velocity. The magnitude of acoustic pressure normalized with respect to ρc, computed at points along the X-axis within the duct using the newly developed integral equations is compared to the analytical solution in Figure 2. In the figure, Z represents the normalized impedance $z/\rho c$ and the distance is normalized with respect to the length of the duct. The excellent agreement between the theoretical and computed solutions validate the process developed here.

To validate the UNIQUE method for exterior problems, consider the problem of a vibrating sphere. The assumed values for the material properties of the acoustic domains were $\rho = 1.21$ kg/m^3 and $c = 343$ m/s. Initially the problem was solved by using regular indirect boundary element method (IBEM). The sound pressure level (SPL) at a normalized radial distance of $r/a = 2$, where r is the radial distance and a is the radius of the sphere, is compared to the analytical solution in Figure 3. The analytical solution for the acoustic pressure at a radial distance of r is given by the expression [26],

$$p(r) = k\rho c V_o \frac{a^2}{r(1 + k^2 a^2)} (ka + j) e^{jk(r-a)},$$

where V_o is the acoustic normal velocity applied at the surface of the sphere. Although the results are in agreement at most frequencies, the computed results exhibit oscillatory nature at frequencies correspond to wave numbers in the vicinity of $k = 4.493$ ($f = 245.3$ Hz) and $k = 7.725$ ($f = 421.7$ Hz).

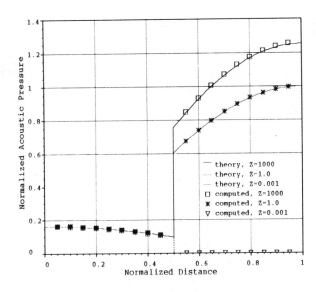

Figure 2: Duct with Unequal Impedance

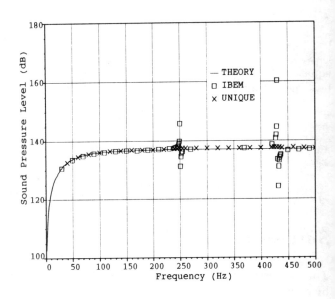

Figure 3: SPL for Sphere

These wave numbers are precisely the first two eigenfrequencies of the the corresponding interior problem. To see the applicability of the currently developed procedure, a large value of impedance was specified on the exterior side and a zero impedance value was specified on the interior side of the acoustic boundary surface. These impedance values force the acoustic velocity to be equal to the structural velocity on the exterior side and simulate a pressure release situation on the interior side. The imposed pressure release condition constrains the influence of the interior acoustic space on the exterior solution. As a consequence, the solution using the UNIQUE method, also plotted in Figure 3, agrees well with the theoretical solution at all frequencies. This confirms the validity of the UNIQUE method for the solution of exterior acoustic problems at all frequencies.

In the next example, the radiation from a rectangular box is studied. The dimensions of the box were taken as $L_x = 0.5$m, $L_y = 0.1$m and $L_z = 0.1$m (see Figure 4). The same material properties used in the previous example were used in this example as well. A uniform velocity of V_o was assumed on one surface of the box along $X = 0$ plane. As before, the problem was solved by using the regular indirect boundary element procedure as well as by using the newly developed UNIQUE method. Figure 5 shows a comparison of the sound pressure level at a point located at a distance of 1 m along the X-axis using the regular indirect boundary element method and UNIQUE procedure. While the behavior of the computed solutions using the UNIQUE procedure is acceptable at all frequencies, the regular BEM solutions failed at frequencies near $f = 343$ Hz and $f = 686$ Hz. These are the eigenfrequencies of the corresponding interior problem.

Finally, the applicability of the UNIQUE method for the solution of realistic problems with complex geometries was investigated by solving the radiation problem of a generic

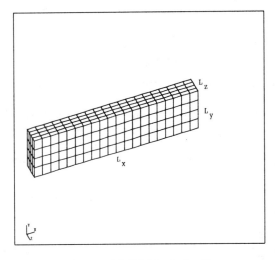

Figure 4: BEM Mesh for Box

submarine. The boundary element model of the submarine is shown in Figure 6. The example was solved by using the regular indirect BEM method as well as by using the UNIQUE method and the CHIEF method based on direct boundary element technique. The values of sound pressure levels computed at an arbitrarily chosen far field point are shown in Figure 7. The figure shows that the results obtained by using the regular indirect BEM are not valid at all frequencies, however, the results using the UNIQUE method are valid at all frequencies and these results correlate well with the CHIEF method solutions. These results validate the applicability of the UNIQUE method for the solution of realistic exterior acoustic problems.

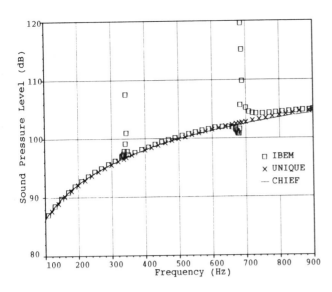

Figure 5: SPL for Box

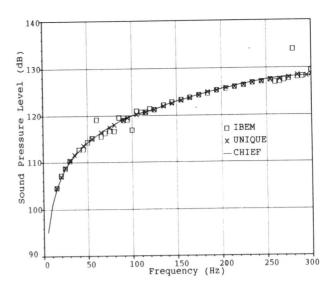

Figure 7: Comparison of SPL for Submarine

neously approximate a wide range of situations ranging from acoustically rigid to pressure release conditions are formulated. The validity and applicability of the proposed process is confirmed by solving a series of acoustic radiation problems.

References

[1] *Numerical Techniques in Acoustic Radiation*, edited by R. J. Bernhard and R. F. Keltie, The American Society of Mechanical Engineers, New York, 1989.

[2] *Boundary Element Methods in Acoustics*, edited by R. D. Ciskowski and C. A. Brebbia, Computational Mechanics Publications, Southampton, Elsevier Applied Science, London, 1991.

[3] Chertock, G. "Sound Radiation from Vibrating Surfaces," *Journal of the Acoustical Society of America*, 1964, 36(7), pp. 1305-1313.

[4] Schenck, H. A. "Improved Integral Formulation for Acoustic Radiation Problems,", *Journal of the Acoustical Society of America*, 1968, **44**, pp. 41-58.

[5] Burton, A. and Miller, G. "The Application of Integral Equation Methods to the Numerical Solution of Some Exterior Boundary Value Problems," *Proceeding of the Royal Society of London*, 1971, **323**, pp. 201-210.

[6] Meyer, W. L., Bell, W. A., Zinn, Z. T. and Stallybrass, M. P. "Boundary Integral Solutions of Three Dimensional Acoustic Radiation Problems," *Journal of Sound and Vibration*, 1978, 59(2), pp. 245-262.

[7] Koopmann, G. H. and Benner, H. "Method for Computing the Sound Power of Machines Based on the

Figure 6: BEM Mesh for Generic Submarine

Conclusion

The boundary element method is a well established tool for the solution of acoustic problems. In particular, the indirect BEM is well suited for the modelling of acoustic domains with complex geometric properties. However, the conventional indirect method solution process for exterior acoustic problems fails at frequencies that correspond to the eigenfrequencies of the relevant interior problems. This irregular behavior is overcome by applying a large impedance value that constrains the acoustic velocity to structural velocity at the outer surface and a zero impedance value that corresponds to pressure release condition at the inner surface of the acoustic domains. To facilitate the application of different impedance values concurrently on either side of the same surface, integral equations that can handle impedance conditions that simulta-

Helmholtz Integral," *Journal of the Acoustical Society of America*, 1982, **71**(1), pp. 78-89.

[8] Seybert, A. F., Soenarko, B., Rizzo, F. J. and Shippy, D. J. "An Advanced Computational Method for Radiation and Scattering of Acoustic Waves in Three Dimensions," *Journal of the Acoustical Society of America*, 1985, **77**, pp. 362-368.

[9] Kipp, C. H. and Bernhard, R. J. "Prediction of Acoustical Behavior in Cavities Using an Indirect Boundary Element Method," *ASME Journal of Vibrations, Acoustics, Stress and Reliability in Design*, 1987, **109**(1), pp. 22-28.

[10] Filippi, P. J. T. "Layer Potentials and Acoustic Diffraction," *Journal of Sound and Vibration*, **54**(4), 473-500, (1977).

[11] Hamdi, M. A. *Formulation Variationnelle par Equations Integrales pour le Calcul de Champs Acoustiques Lineaires Proches et Lointains*, Ph. D Thesis, Universite de Technologie de Compiegne, France, 1981.

[12] Coyette, J. P. and Fyfe, K. R. "Solution of Elasto-Acoustic Problems Using A Variational Finite Element/Boundary Element Technique,", *Numerical Techniques in Acoustic Radiation*, (Eds. Bernhard, R. J. and Keltie, R. F), The ASME, New York, 1989.

[13] COMET/Acoustics User's Manual : Version 3.0, (Automated Analysis Corporation, Ann Arbor, Michigan, 1995)

[14] Lamb, H. *Hydrodynamics*, Cambridge University Press, 1924.

[15] Banerjee, P. K. *Boundary Element Methods in Engineering*, McGraw Hill, London, 1994.

[16] Brebbia, C. A., Telles, J. C. F. and Wrobel, L. C. *Boundary Element Techniques*, Springer-Verlag, Berlin Heidelberg, 1984.

[17] Jones, D. S. "Integral Equations for the Exterior Acoustic Problems," *Q. J. Mech. Appl. Math.*, 1974, **26**, pp. 129-142.

[18] Seybert, A. F. and Rangarajan, T. K. "The Use of CHIEF to Obtain Unique Solutions for Acoustic Radiation Using Boundary Integral Equations," *Journal of the Acoustical Society of America*, 1987, **81**, pp. 1299-1306.

[19] Cunefare, K. A., Koopmann, G. and Brod, K. "A Boundary Element Method for Acoustic Radiation Valid for All Wavenumbers," *Journal of the Acoustical Society of America*, 1989, **85**, pp. 39-48.

[20] Segalman, D. J. and Lobitz, D. W. "A Method to Overcome Computational Difficulties in the Exterior Acoustic Problems,", *Journal of the Acoustical Society of America*, 1992, **9**, pp. 323-329.

[21] Ingber, M. S. and Hickox, C. E. "A Modified Burton-Miller Algorithm for Treating the Uniqueness of Representation Problem for Exterior Acoustic Radiation and Scattering Problems," *Engineering Analysis with Boundary Elements*, 1992, **9**, pp. 323-329.

[22] Dargush, G. F., Raveendra, S. T. and Banerjee, P. K. "Boundary Element Formulation for Structural Acoustics Including Mean Flow Effects," *Computational Methods in Fluid/Structure Interaction*, 1993, AMD-Vol. 178, pp. 39-50.

[23] Mabarek, L. and Hamdi, M. A. "Resolution du Probleme des Frequences Irregulieres dans la Methode des Equations Integrales," Communication presented at $10^e me$ Colloque d'Acoustique Aeronautique et Navale, Marseille, 1986.

[24] Mikhlin, S. G. *Variational Methods in Mathematical Physics*, Macmillan Company, New York, 1964.

[25] Raveendra, S. T. Vlahopoulos, N. and Glaves, A. "An Indirect Boundary Element Formulation for Multi-Valued Impedance Simulation in Structural Acoustics," submitted for publication, 1997.

[26] Pierce, A. D. *Acoustics : An Introduction to its Physical Principles and Applications*, McGraw Hill, New York, 1981.

971948

Modal Content of Heavy-Duty Diesel Engine Block Vibration

Deanna M. Winton and David R. Dowling
University of Michigan

Copyright 1997 Society of Automotive Engineers, Inc.

ABSTRACT

High-fidelity overall vehicle simulations require efficient computational routines for the various vehicle subsystems. Typically, these simulations blend theoretical dynamic system models with empirical results to produce computer models which execute efficiently. Provided that the internal combustion engine is a dominant source of vehicle vibration, knowledge of its dynamic characteristics throughout its operating envelope is essential to effectively predict vehicle response.

The present experimental study was undertaken to determine the rigid body modal content of engine block vibration of a modern, heavy-duty Diesel engine. Experiments were conducted on an in-line six-cylinder Diesel engine (nominally rated at 470 BHP) which is used in both commercial Class-VIII trucks, and on/off-road military applications. The engine was mounted on multi-axis force transducers in a dynamometer test cell in the standard three-point configuration. Standard modal analysis techniques were exploited to determine i) the rigid body modal characteristics of the engine block, and ii) the engine mount force signatures of the six rigid body vibration modes of the engine block. Modal content of the firing engine and its relevance to vehicle simulation is discussed.

INTRODUCTION AND MOTIVATION

To reduce new vehicle development time and cost, full-vehicle computer simulations are both in use and under continuing development for both commercial and military applications. To be effective, these simulations must accurately predict design sensitivities for improvement of existing products and, in military applications, provide sufficient realism for training vehicle operators.

The inherent complexity of a modern ground vehicle, from in-cylinder combustion processes to the suspension system for the wheels and/or tracks, renders direct numerical simulation of all the vehicle's moving parts and fluids beyond the capabilities of the fastest computers. In addition, full simulations are never possible in the early design phase because the level of detail necessary for their implementation has not been reached. Therefore, useful full-vehicle simulations must be both computationally efficient and capable of handling some design uncertainty. This means that modeling decisions must be made on how to relinquish fidelity while including experimental inputs and incorporating only a coarse level of design information. For complex vehicles and vehicle subsystems, the trade-off between computational efficiency and accuracy is seldom known in advance; therefore, experimental testing of simulation predictions are required as well (for additional information see Bretl 1995). Hence, dynamic experimental measurements of vehicles and vehicle components can influence simulations in two ways: i) by providing empirical models and correlations on which to base higher level models, and ii) by providing test results from which simulation accuracy can be assessed.

Realistic simulation of both on- and off-road vehicles requires accurate modeling of the engine and powertrain. The most direct interaction between the vehicle frame and the engine occurs at the motor mounts through engine weight and vibration loads. Given the dynamic complexity of engine phenomena, an overall vehicle simulation would clearly benefit from a reduced-order description of the vibratory motor mount loads produced by a running engine. This paper describes an experimental study that seeks to determine if such an engine vibration model, based on the six rigid body vibration modes of the mounted engine, is possible and, if so, the model's ultimate accuracy. All results are based on experiments conducted with the engine on a test stand in an operable condition. Natural vibration modes were determined via impact testing while the engine was not running. The assessments of the modal formulation's utility are based on measurements of the fired engine near idle, and at a typical operating speed and load.

Most previous engine vibration studies address the isolation problem in which the mounting scheme is chosen to minimize some combination of cost and transmitted loads (see Hata et al. 1987; Johnson et al. 1979; Snyman et al. 1995). Typically, a matrix analysis of a multi-point mounting system is carried out for a single excitation frequency, with invocation of linear superposition for multiple frequency excitation (see Ashrafiuon et al. 1992; Spiekermann 1982). While successful for choosing motor mount stiffness and damping properties, this approach assumes that the excitation from the engine is known and is, therefore, not directly applicable to situations where it is not. The results from other studies of the dynamics of the piston-rod-crankshaft system could be used to predict engine excitation levels, but these models have not been coupled to the influence of vibration-isolating motor mounts to predict realistic vibration loads. Piston-rod-crankshaft dynamic models have either been developed entirely for crankshaft torsion analysis (see Zhengchang et al. 1988), or are multi-cylinder extensions from single cylinder models which simulate the kinematics or dynamics of an assumed rigid crankshaft (see Norling 1978; Shiao 1994). A few previous studies have addressed rigid-body modal vibration of multi-cylinder engines (see Muller et al. 1995; Radcliffe et al. 1983; Schmitt et al. 1976); however, these studies have either been conducted on passenger car spark-ignition engines, or they have not included the vibration state of the running engine. In addition, the experiments described in these previous studies typically infer vibration response from accelerometer signals.

This current study addresses the rigid body modal content of a running heavy-duty Diesel engine on typical vibration isolation mounts based on direct three-component measurements of motor mount loads. In practice, for vehicle design simulation, both the rigid body modes of the mounted engine and the running engine's vibration excitation would have to be predicted apriori from generic cylinder pressure histories, inertial properties of the engine and the engine components, the location and orientation of the engine mounts, and estimates of the directional stiffness and damping coefficients of the mounts. In this study, the rigid body modes of the engine on its mounts and the running engine's vibration excitation are not predicted but are instead determined experimentally using modal analysis techniues. Hence, the results obtained here represent a "best-case" scenario since the modes and excitation predicted by any simulation technique are essentially guaranteed to be less accurate than the experimentally determined ones. Therefore, the main contribution of this paper is the assessment that a formulation of heavy-duty Diesel engine vibration based on rigid body modes is attractive at low engine load and speed, as long as the modes are known accurately. The current findings suggest that the rigid body modes of the quiescent (stationary) engine can account for approximately 90% of engine vibration when the engine is operated near its idle speed at low load (approximately 1% of full load torque). At higher engine load (40% of full load torque) and speed, the formulation is not as accurate, but may still provide acceptable accuracy for vehicle simulations where speed of execution is paramount. Extension of this investigation to include predicted rigid body modes and predicted engine excitation is currently underway.

The remainder of the paper is organized into three sections. The next section describes the experimental step and the instrumentation used in this investigation. The third section provides the experimental results and is split into two parts: i) quiescent engine impact testing modal analysis, and ii) fired (running) engine modal content results. The final section summarizes the conclusions drawn from this study.

EXPERIMENTAL TECHNIQUE

EXPERIMENTAL SET-UP - Modal parameters were determined for the Detroit Diesel Series 60 Engine which was mounted in a dynamometer test cell. The test set-up is shown in Figure 1. A flexible driveshaft, equipped with universal joints on each end, connects the engine to the dynamometer. A torquemeter was installed between the engine flywheel and driveshaft to measure instantaneous torque of the engine.

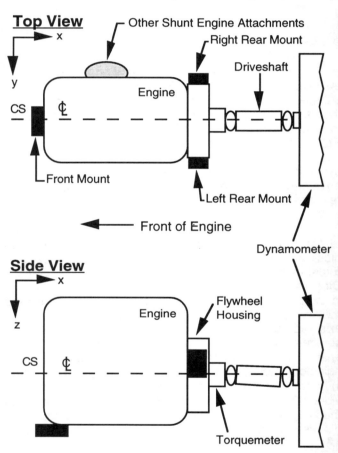

Figure 1. Schematic of experimental set-up
(Dark blocks correspond to mount locations)

The engine was mounted on rubber isolation mounts typically used for this engine type. Three-component load cells were installed beneath each of these three mounting points. These mounting points are indicated as the dark blocks in Figure 1. Load cells were used to measure the instantaneous load transmitted between the rubber engine mount and supporting structure. In vehicle applications the supporting structure would be the vehicle frame or subframe. In this experiment the supporting structure is limited to rigid "feet" attaching the engine to the bedplate of the dynamometer test cell. Because this transmitted load is the desired engine output measurement, load cells were installed in contrast to accelerometers which are typically used in engine vibration measurement. Accelerometers simply measure the kinematics of particular points on an engine. Once the kinematics of the engine mount are measured by an accelerometer, the engine mount stiffness characteristics must be known, or in many cases estimated, in order to predict the ultimate transmitted engine load to the supporting structure. The engine mount stiffness is very difficult to estimate due to inherent nonlinearities in the elastomeric material of the mounts; therefore, the transmitted load was directly measured through use of load cells.

Engine Specifications - The Detroit Diesel Series 60 Engine used is a 470 BHP, 12.7 liter, in-line, six-cylinder, four-stroke heavy-duty engine. It is used in both commercial Class-VIII trucks as well as military applications. Table 1 lists engine specifications.

Table 1. Detroit Diesel Series 60 General Engine Specifications (12.7 liter Automotive; 470 BHP)

Engine Type:	in-line 6; 4 stroke; DI Diesel
Air System (test set-up):	turbocharged water-to-air charge cooling
Displacement:	774 cu. in. (12.7 liters)
Rating (maximum torque @ rpm):	1550 lbf-ft (2101 N-m) @ 1200 rpm
Approximate Engine Idle:	600 rpm
Weight:	
Nom. Weight (dry) - lbf (kg):	2630 (1193)
Nom. Weight(wet) - lbf (kg):	2752 (1248)
Exp. Weight (wet) - lbf (kg): (as mounted in test set-up)	3386 (1536)
Size:	
Length - in (mm):	57 (1448)
Width - in (mm):	34 (864)
Height - in (mm):	50.1 (1273)

In these experiments, the engine was equipped and instrumented for full running conditions; including a laminar air flow intake, an air-to-water charge-air cooler, an exhaust system, and a driveshaft coupling to the DC dynamometer. Most of these devices were attached to the engine according to the top view of Figure 1. They are each sources for shunt force attachments beyond the standard three mounting points. These attachments also contributed to the vibrating dynamics of the engine.

Dynamometer and Driveshaft Specifications - The dynamometer used in the experimental set-up was a General Electric Model 26G26 DC electric dynamometer capable of absorbing 600 HP and supplying 500 HP motoring. This dynamometer was controlled by a Dyne-Systems Dyne-Loc IV controller which was programmed to hold engine speed constant during these experiments.

The driveshaft was a Dana Spicer Series 1810 rubber element driveshaft assembly (Part No. 907912), and was rated up to 1600 lbf-ft torque capacity with a torsional flexibility of 0.33 degree/(100 lbf-ft). This driveshaft was chosen in order to ensure that its rotational resonance was below the engine idle frequency. The driveshaft was installed at a 3 degree rotation from both the crankshaft and dynamometer axes, as shown in Figure 1, in order to ensure limited frictional wear and provide adequate movement of the universal joints during engine operation.

Engine Mount Specifications - The engine was mounted on rubber elements manufactured by Lord Corporation. All mounts were oriented symmetrically with the global coordinate system of the engine (indicated in Figure 1). Each of the two rear mounts (Part No. CB-2204-2) have a center bolt which, upon installation, compresses the washer-like rubber element to a specified width. This specified width along with the material properties of the mount determines the dynamic stiffness of the mount. The single front mounting point incorporates two side-by-side rubber mounts (Part No. SSB33-1000-4). These mounts are similar in installation procedures to the rear mounts; however, they have different material properties and different stiffness characteristics.

The rear mounts have the same stiffness characteristics in the X and Y directions and symmetry around the Z axis (see Figure 1 for coordinate system convention). These mounts are more flexible in the Z direction than in the X or Y directions. Similarly, each of the front mounts have the same stiffness characteristics in the X and Y directions; however, these mounts are stiffer in the Z direction. This stiffness configuration ensured that the rolling resonance of the engine (rotation around the X axis) was below the firing frequency of the engine at nominal idle. The firing frequency of this engine is three times the engine speed (three torque pulses per revolution for an in-line six-cylinder engine).

Load Cell Specifications - Beneath each of the three mounting points, a three-component load cell was installed. These load cells were manufactured by Advanced Mechanical Technology, Inc. (AMTI). Each of these load cells is a strain gauge type device enclosed in a five inch diameter cylindrical-shaped aluminum casing. The strain-gauged, precision elements inside this casing isolate the three orthogonal forces applied between the two axisymmetric, circular, flat plates incorporating the cylindrical casing. These load cells were also oriented symmetrically with the global coordinate system of the engine and engine mounts.

The rear mounting points used the MC5-3-5000 load cell which has a 5000 lbf vertical (Z) capacity and a

2500 lbf horizontal (X and Y) capacity. Each of these two load cells was installed beneath a rear rubber mount with the Z (global) axis corresponding to the symmetrical axis of the load cell. The front mounting point used the MC5-4-10000 load cell which has a 10,000 lbf vertical (Z) capacity and a 5000 lbf horizontal (X and Y) capacity. This load cell was similarly installed beneath the front mounting point (beneath the two front rubber elements) with the Z (global) axis corresponding to the symmetrical axis of the load cell.

The strain gauge outputs of the three load cells were appropriately conditioned and amplified with AMTI signal conditioners. The nine amplified outputs (three forces per mount) were then digitized using Tektronix data acquisition hardware and a computer. The digitized signals were then manipulated with the LabVIEW software package (available through National Instruments Co.) before storage and subsequent modal analysis data reduction.

EXPERIMENTAL PROCEDURE - The experimental procedure was a modified form of a typical modal analysis procedure. In this procedure the system was assumed to be linear; and the modal parameters (frequencies, dampings, and signatures) were experimentally determined from an impact input to the quiescent engine. In this modal analysis, nine loads were measured simultaneously for a single impact (input). In assuming reciprocity of those measurements, one test supplies sufficient data for the curve fitting analysis to extract the modal parameters. However, more tests were taken to aid in extracting accurate curve fitted results. The running engine was then decomposed into a linear combination of the modal signatures found from the curve fitting analysis.

The obvious modification to the modal analysis procedure was that the output was measured as a load and not as an acceleration, a velocity, or a displacement. Therefore, the spectral output could not be cast into the typical modal vectors associated with displacement. Alternatively, the modes found through this study are represented through "modal signatures". These modal signatures represent the character of the loading which the engine applies to the supporting structure through its mounts. At each modal frequency, a modal signature was determined. It is a vector where each element can be essentially defined as the relative amount of load contributing to the total loading applied to the supporting structure. Therefore, each modal signature naturally occurs at its corresponding modal frequency. The running engine was then characterized by these modal signatures. The motor mount forces generated by the running engine were decomposed into a linear combination of modal signatures. This linear combination was represented by the linear weighting coefficients or modal content vector.

The engine was impacted with a rubber hammer near the left rear engine mount. This impact ensured a broad frequency input force typical of modal analysis procedures. The impact hammer, however, was not instrumented with a load cell or an accelerometer. It was simply used as an excitation device which provided adequate broad band frequency input in the frequency range of interest.

Once the engine was impacted with the hammer on the left rear mount, the digitizer and data manipulating software were triggered to simultaneously retrieve and store the nine channels of amplified load cell data at a sampling rate of 2.1 kHz for 5 seconds. In order to facilitate the appropriate trigger, an accelerometer was installed on the left rear mount. The accelerometer voltage output, which instantaneously measured any motion of the engine, was used to trigger the digitizing system. In the post-processing stage of manipulation, the software then computed an amplitude and phase spectrum of the nine temporal outputs via the Fast Fourier Transform (FFT) routine within LabVIEW. Ultimately, three impact tests were used in the following analysis. The left rear mount was impacted in the X, Y, and Z directions.

Upon completion of the impact testing, the engine was run at two different operating points (load and speed). The same digitizing and data acquisition procedure was used. The two operating points used in this study are [750 rpm; 20 lbf-ft] (which is characteristic of the engine idling condition) and [1200 rpm; 615 lbf-ft] (which corresponds to 40% load at 1200 rpm).

DATA REDUCTION METHOD - Data reduction began with extracting the modal parameters from the quiescent engine impact tests through advanced modal curve fitting. Once the modal signatures (more specifically, residues) were determined numerically, the running engine spectra were decomposed into a linear combination of these modal signatures through a least squares analysis. The result is a modal content representation of the running engine.

The spectral data taken from the aforementioned experiments were stored into ASCII format files suitable for spreadsheet manipulation. The nine spectral outputs (each with both magnitude and phase information) were translated into the appropriate format required by the STAR System software (available through Spectral Dynamics, Inc.). This software was used to extract the modal parameters from three impact tests through the software's advanced curve fitting capability. First, the six modal frequencies (natural frequencies) and modal dampings were found and selected. In this step of the curve fitting analysis, the computer simply finds the potential modal frequencies. The user must chose the best set of modal frequencies (and the resulting damping values) for the data. This process consists of many iterations and selective elimination until the appropriate modal frequencies are determined which closely fit the actual measurements. Then the modal signatures corresponding to each of these modal frequencies were calculated. Each modal signature is a nine element complex vector associated with the load contribution of each of the nine outputs to each particular mode.

The modal signatures have units of force per second (lbf/s) because they represent the residues of the frequency response functions (FRF) involved in the urve fitting algorithm. The residues computed by the curve fitting algorithm are not normalized and have the units of the input frequency response function divided by time. In this particular experiment the frequency response functions are the input load spectra which have the units of force (lbf). The input force from the impact hammer was not measured; therefore, the spectra were not normalized with respect to the magnitude of the input force. (A strict frequency response function would have the units of acceleration, velocity, or displacement per unit input force.) The complex modal signatures were fit into a complex modal matrix with each column corresponding to each modal signature (ordered from the corresponding lowest modal frequency to highest modal frequency). Because six modes were found, the resulting modal matrix had six columns (corresponding to each mode) and nine rows (corresponding to the nine load outputs from the engine).

Once the modal matrix was determined, the spectra from the running engine were decomposed into a linear combination of these modal signatures. This decomposition was performed with the MATLAB software (available through The Math Works, Inc.). A linear weighting of these modal signatures results in a modal content vector associated with each discretized frequency value of the input spectrum. Because only six modes were extracted through the curve fitting software and nine spectra were measured, the linear weighting coefficients or modal content vectors were over-determined and were numerically determined in the least square sense for every discretized spectral frequency of the input. The result is essentially a representation of the running engine through the modal content vectors.

The modal content units are governed by the units of the modal signatures. Because the modal signatures have the units of force per unit time (lbf/s), the modal content has the units of time (s). The modal signatures could also be normalized to units of force by dividing out the modal damping (expressed in units of frequency) for each mode. The modal content would then be unitless.

The rigid body modal signatures of an engine are theoretically constant. They can be viewed as fundamental properties of the engine. Modal analysis implicitly includes the assumption that the running engine has the same modal signatures and that the running engine can be decomposed into these modal signatures. Therefore, a set of output load spectra for the running engine can be cast into modal content spectra which represent the relative contribution of each rigid body mode to the response of the running engine.

The underlying linearization assumption is analyzed in this study along with the assumption that the modal signatures, defined by the quiescent engine, can capture most of the character of the running engine. Therefore, a relative error level was calculated (also in MATLAB). This root mean square relative error (RMS relative error) was determined at each discretized frequency of the modal content calculations. It can be explained as the root mean square of the raw error summed over the nine outputs divided by the root mean square of the measured force levels summed over the nine outputs. The raw error is defined as the nine measured force outputs at the discretized frequency minus the predicted nine force outputs at the same frequency. In other words, the raw error is the direct difference between the actual nine spectral magnitudes at a frequency and the re-calculated nine spectral magnitudes at the same frequency based on the least square sense solution for the modal content vector. Because the RMS relative error is inappropriately magnified to outrageous values when the modal content is within the noise range, the RMS relative error is filtered only to show appropriate error when the modal content is out of this noise range.

EXPERIMENTAL RESULTS

MODAL ANALYSIS OF THE STATIONARY ENGINE - Three modal analysis tests were performed. The left rear mount was impacted in the three orthogonal directions (X, Y, and Z). With each impact all nine outputs were recorded simultaneously. The results of these impacts are shown in this section.

Impact Testing Signature Spectra - The spectral results for one impact test are shown in Figures 2, 3, and 4. Only the magnitude information is shown. The impact was applied in the Z direction on the left rear mount. All spectrum have 0.2 Hz discretizing accuracy.

All of the natural frequencies (modal frequencies) indicated by the impact tests are between 0 and 30 Hz. Natural frequencies occur where the response magnitude has a local maximum. These frequencies are assumed to be rigid body natural frequencies. The natural frequencies are noticeably damped and two are coupled. Although only five natural frequencies may be obvious at first glance, a curve fitting analysis finds that two natural frequencies lie close together around 13 Hz.

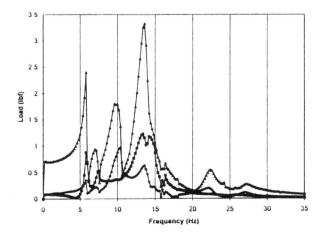

Figure 2. Left rear mount magnitude spectra for left rear (Z) mount impact
-◆- (X) direction -■- (Y) direction -▲- (Z) direction

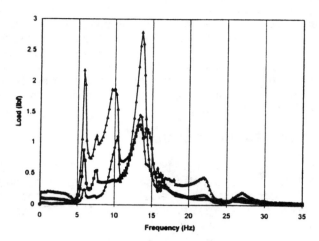

Figure 3. Right rear mount magnitude spectra for left rear (Z) mount impact
-◆- (X) direction -■- (Y) direction -▲- (Z) direction

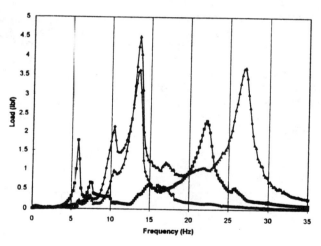

Figure 4. Front mount magnitude spectra for left rear (Z) mount impact
-◆- (X) direction -■- (Y) direction -▲- (Z) direction

Modal Parameter Estimation - This set of spectra, along with the other two sets of impact test spectra, were evaluated in the STAR System software. Six modal frequencies and modal dampings were determined from the (27) frequency response spectra used in the analysis. These frequencies (Hz) and dampings (% of critical) are listed in Table 2. Once these frequencies were determined, a curve fitting analysis was completed on the spectra excited with the left rear Z impact to determine the modal signatures at each modal frequency. A typical signature response (frequency response) curve fitting is shown in Figure 5. The curve fitting analysis closely fits the measured data in all nine spectra. In Figure 5 the magnitude of the right rear response is plotted in units of force (lbf).

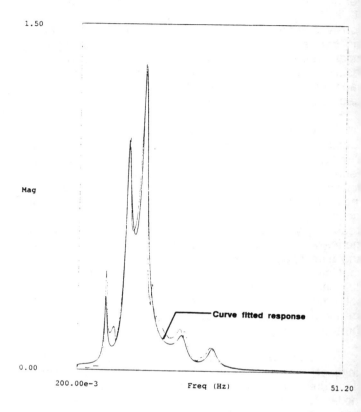

Figure 5. Typical magnitude response curve with modal analysis least square curve fit to extract modal parameters for right rear (X) mount output and left rear (Z) mount impact input

The nature of the modes can be extracted from Table 2. They could also be animated in the STAR System software for visual verification. The first mode is primarily a roll motion (rotation about the X axis) along with some longitudinal sway (Y translation). The second mode has primarily thrust motion (X translation). The third and fourth modes were found very close together around 13 Hz. The nature of these modes are very different. The third mode is primarily a heave motion (Z translation). The fourth mode may be attributed to the extra "shunt" attachments from the engine as indicated in Figure 1. The fourth mode motion includes rotation around the Z axis centered at the exhaust line attachment point. Therefore, shunt attachments must be considered in future analyses. The fifth mode has a much more complex behavior involving pitch and yaw. The sixth mode is primarily a pitching motion (rotation about the Y axis).

Table 2. Experimentally determined rigid body modal frequencies, modal dampings, and modal signatures from a modified modal analysis of a left rear (Z) engine mount impact

	Mode #1	Mode #2	Mode #3	Mode #4	Mode #5	Mode #6
Frequency (Hz)	5.60	9.71	12.95	13.65	20.57	26.16
Damping (% of Critical)	3.73	4.59	3.36	5.53	4.34	2.78
Left Rear:						
X- Magnitude (lbf/s)	1.11	4.93	4.32	2.40	2.55	1.04
X- Phase (degrees)	10.46	-18.97	-169.72	53.25	159.38	144.66
Y- Magnitude	2.99	0.24669	4.38	4.40	0.37674	0.20555
Y- Phase	-178.75	156.13	-42.60	49.50	127.45	122.71
Z- Magnitude	7.42	8.02	17.85	0.06178	5.81	2.11
Z- Phase	3.46	-7.99	-13.85	-144.79	-17.70	-29.44
Right Rear:						
X- Magnitude	0.89247	5.44	7.40	0.80067	1.92	0.91283
X- Phase	164.98	-17.59	166.77	-137.47	-13.93	138.60
Y- Magnitude	3.19	0.09193	5.32	4.90	0.36236	0.16652
Y- Phase	-177.65	123.93	-40.24	50.10	123.49	127.09
Z- Magnitude	7.57	9.49	16.27	3.94	4.07	2.12
Z- Phase	177.36	-5.88	-8.91	-125.97	169.30	-38.92
Front:						
X- Magnitude	0.47030	9.25	18.75	2.62	1.55	0.74021
X- Phase	70.80	-19.59	174.92	57.97	134.01	120.15
Y- Magnitude	6.08	0.72248	1.91	1.55	24.38	0.97114
Y- Phase	-179.21	16.13	116.94	-123.35	-14.08	16.84
Z- Magnitude	0.65908	3.79	25.62	3.27	5.75	28.24
Z- Phase	-70.51	143.66	-9.69	-137.07	-100.63	148.32

MODAL CONTENT OF THE RUNNING ENGINE - The modal signatures determined for the stationary engine are now applied to the running engine. The modal content of the running engine is determined at each discretized frequency of the measured spectra.

Running Engine Signature Spectra - With the engine running at two different operating conditions, the same data acquisition system was used to record the spectrum of the running engine. Figures 6, 7, and 8 show the magnitude spectrum of the engine running at 750 rpm and 20 ft-lbf, which is near the idling condition. Notice that large magnitudes are recorded at harmonics of the first engine order (12.5 Hz). The response is dominated by the half, first, and third engine order harmonics. The large half order harmonics are attributed to the four cycle nature of the engine where a cylinder fires once per two crankshaft revolutions. Figures 9, 10, and 11 show the magnitude spectrum of the engine running at 1200 rpm and 615 ft-lbf. The response at this condition is dominated by the first engine order harmonic at 20 Hz.

Modal Content of the Running Engine - Lastly, these magnitude spectra were decomposed into the modal content of the running engine. Figures 12 and 14 display the modal content of the running engine throughout a frequency range up to 200 Hz.

If the engine excites a mode near its natural frequency (modal frequency), a small input is greatly magnified. Near engine idle, the first mode is contributing much to the response because the half engine order at 6.25 Hz is close the first mode natural frequency at 5.60 Hz. The fourth mode is also largely contributing to the response because the fourth mode natural frequency at 13.65 Hz is near the first engine order frequency at 12.5 Hz.

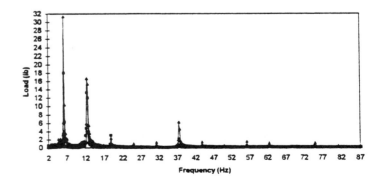

Figure 6. Left rear mount magnitude spectra for running engine at 750 rpm engine speed and 20 ft-lb load
-◆- (X) direction -■- (Y) direction -▲- (Z) direction

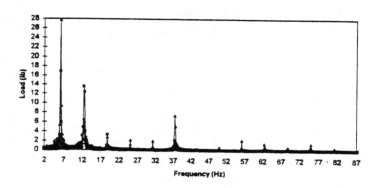

Figure 7. Right rear mount magnitude spectra for running engine at 750 rpm engine speed and 20 ft-lb load
-◆- (X) direction -■- (Y) direction -▲- (Z) direction

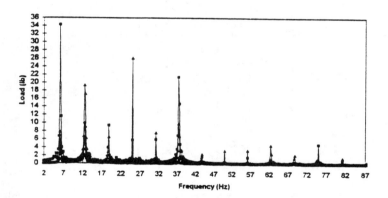

Figure 8. Front mount magnitude spectra for running engine at 750 rpm engine speed and 20 ft-lb load
-◆- (X) direction -■- (Y) direction -▲- (Z) direction

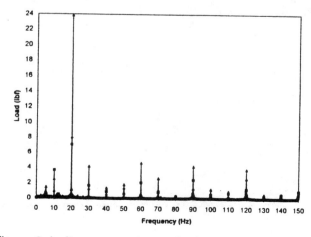

Figure 9. Left rear mount magnitude spectra for running engine at 1200 rpm engine speed and 615 ft-lb load
-◆- (X) direction -■- (Y) direction -▲- (Z) direction

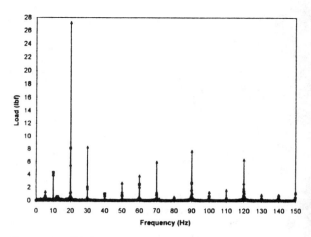

Figure 10. Right rear mount magnitude spectra for running engine at 1200 rpm engine speed; 615 ft-lb load
-◆- (X) direction -■- (Y) direction -▲- (Z) direction

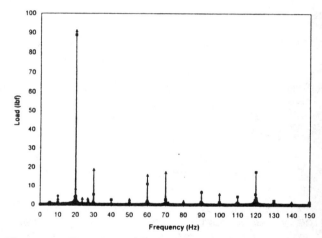

Figure 11. Front mount magnitude spectra for running engine at 1200 rpm engine speed and 615 ft-lb load
-◆- (X) direction -■- (Y) direction -▲- (Z) direction

At the higher speed and load the first mode is no longer dominant. The first engine order frequency is now at 20 Hz which is much higher than the first natural frequency. The fifth and sixth modes are dominating the response. The fifth mode natural frequency is 20.57 Hz, and the sixth mode natural frequency is 26.16 Hz. Both of these natural frequencies are near the first engine order frequency. Additionally, the modal content is spread to much higher frequencies compared to when the engine is idling.

Therefore, by immediate analysis the modal content results intuitively adhere to the modal parameters determined from the quiescent engine. However, in order to quantify the error involved with casting the response of the running engine into a rigid body modal content representation, the RMS relative error was calculated and plotted in Figures 13 and 15. The RMS relative error was filtered such that when the modal content was less than 0.3 (s) in modal weight, the error was filtered to zero. This filtering prevents magnification of the error in unwanted noise of low level inputs. Close to idle, the RMS error is relatively small. With low engine speed and light load, the modal signatures can be used with approximately 90% accuracy to characterize the running engine across the entire response spectrum. With increased engine speed and load, the modal signatures can be used with 80% accuracy up to the fourth engine order. At 4.5 and 6 times the first engine order (corresponding to 90 and 120 Hz), the modal signatures can only be used with 60% accuracy.

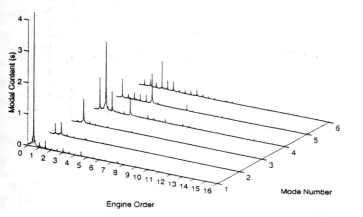

Figure 12. Rigid body modal content of the running engine at 750 rpm engine speed and 20 ft-lb load (1st engine order equals 12.5 Hz)

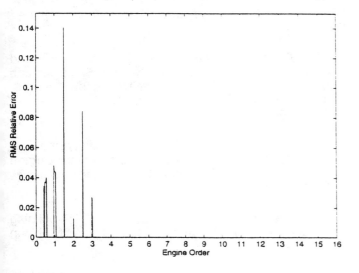

Figure 13. Filtered root mean square relative error of rigid body modal content of the running engine at 750 rpm engine speed and 20 ft-lb load (1st engine order equals 12.5 Hz)

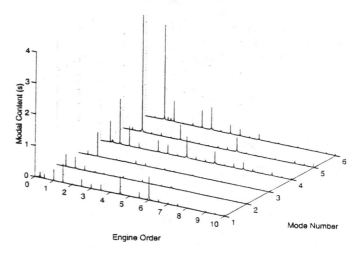

Figure 14. Rigid body modal content of the running engine at 1200 rpm engine speed and 615 ft-lb load (1st engine order equals 20 Hz)

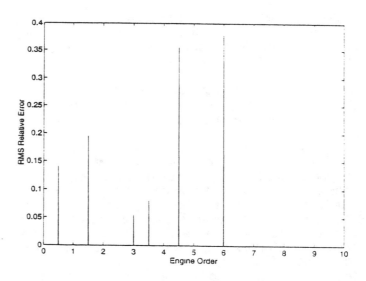

Figure 15. Filtered root mean square relative error of rigid body modal content of the running engine at 1200 rpm engine speed and 615 ft-lb load (1st engine order equals 20 Hz)

Therefore, the modal signatures determined from a quiescent engine modal analysis have limitations when being applied to the heavily-loaded and higher-speed engine. There are many possible reasons for this accuracy limitation. First of all, the running engine provides a change to the assumed stationary state of the engine used to determine modal signatures. The engine has moving internal mechanisms, which typically lie in different positions after every engine shut down, and may provide different reaction forces when they are moving or spinning compared to when they are stationary. For example, the rotational inertia of the crankshaft and flywheel will alter the engine's response in pitch and yaw when it is running. Secondly, the flexural motions of the engine block are not considered in this study and could account for some of the higher frequency error at higher load and speed. In addition, some portions of the engine are not at all rigid. The dynamics of coolant flow and lubricating oil flow through the engine during firing modify the dynamics of the vibrating engine from the quiescent state. Any change in temperature between experiments also deteriorates modal content accuracy since the stiffness and damping of the elastomeric isolation elements in the motor mounts are temperature dependent. However, the current results are encouraging enough to warrant further work.

CONCLUSIONS & DISCUSSION

A generalized modal vibration approach to running engine vibration holds the promise of efficient reduced degree-of-freedom vibration modeling. The feasibility and accuracy of utilizing experimentally determined rigid body modes of a mounted engine as a basis for describing running engine vibration levels have been studied. Two conclusions can be drawn from this effort.

1.) The rigid body modes measured from a hammer impact on a quiescent heavy-duty Diesel engine can effectively describe the character of the vibrating engine with 90% or better accuracy when the engine is running at low load and low speed.

2.) The capability of this modal decomposition to properly describe higher frequency vibrations is degraded with increasing engine speed and load.

However, the modal content results for the loaded engine account for far more than half of the engine's vibration loads. The residual vibration levels, which cannot be attributed to rigid body block motions, are likely to be caused by flexural modes of the engine block, dynamic component-dependent reactions, and/or the influences of the non-solid parts of the engine. Future work will determine which of these possibilities dominates the residual vibrations so that a proper extension to the rigid body modal basis can be identified.

At a more empirical level, a complete modal content study that maps all engine operation points and, therefore, characterizes the dominant rigid body modes across the operating range of the engine, could be used to determine the dominant rigid body modes of the vibrating engine at any speed and load. These dominant modes could then be utilized in a low degree-of-freedom engine vibration model which executes efficiently.

ACKNOWLEDGMENTS

This research was supported by the U.S. Army Tank Automotive Research Development and Engineering Center (TARDEC) through the Automotive Research Center (Contract # DAAE07-94-C-R094) at The University of Michigan.

The authors would also like to acknowledge the contribution of Steve Hoffman and Kevin Morrison for their efforts in setting up and maintaining the test engine, and for their assistance while running the engine tests.

REFERENCES

Ashrafiuon, H., (1992) "Dynamic Analysis of Engine-Mount Systems", *Journal of Vibration and Acoustics*, Vol. 114, pp. 79-83, January 1992.

Bretl, John, (1995) "Advancements in Computer Simulation Methods for Vehicle Noise and Vibration", SAE #951255.

Hata, H. and Tanaka, H., (1987) "Experimental Method to Derive Optimum Engine Mount System for Idle Shake", SAE #870961.

Johnson, Stephen R., and Subhedar, Jay W., (1979) "Computer Optimization of Engine Mounting Systems", SAE #790974.

Muller, Michael, Siebler, Thomas W., and Gartner, Hanno, (1995) "Simulation of Vibrating Vehicle Structures as Part of the Design Process of Engine Mount Systems and Vibration Absorbers", SAE #95221.

Norling, Richard L., (1978) "Continuous Time Simulation of Forces and Motion within an Automotive Engine", SAE #780665.

Radcliffe, Clark J., Picklemann, Mark N., Spiekermann, Charles E., and Hine, Donald S., (1983) "Simulation of Engine Idle Shake Vibration", SAE #830259.

Schmitt, Regis V., and Leingang, Charles J., (1976) "Design of Elastomeric Vibration Isolation Mounting Systems for Internal Combustion Engines", SAE #760431.

Shiao, Yaojung, Pan, Chung-hung, and Moskwa, JJ., (1994) "Advanced Dynamic Spark Ignition Engine Modelling for Diagnostics and Control", *International Journal of Vehicle Design*, Vol. 15, No. 6, pp. 578-596.

Snyman, J.A., Heyns, P.S., and Vermeulen, P.J., (1995) "Vibration Isolation of a Mounted Engine Through Optimization", *Mechanism & Machine Theory*, Vol. 30, No. 1, pp. 109-118.

Spiekermann, Charles E., (1982) "Simulating Rigid Body Engine Dynamics", M.S. Thesis, Michigan State University.

Zhengchang, Xu, and Anderson, R.J., (1988) "A New Method for Estimating Amplitudes of Torsional Vibration for Engine Crankshafts", *International Journal of Vehicle Design*, Vol. 9, No. 2, pp. 252-261.

971949

H∞ Control Design of Experimental State-Space Modeling for Vehicle Vibration Suppression

Zhongyang Guo, Itsuro Kajiwara, and Akio Nagamatsu
Tokyo Institute of Technology

Tsutomu Sonehara
Isuzu Advanced Engineering Center, Ltd.

Copyright 1997 Society of Automotive Engineers, Inc.

ABSTRACT

State-space solutions of $H\infty$ controller have been well developed. Hence to a real structure control design, the first step is to get a state space model of the structure. There are analytical and experimental dynamic modeling methods. As we know, it is hard to obtain an accurate model for a flexible and complex structure by FEM(Finite Element Method). Then the experimental modeling methods are used. In this paper, we use frequency domain modal analysis technique based on system FRF(Frequency Response Function) data and ERA(Eigensystem Realization Algorithm) time domain method based on system impulse response data to establish state-space model in order to design $H\infty$ control law for the purpose of vibration suppression. The robust control implementation is exerted on a testbed (truck cab model device) with three degrees of freedom. The validity of experimental state-space modeling is testified and the obvious vibration control performances are achieved.

INTRODUCTION

In recent years, the implementation of robust stabilization and performance control based on $H\infty$ control theory have been investigated for different structures[1]. A common procedure first involves the generation of a finite element model of the structure for the purpose of control design. But it often results in a nominal plant model with large error to true plant. As a matter of fact, the closed-loop plant response characteristics are sensitive to the plant model used in $H\infty$ control design. Therefore, if significant error exists between the nominal model and the actual system, then experimental results could drastically differ from analytical simulations. Recent experimental verification of applying an $H\infty$ design for active vibration suppression[2,3] has shown

more practical solutions to the control design problem by the way of experimental state-space modeling. With the development of system identification technique, state-space modeling of structures has become sophisticated using experiment data from the structural testing of a physical system. Here in this paper, we handle two methods of structure experimental modeling based on modal identification techniques. One is ERA method[4,5,6,7] and the other is the common modal analysis approach. Constructing a suitable mathematical model for a dynamic system is a major task in the field of dynamic analysis and control design for vibration suppression. In the past, system identification techniques such as frequency domain methods and time domain methods, e.g., the Eigensystem Realization Algorithm (ERA), have been developed using experiment data from the modal testing of flexible structures, to construct a mathematical model either for a controller design or for modal parameter identification including frequencies, damping and mode shapes. The conventional modal analysis is based on FRF curve-fitting technique, which is a SISO (Single-Input/Single Output) approach because mode coordinates can be decoupled. And then several such SISO sub-systems are constructed into a MIMO (Multi-Input/Multi-Output) system. The state-space representation of the constructed system is of less numerical deficiency and the system mode is decided by the number of decoupled mode equations. The multi-input-multi-output time domain ERA technique was originally developed for modal parameter identification and successfully applied to the modal testing of flexible structures. The ERA technique uses the singular values of the finite Hankel matrix formed by impulse response functions (Markov parameters) to determine the system order and further reduce the noise effect on the realized system model by truncating some small singular values. The singular value truncation means that the system modes have stronger energy than noise modes. However, it is possible that the weakly excited modes with less observation energy might be truncated. The retained singular values are dominated by the system signals and the order of system is determined by the number of retained singular values. On the other hand, the truncated singular values are dominated by noise and neglected high frequency dynamics, and hence may be used to estimate the model error bound. Defining the model error bound to be truncated singular values is equivalent to defining an $H\infty$ error bound for the realized system model. Once the $H\infty$ error bound is obtained, the $H\infty$ control method is readily applied to derive a robust controller design for the system represented by the realized model. The $H\infty$ design is chosen since: ①it supplies robust stability to model and sensor uncertainties; ②it achieves performance effectively; ③it handles both disturbance and controller saturation easily; ④and it works not only on simply SISO systems but also on MIMO systems. Therefore, frequency response criteria can easily be shaped to desired specification in order to realize the vibration suppression.

This paper is motivated by the strong connection between the state-space representation of a dynamic system with its mode coordinate equation and $H\infty$ control design, and the relation between the ERA system minimum realization and the $H\infty$ control method. It seems natural to integrate the system modal identification methods with the control methods so that a robust controller design can be achieved directly using the experiment data. This paper intends to outline the fundamental concepts from different disciplines and then integrate them together. In the second section, the modal analysis approach and the ERA algorithm are briefly introduced. In the third section, the test structure and the experimental modeling results are included. The robust control design and experiment implementation are included in the fourth section. Finally, the summary remarks conclude this paper.

STATE-SPACE MODELING FOR ROBUST CONTROL DESIGN

1. MODAL ANALYSIS APPROACH

Here a method is described based on modal analysis for robust state space modeling. With this method, it is assumed that a pole-zero representation of the SISO transfer function is identified from frequency domain domain data. A state-space model based on modal analysis technique offers two advantages over other modeling techniques. The modal model retains a simple physical correspondence between the identified model and the test structure, which is lost in many state space identification methods. Another advantage is that by making symmetry assumption about the structure (i.e., the mass, stiffness, and damping matrices are symmetric), the modal parameters can be obtained from a relatively small number of experimental transfer function measurements. The advantage of synthesizing the MIMO model from SISO representation is of great flexibility in matching the model to the experiment data.
The structure can be represented by

$$M\ddot{q} + C\dot{q} + Kq = f \qquad (1)$$

Where $M, C,$ and K are symmetric and positive definite matrices, standing for mass, damping and stiffness; q is a vector of displacements, and f is a vector of inputs. It is also

assumed that the structure is time invariant and a driving point transfer function(collocated sensor and actuator) is available.

Here exists a orthonormal matrix Φ with respect to M, so that substituting mode coordinate transformation $q = \Phi \xi$ into Eq. (1) produces:

$$I\ddot{\xi} + C_{\Phi}\dot{\xi} + \Lambda \xi = \Phi^T f \qquad (2)$$

here $I = \Phi^T M \Phi$, $C_{\Phi} = \Phi^T C \Phi$, $\Lambda = \Phi^T K \Phi$.

Rewrite Eq. (2) into state-space form:

$$\dot{X} = AX + Bf \qquad (3)$$

where $\dot{X} = \begin{pmatrix} \xi \\ \dot{\xi} \end{pmatrix}$, $A = \begin{bmatrix} 0 & I \\ -\Lambda & -C_{\Phi} \end{bmatrix}$, $B = \Phi^T$

Output equation is presented as:

$$y = CX + Df \qquad (4)$$

C depends on sensing position of mode outputs and D is direct transformation matrix of inputs. C and D have close relationship with physical structure under modal testing. Therefore, using this approach to obtain state-space representation needs some artificial treatments based on its evident physical meaning. So far, the dynamic system state-space realization triple $[A,B,C]$ and D are obtained. They can be used in the robust design with distinct physical meaning.

2. ERA APPROACH

For structural analysis and controller designs, experimental data from structural testing are used to either verify an analytical model or directly determine a mathematical model. The ERA is an algorithm which computes a mathematical model directly from experimental data. In this part, the basic ERA formulations are briefly discussed. The ERA minimum realization is a balanced realization. The ERA balanced realization has been frequently used for model reduction, and further in robust control design. The ERA algorithm uses the singular value decomposition of a Hankel matrix to determine system model under test. Consider the discrete-time state-space model as given by Eq. (5) below. Let the input u be the impulse input at an initial time.

$$\begin{aligned} x(k+1) &= \hat{A}x(k) + \hat{B}u(k) \\ Y(k) &= \hat{C}x \end{aligned} \qquad (5)$$

The time domain description is given by Markov parameters, i.e., impulse response functions

$$Y_k = \hat{C}\hat{A}^k\hat{B} \qquad (6)$$

Note that there are many other ways to obtain the impulse response functions experimentally. For example, inverting a transfer function from the frequency domain to the time domain will generate the corresponding impulse response function. The algorithm begins by forming the $r{\times}s$ block matrix (generalized finite-Hankel matrix) :

$$H(k) = \begin{bmatrix} Y_k & Y_{k+1} & \cdots & Y_{k+s-1} \\ Y_{k+1} & Y_{k+2} & \cdots & Y_{k+s} \\ \vdots & \vdots & \ddots & \vdots \\ Y_{k+r-1} & Y_{k+r} & \cdots & Y_{k+r+s-2} \end{bmatrix} \quad (7)$$

The finite-Hankel matrix $H(k)$ can be expressed as:

$$H(k) = V_r \hat{A}^k W_s \quad (8)$$

where $V_r = [\hat{C}\ \hat{C}\hat{A}\ \cdots\ \hat{C}\hat{A}^{r-1}]^T$, $W_s = [\hat{B}\ \hat{A}\hat{B}\ \cdots\ \hat{A}^{s-1}\hat{B}]$.

V_r and W_s are the observability and controllability matrices, respectively. Assume that there exists a matrix H', such that

$$W_s H' V_r = I_n \quad (9)$$

where I_n is an identity matrix of order n. Equation (8) and (9) imply:

$$\begin{aligned} H(0) H' H(0) &= V_r W_s H' V_r W_s \\ &= V_r W_s = H(0) \end{aligned} \quad (10)$$

Thus H' is the pseudo inverse of $H(0)$. Solution for H' can be obtained as follows. Applying singular value decomposition to the matrix $H(0)$

$$H(0) = U \Sigma V^T \quad (11)$$

where the columns of U and V are orthonormal and Σ is diagonal with positive elements $[\sigma_1, \sigma_2, \ldots, \sigma_n]$. The pseudo inverse can be computed as:

$$H' = [V][\Sigma^{-1} U^T] \quad (12)$$

Define 0_m as the null matrix of order m, I_m an identity matrix, then $E_m^T = [I_m, 0_m, \cdots, 0_m]$. From Eqs. (8), (9) and (12), a minimal realization can be obtained from:

$$\begin{aligned} Y(k+1) &= E_q^T H(k) E_p \\ &= E_q^T V_r \hat{A}^k W_s E_p \\ &= E_q^T V_r W_s H' V_r \hat{A}^k W_s H' V_r W_s E_p \\ &= E_q^T U \Sigma^{1/2} [\Sigma^{-1/2} U^T H(1) V \Sigma^{-1/2}]^k \Sigma^{1/2} V^T E_p \\ &= C [A]^k B \\ Y(0) &= D \end{aligned} \quad (13)$$

The triple $[A,B,C]$ is a minimum realization.
This is the basic ERA formulation.
For a finite dimensional and linear time-invariant system, an exact system realization can be obtained by the ERA algorithm from noise-free measurements[8]. The ERA algorithm is accurate and efficient, particularly for low noise levels, and it produces a minimum order realization. It is a powerful identification algorithm. However, if significant noises are present in the measurements, caution must be taken to identify a proper order for the system model. Moreover, It has been concluded that the reduced ERA model, i.e., after singular value truncation, the retained system model, converges to the reduced balanced model[9]. The balanced realization has been widely used for model reduction and robust controller design.

EXPERIMENTAL STRUCTURE AND IDENTIFICATION

A truck cab model device is used as the testbed, shown in Fig. (1). This model targets to the research of improving the ride comfort of commercial vehicle at the position of driver cab through active vibration control. This test model is simplified into a mass-spring-damping system with three degrees of freedom. Between the cab and chassis frame, an electric actuator is mounted and thus is used to execute exciting source in the test measurement process and to exert active vibration control in the robust control process. Three acceleration frequency response function (FRF) data are measured between u and three measurement stations x_1, x_2 and x_3. The inverse of FRF, i.e., impulse response data are recorded. The modal analysis approach uses FRF data as input data, while the ERA uses impulse response data as input data.

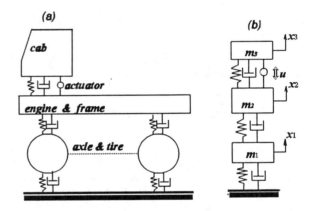

Fig. 1 Truck cab model

1. MODAL ANALYSIS MODELING

Checking Fig. (1), it is clear that there are the following equations in contrast with Eq. (1).

$$f = [0, -u, u]^T \quad q = [x_1, x_2, x_3]^T \quad (14)$$

Considering a coordinate transformation:

$$q = T q_a \quad (15)$$

here $q_a = [x_1, x_2, x_3 - x_2]$, $T = \begin{bmatrix} 1 & 0 & 0 \\ 0 & 1 & 0 \\ 0 & 1 & 1 \end{bmatrix}$.

Then a single exciting dynamic system is presented as the following.

$$\bar{M}\ddot{q}_a + \bar{C}\dot{q}_a + \bar{K}q_a = T^T f = f_a \qquad (16)$$

here $\bar{M} = T^T M T$, $\bar{C} = T^T C T$, $\bar{K} = T^T K T$, $f_a = [0, 0, u]^T$.

Based on FRF test data, the modal parameters are identified as:

$$\begin{aligned} \text{system eigenvalue}: \Lambda &= \text{diag}(\Omega_1, \Omega_2, \Omega_3) \\ \text{mode damping ratio}: \zeta &= \text{diag}(\zeta_1, \zeta_2, \zeta_3) \\ \text{mode shape matrix}: \Phi &= [\phi_1, \phi_2, \phi_3] \end{aligned} \qquad (17)$$

Performing mode decoupling transformation $q_a = \Phi \xi$ and rewriting Eq.(16) result in:

$$I\ddot{\xi} + C_\Phi \dot{\xi} + \Lambda \xi = \Phi^T w + \Phi^T u$$
$$\text{here} \quad C_\Phi = \text{diag}(2\zeta_1\Omega_1, 2\zeta_2\Omega_2, 2\zeta_3\Omega_3) \qquad (18)$$

w is disturbance input vector and u is control input. Eq.(18) is written as state-space form:

$$\dot{q} = Aq + B_1 w + B_2 u \qquad (19)$$

where $\dot{q} = \begin{bmatrix} \xi \\ \dot{\xi} \end{bmatrix}$, $A = \begin{bmatrix} 0 & I \\ -\Lambda & -C_\Phi \end{bmatrix}$, $B_1 = [0 \ \Phi^T]^T$, $B_2 = [0 \ \Phi^T]^T$

Synthesizing the three acceleration outputs, system output equation is presented as:

$$\begin{aligned} y_a &= Cq + D_{21} w + D_{22} u \\ C &= [-\Phi\Lambda, \ -\Phi C_\Phi], \quad D_{21} = \Phi \Phi^T, \quad D_{22} = \Phi \Phi^T \end{aligned} \qquad (20)$$

For the system in Fig. (1), its state-space model parameters are A, $[B_1 \ B_2]$, C, $[D_{21} \ D_{22}]$. The Bode plot of x_3 measurement station due to above state-space model is shown in Fig. (2), and this model nicely correlates test data.

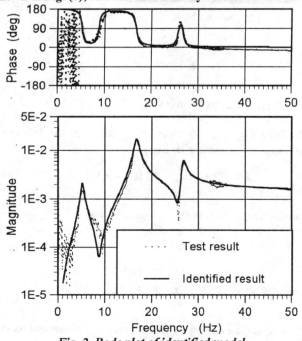

Fig. 2 Bode plot of identified model

2. ERA Modeling

ERA finite Hankel matrix size is taken to be 90×90. The results of state-space model identification with ERA method is effectively obtained and its Bode plot of station x_3 is demonstrated in Fig. (3).

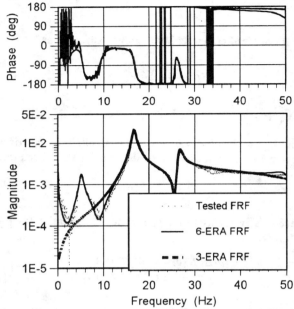

Fig. 3 Bode plot of model identified by ERA

The 6-order ERA state-space minimum realization model which better correlates the experiment data, in Fig. (3), is a overestimated model since the higher level of measurement noise. For the 3-order ERA result, in Fig. (3), the first system mode is truncated because of its relative low signal-to-noise ratio and low observation energy. Now it is clear that in order to get system state-space representation for the robust control design, the modal analysis approach needs some artificial treatments and ERA method can automatically produce system state-space representation. Taking the two identified state-space representations as the following nominal models of robust control design, the developed $H\infty$ robust control design technique is readily available to design the robust controllers.

$H\infty$ CONTROLLER DESIGN AND IMPLEMENTATION

In this section, we apply $H\infty$ control algorithm to design controllers for the problem in this paper. Generally speaking, the $H\infty$ control methods are optimal frequency

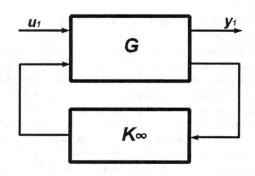

Fig. 4 $H\infty$ control problem

domain algorithms which compute the feedback control law by minimizing the maximum singular value of the system transfer function. It is known that $H\infty$ control methods address the full range of stability margin, sensitivity, and robust optimization. The $H\infty$ control strategy, as compared to classical control technique, provided new techniques and effective results in design control system. That is accomplished by shaping the frequency response characteristics of a plant according to pre-specified performance specification in the form of weighting functions and to keep the robust stability under the meaning of $H\infty$ norm. The basic $H\infty$ control problem is formed in Fig.(4). $K\infty$ is the $H\infty$ controller and G is the plant augmented by robust weighting functions. The optimal $H\infty$ closed-loop control results in $\|T_{y_1 u_1}\| \leq 1$.

As we know, in the above section, the approximations of physical system model by different approaches are obtained. There are always some uncertainties present even when the underlying process is essentially linear. This uncertainty may be due to incomplete knowledge of the physical parameters, neglected high frequency dynamics or invalid assumptions made in the model formulation. Descriptions of these uncertainties determine the tradeoff between achievable performance and robustness of the control design[10]. An improper selection of uncertainty weighting functions may lead to unstable or poor performing controllers on the actual system. In contrast, if descriptions of uncertainty are overly conservative, performance of the closed-loop system may be severely limited. Therefore, tight uncertainty bounds are required to provide robust control design which achieves high performance when implemented on actual system. The appropriate selection of weighting functions[11,12] over the desired frequency range has some relation but is not very explicit in a direct manner. Numerous trial selections are usually required in order to obtain desired performance and/or stability objectives. The goal of $H\infty$ design is to reshape the system open-loop dynamics in order to provide vibration suppression in interested frequency region. Therefore, in this case, performance or sensitivity weighting function is chosen as

$$W_1 = \frac{3.28356 \times 10^8}{s^4 + 49s^3 + 4.0732 \times 10^4 s^2 + 1.1158 \times 10^6 s + 3.28356 \times 10^8} \quad (21)$$

seeing Fig. (5). The suppression of the 2nd and 3rd order modes is pursued.

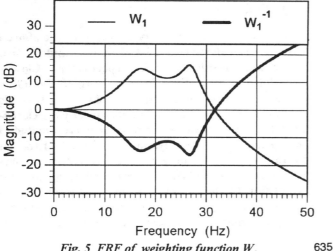

Fig. 5 FRF of weighting function W_1

1. CONTROL IMPLEMENTATION OF MODAL ANALYSIS CASE

The control design block diagram is shown in Fig. (6). Robust control design methods optimize control laws based on knowledge of how model disturbances enter into the problem description. Since controller optimization is based on mathematical system descriptions, accurate accounting and characterization of variations between real structure and their mathematical models is essential. Here the disturbance vector w, in Eq. (19) and (20), consists of d_r disturbance from road surface input, d_u disturbance induced by robust stability evaluation due to the plant multiplicative uncertainty, and d_n acceleration sensor noise induced for the controller gain shaping through a small weighting ε_n.

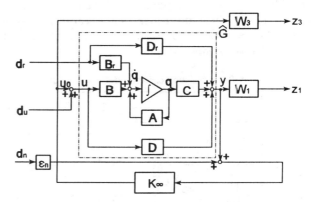

Fig. 6 Augmented Plant for robust control design

The model multiplicative uncertainty is selected as:

$$\Delta_m = |(\hat{G} - G)\hat{G}^{-1}| \quad (22)$$

In Eq. (21) and Fig. (5), $\hat{G}$ is the identified nominal model, y is the feedback signal from x_3, and u the control input. According to above uncertainty description, its weighting function is selected as:

$$W_3 = \frac{s^2 + 75.36s + 35494.56}{s^2 + 1884s + 3549456} \times 100 \quad (23)$$

Robust performance weighting function is chosen as W_1 in Eq. (21). Hence based on Fig. (6), the system plant state-space equation is listed as:

$$\dot{X} = AX + b_1 \omega + b_2 u \quad (24)$$

where $b_1 = [B_r \ B]$, $b_2 = B$,
and observation output equation

$$y = CX + d_{21}\omega + d_{22}u \quad (25)$$

where $d_{21} = [D_r, D, \varepsilon_n]$, $d_{22} = D$.

Control variables z_1 and z_3, as well as their corresponding weighting functions W_1 and W_3, have state-space representation matrices $[A_{w1}, B_{w1}, C_{w1}, D_{w1}]$ and $[A_{w3}, B_{w3}, C_{w3}, D_{w3}]$, respectively. Concluding above analysis results in generalized augment plant G:

$$G \triangleq \left[\begin{array}{ccc|c|c} A & 0 & 0 & b_1 & b_2 \\ B_{w1}C & A_{w1} & 0 & B_{w1}d_{21} & B_{w1}d_{21} \\ 0 & 0 & A_{w3} & 0 & B_{w3} \\ \hline D_{w1}C & C_{w1} & 0 & D_{w1}d_{21} & D_{w1}d_{22} \\ 0 & 0 & C_{w3} & 0 & D_{w3} \\ \hline C & 0 & 0 & d_{21} & d_{22} \end{array}\right] \quad (26)$$

Hereafter, a 12-order proper and rational $H\infty$ controller can be obtained. Then the experimental control result and calculation result can be seen in Fig. (7). The vibration suppression is obviously achieved.

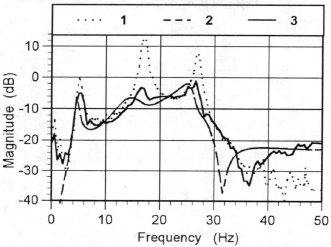

Fig. 7 *Experimental implementation of robust control*

2. $H\infty$ CONTROL IMPLEMENTATION OF ERA CASE

In this part, the mixed sensitivity approach of $H\infty$ control design is used, Fig. (8).

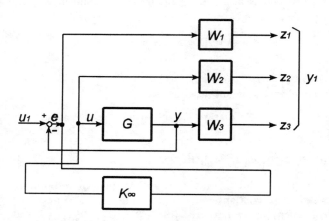

Fig. 8 *Mixed sensitivity control design*

The closed-loop transfer function matrices, from u_1 to e and from u_1 to y, are expressed as:

$$S(s) = [I + L(s)]^{-1}$$
$$T(s) = L(s)[I + L(s)]^{-1} \quad (27)$$

where $L(s) = G(s)K_\infty(s)$.

The matrices $S(s)$ and $T(s)$ are known as the sensitivity function and complementary sensitivity function. The singular values of these matrices can be used to quantify the stability margins and performance of the system. Here W_1 is chosen as the same as that in Eq. (21), aimed to depress the sensitivity of the *2nd* and *3rd* mode frequencies; $W_2 = 0.03$, sometimes W_2 is used for the shaping of controller gain. W_3 is multiplicative uncertainty weighting function. The meanings of other symbols are the same as what mentioned above. The $H\infty$ control design can be expressed as:

$$\left\| \begin{array}{c} \gamma W_1 S \\ W_2 K_\infty S \\ W_3 T \end{array} \right\|_\infty \le 1 \quad i.e. \quad \|T_{y_1 u_1}\|_\infty \le 1 \quad (28)$$

In ERA algorithm, because the truncated singular values can be considered to formulate the error bound of nominal model, this error bound can be defined as the model uncertainty $H\infty$ norm. It is known that ERA algorithm for system realization is a balanced realization. Therefore, the output multiplicative uncertainty is presented as[13]:

$$\|\Delta_m\|_\infty = \|G^{-1}(G - G_k)\|_\infty \le \prod_{i=r+1}^{n} (1 + 2\sigma_i\sqrt{1+\sigma_i^2} + \sigma_i) - 1 \quad (29)$$

σ_i presents the truncated singular value. G_k is the non-reduced ERA realization model and G is the reduced nominal ERA model.

Based on the uncertainty description, 3-order and 6-order ERA state-space model uncertainty weighting functions are chosen as:

$$W_3(3) = \frac{0.5s + 110}{s + 1} \quad \text{and} \quad W_3(6) = \frac{s + 110}{5s + 100} \quad (30)$$

respectively.

The $H\infty$ controllers are easily obtained for the above ERA cases and calculated controller order is 11 and 17 for the 3-order and 6-order model, respectively. Closed-loop experimental control results are shown in Fig. (9). The vibration suppression effectiveness of 6-order ERA closed-loop model is not so fine as that of 3-order ERA closed-loop model. This is because the 6-order overestimated ERA model has non-system dynamics which affect the control performance. On the whole, the vibration suppression effectiveness is explicit in considered frequency region. According to these control results, $H\infty$ controller adds damping to the dynamics of closed-loop system. It is the $H\infty$ damping principle that attenuates vibration.

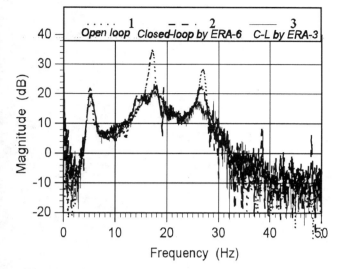

Fig. 9 Experimental implementation of robust control

CONCLUDING REMARKS

The experimental state-space modeling of structure by the modal analysis approach and ERA method establishes the basis of $H\infty$ robust control design. Since experimental data are used, actuator dynamics and sensor dynamics as well as measurement noise are included in the identification modeling process and thus in the contr. l designs. Therefore, the used two approaches are believed to provide the nominal model with fidelity. The modal analysis needs some artificial treatments to derive the system state-space representation, in which, controllability matrix and observability matrix have some physical significance to some extent. This approach can properly correlate the system order with the number of system test modes. While ERA method gives out sate-space representation directly by its reduced and balanced minimum realization. However, if the measurements have high level of noise, one must pay enough attention to the determination of system order. When the system is overestimated, resulting in a non-minimum state-space model, the nominal model can produce better open-loop correlation with experimental data, but results in poor closed-loop control performance as we demonstrated in this paper. This is because with a non-minimum realized model, non-system dynamics are in the nominal model that do not exist in the actual structure. $H\infty$ control method provides effective robust performance and robust stability through careful selections of weighting functions. There is no doubt that the nominal model used in $H\infty$ control design affects the closed-loop control performance. Therefore the accurate measurements of test objects, with as possible as low level of noise, are good for obtaining an accurate model by the experimental state-space modeling approaches.

REFERENCES

[1] Reichert R. T., "$H\infty$ Control Theory for Missile Autopilots", *AIAA Guidance Navigation and Control Conference*, 1989.

[2] Jeffrey Dosch, Dodald Leo and Daniel Inman, "Modeling and Control for Vibration Suppression of a flexible Active Structure", *Journal of Guidance, Control, and Dynamics*, Vol. 18, No. 2, 1995, pp. 340-346.

[3] John L. Crassidis and D. Joseph Mook, "Robust Identification and Vibration Suppression of a Flexible Structure", *Journal of Guidance, Control, and Dynamics*, Vol. 17, No. 5, 1994, pp. 921-928.

[4] Jer-Nan Juang, and Richard S. Pappa, "An Eigensystem Realization Algorithm for Modal Parameter Identification and Model Reduction", *Journal of guidance, Control, and Dynamics*, Vol. 8, No. 5, 1985, pp. 620-627.

[5] Jer-Nan Juang, and H. Suzuki, "An Eigensystem Realization Algorithm in Frequency Domain for Modal Parameter Identification ", *Journal of Vibration, Acoustics, Stress and Reliability in Design*, Vol. 110, No. 1, Jan.,1988, pp. 24-29.

[6] Jer-Nan Juang, J. E. Cooper, and J. R. Wright, "An Eigensystem Realization Algorithm Using Data Correlations(ERA/DC) for Modal Parameter Identification ", *Journal of Control Theory and Advanced Technology*, Vol. 4, No. 1, March 1988, pp. 5-14.

[7] Ralph Quan, "System Identification Using Frequency Scanning and the Eigensystem Realization Algorithm", *Journal of Guidance, Control, and Dynamics*, Vol. 17, No. 4, 1994, pp. 670-675.

[8] Jer-Nan Juang, and Richard S. Pappa, "Effect of Noise on Modal Parameter Identification by the Eigensystem Realization Algorithm", *Journal of Guidance, Control, and Dynamics*, Vol. 9, No. 3, 1986, pp. 294-303.

[9] Jer-Nan Juang, and Jiann-Shiun Lew, "Integration of System Identification and Robust Controller Designs for Flexible Structures in Space", *AIAA Guidance Navigation and Control Conference*, 1990.

[10] Gary J. Balas, and John C. Doyle, "Robustness and Performance tradeoffs in Control Design for Flexible Structures", *IEEE, Proceedings of the 29th Conference on Decision and Control*, Honolulu, Hawaii, Dec., 1990.

[11] Petter Lundström, Sigurd Skogestad and Zi-Qin Wang, "Uncertainty Weight Selection for $H\infty$ and μ Control methods", *IEEE, Proceedings of the 30th Conference on Decision and Control*, Brighton, England, Dec., 1991.

[12] John E. Bible and D. Stephen Malyevac, "Guideline for the Selection of Weighting Functions for $H\infty$ Control", AD-A252 781, NSWCDD/MP-92/43.

[13] Kemin Zhou with Jhon Doyle and Keith Glover, *Robust and Optimal Control*, Prentice-Hall Inc., Englewood Cliffs, NJ, 1996.

971950

Acoustic Modeling and Optimization of Seat for Boom Noise

Yang Qian and Jeff VanBuskirk
Rieter Automotive North America, Inc.

Copyright 1997 Society of Automotive Engineers, Inc.

ABSTRACT

Results of acoustical simulation of a vehicle with seats is presented in this paper, providing some basic understanding how the geometry of the seats as well as the acoustical properties of the seat material can affect the acoustical behavior of the interior. Both a finite element model and a boundary element acoustical model for a minivan with seats are generated. The influence of a change of seat geometry on modes and response is calculated first. In addition, the effects of acoustical properties of the seat material, i.e. airflow resistivity, on absorption respectively boom reduction is investigated.

The simulation results have shown that the geometry of the seats has to be modeled quite accurately in order to achieve good simulation results. It has been found that rather small changes of the seat model may cause noticeable changes in modal behavior and acoustical response. Moreover, it is demonstrated how the airflow resistance of the seat material can have a large effect on the acoustical response at low frequency in a vehicle, especially by modifying the natural frequencies and loss factors of longitudinal modes.

Following these simulation results, a vehicle has been tested with seats made of different materials and also with different geometrical modifications. The test results are in good agreement with the simulation results.

INTRODUCTION

Vehicle cavity boom noise (<150 Hz) is often found in cars or vans, being perceived as acoustical resonances in the passenger compartment. These acoustic resonances of the cavity interior modes are excited by vibrations of the surrounding body resp. trim panels which in turn are generated due to road or engine excitation. A favorable condition for the generation of boom noise is the coincidence of structural body/panel modes and acoustical cavity modes.

Both the natural frequencies and the shapes of the cavity modes are primarily determined by the dimensions of the passenger compartment. Usually these dimensions are hardly changed after the design of the vehicle is finished and it is therefore difficult to modify these modal parameter. Nevertheless, seats are part of the design and they may -simplifying speaking- "obstruct" the acoustical velocity of longitudinal modes, especially when the seats are placed close to the pressure nodes of such modes. Changing the seat geometry may therefore modify acoustical modes to some extent as will be outlined in this paper.

A second way to reduce boom noise is by increasing the modal damping of the acoustical modes by adding absorption to the cavity surface or to other relevant design elements. This measure is equivalent to an increase of damping in structural vibrations and will decrease the resonance peaks in the acoustical response.

In the frequency range from 20-100 Hz, little absorption exists for most of the interior sound absorption parts. This is mainly due to the combination of the two facts that (i) absorbing layers, mounted on the cavity surface, are quite thin compared with the sound wavelength and (ii) the acoustical velocity vanishes close to the surface. Comparing seats with trim parts, it becomes immediately clear that seats offer potential for the increase of absorption. Firstly, they occupy a relative large volume respectively large surface area compared with other interior parts. And they are moreover located in advantageous positions in the interior center.

Having these opportunities in mind, research work is carried out by car manufacturers as well as by Rieter Automotive in order to derive know-how and procedures for the optimization of seats with respect to the reduction of boom noise [1]. Acoustical simulation is intensively used as tool in this work, either by executing boundary element simulations (BEM) or by applying the finite element approach (FEM).

Several sub-variants exist again for the representation of the seats in numerical simulation. When the interest is focused toward the calculation of the natural frequencies and shapes of the acoustical modes, the FEM method is usually applied. In this case, seats may be modeled by appropriate impedances of their surfaces, by "dense air", by porous elements or by even more complex approaches. The most simple approach is the impedance approach which allows to represent both acoustical reflection and absorption. This method is particularly appropriate for rear seats, mounted on a rigid rear seat back. But it has the disadvantage that sound transmission through the front seats cannot be taken into account correctly. In the second approach, the seat interior volumes are modeled as dense air, where the density of the air is increased by a factor of about 10 to 20 compared with normal air. This representation allows to take both reflection and transmission into account, but here absorption is not represented well. This is why a third approach, i.e. modeling of the seats by porous elements will be considered in this paper too.

Integrating porous elements into the modeling introduces unevenly distributed dissipation which leads to complex modes. Only very recently, complex modes can be numerically evaluated by commercial software. This is why numerical simulations with porous seats are applied in this paper only for the prediction of acoustical response functions.

As a second class of approaches for numerical simulations, boundary element modeling can be applied. BEM has the advantage of reduced modeling work (only the boundary surfaces have to be discretized, not the volume itself). But this method again is limited to the calculation of forced response functions; mode calculations are more or less impossible. And the seats can only be modeled by impedance, not by porous elements.

Based upon the advantages and disadvantages of the different simulation approaches, both a boundary element model and a finite element model of a minivan were generated. The FEM model was used for an investigation of the influence of geometrical aspects of seat design on the natural frequencies and shapes of acoustical modes. The BEM model was used for a comparison of acoustical response with/without absorption on the surface around the passenger compartment. And the FEM model was used again for the prediction of the effects of seats made out of porous material respectively for an investigation about the optimization of the flow resistivity of the seat material.

INFLUENCE OF SEATS GEOMETRY

As mentioned above, the cavity mode shapes and cavity mode frequencies are influenced by the seats due to their positions in passenger compartments and dependent on the material. In this section some results of investigations about the geometry of the seats will be presented and discussed.

In this context, it is appropriate to have a closer look to the basic appearance of acoustical modes of a vehicle interior. Figure 1. shows some longitudinal modes of a passenger car and a minivan in a simplified representation. The sine-type curves represent schematically the pressure distributions. As acoustical velocity is related to the gradient of the pressure, the acoustical velocity will exhibit its maximum in the region of the pressure nodes. Placing an obstructing element will probably have the biggest impact on an acoustical mode when the location coincides with a high acoustical velocity. This means that in a passenger car (Figure 1a.) the front seats will probably strongly influence the fundamental longitudinal mode. In a minivan on the other hand, the second row will presumably influence the fundamental mode more whereas both front and rear row may rather show an impact on the second longitudinal mode.

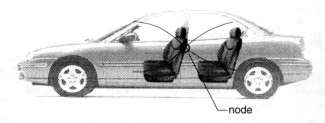

Figure 1a. First longitudinal acoustical mode in a passenger car

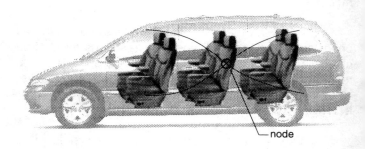

Figure 1b. First longitudinal acoustical mode in a minivan

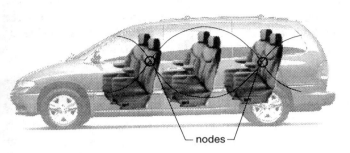

Figure 1c. Second longitudinal acoustical mode in a minivan.

In order to check the validity of this phenomenological vies ("hypothesis"), different seat models were compared with respect to the natural frequencies of the interior modes:

- "down to floor seat" with no air under cushion
- "floating-over-floor" seats with air under them
- "dense-air" seats (same as down-to-floor but represented by air having a density of 5 resp. 10 times the usual value)
- omitting all seats

Table-1 Comparison of first three longitudinal mode frequencies for different seat models

Seat Model	1st Mode (Hz)	2nd Mode (Hz)	3rd Mode (Hz)
1. All seats to floor	48.42	85.44	126.46
2. All seats floating	50.31	91.76	132.24
3. Front seat to floor	50.07	88.84	129.72
4. Middle seat to floor	49.04	92.45	131.99
5. All air seat to floor (5 times of air density)	51.21	93.61	136.43
6. All air seat to floor (10 times of air density)	51.21	93.61	136.43
7. No seat	53.43	101.44	139.53

Table-1 shows the results for some longitudinal mode frequencies for the minivan, calculated with the FEM model. The presence of seats in the passenger compartment reduces the cavity modal frequencies for longitudinal modes since, simplified speaking, the waves have to find their way around the seats, thus traveling virtually longer.

The results in Table 1. Show that the first and third longitudinal mode frequencies are relatively insensitive with respect to the different seat models. The second natural modal frequency is much more sensitive. The high sensitivity of the second mode verifies the hypothesis given above by notifying that the second mode has two nodal planes interfering with seats (see Figure 1c).

In addition to the shift of modal frequencies, there is also a slight modification in mode shapes. As an example demonstrating these modifications, Figures 2a and 2b (see **Appendix**) show the pressure contours of the minivan interior second longitudinal mode for seat models 1 and 2. The differences are most pronounced around the second row of seats.

A further support of the hypothesis given above is given by the fact that the connection of the font seats to the floor is most relevant to the second mode whereas doing the same with the second row shows more effect on the first mode. This shows that connecting seats to the floor influences preferably modes which have a high acoustical velocity at the seat location.

For "dense air" seats, the acoustical wavelength increases which leads to an increase of modal frequencies (models 5 and 6). The longitudinal modal frequencies are somewhat overestimated compared with the rigid seat (model 1) which underlines that these models are to be handled with precaution.

Summarizing, it is found that acoustical modes can be shifted by changing the geometry of the seats. This can yield solutions against booming noise when the boom is due to a coincidence of the natural frequency of a structural body panel mode and the natural frequency of an interior acoustical mode.

BASIC EFFECT OF POROUS SEATS

The modeling variants 6 and 7 in the previous chapter were rather representing changes in material than changes in geometry. It was discussed, how this change modifies the modal frequencies. The subsequent question arises about how a change towards porous materials would modify the response. It is obvious that porous material may lead to dissipation. The principal advantage of dissipation is displayed in Figure 3. Two BEM calculations are done in a usual passenger car. One case with absorption of 5% on seats and the other case without this absorption. The decrease of the response peaks at the resonance conditions is clearly noticeable although there is not much difference in response at intermediate frequencies. Basically, these response peaks represent the well known booming noise and a potential capability of reduction of booms is of course highly welcome. It is for this reason that it is of interest to investigate the potential of porous seats.

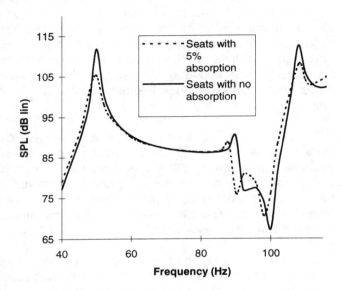

Figure 3. Influence of absorption: SPL in a vehicle for the seats with different absorption coefficients.

The relevant acoustical properties of porous materials can be traced back to two physical phenomena. Firstly, the flow resistance of the material which causes "friction" between the sound wave (i.e. the acoustical velocity) and the skeleton and, secondly, heat transfer mechanisms between air and skeleton. Both effects produce dissipation respectively absorption of the sound wave traveling through the material. This dissipation may strongly reduce the high sound pressure levels at resonance conditions as described. The heat transfer mechanism has the additional effect that the compression modulus of the air shifts partially from the adiabatic extreme ($K \cong 144'000$ N/m^2) towards the isothermal value ($K \cong 100'000$ N/m^2) by which means the sound velocity is reduced.

Energy is typically always the product of a force-type and a velocity-type quantity. Based on that, it becomes obvious that the friction forces in the porous materials will become highly efficient loss generators when they act on high acoustical velocity of the sound wave. Porous materials therefore become highly efficient when they are placed in the vicinity of the nodal planes of the modal pressure distributions where the acoustical velocity is high (this explains also why low frequency absorption is difficult to generate on cavity surfaces). This happens to be valid for seats in vehicles. The front seats are usually located at a place near the maximum particle velocity for the first longitudinal mode in a car (Figure 1a). And, in a minivan, the second seat is located at a node of the first longitudinal mode while the front and third row seats are located at nodes of second longitudinal mode (Figures 1b and 1c).

With seats at a position where the particle velocity of the sound waves exhibits a maximum, the porous materials still need to be carefully selected to provide good absorption. The main reason for this is that when the flow resistance of the material is too high, the acoustical flow through the seats is prevented by which means the absorption drops again. It is thus an optimization task to select the material in such a way that a maximum product of flow resistance and sound velocity is obtained.

The design of seats using a properly selected material thus brings us to a situation which is somewhere in between of the case where no seats are existing (zero flow resistance) and the other extreme case with impervious seats (zero sound velocity). This has some effects on modal frequencies and mode shapes as well as can be imagined by executing two acoustical analysis; one without seats and one with rigid seats. The modal frequencies and shapes exhibit consequently intermediate results.

MODELING OF POROUS MATERIALS

Porous materials consist of two phases: the solid phase (skeleton or frame) and the fluid phase (air). Longitudinal waves can principally occur in both phases. A transversal wave also appears when the skeleton possesses shear stiffness. Porous materials are often differentiated in ones having rigid skeletons and others with a flexible skeleton.

Different models exit for the representation of porous materials. The most simple model is probably the model of Delany and Bazley [2], in which the skeleton is rigid and highly porous. Based on large amounts of experimental work with fiber glass, Delany and Bazley have shown that the airflow resistance is the only significant parameter in these cased for the calculation of material impedance.

A more advanced model, usually applied in cases with a rigid non-vibrating skeleton, is the generalized Rayleigh model [3]. For a harmonic wave, the momentum and continuity equations can be written as:

$$(i\rho_a \omega + R)v + \nabla p = 0$$

$$div\ v = \frac{i\omega\Omega}{\rho c_a^2} \cdot p$$

Where R is airflow resistivity, c_a is the speed of wave propagation and ρ_a is the effective density related to air density ρ through the structure factor K_s. The relation can be written as:

$$\rho_a = K_s \rho$$

where the structure factor K_s, accounts for inertia effects of the material. This model thus needs only the fluid parameters airflow resistance, porosity and structural factor as well as the sound velocity to describe its behavior.

A finite element formulation developed on the basis of this generalized Rayleigh model is available in commercial software for the calculation of frequency response functions [4]. The results following later in this paper for porous seats are accordingly obtained.

More advanced models allow for compression waves through the skeleton too, in addition to the wave through the air. The flow resistance term consequently acts on the difference in velocities of the air and the skeleton. Two cases are again distinguished:

Limp porous materials: the skeleton does vibrate but the stiffness of the skeleton is significantly less than that of air. The airborne wave is the wave of primary importance but the modeling needs the density of the frame in addition to the acoustic/fluid parameters in order to describe its behavior.

Elastic porous materials: the frame does vibrate and the stiffness of the frame is of the same order as the air stiffness. Airborne waves and structureborne waves are superimposed which make this case the most diffcult case to model. In addition to the fluid parameters, the Young's and shear moduli of the skeleton as well as their loss factor are needed in order to predict its acoustical performance.

For limp and elastic porous materials, mechanical and kinematical interactions between solid and fluid phase can be described either by an extended Rayleigh model or by the quite general theory of elastic porous material by Biot [5]. The Biot model is efficient and is becoming the predominant model, taking into account both fundamental fluid structural coupling effects, i.e. mass coupling and viscous forces. Based on Biot's theory, Y. J. Kang and J. S. Bolton [6] developed formulations of foam finite element. And, J.P. Coyette *et al.* [7] also developed formulations for finite element method (VIOLINS), applied for layered material systems.

CASE STUDIES WITH POROUS SEATS

After showing the influence of the seat geometry in the previous section, the effect of the material choice will now be demonstrated by applying materials with varying porosity resp. flow resistivity to the seats of a minivan. The potential of porous seats may be especially high in a minivan because three rows of seats are positioned in the open cavity volume instead of only one row as in case of a usual passenger car.

All seats were modeled as rigid porous elements thus neither considering structural waves through the seat material nor structural vibrations of the seats as a whole. The boundary of the interior compartment is moreover assumed to be rigid and fully reflecting. This interior volume is excited by a piston-like vibration of the driver's toepan. The frequency response functions of the sound pressure level is computed while varying the air flow resistance value of the seat material.

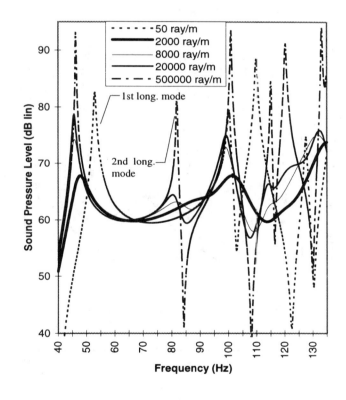

Figure 4a. SPL at driver's inner ear with different seat airflow resistance in a minvan. Vehicle is excited by velocity excitation on driver's toepan.

Figure 4a shows SPL at driver's inner ear. The first peak at the first modal frequency is very high for seats having very high resistivity (quasi-rigid seats), then the value of peak gradually reduces with decreasing of resistivity. As the resistivity decreases to very low value (the seat transfers so-to-speak to air), the peak increases again at a slightly increased frequency. Both extremes (rigid and air) are lacking of dissipation (no flow respectively no resistivity forces). But in between a maximum of dissipation at an optimal value of resistivity can be achieved which greatly reduces the response peaks at the resonance frequencies of the longitudinal modes. It means that an optimal value can be achieved which greatly reduces the peaks of longitudinal modes. Accompanied with the change in dissipation, there is a shift of modal frequencies towards higher values when going from the high resistivity to low values. This is related to the virtual shortening of the car interior. It should be emphasized that the value of the optimal flow resistivity depends on the geometrical details of the car under investigation.

It can be observed that the peak values of first mode are higher than those of second modes at the driver's ear. This is because of the mode shapes where the first mode has its maximum amplitude close to the driver's ear. The second mode, on the other hand, exhibits a maximum amplitude close to the second row of seats (see Figures 1b and 1c), and this becomes accordingly visible when displaying forced response results for that location (Figure 4b).

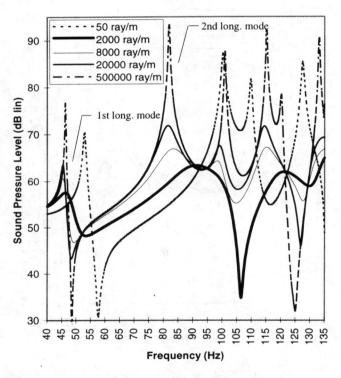

Figure 4b SPL at center position of the second seat with different seat airflow resistance in a minvan. Vehicle is excited by velocity excitation on driver's toepan

VALIDATION MEASUREMENTS

According to the simulation results above, a minvan with different seats, having different airflow resistivities, was tested on a rough road. Figure 5 shows the SPL versus frequency at the driver's inner ear location. The results confirm that the sound pressure level with seat "A" (optimized resistivity) has been significantly reduced at the resonances of the first three longitudinal modes compared with seat "B" which exhibits a too high resistivity.

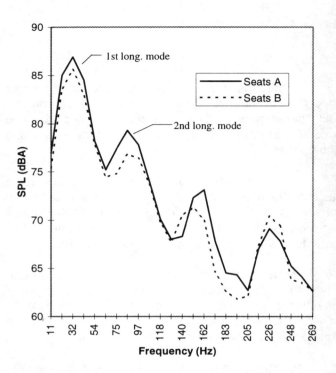

Figure 5. Influence of seat airflow resistance. SPL at driver's ear tested on rough road.

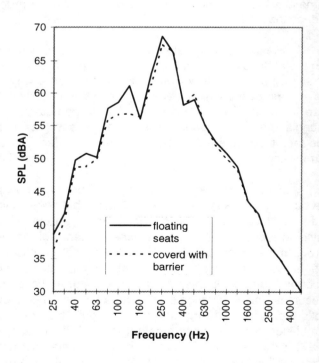

Figure 6. Influence of seat geometry change. SPL at 2nd seat center position tested on rough road.

A further experiment was carried out in order to verify the effect of geometrical modifications by closing the room under the seats. From modal investigations as shown in Figure 2, it was expected that the pressure level at the location of the 2nd row of seats might be reduced by closing the space underneath the 2nd and 3rd row seats. This space was therefore covered with a barrier. Figure 6 shows the measured SPL at the second seat center position for a run-up from 20 to 60 mph on a rough road. The SPL was reduced significantly around the lower 3 modes frequencies. An additional reason for the significant reduction at 125 Hz is the high acoustical sensitivity and high panel contribution of the floor under the second row of seats. This panel excites some of the modes at a pressure maximum when the seats are "floating". The covers under the seat cushion do thus not only tune the longitudinal acoustical mode shapes but also reduce the coupling between structural floor panel modes and acoustical cavity modes.

CONCLUSION

From the results of simulation and test, the influence of seats can be summarized as:

1. A quite small geometrical change of the seats may cause a noticeable change in shapes and natural frequencies of low longitudinal modes and therefore influence the acoustical response. Quite an accurate 3D seat model is necessary for reliable simulations of these effects.

2. The acoustical properties of the material of the seats do have a significant effect on longitudinal modes at low frequencies. The mode shape and mode frequencies and, especially the modal loss factors, can be changed by modifying the airflow resistance even without actually changing the geometry of the seat.

3. An optimized airflow resistance can significantly reduce the resonance peaks at the natural frequencies of the longitudinal modes due to a maximization of the dissipation.

4. More correlated studies are desirable with an improved FEM-representation of the porous material.

REFERENCES

1. R. Frosio, Unikeller conference 1993. Some aspects of numerical simulation of car interior noise.

2. M. E. Delany and E. n. Bazley, 1969 national physical laboratory, aerodynamics division report ac 37. Acoutical Charicteristics of Fibrous Absorbent Materials.

3. K. Attenborough, 1982 Physics Reports 82 (3), 179-227. Acoustical Charicteristics of Porous Materials.

4. SYSNOISE reference manual, theoritical manual, 1995, LMS

5. M. A. Biot, 1956 JASA 28, 168-191. Theory of Propagation of Elastic Waves in a Fluid-Saturated Porous Solid. I. Low Frequency Range. II. Higher Frequency Range.

6. Y. J. Kang and J. S. Bolton, JASA 98, July 1995. Finite element modeling of isotropic elastic porous materials coupled with acoustic finite elements.

7. J. P. Coyette *et al.*, **VIOLINS** (Simulation of **Vi**brations **O**f multi-**L**ayer **In**sulation **S**ystems) reference manual, Numerical Integration Technologies and also, Proc. Inter-Noise 95, vol. 2, 1995, A finite element model for predicting the acoustic transmission characteristics of layered structure

8. R. Frosio, Unikeller conference 1995. Numerical investigation of the acoustical potential of parcel shelf design.

9. Y. Qian and J. Vanbuskirk, Proceedings of the 1995 Noise and Vibration Conference, 951244 Vol. 1. Sound Absorption Composites and Their Use in Automotive Interior Sound control.

10. J. F. Allard, J. F., Propagation of Sound in Porous Media, 1993 Elsevier Science Publishes LTD.

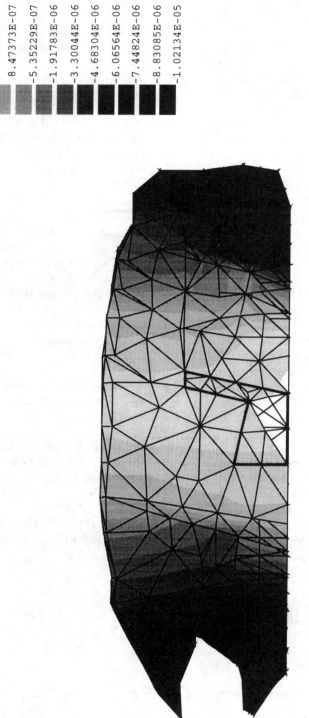

Figure 2a   Contour plot of the 2nd longitudinal mode with seat model 1 (all seats to floor)

Database : rigid.cdb
Results : rigid.rslt

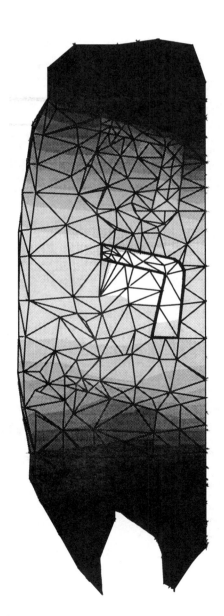

Figure 2b Contour plot of longitudinal mode with seat model 2 (all seats floating)

3   9.1756963E+01 Hz Mode Shape
COMET/Vision Ver 3.1.0 Tue Sep 10 11:36:05 1996

971951

Experimental Verification of Design Charts for Acoustic Absorbers

Mardi C. Hastings and Richard D. Godfrey
Ohio State Univ.

Copyright 1997 Society of Automotive Engineers, Inc.

ABSTRACT

Design charts which predict acoustic absorption of porous insulators were verified experimentally using the two-microphone technique to measure the normal incidence absorption coefficient of three glass fiber materials in two different arrangements — a single-layer sample and a single layer in front of an air space, each backed by a rigid termination. The specific flow resistivities of the materials ranged from 2,000 to 52,000 mks rayls/m. Experimentally determined absorption coefficients were in agreement with those predicted by the design charts. The results indicate that these charts could be a useful tool in designing sound absorbers for practical applications.

INTRODUCTION

The design chart format proposed by Mechel [1] is a useful means for summarizing the sound absorption characteristics of an array of insulators. Figure 1 shows a design chart for the normal incidence absorption coefficient ($\alpha_\perp$) of a single-layer absorber backed by a rigid termination. The chart format is a contour plot of the theoretical sound absorptivity for a specific arrangement of absorbers plotted on two nondimensional axes: $R = R1*d/Z_o$ and $F = f*d/c$, where $R1$ is the airflow resistivity, d the absorber thickness, Z_o the free field acoustic impedance, f the frequency, and c the free field sound speed. Figure 1 indicates that for relatively small airflow resistivity ($R \approx 1.0 - 1.5$), the maximum absorption coefficient occurs at $F \approx 0.25$ which corresponds to a thickness d equal to a quarter of the free field wavelength, $\lambda = c/f$. As would be expected, this resonance condition producing a peak in absorptivity also occurs at odd multiples of $\lambda/4$ (i.e., $F = 0.75$, $F = 1.25$, etc.). The design chart format easily shows the effects of basic parameters used in the design of sound insulators on absorptivity.

The charts are based on the airflow resistivity (i.e., specific airflow resistance per unit thickness) which is either directly measured or calculated using an empirical formula. The sound absorption coefficient is calculated from the propagation constant and characteristic impedance of the material. The calculation is based on a one-dimensional plane wave analysis, and

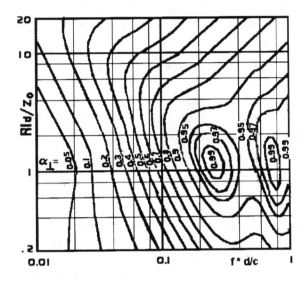

Figure 1. Design chart showing constant absorption lines for a bulk absorber in a normal incidence sound field as a function of nondimensional parameters as proposed by Mechel [1].

the propagation constant and characteristic impedance used are determined from empirical formulas. Since the design charts are based on a combination of empirical formulas, they should be experimentally verified.

In this study, experimental validation of the normal incidence design charts for a rigidly terminated monolayer and a simple multilayer arrangement (an absorber layer and air gap) was completed using glass fiber materials. The experimental measurements for normal incidence were done using the two-microphone technique as specified by ASTM E 1050-90 [2].

ANALYSIS

Much work has been done on the calculation of the acoustic variables — airflow resistivity, propagation constant and acoustic impedance — for a given material by Beranek [3], Beranek and Work [4], Delaney and Bazley [5], Bies [6], and Mechel [1]. Their research resulted in empirical formulas defining these acoustic material properties. They also formulated impedance analyses to determine the performance of monolayer and multilayer sound absorbers.

The airflow resistivity, $R1$, is defined as:

$$R1 = -\Delta p / (\Delta x \times v) \text{ mks rayls/m (Pa-s/m}^2) \qquad (1)$$

where Δp is the static pressure difference across the material thickness Δx, and v is the velocity of flow through the material. In practice, these three parameters are measured according to ASTM C 522 - 87 [7] and then used to calculate $R1$. For glass fiber absorbers, Bies [6] reported an empirical equation for airflow resistivity based on regression analysis of experimental data:

$$R1 = 8750 \times 393.7 (\sigma_m)^{1.53} / d_f^2 \qquad (2)$$

where σ_m is the bulk density of the absorber (lb/ft^3); d_f is the fiber diameter (HT, hundredth thousandths of an inch); and 393.7 is a conversion factor to give $R1$ in mks rayls/m.

Calculation of the propagation constant and acoustic impedance is similar to that of the airflow resistivity because even though they both have theoretical definitions, they are calculated using empirical formulas. The empirical formulas used were developed by Mechel [1]. The characteristic impedance is represented by Z_a, the propagation constant by Γ_a, and normalized values Z_a/Z_o and Γ_a/k_o, are subscripted with an additional "n." ($k_o = \omega/c$ is the free field wave number.) Mechel [1] used the following empirical relationships to develop the design charts:

$$E = \rho_0(f)/(R1) \qquad (3)$$

For $E \le E_x$:

$$Z_{an} = \Gamma_{an} / (j\gamma h) \qquad (4)$$

$$\Gamma_{an} = j\left(1 - j(\gamma / 2\pi E)\right)^{1/2} \qquad (5)$$

For $E > E_x$:

$$Z_{an} = (1 + 0.06082 / E^{0.717}) - j0.1323 / E^{0.6601} \qquad (6)$$

$$\Gamma_{an} = 0.2082 / E^{0.6193} + j(1 + 0.1807 / E^{0.6731}) \qquad (7)$$

where $R1$ is determined using Equation (2), ρ_o is the density of air, $j = (-1)^{1/2}$, $\gamma = 1.403$ is the adiabatic exponent for air, and h the porosity of the absorber. The porosity is defined as one minus the ratio of the absorber density to the fiber density, but can be approximated using $h = 0.95$ for glass fiber absorbers. As explained by

Mechel [1], the transition point, E_x, is different for the real and imaginary parts of each equation. E_x is 0.04 and 0.006 for the real parts of Γ_{an} and Z_{an}, respectively, and 0.008 and 0.02 for the imaginary parts of Γ_{an} and Z_{an}, respectively.

The analyses of monolayer and multilayer absorbers were done for normal incidence with a rigid termination. The sound absorption coefficient, $\alpha_\perp$, for monolayer bulk absorbers with these conditions is determined using:

$$\alpha_\perp = 1 - |r|^2 \qquad (8)$$

where the reflection coefficient, r, is defined as:

$$r = (z - 1) / (z + 1) \qquad (9)$$

and the normalized wall impedance, z, is given by:

$$z = Z_1 / Z_0 = Z_{an} / \tanh(k_0 d\, \Gamma_{an}) \qquad (10)$$

The multilayered absorbers analyzed consisted of a layer of glass fiber material followed by an air space backed by a rigid termination. For a multilayered absorber an input impedance must be calculated. This impedance represents the effective impedance of all the layers combined. Once this impedance is calculated, it is substituted into Equations (8) - (10) to calculate the sound absorption coefficient. For normal incidence, the equations for a bulk absorber with a bulk reacting air gap are:

$$Z_1 = Z_2 \frac{1 + (Z_a \tanh(D_1)) / Z_2}{1 + (Z_2 \tanh(D_1)) / Z_a} \qquad (11)$$

where $D_1 = \Gamma_a d$, $Z_2 = -jZ_o\cot(k_o t_a)$, t_a is the thickness of the air gap, $\Gamma_a = \Gamma_{an} k_o$ the propagation constant for the material, and $Z_a = Z_{an} Z_o$ the impedance constant for the material.

EXPERIMENTAL MEASUREMENTS

The normal incidence absorption coefficients for each of the six configurations were determined according to ASTM E 1050 [2] using a Brüel and Kjær (B&K) Type 4206 Impedance Tube and Type 2032 Digital Signal Analyzer. Measurements were made in one–third octave bands from 50 to 6300 Hz. A tube with a 100–mm internal diameter (large tube) was used for 50 to 1600 Hz, and a tube with a 29–mm internal diameter (small tube) was used for 500 to 6300 Hz. Figure 2 illustrates the setup.

MATERIALS TESTED - Three absorbers made of glass fibers were tested. Table 1 summarizes their properties needed for calculation of the acoustic parameters using the equations given in the previous section. The material samples were 41 mm thick. The total thickness with a 41-mm material layer and air gap was 400 mm. The flow resistivities of the samples resulted in parameter R values that spanned the scale of the design chart shown in Figure 1.

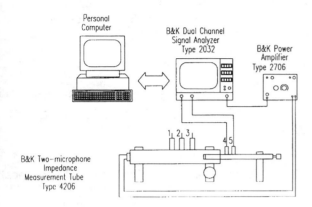

Figure 2. Experimental setup for measurement of normal incidence absorptivity.

Table 1. Absorber Material Properties

Material	Effective Fiber Dia. (HT)	Bulk Density (kg/m^3)	Airflow Resistivity (Pa-s/m^2)	R (dimensionless)
A	55.7	26.8	2,248	0.23
B	37.9	43.2	12,815	1.31
C	39.2	106.7	52,082	5.34

RESULTS

The measured normal incidence absorption coefficients were correlated with Mechel's [1] design chart model summarized by Equations (2) - (11). Figures 3 and 4 display measured and calculated results for the monolayer cases and multilayer cases (single layer in front of an air gap), respectively. All six configurations had a rigid termination. The measured values are points labeled "Lg. Tube" and "Sm. Tube," and the calculated values are curves labeled "Model." The design chart shown in Figure 1 corresponds to the values plotted in Figure 3 for a single-layer glass fiber insulator backed by a rigid termination.

The measured normal incidence absorption coefficients are in good agreement with the model.

Mechel [1] estimates that experimental data should fall within ±20% of the predicted values. Even the additional resonances created by the air gap are tracked fairly well by the model. Figures 3 shows that Material C, which is the most dense and has the highest airflow resistivity, does not provide the best absorption at higher frequencies. This behavior is clearly apparent in the design chart shown in Figure 1. Above the "islands" of high absorptivity for $R \approx 1.0 - 1.5$, at high values of F for $R = 5.34$ the absorption coefficient contours have little, if any, change in value as F increases.

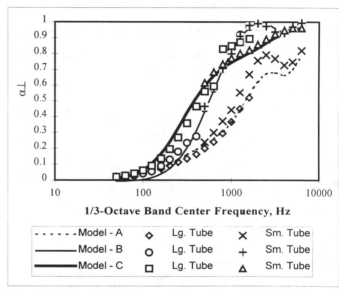

Figure 3. Normal incidence absorption coefficients for three different 41-mm thick glass fiber absorbers.

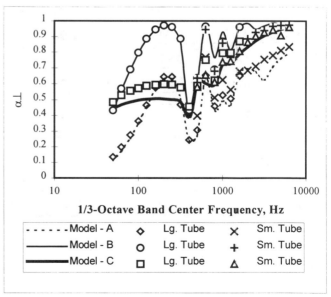

Figure 4. Normal incidence absorption coefficients for three different 41-mm thick glass fiber absorbers in front of an air gap (total thickness of 400 mm).

For a direct comparison with Figure 1, Table 2 summarizes measured and calculated values of the absorption coefficient for the monolayer case of Material B in terms of the dimensionless parameter, F. The values of the absorption coefficients in the design chart would then fall on a horizontal line through $R = 1.31$. For this case, the resonance condition described earlier which results in high absorptivity when the single–layer thickness is equivalent to an odd multiple of $\lambda/4$ ($F \approx 0.25$ and $F \approx 0.75$) is clearly apparent in both measured and calculated absorption coefficients.

Table 2. Summary of Absorption Coefficients for Material B

f (Hz)	F	$\alpha_\perp$ Measured (ASTM E1050-90)	$\alpha_\perp$ Calculated (Mechel Model)
80	0.010	0.03	0.00
100	0.012	0.04	0.00
125	0.015	0.06	0.02
160	0.019	0.09	0.04
200	0.024	0.13	0.07
250	0.030	0.18	0.13
315	0.038	0.24	0.20
400	0.048	0.28	0.31
500	0.060	0.45	0.43
630	0.075	0.58	0.57
800	0.096	0.71	0.70
1000	0.120	0.80	0.81
1250	0.149	0.91	0.90
1600	0.191	0.98	0.96
2000	0.239	0.99	0.99
2500	0.299	0.97	0.99
3150	0.377	0.94	0.96
4000	0.478	0.93	0.94
5000	0.598	0.97	0.97
6300	0.753	1.00	0.99

CONCLUSION

Design charts developed by Mechel [1] were experimentally verified for glass fiber absorbers in a normally incident sound field. The measured absorption coefficients were in good agreement with those predicted by the model. These design charts which summarize the absorption characteristics of an array of acoustic insulators as contour plots of two dimensionless variables provide a useful tool in designing sound absorbers for practical applications.

REFERENCES

1. Mechel, F. P. (1988), "Design Charts for Sound Absorber Layers," The Journal of the Acoustical Society of America, Vol. 83, No. 3, pp. 1002-1013.

2. ASTM Designation: E 1050 - 90 (1990), "Standard Test Method for Impedance and Absorption of Acoustical Materials Using a Tube, Two Microphones, and a Digital Frequency Analysis System," American Society for Testing and Materials, Philadelphia, PA.

3. Beranek, L. L. (1947), "Acoustical Properties of Homogenous, Isotropic Rigid Tiles and Flexible Blankets," The Journal of the Acoustical Society of America, Vol. 19, No.4, pp. 556-568.

4. Beranek, L. L. And Work, G. A. (1949), "Sound Transmission through Multiple Structures Containing Flexible Blankets," The Journal of the Acoustical Society of America, Vol. 21, No. 4, pp. 419-428.

5. Delaney, M. E. And Bazley, E. N. (1970), "Acoustical Properties of Fibrous Absorbent Materials," Applied Acoustics, Vol. 3, pp. 105-116.

6. Bies, D. A. (1971), "Soft Acoustical Blankets," in Noise and Vibration Control (L. L. Beranek et al., eds.), McGraw-Hill, pp. 245-269.

7. ASTM Designation: C 522 - 87 (1993), "Standard Test Method for Airflow Resistance of Acoustical Materials," American Society for Testing and Materials, Philadelphia, PA.

971953

Panel Contribution Study: Results, Correlation and Optimal Bead Pattern for Powertrain Noise Reduction

Anbarasu Nachimuthu and Karen M. Carnago
Chrysler Corp.

Farshid Haste
Automated Analysis Corp.

Copyright 1997 Society of Automotive Engineers, Inc.

ABSTRACT

To understand how the passenger compartment cavity interacts with the surrounding panels (roof, windshield, dash panel, etc) a numerical panel contribution analysis was performed using FEA and BEA techniques. An experimental panel contribution analysis was conducted by Reiter Automotive Systems. Test results showed good correlation with the simulation results.

After gaining some insight into panel contributions for power train noise, an attempt was made to introduce beads in panels to reduce vibration levels. A fully trimmed body structural-acoustic FEA model was used in this analysis. A network of massless beam elements was created in the model. This full structural-acoustic FEA model was then used to determine the optimal location for the beads, using the added beams as optimization variables.

INTRODUCTION

To reduce the structure-borne power train noise levels in the vehicle, different areas of the car have been studied for improvement: power train, engine mounts, engine box, suspension cross member, body panels, instrument panel and passenger cavity panels. This paper highlights the effect of panel contributions, their interaction with the acoustic cavity and the optimum geometry of panel beads.

Finite Element and Boundary Element Analysis (FEA and BEA) were performed to predict and rank panels contributing to powertrian noise. An experiment was then conducted to correlate with the analysis.

One common technique to control the vibrations of body panels is to locally stiffen them by adding beads. However, design engineers often face the challenge of determining the appropriate location and length of these beads. It is important to develop a methodology to optimally locate these beads and to control their effect on different performance functions. Another difficulty in finding the optimal bead pattern is the number of constraints the optimizer faces. For example an improvement in low frequency noise due to beading might cause a degradation in higher frequencies (due to the panel resonance shift from the stiffening effect of these beads).

METHODOLOGY

Structural Model -To study power train noise, a detailed finite element trimmed body model of the vehicle was developed. The model included BIW and non-structural masses. Figure 1 shows the finite element model of the trimmed body.

To reduce the model size, the structural model didn't include all the details of the moveable panels (doors, hood and decklid). Instead, their mass and inertia properties were represented at their center of gravities and their effect was transferred to these panels on the acoustic cavity.

Acoustic Model - Two different techniques were used to represent the acoustic cavity. A Boundary Element Analysis (BEA) model was created to study the acoustic response and panel contributions. The cavity model is shown in Figure 2. This model was made of a 2D shell around the boundary of the interior car cavity. For optimal bead location analysis, using 3D solid elements, an FEA acoustic cavity model was built. Both models were built with the same size elements.

<u>Loads</u> - The input power train load was obtained from wide open throttle test results. While running the engine through a frequency sweep (1500 to 6000 RPM), the corresponding body side and engine side accelerations were measured and the second order component of the signal was filtered out. The dynamic engine mount rates were measured separately. Finally the loads at each engine mount were obtained by multiplying these rates and the measured acceleration values at each mount. These loads were imported into the FEA environment as a set of complex frequency dependent functions.

<u>Frequency Range</u> - The study focused on the second order structure-borne noise caused by the load transmitted through the engine mounts. The engine operating range was 1500 to 6000 RPM (50 Hz to 200 Hz second order). To avoid modal truncation error in the higher end frequencies, a modal search was performed up to 325 Hz (more than 1.5 times the maximum output frequency of 200 Hz). Local A/F and P/F (acceleration and acoustic pressure frequency response to a unit input force) were performed at all body-chassis attachment locations and compared to test A/F and P/F results. The correlation results were satisfactory.

PANEL CONTRIBUTION - Panel contributions were studied both analytically and experimentally.

<u>Analysis</u> - A panel contribution study was performed in two phases: 1-FEA and 2-BEA. MSC/ NASTRAN was used to compute the velocity responses at the cavity boundary between 50 to 200 Hz. The measured power train loads were used as input in this FEA analysis. The velocity values were then transferred to a BEA code (SYSNOISE).

The cavity surfaces of the BEA model were divided into seven subgroups or panels : roof, floor, dash panel, backlite, rear floor, shelf and windshield.

To calculate the contribution of a panel to the Sound Pressure Level (SPL) at a certain point A (driver's ear location in this case), the BEA code calculates the following integral over the panel area:

$$c(S_p) = \int_{S_p} (p \partial_n G + i \rho \omega V_n G) dS$$

S_p : **Panel Surface**

G : **Green's Function**

$c(S_p)$: **Panel Contribution**

p : **Pressure**

ρ : **Density**

V_n : **Normal Velocity**

ω : **Frequency**

This contribution of a group of elements (lumped as one panel) is a complex vector. The summation of all panel contribution vectors at a point A and for a certain frequency is equal to the total sound pressure level at point A. Thus the calculated panel contribution determines how much a specific panel "contributes" to the total SPL at a certain frequency. Depending on the orientation of each panel contribution vector, their contribution could be either positive, zero or even negative. In other words, reducing the vibration level in a certain panel might increase or decrease the sound pressure level at a certain point and at a certain frequency.

<u>Experimental</u> - Experimental panel contributions were performed by Reiter Automotive Systems. Similar to the analysis, the cavity surface was divided into seven panels: rear floor, rear roof, front roof, front floor, top of I/P, front floor and windshield.

Measurement was taken on the smooth roll chassis dynamometer under a wide open throttle acceleration sweep from 1500 RPM to 5600 RPM.

The Panel Radiation Determination (PARADE) technique was used to compute the panel contribution vectors.

OPTIMAL PANEL BEADING - To optimally locate beads on the major cavity panels, a network of FEA massless beam elements was created. This network was connected to the grid pattern of the shell elements in each panel. Each beam element was defined as a variable in the optimization operation.

The combined structural-acoustic FEA model was used in the optimization. Acoustic pressure at the driver's ear location was used as the objective function. The optimizer tried to minimize the objective function while changing the added beam inertia properties.

Based on the optimization results, the beam elements were modified to represent those elements whose inertia properties increased. The non-contributing elements were removed from the network.

RESULTS

The acoustic response at the driver's ear was calculated using the BEA code SYSNOISE. Surface pressures were also obtained for the cavity. The acoustic response at driver's ear is shown in Figure 3. It is clear that undesirable peak responses exist at 100, 140 & 190 Hz. Modal analysis results indicate structural and cavity modes exist at some of these frequencies. The acoustic response calculated from the model had previously been correlated to test data. The overall trend of the response and magnitude of the response agreed with test data.

Panel contribution analysis was then performed using SYSNOISE. The contribution vector for the driver's ear location at each frequency was calculated. Polar diagram and curve plots were constructed for each panel. The contribution vector for the roof, floor and dash panel is shown Figures 4, 5 and 6 respectively. The major contributing panels identified were the roof, windshield, front and rear floors. These analysis results were then compared with experimental data obtained from REITER. The panel ranking obtained using BEA method correlated well with the test data, as shown in Table 1.

CONCLUSION

A finite element model of a trimmed body and the corresponding acoustic cavity were developed. The driver's ear acoustic response and panel contributions were obtained using the BEA code. These results were then compared with test data. A good correlation between analysis and experimental data was established.

Various design proposals were studied using these application methods to reduce power train noise levels.

ACKNOWLEDGMENTS

The authors would like to thank Reiter Automotive Systems for performing and providing the test information and Small Car Platform NVH team members for their active support in this project.

REFERENCES

[1]   Numerical Integration Technologies NV, SYSNPOISE Rev 5.2 User's Manual, Leuven, August 1994.

[2]   The Macneal-Schwendler Corporation, MSC/NASTRAN User's Manual V68, Los Angeles, August 1994

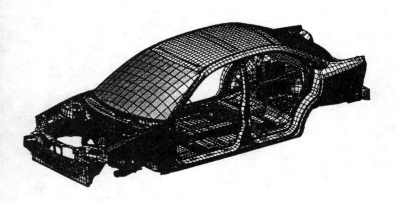 

Figure 1: Structural FEA Model of the Trimmed Body

Figure 2: BEA Cavity Model

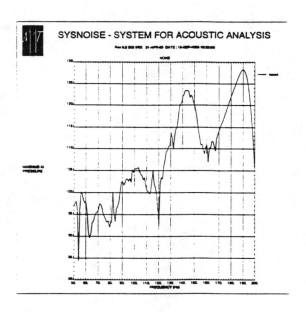 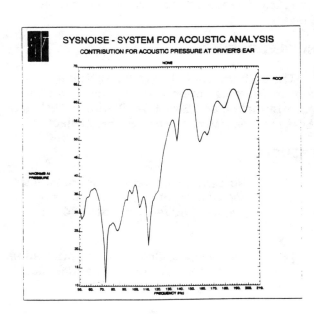

Figure 3: Acoustic Response at Driver's Ear for Powertrain Second Order WOT Load

Figure 4: Roof Panel Contribution

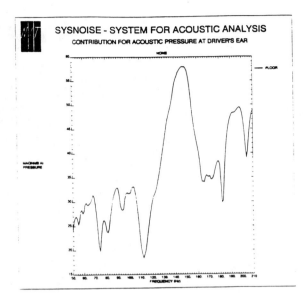

Figure 5: Floor Panel Contribution

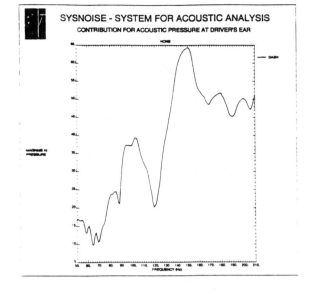

Figure 6: Dash Panel Contribution

CONTRIBUTING PANELS

Engine rpm	Analysis	Test - Reiter
3700	Roof	Front Roof
4500	Roof Dash panel Windshield Floor	Rear Floor Front Roof Top of IP Windshield
5000	Windshield Dash Panel Roof	Windshield Top of IP Front Roof Rear Floor
5700	Roof Windshield Rear Floor Front Floor	Front Roof Front Floor Rear Floor
6200	Roof Rear Floor Windshield	Rear Roof Rear Floor

Table 1: Panel Contribution Results Comparison

971954

Prediction of Radiated Noise from Engine Components Using the BEM and the Rayleigh Integral

A. F. Seybert
University of Kentucky

D. A. Hamilton and P. A. Hayes
Cummins Engine Co.

Copyright 1997 Society of Automotive Engineers, Inc.

ABSTRACT

This paper examines the feasibility of using the boundary element method (BEM) and the Rayleigh integral to assess the sound radiation from engine components such as oil pans. Two oil pans, one cast aluminum and the other stamped steel, are used in the study. All numerical results are compared to running engine data obtained for each of these oil pans on a Cummins engine. Measured running-engine surface velocity data are used as input to the BEM calculations. The BEM models of the oil pans are baffled in various ways to determine the feasibility of analyzing the sound radiated from the oil pan in isolation of the engine. Two baffling conditions are considered: an infinite baffle in which the edge of the oil pan are attached to an infinite, flat surface; and a closed baffle in which the edge of the oil pan is sealed with a rigid structure. It is shown that either of these methods gives satisfactory results when compared to experiment. It is also demonstrated that the Rayleigh integral gives results comparable to the BEM.

INTRODUCTION

The diesel engine is an important source of power in the world today, particularly for industrial and commercial uses. The diesel's popularity stems in large part from its superior efficiency compared to gasoline engines, while sharing the gasoline engine's advantages of portability, widespread fuel availability, reliability, and high power-to-weight ratio.

One of the principal applications for direct injection diesel engines is in medium and heavy trucks, whose noise levels are regulated by the U.S. Environmental Protection Agency. The engine is typically responsible for approximately 70 percent of the total truck noise as measured by a passby test. Of the various engine components, the oil pan is one of the most significant radiators of sound because of its large area and the fact that, in a truck, it often hangs below the truck's frame rails giving line-of-sight exposure to the microphones in a passby noise test.

The noise radiated from engine components such as the oil pan, valve covers, etc., has traditionally been evaluated using production or prototype tests in engine test cells or on trucks. The objective of this paper is to examine the use of numerical methods for predicting noise radiated from two oil pans for the Cummins M11 engine. One oil pan is an aluminum casting, and the other is a steel stamping.

Specifically, it is of interest to determine how well the boundary element method (BEM) can predict the radiated noise of engine components without modeling the complete engine. That is, is it possible to calculate the sound radiated from the surfaces of a component, such as the oil pan, without considering the baffling effects of the remaining surfaces of the engine on the oil pan radiation? If this approach is successful, significant engineering effort can be saved because only the oil pan geometry needs to be modeled. However, the major advantage of such an approach is the considerable reduction in turn-around time for executing the BEM model, which is roughly proportional to N^2 or N^3, where N is the number of nodes in the model.

To answer this question, vibration measurements were made at rated engine operating conditions on a grid of points for the two oil pans under study. The vibration data were used in BEM models of the two oil pans, and the predicted noise was compared to measured values. Two types of baffling were used in the BEM models. In the first case, the BEM model of the oil pan was closed at the upper edge by a set of non-radiating (rigid) elements. In the second case, an infinite, rigid surface was attached to the upper edge of the BEM model of the oil pan. Both types of baffling eliminated the problem of noise radiated from the inner surfaces of the oil pan, which, of course, does not contribute to the exterior noise radiation.

The Rayleigh integral was also used to predict the noise radiated by the oil pans. The Rayleigh integral is an approximate radiation integral in which it is assumed that the radiating surface is flat and lies in an infinite, flat baffle. Even though most engine components do no satisfy theses assumptions in a strict sense, the errors introduced in the calculations may be acceptable for many situations. Further, because the CPU time for the Rayleigh integral is much less than the BEM, the Rayleigh integral may be an acceptable method for the rapid evaluation of component designs.

The current work is not the first time that the BEM has been used to evaluate engine or engine component noise. The BEM has been evaluated in a number of situations and found to yield accurate results if the vibration levels are accurate [1,2]. (It is for this reason that in the present paper measured values of vibration, rather than those obtained from a finite element model, are used as input to the BEM.) Smith and Bernhard [3] used the BEM and the Rayleigh integral to predict the noise radiated by a valve cover mounted on a concrete floor and excited by a mechanical shaker. In their carefully controlled test, it was found that the sound pressure level predicted by the BEM agreed well with measured values. However, the sound pressure level predicted by the Rayleigh integral varied up to 20 dB from that measured, and the authors recommended that the Rayleigh integral be used with caution. On the other hand, Hayes and Quantz [4] used the Rayleigh integral to calculate the radiated noise and the radiation efficiency of valve covers and found that the overall measured and predicted sound power levels agreed to within 1 dB.

EXPERIMENT

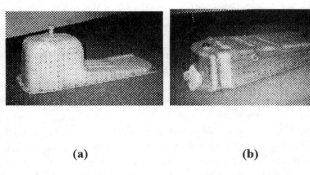

Figure 1: Photographs showing (a) steel and (b) cast aluminum oil pans used in the current investigation.

The two oil pans used in the present paper are shown in Fig. 1. One is made of low-carbon steel approximately 2.1 mm thick and deep-drawn into a mold. The other is an aluminum sand-casting with a wall thickness of approximately 10.7 mm. Each oil pan has a mass of approximately 7 kg.

The oil pans were tested on a Cummins M11 direct-injection diesel engine operating at full load at 1800 rpm (370 h.p.) in a hemi-anechoic engine test cell at Cummins' Walesboro Noise Facility. At this operating condition, oil pan vibration, sound intensity data, and radiated sound pressure level data were obtained for each oil pan.

The vibration data were obtained over a grid consisting of 542 and 512 points, respectively, for the steel and aluminum oil pans. The grid spacing was approximately 38 mm. A moving accelerometer was used to measure the vibration amplitude at each grid point, and the phase was measured with respect to a second (reference) accelerometer fixed at one of the grid points. At each grid point, the moving accelerometer was held in place by a high-strength magnetic mounting base which attached directly to the steel oil pan. For the aluminum oil pan, flat steel washers were attached to the oil pan using a high-temperature structural adhesive, and the magnetic mounting base was attached to the steel washers.

Sound intensity data were measured using a standard intensity probe. The intensity probe was moved from point-to-point and held stationary during the measurement by a fixture. The distance from the surface of the oil pan to the front of the sound intensity probe was about 75 mm.

At 1800 rpm, the fundamental engine firing frequency is 15 Hz. The vibration data, as well as all sound data, were analyzed using an 801-line spectrum from 0 to 4 kHz, i.e., a frequency resolution of 5 Hz. Ideally, therefore, there were two frequency data points between each harmonic of the engine noise or vibration. The data above 3 kHz were discarded because the measurement grid used for the vibration data was not small enough to resolve vibrational modes above this frequency. Data below 100 Hz were also discarded because of the low-frequency cut-off of the test cell.

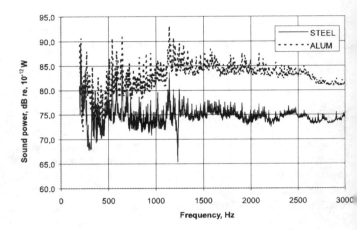

Figure 2: Measured sound power spectra of the aluminum and steel oil pans.

Figure 2 shows the sound power levels of the two oil pans determined from the sound intensity measurements. The overall sound power level of the aluminum oil pan is approximately 8 dB(A) higher than that of the steel oil pan, even though the vibration level of the aluminum oil pan is actually less than that of the steel oil pan [5]. This anomaly is due to the fact that the steel oil pan's response is dominated by a large number of acoustically slow modes which have low radiation efficiency compared to the modes of the aluminum oil pan.

NUMERICAL MODELING

The sound radiated from a vibrating surface may be determined by using the BEM which solves the Helmholtz integral equation [6]

$$C(P)p(P) = -\int_s [p(Q)G'(Q,P) + jkz_0 v_n(Q)G(Q,P)]dS(Q) \quad (1)$$

where $G(Q,P) = exp(-jkr)/r$ is the free-space Green's function between a point Q on the vibrating surface S and a receiver point P, p is the sound pressure, v_n is the velocity normal to the surface S, and $C(P)$ is a constant that depends on the location of P. In Eq. (1), $G'(Q,P)$ is the normal derivative of G, $z_0 = \rho_0 c$ is the characteristic impedance of the medium, ρ_0 is the density of the medium, $k = \tilde{\omega}/c$ is the wavenumber, $\tilde{\omega}$ is the angular frequency, and c is the speed of sound.

In the BEM, Eq. (1) is solved by discretizing the radiating surface into small elements. Using standard element technology, a system of equations is formed:

$$[A(\omega)]\{p\} = [B(\omega)]\{v_n\} \quad (2)$$

relating the normal velocity of the surface S to the sound pressure. In Eq. (2), the A and B matrices depend on the geometry of the surface and frequency. Once the normal velocity is provided as input, the sound pressure on the surface may be found from Eq. (2). The sound intensity may then be calculated, and the sound power is determined by integrating the sound intensity over the surface S.

If the vibrating surface S is flat and part of an infinite, rigid baffle, the Rayleigh integral [6] may be used to compute the sound pressure at a point P.

$$p(P) = -(jkz_0/2\pi) \int v_n(Q)G(Q,P)ds(Q) \quad (3)$$

The Rayleigh integral is much faster to calculate sound pressure than Eq. (1) because it does not require that a system of equations be assembled and solved.

BEM MODELS

BEM models were constructed from the vibration measurement grids of each oil pan, as shown in Fig. 3. Each group of four grid points formed a quadrilateral boundary element. In a few places, triangular elements were used to complete the model. Figure 3 shows that each BEM model was closed by defining a number of additional nodes and elements connecting the upper edge of the oil pan (where it attaches to the engine). The purpose of the additional elements is to simulate the engine which blocks the radiation of sound from the inside surfaces of the oil pan. These elements are assumed rigid and contribute no noise to the calculated sound. The total number of nodes for the aluminum and steel oil pan models was 608 and 615, respectively.

One potential problem with the models in Fig. 3 is that the baffling effect of the engine may not be correctly accounted for by simply closing off the upper edge of the oil pan models. In reality, the sides of the engine reflect sound radiated from the oil pan below and prevent the cancellation of sound to a certain extent from the two sides of the oil pan. Both of the models in Fig. 3 allow sound to diffract over the top of the oil pans, thereby creating the opportunity for sound cancellation, particularly at lower frequencies, that does not exist in reality.

To simulate correctly the baffling effect of the engine on the oil pan, the BEM mesh would have to include the entire engine. This would increase the number of nodes in the model several-fold, and increase the CPU and turn-around time by at least an order of magnitude. This is unacceptable for most development projects in which it is desirable to investigate a number of design options in a short period of time. Consequently, there is considerable motivation for looking at alternatives such as shown in Fig. 3.

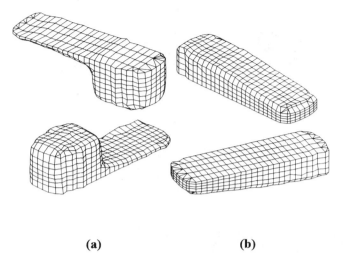

(a) **(b)**

Figure 3: BEM models for the (a) steel and (b) aluminum oil pans

The importance of engine baffling was examined by attaching the oil pan model to an infinite baffle, as shown in Fig. 4 for the aluminum oil pan model. The two approaches, Fig. 3b and Fig. 4, represent two extremes of baffling: in Fig. 3 there is no baffling and in Fig. 4 there is complete baffling. The actual baffling by the engine is between these two extreme situations.

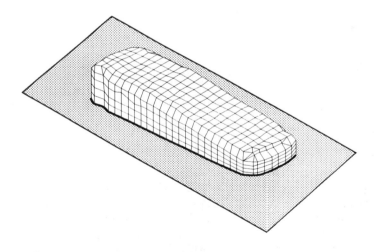

Figure 4: BEM model for the aluminum oil pan with an infinite baffle.

BEM RESULTS

Figures 5 and 6 show the comparison between the measured and calculated sound power level for the aluminum and steel oil pans, respectively, using the BEM models in Fig. 3. The BEM calculations were carried out using our in-house program BEMAP.

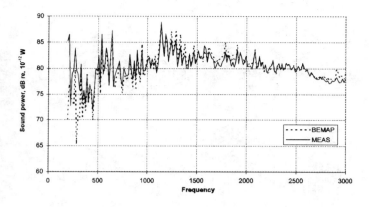

Figure 5: Comparison between the measured and BEM sound power for the aluminum oil pan.

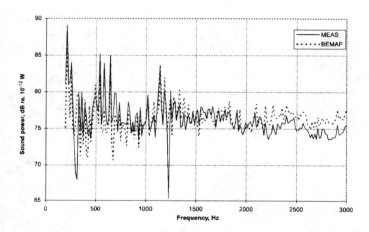

Figure 6: Comparison between the measured and BEM sound power for the steel oil pan.

The overall measured and calculated sound power levels for each oil pan agreed to within 0.1 dB. However, in Fig. 6 there is a discrepancy between the measured and calculated sound power of approximately 1.5 dB above 2200 Hz. This is likely due to insufficient resolution in the measurement grids used to measure vibration or sound intensity, or both, for the steel oil pan. The steel oil pan, because it is thinner, has a modal density much higher than that of the aluminum oil pan. Because the measurement grid size was the same for both oil pans, the effect on resolution errors will be noticed first in the results for the steel oil pan.

In both Figs. 5 and 6, the BEM calculation of the sound power below 1 kHz is approximately 1.5 dB less than the measured values. This is presumably due to the lack of baffling in the BEM models in Fig. 3. To test this hypothesis, the BEM model of the aluminum oil pan was analyzed again using the infinite baffle arrangement shown in Fig. 4. Figure 7 shows the comparison of the BEM results for the models in Fig. 3b and Fig. 4. It may be seen from Fig. 7 that the infinite baffle increases the sound power by about 1-2 dB in the low frequency region.

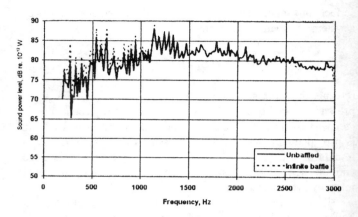

Figure 7: Sound power calculated by the BEM for the aluminum oil pan with and without infinite baffle.

RAYLEIGH INTEGRAL RESULTS

An option in BEMAP allows the computation of sound power and sound pressure using the Rayleigh integral in Eq. (3). Because the Rayleigh integral assumes the radiating surface is flat and that it is part of an infinite, rigid baffle, we would expect the Rayleigh integral to yield results consistent with the BEM when the radiating surface conforms to such assumptions. In Figure 8, the results from the Rayleigh integral are compared to the BEM and experimental results for the aluminum oil pan (from Fig. 5). It can be seen from Fig. 8 that the Rayleigh integral gives excellent agreement with the BEM and measured results.

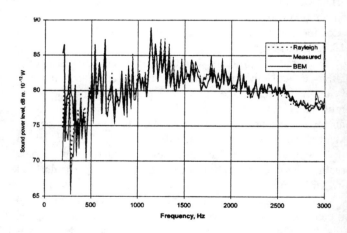

Figure 8: Comparison of the measured sound power for the aluminum oil pan with the sound power calculated by the BEM and the Rayleigh Integral.

SUMMARY AND DISCUSSION

The objective of this paper has been to determine if numerical methods such as the BEM and the Rayleigh integral can be used to calculate noise radiated by engine components without modeling the complete engine. Running-engine noise and vibration data for two oil pans were used to evaluate the numerical methods. It was found that the baffling effect of the engine could be simulated effectively by attaching the BEM model of the oil pan to an infinite baffle, thereby preventing cancellation of sound over the top of the oil pan model. The use of the infinite baffle to simulate the engine does not add any engineering effort or CPU time. Closing the top of the oil pan with non-radiating boundary elements was also effective, but resulted in an error of approximately 1.5 dB below 1 kHz. Closing the top of the oil pan also added to the modeling effort and the CPU time for execution of the BEM.

In this paper, all of the data comparing BEM with experiment were sound power data. The BEM was also used to calculate the sound pressure level in the near field of the oil pans used in this study, and these levels agreed quite well with the experimental results [5]. As a practical matter, however, sound pressure data from the engine used in the current study were not conclusive in the evaluation of oil pan noise in the far field because of sound radiation from the other parts of the engine. This was particularly true for the steel oil pan which was significantly quieter than the aluminum pan.

The Rayleigh integral was shown to be in excellent agreement with the BEM results. The success of the Rayleigh integral is probably due to the fact that the oil pan surfaces are reasonably flat radiators of sound. It is likely that the Rayleigh integral would also predict accurately the sound power of other engine components such as valve covers, front covers, and the sides of the engine.

The Rayleigh integral is much faster than the BEM because it is not necessary to assemble and solve a system of equations. For the same reason, the Rayleigh integral also requires only a fraction of the RAM that is needed for a BEM solution.

How well the Rayleigh integral will predict the sound pressure level at specific points was not examined in this paper. However, based on other work [3], it appears that the BEM is better suited for the calculation of sound pressure levels than is the Rayleigh integral. This is not a major shortcoming of the Rayleigh integral, because, in many engineering development projects, sound power is the preferred basis for comparing alternative designs.

It was noted earlier that the average vibration level of the aluminum oil pan was less than that of the steel pan even though the sound power was greater, as shown in Fig.2. This phenomenon can be explained by examining the radiation efficiency of the two oil pans, calculated from the BEM and shown in Fig. 9

It may be seen from Fig. 9 that the sound radiated by the aluminum oil pan is due to acoustically fast modes [6] with relatively high radiation efficiency. On the other hand, the radiation of sound from the steel oil pan is dominated by acoustically slow modes. Acoustically slow modes exhibit considerable cancellation between adjacent regions of positive and negative volume velocity, resulting in less sound power radiated when compared to acoustically fast modes.

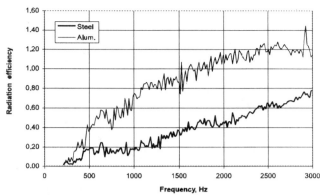

Figure 9: Sound radiation efficiency of the aluminum and steel oil pans, as calculated by the BEM.

REFERENCES

1. Seybert, A.F., Chorine, R., Calculation of the Sound Intensity and Sound Radiation Efficiency of Structures from Vibration Data, *Proceedings*, Noise-Con 88, June 20-22, 1988, pp. 609-614.

2. Seybert, A. F., and Oswald, B., Experimental Validation of Boundary Element Methods for Noise Prediction, *Proceedings*, Inter-Noise 92, Toronto, Canada, pp. 1137-1142.

3. Smith, D. C., and Bernhard, R. J., Verification of Numerical Acoustic Radiation Predictions, paper 891171, SAE Noise and Vibration Conference, Traverse City, Michigan, May, 1989

4. Hayes,, P.A., and Quantz, C. A., Determining Vibration, Radiation Efficiency, and Noise Characteristics of Structural Designs Using Analytical Techniques, paper 820440, SAE Conference, Detroit, Michigan, February, 1982.

5. Hamilton, D. A., A Comparison of the Boundary Element and Rayleigh Integral Methods for Calculating Noise from Vibrating Bodies, M.S. thesis, University of Kentucky, 1996.

6. Fahy, Frank, *Sound and Structural Vibration*, Academic Press, Inc., London, 1985.

971955

Interior Noise Prediction Process for Heavy Equipment Cabs

Andrew F. Seybert, Tiemin Hu, David W. Herrin, and Robert S. Ballinger
University of Kentucky

Copyright 1997 Society of Automotive Engineers, Inc.

ABSTRACT

This paper is concerned with the prediction and experimental verification of the interior noise of cabs used on construction, highway, and farm equipment. The typical heavy equipment cab is totally enclosed and partially lined with absorbing materials but is much stiffer and more massive than automobile passenger compartments. The process to analyze a construction cab is explained in detail. Selected results are also presented to show the value of the method.

INTRODUCTION

Reducing noise has become a priority in many industries including the automotive and construction industries [1]. The interior noise level of vehicles and cabs, for example, has a major effect on the acceptance and sales potential of automobiles and construction equipment. Additionally, if interior noise levels are too high, a vehicle may fail to meet domestic or foreign noise regulations, and may also make it difficult for the operator to hear desirable sounds such as warning signals. Consequently, considerable effort is spent minimizing the interior noise in vehicles and cabs.

Traditionally, noise reduction has been achieved by reducing structural vibration or by adding sound absorbing material (foam and carpet) to the interior surfaces to dissipate acoustic energy [2]. However, determining which panels are the major noise contributors is a time-consuming experimental process. In addition, changing the structural design after a prototype has been developed is costly, and adding sound absorbing material increases the cost and complicates the manufacturing. Due to the complexity of the problem, time constraints, and the cost of design changes, numerical methods are being used increasingly to predict the impact of design changes on the interior sound levels of vehicles or cabs [3-5].

The capability has been developed to model and experimentally validate the vibration and interior noise of vehicles and cabs. This capability includes the application of finite element and boundary element technology for modeling and the use of experimental modal analysis for model verification. This capability also includes solid modeling, characterization and measurement of acoustic material properties, and sound intensity measurements. The present paper outlines a process for modeling and predicting the interior noise of heavy equipment cabs using the finite element method for the structural modeling and the boundary element method for the acoustic modeling.

OVERVIEW OF THE NOISE MODELING PROCESS

Sources and Paths

Figure 1 shows a typical cab structure isolated from the rest of the vehicle and auxiliary equipment. Most interior noise is either structureborne or airborne, caused by sound impinging on the exterior of the cab or by vibration introduced at some point on the cab structure. Sound sources exterior to the cab consist of the engine, exhaust stack opening, and auxiliary sources such as pumps and compressors driven by the engine. These sound sources introduce vibrations on the exterior of the cab which transmit sound to the interior. Airborne noise may also be introduced

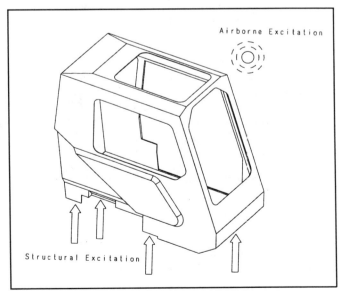

Figure 1: Typical cab showing airborne and structure-borne noise paths

through openings or gaps in the cab through which sound energy may flow without interacting with the structure. In a well-constructed, totally-enclosed cab, this secondary path should be negligible.

The structureborne noise is produced by vibration introduced directly to cab. Normally, these vibration sources excite the cab through the frame which is coupled to the cab by rubber isolators. Additional structureborne excitation may occur due to auxiliary equipment that is attached to the cab structure. However, these excitations normally have minimal impact compared to the excitation at the cab isolator attachment points.

A Process Model for Interior Noise Prediction

A process model for modeling and prediction of cab interior noise is shown in Figure 2. The process model consists of two parallel efforts corresponding to the modeling of the airborne and the structureborne paths, as discussed above. The finite element and boundary element models used in the prediction of the interior noise will be discussed below. As seen from Figure 2, it is necessary to provide estimates of damping to the finite element model. It is also necessary to supply to the boundary element model estimates of the acoustic impedance of the seat, headliner, and other sound absorbing materials inside the cab. Both damping and acoustic impedance estimates come either from direct measurement or from the experience of the user.

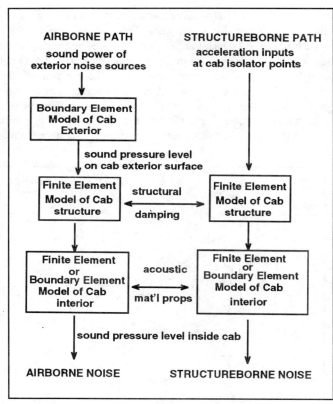

Figure 2: Interior Noise Prediction Process Model.

The interior noise may be predicted using either an acoustic finite element or a boundary element model. The latter is faster to build than the former, because it is essentially a skin of the cab structure, but runs much slower for multiple frequencies. The interior boundary element model is very similar to the exterior one, except the user must account for the seat and the acoustic impedances of other absorbing surfaces.

STRUCTURAL FINITE ELEMENT ANALYSIS

An integral step in the prediction of interior noise levels is to perform a finite element analysis of the cab structure to determine the dynamic displacements. A finite element model of the cab must be developed using a pre-processor program. Since the dynamic displacements rather than the static stresses are of interest, the mesh can be of uniform coarseness. The density of the finite element mesh must, at a minimum, represent the mode shapes in the frequency range of interest and should preferably be of sufficient density to represent modes well beyond the frequency range of interest.

Figure 3 shows the structural finite element model of the heavy equipment cab in Figure 1. The model shown has approximately 50,000 degrees of freedom, though many cab models exceed 100,000 degrees of freedom. Shell elements are used to model the panels on the cab and beam elements are used to model the roll-over-protection-system and the frame of the cab. Concentrated masses are used to model the steering column and the seat since each of these adds significant mass but little stiffness to the structure of the cab.

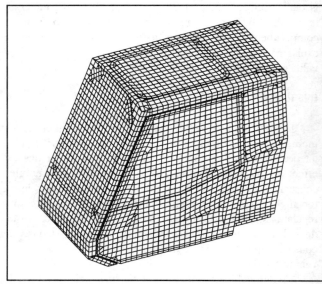

Figure 3: Finite element model of the cab.

Structural finite element runs are made for both the airborne and the structureborne inputs. The airborne inputs are modeled as dynamic forces on the exterior nodes of the cab. These dynamic forces are determined from the sound pressures calculated by the external boundary element run (discussed below) using a file translator to convert the sound pressures on the surface to equivalent concentrated forces at the nodes.

The effects of structureborne sources are usually determined experimentally by measuring the acceleration levels at the cab isolation mounts and at other locations where structural sources are present. These accelerations are enforced on the finite element model, and the response due to these inputs can be calculated. Since multiple accelerations are being enforced simultaneously, the large-mass approach to handle enforced accelerations produces erroneous results because a large mass at each input point significantly changes the nature of the system. Instead, the

Lagrange Multiplier Technique is chosen over the large-mass approach. Using this technique, multiple enforced acceleration inputs may be applied in a single finite element run.

A modal frequency response analysis is used to determine the forced response. This is preferable to using the direct frequency response approach because one is usually interested in analyzing the model at numerous frequencies and using different load sets. Traditionally, models of this size have been solved using disk- and time-saving techniques such as component mode synthesis and Guyan reduction. However, as the speed of workstations increases and more efficient eigenvalue solvers such as the Lanczos method are developed, complete systems can be solved in a single run without having to resort to such techniques.

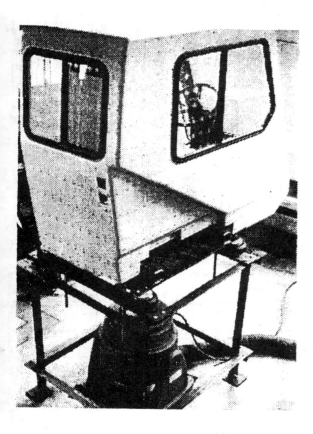

Figure 4: Experimental setup to measure cab modes and damping.

EXPERIMENTAL CORRELATION

The structural finite element model is correlated with experimental results to both validate the model and to characterize the damping for the modal frequency response analysis. Figure 4 illustrates a typical experimental setup for a cab used to validate modes and determine damping. The cab is mounted on a simulated vehicle frame connected to the isolation mounts and excited using a 1000-pound electromechanical shaker. The finite element model is verified by comparing the experimental and computed modes of vibration. Figure 5 shows a comparison of experimental and finite element modes for a door at 150 Hz. From the experimental curve-fit, the damping for each mode may be estimated for use in the finite element analysis.

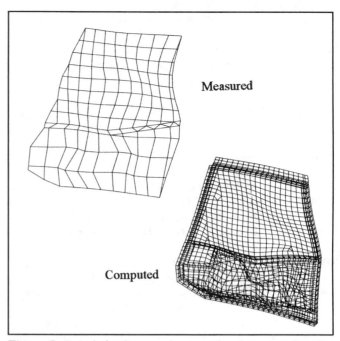

Figure 5: Correlation between measured and computed mode of the cab door.

BOUNDARY ELEMENT ANALYSIS

Boundary element analyses are required to predict both the sound pressure on the exterior of the cab and the interior noise. The meshes used for both exterior and interior boundary element analyses are quite similar, and can be created by coarsening the finite element mesh using either an automatic mesh coarsener or manually using the finite element mesh as a guide. Because the boundary element mesh is much coarser, developing the boundary element mesh takes only a fraction of the time required to create the finite element mesh. Figure 6 shows the boundary element mesh created from the finite element mesh in Figure 3.

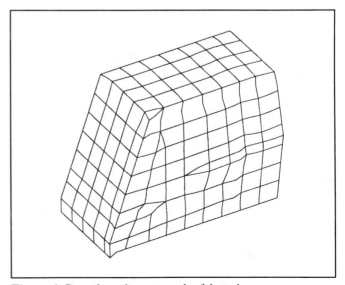

Figure 6: Boundary element mesh of the cab.

The first step in predicting the interior noise due to the airborne path is to quantify the sound power of the exterior noise sources. This may be done experimentally, such as by using sound intensity measurements, or may be the result of radiation models of engines and engine components. It is important to determine the sound power as accurately as possible over the frequency range of interest.

The sound pressure levels on the exterior surface of the cab are determined from the boundary element analysis and the sound power of the airborne sources. For this prediction, the cab is assumed to be rigid (this is a good approximation for cabs because they are generally very stiff). The exterior sources may be modeled as individual point sources or as distributions of point sources, depending on how compact they are. For example, an exhaust stack opening may be modeled quite accurately as a single point source, but an engine may require a distribution of several sources if it is large compared to the wavelength of sound at the highest frequency of interest.

An interior boundary element analysis is conducted to estimate the sound pressure levels inside the cab from either the airborne or structureborne sources. The mesh is similar to that used for the exterior analysis except the seat geometry and other interior details need to be included. The boundary conditions for the interior boundary element analysis consist of the velocities obtained from the structural finite element analysis and impedances for sound absorbing surfaces such as the seat and the headliner. These impedances are usually measured using the ASTM E 1050 test method. Figure 7 shows results from an ASTM E 1050 test for a seat.

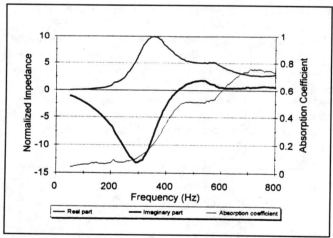

Figure 7: Impedance of seat used in boundary element model of cab interior.

Figure 8 shows a comparison between the results of an interior boundary element analysis and experimental results for a point in the interior of the cab. For this analysis, the excitation was provided by the mechanical shaker system in Figure 4. As the graph shows, the correlation between the predicted and experimental results is quite good, particularly below 1100 Hz. Above 1100 Hz, the experimental and predicted results have a similar trend, but at specific frequencies differences are clearly seen.

Figure 9 shows the mode shape of the cab roof at 1200 Hz. From this figure, it is clear the finite element model of the cab (Figure 3) or the boundary element model (Figure 6) is not fine enough to represent structural modes at high frequencies. This is why the correlation between the measured SPL and the predicted SPL inside the cab is not so good above 1100 Hz.

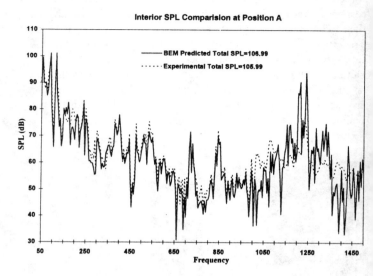

Figure 8: Comparison between measured and predicted sound pressure level inside of the cab.

Table 1 shows a comparison between sound levels measured and predicted at several points in the cab using the process described above.

Position	Calculated	Measured	Difference
A	107.0	106.0	+1.0
B	109.4	108.5	+0.9
C	105.8	104.0	+1.8
D	108.2	108.4	-0.2
E	105.5	104.4	+1.1

Table 1: Comparison of measured and predicted sound levels at several points in the cab.

Using the results from the boundary element analysis, panel participations can be estimated and utilized to drive design changes. Certain panels can be stiffened using ribs or beams. Additionally, absorptive materials can be added to strategic locations. If structural changes are made to the cab, then the structural finite element and interior boundary element analyses would need to be rerun. However, if absorptive material is added, only the interior boundary element analysis would need to be rerun in that case. Unfortunately, identifying which design changes to make can be a time consuming process. Thus, methodologies need to be developed to aid in making design decisions.

SUMMARY AND CONCLUSIONS

Using finite element and boundary element analysis, it is now feasible with engineering workstations to predict the interior noise of vehicle cabs. Compared to automobile passenger compartments which are relatively compliant, cabs used for construction and farm applications are quite stiff, resulting in lower modal density. The lower modal density of cabs compared to automobile passenger compartments makes it possible to predict interior noise at higher frequencies (up to 1500 Hz) using finite element and boundary element technology.

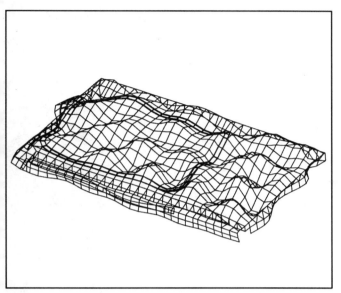

Figure 9: The mode shape of the cab roof at 1200 Hz.

The effort (both human and computer) required to conduct interior noise predictions can be significant, depending on the complexity of the cab structure and the frequency range of interest. Most of the human effort is expended in the finite element modeling of the structure; the boundary element modeling effort is small in comparison, as shown in Fig. 10. By contrast, most of the computing resources are used for conducting the interior and exterior boundary element analyses, as illustrated in Fig. 10. The solve time needs to be reduced since it effects how quickly design iterations can be made. However, as computing speeds increase and as software becomes more efficient, it should feasible to use noise prediction technology and the process described in the paper to design quieter vehicle cabs.

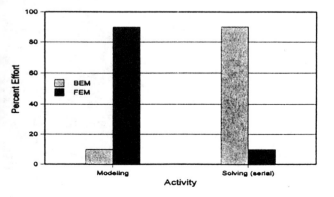

Figure 10: Comparison of effort required for various activities associated with interior noise predictions.

REFERENCES

1. Nefske, D. J., Wolf, Jr, J. A., and Howell, L.J., "Structural-Acoustic Finite Element Analysis of the Automobile Passenger Compartment: A Review of Current Practice", J. of Sound and vibration, 80(2), 1982.
2. Kawenski, T. and Brooks, M., "Interior Noise Reduction Methods for Heavy Trucks Using Acoustical Materials", SAE Proceedings of the 1989 Noise and Vibration Conference, Traverse City, Michigan, 1989.
3. Yang, T. and Tseng C., "On Implementation of Optimization and Boundary Element Method for Noise Control In Cavities", Third International Congress on Air- and Structure-Borne Sound and Vibration, Montreal, Canada, 1994.
4. Utsuno, H., Wu, T.W., and Seybert, A.F., "Prediction of Sound Fields in Cavities with Sound Absorbing Materials", AIAA Journal, Vol.28, No.11, 1990.
5. Huff, Jr, J.E. and Bernhard, R.J., "Acoustic Shape Optimization Using Parametric Finite Elements", 1995 Design Engineering Technical Conferences, Vol. 3-B, ASME, 1995.

971956

Noise Source Identification in a Highly Reverberant Enclosure by Inverse Frequency Response Function Method: Numerical Feasibility Study

Drew A. Crafton, Jingdong Ding, and Jay H. Kim
University of Cincinnati

Copyright 1997 Society of Automotive Engineers, Inc.

ABSTRACT

In highly reverberant enclosures, the identification of noise sources is a difficult and time consuming task. One effective approach is the Inverse Frequency Response Function (IFRF) method. This technique uses the inverse of an acoustic FRF matrix, that when multiplied by operating pressure response data reveals the noise source locations. Under highly reverberant conditions the deployment of a sound absorbing body is especially useful in reducing the effects of resonant modes that obscure important information in the FRFs. Without the absorption, the IFRF method becomes practically difficult to perform in these environments due to poor conditioning of the FRF matrix. This study investigates the feasibility of using Boundary Element and Finite Element Methods to establish the frequency response functions between selected panel points and microphones in the array. The advantage of numerical FRF development is a practical savings of time and physical effort by reducing the amount of data that must be acquired by experimental testing. By providing additional flexibility in the design of the test, the material selection for the absorptive center body, as well as the strategy of microphone deployment, could be made through use of numerical analyses without additional physical setup demands.

NOMENCLATURE

BEM   Boundary Element Method
FEM   Finite Element Method
FRF   Frequency Response Function
IFRF   Inverse Frequency Response Function
$[H(\omega)]$ Frequency Response Function Matrix
$[H(\omega)]^+$ Pseudo-inverse of FRF matrix
$\{P(\omega)\}$ Output vector - Sound Pressure Level
$\{Q(\omega)\}$ Input vector - Acoustic source strength

1. INTRODUCTION

Acoustic array techniques including Near-Field Acoustic Holography (NAH) and IFRF methods have been developed and used in recent years for noise source imaging [1,2]. NAH and IFRF methods are most easily applied to the identification of noise sources under free-field conditions. While NAH has been applied to an interior noise source identification problem using a double plane deployment of microphone arrays [3], this technique has the limitation of being restricted to relatively simple geometries.

With these considerations, the IFRF method is an attractive option for the identification of noise sources in a reverberant enclosed space. There are no restrictions on the geometry or physical requirements dictated by the free field assumption, as these factors are incorporated in the procedure. However, of practical importance when applying the IFRF method to a highly reverberant system is the existence of strong standing waves. Since these resonant modes dominate the information in the FRFs near the natural frequencies, the FRF matrix becomes ill-conditioned at these frequencies and therefore difficult to invert. To overcome this difficulty, the application of the IFRF method in conjunction with the insertion of a center body made of a sound absorptive material has been discussed in Ref [4]. The practical advantage of this approach has been clearly demonstrated in the experimental application of the IFRF method under these reflective conditions [5]. In this work, a fundamental study is done to examine the feasibility of using numerical methods in the design or implementation of such a scheme.

In particular, the use of numerical methods to advance the concept developed in [4,5] and to aid in the further refinement of the experimental measurement technique are investigated.

2. IFRF METHOD APPLIED TO NOISE SOURCE IDENTIFICATION

The acoustic frequency response function (FRF) relationship between the input vector $\{Q(\omega)\}$ and the output vector $\{P(\omega)\}$ is described as:

$$\{P(\omega)\}=[H(\omega)]\{Q(\omega)\} \qquad (1)$$

Equation (1) may be applied repeatedly using known inputs $\{Q(\omega)\}$ and measuring the resulting responses $\{P(\omega)\}$ to sufficiently define the FRF matrix $[H(\omega)]$. Once $[H(\omega)]$ is determined, unknown inputs to the system can be identified from the measured operating pressure responses $\{P(\omega)\}$ as follows:

$$\{Q(\omega)\}=[H(\omega)]^+ \{P(\omega)\} \qquad (2)$$

where, $\{Q(\omega)\}$ is a vector representing acoustic inputs or noise sources to be identified, $[H(\omega)]^+$ is the pseudo-inverse of the FRF matrix $[H(\omega)]$ and $\{P(\omega)\}$ is the vector of pressure responses at the observation points. Therefore, if the FRF matrix $[H(\omega)]$ is available and is invertible, the input noise sources $\{Q(\omega)\}$ can be identified by the IFRF relation described in equation (2) from the measured pressure responses $\{P(\omega)\}$.

While the IFRF method procedure can be applied to the identification of inputs to any linear dynamic system, its application to the noise source identification in a highly reverberant enclosure becomes exceedingly difficult due to the presence of standing waves. At frequencies near their natural frequencies, the FRF matrix becomes ill-conditioned as the response is dominated by the resonant modes associated with these standing waves. Effectively, the true noise sources cannot be distinguished from the apparent sources associated with the reflections. To lower the effects of these standing waves, a sound absorbing structure consisting of a fibrous absorber material may be inserted inside the reverberant enclosure to attenuate the reflected waves. This method is especially well suited for noise source identification purposes when the noise sources are located on or near the boundaries of the enclosure as there are no treatments to the structure itself which would affect the characteristics of the native enclosure being tested.

3. SYSTEM CONFIGURATION

For the purpose of fundamental study, the IFRF method is applied to a rectangular shaped enclosure. As shown in Figure 1, this box consists of Plexiglas walls attached to a wooden frame. Figure 2 illustrates the dimensions of this box and the absorptive center body insertion used in this study. Figure 3 shows the complete arrangement for experimental IFRF implementation including the absorptive material.

Figure 1 - Rectangular Test Enclosure

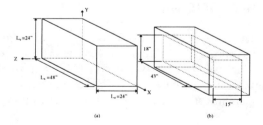

Figure 2 Geometry and coordinate system of the rectangular enclosure, (a) without absorptive insertion, (b) rectangular absorptive insertion present

Figure 3 - Enclosure with microphone array and absorptive material

For comparison purposes, the frequency responses to a point source (monopole) are established for both the empty box and the box with the absorptive insertion at several response locations using the experimental and numerical methods. Experimentally, a baffled speaker representing a simple monopole source (Figure 4) is placed at the selected source point and FRFs are obtained between that source location and each of the microphone points (Figure 5). Numerically, an FRF calculated between the monopole source and the each of recovery points which correspond to microphone locations.

Figure 4 - Source Location

Figure 5 - Response Locations

4. NATURAL MODE ANALYSIS

Natural frequencies and modes of the rectangular cavity considered are obtained by FEM and analytic methods to check the validity of the acoustic model's parameters including the selected mesh size over the frequency range of interest. All outside walls were considered to be rigid. Analytically, the natural frequencies of the empty box are readily available. Natural frequencies are given by Equation (3).

$$f_{lmn} = \frac{c_o}{2}\sqrt{\left(\frac{l}{a}\right)^2 + \left(\frac{m}{b}\right)^2 + \left(\frac{n}{c}\right)^2} \quad (3)$$
$$l, m, n = 0, 1, 2, \ldots$$

where a, b, and c are the length, width, and height of the enclosed rectangular cavity, l, m, n are mode numbers, and c_o is the speed of sound [6]. The modal analysis results from two commercial FEM packages used in this work are shown up to 660 Hz in Table 1.

The three cases provided comparable results in the low frequency range. This suggests that discrepancies between the experimental results and numerical results in this frequency range occur from approximations other than the meshing.

Table 1: Natural Frequencies of Box without Absorber (Hz)			
#	FEM 1	FEM 2	Analytic	Exper.
1	141.03	139.83	151.39	158
2	291.72	282.07	302.78	298
3	291.72	282.07	302.78	298
4	291.72	282.07	302.78	298
5	324.02	314.83	338.52	334
6	324.02	314.83	338.52	334
7	412.55	398.90	428.20	416
8	412.55	398.90	428.20	416
9	412.55	398.90	428.20	416
10	435.99	422.70	454.17	440
11	461.25	429.12	454.17	440
12	505.27	488.55	524.43	520
13	545.76	513.53	545.84	574
14	545.76	513.53	545.84	574
15	618.83	583.44	605.56	594
16	652.31	583.44	605.56	594
17	652.31	583.44	605.56	594
18	652.31	585.89	624.20	640
19	667.38	599.96	624.20	640
20	667.38	599.96	624.20	640

5. EXPERIMENTAL STUDY

An experimental analysis was performed in order to compare the actual FRFs between selected source/microphone locations and the numerically generated FRFs from FEM/BEM analyses. The test chamber shown in Figure 1 was tested using a random excitation from a baffled speaker driver that approximates a monopole source. The range of frequencies from 0 - 800 Hz was studied. The input strength of the speaker driver was monitored using an accelerometer mounted on the surface of the cone. Representative examples of the measured results are shown in Figures 6 through 9. Comparing frequency responses and coherence for the case of the empty box (Figure 6) and the box with the absorptive insertion (Figure 7), the expected improvement in measurement quality is easily noticed. Sharp drops in coherence result from electrical noise present in the data at harmonics of 60 Hz. Shown clearly in Figure 8, the insertion of the absorptive body substantially reduces the magnitude of the resonance peaks caused by standing waves. This would improve the numerical conditioning of the FRF matrix $[H(\omega)]$ which must be inverted in the IFRF procedure. Therefore, identification of noise sources in reverberant environments will be improved by use of this process with absorption.

Additionally, the experimental study provided for a check of the validity of the rigid wall assumption used in the numerical model. To test this assumption, an accelerometer was mounted on one of the sidewall panels. The coherence between the panel motion and the measured sound pressure was examined. Since the coherence in this measurement was low, this indicated that there was little causality between the panel movement and the measured sound pressure in the frequency range of study. This shows that the structural coupling effect does not have to be considered in the numerical analyses for this test setup.

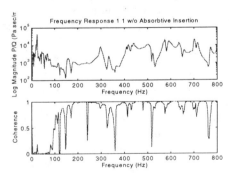

Figure 6 - Experimental FRF 1, without Absorption

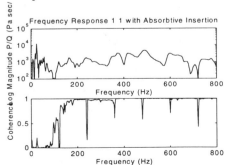

Figure 7 - Experimental FRF 1, with Absorption

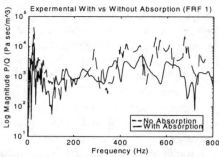

Figure 8 - Comparison of Experimental FRF 1 Magnitudes for Cases With and Without the Absorptive Center Body

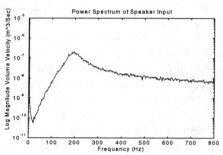

Figure 9 - Power Spectrum of Input Source (Speaker Driver)

6. NUMERICAL STUDY

The FRFs between the monopole location and the response points were performed for the case without absorption and for two cases with absorption present. These cases, used different assumptions as to the impedance characteristics of the absorptive material. One set of FRFs was calculated using a constant impedance value over the frequency range. The value used was the surface normal impedance taken at 500 Hz (Z_a=166.4 - 688.5j). It was based on an experimental measurement result using a 5 cm thickness of the material tested in a standing wave tube. The acoustic characteristics of the absorption body are assumed to be represented by this local surface impedance. Another FRF set was computed using an analytical estimate of the impedance as shown in figure 10. This estimate depends on the parameters such as fiber diameter and bulk density for the absorptive material used in testing. These parameters are not always easy to measure. A two microphone method to determine experimentally impedance as a function of frequency or more elaborate numerical modeling may be an improvement [7].

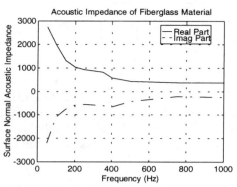

Figure 10 - Impedance of absorptive material as a function of frequency used in BEM analysis

Figures 11 and 12 compare the frequency response results from the BEM analysis between response point 1 and the source for the cases with and without the absorptive insertion. Consistent with the measured cases in Figures 6 and 7, the inclusion of the absorptive material lowers the resonance peaks of the standing wave modes. Compared with the experimental frequency response functions, the BEM analysis produced good results the empty box case, shown in Figure 13. Also the BEM analysis produced results which are qualitatively reasonable in the absorptive case, as seen in Figure 14.

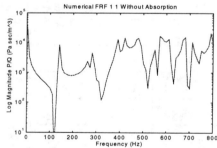

Figure 11 - BEM FRF 1, without Absorption

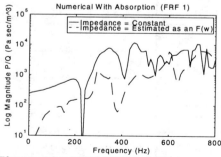

Figure 12 - BEM FRF 1, with two Absorption characteristics.

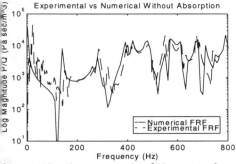

Figure 13 - Comparison of Numerical and Experimental Results for the Cases Without Absorption

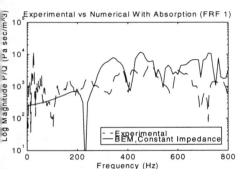

Figure 14 - Comparison of Numerical and Experimental Results for the Cases With Absorption

CONCLUSIONS

A fundamental study has been performed to investigate the feasibility of replacing experimentally measured FRFs with numerically generated FRFs when using the IFRF method acoustic array technique for noise source identification in a reverberant enclosure. An example of a practical application would be noise path analysis with an untrimmed auto body. The numerical FRFs could be computed from a boundary element model and assembled prior to the development of a prototype. This would leave only the acquisition of operating data and application of the pseudo-inverse of the FRF matrix to determine the noise sources and their strengths.

A rectangular box was used as the example for both experimental and numerical study. Acoustic BEM analyses were performed to obtain frequency response functions between the source point and the measurement points and compared with the equivalent experimental measurements. This comparison served to check the BEM modeling accuracy including mesh size. The experimental study also served to verify some of the basic assumptions such as the rigid wall assumption.

Refinement in modeling of the absorptive body is considered to be the most important improvement necessary for the full implementation of the IFRF method for the intended purpose of noise source identification in a reverberant enclosure. This has yet to be realized by numerical methods. The numerical IFRF procedure may be best suited for design purposes, such as the optimum sizing of the clearance between the wall and the array surface, the geometrical deployment of the arrayed microphones, and the effects of the surface geometry, etc.

References

[1] Dumbacher, S.M., Blough, J.R., Hallman, D.L., and Wang, P.F., "Source Identification Using Acoustic Array Techniques", Proceedings of the 1995 Noise and Vibration Conference, Traverse City, Michigan, SAE Paper 951360, pp.1023-1035.

[2] Dumbacher, S. M., and Brown, D. L., "Source Imaging Using Acoustic Inverse FRF Array Technique", Proceedings of the 14th International Modal Analysis Conference, Detroit, Michigan, 1996.

[3] Hallman, D., and Bolton, J.S., "A Technique for Performing Source Identification in a Reflective Environment by Using Nearfield Acoustic Holography", Proceedings of Noise-Con 1993, pp 479-484.

[4] Dumbacher, S. M., and Brown, D. L., "Identifying Structure and Airborne Noise Sources in Interiors Using Inverse FRF Method", Noise-Con96, Seattle, Washington, 1996

[5] Dumbacher, S. M., and Brown, D. L., "Practical Considerations of the IFRF Technique as Applied to Noise Path Analysis and Acoustical Imaging", Proceedings of the 15th International Modal Analysis Conference, Orlando, FL 1997

[6] Kinsler, A. R. Frey, A. B. Coppens, and J. V. Sanders, " Fundamentals of acoustics", John Wiley & Sons, Third edition (1982).

[7] Semeniuk, B.P., Morfey, C.L., and Petyt, M., "Acoustic and structural wave propagation through a fibrous, highly porous thermal insulation blanket", Proceedings of ISMA21, Leuven, Belgium, 1996.

971957

Estimation of a Structure's Inertia Properties Using a Six-Axis Load Cell

Mark Stebbins, Jason Blough, Stuart Shelley, and David Brown
University of Cincinnati

Copyright 1997 Society of Automotive Engineers, Inc.

ABSTRACT

A new method to estimate a structure's inertia properties using a prototype load cell designed to measure all loads and moments applied to a structure is presented. This prototype six-axis transducer approach employs 32 piezoelectric sensing elements which are arranged to form the load cell. These redundant measurements are used to determine the principal forces and moments from an overdetermined set of equations. Calibration of this multi-crystal load cell is performed with a fixture that utilizes a calibration mass and quasi-free-free boundary conditions. The resulting calibration matrix is a 6x32 transformation from the coupled measurements to a decoupled set of pseudo measurements consisting of the forces acting on a structure. With this transducer and its calibration matrix, a system's inertia properties can be estimated. A thorough discussion of both the calibration and inertia estimation procedure with a experimental test case is presented.

INTRODUCTION

The determination of inertia properties (mass, center of gravity location, and inertia tensor) of a structure is a very important aspect of the design process. For complicated structures these properties can be estimated using analytical approaches, such as solid or finite element modeling. When a model of the structure exists from the initial design process, very little additional time is required to obtain the inertia properties.

In the case where a model does not exist, or when modifications are made to the original structure for use in different applications, building or modifying an accurate model can be time consuming. A good example of this is in the case of an automobile powertrain. For a given engine different transmissions and accessories may be used which will modify the inertia properties of the complete powertrain. It may not be cost effective to build or modify a computer model for each case.

To address this problem, methods presented by various authors [1-12], have been developed to experimentally determine these properties. Other common experimental approaches include the use of a bifilar or trifilar pendulum. The trifilar method sometimes take days to properly fixture and obtain enough measurements to estimate inertia properties. The method presented in this paper employs a prototype 6 degree of freedom load cell. The experimental approach is different from the previously developed methods, while the theory remains basically unchanged.

This method uses rigid body dynamics to estimate the structure's inertia properties. The prototype 6 degree of freedom load cell is used to measure the external forces which act on a structure. Simultaneously, translational accelerometers are used to measure the rigid body response of the structure. A digitizer can be used to quickly and accurately measure the physical geometry of the structure which is utilized during the estimation procedure. Depending on the structure and the number of triaxial acceleration measurements used, the proposed method may take between 2 to 4 hours from beginning of setup to the estimation on inertia properties.

THEORY

First, consider a free body diagram of the rigid body shown in Figure 1. Newton's second law of motion can be used to describe its rigid body dynamics. This law states that the summation of the external forces which act on the body equal the product of its mass and acceleration at its center of gravity, as in Equation 1.

$$\sum_{i=1}^{n} \bar{\mathbf{F}}_i = \mathbf{m}\bar{\mathbf{a}}_{cg} \tag{1}$$

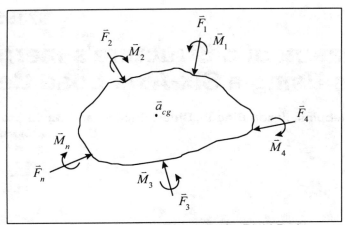

Figure 1. Free Body Diagram of a Rigid Body

Even if the center of gravity were known, its location may not allow for the direct measurement of accelerations. The center of gravity of a structure can be located in an inaccessible location as is the case with a solid object. For a structure like an automobile frame its location may actually be at a location in space where an accelerometer cannot be attached to the structure. Rigid body dynamics can be used to indirectly measure the accelerations at the center of gravity, or any location on the structure.

A reference point P on the structure is chosen with respect to which the rigid body translations and rotations are calculated. This is also the point where the 6 degree of freedom (DOF) load cell is attached to the structure. The structure of interest is excited through the 6 DOF load cell allowing the amplitude and direction of the excitation to be measured. Figure 2 shows a special case of the previous free body diagram, where all of the external forces are applied through and measured by the 6 DOF load cell.

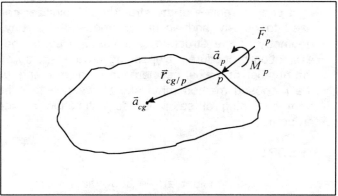

Figure 2. Rigid Body with all External Forces Applied at P

The following expression describes the acceleration of the center of gravity in terms of the acceleration of reference point P and the relative acceleration of the center of gravity relative to point P.

$$\vec{a}_{cg} = \vec{a}_p + \vec{a}_{cg/p} \tag{2}$$

Equation 2 can be rewritten to separate the normal and tangential acceleration components that make up the total relative acceleration term.

$$\vec{a}_{cg} = \vec{a}_p + \dot{\vec{\omega}} \times \vec{r}_{cg/p} + \vec{\omega} \times (\vec{\omega} \times \vec{r}_{cg/p}) \tag{3}$$

The previous expression can be linearized by neglecting the normal acceleration term. This introduces negligible error since the angular velocity of the rigid body during testing is very small.

$$\vec{a}_{cg} = \vec{a}_p + \dot{\vec{\omega}} \times \vec{r}_{cg/p} \tag{4}$$

Expressing Equation 4 in vector matrix notation

$$\begin{Bmatrix} \ddot{x}_{cg} \\ \ddot{y}_{cg} \\ \ddot{z}_{cg} \end{Bmatrix} = \begin{bmatrix} 1 & 0 & 0 & 0 & z_{cg/p} & -y_{cg/p} \\ 0 & 1 & 0 & -z_{cg/p} & 0 & x_{cg/p} \\ 0 & 0 & 1 & y_{cg/p} & -x_{cg/p} & 0 \end{bmatrix} \begin{Bmatrix} \ddot{x}_p \\ \ddot{y}_p \\ \ddot{z}_p \\ \ddot{\theta}_{x_p} \\ \ddot{\theta}_{y_p} \\ \ddot{\theta}_{z_p} \end{Bmatrix}$$

$$\{a_{cg}\} = [\Psi_{cg/p}]\{a_p\} \tag{4}$$

The result from Equation 4 can then be substituted into Equation 1 to yield:

$$\sum_{i=1}^{n} \vec{F}_i = m\left(\vec{a}_p + \ddot{\vec{\theta}}_p \times \vec{r}_{cg/p} \right) \tag{5}$$

Newton's second law of motion can be applied once again to write an expression that represents the sum of the moments about point P.

$$\sum_{i=1}^{m} \vec{M}_{p_i} = I_p \ddot{\vec{\theta}}_p + \vec{r}_{cg/p} \times m\vec{a}_p \tag{6}$$

Writing Equations 5 and 6 in vector matrix notation yields:

$$\begin{Bmatrix} F_{x_p} \\ F_{y_p} \\ F_{y_p} \\ M_{x_p} \\ M_{y_p} \\ M_{z_p} \end{Bmatrix} = \begin{bmatrix} m & 0 & 0 & 0 & m \cdot z_{cg/p} & -m \cdot y_{cg/p} \\ 0 & m & 0 & -m \cdot z_{cg/p} & 0 & m \cdot x_{cg/p} \\ 0 & 0 & m & m \cdot y_{cg/p} & -m \cdot x_{cg/p} & 0 \\ 0 & -m \cdot z_{cg/p} & m \cdot y_{cg/p} & I_{xx_p} & I_{xy_p} & I_{xz_p} \\ m \cdot z_{cg/p} & 0 & -m \cdot x_{cg/p} & I_{yx_p} & I_{yy_p} & I_{yz_p} \\ -m \cdot y_{cg/p} & m \cdot x_{cg/p} & 0 & I_{zx_p} & I_{zy_p} & I_{zz_p} \end{bmatrix} \begin{Bmatrix} \ddot{x}_p \\ \ddot{y}_p \\ \ddot{z}_p \\ \ddot{\theta}_{x_p} \\ \ddot{\theta}_{y_p} \\ \ddot{\theta}_{z_p} \end{Bmatrix}$$

$$\{F_{t_p}\} = [I_p]\{a_p\} \tag{7}$$

Equation 7 can be reformulated to obtain a solution for the inertia values. The symmetric property of the inertia tensor can be utilized in the reformulation of Equation 7 to yield:

$$\begin{Bmatrix} F_{x_p} \\ F_{y_p} \\ F_{z_p} \\ M_{x_p} \\ M_{y_p} \\ M_{z_p} \end{Bmatrix} = \begin{bmatrix} \ddot{x}_p & 0 & -\ddot{\theta}_{z_p} & \ddot{\theta}_{y_p} & 0 & 0 & 0 & 0 & 0 & 0 \\ \ddot{y}_p & \ddot{\theta}_{z_p} & 0 & -\ddot{\theta}_{x_p} & 0 & 0 & 0 & 0 & 0 & 0 \\ \ddot{z}_p & -\ddot{\theta}_{y_p} & \ddot{\theta}_{x_p} & 0 & 0 & 0 & 0 & 0 & 0 & 0 \\ 0 & 0 & \ddot{z}_p & -\ddot{y}_p & \ddot{\theta}_{x_p} & \ddot{\theta}_{y_p} & \ddot{\theta}_{z_p} & 0 & 0 & 0 \\ 0 & -\ddot{z}_p & 0 & \ddot{x}_p & 0 & \ddot{\theta}_{x_p} & 0 & \ddot{\theta}_{y_p} & \ddot{\theta}_{z_p} & 0 \\ 0 & \ddot{y}_p & -\ddot{x}_p & 0 & 0 & 0 & \ddot{\theta}_{x_p} & 0 & \ddot{\theta}_{y_p} & \ddot{\theta}_{z_p} \end{bmatrix} \begin{Bmatrix} m \\ m \cdot x_{cg/p} \\ m \cdot y_{cg/p} \\ m \cdot z_{cg/p} \\ I_{xx_p} \\ I_{xy_p} \\ I_{xz_p} \\ I_{yy_p} \\ I_{yz_p} \\ I_{zz_p} \end{Bmatrix}$$

$$\left\{ F_{t_p} \right\} = \left[A_p \right] \left\{ I_{v_p} \right\} \tag{8}$$

At this point the inertia properties can not be determined because there are 10 unknowns with only 6 equations. The number of equations can be doubled by separating the force and acceleration data into it's real and imaginary parts. This not only gives a sufficient number of equations to solve this problem, but it also forces the solution to be real.

$$\begin{Bmatrix} \mathbf{real}\left(\left\{ F_{t_p} \right\}\right) \\ \mathbf{imag}\left(\left\{ F_{t_p} \right\}\right) \end{Bmatrix} = \begin{bmatrix} \mathbf{real}\left(\left[A_p \right]\right) \\ \mathbf{imag}\left(\left[A_p \right]\right) \end{bmatrix} \left\{ I_{v_p} \right\} \tag{9}$$

Translational accelerometers must be placed at different locations on the structure so that the rigid body modes can be completely described. Rigid body dynamics can be used again to relate the translational accelerations at points j, r, etc. and the translational and rotational accelerations at reference point P.

$$\begin{Bmatrix} \{a_j\} \\ \{a_r\} \\ \vdots \end{Bmatrix} = \begin{bmatrix} \left[\Psi_{j/p}\right] \\ \left[\Psi_{r/p}\right] \\ \vdots \end{bmatrix} \{a_p\} \tag{10}$$

$$\{a_v\} = \left[\Psi\right]\{a_p\} \tag{11}$$

Weighting matrices can be applied on both sides of Equation 11 to select a subset of the measured accelerations. The least squares estimate of the acceleration at point P can be obtained by taking the pseudo-inverse of the product of the weighting matrix and the rigid body transformation matrix:

$$\{a_p\} = \left[\left[\mathbf{W}\right]\left[\Psi\right]\right]^+ \left[\mathbf{W}\right]\{a_v\} \tag{12}$$

The weighting matrices and the rigid body transformation matrices can be combined to form the transformation matrix $\left[T_{AA}\right]$, which transforms translational accelerations of different points on the structure to the translational and rotational rigid body accelerations referenced to point P.

$$\left[T_{AA}\right] = \left[\left[\mathbf{W}\right]\left[\Psi\right]\right]^+ \left[\mathbf{W}\right] \tag{13}$$

$$\{a_p\} = \left[T_{AA}\right]\{a_v\} \tag{14}$$

Substituting Equation 14 in 7 yields:

$$\left\{ F_{t_p} \right\} = \left[I_p\right]\left[T_{AA}\right]\{a_v\} \tag{15}$$

Fourier transforming Equation 15 into the frequency domain yields:

$$\left\{ S_{F_{t_p}} \right\} = \left[I_p\right]\left[T_{AA}\right]\{S_{a_v}\} \tag{16}$$

Equation 16 can be modified to use auto and cross power measurements that are acquired during the experimental procedure. One way to do this is to postmultiply both sides of the previous expression by the hermitian of the product of the rigid body transformation matrix $\left[T_{AA}\right]$ and the acceleration vector $\{S_{a_v}\}$:

$$\left\{ S_{F_{t_p}} \right\}\left(\left[T_{AA}\right]\{S_{a_v}\}\right)^H = \left[I_p\right]\left[T_{AA}\right]\{S_{a_v}\}\left(\left[T_{AA}\right]\{S_{a_v}\}\right)^H \tag{17}$$

The Fourier transformed vector of the forces at point P can be expressed as the product of the load cell's calibration matrix and the raw output from the 6 DOF load cell [13]:

$$\left\{ S_{F_{t_p}} \right\} = \left[T_{FS}\right]\{S_F\} \tag{18}$$

Substituting Equation 18 into 17 yields:

$$\left[T_{FS}\right]\{S_F\}\left(\left[T_{AA}\right]\{S_{a_v}\}\right)^H = \left[I_p\right]\left[T_{AA}\right]\{S_{a_v}\}\left(\left[T_{AA}\right]\{S_{a_v}\}\right)^H \tag{19}$$

Explicitly showing the cross spectral average summations (N averages), Equation 19 becomes:

$$\sum_{i=1}^{N}\left(\left[T_{FS}\right]\{S_F\}\{S_{a_v}\}^H\left[T_{AA}\right]^H\right) = \sum_{i=1}^{N}\left(\left[I_p\right]\left[T_{AA}\right]\{S_{a_v}\}\{S_{a_v}\}^H\left[T_{AA}\right]^H\right) \tag{20}$$

The constant matrices in the previous expression can be moved outside the summation.

$$\left[T_{FS}\right]\sum_{i=1}^{N}\left(\{S_F\}\{S_{a_v}\}^H\right)\left[T_{AA}\right]^H = \left[I_p\right]\left[T_{AA}\right]\sum_{i=1}^{N}\left(\{S_{a_v}\}\{S_{a_v}\}^H\right)\left[T_{AA}\right]^H \tag{21}$$

Conducting the necessary averages Equation 21 becomes:

$$\left[T_{FS}\right]\left[G_{S_F S_{A_v}}\right]\left[T_{AA}\right]^H = \left[I_p\right]\left[T_{AA}\right]\left[G_{S_{A_v} S_{A_v}}\right]\left[T_{AA}\right]^H \tag{22}$$

The cross-spectral matrices in Equation 22 are the measurements that are made during the experimental procedure. Applying the load cell calibration to the raw load cell outputs and rigid body transformations in Equation 22 yields:

$$[G_{F_{t_p} A_p}] = [I_p][G_{A_p A_p}] \qquad (23)$$

If only the first columns of $[G_{F_{t_p} A_p}]$ and $[G_{A_p A_p}]$ are considered, this information can be substituted into Equation 8 yielding:

$$\{F_{t_p}\}_{col1} = [A_p]_{col1}\{I_{v_p}\} \qquad (24)$$

A solution cannot yet be obtained because there are 10 unknowns with only 6 equations. Additional equations can be obtained from the subsequent columns of $[G_{F_{t_p} A_p}]$ and $[G_{A_p A_p}]$. These additional equations are appended to the bottom of Equation 24 to obtain Equation 25.

$$\begin{Bmatrix}\{F_{t_p}\}_{col1}\\ \{F_{t_p}\}_{col2}\\ \vdots \\ \{F_{t_p}\}_{col6}\end{Bmatrix} = \begin{bmatrix}[A_p]_{col1}\\ [A_p]_{col2}\\ \vdots \\ [A_p]_{col6}\end{bmatrix}\{I_{v_p}\} \qquad (25)$$

$$\{F_{t_p}\}_T = [A_p]_T\{I_{v_p}\} \qquad (26)$$

In the event that $\{F_{t_p}\}_T$ and $[A_p]_T$ are complex numbers, a real solution for the inertia properties can be obtained by separating real and imaginary parts:

$$\begin{Bmatrix}\text{real}(\{F_{t_p}\}_T)\\ \text{imag}(\{F_{t_p}\}_T)\end{Bmatrix} = \begin{bmatrix}\text{real}([A_p]_T)\\ \text{imag}([A_p]_T)\end{bmatrix}\{I_{v_p}\} \qquad (27)$$

Finally an estimate for the inertia properties can be obtained by taking the pseudo inverse of the previous expression.

$$\{I_{v_p}\} = \begin{bmatrix}\text{real}([A_p]_T)\\ \text{imag}([A_p]_T)\end{bmatrix}^H \begin{Bmatrix}\text{real}(\{F_{t_p}\}_T)\\ \text{imag}(\{F_{t_p}\}_T)\end{Bmatrix} \qquad (2$$

Another formulation similar to that used to deri Equation 17 can be obtained by postmultiplying bo sides of Equation 16 by the hermitian of the $\{S_{F_{t_p}}$ vector. These two formulations may be used separate or together depending on the data under consideration.

6 DOF LOAD CELL

DESCRIPTION - The prototype 6 DOF load c consists of 32 piezoelectric sensing elements. Half the sensing elements are shear while the other half a compression crystals. Each of the individual elemen were assembled in the load cell so their output would b sensitive to some degree to the three translational force and three moments. The cross-axis sensitivity of th elements are not only affected by each element individual characteristics, but by their location in the loa cell. Each of the 32 coupled outputs are uncoupled in the 6 fundamental forces and moments by means of calibration matrix.

CALIBRATION - A 6 DOF calibration method used to determine the calibration matrix that uncouple the output from the load cell. Figure 3 shows the 6 DC load cell preloaded by a center stud between a excitation mass and a rigid calibration mass with know inertia properties. Eight triaxial accelerometers a mounted on the calibration mass to measure its rig body motion. The assembled calibration fixture is the suspended from an engine hoist to simulate pseudo fre free boundary conditions.

Figure 3. 6 DOF Load Cell Calibration Fixture

An impact hammer is used to randomly impact i all directions on the excitation mass. The signal from th impact hammer is used as a trigger for each of th acquisition averages. The 6 DOF load cell measure and transmits the six components of force to th excitation mass. The voltage outputs from the load cell and the measured rigid body acceleration of the calibration mass are used with Equation 7 to obtain a calibration matrix. This calibration matrix is stored for later use to uncouple the outputs from the load cell.

INERTIA PROPERTY ESTIMATION

EXPERIMENTAL SETUP - A simple structure was considered in the application of this inertia estimation procedure. This structure consisted of an adapter plate which was bolted on top of a large cylindrical mass. The purpose of the adapter plate is to provide a smooth mounting surface for the load cell. Additionally, a smaller cylindrical mass was attached to the bottom of the large cylindrical mass. Because of the simplicity of this structure, hand calculations could be used to calculate its inertia properties.

Once again, a center stud is used to attach the excitation mass and calibrated 6 DOF load cell at the point of interest on the structure. The assembled structure with the attached excitation mass and load cell is shown in Figure 4. Once the fixture is assembled it is suspended by a flexible cable as shown in Figure 5.

Figure 4. Assembled Structure with Attached Excitation Mass and 6 DOF Load Cell

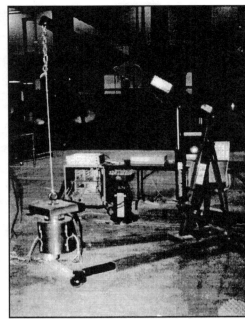

Figure 5. Suspension of the Assembled Fixture from an Engine Hoist

EXPERIMENTAL PROCEDURE - An impact hammer was used to randomly excite in all directions on the excitation mass. Each of the averages was triggered off of the signal from the impact hammer. The channels of the load cell and the accelerometers were simultaneously acquired using a HP3565 data acquisition system. An Exponential window was applied to all of the measurement channels to reduce leakage on the acquired data. A total of 50 random impacts on the excitation mass were used in the averaging of cross-power measurements of all the channels. The data was saved and later post-processed to obtain the inertia property estimates for the structure.

POST-PROCESSING - MATLAB was used with user programmed mfiles to process the experimental data to obtain the inertia property estimates. The necessary cross-power measurements $\left[G_{S_{F_{tp}} S_{AV}} \right]$ and $\left[G_{S_{AV} S_{AV}} \right]$ outlined in Equation 22 were calculated from the experimental data. The 6 DOF load cell calibration matrix $\left[T_{FS} \right]$ and the rigid body transformation matrix $\left[T_{AA} \right]$ are applied to the experimental data. Finally, this modified experimental data can be inserted into Equation 28 where an estimate of the inertia properties can be obtained.

RESULTS - The computed and experimentally estimated results are compared in Table 1. NaN in the table represents a division by zero. The experimental results agree very well with the hand calculations. The largest difference in the center of gravity location between the two methods is 43 thousands of an inch.

	Actual	Estimated	Difference	Error %
mass (Kg)	187.7370	185.1176	2.6194	1.3953
$x_{cg/p}$ (m)	0.0036	0.0038	-0.0002	-6.0280
$y_{cg/p}$ (m)	0.0000	-0.0002	0.0002	NaN
$z_{cg/p}$ (m)	-0.1825	-0.1836	0.0011	-0.6059
Ixx (Kg*m^2)	9.0740	9.0953	-0.0213	-0.2343
Ixy (Kg*m^2)	0.0000	0.0056	-0.0056	NaN
Ixz (Kg*m^2)	0.2481	0.2744	-0.0263	-10.5903
Iyy (Kg*m^2)	9.1297	9.0989	0.0308	0.3368
Iyz (Kg*m^2)	0.0000	-0.0026	0.0026	NaN
Izz (Kg*m^2)	2.1249	2.1023	0.0225	1.0606
Ipp1 (Kg*m^2)	9.1297	9.1090	0.0206	0.2262
Ipp2 (Kg*m^2)	9.0828	9.0959	-0.0131	-0.1438
Ipp3 (Kg*m^2)	2.1160	2.0916	0.0244	1.1550

Table 1. Inertia Property Estimation Results

CONCLUSION

A method has been presented to experimentally estimate the inertia properties of a rigid structure. This method utilizes a calibrated prototype 6 DOF load cell to measure the external forces and moments which are used to excite the structure. Triaxial accelerometers are mounted on the structure to determine its rigid body motion. These measurements can then be inserted into formulations derived by applying Newton's second law.

Comparing the experimental results with hand calculations it's clear this method worked well in the example presented. Perturbed masses added with known inertia properties at different locations can also be used to obtain redundant estimates of the inertia properties. These results may then be averaged to improve all estimates. Inertia properties estimated by attaching the 6 DOF load cell at different locations on the structure can be compared by applying the parallel axis theorem.

The results from this initial investigation are encouraging. Further testing needs to be conducted to determine the repeatability of this method. Different center stud materials could be used to resolve how the stiffness of the center stud influences the results. Another area that may need to be further investigated is the gravitational affects on the acceleration measurements.

.. This method will be further validated with more complicated structures by comparing experimental results to the results from a high fidelity solid model. For large structures where a single application point would not be sufficient to suspend the structure, multiple 6 DOF load cells could be used to support and excite the structure. .

Care has to be taken when applying this method to flexible structures. This is because the theory assumes the structure of interest is rigid. If the first flexible mode is much higher in frequency than the last rigid body mode then this method may still be used by ensuring the data that is used contains very little if any effects from the flexible modes. Modal analysis may have to be used to remove the effects of flexible modes in the case where the first flexible and last rigid body modes are close in frequency.

A second prototype 6 DOF load cell has been developed and will be evaluated. This prototype has only nine sensors. Three of the sensing elements are compression while the other six are shear crystals. The number of sensing elements still allows for the measurement of the six components of force even in the event of a faulty sensor. The reduced number of outputs greatly reduces the number of acquisition channels needed to perform the inertia property estimation procedure.

REFERENCES

[1] Okubo, N., Furukawa, T., "Measurement of Rigid Body Modes for Dynamic Design", Proceedings of the 2nd International Modal Analysis Conference, 1984.

[2] Bretl, J., Conti, P., "Rigid Body Properties from Test Data", Proceedings of the 5th International Modal Analysis Conference, 1987.

[3] Wei, Y. S., Reis, J., "Experimental Determination of Rigid Body Inertia Properties", Proceedings of the 7th International Modal Analysis Conference, 1989.

[4] Furusawa, M., "A Method of Determining Rigid Body Inertia Properties", Proceedings of the 7th International Modal Analysis Conference, 1989.

[5] Pandit, S. M., Yao, Y. X., Hu, Z. Q., "Dynamic Properties of Rigid Body and Supports from Vibration Measurements", Structural Vibration and Acoustics; American Society of Mechanical Engineers DE-Vol. 34, 1991.

[6] Fregolent, A., Sestieri, A., Falzetti, M., "Identification of Rigid Body Inertia Properties from Experimental Frequency Response", Proceedings of the 10th International Modal Analysis Conference, 1992.

[7] Pandit, S. M., Hu, Z. Q., Yao, Y. X., "Experimental Technique for Accurate Determination of Rigid Body Characteristics", Proceedings of the 10th International Modal Analysis Conference, 1992.

[8] Mangus, J. A., Passerello, C., Van Karsen, C., "Direct Estimation of Rigid Body Properties from Frequency Response Functions", Proceedings of the 10th International Modal Analysis Conference, 1992.

[9] Mangus, J. A., Passerello, C., Van Karsen, C., "Estimating Rigid Body Properties from Force Reaction Measurements", Proceedings of the 11th International Modal Analysis Conference, 1993.

[10] Toivola, J., Nuutila, O., "Comparison of Three Methods for Determining Rigid Body Inertia Properties from Frequency Response Functions", Proceedings of the 11th International Modal Analysis Conference, 1993.

[11] Huang, S. J., Cogan, S., Lallement, G., "Experimental Identification of the Characteristics of a Rigid Structure on an Elastic Suspention", Proceedings of the 13th International Modal Analysis Conference, 1995.

[12] Urgueira, A. P., "On the Rigid Body Properties Estimation from Modal Testing", Proceedings of the 13th International Modal Analysis Conference, 1995.

[13] Stebbins, M. A., Blough, J. R., Shelley, S. J., Brown, D. L., "Multi-Axis Load Cell Calibration and Determination of Sensitivities to Forces and Moments", Proceedings of the 15th International Modal Analysis Conference, 1997.

971958

Practical Aspects of Perturbed Boundry Condition (PBC) Finite Element Model Updating Techniques

Michael Yang, Randall Allemang, and David Brown
University of Cincinnati

Copyright 1997 Society of Automotive Engineers, Inc.

ABSTRACT

The perturbed boundary condition (PBC) model updating procedure has been developed to correct the finite element model [1]. The use of additional structural configurations adds more experimental information about the system and so better updating results can be expected. While it works well for simulated examples, practical limitations and additional requirements arise when it is used to update engineering structures. In this paper, the merits and the practical limitations of the techniques will be discussed in depth through the updating of a simulated system where the "measured" data is generated by computer and a real test structure where the experimentally measured data is noisy and distorted due to leakage. Useful suggestions and recommendations are drawn to guide the model updating of practical engineering structures.

INTRODUCTION

In today's industry, an accurate finite element model is important to conduct the dynamic simulations of a structure and machine. To obtain such an accurate model, a model updating procedure is always necessary that localizes and corrects the errors with the finite element model by minimizing the difference of some quantities between the analytical results and those obtained through experiment. Among various model updating techniques, the ones that update the physical parameters and use measured frequency response functions have been proved more promising [1,2].

Most updating techniques investigate a structure under single structural configuration only. Since some structural parameters may not be well tested within limited frequency band under single configuration, the updated model will be unable to predict the dynamic properties well. This is especially true if the model errors are from some local components and if the in-situ structure is placed under different boundary conditions from the one under test. As a remedy, the perturbed boundary condition model updating techniques are developed which construct the finite element models and test the structures under several structural configurations. The use of additional structural configurations adds more information about the system and so will generally give better updating results. In reference [1], the theory and procedure of the PBC model updating techniques is presented.

While the PBC model updating techniques work well for academic examples, practical limitations and additional requirements arise when it is used to update the models of engineering structures. In this paper, these problems will be discussed in depth through the updating of both a simulated example and a real test structure, where the measured data for the former is generated by computer and that for the latter contains measurement noise and suffers severe signal processing leakage. Suggestions and recommendations are drawn which are useful to guide the model updating for practical engineering structures.

THEORETICAL BACKGROUND

PBC MODEL UPDATING EQUATIONS - The finite element model of a structure is usually represented by the system matrices $[M]$, $[K]$, and $[C]$. Assuming that the differences between the original finite element model and the updated one can be expressed by a first order linear Taylor expansion, we have

$$[M] = [M_A] + \sum_i \alpha_i [M]^i \tag{1}$$

$$[K] = [K_A] + \sum_j \beta_j [K]^j \tag{2}$$

$$[C] = [C_A] + \sum_k \gamma_k [C]^k \tag{3}$$

where α_i, β_j, and γ_k are the relative updating parameters, and $[M]^i$, $[K]^j$, and $[C]^k$ are the derivative matrices with respect to specific element in the global coordinate system. The purpose of model updating is to determine the values of these parameters by comparing the dynamic properties between the

$$[A]\{X\} = \{B\} \tag{4}$$

of where the vector $\{X\}$ is a subset of all three types of updating parameters. In the PBC model updating procedure, the data matrices corresponding to each of the S configurations can be constructed separately and then stacked together to get the matrix $[A]$ and the vector $\{B\}$,

$$[A] = \begin{bmatrix} A_1 \\ A_2 \\ \cdots \\ A_s \end{bmatrix}, \quad \{B\} = \begin{Bmatrix} B_1 \\ B_2 \\ \cdots \\ B_s \end{Bmatrix} \tag{5}$$

If Configuration 1 is defined as the original structure, the data matrices and the observation vectors for the perturbed configurations are respectively

$$\begin{aligned} [A_s] &= [A_1] + [\delta A_s] \\ \{B_s\} &= \{B_1\} + \{\delta B_s\} \end{aligned} \quad 2,3,\ldots S \tag{6}$$

Depending on the algorithm that is used, $[A_s]$, $[\delta A_s]$, $\{B_s\}$, and $\{\delta B_s\}$ will have different expressions. For current study, the expressions developed in reference [1] are used and are not presented here for brevity.

CONFIGURATION SELECTION PROCEDURE - By perturbing a structure, the parameters that are not well tested under the original configuration may be well tested under the perturbed configuration. The combination of the data thus adds more experimental information about the test structure. In mathematical term, it implies that the test data base of the perturbed structure differs from that of the original structure, and the combination of both results in a larger test data base. In reference [3], a PBC Configuration Selection Procedure (CSP) is proposed to evaluate if a perturbation is effective or if two perturbations are similar to each other. This configuration selection procedure judges the principal angles between the data space of the original structure and that introduced by perturbation. It starts to find the base orthogonal matrices $[Q_1]$ and $[Q_s]$ for these two data spaces by QR permutation decomposition,

$$[A_1][P_1] = [[Q_1][R_1] \quad [O_1]] \tag{7}$$

$$[\delta A_s][P_s] = [[Q_s][R_s] \quad [O_s]] \quad s=2,3,\ldots,S \tag{8}$$

where $[O_1]$ and $[O_s]$ represent the non-significant parts that are set by some cut-off criterion. And, a principal matrix is formed by

$$[C_s] = [Q_1]^H [Q_s] \quad s=2,3,\ldots,S \tag{9}$$

which has the size of N_1 by N_s and all of its singular values fall into the range between zero and one. The principal angles

for the sth configuration are therefore defined as the arc-cosines of the singular values,

$$0 \le \theta_1 \le \cdots \le \theta_i = \cos^{-1}(\sigma_i) \le \cdots \le \frac{\pi}{2} \tag{10}$$

If all principal angles have zero degrees, it implies that these two spaces are parallel to each other and so no additional information is introduced by perturbation. On the contrary, if all principal angles are 90 degrees, these two spaces are orthogonal to each other and so maximal information is introduced by perturbation. In practice, most of the principal angles fall between zero and 90 degrees. By common practice, the principal angles that are greater than 60 degrees are considered significant. Therefore, the number of large principal angles indicates the degree of perturbation effectiveness.

EXAMPLES

UPDATING OF A 15 DOF SYSTEM - The 15 degree-of-freedom lumped mass system was updated in reference [1] and it is shown in Figure 1. It characterizes five local modes and two repeated modes. In total, three different perturbations are made. Together with the original model, there are in total four structural configurations. Configuration 1 is the original system; Configuration 2 is constructed by multiplying masses 11-15 by a factor of 10; Configuration 3 by dividing springs 18-22 by a factor of 5; and Configuration 4 by dividing springs 13-17 by a factor of 5. The "measured" frequency response functions for all four configurations are generated in the frequency range from 0 to 200 Hz with a frequency resolution of 0.5 Hz, and then contaminated with 1% (or down 40 dB) proportional and 2% (or down 34 dB) background noise. The damping level is about 1% for all modes. The finite element modeling errors are made on masses 11, 12, 13, 14, and 15 with a factor of four, and on springs 1 and 6 with a factor of one. The existence of these modeling errors causes the frequency differences between the original finite element models and the "measured" models of all four configurations at a level about -10% for the first 10 modes and about -45% for the last five local modes. MAC analysis shows that very poor correlation exists between the analytical modes and the measured modes, especially for the last five local modes and the pairs of repeated modes where no correlation can be figured out at all. It therefore puzzles the updating procedures that are based on the modal parameters that require correct pairing of modal parameters.

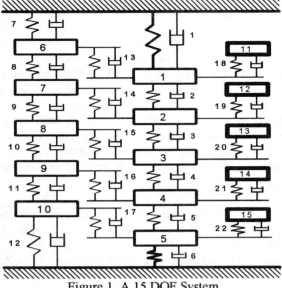

Figure 1. A 15 DOF System

Test data is collected from coordinates 1, 6, 11, 12, 13, 14, and 15 for all four configurations. The number of degrees of freedom of the analytical models is reduced from 15 to 7 by a dynamic model reduction method [3]. The frequency band of interest is defined in the range from 0 to 80 Hz and the updating frequencies are only selected within this range. For the updating frequencies, each frequency is selected at each of the resonant peaks within the frequency band of interest, which means that 10 updating frequencies are chosen respectively for Configurations 1 and 4, and 15 updating frequencies are chosen respectively for Configurations 2 and 3.

In the updating, updating parameters are first localized using the SVD/QR based model error indicator function [4]. All three perturbed configurations are checked using the configuration selection procedure (CSP) mentioned above for their effectiveness of perturbation. Their principal angles are calculated and plotted in Figure 2 which shows that Configuration 3 posses the largest number of large principal angles and is thus considered the best one. In the same sense, Configuration 4 is better than Configuration 2 since the number of significant principal angles is four for the former and is only one for the latter.

Updating is then conducted and the results corresponding to different PBC configuration combinations are calculated. Figure 3 plots and lists the root mean square (RMS) values of the updated parametric factors with respect to five different combinations of PBC configurations. It shows that when the most effective perturbation 3 is involved and when more configurations are used, smaller variance of, i.e. better, updated parameter factors is obtained. As a result, cases d) and e) are better than case c), and case c) is better than cases a) and b). Since case a) uses only one configuration and case b) includes a less effective perturbation, the results from these two cases is not as good as other cases.

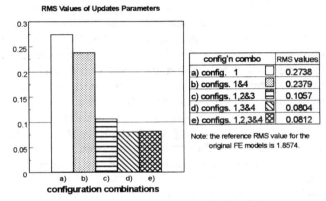

Figure 3. RMS Values of the Updated Factors

In all five PBC updating cases, the updated models can all well predict the modal frequencies inside the band of interest. The relative errors for these frequencies all fall in the levels of less than 3%, down from the original 10% levels. For the modal frequencies outside the band of interest, the predictions in cases a) and b) are very poor - the relative errors are as high as 20%. On the other hand, in cases c), d), and e) where proper number of PBC configurations are used, those errors for the out-of-band modes are dramatically reduced to the levels about 3%, down from the original 47%. Therefore, the use of properly selected PBC configurations can improve not only the in-band but also the out-of-band frequency properties. This is because the updated parameters are more accurate (with smaller RMS value). Figure 4 plots the frequency response functions of Configuration 1 at coordinate (12,11). The frequency response functions are compared among the measured one, the one before correction, and the ones reconstructed from the results by single configuration updating and by PBC updating respectively. The reconstruction is very good (bold solid line) across the whole frequency range when three PBC configurations are used while it is only good within the band of interest when only single configuration is used (dash line).

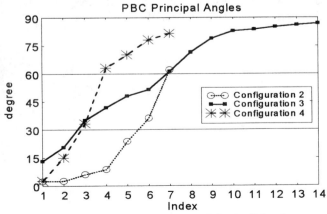

Figure 2. 15 DOF PBC System Principal Angles

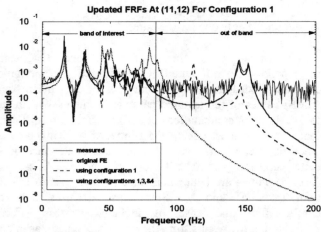

Figure 4. Comparison of FRFs at Position (11,12)

UPDATING OF AN H-FRAME STRUCTURE - The H-frame structure is constructed by two 6-foot-long and one 3-foot-long steel box beams, welded together in H-shape. Four small steel plates of 6½"×4"×½" are then welded at the four ends to which four aluminum plates of 12"×12"×1" are connected by bolts. A finite element model using shell or solid finite elements plus very fine mesh sizes is more accurate but the model size becomes unreasonably large. A finite element model using beam elements is much simple and is preferred for this structure but it results in too higher modal frequencies since it treats the joints between the box beams as rigidity. In this example, the three box beams are modeled using coarse two dimensional beam elements. The connections between the box beams are modeled by three springs. The end metal plates and all perturbation mass plates are treated as lumped parameters. The equivalent lumped mass and mass moment of inertia are respectively 5.63 kgs (14.64 lbm) and 184.69×10^3 kg.m^2 (118.85 lbm.in^2) for the end metal plates at each of the four ends. The model is shown in Figure 5.

Two perturbed structural configurations are constructed by bolting mass plates to the selected ends of the frame. Two sets of mass plates were made, each consisting of one 10⅞"×10⅞"×1" and two 12"×4"×1" steel plates. The equivalent mass and mass moment of inertia for one mass set is 27.3 kg (60.3 lbm) and 1.938×10^6 kg.m^2 (1250 lbm.in^2) respectively. Configuration 1 is defined as the original configuration; Configuration 2 is constructed by attaching one perturbation mass set to the end at node 13 and another set to the end at node 26; and Configuration 3 is constructed by attaching one set to the end at node 13 and another set to the end at node 14.

The finite element models of the frame are then modeled using 30 2-D beam elements with a mesh size of 6 inches, together with four lumped masses, four mass moments of inertia, and six isolated springs. In total, 33 nodes are selected that defines analytical models of 99 degrees-of-freedom. The values of the six isolated springs are referred by the stiffness of the jointing beam elements such as elements 25 and 30 in Figure 5. Thus, the translation stiffness has a value of 100.0×10^6 lb.in^{-1} and the rotational stiffness a value of 17.75×10^6 lb.in.rad^{-1}.

In the tests, 37 transverse acceleration responses with respect to 2 references were taken, separately for all three structural configurations. The data acquisition system includes an HP VME mainframe and LMS CADA-X software. Two independent burst random (50%) excitation signals were applied at coordinates 26:+X and 29:+Y respectively and the flattop windows (which may not be appropriate) were applied for handling leakage. The frequency band of interest was set in the range of 0 and 200 Hz with a frequency resolution of 0.5 Hz. This range covers six lower vibrating modes in the X-Y plane.

The coherence and the FRF reciprocity were checked before measurements started. Figure 6a) is the coherence at coordinate 26:+X, and significant drop at all resonant peaks shows severe leakage existing because the structure is very lightly damped and the improper windows were used. Figure 6b) plots the FRF reciprocity between the two references. Except at all resonant peaks and in the lower frequency range from 0 to 10 Hz, the reciprocity is very good since the differences between the two reciprocal frequency response functions are mostly less than 5%.

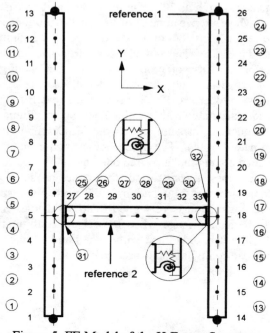

Figure 5. FE Model of the H-Frame Structure

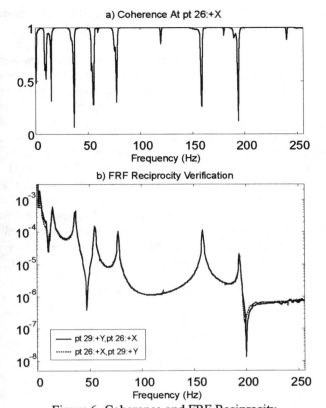

Figure 6. Coherence and FRF Reciprocity

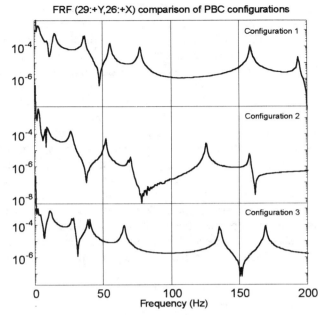

Figure 7. Typical Measured FRFs

Figure 7 compares the measured frequency response functions at position (29:+Y, 26:+X), for all three configurations. It shows that the change of frequency properties due to the structural perturbations is significant. As the result of large signal processing leakage, however, the measured frequency response functions are severely distorted at all resonant peaks and so the data at these frequencies are unreliable. The distortion is even more severe for mode 2 of Configuration 1, modes 1 and 4 of Configuration 2, and the first three modes of Configuration 3, where the peaks are split apart!

The experimental modal parameters are then estimated using the X-Modal modal parameter estimation software that is developed at the UC-SDRL. The estimated modal parameters are generally not accurate due to the severe data distortion at the resonant peaks. It is for sure that, if an modal model updating procedure is used, good updating results can not be expected. For the updating procedure that uses the measured frequency response functions, however, this is not a problem since one can choose the updating frequencies from where the data shows good quality.

Since the number of degrees of freedom of the analytical models is 99 while that for the test models is 37, a model reduction procedure is applied to match the analytical and the test models. The analytical modal parameters are then calculated from the analytical finite element system matrices. The MAC values between the analytical results and the test parameters range from 66% to 97%. The differences of frequencies between the measured and the tested reach at the levels as high as 20% for all modes for all three configurations. It becomes clear that significant modeling errors exist and the model must be updated.

The property parameters of the H-frame structure include the bending stiffness EI_{zz}, the axial stiffness EA and the distributive mass density ρA for the beam elements plus eight lumped mass elements and six isolated spring elements that results in 104 elements in total. On the other hand, the number of modes in the band of interest is only six, which set certain limitation on the maximum number of updating parameters. Since the three box beams have the same cross sectional properties, we can group all of beam elements and use the three cross-sectional parameters for this group. Thus, the total number of updating parameters is reduced to 17.

The updating procedure using frequency response functions provides a flexible way to choose the updating frequencies. The frequencies below 10 Hz and at all resonant peaks should be avoided because they are either noisy or distorted. In addition, frequencies at all resonance of the finite element models should not be selected because they have infinite values and will create numerical problems when the finite element damping is not modeled. With this consideration, 48 updating frequencies are selected for Configuration 1, 65 for Configuration 2 and 66 for Configuration 3.

dominating error sources of the finite element model. The error localization equation is then constructed using the impedance updating method presented in reference [1]. The singular values of the data matrix is checked against the information content represented by the data matrix. Figure 8 plots such singular values of a typical data matrix with all 17 updating parameters. It shows that the significant number of singular values is nine only. This number is therefore used to conduct QR permutation decomposition with column pivoting. In such a way, the parameters that are detected include the two rotational springs (parameter numbers 14 and 17), the four lumped masses (parameter numbers 1, 3, 5, and 7), and the mass density and the bending stiffness of the box beams (parameter numbers 9 and 10). The use of these eight parameters reduces the equation error residual norm to 4%, as shown in Figure 9. The remaining nine parameters are less significant and so can be excluded from updating.

impedance model updating procedures over the modal model updating procedures.

To check the effectiveness of the perturbed structural configurations, the PBC configuration selection procedure is used. The principal angles for Configurations 2 and 3 are calculated and plotted in Figure 10. Configuration 2 has two principal angles that are greater than 60 degrees and Configuration 3 has nine such principal angles. With respect to the number of significant principal angles, Configuration 3 is better than Configuration 2. Therefore, Configuration 1 and Configuration 3 are selected as the optimal set of PBC configurations. The structure is then updated using the data from these two configurations. The results from the PBC updating is also compared with those when only Configuration 1 is used. Updating solutions are achieved through an iterative approach.

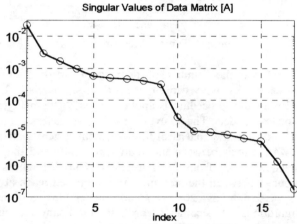

Figure 8. Singular Values of the Data Matrix

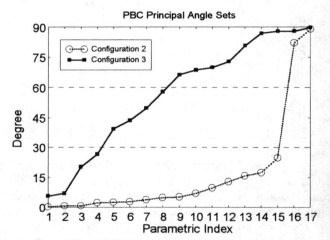

Figure 10. PBC Principal Angles of the H-Frame

One fact to be noticed is that in the frequency range below 200 Hz there exist only six modes while the data matrix has a rank of nine. This is because some of the out-of-band modes still have influence within the band of interest when the frequency response functions are used. As a result, the rank of the data matrix is usually couple of orders higher than the number of modes. This is an advantage of the

For the updating using data from single configuration, Figure 11 shows the updated parameters in iterative form for Configuration 1. After about four iterative steps, convergent solution is achieved. The updated parameters are then used to correct the finite element model, from which the modal frequencies are calculated. The MAC values between the updated model and the measured structure are also calculated for the first six modes, as shown in Table 1. It shows that the modal frequency errors between the finite element model and the test structure all fall into the levels of less than 1% after correction, down from those up to 20% before correction. The results are really satisfactory.

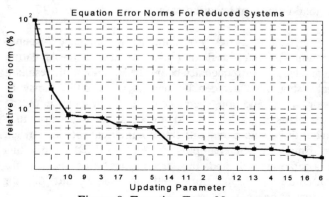

Figure 9. Equation Error Norms

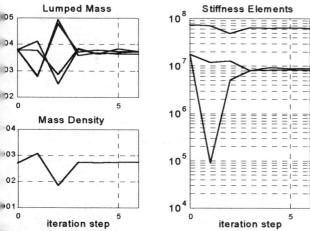

Figure 11. Updated Parameters in Iterative Form Using Configuration 1 Only

Mode No.	Measured (Hz)	Updated (Hz)	Error (%)	MAC (%)
1	14.77	14.77	0.02	96.8
2	36.67	36.94	0.80	83.7
3	55.39	55.44	0.14	88.6
4	77.56	76.89	-0.88	89.3
5	158.7	158.93	0.15	92.5
6	194.0	193.42	-0.07	87.3

Table 1. Modal Properties of Configuration 1 Using the Data from Configuration 1 Only

Mode No.	Measured (Hz)	Updated (Hz)	Error (%)	MAC (%)
1	10.80	10.99	0.18	95.1
2	27.69	27.90	0.78	63.9
3	39.53	39.33	-0.38	73.4
4	65.14	64.51	-0.90	70.1
5	135.6	135.48	-0.30	94.6
6	169.4	169.95	0.33	97.5

Table 2. Modal Properties of Configuration 3 Predicted by PBC Updating

Next, PBC model updating is conducted where the data from both Configurations 1 and 3 are used. Figure 12 shows the updated parameters in iterative form. One may notice that there are fluctuations with the iterative results. Recall that Configuration 3 is constructed by adding about 60 gm (or 0.155 lb) mass plates respectively at the nodes 13 and 14 of Configuration 1, and the end mass plates are then approximated as lumped parameters. This rough treatment of the perturbed masses will definitely introduce additional errors into the finite element models. The errors are different between the models of the two configurations. When a PBC updating procedure is used, however, it treats the end mass as the updating parameter for both configurations uniquely rather than separately. The fluctuations of the updated results just reflect this inconsistency of the updating parameters. We also find that the fluctuations of the updated lumped masses of Configuration 3 are smaller when compared with those of Configuration 1. This is because the amounts of lumped masses of Configuration 3 is about seven times of those of Configuration 1, and thus they apply larger weighting on them during a least squares estimation procedure. The dotted lines in the figure represent the updated masses on the perturbed ends and are more stable due to larger weightings. In this case, the fluctuations of updated parameters, when comparing with their original values, are small and so can be neglected. As long as the modeling errors of the perturbed masses are relatively small, the fluctuations of the updated parameters will be small.

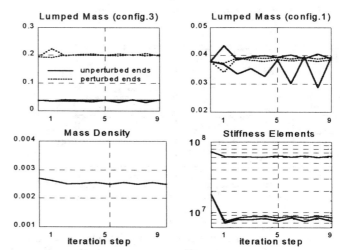

Figure 12. PBC Updating - Updated Parameters of Both Configurations 1 and 3

Table 2 shows the modal parameters for updated Configuration 3. The frequency errors for the first six modes between the finite element models and the test structures all fall into the levels of less than 1% after correction, from those up to 20% before correction. At this point, it seems that both single configuration updating and PBC updating can give very good updating results for the configurations that are used. When the updating results from only Configuration 1 are used to correct other configuration that is not used in updating, such as Configuration 3, however, the predicted parameters are not so exciting, as shown in Table 3, when comparing with the results in Table 2 where PBC updating is used.

In Table 4, we list some of the property parameters of the H-frame structure, updated using the data from different combination of PBC configurations. The corrected values differ a little from the initial ones for the distributed mass and a little more for the bending stiffness. The differences are, however, large (more than -50%) for the rotational springs. This verifies that the dominating modeling errors of the H-frame are due to the inaccurate treatment of the rotational stiffness at the two beam connection joints. Table 4 also shows that the results when using the data from more configurations as in the last two cases are less variant than those when using the data from single configurations as in the first two cases.

Mode No.	Measured (Hz)	Updated (Hz)	Error (%)	MAC (%)
1	10.80	11.18	1.88	96.5
2	27.69	28.28	2.15	63.9
3	39.53	39.63	0.38	73.1
4	65.14	64.71	-0.61	73.4
5	135.6	132.96	-2.15	94.2
6	169.4	166.24	-1.86	97.3

Table 3. Modal Properties of Configuration 3 Predicted by Single Configuration Updating

PBC Config Combo	ρA ($\times 10^{-3}$ lb.in^{-1})	EI_{ZZ} ($\times 10^6$ lb.in^2)	$K_{\theta\,31}$ ($\times 10^6$ lb.in.rad^{-1})	$K_{\theta\,32}$ ($\times 10^6$ lb.in.rad^{-1})
Initial	2.710	73.59	17.75	17.75
1	2.754	64.32	8.90	8.43
3	2.441	61.57	7.37	7.60
1 & 3	2.560	60.00	8.38	8.91
1, 2 & 3	2.570	61.55	7.97	8.40

Table 4. Updated Parameters of the H-Frame

As a visual estimation, the frequency response functions at coordinate (26:+X,29:+Y) of the corrected models of Configurations 1 and 3 are constructed and compared in Figure 13. Excellent matches are found between the updated and the measured frequency response functions across the whole frequency band in Figure 13a) and in Figure 13c) where the data of the configurations themselves is used in updating. Less accurate, but still good, match is also found in Figure 13b) for Configuration 3 where only the data from Configuration 1 is used in updating.

While the reconstructed frequency response functions match well with the measured ones across the whole frequency range, the MAC values listed in Tables 1, 2, and 3 are not as good as expected. Some of the MAC values are even as low as 64%! The major reason is because the quality of the estimated mode shapes or the residues are questionable due to severe distortion of the measured frequency response functions at the resonant peaks due to leakage. This is one of the advantages of using frequency response functions in the updating procedure. Distortion errors near the resonance peaks in the measured frequency response functions can be minimized.

CONCLUSION

In this paper, the PBC model updating technique that uses measured frequency response functions from PBC structural configurations [1] is used to update one simulated system and one real H-frame structure. Valuable experience has been gained through the investigation conducted in the reported research.

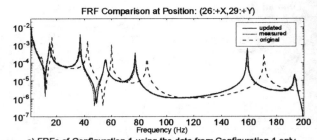
a) FRFs of Configuration 1 using the data from Configuration 1 only

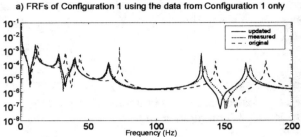
b) FRFs of Configuration 3 using the data from Configuration 1 only

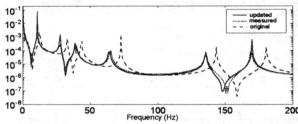
c) FRFs of Configuration 3 using the data from both Configurations 1 and 3

Figure 13. FRFs of the Updated H-Frame

Since the technique uses the frequency response function instead of the estimated modal parameters, it is free of the inaccuracy of the estimated modal parameters. Especially, there is not requirement for pairing the modal parameters between the analytical model and the test structure which may become very difficult when the analytical-test correlation is low and when closed modes are present.

The technique adds large flexibility in choosing the updating frequencies. The updating frequencies can be selected only from the regions which the measured data is in good quality. Therefore, the technique can be used in the situations when the test data contains severe leakage, when nonlinearity exists, or when the data is noisy in some regions. In these cases, accurate estimation of modal parameters becomes impossible.

For the selected frequencies to well represent all the modal information inside the frequency band of interest, at least one frequency should be selected around each mode for all modes in the band, and the number of selected frequencies must not be less than the number of modes. An oversized number of updating frequencies is always useful to average out the measurement noise. An overlapped plot of all frequency response functions and that of coherence are very helpful in determining the good frequency band.

Since at the resonant frequencies of a finite element model, the analytical frequency response functions reach infinite values, those frequencies should always be avoided. In an iterative updating procedure, however, initially well selected frequencies may become bad frequencies for some intermediately-updated finite element models and so the procedure may become divergent. If it happens, a re-adjustment of updating frequencies is necessary. An automatic adjustment of updating frequencies embedded in the algorithm is possible to fix this problem.

The number of updating parameters is dependent on the number of significant singular values of the resultant least squares data matrix. This number is usually a couple of orders higher than the number of modes in the frequency band of interest. An error indicator function is useful to localize the erroneous parameters. If there are too many parameters to be updated, grouping updating parameters and updating in several levels are always useful.

In general, a PBC model updating procedure, where the data from more than one structural configuration is used, will generate better, or more accurate, updating results since the PBC testing provides more useful information about the system. The ability of accurately modeling the perturbations, however, is very critical to the success of a PBC model updating procedure. If the perturbation causes additional modeling errors, the errors will be inconsistent with the original model errors and so the updated parameters become inconsistent. The potential modeling errors in the frequency range of interest due to the introduction of PBC perturbation should be carefully checked. The PBC configuration selection procedure is useful to determine the optimal set of PBC configurations. Theoretically, more PBC configurations will definitely provide more information about the system. Practically, more PBC configurations may introduce more uncertainties of the inconsistent modeling error sources and additional experimental and computational efforts. There must be a trade-off among the number of PBC configurations, the experimental cost and feasibility, the computational cost, and the modeling ability of the perturbed components.

REFERENCES

[1] Yang, M. and D.L. Brown, "Model Updating Techniques Using Perturbed Boundary Condition (PBC) Testing Data," *Proceedings of the 14th International Modal Analysis Conference*, pp 776-782, 1996.

[2] Lammens, S., M. Brughmans, J. Leuridan and P. Sas, "Updating of Dynamic Finite Element Models Based on Experimental Receptances and the Reduced Analytical Dynamic Stiffness Matrix," Proceedings of 1995 SAE N&V Conference, Traverse City, MI, pp 117-126, 1995.

[3] Yang, M. and D.L. Brown, "An Improved Procedure for Handling Damping During Finite Element Model Updating," *Proceedings of the 14th International Modal Analysis Conference*, pp 576-584, 1996.

[4] Yang, M. and D.L. Brown, "SVD/QR Based Model Error Indicator Function," *Proceedings of the 15th International Modal Analysis Conference*, Orlando, FL, Feb., 1996.

NOMENCALTURE

$[A]$, $[As]$	assembled PBC and individual data matrices
$\{B\}$, $\{Bs\}$	assembled PBC and individual error vectors
$[\delta As]$, $\{\delta Bs\}$	quantities introduced by perturbations
$[C]$, $[CA]$	updated and original damping matrices
$[Cs]$	PBC principal matrix for Configuration s
$[K]$, $[KA]$	updated and original stiffness matrices
$[M]$, $[MA]$	updated and original mass matrices
$[M]i$, $[K]j$, $[C]k$	updating element derivative matrices
N1	dimension of original data space
Ns	dimension of perturbed data space
S	number of PBC configurations
$[O1]$, $[Os]$	less significant parts of data matrices
$[P1]$, $[Ps]$	permutation matrices
$[Q1]$, $[Qs]$	significant data base orthogonal matrices
$[R1]$, $[Rs]$	significant upper triangular matrices
$\{X\}$	vector consisting of updating parameters
αi	mass updating parameter factor
βj	stiffness updating parameter factor
γk	damping updating parameter factor singular value
θ	principal angle

971959

Prediction of Powerplant Vibration Using FRF Data of FE Model

Yukitaka Takahashi, Toshibumi Suzuki, and Masayoshi Tsukahara
Honda Research and Development Co., Ltd.

Copyright 1997 Society of Automotive Engineers, Inc.

ABSTRACT

Recently, for the purposes of shortening the development period, the estimation of powerplant vibration has become more important in the early design stage, and eigenvalue analysis by FEM is commonly used to solve this problem. Eigenvalue analysis cannot directly predict vibration levels that affect the durability of each component and the vibration of a car body, however it is necessary to predict powerplant vibration in order to estimate exciting force under running conditions. Another factor adding to the difficulty of prediction is the instability of exciting force and various other non-linear characteristics. This paper presents a new approach using FRF data from FE models for accurate prediction of engine vibration under running conditions. By applying this approach to an in-line four cylinder engine, the predicted vibration is reasonably comparable with experimental results.

INTRODUCTION

In recent years, the performance required of automobiles has been highly diversified in accordance with the times. As a result, in addition to the primary needs of transportation and logistics, there has been strong demand for faster, safer, more comfortable, more cost-effective, lighter, more fuel-efficient and quieter means of transportation with higher power generation.
Because of environmental considerations, as well as changes in consumer preferences regarding noise and vibration, powerplant vibration affecting external and internal noise of the vehicle must be reduce.
Fierce competition among manufacturers and a matured market encourage more reliable development and swifter response to market trends in spite of shorter research and development period. To cope with this, more precise analytical prediction technologies are in desperately needed.

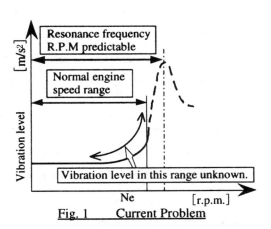

Fig. 1   Current Problem

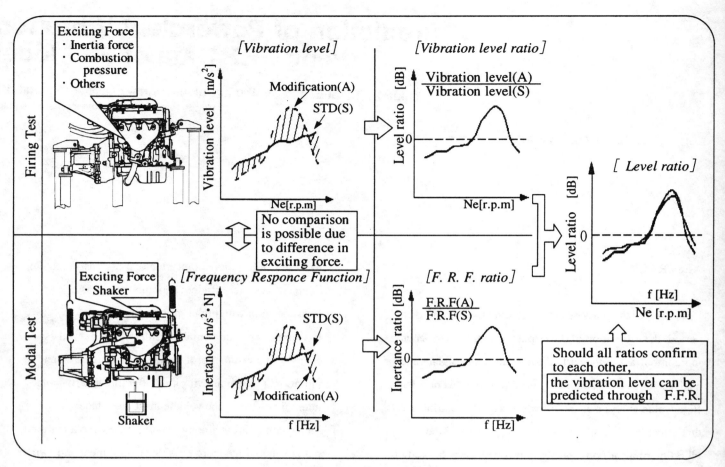

Fig.2 Basic concept

In the current design stage of evaluating the basic configuration of a powerplant, eigenvalue analysis using an FE model is one of the most popular methods. With this method, it is possible to predict the exact resonance frequencies. Little achievement, however, has been experienced when it comes to the analysis of vibration level, as shown in Fig. 1.

This paper reports a method of predicting the vibration level of an engine during the firing test based on actual vibration data, as well as the difference in frequency response functions (FRF) of the FE models before and after design modification. Furthermore, the method was applied to a transversally-mounted in-line four cylinder engine in a front-wheel drive layout.

DESCRIPTION OF ANALYSIS METHOD

BASIC CONCEPT

The basic concept of the method is shown in Fig. 2. The upper column contains the data of the firing test whereas the modal test is depicted in the lower column.

In the firing test, engine vibration occurs as a result of several exciting forces, namely inertia and combustion pressure. In the modal test, the engine is forcibly excited by a shaker. The inertance corresponding to each frequency can be measured. Obviously, no direct comparison between the data is possible.

Focus was given to the ratio of each measurement parameter (i.e., vibration level and FRF). The ratio of each parameter is calculated according to data taken from powerplant with different auxiliary equipment specifications. Then, should all the ratios conform, it would be possible to predict the actual vibration level of the modified powerplant based on the ratio of FRF.
This ratio method was then applied to an actual evaluation using two in-line four-cylinder engines with different specifications of auxiliaries. The result is shown in Fig. 3. That the vibration level ratio and FRF ratio showed a similar tendency was confirmed.

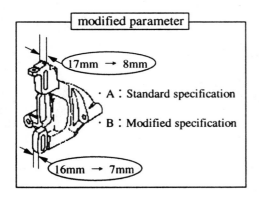

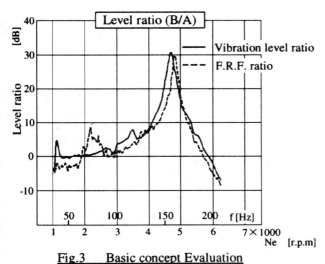

Fig.3   Basic concept Evaluation

ANALYSIS PROCEDURE

Figure 4 shows the flow chart of the analysis. The procedure is as follows;
1) FEM model production:
    The FEM model of each powerplant component are produced. They include the cylinder block, clutch casing, transmission casing and cylinder head.
2) Joint model production:
    The joint properties between the two adjacent components are identified ([K], [C]) based on the method described in the next section. Using the identifications, the connection area model is produced.
3) FE model assembly:
    The models produced in step 1) and 2) are assembled to produce the FE model corresponding to the state of the actual powerplant.
4) Frequency response analysis:
    The transfer functions of engine model before and after the design modification are calculated.
5) Vibration levels are measured during firing test.
6) Vibration levels predicted after design modification:
    The vibration level during the firing test, using the modified powerplant, is predicted based on the achievements in steps 4) and 5).

IDENTIFICATION OF JOINT PROPERTIES

The simulation results tend to be easily affected by the joint properties in the case of a complicated mechanism like an engine where a number of parts are assembled together. This must be considered for precise performance prediction. The joint properties were identified and applied to the FE powerplant model in the case.

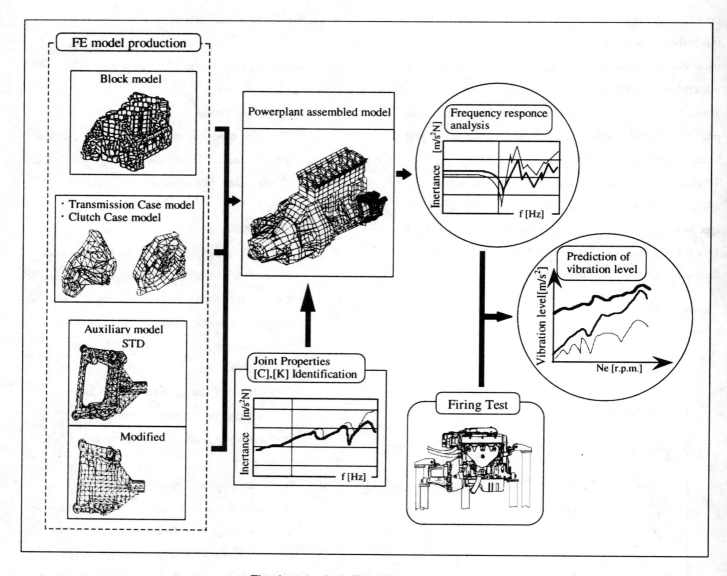

Fig. 4   Analysis Flow Chart

The motion equation of a system featuring viscous damping is defined as follows assuming that two components A and B are connected via a joint D, as shown in Fig. 5;

$$[M]\{\ddot{X}\}+[C]\{\dot{X}\}+[K]\{X\}=\{F\}$$
$$\{-\omega^2[M]+j\omega[C]+[K]\}\{X/F\}=\{I\}$$

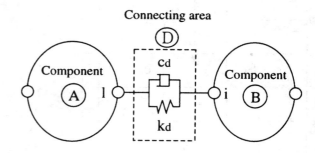

Fig.5   System Diagram with Connecting area

Then, assuming that;

$$\{X/F\} = H(\omega) = R(\omega) + jI(\omega)$$

Simplifying the imaginary parts leads to;

$$\omega[C] = -\{-\omega^2[M]+[K]\}[I(\omega)][R(\omega)]^{-1}$$

Thus, the imaginary part of the system where components A and B are connected is as follows;

$$\begin{bmatrix} -\omega[C_{A1}] & -\omega[C_{A2}] & [0] & \\ -\omega[C_{A3}] & -\omega[C_{a}]+\omega[C_d] & & -\omega[C_d] \\ [0] & & -\omega[C_{B1}] & -\omega[C_{B2}] \\ & -\omega[C_d] & -\omega[C_{B3}] & -\omega[C_{bi}]+\omega[C_d] \end{bmatrix}$$

$$= -\{-\omega^2[M]+[K]\}[I(\omega)][R(\omega)]^{-1} \quad \cdots (1)$$

Values of [M], [K], R(ω) and I(ω) can be obtained from the FE model and FRF from modal test. A value [Cd] that constantly satisfies equation (1) can be obtained by convergent calculation. Also, a value [Kd] can be obtained through the method similar to that for [Cd]. As a result, the joint properties are successfully identified. The method previously mentioned was then evaluated using a simple box-shape rig as shown in Fig.6. The result is shown in Fig. 7. The model consists of a dummy engine block and transmission connected to each other with four bolts.

The joint properties were taken into the model for analysis, which revealed that transfer function prediction was possible with an accuracy of ±2dB in the range between 0-500Hz.

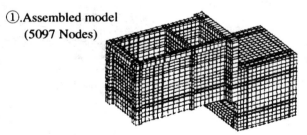

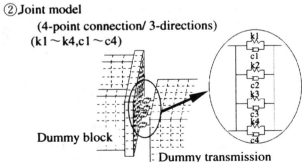

Fig.6   Test Rig model

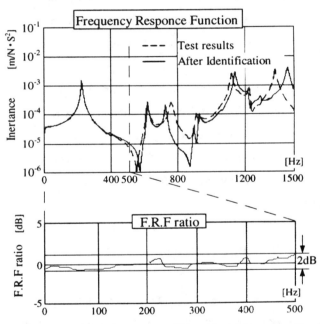

Fig.7   Verification result of identified joint properties

APPLICATION TO ENGINE DESIGN EVALUATION

In the following paragraphs, an example of application of this method to the evaluation of basic configuration for a transversally-mounted in-line four cylinder engine in a front-wheel drive layout.

EVALUATION OF VIBRATION LEVEL REDUCTION

The vibration level and resonance frequency of the powerplant are well known to be factors of paramount importance because they often cause booming noise and decrease engine durability.

The prototype powerplant could not achieve the target resonance frequency that affected booming noise, thus compliance with the target vibration level under running conditions was an immediate concern. To make the best of this situation, the prototype was used as a base for evaluation of the newly developed method.

Because the powerplant could not achieve the target, a drastic improvements to the powerplant was necessary. Various simulations were carried out to improve engine stiffness using bearing cap(B/CAP) beams, gusset stiffeners, etc.

The vibration level reduction effect was predicted using the new method. The predictions were then compared with the actual measurements of the prototype engine, the results of which are shown in Fig. 8. The difference was kept within ±2dB, showing that the method has a practical level of accuracy.

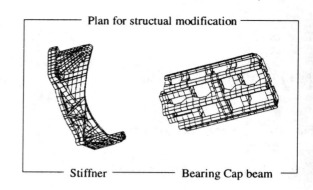
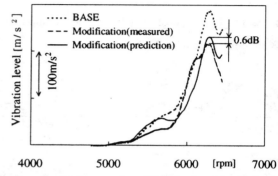

Fig.8 Application result to Vibration level Reduction

WEIGHT REDUCTION EVALUATION

The method was also applied to the weight reduction process for auxiliary brackets. The prototype powerplant featured cast iron brackets. The use of aluminum for weight reduction was considered at one stage for the purposes of improved fuel economy and reduced costs. Simulations for all brackets were conducted based on the concept that weight reduction should be ensured while maintaining the vibration level at the present level so as to

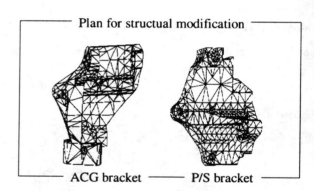
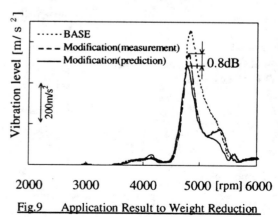

Fig.9 Application Result to Weight Reduction

prevent the design changes from leading to more severe vibration and decreased durability.

A comparison between the simulation and actual measurements was performed and the result is shown in Fig. 9. The difference between the two was a mere ± 2dB, which confirmed the practical accuracy of the method.

The other significant achievement lay with the bracket fixtures where an additional bolt maintained the vibration level at that of the cast iron bracket. This suggested a successful weight reduction was possible by altering the material.

At this stage, it was concluded that the use of the method makes it possible to promote smoother development flow, function evaluation, and prediction of possible results due to design changes. Therefore, it effectively reduced the time required for prototype development and analysis.

CONCLUSION

1) A vibration prediction method was developed as an evaluation tool for basic configuration work of powerplants. This was achieved by using FRF data of FE powerplant model where the joint properties were taken into account.

The method allows prediction of the degree of vibration level after the design modification through FEM frequency response analysis, provided that the actual measurement data of the level of the powerplant to be modified is available.

2) The application of the method to the front-wheel drive transversally-mounted in-line four cylinder powerplant proved the method is accurate enough for practical use.

REFERENCE

(1). K. Semba, T. Takahashi et al., 'Finite Element Eigenanalysis of Frontwheel Drive Powertrain' 7th IMAC Vol. 9 pp.1116-1122, 1989.

(2). T. Uchida, et al., 'Honda New In-Line Five Cylinder Engine---Noise and Vibration Reduction' SAE Technical Paper 900389.

(3). K. Semba, Y. Takahashi, et al., 'Powerplant Vibration Analysis using Super Computer' Proc. of JSAE, Vol. 901, pp.65-68, 1990 (Japanese).

(4). Jimin He, D.J. Ewins, 'Identification of damping properties in vibrating structures.' Proc. 1st International Conf. on Motion and Vibration Control, Yokohama, Japan, pp.982-987, 1992.

(5). J.Ouerengässer, J.Meyer, et al., 'NVH optimization of an In-line 4-cylinder powertrain.' SAE Technical Paper 951294.

971960

Heavy Duty Diesel Engine Noise Reduction Using Torsional Dampers on Fuel Pump Shafts

Neil Hutton
Holset Engineering Co., Ltd.

Copyright 1997 Society of Automotive Engineers, Inc.

ABSTRACT

An experimental study has been carried out to investigate the effect of the level of fuel pump shaft Torsional Vibration (T.V.) on engine noise. The study was carried out on several 6 cylinder automotive DI heavy duty diesel engines. The level of fuel pump T.V. was changed by fitting torsional dampers to the fuel pump shaft. A strong correlation was found between the T.V. level at 12th order of fuel pump rotation and the overall engine noise. Reducing the 12th order of fuel pump T.V. gave significant reductions in both overall and broad-band engine noise.

INTRODUCTION

Ever more stringent emission legislation and increased truck operator expectations are driving the search for lower emissions, better fuel economy and increased power from heavy duty truck DI diesel engines. The emissions benefits of high pressure fuel injection with variable timing control have been shown [1,2,3] and have led to the trend for fuel injection pressures to increase on recent diesel engine designs. 1000 bar injection pressure is now common, 1600 bar is in production [3] and 2000 bar is a future possibility.

Unfortunately, higher fuel injection pressure often leads to increased engine noise due to a combination of :
- higher levels of combustion noise
- increased fuel pump noise
- increased timing gear noise.

While the first two can be significant, this paper deals only with the latter cause of increased engine noise. It has been shown that on engines with high injection pressures, timing gear noise can be the dominant noise source on the engine [4]. On these engines, the reduction of timing gear noise is fundamental to achieving engine noise reduction.

BACKGROUND

The increasing significance of timing gear noise as a major factor on the overall level of truck engine noise has led to both practical [4-7] and theoretical [8,9] studies being carried out. These have determined the mechanism by which timing gear noise is generated and put forward methods of reducing it.

Most current heavy duty diesel engines use fuel systems in which the fuel injection pressure is generated by a pump using a cam/plunger mechanism. The level of instantaneous torque on the fuel injection pump shaft is negative when injection pressure is being generated and positive at the end of injection. The higher the injection pressure required, the larger the torque loadings on the shaft. For a 6 cylinder engine, this load and then unload sequence is applied to the fuel pump shaft 6 times per fuel pump shaft revolution (i.e. at 3rd order of crankshaft rotation). On a high injection pressure fuel system, the torque reversals associated with the generation of injection pressure are large enough to drive the fuel pump shaft and its timing gear backwards and forwards through the backlash available between it and the next gear in the timing gear train. These gear tooth separations and subsequent tooth impacts are the main mechanism for timing gear noise generation [4,5]. The energy in the tooth impacts is proportional to the impact velocity, as is the noise generated by the impact. Thus gear noise can be reduced by controlling the relative motion of the timing gears to reduce this impact energy. Various methods have been put forward for reducing gear noise. These can be divided into two groups :

REDUCED INPUT FORCE - If the forces input to the timing gear system are reduced, the gear motion will also be reduced. For a six cylinder engine, the most significant torsional forces input to the timing gear train are from the crankshaft and the fuel pump [4,9]. Methods of reducing the input forces are:
- Nodal Drive -Driving the timing gear drive from the flywheel end of the engine (adjacent to the crank T.V. node) is referred to as a nodal drive. This method has been reported to give up to 2 dBA noise reduction [3,7,10].
- Oversize Crank Damper - Fitting a larger damper than is required for crank stress reduction can minimize crank nose T.V. [4,6,7]
- The input forces from the fuel pump can be reduced by changing to a different type of fuel injection system. Common rail fuel systems probably offer the best long term prospect for reducing timing gear noise on diesel engines since they have a near constant torque requirement. However they are likely to emerge with the next generation of engines rather than as a retrofit to current engines.

REDUCED GEAR SYSTEM RESPONSE - For a given torsional force input, the torsional motion of the timing gear system can be reduced by :
- Reduced backlash - Anti-backlash gears [4-7] have been shown to give up to 1.5 dBA noise reduction.
- Gears Tooth Form - Maximum conjugacy gears [4-7, 10, 11] and precision honed gears [12] have been reported to give noise reductions in timing gear drives.
- Pendulum Absorber - These remove energy from a particular order of rotation. Pendulum absorbers have been applied to a 'lively' fueling cam [13].
- Increased Inertia - Increased timing gear inertia will reduce acceleration for a given input force. Experimental tests showed gear acceleration reduced [4-7] but modeling [8,9] predicted little effect on rattle/hammering.
- Load by constant torque devices - Adding a hydraulic pump to the end of the timing gear train significantly reduced gear acceleration pulses as the pump was loaded [5-7]. Croker et al. [8] predicted that realistic loads were sufficient to have a significant effect on gear rattle.

This paper proposes the adoption of an alternative method of gear noise reduction which falls into the second category, namely the absorption of energy by the addition of a torsional damper to the fuel injection pump shaft. This approach has been suggested before [9,14], but this paper examines a damper optimization method in detail.

MEASUREMENT

The T.V. level of the various engine shafts was measured using the signal from an inductive pick-up adjacent to a toothed wheel on the shaft. By measuring the time between consecutive zero volt crossing points in the signal, instantaneous shaft speed was calculated for every tooth pass. Further processing using FFT and integration in the frequency domain gave angular displacement in degrees.

This measurement method has two major drawbacks. Firstly, it can give no information about orders higher than half the number of teeth. This was not a problem for the work reported in this paper since it was mainly concerned with 12th order T.V., easily within the measurement range for the gears used. A second problem with the measurement method is that it fails to differentiate between true torsional vibration and lateral or axial movement of the gear being measured. For example, if a Ø150 mm gear with a 15° helical tooth form moves axially 0.1 mm, it will appear to the sensor as a torsional motion of 0.02°. If the same gear wheel moves laterally 0.1 mm in a direction 90° to the inductive pickup, it will appear to the sensor as a torsional motion of 0.04°. While errors in T.V. measurement due to these shaft movements could be significant, the T.V. levels being measured in this study were considered to be generally high enough to render the effect of the likely errors secondary. In addition, a more rigorous measurement technique would require increased installation complexity, losing the advantages of a simple portable measurement system that could quickly be applied to engines. The T.V. levels in this paper are given as 0-peak degrees displacement.

The four engines studied in this paper were six cylinder in-line DI diesel engines for truck applications. All were turbocharged, with the timing gear drives located at the front of the engine i.e. driven from the free end of the crankshaft.

The engines were installed in semi-anechoic test cells. Noise levels were measured using microphones positioned 1m from the engine. Levels are given as A weighted Sound Pressure Level (SPL) in dBs, either for an individual microphone or for an average of several microphones.

Measurements reported in this paper were all with the engines at full load. Fuel pump dampers were found to give the largest noise reductions when timing gear noise is most dominant. This usually occurs when the fuel injection loads are highest i.e. at full load.

Measurements were taken during two types of test : steady speed and ramped speed. The noise measurements were taken at steady speeds to allow averaging to take place. The T.V. measurements were all taken during ramp tests. These are a quasi steady-state test with the engine speed being slowly ramped through the operating range by adjusting the brake load. The rate of speed changes during these ramped speed tests was set to cover the engine operating speed range in between 1 and 2 minutes.

All of the rubber fuel pump dampers used in the tests described in this paper were of a conventional rubber in compression design. By changing the rubber stiffness, the resonant frequency of the damper could be changed to matched the installation.

ENGINE 1

The first engine tested was fitted with a Bosch in-line fuel pump and had a rated power of 170 kW. The initial tests on this engine were reported in [14].

For ease of testing, the fuel pump dampers were fitted onto an extension of the fuel pump shaft which passed forwards through the front cover, i.e. the damper was outside the timing gear case (figure 1). A secondary cover attached to the engine front cover was fitted over the damper and shaft extension. This controlled the direct radiation of noise from the damper itself. The rubber fuel pump damper was Ø130 diameter with an active inertia of 0.005 kg m².

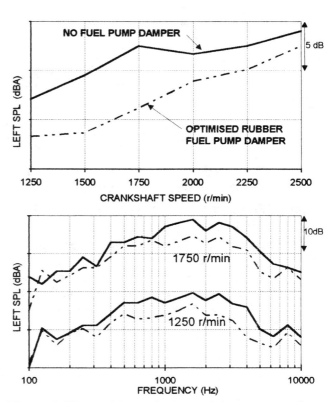

Figure 2 Effect of fitting fuel pump damper on noise of Engine 1

The addition of the fuel pump damper had a dramatic effect on the noise of engine 1. Figure 2 shows the noise reductions seen at the front microphone when the optimised rubber fuel pump damper was fitted. It shows overall noise through the speed range and 1/3 octave noise at two steady speeds. The overall noise level was significantly reduced throughout the speed range, the largest reduction being more than 6 dBA at 1750 r/min. The broad band nature of the noise reduction can be seen in the 1/3 octave spectra, levels being lowered across the whole audible frequency spectrum.

The results shown in figure 2 are for the front of the engine. The noise levels reductions measured to the left (fuel pump) and right sides of the engine were 6.4 and 4.9 dBA respectively at 1750 r/min. It was clear from the size of these noise reductions that the fuel pump damper was having a major effect on the dominant noise generation mechanism on this engine, which indicated that gear noise was by far the largest source of noise.

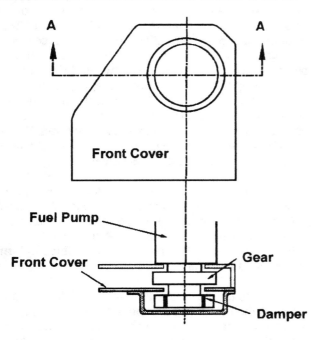

Section on A-A
Figure 1 Installation of prototype "External" fuel pump damper on Engine 1

Figure 3 shows the fuel pump T.V. with no fuel pump fitted. The fuel pump shaft has an undamped torsional resonance at 165 Hz. This is

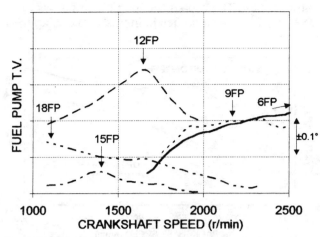

Figure 3 Fuel pump T.V., Engine 1, no fuel pump damper.

most clearly visible at 1650 r/min in the 12FP order (6E) but can also be seen at 2200 r/min in the 9FP order and at 1320 r/min in 15FP. The resonance was above maximum speed for the 6FP order and at the lowest measured speed for the 18 FP order as indicated.

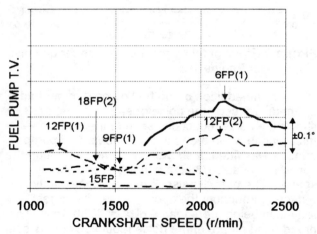

Figure 4 Fuel pump T.V., Engine 1, optimized fuel pump damper.

The torsional effect of adding the damper to the fuel pump system can be seen by comparing figure 4 with figure 3. Adding the fuel pump damper modified the system to give two torsional resonant modes: one at lower frequency and one at higher frequency than the undamped resonance. These damped modes were at 110 and 210 Hz. The lower can be seen in figure 4 at 2150 r/min in the 6FP order, 1500 r/min in the 9FP order and 1050 in the 12FP order. The higher frequency resonance can be clearly seen at 2100 r/min in the 12FP order and more faintly at 1400 r/min in the 18FP order. The relative amplitude of the two damped resonant modes was modified by changing the stiffness of the rubber used in the damper elastic member. This changed the resonant frequency of the fuel pump damper, shifting energy from one fuel pump shaft resonant mode to the other. The optimum damper tune for minimizing T.V. is when the level of the two damped peaks are the same. This was achieved on Engine 1 with a damper whose frequency was circa 160 Hz.

Figure 4 shows that the fuel pump damper was acting as a tuned device with a resonant frequency of 160 Hz, yet figure 2 shows that adding the damper gave a broad band noise reduction from 300 Hz right up to 10 kHz. The best way to explain this apparently paradoxical result is to note that the major component of timing gear noise is generated by tooth impacts which give rise to broad-band noise. By reducing the torsional motion of the fuel pump shaft the energy in the impacts between the fuel pump gear teeth and the driving gear teeth was reduced, leading to less broad band noise being generated.

Listening to the engine showed that a noise quality change was associated with the overall noise reduction. Phrases like "smoother", "less harsh" and "reduced hammering" were used to describe the noise with the fuel pump damper fitted. This supports the evidence for a reduction in tooth impact noise.

Comparing the change in overall noise shown in figure 2 with the change in single orders of fuel pump T.V. plotted in figure 3 and 4, it can be seen that there is a similarity between the change in overall noise level and the change in the 12FP order of T.V. Both overall noise and 12FP T.V. were reduced throughout the speed range by the addition of a fuel pump damper, the difference between the damped and undamped levels generally tapering as the engine speed increased. The maximum reduction occurred at around 1700 r/min. None of the other orders of fuel pump T.V. show these features. From this we can infer some relationship between the level of 12 FP T.V. and overall noise level.

A further demonstration of the relationship between 12FP order and overall noise level is shown in figure 5. This shows the engine noise levels and 12FP T.V. levels for an inertia disk and two non-optimum frequency fuel pump dampers fitted in turn to the fuel pump shaft.

The inertia of the disk was 0.005 kg m², the same as the fuel pump dampers. Its effect on fuel pump T.V. was to reduce the resonant frequency of the system. This is seen in figure 5 by the peak level of the 12FP order occurring at 1250 r/min with the inertia fitted (c.f. 1650 r/min undamped in figure 4). Reducing the speed at which the peak occurred also moved the flanks of the peak. This increased 12FP

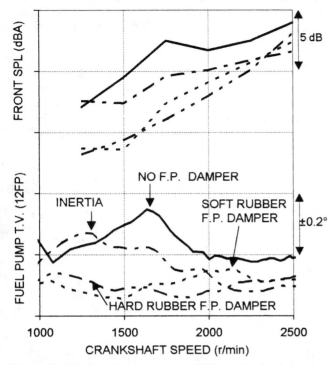

Figure 5 Comparison between 12th order fuel pump T.V. and overall noise level (Engine 1)

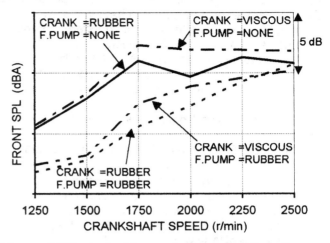

Figure 6 Effect of rubber and viscous crankshaft dampers on overall engine noise with and without fuel pump damper (Engine 1)

level at low speed and reduced it above about 1300 r/min. This low speed increase / high speed decrease characteristic was also seen in the overall noise level.

The non-optimum fuel pump dampers were fitted with elastic members made with rubber stiffness either side of optimum. This changed the relative levels of the two resonant modes. Figure 5 shows that the softer rubber (lower frequency) damper lowered the 12FP T.V. level at 1050 r/min while increasing it 2100 r/min. The harder rubber damper had the opposite effect, raising the lower speed T.V. level and reducing the higher speed peak level. Once again, the overall noise levels closely followed the trend of the 12FP T.V. This important result confirmed that there was a relationship between 12FP order T.V. and overall noise level.

Despite the large changes to the fuel pump T.V. levels achieved by fitting the different fuel pump dampers, only small changes were seen in the crankshaft T.V. levels. This was also an important result since it implied that the fuel pump loading was the dominant loading in the timing drive system. As a first approximation, the fuel pump was acting as a separate torsional system to the crankshaft.

The levels of crankshaft T.V. on Engine 1 were changed by fitting a viscous crank damper in place of the standard rubber damper. This change in crank damper type increased the high speed engine noise levels by up to 3.5 dBA (figure 6, upper curves where F.PUMP=NONE). The noise reduction due to fitting the optimized fuel pump damper was then compared for both the rubber and the viscous crankshaft damper. This is shown in figure 6 as the difference between the pair of curves for each crank damper type. Except at maximum speed, the noise reduction due to the F.P. damper is within 0.6 dBA for both types of crank damper. This means that the noise changes due to the crankshaft damper and due to the fuel pump damper are additive. This result adds further weight to the conclusion reached above that the crankshaft and the fuel pump shaft were acting as independent torsional systems.

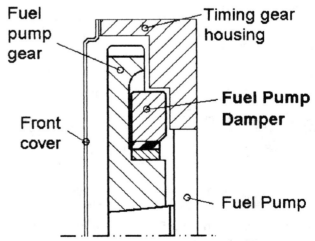

Figure 7 Installation of "Internal" fuel pump damper on Engine 1

A new rubber fuel pump damper was designed to fit inside the front cover of the engine (figure 7). It was mounted directly onto the fuel pump gear wheel, between the gear and the fuel pump. This internal damper was designed to have the same resonant frequency as the optimized external damper although the limited available space meant that the inertia had to be reduced by 20% to 0.004 kg m². When fitted to the engine, the internal damper gave the same noise reduction as the external damper.

The tests on Engine 1 led to the following conclusions:
- adding a fuel pump damper gave large broad band noise reductions.
- there was a strong relationship between the level of 12FP fuel pump T.V. and overall noise.
- the resonant frequency of the fuel pump damper was critical, e.g. a change of 25% in frequency gave 3.5 dBA less noise reduction.
- the crankshaft and fuel pump were acting as two separate torsional systems.
- the inertia of the damper was relatively unimportant, similar noise reductions being achieved with 60 and 150% inertia at constant damper frequency.

ENGINE 2

The extensive test programme that had been carried out on Engine 1 showed that large engine noise reductions could be achieved by using a fuel pump damper to minimize 12th order of fuel pump T.V. A second engine was investigated to see if this result was specific to Engine 1 or was more general. Engine 2 was a similar size to engine 1, and was again fitted with a Bosch in-line fuel pump.

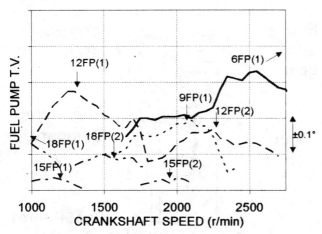

Figure 8 Fuel pump T.V., Engine 2, no fuel pump damper

Figure 8 shows the undamped fuel pump T.V. for Engine 2. There is a resonance at about 145 Hz, visible at 2000 r/min in the 9FP order, 1350 r/min in the 12FP and 1200 r/min in the 15FP order. There is also a second, higher frequency resonance at about 240 Hz, seen at 2300 r/min in the 12FP order, 1950 r/min in the 15FP order and 1600 r/min in the 18FP order.

Using the lessons learned on Engine 1, an internal fuel pump damper was added to the fuel pump gear wheel. The damper was tuned to act on the 12FP peak at 1300 r/min. Figure 9 shows the effect that this "soft" damper had on the 12FP order of

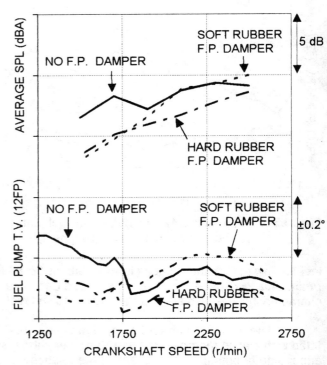

Figure 9 Effect of fuel pump dampers on 12FP T.V. and overall noise (Engine 2)

fuel pump T.V. and on the engine overall noise. The level of 12FP was reduced in the low speed range as expected. At the lowest steady speed measured, this reduction in 12FP level due to the "soft" damper also gave an overall noise reduction of 3.5 dBA. However, at speeds above 1750 r/min the level of 12FP with the fuel pump damper fitted was above the level with no damper. While the engine noise level does not exactly follow this characteristic, the trend in the overall noise level is similar to that for the 12FP T.V.

A "hard" rubber damper with a nominal frequency of 280 Hz was then tested. As would be expected, this higher frequency damper was not as effective at reducing the 12FP level around the lower resonance as the "soft" damper, but still gave lower levels than with no damper. Unlike with the "soft" damper, the T.V. level around the higher frequency resonance was also reduced. The "hard" damper gave a lower level of 12FP throughout the speed range than with no fuel pump damper. This characteristic was also seen in the overall noise level, the "hard" damper giving between 1 and 3 dBA noise reduction through the speed range except at 2500 r/min.

The results from Engine 2 were encouraging in that they backed up the main conclusion from the tests on Engine 1, namely that there is a strong link between the level of 12FP order of fuel pump T.V. and overall engine noise. This was despite the fact that Engine 2 had a more complex fuel pump torsional system with two resonances being present.

ENGINE 3

Following the tests on Engine 1 which had a strong single fuel pump shaft resonance and on Engine 2 which had two weaker resonances, a third engine was tested. Engine 3 was slightly larger than engines 1 and 2, with a rated power of 190 kW.

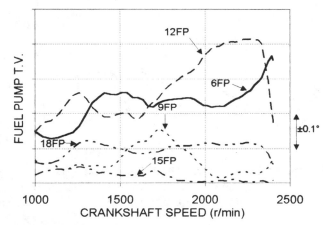

Figure 10 Fuel pump T.V., Engine 3, no fuel pump damper

The undamped fuel pump T.V. for Engine 3 is shown in Figure 10. A resonance was present at 125 Hz, seen as the rising flank in the 6FP order above 2200 r/min and as the peaks seen at 1650 r/min in the 9FP order and 1250 r/min in the 12FP order. However, examining the crank T.V. showed a crankshaft torsional resonance at 122 Hz. The dominant orders of crank T.V. were 3E, 4.5E and 6E, and it was concluded that the features noted above in the fuel pump T.V. were due to the timing gear being driven by the crankshaft motion (3E = 6FP, 4.5E=9FP, 6E=12FP). Apart from the features mentioned above which were forced by the crankshaft resonance, the fuel pump T.V. for Engine 3 did not exhibit any strong resonances.

Two different types of fuel pump damper were tested on Engine 3: a rubber damper and a viscous damper. A rubber damper is a tuned device whose damping effect is greatest at its resonant frequency. A viscous damper produces damping by shearing silicon fluid, and provides broad band damping. The size of the viscous and rubber fuel pump dampers was similar, both being fitted inside the timing gear case (similar to figure 7). The space available between the fuel pump gear and the fuel pump restricted the active inertia of the rubber damper to 0.002 kg m², and that of the viscous damper to 0.0012 kg m².

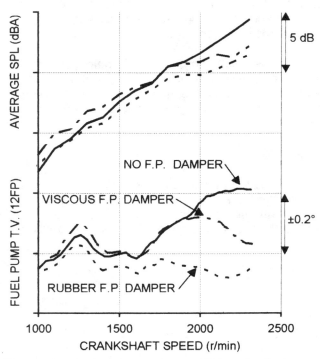

Figure 11 Comparison between 12th order fuel pump T.V. and overall noise level (Engine 3)

Figure 11 shows the comparison between 12th order of fuel pump T.V. and the overall average noise level for Engine 3 with different fuel pump dampers fitted. The curves are for : no fuel pump damper; a viscous fuel pump damper and an optimized rubber fuel pump damper. Once again a relationship was seen between the 12FP order and overall noise level.

Adding the viscous fuel pump damper increased the level of 12FP order of fuel pump T.V. up to 1500 r/min and reduced it above 1900 r/min. A similar characteristic was seen in the overall noise level. Adding the rubber damper reduced the 12FP level throughout the speed range, and again a similar characteristics was seen in the overall noise level. Thus Engine 3 also displayed the strong link between the level of 12FP order of fuel pump T.V. and the overall level engine noise found on the first two engines.

As mentioned above, the fuel pump T.V. peak in the 12FP order at 1250 r/min was forced by the 6E crankshaft T.V. level. No large change in the engine noise level could be seen at this speed, indicating that gear noise was not dominating the engine noise at this low speed.

Comparing the results for Engine 3 (figure 11) with those for Engine 1 (figures 2 and 3) showed that the relationship between the amount of reduction in 12FP order T.V. and overall noise reduction is not consistent. The amount of noise reduction will depend on the dominance of timing gear noise over other engine noise sources.

The results from Engine 3 again showed the strong link between the level of 12FP order of fuel pump T.V. and overall engine noise, despite the fact that the fuel pump shaft on Engine 3 was non-resonant.

ENGINE 4

The fourth engine tested was different from the first three engines in that it had a unit injector fuel system with the injectors were driven from the camshaft. This engine had a rated power of 260 kW.

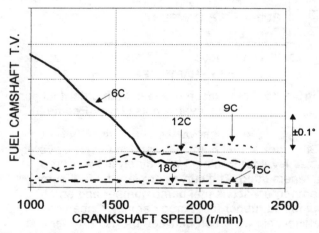

Figure 12 Camshaft T.V., Engine 4, no camshaft damper

Figure 12 shows the camshaft T.V. for Engine 4 in standard undamped condition. There appeared to be a camshaft resonance at about 170 Hz, visible in the 9FP and 12FP orders. However, the crankshaft had a resonance at approximately the same frequency, and comparing the 3E, 4.5E and 6E orders of crankshaft T.V. with the 6FP, 9FP and 12FP orders showed close similarities. This indicated that the camshaft motion on Engine 4 was being forced by the crankshaft.

The amplitude of shaft motion was similar for both shafts. Since there is a 2:1 gear ratio between the crank and camshaft, one degree of crankshaft rotation equals half a degree on the camshaft. The similar levels of T.V. indicated that the camshaft was amplifying the crankshaft motion.

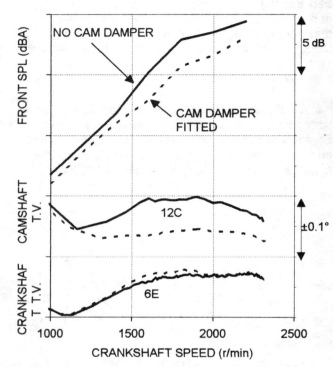

Figure 13 Effect of camshaft damper on 12C camshaft T.V., 6E crankshaft T.V. and overall noise (Engine 4)

Figure 13 shows camshaft T.V. when a viscous damper was attached to the front of the camshaft gear inside the front cover of the engine. This camshaft damper reduced the level of the 12C camshaft T.V. above 1200 r/min. The magnitude of the reduction increased as the speed rose to 1600 r/min, was relatively constant at approximately 50% from there to 2000 r/min and then tapered off.

The 6E crankshaft T.V. order is also shown in Figure 13 as the undamped motion of the camshaft was being forced by the crankshaft (6E=12C). The addition of the camshaft damper slightly increased the level of the crankshaft 6E T.V. around 1600 r/min. This increase was due to the crankshaft resonant frequency being slightly reduced by the addition of the inertia to the camshaft.

The major decrease in camshaft 12C level at the same time as a slight increase in crankshaft 6C level showed that although the camshaft motion was being forced by the crankshaft, the magnitude of the camshaft response could be reduced.

The overall noise level showed a significant reduction above 1500 r/min. The trend of the reduction was similar to reduction in the 12C order of camshaft T.V. Although the crankshaft on Engine 4 was forcing the camshaft motion, it was the magnitude of the camshaft motion and not of the crankshaft motion that was controlling the engine noise level. This again demonstrated the relationship between the twelfth order of camshaft (fuel pump shaft) and overall noise level seen on the previous three engines.

CONCLUSIONS

A practical investigation was carried out on several different six cylinder DI diesel engines. For each engine, the torsional vibration of the fuel pump shaft was modified by the addition of a torsional vibration damper. Significant noise reductions were achieved.

A strong link was seen between the level of 12th order of fuel pump torsional vibration and the overall noise level on all of the engines.

A rubber fuel pump dampers is a tuned device with a single torsional resonant frequency. The engine noise reductions achieved by fitting rubber fuel pump dampers were broad band. This indicated that the fuel pump damper reduced the impacts between the timing gear teeth.

The most important parameter for a rubber fuel pump damper was its resonant frequency. The size of the inertia ring and level of damping in the rubber were secondary factors.

For a given size, a correctly tuned rubber fuel pump damper was more effective than a viscous fuel pump damper.

Adding a viscous camshaft damper to an engine with a unit injector fuel system demonstrated similar noise reduction features to those seen on the engines with in-line fuel pump fuel systems.

Fuel pump shaft dampers with a relatively small inertia gave significant noise reductions. Their small size made it easy to engineer an installation with a minimum of engine design changes. Fuel pump shaft dampers can thus offer a cost effective solution to engine noise reduction.

Further studies are ongoing to determine why the 12FP/6E order of T.V. is important to the generation of timing gear noise on a 6 cylinder engine, and the mechanism which gives rise to it.

ACKNOWLEDGMENTS

Special thanks are due to Jeff Baker at ADAU, ISVR, Southampton who first suggested fuel pump damping to us and who carried out most of the initial test work. The author also wishes to thank his colleagues at Holset for their support in carrying out this work.

REFERENCES

[1] Zelenka P., Kriegler W., Herzog P.L. and Cartellieri W.P. "Ways Towards the Clean Heavy-Duty Diesel", SAE paper 900602

[2] Schulte H., Dürnholz M. and Wübbeke K. "The Contribution of the Fuel Injection System to Meeting Future Demands on Truck Diesel Engines", SAE paper 900822

[3] Schittler M., Bergmann H & Flathmann K. "The New Mercedes-Benz OM 904 LA Light Heavy-Duty Diesel Engine for Class 6 Trucks", SAE paper 960057

[4] Spessert B. and Ponsa R. "Investigation in the Noise from Main Running Gear, Timing Gears and Injection Pump of DI Diesel Engines" SAE paper 900012

[5] Wilhelm M., Laurin S., Schmillen K. and Spessart B. "Structure Vibration Excitation by Timing Gear Impacts" SAE paper 900011

[6] Spessert, Laurin, Schmillen and Wilhelm "Noise Excitation by the Timing Gear Train (Especially of Six Cylinder Diesel Engine", CIMAC, 19th International Congress on Combustion Engines, Florence 1991

[7] Wilhelm M. and Spessert B. "Vibration and noise excitation in the timing gear train of diesel engines", I.Mech.E. paper C432/122, 1992

[8] Croker D.M., Amphlett S.A., and Barnard A.I. "Heavy Duty Diesel Engine Gear Train Modeling to Reduce Radiated Noise", SAE paper 951315, SAE 1995 Noise and Vibration Conference.

[9] Pfeiffer F. and Prestl W. "Hammering in Diesel-Engine Driveline Systems", Nonlinear Dynamics 5 1994 (Kluwer Academic Publishers,Netherlands), Vol 5 p477-492.

[10] Spessert B., Esche D. and Giebel G. "The new Deutz FM 1012/1013 engines: Vibration, noise and cooling system concept", 4th International EAEC Conference: Vehicle and Traffic Systems Technology, June 1993, Vol. 2 p133-153.

[11] Pettitt R. and Towch B. "Noise reduction of a four litre direct injection diesel engine", I.Mech.E. paper C22/88 1988.

[12] "Honed Gears Speed Quiet Engine Assembly", OEM Design, June 1996, p51

[13] Balek S.J. and Heitzman R.C. "Caterpillar 3406E Heavy Duty Diesel Engines", SAE 932969

[14] Baker J.M., Bazeley G., Harding R., Hutton D.N. and Needham P. "Refinement Benefits of Engine Ancillary Dampers". I.Mech.E. paper C487/041/94

ABBREVIATIONS

Throughout the paper orders of engine rotation (crankshaft speed) are denoted by "E" after a number e.g. 6E is the sixth order of engine rotation. Orders of fuel pump rotation are denoted by "FP" after a number, e.g. 12FP is the twelfth order of fuel pump rotation. Similarly, orders of camshaft rotation are denoted by "C" after a number e.g. 12C is the twelfth order of camshaft rotation.

Since both the fuel pump and the camshaft of a four stroke engine rotate at half engine speed, for any given crankshaft speed the frequency of the 6E order is equal to the 12FP order and the 12C order.

971961

Attenuation of Engine Torsional Vibrations Using Tuned Pendulum Absorbers

Steven W. Shaw, Vishal Garg, and Chang-Po Chao
Michigan State Univ.

Copyright 1997 Society of Automotive Engineers, Inc.

ABSTRACT

In this paper results are presented from a study that investigates the use of centrifugally driven pendulum vibration absorbers for the attenuation of engine torsional vibrations. Such absorbers consist essentially of movable counterweights whose center of mass is restricted to move along a specified path relative to the rotational frame of reference. These devices are commonly used in light aircraft engines and helicopter rotors. The most common designs use a circular path for the absorber, tuned to a particular order of rotor disturbance, although more recent developments offer a wider variety of paths. Our goal here is to evaluate the system performance for a range of path types with different types of tuning. This analytical study is carried out for a simple mechanical model that includes a rotor and an absorber riding along a quite general path. Approximate solutions are obtained using a perturbation scheme and compared with detailed computational results. These results provide some valuable guidelines for path design considerations.

1 INTRODUCTION

Torsional vibrations in IC engines are induced primarily by torques transmitted to the crankshaft from cylinder gas pressure and from inertial loading from pistons and connecting-rods. These vibrations can propagate throughout the drivetrain and often cause fatigue and NVH difficulties. They can be addressed in many ways, the most common of which for automotive applications is the application of tuned dampers attached at the end of the crankshaft. These dampers are tuned for troublesome frequencies and reduce resonance amplification curves in the classic manner of a tuned damper (Den-Hartog [1], pp. 93-106).

In light aircraft engines it is common to employ movable counterweights as absorbers in order to address a given *order* of vibration, as opposed to a given frequency. In this way the source of vibration is counteracted at *all* engine operating speeds, not just at resonances. These devices smooth out torsional vibrations of the crankshaft by oscillating in such a manner that they produce a torque on the crankshaft that opposes the applied oscillatory torques. In order to do so they must be properly tuned such that their natural vibration frequency is the same (or nearly the same) as that of the applied torque harmonic being addressed. This is accomplished by specifying the path that the absorber center of mass must follow (rel-

ative to a rotational frame of reference attached to the crankshaft). There exist many kinematic arrangements to accomplish this tuning, as described in detail in KerWilson [2].

These dynamic balancers, known by many names and herein called *centrifugal pendulum vibration absorbers*, or CPVA's for short, have a long and successful history. The flurry of development during WWII lead to an improved understanding of engine vibration and many innovations resulted, including the wide application of CPVA's [3]. Without these devices many of the most popular engines of that era could never have been put into service. The technology developed during that time is still in use, as for example, in a recent racing engine application [4].

It should be pointed out that until around 1980 all implementations used circular absorber paths designed by using linear theory with a slight tuning adjustment to avoid certain undesirable nonlinear behaviors that can occur at moderate amplitudes of absorber motion [5]. There have been some significant developments in recent years, and these fall into two basic categories: The first group deals with improved mechanical implementations and the second group, which is of interest here, considers varied absorber paths that offer improved performance. In particular, cycloidal path absorbers have been put into use for helicopter rotors [6], and epicycloidal path absorbers have been tried in automotive engines [7, 8, 9].

In this paper we consider a systematic treatment of a wide variety of absorber paths and offer performance assessments and some design guidelines. For this purpose, an idealized model is considered consisting of a simple rotating inertia subjected to a periodic torque and fitted with a set of CPVA's tuned to a given order. In this study we assume that all absorbers move in a synchronous manner, thereby creating the effect of a single absorber mass. Using a perturbation analysis, results are obtained that provide valuable guidelines for initial designs of absorber systems, regardless of the kinematic configuration adopted, and extensive numerical results for a specific set of parameter values are given for a sample system. It should be noted that this work extends that of Lee and Shaw [10] to include the important effects of absorber mistuning.

The paper is organized as follows. Section two describes the basic system model considered. In section three the formulation, geometry and characteristics for the absorber paths under consideration are given. In section four the dynamic equations of motion are derived and results from the perturbation analysis and computations are given. In section five, based on the results of the foregoing analysis, some design guidelines for absorber paths are offered. Finally, some conclusions and directions for future work are given in section six.

2 THE SYSTEM

In this section the system configuration, which will be used as the basis for absorber path designs in the next section, is first described. A schematic diagram of the dynamical system being considered is shown in Figure 1. This idealized model consists of a simple rotating inertia with a single absorber riding along a path that is fixed relative to the rotor. The rotor has a moment of inertia I_d, while the absorber is taken to be a point mass of mass m_a (for a bifilar absorber, the absorber's rotational inertia can be simply included in I_d, as it undergoes pure translation relative to the rotor). The rotor spins at a nominal rate Ω, but undergoes torsional oscillations due to a disturbing torque $T(\theta)$, which models the fluctuating loads acting on the rotor. For present purposes it is sufficient to consider this torque to be of the form $T(\theta) = T_n \sin(n\theta)$. This captures the dominant harmonic that is encountered in many applications (e.g., $n = N/2$ for a four-stroke, N-cylinder engine).

To specify the absorber path we first define the vertex to be the point on the absorber path which is furthest from point O in Figure 1, where O represents the center of rotation. From the vertex, an arc-length variable S is assigned to specify the location of the absorber center of mass along its path; see Figure 1. (Note that for circular paths, it is simpler to use the pendulum angle, whereas S is convenient for more general paths.)

The distance from a point on the path to point O is denoted as R, and R is expressed as a function of S, that is, $R = R(S)$. The path can be uniquely determined once $R(S)$ is chosen. The value of R at the vertex is denoted as R_o, that is, $R_o = R(0)$. Thus the absorber has a nominal moment of inertia of $I_1 = m_a R_o^2$ about point O. It is assumed that the absorber path is symmetric about its vertex, so that $R(-S) = R(S)$.

With the basic system and absorber path formulation in hand, we now turn to determining the functions $R(S)$ for various types of paths.

3 ABSORBER PATHS: GEOMETRY AND CHARACTERISTICS

Circular paths are the most easily manufactured and commonly used paths. However, a circular path absorber experiences an amplitude-dependent frequency, just as does a simple swinging pendulum. In particular, the pendulum frequency of free vibration decreases as the amplitude of vibration increases.

In the area of absorber path designs, a recent developmental theme has been to move away from the circular path in order to avoid this shortcoming. The first well-reasoned use of a non-circular path appears to be the cycloidal path offered by Madden [6], which was proposed and implemented in helicopter rotors for the suppression of rotor shake. Along the same lines, the epicycloidal path of Denman [7] offered an even better solution since this type of path maintains a constant absorber oscillating frequency at all amplitudes of absorber motion. The epicycloidal path was implemented in an experimental study at Ford Motor Company, the results of which are summarized in the paper by Borowski et al. [8].

In order to account for all these path types we adopt a formulation from Denman [7] that captures all such paths in a single mathematical formulation. This is described first, after which we give the specific forms for three commonly considered paths. In each case we will take the linear tuning constant to be a free parameter in the designs considered.

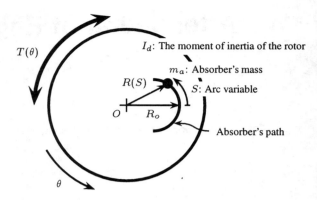

Figure 1: Schematic diagram for the CPVA and the rotor from the cross-section view.

3.1 The General Path Geometry

H.H. Denman [7], acting on a suggestion from V. J. Borowski, proposed a one-parameter family of absorber paths specified by

$$\rho^2(S) = \rho_o^2 - \lambda^2 S^2 \qquad (1)$$

where $\rho(S)$ is the local radius of curvature for the path at the location on the path specified by the variable S, and ρ_o is the radius of curvature at the vertex of the path. The path is designed to be symmetric with respect to $S = 0$; thus, $\rho^2(S)$ is an even function of S. It is known that ρ_o must be selected to be

$$\rho_o = \frac{R_o}{m^2 + 1} \qquad (2)$$

in order to tune the small-amplitude absorber oscillation frequency to be equal to $m\,\Omega$ [7]. As the amplitude of absorber motion increases, the frequency begins to be influenced by the nonlinearity parameter λ, which can take any value between zero and one. It can be shown that in the limiting cases where $\lambda = 0$ and $\lambda = 1$, this formulation describes a circle and a cycloid, respectively. In the special case when $\lambda^2 = m^2/(m^2 + 1)$, the formulation (1) describes an epicycloid with its base circle of radius $(R_o - \rho_o)$ centered at O. This is the "standard epicycloid" offered by Denman for which, as we show below, the frequency of absorber oscillation is equal to $m\,\Omega$ *at all amplitudes* when the rotor speed is constant.

3.2 Mistuning

In order to overcome the shortcomings of the circular path and expose a larger class of absorber paths, it is worthwhile to examine the effect of intentionally mistuning the linear frequency of the absorber relative to that of the disturbing torque. To account for such mistuning we assume that the absorber is tuned to order m, where

$$m = n(1 + \sigma), \qquad (3)$$

n is the order of the applied torque and σ is a measure of the mistuning built into the small amplitude absorber dynamics. Note that in practice such a path can be achieved by fixing the following ratio: $\rho_o/R_o = 1/(m^2 + 1)$.

3.3 Path Representations

Before proceeding with the derivation of the equations of motion, the absorber paths need to be represented by the function $R(S)$. This can be done by geometric calculations using the local curvature $\rho(S)$ in equation (1), as presented by Denman [7]. Since we are primarily interested in leading-order nonlinear effects, we nondimensionalize $R(S)$ and expand it in terms of S, as follows,

$$x(s) = 1 - m^2 s^2 + K_4 s^4 + \mathcal{O}(s^6) \tag{4}$$

where

$$
\begin{aligned}
K_4 &= \frac{(m^2 + 1)^2 (m^2 - \lambda^2 (1 + m^2))}{12}, \\
s &= \frac{S}{R_o}, \quad \text{and} \\
x(s) &= \left(\frac{R(R_o s)}{R_o} \right)^2.
\end{aligned}
$$

It is shown in the next section that the introduction of the parameter σ provides the desired linear mistuning by the presence of $m = n(1 + \sigma)$ in the quadratic term, and that the nonlinear character of the path is captured by K_4, which is a function of λ and m.

Note that for different values of λ between zero and one, the formulation (4) give various smooth paths including the following three commonly considered paths:

Circles:

$$
\begin{aligned}
x(s) &= 1 - \frac{2m^2 \{1 - \cos\left[(m^2 + 1) s\right]\}}{(m^2 + 1)^2} \\
&= 1 - m^2 s^2 + \frac{m^2 (m^2 + 1)^2 s^4}{12} + \mathcal{O}(s^6). \tag{5}
\end{aligned}
$$

Cycloids:

$$
\begin{aligned}
x(s) &= 1 - \left(m^2 + \frac{3}{4}\right) s^2 + \frac{\{\sin^{-1}\left[(m^2 + 1) s\right]\}^2}{4(m^2 + 1)^2} \\
&\quad + \frac{\sin^{-1}\left[(m^2 + 1) s\right] \sin\left[2\sin^{-1}(m^2 + 1) s\right]}{4(m^2 + 1)^2} \\
&= 1 - m^2 s^2 - \frac{(m^2 + 1)^2 s^4}{12} + \mathcal{O}(s^6). \tag{6}
\end{aligned}
$$

Standard Epicycloids:

$$x(s) = 1 - m^2 s^2. \tag{7}$$

4   TORSIONAL VIBRATION RESPONSE

4.1   Equations of Motion

The equations of motion are determined by Lagrange's method, under the assumptions that there is no friction or damping, gravitational effects are small compared to centrifugal effects, and the nominal speed of the rotor is Ω. Following the application of Lagrange's method and a straightforward nondimensionalization procedure we obtain the following equations of motion:

$$\ddot{s} + g(s)\ddot{\theta} - \frac{1}{2}\frac{dx}{ds}(s)\dot{\theta}^2 = 0, \tag{8}$$

$$
b_o\ddot{\theta} + \left[\frac{dx}{ds}(s)\dot{s}\dot{\theta} + x(s)\ddot{\theta} + g(s)\ddot{s} + \frac{dg}{ds}(s)\dot{s}^2 \right]
$$
$$= \Gamma_n \sin(n\theta), \tag{9}$$

where θ is the rotor angle, $\tau = \Omega t$ is the dimensionless time, $(\dot{\cdot}) = \frac{d(\cdot)}{d\tau}$ is the derivative with respect to τ, $b_o = \frac{I_d}{I_1}$ is the ratio of rotor inertia to nominal absorber inertia (typically much larger than one), $\Gamma_n = \frac{T_n}{I_1 \Omega^2}$ is the dimensionless magnitude of the applied torque, and $x(s)$ and $g(s) = \sqrt{x(s) - \frac{1}{4}\left(\frac{dx}{ds}(s)\right)^2}$ are functions specified by the path.

Note that in the unforced case, $\Gamma_n = 0$, the rotor can run at a constant speed, $\dot{\theta} = 1$ in this nondimensional form, with the absorber stuck at its vertex, $s = 0$. This is the usual solution about which one expands for deriving the response of the system.

Equations (8) and (9) represent an autonomous dynamical system (that is, it contains no terms that explicitly depend on time) in s and θ. Note that this system has several nonlinear terms, including the torque, $\Gamma_n \sin(n\theta)$. The expansion of $\sin(n\theta)$ in a perturbation analysis will involve certain technical difficulties, since θ is not restricted to be small. However, since the rotor never reverses its direction, *i.e.*, $\dot{\theta} > 0$ always, functions of τ can be expressed as functions of θ, which makes possible a switch of the independent variable from τ to θ. As a result, the system becomes non-autonomous, as the new independent variable θ appears explicitly in equation (9), and the originally highly nonlinear term $\sin(n\theta)$ is transformed to a simple external excitation term. Note that with this change of variables the angular speed of the rotor, $\dot{\theta}$, is treated as a separate dependent variable, denoted here as y. In addition, the equations of motion (8) and (9) transform to the following,

$$ys'' + \left[s' + g(s) \right] y' - \frac{1}{2}\frac{dx}{ds}(s)y = 0, \tag{10}$$

$$g(s)y^2 s'' + \left[b_o + x(s) + g(s)s' \right] yy' + \frac{dx}{ds}(s)s'y^2$$
$$+ \frac{dg}{ds}(s)s'^2 y^2 = \Gamma_n \sin(n\theta), \tag{11}$$

where $y = \dot{\theta}$ and $(\cdot)'$ denotes $\frac{d(\cdot)}{d\theta}$. Note that in this formulation the rotor angular acceleration $\ddot{\theta}$, the quantity which is a key measure of torsional vibration levels, is equal to yy', and the zero-torque base operating point is $y = 1$ and $s = 0$.

Note that these equations of motion show why the standard epicycloidal path is desirable. Consider the case of constant rotation rate, $y = 1$, (say, for a huge flywheel). For the epicycloidal path of equation (7), the resulting equation of motion for the absorber mass becomes, $s'' + m^2 s = 0$, which is simply the linear oscillator with frequency m. Note that this implies that the response of the absorber is purely harmonic of frequency m, *over all amplitudes*. This is good, but not completely satisfactory, since the harmonic absorber motion does not, unfortunately, induce a purely harmonic torque on the rotor. The absorber motion in fact generates higher-order harmonics on the rotor that may excite higher-order resonances of the system. In order to avoid these problems, one can use additional absorbers tuned to higher harmonics, or one can use a newly- developed nonlinear absorber system that generates no additional harmonics (see [11] for details).

The design variables σ and λ appear in the equations of motion through the functions $x(s)$ and $g(s)$. In order to evaluate the effectiveness of a proposed design, one must determine the steady-state solution of these dynamic equations. From these solutions one can distill measures of performance, such as the levels of torsional vibration of the rotor and the corresponding amplitudes of absorber response. While it is currently popular to attack such problems by directly using numerical methods on a computer, such a tact will not be effective here, as the system has no dissipation and will not settle down into a steady state in standard simulations. Furthermore, analytical solutions that contain the design parameters, even if only approximate, are much more convenient for design evaluations.

4.2 Method of Analysis

With the equations of motion and the path functions now available, we derive approximate solutions for the steady-state response. As no resonances or instabilities are expected to occur, a regular perturbation method is employed. To prepare for the procedure, the nondimensional torque level Γ_n is assumed to be small, and the response variables s and y are expanded in Taylor series in terms of Γ_n, as follows:

$$s(\theta) = \Gamma_n s_1(\theta) + \Gamma_n^2 s_2(\theta) + \Gamma_n^3 s_3(\theta) + \cdots, \quad (12)$$
$$y(\theta) = 1 + \Gamma_n y_1(\theta) + \Gamma_n^2 y_2(\theta) + \Gamma_n^3 y_3(\theta) + \cdots. \quad (13)$$

Note that the leading term for y is unity since $y \to 1$ as $\Gamma_n \to 0$. The y_j's for $j = 1, 2, \ldots$ represent the torsional oscillations. Likewise, note that $s \to 0$ as $\Gamma_n \to 0$. The solution is then achieved by substituting these expansions for s and y into the equations of motion, (10) and (11), expanding in Γ_n, collecting terms of the same order in Γ_n, and solving the resulting linear differential equations for the periodic solutions of the s_j's and y_j's for $j = 1, 2, 3, \ldots$. This can be done for a general path, yielding results that depend on the design parameters. The computer-assisted symbolic manipulation program $Mathematica$ was used extensively in carrying out these calculations.

4.3 Analytical Results

We take the absorber path to have the general form described by equation (4) and then specialize to the cases of interest by assigning specific values for σ and K_4. Following the methodology described above, the leading order terms in the Γ_n-expansion, that is, the linearized system equations of motion, are found to be

$$s_1''(\theta) + n^2 s_1(\theta) + y_1'(\theta) = 0, \quad (14)$$
$$s_1''(\theta) + (b_o + 1)y_1'(\theta) = \sin(n\theta), \quad (15)$$

which render the following linearized system response,

$$s_1(\theta) = -\left(\frac{\sin(n\theta)}{m^2(1+b_0) - b_0 n^2} \right)$$

$$y_1(\theta) = -\left(\frac{(m^2 - n^2)\cos(n\theta)}{n(m^2(1+b_0) - b_0 n^2)} \right). \quad (16)$$

Note that when the absorber path is tuned exactly at $m = n$, this solution reduces to

$$s_1(\theta) = -\frac{\sin(n\theta)}{n^2}, \quad y_1(\theta) = 0, \quad (17)$$

which is precisely the desired system behavior since the rotor runs at a constant speed, even in the presence of the applied torque. At this level of approximation the absorber dynamics exactly counteract the applied torque. If one can ensure that vibrations amplitudes remain small, the above design without mistuning (that is, $m = n$ or, equivalently, $\sigma = 0$) represents a good design, and it is independent of λ (or, equivalently, of K_4) since nonlinearities do not come into play. In such a case, circular paths can be used as well as any other, so long as the curvature at the vertex is properly chosen. In many applications, however, the rotor is subjected to a wide range of torques and operating speeds and it is impossible to keep amplitudes small in all cases. To do so would require a small dimensionless disturbance torque, implying one of the following situations: a high rotation speed — which is not always possible; or a large I_1, which implies either a large absorber mass m_a or a large moment arm R_o, which are undesirable due to weight, responsiveness, and spatial considerations. (These results follow from the fact that the normalized torque amplitude is given by $\Gamma_n = T_n/(m_a R_o^2 \Omega^2)$.) Based on these shortcomings, the system dynamics for moderate amplitudes of vibration are investigated in the following in order to find paths that offer desirable performance over a wide range of amplitudes.

Higher order terms for s and y are determined by straightforward application of the perturbation procedure, yielding

$$\begin{aligned} s = \ & \Gamma_n s_1(\theta) + \Gamma_n^2 \alpha_0 \sin(2n\theta) \\ & + \Gamma_n^3(\alpha_1 \sin(n\theta) + \alpha_2 \sin(3n\theta)) + \cdots, \quad (18) \\ y = \ & 1 + \Gamma_n y_1(\theta) + \Gamma_n^2 \alpha_3 \cos(2n\theta) \\ & + \Gamma_n^3(\alpha_4 \cos(n\theta) + \alpha_5 \cos(3n\theta)) + \cdots, \quad (19) \end{aligned}$$

where the α's are listed in the Appendix. Using these results, the approximate form for the the level of torsional vibration, as expressed by the rotor angular acceleration $\ddot{\theta}$, is found to be

$$\begin{aligned} \ddot{\theta} = \ & yy' \\ = \ & \frac{\Gamma_n(m^2 - n^2)\sin(n\theta)}{m^2 + b_0 m^2 - b_0 n^2} \\ & + \frac{\Gamma_n^2(-3(m^4 n + m^2 n^3))\sin(2n\theta)}{2\alpha_{01}} \\ & - n\Gamma_n^3(\alpha_4 \sin(n\theta) + 3\alpha_5 \sin(3n\theta)) + \cdots. \quad (20) \end{aligned}$$

Note that the n and $2n$ harmonics in $\ddot{\theta}$ are independent of K_4, that is, they are not affected by the nonlinear nature of the path. This has the interesting implication that one cannot avoid the generation of secondary torque harmonics, even by clever selection of the nonlinear nature of the path. Also note that, as expected, all orders of $\ddot{\theta}$ are affected by m.

4.4 Compuational Results

Results from numerical computations are obtained in order to verify the analytical results and determine thier limitations. As stated above, straightforward simulations will not work in the present case

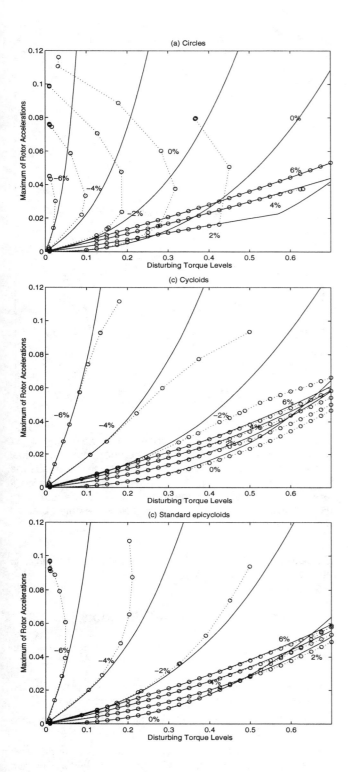

Figure 2: The amplitude of rotor acceleration, $\max(\ddot{\theta})$, versus the disturbing torque level, Γ_n, for $n = 2$ and $b_o = 6.017$: (a) circles, (b) cycloids, and (c) epicycloids; with mistuning levels ranging from -6% to +6%; solid lines are from equation(20); "o" are from the numerical results.

since no system dissipation is present and thus the system will never settle into a steady-state. Therefore, an approach is employed wherein numerical root solving methods are used to determine the initial conditions corresponding to periodic solutions of the differential equations of motion, (10) and (11). This is implemented using a standard optimization routine programmed to carry out the desired task. In this way, numerical results for the periodic steady-state responses were obtained over a wide variety of parameter conditions. The output of these results can be used to determine torsional vibration levels, absorber amplitudes, and to plot response traces versus the rotor angle. This method and results from more examples can be found in [12].

For some parameter conditions the convergence of this method was not reliable. This appears in some of the response curves to follow, where isolated numerical data points are slightly irregular. The overall response curves are quite smooth in spite of these points, and the trends of interest are obvious.

4.5 An Example

The analytical results are shown and compared with numerical results obtained for a sample system, a four-stroke, four-cylinder, in-line engine, for which the primary excitation torque is second order, $n = 2$, and the inertia ratio has a value of $b_o = 6.017$ (these data are borrowed from [7]).

Figure 2 shows results for the peak value of angular acceleration, $\max(\ddot{\theta})$, as a function of torque amplitude for three different paths and a variety of σ values. (Note that here σ is given in terms of percentages.) In this figure, results from the analysis are shown as solid curves while the numerically obtained data points are shown as open circles. In a given design one hopes to achieve a low value of $\max(\ddot{\theta})$ over the torque range of interest, while maintaining a moderate amplitude for the absorber motion.

The following general observations are made regarding the results shown in Figure 2. First, note that the analytical results are generally very good, except when the response curves bend back on themselves. This is an important point to understand. As the torque level is increased, the system response initially grows smoothly, but if such a bend is encountered, a sudden increase in the response level will occur, in fact to a branch of solutions well above those shown in the figures. This nonlinear *jump* in the response is the reason that one cannot use perfectly tuned circular path absorbers, and it is why mistuning is intentionally built into the absorber path [5]. Note that the perfectly tuned circle fails by a jump at a torque level of about 0.3, whereas a circle with +6% mistuning is well behaved out to at least 0.7.

It is also important to note in Figure 2 that for those cases in which the response is nearly a straight line, the linear response is satisfactory. However, it is not possible using linear theory to determine where the linear result breaks down, and the breakdown can be dramatic.

Figures 3 and 4 show analytical and numerical system responses in terms of absorber motions and the corresponding rotor accelerations over one excitation period for a torque level of $\Gamma_n = 0.5$. The samples shown are for a circular path with mistuning of $\sigma = 4\%$, and for a perfectly tuned cycloidal path. These paths are among those corresponding to the satisfactory designs discussed in section 5.1 below.

It can be seen from Figures 3(a) and 4(a) that the absorber responses are all quite close to pure harmonic motion. It is also seen from Figure 4 that the perturbation approach gives satisfactory pre-

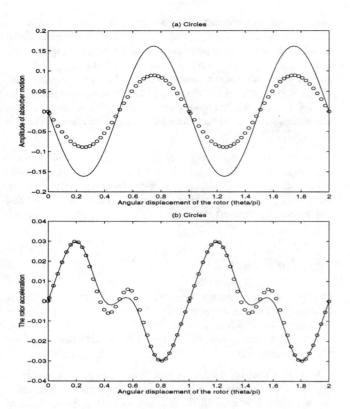
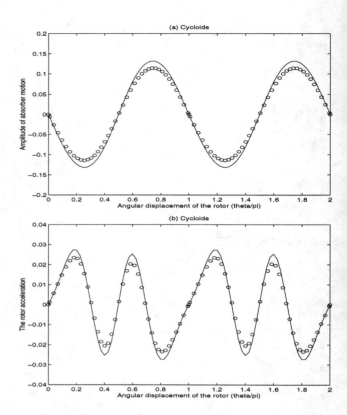

Figure 3: A sample system response for $n = 2$ and $b_o = 6.017$ at a torque level of $\Gamma_n = 0.5$. The absorber path is circular with mistuning of $\sigma = +4\%$. (a) The absorber motion $s(\theta)$ versus the displacement of the rotor, θ/π. (b) The rotor acceleration $\ddot{\theta}$ versus the displacement of the rotor, θ/π. Solid lines are from the analytical perturbation results; "o" are data from the numerical method.

Figure 4: A sample system response for $n = 2$ and $b_o = 6.017$ at a torque level of $\Gamma_n = 0.5$. The absorber path is cycloidal with zero mistuning. (a) The absorber motion $s(\theta)$ versus the displacement of the rotor, θ/π. (b) The rotor acceleration $\ddot{\theta}$ versus the displacement of the rotor, θ/π. Solid lines are from the analytical perturbation results; "o" points are data obtained from the numerical method.

dictions for the absorber motion and the rotor acceleration for the cycloidal path. However, for the circular path, Figure 3(a) shows that the absorber response is not well predicted by the perturbation approach. This is due to the fact that the perturbation method breaks down when the system is close to the point where a "*jump*" occurs. To solve this problem, a method will be introduced in section 5.2 that provides a better prediction for the amplitudes of the absorber motions. Figure 3(b) shows the rotor acceleration for the circular path, where it can be seen that despite the aforementioned shortcomings, the perturbation approach still renders an accurate peak value for $\ddot{\theta}$, even at this large torque level where the absorber response is poorly approximated.

5 DESIGN GUIDELINES

The selection of the path parameters (m, K_4) (or, equivalently, (σ, λ)) in order to best reduce torsional vibration levels over a range of torque amplitudes is discussed in section 5.1. The limitations imposed by the jump phenomenon are then considered in section 5.2.

Note that the results and conclusions drawn here are for the particular values of n and b_o being considered. However, the analytical results are quite reliable and are formulated in terms of general parameter values. Therefore, they can be used in design studies to select the best path for any values of b_o and n.

5.1 Minimizing Torsional Vibration

Based on the results obtained, the following general observations can be made:

- Absorbers generally perform better (*i.e.*, $\max(\ddot{\theta})$ is smaller) with positive mistuning than with negative mistuning.
- For very small torque levels, the results are best for zero mistuning, and are essentially independent of the path type.
- However, with zero mistuning, the circle fails at a torque level of about 0.3, whereas the cycloid and epicycloid survive and perform well out beyond 0.7.
- A satisfactory overall performance is offered for a circular path with $= 4\%$ mistuning.
- The other paths also perform well for this level of mistuning, but will require more sophisticated machining methods to implement.

Based on the relationship between angular acceleration, $\ddot{\theta}$, and the path parameters, (m, K_4) in equation (20), the goal of minimizing the level of $\ddot{\theta}$ over a range of torque levels could be pursued using an optimization approach, as follows. For the values of b_o and n of interest, an objective function is formulated that involves a measure of angular acceleration weighted in some manner over the range of torque levels of interest. Using this objective function and the two path parameters as design variables, a minimum is sought. Such an approach was carried out in [10] for the limited case of perfectly tuned paths ($\sigma = 0$) at a fixed torque amplitude. However, it must be noted that Figure 2 indicates that the sensitivity of the results to changes in the design parameters near a desirable solution may not be significant. For the present example several paths offer performance that is close to the optimal. Thus, such a detailed optimization study may not be worth the effort involved.

5.2 Avoidance of Jumps

The jump phenomenon has been observed and documented in analytical studies and in real applications for absorbers using circular paths [5].

This phenomenon occurs when the amplitude of the absorber motion grows rapidly (even discontinuously) with respect to the torque amplitude Γ_n. Such jumps are encountered as the torque level is increased beyond the points of vertical tangency in the nonlinear response curves. For example, refer to the zero mistuned circular paths at around $\Gamma_n = 0.33$ in Figure 5. These jumps can spell disaster for a system, since the response beyond such points results in a torque generated by the absorber that is in phase with the applied torque, causing a dramatic increase in torsional vibration amplitude.

The following jump analysis also provides additional important information relative to the perturbation results, and this is crucial if one is to use the analytical results in a design process. Specifically, the regular perturbation results as derived above will generally give smooth, single-valued response curves as a function of torque amplitude. However, these have a limited range of validity, and the most important limitation is the jump encountered. (Recall that the approximations become less valid as the jump is approached.) A bound on the torque range over which the steady-state perturbation results are valid can be approximated by determining the torque level at which a jump occurs.

Mathematically speaking, a jump occurs when $\frac{ds_a}{d\Gamma_n} \to \infty$, where s_a is the amplitude of the absorber motion. However, jumps do not occur for all paths, as the absorber amplitude is limited by the fact that $g(s)$ must be real. (At points where $g(s) = 0$, the absorber point mass reaches a singular point, specifically a cusp, in the path. In practice, absorbers can not achieve such operating levels.) Let s_J and s_g be the absorber amplitude at which the jump occurs and the amplitude at which $g(s) = 0$, respectively. Then a jump can occur only if $s_J < s_g$. This condition is very difficult to check, as estimates of s_J are not easily obtained.

We can, however, estimate s_a by noting that in this undamped case s is a series of sine harmonics only, and since the order n sine harmonic is dominant, s reaches its maximum (or minimum) at $\theta = \frac{\pi}{2n}$. This assumption is reasonable since the fundamental harmonic for s, which corresponds to order n, is quite close to the linear solution of the system, while the remaining harmonics are generated by nonlinear terms, and these remain relatively small. (This is verified by observations of the simulations.) With this assumption, the amplitude s_a is affected only by the coefficients of the *odd* order sine harmonics, since all even order sines vanish at $\theta = \frac{\pi}{2n}$. With the

order m absorber paths given, the amplitude s_a achieved in this way is a polynomial in Γ_n and contains only odd powers of Γ_n (as seen from equation (21) below). In order to calculate s_J, the relationship between s_a and Γ_n is inverted so that Γ_n is approximated by a series expansion in terms of s_a. Then, the value of s_J is taken as the value of s_a at which $\frac{d\Gamma_n}{ds_a} = 0$. At this jump point, the corresponding torque amplitude, denoted as $(\Gamma_n)_J$, is obtained by substituting s_J into the relationship between Γ_n and s_a. This calculation is outlined below.

Using the assumptions described above, we first obtain the approximation

$$s_a = \frac{\Gamma_n}{m^2 (1 + b_0) - b_0 n^2} + \alpha_5 \Gamma_n^3 + \cdots, \qquad (21)$$

where

$$\alpha_5 = \alpha_2 - \alpha_1. \qquad (22)$$

The inverse relationship between s_a and Γ_n is obtained by expressing Γ_n as a Taylor series expansion in terms of s_a with unknown coefficients, expanding in terms of s_a, matching terms of equal power in s_a, and solving for the unknown coefficients. This yields

$$\begin{aligned} \Gamma_n &= \left(m^2 (1 + b_0) - b_0 n^2 \right) s_a \\ &\quad - \left(m^2 (1 + b_0) - b_0 n^2 \right)^4 \alpha_5 s_a^3 + \cdots. \end{aligned} \qquad (23)$$

The jump occurs when $\frac{d\Gamma_n}{ds_a} = 0$, that is, $\frac{d\Gamma_n}{ds_a}(s_J) = 0$. It is thus determined that the jump will occur when the absorber amplitude reaches a magnitude given approximately as follows,

$$s_J \approx \frac{1}{\sqrt{3\alpha_5 \left(m^2 (1 + b_0) - b_0 n^2 \right)^3}}. \qquad (24)$$

The corresponding jump torque $(\Gamma_n)_J$ is estimated by using this amplitude in the expression for the torque amplitude given above, yielding

$$(\Gamma_n)_J \approx \frac{2}{3\sqrt{3\alpha_5 \left(m^2 (1 + b_0) - b_0 n^2 \right)}}. \qquad (25)$$

The sign of the parameter α_5 is of particular interest in determining whether or not a jump will occur, since it appears in a square root in the above expressions. Note that in order for a jump to occur, both $g(s_J)$ and s_J must be real, where

$$g(s) = \sqrt{1 - m^2 (m^2 + 1) s^2 + K_4 (4m^2 + 1) s^4 + \cdots}. \qquad (26)$$

These results are useful in determining whether a given path is susceptible to the jump behavior over a given torque range, a crucial issue when evaluating proposed designs.

Specific results for each of the paths under consideration for the system with parameter values $b_o = 6.017$ and $n = 2$ are shown in Figure 5. This demonstrates the accuracy of the modified analytical solutions when compared against those obtained by accurate numerical techniques. It is also important to note that these results do not break down at the jump points, and are thus valid over a larger operating range. It should also be noted that the angular acceleration levels predicted by this modified analytical approach similarly match the numerical results shown in Figure 2.

Some specific conclusions regarding jumps for the three path types are now described.

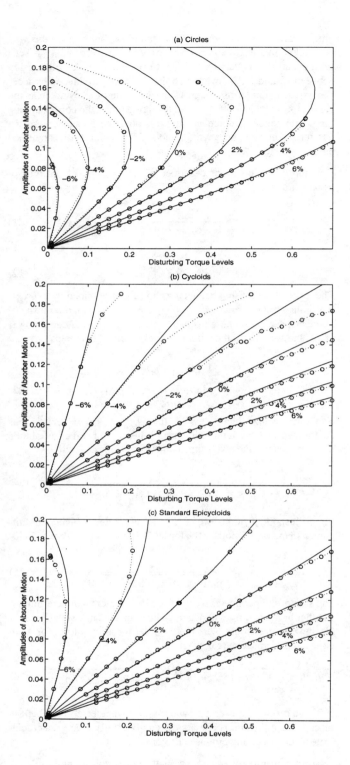

Figure 5: The amplitude of absorber motion s_a versus the disturbing torque level Γ_n for $n = 2$ and $b_o = 6.017$: (a) circles, (b) cycloids, and (c) standard epicycloids; with mistuning values from from -6% to +6%; solid lines are from equation (23); "o" are data from numerical results.

Circles:

With $K_4 = \frac{m^2(m^2+1)^2}{12}$, the critical coefficient α_5 can be computed using the expressions given in the Appendix. The relationship between Γ_n and s_a for various mistuning levels can be then predicted using equation (23). Figure 5(a) shows this relationship and a comparison with numerically obtained results. In each case the jump torque level $(\Gamma_n)_J$ is the level at which the curve has a vertical tangency, and s_J is the corresponding value of s_a. It is seen that for 0% mistuning, the analysis predicts $s_J = 0.124$ and $(\Gamma_n)_J = 0.330$, which are very close to the values obtained by numerical methods. One can also easily confirm that $g(s_J) > 0$ for this case. Thus, as is well known, a jump occurs for the circular path with zero mistuning. In fact, using the analysis presented here, one can estimate whether or not a jump will occur for any level of mistuning, and the jump points can be accurately approximated.

Cycloids:

With $K_4 = -\frac{(m^2+1)^2}{12}$, the parameter α_5 can be computed using the expressions given in the Appendix. For $-6\% \leq \sigma \leq +6\%$, jumps do not occur, as α_5 is either negative or small positive, which renders s_J or $g(s_J)$ not real. This result is confirmed in Figure 5(b), where no vertical tangencies occur for any of the response curves. This is backed up by the numerically obtained results.

Standard Epicycloids:

With $K_4 = 0$, the parameter α_5 can be computed using the expressions given in the Appendix. Figure 5(c) shows that jumps occur only when the mistuning is significantly negative. In light of the poor performance they offer, such paths would never be used in practice; see Figure 2.

6 CONCLUSIONS

6.1 Summary

First, one can see from Figures 2 and 5 that no matter which path is considered, negative mistuning leads to larger rotor accelerations and are more likely to produce jumps. Therefore, negative mistuning should be disregarded while designing absorber paths. Second, in the range of small disturbing torque levels ($\Gamma_n \leq 0.4$), cycloids and epicycloids with zero mistuning offer excellent absorber performance. However, in the range of larger torques, $\Gamma_n \geq 0.4$, some level of positive mistuning is required in order to achieve good performance for some paths. This is absolutely crucial for circles, and mildly so for epicycloids.

6.2 Limitations and Future Work

A preliminary analysis is presented in this study in order to predict absorber performance over a range of applied torques by utilizing an idealized mechanical model. This model has many limitations that need to be addressed, as the ignored effects may be important. In this light, one should realize that the results presented here should be used as rough guidelines, and one should not use them for fine tuning of the path parameters. Here we describe just a few of the model limitations and what is known about them. We also offer some additional comments about another absorber system and some speculation about some potential added benefits of using such absorbers.

Dissipation is neglected in the model. Damping generally reduces the performance of absorbers in terms of rotor acceleration, but it also generally causes a slightly increased torque range for acceptable performance. An improved model that accounts for this could be used to study these aspects and to compute the dynamic stability of the steady state response of the system.

Also, the system model should be generalized to include the dynamics of a set of N absorbers, even if they are identical. Such sets of absorbers are always used for balancing and/or due to the restricted space around the rotor. A recent study has shown that multi-absorber systems can behave quite differently than this idealized model. In particular, the absorbers do not always respond in a synchronous manner and the amplitude of one absorber in the set may become quite large beyond a certain torque level. This localized motion limits the torque range in a way not predicted in the current work, or by linear theory. See the recent papers [13, 14] for details.

Bifilar absorbers employ rollers whose dynamic effects are not captured by the present model. They can roll and/or slip, depending on operating and lubrication conditions, and have a rather unpredictable effect on system performance. However, their effects will be small if their inertia is small relative to that of the absorber mass. Denman includes a detailed analysis of roller dynamics in his study [7].

Since a rotor is not rigid, future models should account for torsional flexibility. While the rotor is generally quite stiff when compared to the absorbers, even a small amount of relative motion between absorber paths may have a significant effect on the system performance. This may be especially important when several absorbers, and the applied torques, are spaced out along a crankshaft.

Likewise, in practice the applied torque is much more complex than the simple model considered here. When multiple harmonics are encountered, one can often get satisfactory results by simply designing an absorber path for each individual harmonic, ignoring in turn the other harmonics. The paper by Borowski et al.[8] describes the success of such an approach for a four-cylinder engine. Similarly, the paper of Lee and Shaw [9] offers a more comprehensive dynamic model for an IC engine. It includes detailed effects for the inertia of engine components, friction in bearings, and gas pressure in cylinders. One could (and should) test proposed absorber paths using such a model before building hardware.

Systematic experiments need to be conducted to explore the utility of these path types. In the list of references below, only the work of Albright et al. [4] offers quantitative results from experiments, and this was done only for a specific engine and a particular set of circular path absorbers.

Since these absorbers smooth out irregularities in the torsional system response, one can reconsider the use of flywheels that are currently used for this purpose. While flywheels are mechanically simpler, they significantly reduce system responsiveness, and this may be a critical consideration for high-performance applications.

It should be noted that the recently designed subharmonic absorber system described in Lee et al. [11] offers *absolutely perfect performance* by the measures used in this study. Its main limitation is that it does require a bit more space for absorber movement, and it has yet to be experimentally tested. However, for a system model with no damping, it renders the rotor acceleration to be *absolutely zero* over a large torque range using only two absorbers.

ACKNOWLEDGEMENT

The first author would like to acknowledge the fruitful and stimulating interactions he has had with several people over the years on this topic, including Professor Harry Denman of Wayne State University, Mrs. Victor Borowski and Al Berger of Ford Motor Company, Professor Don Cronin of the University of Missouri-Rolla, Dr. Cheng-Tang Lee, now working in Taiwan, and the current co-authors. This work has been partially supported by the National Science Foundation.

References

[1] J. P. DEN HARTOG. Tuned pendulums as torsional vibration eliminators. In *Stephen Timoshenko 60th Anniversary Volume*, pages 17–26. The Macmillan Company, 1938.

[2] W. KER WILSON. *Practical Solution of Torsional Vibration Problems*, chapter XXX. Volume IV, Chapman and Hall Ltd, London, London, 3rd edition, 1968.

[3] G. WHITE, 1995. *Allied Aircraft Piston Engines of World War II*. SAE Inc., Warrendale, PA.

[4] M. ALBRIGHT, T. CRAWFORD, and F. SPECKHART. Dynamic testing and evaluation of the torsional vibration absorber. volume II, Engines and Drivetrains, P-288, pages 185–192. Motor Sports Engineering Conference Proceedings, 1994.

[5] D. E. NEWLAND, 1964, *ASME Journal of Engineering for Industry* **86**, 257–263. Nonlinear aspects of the performance of centrifugal pendunlum vibration absorbers.

[6] J. F. MADDEN, 1980, *United States Patent No. 4218187*. Constant frequency bifilar vibration absorber.

[7] H. H. DENMAN, 1992, *Journal of Sound and Vibration* **159**, 251–277. Tautochronic bifilar pendulum torsion absorbers for reciprocating engines.

[8] V. J. BOROWSKI, H. H. DENMAN, D. L. CRONIN, S. SHAW, J. P. HANISKO, L. T. BROOKS, D. A. MILULEC, W. B. CRUM, and M. P. ANDERSON. Reducing vibration of reciprocating engines with crankshaft pendulum vibration absorbers. The Engineeering Society for Advanced Mobility Land, Sea, Air and Space, 1991. SAE Technical Paper Series 911876.

[9] C.-T. LEE and S. W. SHAW. Torsional vibration reduction in internal combustion engines using centrifugal pendulums. In *ASME Design Engineering Technical Conference*, volume DE-Vol. 84-1, Volume 3-Part A, pages 487–492, 1995.

[10] C.-T. LEE and S. W. SHAW. A Comparative Study of Nonlinear Centrifugal Pendulum Vibration Absorbers. In *Nonlinear and Stochastic Dynamics, ASME*, volume AMD-Vol. 192/DE-Vol. 78, pages 91–98, 1994.

[11] C.-T. LEE, S. W. SHAW, and V. T. COPPOLA, 1997, to appear, *ASME Journal of Vibration and Acoustics* . A subharmonic vibration absorber for rotating machinary.

[12] V. K. GARG, 1996. Effects of mistuning on the performance of centrifugal pendulum vibration absorbers. Master's thesis, Department of Mechanical Engineering, Michigan State University, East Lansing.

[13] C.-P. CHAO, C.-T. LEE, and S. W. SHAW, 1997, to appear, *Journal of Sound and Vibration* . Non-unison dynamics of multiple centrifugal pendunlum vibration absorbers.

[14] C.-P. CHAO, S. W. SHAW, and C.-T. LEE, 1997, to appear, *ASME Journal of Applied Mechanics* . Stability of the unison response for a rotating system with multiple centrifugal pendunlum vibration absorbers.

APPENDIX

The α's are given by

$$\alpha_0 = \frac{-2m^4 - 2b_0 m^4 + 5b_0 m^2 n^2 - 3b_0 n^4}{2n\alpha_{01}}$$

$$\alpha_3 = \frac{-m^6(1+b_0) + m^4 n^2 (5+6b_0)}{4n^2 \alpha_{01}}$$
$$+ \frac{-m^2 n^4 (-2+9b_0) + 4b_0 n^6}{4n^2 \alpha_{01}}$$

$$\alpha_{01} = \left(m^2 + b_0 m^2 - 4b_0 n^2\right)\left(m^2 + b_0 m^2 - b_0 n^2\right)^2$$

$$\alpha_{11} = 1 + 2b_0 + b_0^2$$

$$\begin{aligned}
\alpha_1 = & \left((-4\alpha_{11})m^8 + (-12K_4\alpha_{11})m^2 n^2\right. \\
& + \left(7 + 34b_0 + 27b_0^2\right)m^6 n^2 + (3\alpha_{11})m^8 n^2 \\
& + \left(48b_0 K_4 + 48b_0^2 K_4\right)n^4 \\
& + \left(-8 - 63b_0 - 71b_0^2\right)m^4 n^4 \\
& + \left(-4 - 19b_0 - 15b_0^2\right)m^6 n^4 \\
& + \left(28b_0 + 68b_0^2\right)m^2 n^6 + \\
& \left. \left(16b_0 + 12b_0^2\right)m^4 n^6 - 20b_0^2 n^8\right) / \\
& \left(8n^2 \alpha_{01}\left(m^2 + b_0 m^2 - b_0 n^2\right)^2\right)
\end{aligned}$$

$$\begin{aligned}
\alpha_2 = & \left((-8\alpha_{11})m^8 + (4K_4\alpha_{11})m^2 n^2 + \right. \\
& \left(-1 + 66b_0 + 67b_0^2\right)m^6 n^2 \\
& + (-\alpha_{11})m^8 n^2 + \left(-16b_0 K_4 - 16b_0^2 K_4\right)n^4 \\
& + \left(-59b_0 - 167b_0^2\right)m^4 n^4 + \left(4 + 9b_0 + 5b_0^2\right)m^6 n^4 \\
& \left. + \left(-12b_0 + 168b_0^2\right)m^2 n^6 \left(-16b_0 - 4b_0^2\right) - 60b_0^2 n^8\right) / \\
& \left(8n^2 \alpha_{01}\left(m^2 + b_0 m^2 - b_0 n^2\right)\left(m^2 + b_0 m^2 - 9b_0 n^2\right)\right)
\end{aligned}$$

$$\begin{aligned}
\alpha_4 = & \left((-\alpha_{11})m^{10} + \left(8b_0 + 8b_0^2\right)m^8 n^2\right. \\
& + \left(12K_4 + 12K_4 b_0\right)m^2 n^4 \\
& + \left(3 - 12b_0 - 22b_0^2\right)m^6 n^4 \\
& + (-4 - 4b_0)m^8 n^4 - 48b_0 K_4 n^6 \\
& + \left(-10 + 20b_0 + 28b_0^2\right)m^4 n^6 + (4 + 20b_0)m^6 n^6 \\
& \left. + \left(-14b_0 - 17b_0^2\right)m^2 n^8 - 16b_0 m^4 n^8 + 4b_0^2 n^{10}\right) / \\
& \left(8n^3 \alpha_{01}\left(m^2 + b_0 m^2 - b_0 n^2\right)^2\right)
\end{aligned}$$

$$\begin{aligned}
\alpha_5 = & \left((-\alpha_{11})m^{10} + \left(12 + 28b_0 + 16b_0^2\right)m^8 n^2\right. \\
& + \left(-12K_4 - 12b_0 K_4\right)m^2 n^4 \\
& + \left(15 - 60b_0 - 78b_0^2\right)m^6 n^4 \\
& + (4 + 4b_0)m^8 n^4 + 48b_0 K_4 n^6 \\
& + \left(2 + 24b_0 + 148b_0^2\right)m^4 n^6 + (-12 - 28b_0)m^6 n^6 \\
& \left. + \left(\left(34b_0 - 121b_0^2\right)m^2 n^8 + 48b_0 m^4 n^8 + 36b_0^2 n^{10}\right)\right/ \\
& \left(8n^3 \alpha_{01}\left(m^2 + b_0 m^2 - b_0 n^2\right)\left(m^2 + b_0 m^2 - 9b_0 n^2\right)\right)
\end{aligned}$$

971962

Influence of Tensioner Friction on Accessory Drive Dynamics

M. J. Leamy, N. C. Perkins, and J. R. Barber
University of Michigan

R. J. Meckstroth
Ford Motor Co.

Copyright 1997 Society of Automotive Engineers, Inc.

1   ABSTRACT

Belt drives have long been utilized in engine applications to power accessories such as alternators, pumps, compressors and fans. The first belt drives consisted of one or more V-belts powering fixed-centered pulleys and were pretensioned by statically adjusting the pulley center separation distances. In recent years, such drives have been replaced by a single, flat, 'serpentine belt' tensioned by an 'automatic tensioner.' The automatic tensioner consists of a spring-loaded, dry friction damped, tensioner arm that contacts the belt through an idler pulley. The tensioner's major function is to maintain constant belt tension in the presence of changing engine speeds and accessory loads. The engine crankshaft supplies both the requisite power to drive the accessories as well as the (unwanted) dynamic excitation that can adversely affect the accessories and the noise and vibration performance of the belt.

The objective of this study is to model the rotational response of each accessory element to harmonic excitation from the crankshaft. This system model includes a nonlinear component model of the tensioner that captures the effect of tensioner arm dry friction. A numerical scheme is created to subsequently integrate the equations of motion and solve for the response of each pulley and the tensioner arm. Computed results illustrate tensioner stick/slip motions, sub- and super-harmonic responses, and secondary resonances; phenomena that cannot be captured in previous linear models of accessory drive dynamics.

2   INTRODUCTION

Automotive applications of belt drives have evolved in the past ten to fifteen years from drives employing multiple V-belts driving fixed-centered pulleys to a single drive employing a flat *serpentine* belt and an automatic tensioning device, as illustrated in Figure 1. The serpentine belt couples the entire front end accessory drive (FEAD) system, and powers various engine accessories including the alternator, water pump, power steering pump, etc. The auto-

matic tensioning device, referred herein as the *tensioner*, consists of an idler pulley pinned to a rigid arm which pivots about a fixed point. The motion of the arm is resisted by a coil spring and a dry friction damper. The tensioner attempts to maintain constant belt tension in the slack side belt span between the crankshaft pulley and the tensioner pulley, over a wide range of belt operating speeds and accessory operating torques [1]. Near constant tension is achieved by tensioner arm rotations which take up belt slack in the system. The frictional damping is realized at two mating surfaces, one stationary with the hub and one rotating with the arm, that are pressed together under the prescribed normal pressure of an adjusting bolt. As the arm rotates, a dry friction torque results which resists the arm motion. This friction provides the major source of vibration energy dissipation in the FEAD.

Dry friction damping represents a strong nonlinearity which can generate rich dynamic response not seen in linear viscous systems. Excitation at one frequency may produce harmonic responses at multiples (super-harmonics) and fractions (sub-harmonics) of the excitation frequency. In addition, this nonlinearity may also generate secondary resonances (sub- and super-harmonic) [2]. Resonant (excessive) belt response can induce belt slip, and thus loss of power transmission, as well as induce unwanted belt noise. This study is motivated by the need to develop a nonlinear FEAD model that is capable of capturing the nonlinear effects of dry friction at the tensioner arm.

Several investigators have modeled the rotational response of the FEAD system to dynamic excitation from the crankshaft. Beikmann [1] studied a weakly (geometric) nonlinear model of a prototype FEAD composed of one accessory, a crankshaft pulley, and a tensioner. Tensioner damping was linear and Beikmann examined steady-state, linear dynamic and nonlinear dynamic responses. Barker et al. [3] developed a FEAD model and studied *transient* rotational response of the accessories due to engine accelerations. Their model treated the combined elastic and friction torque on the tensioner arm as an experimentally known quantity which depended on the direction of the arm motion and its angle from the free position. They assumed the arm always locked, and later possibly unlocked, whenever its angular velocity passed through zero. Hwang et al. [4] developed a model for evaluating the *steady-state* response of the FEAD to crankshaft excitation, the nearness to slip, as well as the rotational natural frequencies and vibration modes of the system. Their model replaced the dry friction damper with an equivalent viscous damper. Most recently, Kraver et al. [5] used a complex modal analysis procedure to re-compute the frequency response of the FEAD. The friction damper was again replaced by an equivalent viscous damper in [5]. This paper specifically focuses on 1) modeling the dry friction in the tensioner as Coulomb damping and, 2) numerically implementing the resulting nonlinear FEAD model. The influence of the dry friction is highlighted by the rich computed responses over a range of steady engine speeds.

3 RESPONSE MODEL

3.1 Modeling Assumptions

The front end accessory drive to be modeled is schematically illustrated in Figure 1. The rotation of each accessory pulley and the tensioner arm represent the (unknown) responses that are excited by prescribed excitation from crankshaft motion and accessory torque fluctuations. The assumptions used in deriving the model are:

(1) The belt is uniform along the axial direction and stretches in a quasi-static manner.

(2) Longitudinal and lateral belt motions remain decoupled [1].

(3) The tensioner spring is linear and the tensioner dry friction is governed by a classical Coulomb law, refer to Section 3.2.

(4) Energy losses incurred as the belt creeps against the pulleys in the pulley *slip zones* are negligible; see Section 3.3.

(5) The slip zone remains a fraction of the entire angle of wrap [6, 7] (no gross slip).

3.2 Equations of Motion

Each fixed center accessory pulley shown in Figure 1 is assigned a single degree of freedom $\theta_i(t)$, which represents the *unsteady* component of the pulley's rotation. The equation of motion of the ith fixed centered accessory pulley is:

$$I_i \ddot{\theta}_i = R_i(T_{i-1} - T_i) - Q_i - C_{num}\dot{\theta}_i, \quad i = 2, 3...6 \quad (1)$$

where T_i denotes the tension in the ith belt span, R_i the ith pulley radius, Q_i the ith pulley external torque, and C_{num} denotes an introduced accessory damping coefficient as discussed in Section 4. The moment of inertia about the fixed center, I_i, accounts for both the ith pulley's moment of inertia and the ith accessory's contributions to the moment of inertia about the fixed center.

The motion of the tensioner arm and the rotation of the tensioner pulley are described by two additional degrees of freedom $\theta_7(t)$ and $\theta_t(t)$, respectively; refer to Figure 1. Following [4], the two equations of motion governing tensioner response are:

$$I_7(\ddot{\theta}_7 - \ddot{\theta}_t) = R_7(T_6 - T_7) - Q_7 - C_{num}\dot{\theta}_7, \quad (2)$$

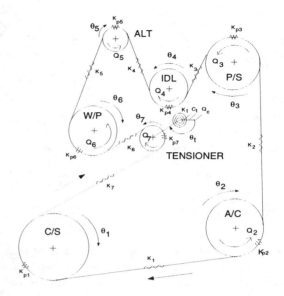

Figure 1: Front End Accessory Drive (FEAD). The crankshaft (C/S) powers a serpentine belt that drives the engine mounted accessories: water pump (W/P), alternator (ALT), idler (IDL), power steering pump (P/S), and air conditioner compressor (A/C). An automatic tensioner is an integral part of the FEAD.

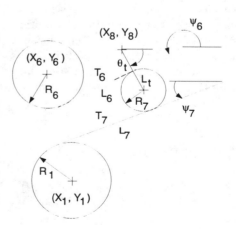

Figure 2: Tensioner Assembly Geometry

$$(I_t + I_{7t})\ddot{\theta}_t - I_7\ddot{\theta}_7 + K_t(\theta_t - \theta_0) + (Q_c - Q_{c0})$$
$$-C_{num}\dot{\theta}_7 = R_7(T_7 - T_6) - Q_t + m_{eff}L_{eff}g\cos(\theta_t)$$
$$-L_t\sin(\Psi_6 + \theta_t)\left[T_6 - \rho_b AV^2\right]$$
$$+L_t\sin(\Psi_7 + \theta_t)\left[T_7 - \rho_b AV^2\right], \quad (3)$$

where Coulombic friction torques (Q_c, Q_{c0}) and viscous damping torque $(C_{num}\dot{\theta}_7)$ are added to capture Coulombic and viscous dissipation. The installed position of the tensioner arm is denoted by θ_0 and is assumed known. Thus $\theta_t = \theta_0$ when the belt is installed on the FEAD at rest (i.e. under conditions of zero engine speed and zero accessory torques). Small differences (due to stretching) in belt speed, V, between the various spans of the belt are neglected, and hence V is based solely on the crankshaft pulley angular speed and radius. The associated frictional torque at installation is denoted Q_{c0}. Q_t is the preload torque of the tensioner arm spring. Here, I_t and I_{7t} are the moments of inertia about the pivot point of the tensioner arm and the tensioner pulley, respectively, and K_t is the stiffness of the hub mounted tensioner spring. The variable m_{eff} represents the total mass of the tensioner arm and pulley assembly, L_{eff} represents the distance between the mass center of the arm/pulley assembly and the arm pivot point, and L_t is the distance from the pivot point to the tensioner pulley center. The gravitational constant is denoted by g, the belt density by ρ_b, and A denotes the belt's cross sectional area. The angles Ψ_6 and Ψ_7, defined in [4] and illustrated in Figure 2, measure the inclination of the sixth and seventh belt spans from the horizontal. Note from Figure 2 that Ψ_6 is measured from the horizontal in the three o-clock position, while Ψ_7 is measured from the horizontal in the nine o-clock position. These angles and span lengths are related through:

$$\Psi_6(\theta_t) = \Pi + tan^{-1}\left[\frac{Y_7 - Y_6}{X_7 - X_6}\right]$$
$$+tan^{-1}\left[\frac{R_6 + R_7}{L_6}\right], \quad (4)$$

$$\Psi_7(\theta_t) = tan^{-1}\left[\frac{Y_7 - Y_1}{X_7 - X_1}\right] - tan^{-1}\left[\frac{R_7 + R_1}{L_7}\right], (5)$$

$$X_7 = X_8 + L_t\cos(\theta_t), \quad (6)$$

$$Y_7 = Y_8 - L_t\sin(\theta_t), \quad (7)$$

$$L_6(\theta_t) = ((X_6 - X_7)^2 + (Y_6 - Y_7)^2$$
$$- (R_6 + R_7)^2)^{\frac{1}{2}}, \quad (8)$$

$$L_7(\theta_t) = ((X_7 - X_1)^2 + (Y_7 - Y_1)^2$$
$$- (R_7 + R_1)^2)^{\frac{1}{2}}, \quad (9)$$

where (X_i, Y_i) locate the center of the ith pulley. The above relationships for $\Psi_6(\theta_t)$ and $\Psi_7(\theta_t)$ are valid when the tensioner pulley and the crankshaft pulley are in the quadrant shown in Figure 2, while similar relationships are readily developed for the remaining three quadrants.

The frictional torque term, Q_c, in equation (3) is defined as follows. The construction of a tensioner arm includes mating surfaces at the hub that are designed to generate dry friction. The frictional model employed herein is the classical Coulomb model. Thus, the frictional torque is a prescribed maximum value ($|Q_c| = Q_m$) when the tensioner arm is moving ($\dot{\theta}_t \neq 0$) and acts in opposition to this motion. However, when the arm is at rest ($\dot{\theta}_t = 0$), the magnitude and direction of the frictional torque are those required to maintain equilibrium, with the magnitude limited by the bounds $\pm Q_m$. In this case, the magnitude and direction of the frictional torque are determined from the equations of motion by setting $\ddot{\theta}_t$ and $\dot{\theta}_t$ to zero and solving for Q_c. In the limiting case when the arm is stationary, but motion is impending ($\ddot{\theta}_t$ is not zero), the magnitude of Q_c is set to its maximum and its direction is opposite the impending motion. Thus,

$$\begin{aligned} Q_c &= +Q_m & \dot{\theta}_t &> 0 \\ -Q_m \leq Q_c &\leq +Q_m & \dot{\theta}_t &= 0 \\ Q_c &= -Q_m & \dot{\theta}_t &< 0 \end{aligned} \quad (10)$$

3.3 Belt Constitutive Relations

The tension in any belt span is the sum of an initial reference belt tension, T_0, and that induced by an additional longitudinal stretch of the belt, Δ_i

$$T_i = T_0 + K_i\Delta_i \quad (11)$$

where $K_i = \frac{EA}{L_i}$ is the axial stiffness of the ith belt span, assuming the belt stretches quasi-statically. Following [3], the belt stretch, Δ_i, consists of a component due to the relative rotation of the adjacent pulleys, $R_i\theta_i - R_j\theta_j$, and a component δ_j, denoting the creep of the belt on the contact arc of the jth pulley.

$$\Delta_i = R_i\theta_i - R_j\theta_j - \delta_j,$$
$$j = i + 1 \text{ except when } i = 7, \text{ then } j = 1 \quad (12)$$

To understand how the component δ_j develops, consider first rigid body rotations of the ith and jth pulleys. For positive rotations ($\theta_i, \theta_j > 0$), the ith pulley reels in belt while the jth pulley pays out belt. As is described in [6, 7], the belt slips in a specific arc on the trailing edge of the pulley only, in what is referred to as a *slip arc*. Thus, an additional amount of belt is released by the jth pulley as the belt creeps in the slip arc at the trailing edge of the belt/pulley contact region. This additional belt length decreases the stretch of the ith span. The jth pulley's slip arc affects the ith span's stretch. The slip arc of the jth pulley is modeled herein by introducing an additional spring of stiffness $K_{pj} = \frac{EA}{r_j\beta_j}$, where β_j represents the angle of wrap of the jth pulley. The use of this stiffness only approximates the more complex mechanics of the belt in the slip zone [6, 7] and is inherited from previous models [3, 4]. The amount of belt released in the jth pulley's slip arc can be estimated using the average tension across the slip arc:

$$\delta_j = \frac{(T_j - T_0) + (T_i - T_0)}{2K_{pj}} \quad (13)$$

Together, equations (11)-(13) relate the span tensions to the rotational degrees of freedom.

Following [4], the reference tension, T_0, is evaluated as a function of the equilibrium tensioner arm angle, θ_e. The equation of tensioner equilibrium is obtained from (3) upon eliminating all time varying terms. Evaluating the resulting equilibrium equation with $T_6 = T_7 = T_0$ and with zero belt speed yields the reference tension for the accessory drive with a Coulombic tensioner:

$$T_0 = \frac{K_t(\theta_0 - \theta_e) - Q_t + m_{eff}L_{eff}g\cos(\theta_e) - Q_{ce}}{L_t\left[\sin(\Psi_6 + \theta_e) - \sin(\Psi_7 + \theta_e)\right]} \quad (14)$$

where Q_{ce} is the tensioner arm frictional torque at equilibrium. This reference tension is required as an initial condition to start the numerical simulation discussed in Section 4. Any value of Q_{ce} in the range $-Q_m \leq Q_{ce} \leq +Q_m$ may be prescribed and this choice affects θ_e, but not T_0. Thus, there exists a range of values of θ_e for which the tensioner arm will remain under equilibrium. The range of θ_e is directly proportional to the maximum exertable Coulomb friction, Q_m. In the following, the value of θ_e is the average value in this range for which $Q_{ce} = 0$.

3.4 Example Excitation

The excitation considered herein models prescribed crankshaft angular acceleration. In particular, the excitation is assumed to be that produced by a six cylinder engine and is dominated by three torque pulses per crankshaft revolution. In this example, the torque pulses produce harmonic oscillations of the crankshaft angular velocity having magnitude of 18 revolutions per minute:

$$\dot{\theta}_1 = \left[18\cos\left(3n\frac{2\Pi}{60}t\right)\right]\frac{2\Pi}{60} \quad (15)$$

Here, n is the steady component of the engine angular velocity measured in revolutions per minute, t is time measured in seconds, and $\dot{\theta}_1$ is measured in radians per second. The selected excitation is considered typical of one engine application and serves to illustrate one possible example.

4 NUMERICAL SIMULATION

Equations (1) - (15) comprise a system of ordinary differential equations and algebraic equations for the solution of the unknown angular displacements of each pulley($\theta_i(t)$, i=2,3,...,7) and the tensioner arm angle $\theta_t(t)$, which are complete once the equilibrium position of the tensioner arm (θ_e) is calculated. As in [4], the equilibrium position is determined by setting the time derivative terms to zero in the equations of motion and, in conjunction with the belt constitutive relations, computing θ_e. The remaining equilibrium angular displacements of each accessory and the equilibrium belt tensions are then calculated based on the obtained θ_e. Next, the equations of motion are cast in first order form to facilitate numerical time integration:

$$\dot{Y} = F(Y) \quad (16)$$

where

$$Y = \begin{bmatrix} \theta_t \\ \theta_2 \\ \theta_3 \\ \theta_4 \\ \theta_5 \\ \theta_6 \\ \theta_7 \\ \dot{\theta}_t \\ \dot{\theta}_2 \\ \dot{\theta}_3 \\ \dot{\theta}_4 \\ \dot{\theta}_5 \\ \dot{\theta}_6 \\ \dot{\theta}_7 \end{bmatrix}$$

is the vector of response coordinates for the FEAD system. The initial values of the displacements are selected to be equal to their equilibrium values, and therefore, the initial angular velocities are set to zero.

An adaptive time step, fourth/fifth order Runge-Kutta explicit integration method from [8] was selected to integrate the system equations (16). A quick overview of this method is required to explain modifications which were necessary. The method includes: 1) a primary routine that advances the solution vector by one Runge-Kutta time step, 2) a stepper routine that calls the primary routine multiple times in order to evaluate the largest time step size (at the current time) which will achieve a user specified error tolerance, and 3) a driver routine that calls the stepper routine to advance the solution vector from the initial time to the final time, one (adaptive) step at a time. The primary routine calls a user supplied subroutine which evaluates the first derivatives of the coordinates (16). The stepper routine provides the adaptive feature of the method and chooses the time step using *step doubling*. Step doubling compares the solution vector after taking one step in time to the solution vector after taking two half steps in time, and measuring the resulting truncation error of each element in the solution vector of the first full step. The truncation error for each element is scaled to approximately the previous value of the element (see [8]), and the maximum scaled truncation error is then used to determine whether the step size should be increased or decreased.

Modifications to the method above were necessary to enable the integration of Equation (3) which contains the Coulomb friction nonlinearity. A separate subroutine was created to calculate the external torque acting on the tensioner arm, excluding the frictional torque. The stepper routine determines the value of Q_c used in the primary routine based on the external torque subroutine and the position and velocity of the tensioner arm $(\theta_t, \dot{\theta}_t)$, as follows. In the primary routine, if the magnitude of the external torque supplied by the stepper routine is less than Q_m, the unknowns $\dot{\theta}_t$ and $\ddot{\theta}_t$ are set to zero. Each time the stepper routine calls the primary routine to calculate a new solution vector, several checks are made on the returned state of the tensioner arm $(\theta_t, \dot{\theta}_t)$. First, if $\dot{\theta}_t$ has changed sign, *and* the magnitude of the computed external torque is *greater* than Q_m, then the solution vector is accepted and Q_c is set to $\pm Q_m$, where the sign is selected opposite that of $\dot{\theta}_t$. However, if the magnitude of the computed external torque is *less* than Q_m, $\dot{\theta}_t$ is set to zero (i.e. the tensioner arm is momentarily stuck) in the solution vector and Q_c is set equal and opposite to the computed external torque. Second, if the tensioner arm changes from a stuck state to a moving state, the magnitude of Q_c is set to Q_m and the direction of Q_c is chosen opposite that of $\dot{\theta}_t$. Finally, if the arm remains stuck, Q_c is updated to oppose the current external torque on the arm.

Without further modifications, the method decreases the time step size ad finitum when the tensioner arm changes from a state of slip to a state of stick. The (infinite) reduction in step size occurs since the stepper routine must compare a solution vector obtained from a full step evaluated with the magnitude of Q_c equal to Q_m, to a solution vector obtained after two half steps which during the first half step, $|\,Q_c\,| = Q_m$, and during the second half step, Q_c equals the value required to hold the arm stationary. The resulting scaled truncation error is rarely small due to this discontinuity in Q_c. The stepper subroutine will then decrease the step size until both half steps use $|\,Q_c\,| = Q_m$, and the resulting truncation error will now be acceptable. However, the solution will then be advanced by the driver routine only to encounter the same problem on the next call to the stepper, etc. In this manner, the time step is continually decreased and, if uncorrected, the simulation does not advance past the time of stick of the tensioner arm.

A second problem due to the discontinuity in Q_c is encountered when the arm changes direction without sticking. This transition is numerically unstable since the net effect of the transition is a large torque applied to the system in a small period of time (equal to the time step).

The above difficulties are surmounted by two modifications. First, the time step is monitored and when an infinitely decreasing sequence of time steps is detected, the time step is set to a prescribed minimum time step. Second, the equations of motion are re-formulated to include viscous damping (of the form $C_{num}\dot{\theta}$); refer to Equations (1) and (2). This added term dampens the computed response following a discontinuity in Q_c. The value of this (numerical) damping is selected to be less than ten percent of critical damping in all modes and stabilizes the numerics during these short time intervals.

The transition from a stuck state of the tensioner arm to a moving state may once again admit discontinuities in Q_c if the time step is not selected appropriately small during this transition. An appropriately small time step is determined by adding an additional term to the truncation error calculation when $\dot{\theta}_t$ equals zero. The maximum scaled truncation error is selected as the maximum of the solution vector scaled truncation errors *and* the change in the external torque scaled by Q_m, which ensures a time step in which the change in Q_c at break away is small enough to be numerically stable.

Table 1: Example FEAD System

PULLEYS

Pulley No. i	Type	Location (X_i, Y_i) (mm)	Radius R_i (mm)	Moment of Inertia, I_i (kgm^2)
1	C/S	(0,0)	81.25	0.122
2	A/C	(261.5,60)	64.50	0.00415
3	P/S	(252,234)	70.60	0.00131
4	IDL	(90.3,251.1)	41.15	0.000263
5	ALT	(86,354)	30.00	0.00421
6	W/P	(0,167.5)	67.50	0.00176
7	TEN	N/A	38.10	0.000208

TENSIONER ARM

(X_8, Y_8) (mm)	θ_0 (deg)	I_7 (kgm^2)	$I_t + I_{7t}$ (kgm^2)
(142.0,207.5)	93	0.000208	.001485

$meff$ (kg)	$Leff$ (mm)	L_t (mm)
0.9163	25	53

Q_t (Nm)	C_{num} $(\frac{Nm\,sec}{rad})$	K_t $(\frac{Nm}{rad})$	Q_m (Nm)
25.19	0.02825	28.25	3.95

BELT

ρ_b (kg/m)	EA (N)
0.1036	80068

5 RESULTS

As discussed in the previous section, the equations of motion for a FEAD system with a Coulombic tensioner are integrated using a fourth/fifth order Runge-Kutta algorithm modified to handle the Coulomb nonlinearity. An example FEAD system is introduced here to illustrate the capabilities of the model. Consider a six cylinder production engine with a FEAD system defined by the parameters listed in Table 1. The maximum frictional torque exerted by the tensioner hub was chosen to be $Q_m = 3.95\ Nm$ based on experimental measurements.

The results of numerical simulations that follow yield new conclusions about the response of the accessory drive which can not be obtained using previous linear models. For example, the simulations reveal that the response of the accessory pulleys due to harmonic crankshaft excita-

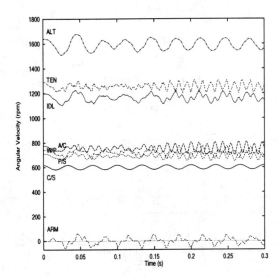

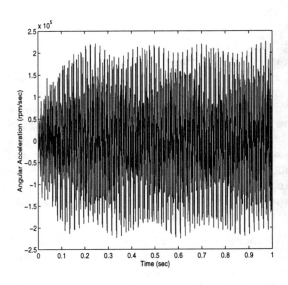

Figure 3: Accessory and tensioner arm velocity at 600 RPM demonstrating 3rd order super-harmonic response and stick-slip motion of the tensioner arm. The notations of each component are provided in Figure 1.

Figure 4: Tensioner pulley response at 3000 RPM demonstrating sub-harmonic and super-harmonic response

tion contains not only the excitation harmonic, but higher and lower order harmonics. The simulations also reveal secondary (sub- and super-harmonic) resonances. These new phenomena are described below.

Figure 3 shows the angular velocity history of each accessory pulley due to crankshaft excitation for the case of relatively low engine speed. In this simulation, the crankshaft rotates with a steady component of 600 revolutions per minute ($n = 600\ RPM$), and with an unsteady component of 18 RPM. The simulation time of 300 milliseconds corresponds to approximately nine excitation cycles. The resulting excitation frequency ($\frac{3n}{60}$ Hz) is 30 Hz. Illustrated are 1) the prescribed harmonic motion of the crankshaft, and the resulting responses of 2) each accessory pulley, and 3) the tensioner arm. The response of the accessory pulleys and the tensioner arm are, in general, not harmonic due to the action of the nonlinear (Coulomb) element. In particular, the response of the power steering, water pump, air conditioner, and idler pulleys contain primarily two harmonics: 1) a (fundamental) harmonic coincident with the excitation frequency (30 Hz), and 2) a harmonic of three times the excitation frequency (90 Hz). This second harmonic is commonly referred to as a *super-harmonic* of order three [2]. Higher order super-harmonics also appear in the responses of the lightest (smallest I) components: 1) the tensioner and 2) the idler pulleys. By contrast, the alternator (largest I) response is well described by the fundamental harmonic alone. Again, all super-harmonic responses exist by virtue of the nonlinearities generated by Coulomb friction acting on the tensioner arm. Note also the regions of stick and slip in the tensioner arm history which ultimately generate the higher harmonics in the accessory responses.

The previous simulation clearly shows the existence of a super-harmonic response. Sub-harmonic responses also exist as illustrated in a second simulation for the engine op-

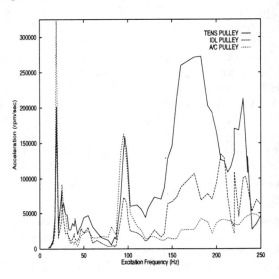

Figure 5: Frequency response of 3 FEAD accessories after 10 excitation cycles, exhibiting secondary resonances

erating at moderate speed (3000 RPM). The crankshaft is again assumed to have an unsteady component of rotation with magnitude 18 RPM. At 3000 RPM, the crankshaft delivers an excitation with frequency 150 Hz. The simulation results illustrated in Figure 4 were computed for 1 second (150 excitation cycles). Observe in the figure that the tensioner pulley response ($\ddot{\theta}_7$) contains a high oscillation frequency and a much lower modulation frequency. This latter frequency component is a probable sub-harmonic of approximately 2.3 Hz. A Fourier analysis (not shown) also reveals the existence of a super-harmonic of frequency 450 Hz (three times the excitation frequency). The 450 Hz frequency nearly aligns with the sixth natural frequency of the system (449 Hz) and is here responsible for the large response magnitude. This is an example of a secondary resonance.

An important effect captured by the new model (which is absent from all previous linear models) is the existence

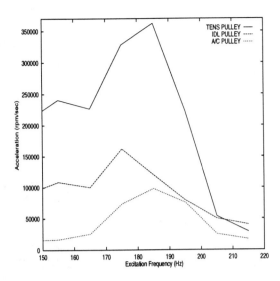

Figure 6: Portion of frequency response curve after 100 excitation cycles resolving two distinct resonances

of secondary resonances. Both primary and secondary resonances are observable in the frequency responses illustrated in Figure 5, for the tensioner arm, the idler pulley, and the air conditioner pulley. These frequency responses were generated by first completing the simulations for ten excitation cycles at each of many excitation frequencies within the range of 10 Hz to 250 Hz. At the conclusion of a simulation, the magnitude of the acceleration response for each pulley was recorded. In Figure 6, a second set of frequency response curves are provided over a limited range of 150 Hz to 220 Hz. This set is generated by running the simulation for 100 excitation cycles. After 100 excitation cycles, the broad response peak in Figure 5 over this frequency range is now resolved into two response peaks. It should be noted that ten excitation cycles, and possibly even 100 excitation cycles, may not be enough cycles to achieve steady-state response at all excitation frequencies, especially near a resonant frequency. Future work includes developing a method to predict the steady-state response of each accessory pulley and automatically generating their associated frequency response curves [9].

The linear model predicts (primary) resonances solely at the natural frequencies for the accessory drive: 19.7, 98.4, 102.7, 189, 240, 449, and 517 Hz; refer to [4]. The first five of these primary resonances fall within the frequency range of the present example. Note the prominent peak appearing at the first natural frequency, a wider combined peak at the second and third natural frequencies, and a sharp peak at the fifth natural frequency. In addition to these peaks are secondary resonances which derive from sub-harmonic and super-harmonic resonant response components. Evident are modest secondary resonances in Figure 5 corresponding to approximately one-third of the second natural frequency, followed by another at approximately one-fifth of the fourth natural frequency, with probable others dispersed throughout the frequency range. The unusually broad peaks between 150 and 200 Hz in Figure 5 are seen to be resolved into two resonances in Figure 6. The resonance at 189 Hz corresponds to the fourth natural frequency, and the resonance at 150 Hz corresponds to one-third of the sixth natural frequency. In this particular example FEAD, the secondary resonances are clearly less pronounced than the primary resonances. Larger secondary resonances and secondary resonances of other orders may very well exist for different designs and/or engine operating conditions.

6 SUMMARY AND CONCLUSIONS

A nonlinear model of the rotational response of the front end accessory drive has been developed together with a numerical solution method. The model and solution method capture the nonlinear effects of Coulomb damping at the tensioner. This nonlinear damping may produce stick-slip motions of the tensioner arm and associated sub- and super-harmonic responses of the accessory pulleys. Furthermore, the nonlinear damping may generate secondary resonances (sub- and super-harmonic resonances) that, in addition to the primary resonances, may lead to excessive vibration, noise, and/or power loss. Linear FEAD models successfully predict primary resonances but can not predict the secondary resonances observed herein. FEAD designers should be aware of these secondary resonances and, in particular, avoid a system design with any (primary or secondary) resonances near idle speed or cruising speed.

ACKNOWLEDGEMENTS

The authors gratefully acknowledge Ford Motor Company for support of this research endeavor, and acknowledge partial support from the University of Michigan Automotive Research Center, funded by the U.S. Army (TACOM). They also acknowledge Professor V. Coppola of the University of Michigan for his fruitful discussions.

7 REFERENCES

1 Beikmann, R. S., 1992, "Static and Dynamic Behavior of Serpentine Belt Drive Systems: Theory and Experiment". Ph.D. Dissertation, University of Michigan, Ann Arbor, MI.

2 Nayfeh, A. H., Mook, D. T., 1979, *Nonlinear Oscillations*, New York, John Wiley & Sons, Inc.

3 Barker, Clark R., Oliver, Larry R., Brieg, William F., 1991, "Dynamic Analysis of Belt Drive Tension Forces During Rapid Engine Acceleration," *SAE Congress*, Detroit, Michigan, 910687, pp. 239-254.

4 Hwang, S.-J., Perkins, N. C., Ulsoy, A. G., Meckstroth, R.J., 1994, "Rotational Response and Slip Prediction of Serpentine Belt Drive Systems," *ASME Journal of Vibration and Acoustics*, Vol. 116, pp. 71-78.

5 Kraver, T. C., Fan, G. W., Shah, J. J., 1996, "Complex Modal Analysis of a Flat Belt Pulley System With Belt Damping and Coulomb-Damped Tensioner," *Journal of Mechanical Design*, Vol 118, pp. 306-311.

6   Gerbert, G. G., 1991 "On Flat Belt Slip," *Vehicle Tribology*, Tribology Series 16, pages 333-339, Amsterdam. Elsevier.

7   Johnson, K. L., 1985, *Contact Mechanics*, London, Cambridge University Press, Chapter 8.

8   Press, W. H., Teukolsky, S. A., Vetterling, W. T., Flannery, B. P., 1986, *Numerical Recipes in Fortran*, 1st. Edition, Cambridge University Press.

9   Leamy, M. J., Perkins, N. C., submitted, "Periodic Response of Front End Accessory Drives with Dry Friction," *ASME Sixteenth Biennial Conference on Mechanical Vibration and Noise*, Sept. 14-17, 1997, Sacramento, CA.

971963

Development of an Isolated Timing Chain Guide System Utilizing Indirect Force Measurement Techniques

Wayne Nowicki
Roush Anatrol

Eric Sheffer
Ford Motor Co.

Copyright 1997 Society of Automotive Engineers, Inc.

ABSTRACT

This paper outlines the development process of a vibration isolation system for the timing chain guides of an internal combustion engine.

It was determined through testing that the timing chain guides are a significant path by which the chain/sprocket impacts are transmitted to other powertrain components. These components radiate the energy as chain mesh order narrow band sound as well as wide band energy. It was found that isolation of the chain guides produced a significant reduction in radiated sound levels, reduced mesh frequency amplitudes, and improved sound quality.

The development process utilized indirect force measurement techniques for simulation of the chain loading and FEA prediction of the resulting chain guide forces and displacements.

The design of the isolation system involved material selection based on dynamic properties, frequency and temperature ranges, the operating environment, FEA geometry optimization, and durability testing.

INTRODUCTION

Sound quality and low noise levels inside vehicles have increasingly become important discriminators to automotive customers [1,2].

Cam chain noise is one of many sources of noise that has been identified as a problem in vehicle passenger compartments [1]. With the move towards high tech multi-cam engines with multiple drive chains, Figure 1, the cam chain noise has a greater potential to become an issue.

The improvements in noise and sound quality of many passenger car engines further unmasks chain noise and provides more of a need to address this issue.

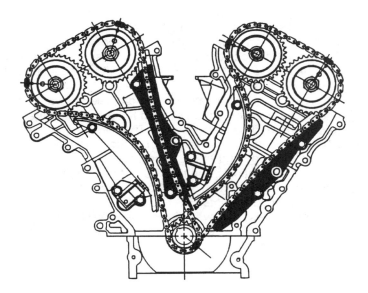

FIGURE 1: Typical Dual Overhead Cam Engine Timing Chain Layout with Chain Guides Highlighted

Chain noise is initiated through several mechanisms, two of the most significant of which are the chain and sprocket meshing, and chordal action [3,4] which results in chain vibration. These mechanisms create tones at chain mesh frequency and also produce broadband noise. The chain mesh frequency is the number of times per second with which the sprocket meshes with the chain. This frequency is equivalent to the number of teeth on the driving sprocket multiplied times the crankshaft speed. For this case study there were 18 teeth on the sprocket which would create 18th order chain noise along with higher order harmonics as shown in Figure 2.

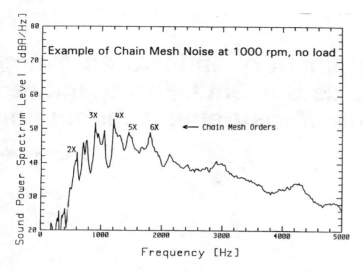

FIGURE 2: Sound Power of Chain Mesh Frequency and Related Harmonics

The magnitude of the chain mesh noise and its harmonics is dependent on the engine speed and the modal alignment with engine components. The broadband noise is dependent on the design of the chain and the contact that the chain makes with the guides and tensioner arms.

The radiation of chain noise is amplified by various engine surface components, most notably the engine front cover and the cam covers. This radiated noise then becomes an issue as it travels from the powertrain into the vehicle body and results in increased interior noise and tactile vibration. Due to its tonality and high frequency nature, chain mesh noise may potentially lead to customer complaints [5,6,7].

The process of developing the isolated timing chain guide system is illustrated in the Flow Diagram, Figure 3.

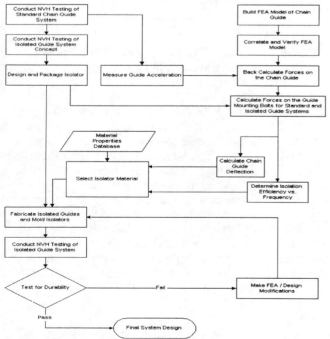

FIGURE 3: Development Process Flow Diagram

EXPERIMENTAL EVALUATION

To determine the effect of chain guide isolation, the NVH level for the production non-isolated chain guide system was first established. The tests performed consisted of near and far field sound pressure, powerplant sound power, and operating vibration on major radiating surfaces determined through near field component intensity measurements. The main components radiating chain mesh noise were the front cover and cam covers, although mesh frequencies were also evident to some degree on all surfaces.

A concept level chain guide isolation system was developed to determine if the guides were an important transmission path for chain noise. The system consisted of standard rubber isolators located at the mounting locations. The first design of an isolation system for the chain guides avoided the long term survivability issues, and was built to give maximum compliance to the guide mountings, while maintaining chain tension and timing. The isolated guide engine was tested in an identical manner to the baseline engine, and the NVH performance was documented. Figure 4 shows the sound power comparison for the accessory side of the baseline and the isolated chain guide engines. A 1.3 dBA overall level reduction, and a 6 dBA reduction at the mesh frequency was achieved with guide isolation. Since significant improvements were realized, a production feasible optimized version of the isolated chain guide system was then developed.

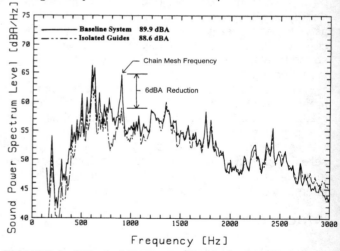

FIGURE 4: Effect of Prototype Guide Isolation on Sound Power of the Accessory Side of the Engine

The process to develop an isolated guide system that optimizes the isolation efficiency while producing a sufficiently rigid system for survivability would require many iterations and test cycles. The development process utilized analytical methods to predict the optimum design and isolator modulus for achieving both results. Experimental data, in the form of acceleration measurements on each guide were used as inputs to the analysis. The maximum number of accelerometers that would package within the guide were used to provide the highest number of system responses. Acceleration data was acquired for 3 different load and speed conditions. The full crosspower acceleration matrix was measured to capture the required amplitude and phase information.

Chain mesh and combustion orders are dominant in chain guide vibration, as expected. The accelerometer locations and a representative autopower and crosspower trace for one location are shown in Figure 5. The tests were repeated with fewer transducers to ensure that mass loading was not apparent.

ANALYTICAL OPTIMIZATION OF ISOLATOR PROPERTIES

The isolation optimization process used indirect force determination methodology to predict the effectiveness of the proposed isolator design and to optimize the isolator stiffness requirements. This method has proven useful in a variety of applications from helicopter rotors to engine mounts [8,9] for which it is difficult or impossible to measure the actual forces without disturbing the system. The timing chain load was predicted based on the vibration responses measured on the chain guides. The force transmitted to the block due to the chain load was then predicted while varying isolator design parameters, to minimize the transmitted load while maintaining sufficient guide rigidity.

The initial step was to model the chain guides using standard finite element methods. The guides consisted of plastic wear strips attached to aluminum castings with three mounting locations as shown in Figure 6.

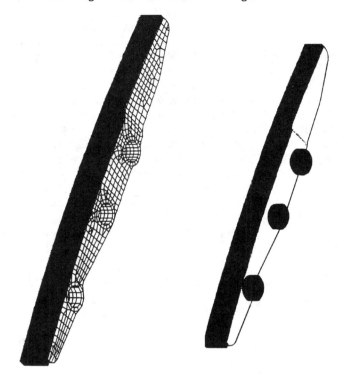

FIGURE 6: Finite Element Model of Left Chain Guide

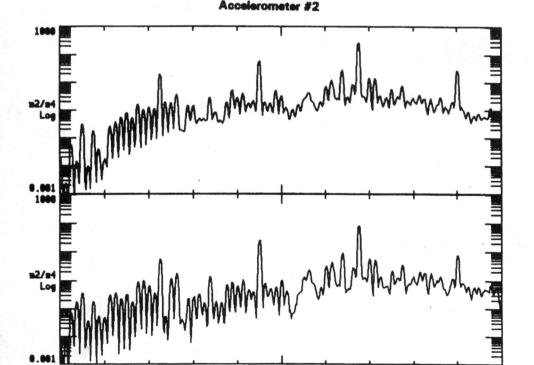

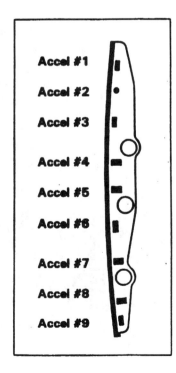

FIGURE 5: Location and Orientation of Accelerometers for the Left Chain Guide Response Measurements and Representative Autopower and Crosspower Measurements at 3000 RPM

733

The flexible modes of the guides were predicted using free-free boundary conditions, for comparison to experimentally obtained data for correlation of the models. Figure 7 compares the analytical and experimental mode frequencies for the left chain guide, with the deviation from the 45 degree line indicating the degree of correlation. The model was modified until the maximum error was less than 5 percent.

The input to this analysis was the guide acceleration responses in the form of a 9x9 crosspower matrix. The number and location of the responses used were determined by the mode predictions and the spatial constraints of the guides. A sufficient number of responses was used to provide the redundancy required to minimize error in the matrix inversion process [8,9]. This is accomplished by having more accelerometers that forces needed to define the chain loading.

The locations for the forces were chosen in an iterative process, starting with a distributed load in the direction normal to the chain guide and in the direction of chain travel. The number of forces must be less than the number of responses for error minimization (as noted above) and less than the number of modes participating to provide a unique solution to the transformation matrix inversion process [10]. The transformation matrix is the frequency response between the forces F to be predicted and the vibration responses X measured.

$$|X| = |H||F| \qquad (1)$$

The H matrix is generated analytically from the finite element model. The forces are then calculated from the acceleration responses and the inverted H matrix as follows:

$$|F| = |H|^{+1}|X| \qquad (2)$$

where the matrix H^{+1} is the pseudo-inverse of the H matrix, a process used for the inversion of a rectangular matrix by least squares solution [8].

The force back-calculation is performed using equation (2), and the results are fed back into equation (1) to calculate the acceleration responses produced by the analytically generated force set. The comparison of the analytically generated acceleration results to the original experimentally measured data is an indication of the validity of the force set. The number of forces and their locations was modified until reasonable correlation was achieved. Figure 8 shows the final force locations and the correlation of the experimental and analytical responses.

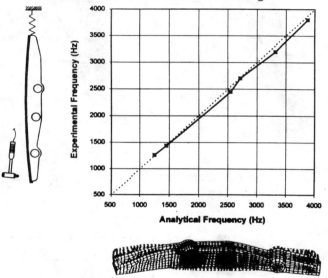

FIGURE 7: Comparison of Analytical and Experimental Modes for Correlation of Model

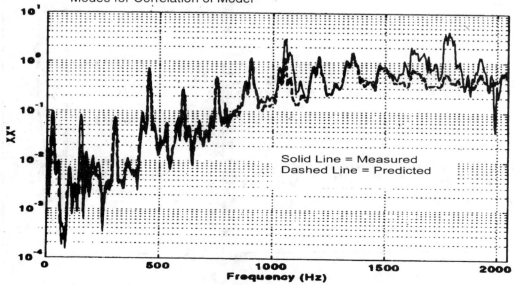
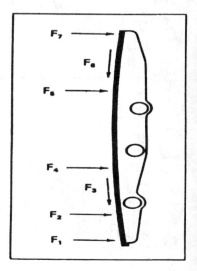

FIGURE 8: Location of Forces for Simulating Chain Load and Resulting Response Prediction Compared to Measured Response Data

After a representative force set was developed to simulate chain loading, the chain guide deflection was predicted. The maximum deflection allowable was based on system integrity and chain control, maintaining proper internal clearances, and safe isolator preload. This provided the lower bounds for the stiffness required for the isolators. Figure 9 shows the deflection for isolators of two different stiffnesses, as quantified by Young's modulus of 6.9 N/mm^2 and 13.8 N/mm^2. The analysis produces conservative results since the system is assumed linear. The deflection is reduced as expected with increased modulus. The analysis was repeated until acceptable deflection was achieved.

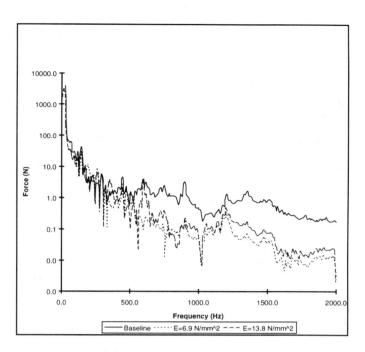

FIGURE 10: Prediction of Force Transmitted Into the Block Due to Chain Mesh Excitation

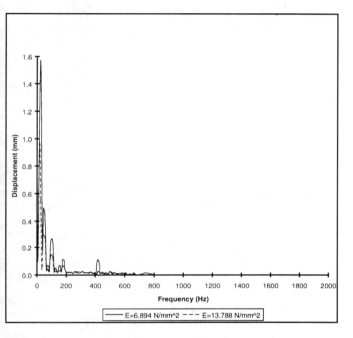

FIGURE 9: Prediction of Isolated Guide Deflection Due to Chain Loading

To optimize the performance of the chain guide isolation, a prediction of the force transmitted to the block through the guide due to the chain loading was performed. This analysis provided the upper bound for isolator stiffness to effectively reduce transmitted chain mesh vibration. Figure 10 is a prediction of the force transmitted to the block for the 6.9 N/mm^2 and 13.8 N/mm^2 modulus isolators, compared to the baseline system with no isolation.

This force prediction shows the degree of vibration reduction, and the frequency range over which isolation occurs. The lower the stiffness, the greater the reduction and the lower the cut-off frequency at which isolation starts to occur. Combining the results of the force predictions with the deflection analysis, the 13.8 N/mm^2 modulus isolator was chosen as the best compromise for vibration reduction and durability.

MATERIAL SELECTION

In choosing a suitable material for use as the guide isolation, only elastomers with excellent resistance to the engine environment, namely oil, solvents, and high temperatures were considered. After developing a list of potential materials, a review of the dynamic material properties was performed to determine the best choice for performance. The goal was to find the material with the most stable modulus and damping properties with respect to temperature and frequency. Stable material properties prevent the isolation system from becoming detuned as the engine temperature varies, thus losing effectiveness. It also ensures similar performance over a wider frequency range.

Using Roush Anatrol's material data base, the material chosen for isolation of the chain guides was a fluorocarbon. As can be seen in Figure 11, the modulus is relatively constant from 55°C to 150°C.

The damping loss factor is also constant over most of that range. The various curves superimposed on the figure indicate that there is some variation of stiffness with frequency, but not sufficient to be of concern since stiffness always increases with frequency [11]. Thus, designing for lower frequency stiffness ensures no compromising of durability due to higher frequency excitation.

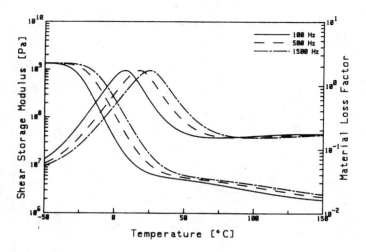

FIGURE 11: Dynamic Properties of the Fluorocarbon Material Chosen for the Final Chain Guide Isolator Design

FINAL NVH TESTING

Upon completion of the material selection process the isolators were molded and the chain guides were machined to accept the finalized design. Testing for quantification of NVH improvement was conducted in the same manner as the initial testing. Significant reductions in chain noise and vibration were realized. Figures 12 and 13 show up to 8 dBA and 6 m/s^2 improvements due to isolation. Reduced stiffness produces the best results up to the point when chain control and clearance limitations cause a softer system to have a negative effect on noise and vibration. However, all chain guide isolation levels were improvements over the non-isolated chain guide system, and the optimized design was verified experimentally.

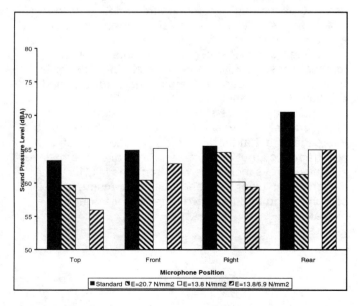

FIGURE 12: Effect of Chain Guide Isolation on Sound Pressure 1 Meter From the Engine

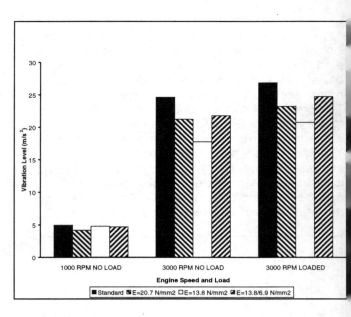

FIGURE 13: Effect of Chain Guide Isolation on Front Cover Vibration

DURABILITY TESTING

Durability testing was conducted to verify the survivability of the isolator design and the integrity of the timing chain system with isolation. The cycle used was a 300 hour cycle at various speeds and loads, with more than 90 percent of the time spent at a wide open throttle condition. Inspection intervals were set at every 50 hours where the front cover was removed and the isolators were visually inspected for separation without removal.

The durability testing uncovered a tendency of the isolators to tear at the flange. Although the system integrity was not compromised, a FE stress analysis was performed to correct the problem. A stress concentration was revealed and corrected through a stress relief in the molded part. Retesting of the improved isolator design met durability requirements.

CONCLUSIONS

In summary, the following conclusions were derived from the project:

- The timing chain guides were identified as a significant path for transmission of timing chain related noise.

- Significant reductions in noise and vibration were realized through the isolation of the timing chain guides of a dual overhead cam engine.

The application of indirect force measurement techniques proved to be useful and fundamental in providing design direction for a situation where force measurement was impractical.

An analytical method to predict the timing chain forcing function was developed.

An isolated timing chain guide system was developed that is easy to manufacture and assemble, is lightweight and durable, and packages in the present engine design.

ACKNOWLEDGMENTS

We would like to thank all those from Ford Motor Co., and Roush Anatrol who contributed to the success of this project.

REFERENCES

[1] "Development of Vehicle Sound Quality - Targets and Methods", Matthias Schneider, Michael Wilhelm, and Norbert Alt. SAE 951283, Traverse City Noise and Vibration Conference, 1995.

[2] "A Study of Noise in Vehicle Passenger Compartments During Acceleration", K. Tsuge, et al SAE 880965.

[3] "On the Noise of Roller Chain Drives", K. Uehara Proceedings of the Fifth World Congress on Theory of Machines and Mechanisms 1979.

[4] "Polygonal Action in Chain Drives", R.A. Morrison Machine Design 1952.

[5] "Psychoacoustics-Facts and Models", E. Zwicker and H. Fastl, Springer, Berlin 1990.

[6] "Effects of Powerplant Vibration on Sound Quality in the Passenger Compartment During Acceleration", H. Aoki, et al SAE 870955.

[7] "Environmental Noise Criteria for Pure Tone Industrial Noise Sources", J. Lilly, Noise-Con 94.

[8] "Modal Verification of Force Determination for Measuring Vibratory Loads", F.D. Bartlett and W.G. Flannelly, Journal of the American Helicopter Society, Vol 24 (2), 1979, PP. 10-18.

[9] "Identification of Forces Generated by a Machine Under Operating Condition", N. Okubo, S. Tanabe, and T. Tatsuno, Proceedings of the 3rd IMAC, 1985.

[10] "Indirect Identification of Excitation Forces by Modal Coordinate Transformation", G. Desanghere and R. Snoeys.

[11] "Vibration Damping", Nashif, Jones, and Henderson, pp 89-90.

971964

Dynamic Analysis of Layshaft Gears in Automotive Transmission

Teik C. Lim and Donald R. Houser
Ohio State Univ.

Copyright 1997 Society of Automotive Engineers, Inc.

ABSTRACT

In this paper, we will present parametric results of performing dynamic analysis of layshaft gear trains typically used in automotive transmissions with emphasis on the vibratory response due to transmission error excitation. A three-dimensional multiple degrees of freedom lumped parameter dynamic model of a generic layshaft type geared rotor system (with three parallel rotating shafts coupled by two sets of gear pairs) has been formulated analytically. The model includes the effects of both rotational and translational displacements of each gears, and bounce and pitch motions of the counter-shaft. The natural frequencies and mode shapes are computed numerically by solving an eigenvalue problem derived from applying harmonic solutions to the equations of motion. The complete set of mode shapes are then used in forced response calculations based on the modal expansion method to predict gear accelerations, dynamic transmission errors, mesh force and bearing loads. Structural resonance prediction is validated by comparison to finite element result of a baseline system. The proposed model is also used to examine the effects of system parameters such as counter-shaft diameter and bearing stiffnesses on the dynamic response due to transmission error excitation. Results of these two case studies are discussed in this article.

INTRODUCTION

The layshaft gear train configuration is currently being used in numerous automotive transmission and transaxle applications. Its basic layout consists of an input shaft driving a counter-shaft that in turns exert motion on the output shaft through a second stage gear pair. There are usually two pairs of gears in mesh simultaneously with one pair between the input shaft and counter-shaft, and a second pair that couples the counter-shaft to the output shaft. Therefore, there are two sets of loaded gears on the counter-shaft that are engaged with their matching gear pairs during the process of power transmission from the engine (power source) to the load (power absorption unit). Figure 1 shows a typical layout of a layshaft gear train system that is being analyzed in this paper.

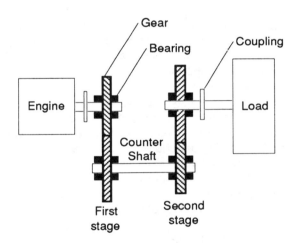

Figure 1. Schematic of a two-stage layshaft geared rotor system.

One of the earliest attempt to model a double reduction gear train system similar to the problem considered in this paper was by Benton and Seireg [1] in 1980. They developed a simple two degrees of freedom torsional model with time-varying mesh stiffness to study dynamic amplifications and instabilities. Later on, Iida et. al. [2-4] formulated a 10 degrees of freedom lumped parameter model of a two-stage geared rotor system. In their formulation, the effects of coupled translational and torsional motions of the counter-shaft were modeled. However, the model did not include translational displacements of the gears on the input and output shafts. In both of these shafts, only rotational degrees of freedom have been modeled. They used this model to compute mode shapes and to study the effects of relative mesh orientation on critical speeds. At about the same period, Umezawa et. al. [5] conducted a series of experiments to evaluate gear vibration response due to spacing between the first and second stage meshes along the axial length of the counter-shaft. They also examined the effects of torque and tooth mesh phase lag between the first and second stage gear pairs. In 1989, Choy et. al [6] used a modal synthesis approach to analyze the dynamics of multi-stage geared rotor systems. Most recently, Velex and Saada [7] in 1991 developed a Ritz-based finite element model of a layshaft gear train system to predict dynamic tooth loads.

It is the objective of this research study to formulate a relatively simple linear lumped parameter system model, but includes both gear rotational and translational displacements, of a generic layshaft gear train system that can be parametrized for analyzing the effects of system characteristics on vibration response due to transmission error excitation. This analytical model is intended for use as a vibration analysis tool at the concept design stage of automotive transmissions or transaxles. Due to the simplicity of its formulation, it can be easily manipulated mathematically to perform a wide range of parametric studies and compute the effects of different layout and structural modifications on gear dynamic response.

In this paper, we have used the proposed lumped parameter model to analyze the dynamic characteristics of generic layshaft gear train systems. The predicted modal characteristic of a baseline system is compared to numerical result of a finite element model. The proposed model is also used to examine 2 parametric case studies: (1) effects of counter-shaft diameter and (2) effects of bearing stiffnesses.

MODELING ASSUMPTION

The layshaft gear train system illustrated in Figure 1 consists of several main components:

- engine (power source)
- first stage pair
- counter-shaft
- second stage gear pair
- load (power absorption unit)
- 4 sets of bearings

The engine and load are modeled as large mass moment of inertias about their rotational axes. They are assumed connected to the layshaft geared rotor system in the rotational direction only through the torsional rigidities of the shafts and flexible couplings. Each gears have four degrees of freedom represented by three orthogonal translational displacements and a rotational motion. The gears can be either helical or spur. The effects of gear pitch motions about the two orthogonal axes perpendicular to the rotational axis are not included. A set of bearings are used to support each gear. The driven gear of the first stage and driving gear of the second stage are rigidly attached to the counter-shaft. The torsional and axial dynamic compliances of the counter-shaft are modeled specifically. It is also allowed to bounce and pitch like a rigid body. The excitation sources are mainly from the transmission errors in both sets of the gear pairs. Only the fundamental mesh harmonic is being considered in the analysis. Other secondary effects such as mesh stiffness tooth-to-tooth variations, non-linearity, gear tooth contact friction forces and gear body compliances are beyond the scope of this study.

MATHEMATICAL FORMULATION

A schematic of a three-dimensional lumped parameter model with 18 independent coordinates to represent a generic layshaft gear train system is shown in Figure 2. Each rotational element is assigned an integer value ranging from numbers 1 to 7 as reference.

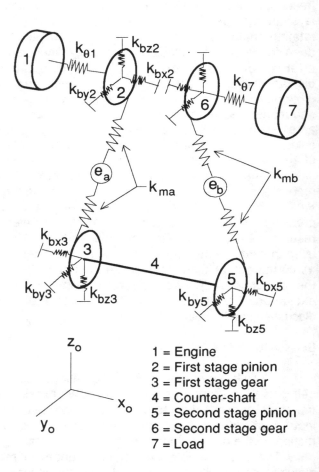

1 = Engine
2 = First stage pinion
3 = First stage gear
4 = Counter-shaft
5 = Second stage pinion
6 = Second stage gear
7 = Load

Figure 2. Lumped parameter model of a layshaft gear train system. The bearing stiffness terms are collinear with the reference axes defined by x_o, y_o and z_o.

The geometrical layout of the drive train and its two mesh orientations are defined with respect to a fixed coordinate system that consists of three orthogonal axes labeled as x_o, y_o and z_o, where x_o is assumed parallel to the axis of rotation. The other two axes y_o and z_o are defined arbitrarily. The relative locations of the three parallel shafts with respect to this fixed coordinate axes and the gear pressure angles of the first ϕ_1 and second ϕ_2 pairs can be used to determine their mesh orientation angles, α_1 and α_2. To do so, consider the shaft layout angles η_1 and η_2 for the first and second stage gear pairs respectively. They are defined by the straight lines between adjacent shaft center points and y_o. Hence, corresponding mesh orientation is given by $\alpha_k = \eta_k - \phi_k$ where subscript k could refer to either first k=1 or second k=2 stage mesh. These angles are used to relate local gear displacement coordinates that are parallel to the line of mesh actions to the fixed reference x_o, y_o and z_o axes.

Next the equations of motion are formulated in the displacement coordinates defined by the x_o, y_o and z_o axes. They are derived from free body diagrams constructed for each components in Figure 2. The undamped torsional vibration equations of motion of the engine (denoted by subscript 1) and load (denoted by subscript 7) are

$$I_{x1}\ddot{\theta}_1 + k_{\theta 1}(\theta_1 - \theta_2) = 0 \qquad (1a)$$

$$I_{x7}\ddot{\theta}_7 + k_{\theta 7}(\theta_7 - \theta_6) = 0 \qquad (1b)$$

where θ_2 is the rotational displacement of first stage driving gear and θ_6 defines the rotational displacement of second stage driven gear. The torsional stiffness terms of the shafts with flexible couplings connected to the engine and load are represented by $k_{\theta 1}$ and $k_{\theta 7}$ respectively. For the first stage gear pair (labeled with subscript a) that includes gears 2 and 3, the undamped equations of motion based on coordinate vector $\{x_2, y_2, z_2, \theta_2, x_3, y_3, z_3, \theta_3\}$ can be shown to be

$$m_2\ddot{x}_2 - k_{ma}\delta_a \sin(\Psi_a) + k_{bx2}x_2 = k_{ma}e_a \sin(\Psi_a) \qquad (2a)$$

$$m_2\ddot{y}_2 + k_{ma}\delta_a \cos(\Psi_a)\sin(\alpha_a) + k_{by2}y_2$$
$$= -k_{ma}e_a \cos(\Psi_a)\sin(\alpha_a) \qquad (2b)$$

$$m_2\ddot{z}_2 - k_{ma}\delta_a \cos(\Psi_a)\cos(\alpha_a) + k_{bz2}z_2$$
$$= k_{ma}e_a \cos(\Psi_a)\cos(\alpha_a) \qquad (2c)$$

$$I_2\ddot{\theta}_2 - r_2k_{ma}\delta_a \cos(\Psi_a) + k_{\theta 1}(\theta_2 - \theta_1) = r_2k_{ma}e_a \cos(\Psi_a) \qquad (2d)$$

$$m_3\ddot{x}_3 + k_{ma}\delta_a \sin(\Psi_a) + k_{x4}(x_3 - x_5) + k_{bx3}x_3$$
$$= -k_{ma}e_a \sin(\Psi_a) \qquad (2e)$$

$$m_3\ddot{y}_3 - k_{ma}\delta_a \cos(\Psi_a)\sin(\alpha_a) + k_{by3}y_3$$
$$= k_{ma}e_a \cos(\Psi_a)\sin(\alpha_a) \qquad (2f)$$

$$m_3\ddot{z}_3 + k_{ma}\delta_a \cos(\Psi_a)\cos(\alpha_a) + k_{bz3}z_3$$
$$= -k_{ma}e_a \cos(\Psi_a)\cos(\alpha_a) \qquad (2g)$$

$$I_3\ddot{\theta}_3 - r_3k_{ma}\delta_a \cos(\Psi_a) + k_{\theta 4}(\theta_3 - \theta_5) = r_3k_{ma}e_a \cos(\Psi_a) \qquad (2h)$$

where r_k is the base circle radius of gear k=2 or 3, ψ_a is the helix angle, k_{ma} is the averaged mesh stiffness, e_a is the transmission error excitation, k_b term refers to bearing stiffness, and δ_a is the elastic deflection along the line of mesh action between gears 2 and 3. The expansion of δ_a in terms of gear displacement vectors will be given later. Also, note that the effects of axial k_{x4} and torsional $k_{\theta 4}$ stiffness terms of the counter-shaft on the dynamics of the driven gear in the first stage are included in Equations (2e) and (2h) respectively.

A similar sets of undamped equations of motion can be derived for the second stage gear pair, which is denoted by subscript b, using the coordinate vector $\{x_5, y_5, z_5, \theta_5, x_6, y_6, z_6, \theta_6\}$ of gears 5 and 6. They are

$$m_5\ddot{x}_5 - k_{mb}\delta_b \sin(\Psi_b) + k_{x4}(x_5 - x_3) + k_{bx5}x_5$$
$$= k_{mb}e_b \sin(\Psi_b) \qquad (3a)$$

$$m_5\ddot{y}_5 + k_{mb}\delta_b \cos(\Psi_b)\sin(\alpha_b) + k_{by5}y_5$$
$$= -k_{mb}e_b \cos(\Psi_b)\sin(\alpha_b) \qquad (3b)$$

$$m_5\ddot{z}_5 - k_{ma}\delta_b \cos(\Psi_b)\cos(\alpha_b) + k_{bz5}z_5$$
$$= k_{ma}e_b \cos(\Psi_b)\cos(\alpha_b) \qquad (3c)$$

$$I_5\ddot{\theta}_5 + r_5k_{mb}\delta_b \cos(\Psi_b) + k_{\theta 4}(\theta_5 - \theta_3) = -r_5k_{mb}e_b \cos(\Psi_b) \qquad (3d)$$

$$m_6\ddot{x}_6 + k_{mb}\delta_b \sin(\Psi_b) + k_{bx6}x_6 = -k_{mb}e_b \sin(\Psi_b) \qquad (3e)$$

$$m_6\ddot{y}_6 - k_{mb}\delta_b \cos(\Psi_b)\sin(\alpha_b) + k_{by6}y_6$$
$$= k_{mb}e_b \cos(\Psi_b)\sin(\alpha_b) \qquad (3f)$$

$$m_6\ddot{z}_6 + k_{mb}\delta_b \cos(\Psi_b)\cos(\alpha_b) + k_{bz6}z_6$$
$$= -k_{mb}e_b \cos(\Psi_b)\cos(\alpha_b) \qquad (3g)$$

$$I_6\ddot{\theta}_6 + r_6k_{mb}\delta_b \cos(\Psi_b) + k_{\theta 7}(\theta_6 - \theta_7) = -r_6k_{mb}e_b \cos(\Psi_b) \qquad (3h)$$

Here, we note that the effects of counter-shaft compliances are formulated into the driving gear equations rather than the driven gear ones given in Equation (2). In both stages, the elastic deflections δ_a and δ_b along the line of actions normal to the tooth surfaces in contact can be derived in terms of gear center displacements and input transmission errors. Consider a pair of gears with helix angle ψ shown in Figure 3 that is oriented in such a way that its mesh orientation is defined by α. From basic gear kinematics, we can easily determine the elastic deflection δ between gear k and k+1 as

$$\delta = e + (x_{k+1} - x_k)\sin(\psi) + [\,(z_{k+1} - z_k)\cos(\alpha) - (y_{k+1} - y_k)\sin(\alpha)$$
$$- (r_{k+1}\theta_{k+1} + r_k\theta_k)\,]\cos(\psi) \qquad (4)$$

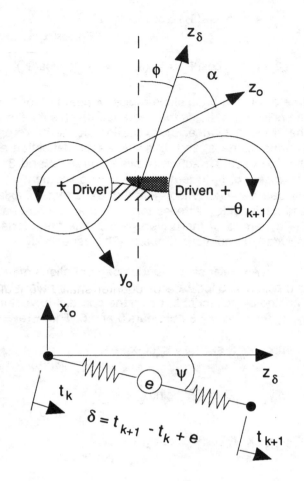

Figure 3. Mesh orientation of a gear pair with respect to x_o, y_o and z_o axes.

This equation is general and applies to both first and second stage meshes. It may be noted that $(\delta-e)$ is known as the dynamic transmission error (DTE). Also, the mesh force f_m can be computed by multiplying δ with mesh stiffness k_m. Substituting Equation (4) into the two sets of Equations (2) and (3) using the appropriate subscripts and combining them with the pair of formulations in Equation (1), we can obtain a stiffness matrix of dimension 18 given by

$$[K] = \begin{bmatrix} k_{11} & k_{12} & & & & & \\ k_{21} & k_{22} & k_{23} & & & & \\ & k_{32} & k_{33} & & k_{35} & & \\ & & & k_{53} & k_{55} & k_{56} & \\ & & & & k_{65} & k_{66} & k_{67} \\ & & & & & k_{76} & k_{77} \end{bmatrix} \quad (5)$$

where k_{11}, k_{12}, k_{22}, ..., k_{77} are the sub-matrices corresponding to the respective components referred to by their two subscripts. For example, k_{35} consists of stiffness terms that couple displacement coordinates of gear 3 to gear 5.

Up to this point, we have not included the effects of counter-shaft pitch and bounce motions. In order to incorporate its effects, we need to introduce additional constraint equations to be able to relate gear displacement vector $\{y_3, y_5, z_3, z_5\}$ to counter-shaft center of gravity displacement vector $\{y_4, z_4, \theta_{y4}, \theta_{z4}\}$. Assuming that the counter-shaft is uniform along its length, its center of gravity location can be determined by

$$L_a = \frac{L_4(m_5+m_4)}{m_3+m_5+2m_4} \quad ; \quad L_b = L_4 - L_a \quad (6)$$

where L_4 is the counter-shaft gear separation distance along the length of the shaft, and L_a and L_b are the distances between the centers of gravity of the counter-shaft and gears 3 and 5 respectively. Given L_a and L_b, we can derive a set of transformation equations from the vectors $\{y_3, y_5, z_3, z_5\}$ to $\{y_4, z_4, \theta_{y4}, \theta_{z4}\}$ assuming that the counter-shaft behaves rigidly relative to these coordinates. The 4 constraint equations are

$$y_3 = y_4 - \theta_{z4} L_a \quad (7a)$$
$$z_3 = z_4 + \theta_{y4} L_a \quad (7b)$$
$$y_5 = y_4 + \theta_{z4} L_b \quad (7c)$$
$$z_5 = z_4 - \theta_{y4} L_b \quad (7d)$$

Accordingly, we can apply these constraint equations to obtain a new set of system stiffness matrix $[K']$ and force vector $\{f'\}$ as follows

$$[K'] = [T]^T [K] [T] \quad ; \quad \{f'\} = [T]^T \{f\} \quad (8a,b)$$

where $[T]$ is the transformation matrix that is equivalent to the relationships given by Equation (7). The effective mass moment of inertia terms in the new mass matrix $[M']$ corresponding to θ_{y4}, and θ_{z4} are

$$I_{y4} = I_{y3} + I_{y5} + \frac{m_3 m_5 L_4^2}{m_3+m_5} + \frac{m_4 L_4^2}{12} + m_4(0.5L_4 - L_a)^2 \quad (9a)$$

$$I_{z4} = I_{z3} + I_{z5} + \frac{m_3 m_5 L_4^2}{m_3+m_5} + \frac{m_4 L_4^2}{12} + m_4(0.5L_4 - L_a)^2 \quad (9b)$$

Hence, the system equations of motion in matrix form is given by

$$[M']\{\ddot{q}\} + [C']\{\dot{q}\} + [K']\{q\} = \{f'\} \quad (10)$$

where $[M']$, $[C']$ and $[K']$ are the mass, damping and stiffness matrices respectively, and

$$\{q\} = \{ \theta_1, x_2, y_2, z_2, \theta_2, x_3, \theta_3, y_4, z_4, \theta_{y4}, \theta_{z4},$$
$$x_5, \theta_5, x_6, y_6, z_6, \theta_6, \theta_7 \}^T. \quad (11)$$

The undamped natural frequencies and their corresponding mode shapes can be computed from the eigenvalue problem $([K'] - \omega^2 [M']) = \{0\}$ that is formulated by applying harmonic solution $\{q\} e^{j\omega t}$ to Equation (10). For each mode, we assume a constant modal damping to represent the effects of $[C']$. The dynamic response of any degree of freedom in the lumped parameter model can be computed from the modal expansion technique by superposition of the modal responses due to all 18 system modes [8].

BASELINE MODEL SIMULATION

The proposed lumped parameter model is applied to a baseline system with design parameters listed in Table 1. This gear train system consists of two identical speed reducer spur gear pairs having gear ratio of 1:2 and $\phi = 20$ degrees pressure angle. Time-averaged mesh stiffness is $k_m = 2.0 \times 10^8$ N/m. The four sets of bearing stiffnesses are assumed to be identical. Also, both the input and output shaft torsional compliances are identical with stiffness value equals to $k_{\theta 1} = k_{\theta 7} = 12.7$ N-m. First the modal predictions for modes of significant interest, implying those that are mostly characterized by coupled torsional and translational motions, are compared to results from finite element calculations which has been performed using a general purpose finite element simulation code [9]. The direct comparison is shown in Table 2. Maximum deviations between the two calculations are found to be within 5 percent error. These discrepancies are primarily due to bending compliance of the counter-shaft which has not been included in the lumped parameter formulation. This factor is consistent with the fact that our lumped

Table 1. Design parameters of a baseline layshaft gear train system.

engine I_1 (kg-m^2)	0.0472
load I_7 (kg-m^2)	0.00472
pinion inertia I_2, I_5 (kg-m^2)	4.72E-5
gear inertia I_3, I_6 (kg-m^2)	3.78E-4
pinion radius r_2, r_5 (mm)	25.0
gear radius r_3, r_6 (mm)	50.0
pressure angle ϕ (degree)	20.0
gear ratio	1:2
mesh stiffness (N/m)	2.0E+8
T.E. e_a, e_b (mm)	0.001
counter-shaft mass m_4 (kg)	1.51
counter-shaft diameter (mm)	40.0
counter-shaft length (mm)	156.25
bearing stiffness k_b (N/m)	2.0E+9
Input shaft stiffness $k_{\theta 1}$ (N-m)	12.7
Output shaft stiffness $k_{\theta 7}$ (N-m)	12.7
Damping ratio	0.06
Helix angle ψ (degree)	0

Table 2. Comparison of predicted resonance frequencies in Hertz and finite element calculations.

Mode	Model	FEM	Error (%)
3	44	44	0.0
4	3,907	3,784	-3.3
5	7,018	6,867	-2.2
6	7,167	7,110	-0.8
7	8,474	8,518	0.5
9	10,928	10,444	-4.6
14	14,586	14,443	-1.0
17	19,505	19,334	-0.9

parameter model calculations are slightly stiffer in most cases. A description of the corresponding mode shapes are given in Figure 4 for modes 4 through 7, 9, 14 and 17. Except for modes 4 (higher order torsional mode) and 6 (pitch and bounce of counter-shaft), we found significant coupling between the torsional and translational motions of the gears as shown in the mode shape results. These modes play very important roles in affecting mesh force amplifications and resonance peaks in the vibration response due to transmission error excitations.

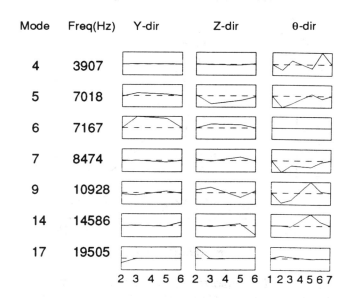

Figure 4. Subset of mode shapes of the baseline layshaft gear train system. Labels: 1=engine, 2=first stage pinion, 3=first stage gear, 4=counter-shaft, 5=second stage pinion, 6=second stage gear, and 7=load.

The results of forced response calculations for the baseline layshaft gear train system due to constant amplitude transmission error input in either the first e_a or second e_b stage gear mesh are discussed next. Figures

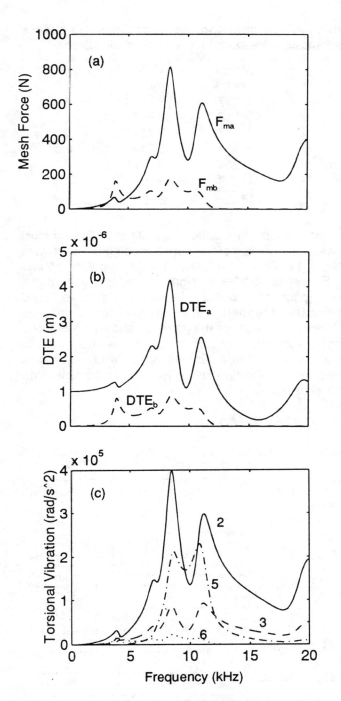

Figures 5a-c. Dynamic mesh and torsional response due to transmission error excitation of $e_a=0.001$mm at the first stage gear mesh.

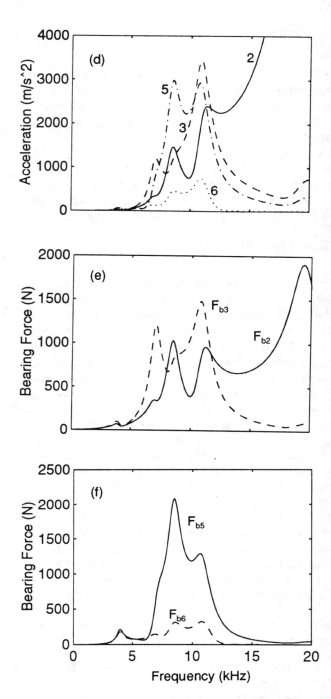

Figures 5d-f. Acceleration response and bearing loads due to transmission error excitation of $e_a=0.001$mm at the first stage gear mesh.

5 and 6 show dynamic response due to $e_a=e_b=0.001$mm at the fundamental mesh frequency. In both figures, we show plots of dynamic mesh force and dynamic transmission errors at both mesh points due to only one transmission error input applied at any one time. Most of the peak responses which correspond to the system resonance frequencies are found to occur within the range of 3-20kHz. The driving-point mesh force and DTE functions can be seen to be generally much higher in amplitude than the cross-point functions at the other mesh point where no transmission error excitation is being applied, as expected. It may be noted the difference in mesh force or DTE peak amplitude at about 8.5 kHz between the driving-point and cross-point mesh points are only significant when the first stage transmission error input is applied. The response amplitude difference in the vicinity of 8.5 kHz is relatively small when we only have transmission error

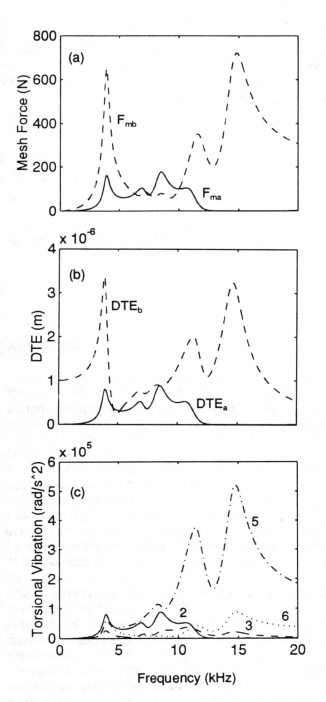

Figures 6a-c. Dynamic mesh and torsional response due to transmission error excitation of $e_b=0.001$mm at the second stage gear mesh.

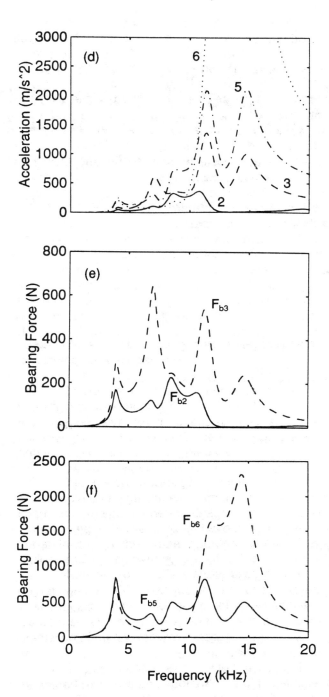

Figures 6d-f. Acceleration response and bearing loads due to transmission error excitation of $e_b=0.001$mm at the second stage gear mesh.

in the second stage gear pair as shown in Figure 6a. This behavior is evident from the mode shape plot in Figure 4 showing high degree of coupling between torsional and translational motions in the first stage gear pair. A similar explanation can be given to mesh force and DTE peaks at about 14.5 kHz where the reverse condition exist. In this situation, the response peak difference is significant when we have transmission error in the second stage gear pair e_b. This mode is not even excited by e_a. On the other hand, the contributions of mode 9 (with the resonance frequency at about 10.9 kHz) to mesh force and DTE functions are equally strong for both cases of transmission error excitations. This is due to the fact that we have near symmetrical pattern of mode shape in the first and second stage gear mesh components.

Results of dynamic response in terms of radial translational and torsional accelerations of each gear pairs and their bearing loads F_b are also given in Figures 5d-f and 6d-f. In the vibration response functions, we concluded that the peak amplitudes are generally higher for gear pair closer to the excitation point. In the bearing force results, we observed that F_b corresponding to smaller gears have higher amplitude of force transmissibilities, especially at higher frequencies as shown in Figures 5e and 6f. This is because lighter gears tend to vibrate more vigorously and therefore causing higher levels of dynamic forces in the bearings. Also, in general bearings that are closer to the excitation point experienced higher level of dynamic loads. For example, when we applied transmission error excitation to the second stage gear pair, the bearings supporting this stage transmit larger dynamic loads compared to bearings at the first stage. Also, we found that F_{b3} of the first stage bearing supporting the counter-shaft is higher in amplitude than F_{b2} of the other first stage bearing that is connected to the input shaft. This behavior is seen to occur over a wide frequency range.

CASE 1: EFFECTS OF COUNTER-SHAFT DIAMETER

The effects of counter-shaft diameter d_4 on resonance frequencies ω_n and dynamic response are investigated in this example case. Using the baseline diameter of $d_{4,ref} = 40$ mm as reference, we vary d_4 such that the ratio $\sigma = d_4/d_{4,ref}$ is between 0.1 to 2.0. Its effect on natural frequencies ω_n are shown in Figure 7. By increasing counter-shaft diameter, we are essentially reducing its torsional compliance, and increasing its mass moment of inertia and total mass. Since both parameters have opposite effects on the natural frequencies, some of the modes could be decreasing or increasing in frequency within certain diameter range and remain constant elsewhere. The inflection points are approximately close to $\sigma=1$. The exact locations depends on the strain and kinetic energy distributions within the counter-shaft component. Also, since the mass is proportional to σ^2, and compliance is inversely proportional to σ^4, the rate of change of mass and compliance values depend on whether $\sigma <1$ or $\sigma >1$. The natural frequencies ω_n shown in Figure 7 are bounded by modes 4 and 17 which are the lowest and highest natural frequencies in this subset group of modes. Modes 5, 6, 7, and 9 are generally decreasing in frequency as counter-shaft diameter increases with the exception of a narrow range of σ values. On the other hand, mode 17 is relatively insensitive to counter-shaft diameter up to $\sigma =1.5$. Beyond this point, it starts to increase rapidly with increasing σ. Modes 4 and 14 have their own unique characteristics. In the case of mode 4, the resonance frequency starts to decrease very rapidly with decreasing σ. This is because mode 4 is a torsional mode that is primarily controlled by counter-shaft torsional compliance. When its compliance increases with decreasing shaft size, the two-stage layshaft gear train system begins to split into two weakly coupled single mesh gear pair sub-systems. Mode 14 depicts a shallow 'u-shape' over most of the lower range of σ. At higher σ values, it basically stabilizes and remain almost constant.

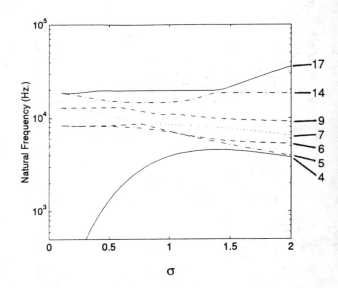

Figure 7. Effect of counter-shaft diameter on the natural frequencies.

For each discrete value of σ between 0.1 and 2.0, a set of forced response calculations were also performed. Both mesh force spectra $F_m(\omega)$ and bearing loads $F_b(\omega)$ were computed. These two response functions are of great interest because they provide a measure of the effects of system dynamics on mesh force amplifications, and the effectiveness of vibration transmissibility from mesh point to bearing locations. The results are shown in Figures 8 and 9 for the mesh force F_{ma} and bearing load F_{b3} respectively. At either very low and high values of σ, the mesh force spectrum is dominated by a single peak. For $\sigma <<1$ mode 5 seems to be dominant, while for $\sigma >>1$ mode 7 becomes the major contributor. In the vicinity of $\sigma =1$, we found two resonance peaks (modes 6 and 7) in $F_{ma}(\omega)$. In the case of $F_{b3}(\omega)$, two force peaks are always present in most cases. At low value of σ, modes 5 and 7 are significant. As σ increases, the contribution from mode 5 reduces, but at the same time, mode 9 becomes a key factor. Mode 7 basically remains as a strong contributor to F_{b3} throughout the entire range of σ used here.

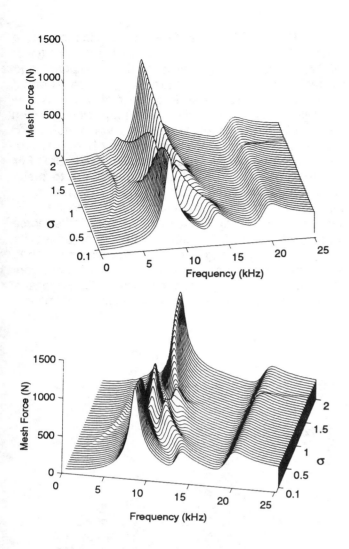

Figure 8. Driving-point mesh force spectra as a function of counter-shaft diameter due to transmission error excitation at the first mesh.

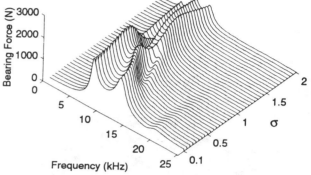

Figure 9. Effect of counter-shaft diameter on its bearing load (first stage). Here we have imposed transmission error excitation at the first mesh only.

CASE 2: EFFECTS OF BEARING STIFFNESS

Next the effects of bearing stiffnesses k_b on the natural frequencies ω_n, mesh force response $F_m(\omega)$ and bearing loads $F_b(\omega)$ are examined. First we parametrized the problem by introducing a dimensionless variable $\gamma = (k_b/k_m)$ similar to case 1. This enable us to perform the parametric studies in a more general manner in order to predict typical dynamic response trends as a function of dimensionless bearing stiffness parameter γ. By varying γ from 10^{-1} to 10^3, we compute the mode shapes and plotted the corresponding resonance frequencies against γ as shown in Figure 10. All the natural frequencies ω_n are found to either increase monotonically for specific range of γ or remain relatively constant. They are also bounded by the lowest and highest modes similar to the previous result. We did not observe any crossing of the functions; hence the mode ordering remains stable.

Modes 4 through 6 first increase up to a certain point and then level off. The transition points are higher for higher mode. On the other hand, modes 7, 9 and 14 are always increasing within the parameter range shown. The highest mode 17 at first remain insensitive to variation in γ up to about $\gamma=30$ and then starts to increase monotonically. It may be noted that these transition points are really indications of fundamental shift in the distribution of strain energy within the layshaft gear train system. When a resonance mode is monotonically increasing in its natural frequencies, the change in modal strain energy in the bearing stiffness element is drastic. The rate of change depends on slope of this natural frequency curve. In the range where the curve remains horizontal, the strain energy in the bearing is essentially stable going from one value of γ to the next.

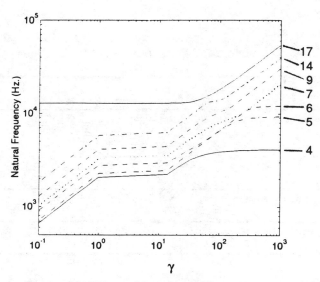

Figure 10. Effect of bearing stiffness on ω_n.

Figures 11 and 12 show variation in mesh force function $F_{ma}(\omega)$ and counter-shaft bearing load $F_{b3}(\omega)$ of the first stage. The peak force levels corresponding to modes 5 and 17 (see Figure 4) can be seen to dominate the overall mesh force for high $\gamma \gg 1$ and low $\gamma \ll 1$ values of bearing stiffnesses respectively. When resonance frequencies are relatively insensitive to change in bearing stiffness values, that is $\Delta\omega_n/\Delta\gamma \approx 0$, the mesh force level is found to be maximum. For example, mode 17 contributes significantly to mesh force peak for $\gamma < 1$ and slowly reduces for $\gamma > 1$ when its frequency starts to increase monotonically. On the other hand, mode 5 initially ($\gamma < 1$) does not affect mesh force response substantially, but gradually increases its contribution in the range of $\gamma > 1$. In the case of F_{b3}, the levels are relatively small when the bearings are very compliant. As bearings become stiffer, F_{b3} increases due to higher stiffness values. At some mid-point near

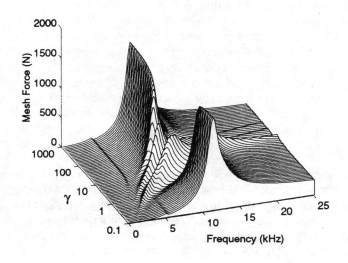

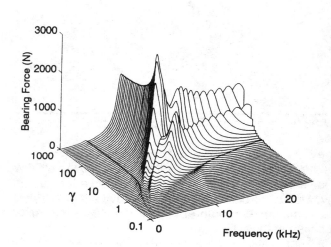

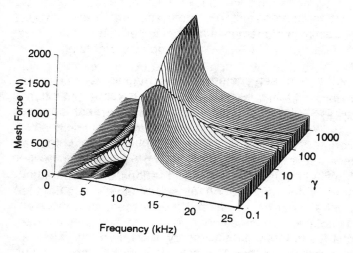

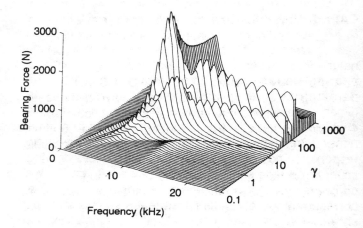

Figure 11. Driving-point mesh force spectra as a function of bearing stiffness due to transmission error excitation at the first mesh.

Figure 12. Effect of bearing stiffness on the counter-shaft bearing load of the first stage. Here, we have imposed only the first mesh transmission error.

the baseline case of $k_b=10k_m$, F_{b3} reaches a maximum. As γ increases further, F_{b3} levels start to reduce. The peaks eventually reach certain lower limits as shown in Figure 12 even though the natural frequency kept increasing. At this point, the bearings are much more rigid compared to the rest of the compliant elements in the gear train system. Hence, continuing to increase γ does not change the magnitude of the F_{b3}.

CONCLUDING REMARKS

A three-dimensional lumped parameter dynamic model of dimension 18 of a generic layshaft gear train system has been derived to study the effects of design parameters on the vibratory response due to transmission error excitation. The formulation was developed to account for coupled translational and torsional motions of the gears. Modal prediction of a baseline system was verified numerically by comparison to finite element calculation. Forced response computation was performed to show the capability of the model to predict mesh force spectrum, dynamic transmission error, bearing loads and gear acceleration response. The model was also used to perform two specific sets of parametric analyses by varying counter-shaft diameter and bearing stiffnesses. Their effects on natural frequencies and peak dynamic responses were examined in detailed. The results presented in this paper suggests that this simple lumped parameter model of a layshaft gear train system can be used to perform numerous calculations to examine dynamic response trends and evaluate new design layout.

ACKNOWLEDGMENTS

The authors wish to acknowledge support from The Ohio State University Gear Dynamics and Gear Noise Research Laboratory to conduct this work. Also, we want to thank Mr. Jonny Harianto for generating some of the plots and results presented in this paper.

REFERENCES

1. M. Benton and A. Seireg, "Factors Influencing Instability and Resonances in Geared Systems," *ASME Journal of Mechanical Design*, Vol. 103, pp. 372-378, 1981.

2. H. Iida and A. Tamura, "Coupled Torsional-flexural Vibration of a Shaft in a Geared System," Proceedings of the IMechE Conference on Vibrations in Rotating Machinery, pp. 67-72, 1984.

3. H. Iida, A. Tamura, M. Oonishi, "Coupled Torsional-flexural Vibration of a Shaft in Geared System, 3rd Report, Dynamic Characteristics of a Counter Shaft in a Gear Train System," *Bulletin of JSME*, Vol. 28, pp. 2694-2698, 1985.

4. H. Iida, A. Tamura and H. Yamamoto, "Dynamic Characteristics of a Gear Train System with Softly Supported Shafts," *Bulletin of JSME*, Vol. 29, pp. 1811-1816, 1986.

5. K. Umezawa, T. Ajima and H. Houjoh, "Vibration of Three Axes Gear System," *Bulletin of JSME*, Vol. 29, pp. 950-957, 1986.

6. F.K. Choy, Y.K. Tu, M. Savage and D.P. Townsend, "Vibration Signature Analysis of Multi-stage Gear Transmission," Proceedings of the ASME International Conference on Gearing and Power Transmission, pp. 383-387, 1989.

7. P. Velex and A. Saada, "A Model for the Dynamic Behavior of Multistage Geared Systems," Proceedings of the 8th World Congress on The Theory of Machines and Mechanisms, Vol. 2, pp. 621-624, 1991.

8. D.J. Ewins, *Modal Testing: Theory and Practise*, Research Studies Press Ltd., England, 1986.

9. The MacNeal-Schwendler Corporation, *MSC/Nastran Finite Element Software Version 68*, 1996.

971965

Numerical Methods to Calculate Gear Transmission Noise

W. Hellinger, H. Ch. Raffel, and G. Ph. Rainer
AVL List GmbH

Copyright 1997 Society of Automotive Engineers, Inc.

ABSTRACT

This report shows the methods, which AVL uses for the calculation of gear box noise.

The analysis of the gear box structure (housing) is done using finite element method (FEM), thereby the natural frequencies are calculated as well as forced vibrations.

As input for the FE calculation of the forced vibrations, the dynamic bearing forces of the shafts in the gear box or the dynamic tooth mesh are used. These forces are determined using the MBS (multi body system) software GTDYN, considering the torsional vibrations as well as axial and bending vibrations.

Several examples of calculation results for the investigation of the gear dynamics are shown within the scope of this report.

INTRODUCTION

With increasing demands on silent running of motor-vehicles, we have to pay increasing attention to the noise generated by transmission units.

With transmission units two principal kinds of noise - gear rattle and gear whine - can be distinguished.

Gear rattle [1] is mainly caused by the changes of contact in the tooth meshes, while they do not transmit torques. The most important influencing parameters for the gear rattle are the speed irregularity at the input side of the transmission unit, the backlash in the tooth meshes and the friction and damping forces acting at the gear transmission components.

The **gear whine** is caused by the tooth meshes between the load transmitting gears (active speed, differential). The frequency of this noise is determined by the meshing frequencies of the corresponding tooth meshes.

The intensity of the noise detected in the vehicle is strongly influenced by the way of transfer of the vibrations caused by the meshing forces via the shafts and supports (bearings) into the gear box structure.

To improve the support of the engineers working at the test beds, when clarifying noise phenomena of transmission units at the one hand and to enable reliable assessment of various variants during gear unit design at the other hand, according methods and tools for the numerical investigation [2, 3, 4, 5] of gear transmission dynamics are required.

This report presents the numerical methods for the investigation and reduction of gear transmission noise used at AVL.

CALCULATION PROCEDURE

The procedure for numerical investigations of gear transmission noise is devided essentially into two parts.

The dynamic behaviour of the gear box structure is analysed with the help of the finite element method (FEM). The excitation forces acting at the structure are determined by using software for the calculation of the dynamic behaviour of multi body systems (MBS).

The sequence of the single tasks and the interfaces between the two calculation parts - FE and MBS calculations - are shown in figure 1.

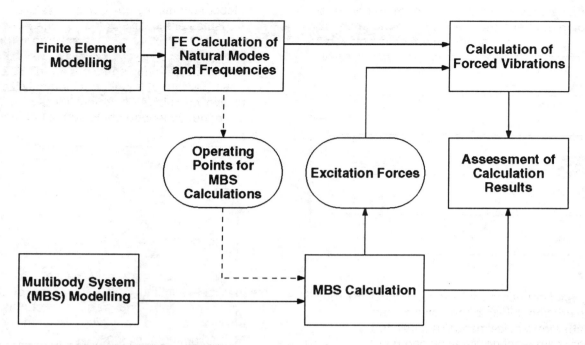

Figure 1: Sequence and data flow for the calculation of gear transmission noise

FE and MBS modelling can be performed in parallel. The most interesting operating points for the simulation of the dynamics of the MBS are determined with respect to the rotational speed irregularity at the clutch, if gear rattle is to be investigated. For gear whine the operating points are chosen by comparison of the natural modes and frequencies of the gear box structure with the meshing frequencies of the single

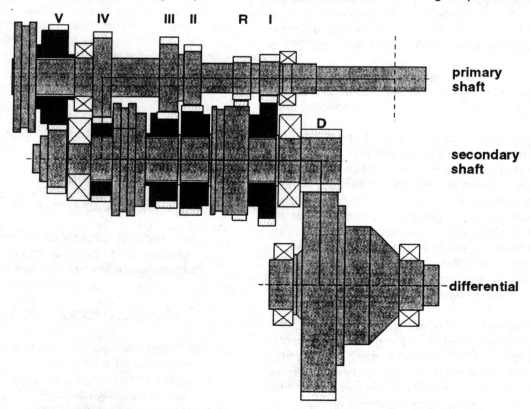

Figure 2: Model of a five-speed passenger car transmission unit (fourth gear operating)

gear meshes.

On the one hand the excitation forces calculated by the MBS software can be used directly for the comparison of different design variants, on the other hand they can be used as excitation forces for subsequent calculations of forced vibrations using FE codes or using special software for nonlinear dynamics as for example the software EXCITE [6].

The next sections shall describe the two main tasks for the calculation of gear transmission noise more detailed.

CALCULATION OF EXCITATION FORCES

For the calculation of excitation forces acting on the transmission unit, the MBS software GTDYN [7] is used. This software contains predefined elements for shafts, gears (including the gear meshes), roller and slider bearings.

The calculations consider torsional degrees of freedom as well as all degrees of freedom for bending and translation (radial and longitudinal) motion of shafts and gears. Thus the influence of the bending deformation of shafts on actual backlashes in the tooth meshes is included in the simulation calculations.

Additionally periodically varying meshing stiffnesses are considered if required. As for example variations of the pitch in a tooth mesh have consequences for the actual tooth mesh stiffness, this influence on the dynamic behaviour of the system and the thereby caused dynamic forces must be considered in simulation calculations.

MBS MODEL: The measured or calculated speed irregularity is prescribed at the clutch position. The model of the primary shaft as shown in figure 2 consists of 16 shaft elements and five gear wheels (for first, second, third, fourth and reverse speed), which are fixed to the shaft.

The gear wheel for the fifth speed is designed as a free running gear. The clearances between the free running gear and the primary shaft are considered in the calculations for axial direction as well as for radial motions.

The primary shaft is supported by two tapered roller bearings, which in the dynamic calculation are represented by their clearance, stiffness and damping values.

At the secondary shaft the gear wheels for fifth speed and differential are fixed. The gears for the remaining speeds are modelled as free running wheels. In the model the pawls, which connect the free running gears with the shaft when the according gears are operating, are fixed at the shaft.

In the results presented in this paper the calculation has been performed for the operation of fourth gear. The according free running gear is supported at the appropriate shaft element in radial and axial direction. The torque is transmitted from the free wheel via a rotational stiffness and damping to the pawl and to the secondary shaft. The secondary shaft is supported by one ball bearing and one roller bearing.

The differential is represented in the calculation model as a short shaft, which is supported by two tapered roller bearings. At the differential the (constant) output torque is applied.

RESULTS OF THE MBS CALCULATIONS: The possibilities, which arise due to the use of the herein presented calculation of excitation forces, are described using the results obtained for several gear unit variants.

The already shown model of the transmission unit was used as baseline for the calculations of variants. The chosen operating point was fourth gear, engine speed 1700 rpm. In practical use of the gear unit in a passenger car, gear whine was detected at the mentioned operating conditions.

Several variants of the model were used to investigate the influence of several parameters on the dynamic loads in the system. The rotational excitation at the clutch and the applied output torque at the differential remained unchanged for all calculated variants.

As the rotational speed irregularity at the clutch was known neither by measurement nor by calculation, a sinusoidal waveform was used to prescribe the speed irregularity. The frequency for the speed fluctuation was set to the second order of the engine speed, the maximum amplitude was assumed to be 3 rad/s.

A constant output torque of 390 Nm was applied at the differential in all investigated variants.

Meshing data: The module used for the gears was between 1.5 and 2.5 mm, the helix angles were around 30 degree.

For the calculations the following variants were considered:

Variant A: Entire model as shown in figure 2 assuming constant meshing stiffnesses and without consideration of radial and axial clearances in the supports of the free running gears.

The constant mesh stiffness was calculated from the theoretical meshing data corresponding to DIN standards. The stiffness value used for the tooth mesh of the fourth speed was about 320 kN/mm, for the gear mesh between secondary shaft and differential it was about 515 kN/mm.

Variant B: Entire model as shown in figure 2

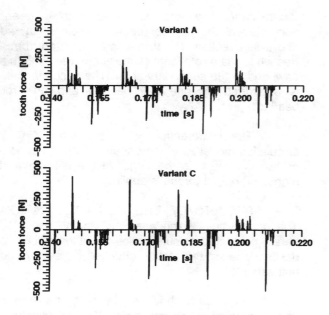

Figure 3: Contact forces in the tooth mesh for the first gear - comparison between the variants A (top) and C (bottom)

with variable meshing stiffnesses in the perfect (no deviations from the theoretical meshing geometry) tooth meshes without consideration of radial and axial clearances in the supports of the free running gears.

The fluctuations of the mesh stiffness were estimated. The stiffness fluctuations used in the calculation varied between 310 and 325 kN/mm (fourth gear) and between 495 and 520 kN/mm (differential). This fluctuation was used to represent the change of the mesh stiffness due to the change of tooth pairs in contact.

<u>Variant C</u>: Same as variant B but with maximum radial and axial clearances in the supports of the free running gears.

<u>Variant D</u>: Same as variant C, but with an uniform deviation from the theoretical tooth geometry for all teeth.

As no information was available about the actual manufactoring errors, this influence was considered by just increasing the variation of the meshing stiffness. The stiffness used in this calculation variant for the gear mesh between secondary shaft and differential changed between 475 and 520 kN/mm.

For each of the specified variants three engine cycles have been calculated. In the following diagrams the calculated results for the third cycle are shown.

GEAR RATTLE: The gear rattle noise is mainly caused by the changes of flank contacts in the tooth meshes between the gears, which currently do not transmit power between the shafts. The main influencing parameters for gear rattle are represented by the moments of inertia of the (free running) gears, the speed irregularity at the clutch, the backlash in the tooth meshes and the clearances in the supports of the free running wheels.

Figure 3 shows the influence of the radial and axial clearances between the shafts and the free running wheels on the changes of contact and the magnitude of the thereby caused meshing forces.

The results indicate an increase of the maximum tooth forces up to 20 percent, if the clearances in the supports of the free running gear wheels are considered. Additionally a comparison of the two variants shows that the times of contacts in variant C show a significant time shift compared to variant A, due to the passing through the radial and axial clearances. Caused by the increased relative velocity between the tooth flanks and the thereby increased energy contents when the impact occurs, variant C shows also higher and shorter peak forces. This leads - due to the transmission of the forces by shafts and bearings - to a significant increase of vibration excitation for the gear box structure.

In our example the calculated contact forces are relatively small. Gear rattle is no problem for the examined gear unit at the considered operating point.

GEAR WHINE: In practical operation gear whine is detected at the examined gear transmission unit at engine speeds of about 1700 rpm with fourth gear operating. As already mentioned, this type of noise usually appears, if natural frequencies of the gear housing coincide with single meshing frequencies. In the example shown (figure 2, fourth gear operating) the meshing frequencies for the chosen operating point are at 1160 Hz in the gear mesh for the fourth gear and at 510 Hz in the tooth mesh between secondary shaft and differential.

As the gear whine was hardly reproducible with the calculation models used up to now (as they use constant meshing stiffnesses and consider only the torsional degree of freedom), the calculations for the presented variants have been performed to investigate, if the mentioned effects can be simulated better by consideration of additional degrees of freedom and by consideration of variable meshing stiffnesses.

The calculated results for the bearing forces have been subject to fast fourier transmission (FFT), to have easier comparisons between the single variants. The simulation results have been compared using these frequency spectra.

The FFT of the bearing force of the tapered roller bearing at the free end of the primary shaft is shown in figure 4. The bearing force is mainly influenced by the meshing forces of the fourth gear, which is next to this bearing. The shaded area marked in figure 4 indicates the frequencies between 1110 and

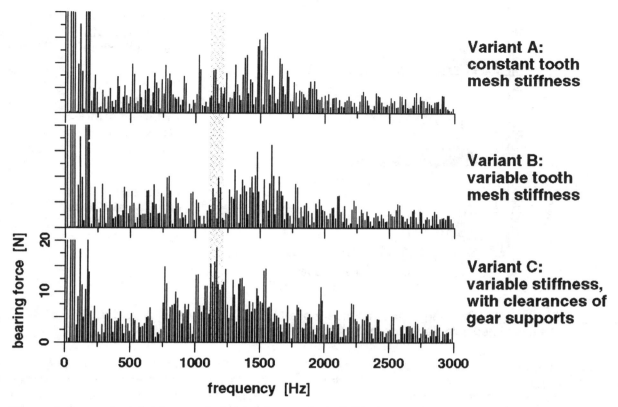

Figure 4: Support at the free end of the primary shaft, FFT of the calculated bearing forces for the calculation variants A, B and C

1210 Hz. This range is equivalent to the meshing frequency of the fourth gear (1160 Hz) +/- 50 Hz.

When looking at the frequency contents, the consideration of angle-dependent variable meshing stiffness (variant B) shows no significant differences compared to the calculation with constant meshing stiffness (variant A). As soon as the clearances between the free running gear wheels and the secondary shaft are considered in the calculations (variant C), clear changes in the frequency spectrum of the calculated bearing force are detected.

Due to the dynamic changes of contact in the clearances additional 'intermediate' frequencies are excited. Viewing the wrapping curves of the frequency shapes in figure 4 shows that significant differences between variant C and the remaining variants appear in the frequency range between 1110 and 1210 Hz only.

As several teeth are assumed to be in contact in the tooth mesh at the same time, the changes in the meshing stiffness are relatively small. For that reason, the influence of the varying meshing stiffness on the watched bearing force is small, too. Only if stiffness

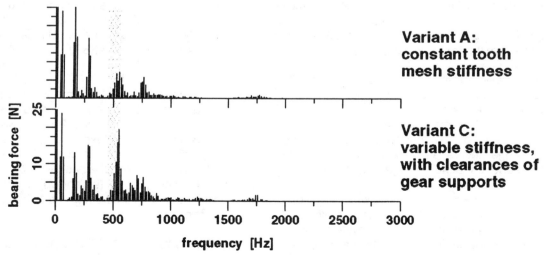

Figure 5: Support of the differential gear (support position below the engine-side bearing of the secondary shaft), FFT of the calculated bearing forces for the calculation variants A and C

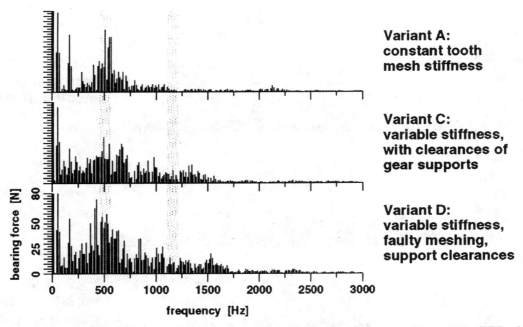

Figure 6: Ball bearing of the secondary shaft (between the gears IV and V), FFT of the calculated axial force for the calculation variants A, C and D

variations interfere with clearances in the supports of the free running gears, the components of the bearing forces around the meshing frequency rise.

As described later on, one natural frequency of the outer wall and the bearing walls of the gear box was found at about 1180 Hz. As the MBS calculations for the mentioned operation conditions resulted in relatively high forces, clear gear whine must be expected in this frequency range.

Figure 5 shows the frequency spectra for the axial force in one of the tapered roller bearings of the differential. The shaded area (460 to 560 Hz) indicates the meshing frequency for the tooth mesh between secondary shaft and differential gear (510 +/- 50 Hz).

The consideration of variable meshing stiffness and additional clearances (variant C) leads to calculated forces with three times the force amplitudes as they have been calculated for variant A (constant meshing stiffness and no consideration of clearances in the supports of the free running gears).

Especially in the range between 500 and 750 Hz variant C shows a significant increase of the dynamic force amplitudes compared to variant A. If parts of the gear box structure next to the watched bearing have natural frequencies in the mentioned frequency range, increased noise (gear whine) must be expected due to the increased excitation forces.

Another calculation variant has been investigated to show the influence of a faulty tooth mesh on the dynamic forces in the system. In this example, an uniform fault in the tooth mesh between secondary shaft and differential gear was assumed. In the calculation this case was described by a short-time decrease in the meshing stiffness (variant D).

The effects of these erratic changes of contact stiffness are represented in figure 6 by showing the FFT of the axial forces in the ball bearing of the secondary shaft.

The comparison of the bearing force components for the variants A and C shows an increase of the force amplitudes between about 200 and 1600 Hz. The only exception is the frequency range around 510 Hz (meshing frequency secondary shaft - differential gear).

Due to the consideration of the axial clearances between the free running gears and the secondary shaft, the pronounced peak between 500 and 550 Hz seen in the FFT for variant A is spread to the wider frequency range between about 300 to 700 Hz with smaller maximum values in variant C.

The consideration of the faulty tooth mesh results in a significant increase of the bearing forces. Strongly increased force amplitudes are located at about 400 to 600 Hz, at 800 Hz and at about 1600 Hz (fig. 6, variant D).

To give a brief summary, we can state that in our case variant C is convenient to describe the real behaviour of the gear unit. Although we do not have any measurement values of dynamic bearing forces, the increased excitation forces at frequencies between 1100 and 1200 Hz and a natural mode of the gear box structure at 1180 Hz (as shown in the next chapter) indicate that gear whine noise is to be expected at this operating point.

ANALYSIS OF THE STRUCTURAL DYNAMICS

The determination of the dynamic behaviour of the gear box structure is performed using the Finite

Element Method (FEM) and can be divided into the following tasks:

- Modelling of the structure
- Analysis of the natural frequencies
- Calculation of forced vibrations

MODELLING OF THE GEAR BOX STRUCTURE: The FE model consists of the gear box with simple models of primary shaft, secondary shaft and differential and of a rough model of the engine structure.

Gear box and engine are represented mainly by shell elements. In regions with mass accumulations (e.g. support positions, ...) solid elements are used. Shafts and bearings are represented by beam elements, gears consist of shell and lumped mass elements. The gear wheels for the fourth gear located in the path of torque transmission and the gears between secondary shaft and differential are connected in the line of action (perpendicular to the tooth flanks) by beam elements. Figure 7 shows the FE model of the entire driving unit.

As a first numerical check of the calculation model an analysis of the static deformation is performed. For this test a torque is applied at the input of the transmission unit (primary shaft). This calculation gives a first overview about the general distribution of stiffness and the flow of torque in the transmission unit. At the differential the output torque appears according to the transmission ratio of the gear unit. The output torque is also used as a check if the model supplies reasonable results.

DETERMINATION OF NATURAL FREQUENCIES: The calculation of the natural frequencies is the first step during the optimization of the dynamic behaviour of gear boxes. All significant structural modes are determined and ranges where

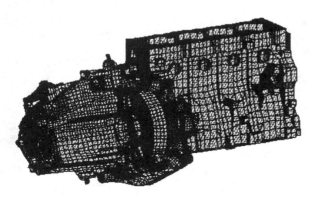

Figure 7: FE model of a driving unit consisting of engine and gear box structure

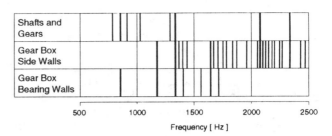

Figure 8: Table of natural frequencies

resonancies are possible (superpositions of vibrations of the shafts in the gear box and the gear box structure) are detected.

Figure 8 shows the table of natural frequencies for primary and secondary shaft, outer walls and bearing walls of the gear box in the frequency range between 500 and 2500 Hz

Bending vibrations of first and second order for the secondary shaft appear in the frequency ranges 790 to 915 Hz and 2100 to 2340 Hz and for the primary shaft between 1030 and 1340 Hz. Local modes of the outer walls of the gear housing have been detected at 1180 Hz, 1340 to 1440 Hz and in the frequency range above 1600 Hz. The bearing walls participate in the frequencies between 860 and 1710 Hz in the global oscillation behaviour.

Possible resonance areas are located at 860 Hz (secondary shaft and bearing walls), at 1180 Hz (gear housing and bearing walls), 1650 Hz (gear housing and bearing walls) and at frequencies of 2075 and 2340 Hz (secondary shaft and gear housing).

The comparison of the resonance areas calculated by FEM with the frequency contents of the calculated bearing forces (MBS) confirms the critical frequency ranges 1160 / 1180 Hz and 1600 to 1650 Hz.

FORCED VIBRATIONS: The target of the calculation of the forced vibrations is the determination of possible excitations of the gear box structure by the gear meshes, which actually transmit the torque.

Principally the way of calculation is structured as follows:

In a first step the actually operating gears are loaded with impact excitations. This calculation is done for each gear separately. Due to this method it is possible, to have a clear relation between excitations and the corresponding structural response. In the second step the dynamic bearing or meshing forces (MBS calculation) are considered as excitation forces. This describes the entire excitation in real operation.

The fig. 9 to 11 show as example the transfer mobilities calculated for an impact excitation at the secondary shaft (fourth gear) for the basic and one modified gear housing.

Figure 9 shows significant raise of the integral level at the gear box external structure near the secondary shaft and at the differential housing.

Afterwards design modifications as local increase of wall thicknesses at the outer walls of the gear box and introduction of additional ribs at the gear box and differential housing have been considered, to shift the natural frequencies of the significant gear housing modes. The goal for the modifications was, to avoid the resonances and to reduce the characteristic raise of integral level. An analysis of natural modes and a calculation of forced vibrations was performed also for the modified structure.

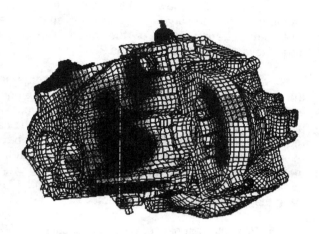

Figure 9: Transfer mobility, basic variant, 4th gear, integral level 0 to 5000 Hz

Figure 10 shows the improvement of the structural response at the gear housing and at the differential housing obtained by the described modifications of the structure.

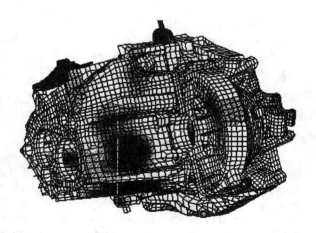

Figure 10: Transfer mobility, modified variant, 4th gear, integral level 0 to 5000 Hz

In figure 11 the transfer mobilities of basic and modified structure are represented. The peaks at frequencies of 1180 Hz, 2075 Hz, 2750 Hz and 3750 Hz are reduced significantly due to the design modifications.

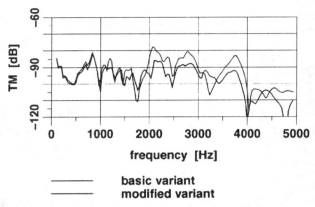

basic variant
modified variant

Figure 11: Transfer mobility, 4th gear, comparison basic variant - modified variant

CONCLUSIONS

With the calculation of gear transmission noise both, the structural dynamics as well as the excitation forces have to be considered.

For the calculation of the excitation forces MBS methods are suitable. The described investigations show that for the excitation forces radial and axial clearances especially in the supports of the free running gears must be considered besides the torsional vibrations, as they have significant influence on the dynamic forces in the system.

Additionally, consideration of variable meshing stiffnesses in the tooth meshes is desirable, as it has decisive influence on level and frequency contents of the dynamic excitation forces.

Structural dynamics is investigated using FEM. Thereby natural frequency analysis as well as calculation of transfer mobilities are used to assess various design variants.

The combination of both mentioned calculation methods (MBS and FEM) and their results bring a valuable support for the designer as well as for the test bed engineer.

REFERENCES

[1] Rust, A.; Brandl, F.K.; Thien, G.E.
"*Investigations into Gear Rattle Phenomena - Key Parameters and their Influence on Gearbox Noise*"
First International Conference on Gearbox Noise and Vibration, Cambridge, 1990

[2] Priebsch, H.H.; Hellinger, W.; Loibnegger, B.; Rainer, G.Ph.
"*Application of Computer Simulation for the*

Prediction of Vibration and Noise in Engines"
ATA 3rd Int. Conference, Firenze, 1992

[3]   Priebsch, H.H.; Affenzeller, J.; Kuipers, G
      "Structure Borne Noise Prediction Techniques"
      SAE-Paper 900019, 1990.

[4]   Rainer, G.Ph.; Loibnegger, B.
      "Design Supporting Analysis for Powertrain Noise and Vibration"
      IiM-Conference, Vienna, 1995.

[5]   Hellinger W.; Priebsch, H.H.; Rainer, G.Ph.; Tzivanopoulos, G.
      "Auslegung von Ventiltrieben für Hochleistungsmotoren - Rechnerische und Experimentelle Verfahren"
      16. Internationales Wiener Motorensymposium, Wien, 1995.

[6]   EXCITE User Manual, Graz, 1996.

[7]   GTDYN User Manual, Graz, 1995.

971966

Gear Noise Reduction of an Automatic Transmission Through Finite Element Dynamic Simulation

Brian Campbell, Wayne Stokes, and Glen Steyer
Structural Dynamics Research Corp.

Mark Clapper, R. Krishnaswami, and Nancy Gagnon
Ford Motor Co.

Copyright 1997 Society of Automotive Engineers, Inc.

ABSTRACT

Numerous authors have previously published on the effects of system dynamics on gear noise in automotive applications [1,2]. It is now widely understood that the torsional compliances and inertias of propeller shafts and pinion gear sets are a controlling factor in final drive gear noise for rear wheel drive vehicles. Considerable progress has been achieved in using finite element simulations of the driveline dynamics to improve the system in regards to gear noise. However very few published results are available showing the application of dynamic simulation methods to automatic transmissions which require considerations of the complications due to epicyclical gear sets.

This paper documents the successful application of finite element dynamics modeling methods to the prediction of gear noise from the gear set in a rear wheel drive automatic transmission. The model was used to investigate the effects of component inertias, stiffnesses, and resonances. Specifically, the ring gear and shaft resonances and the tail stock housing stiffness were found to be significant design factors which influence the gear whine.

Model construction issues will be discussed as well as correlation of predicted gear noise traces with operating measurements. Finally, a comparison will be provided of the noise reduction achieved by incorporating the recommended design modifications into prototype hardware.

INTRODUCTION

As improvements have been made to vehicle NVH characteristics, some concerns that have been masked by other NVH sources are now requiring attention. In some cases, transmission gear noise concerns are now becoming more of an issue with discerning vehicle operators.

In order to reduce gear noise there are typically two ways of approaching the problem. The first involves reducing the gear transmission error. The techniques commonly used are to modify the gear tooth geometry and the gear manufacturing process.

The second approach involves modifying the dynamics of the gear train system through structural changes to the components. Before one can make these changes it is valuable to have a thorough understanding of the dynamics of the system. This understanding can best be gained with a combined test and analysis approach. This paper describes how operating and artificial excitation tests of a Ravineaux gear set transmission can be combined with finite element simulation of the system to yield an accurate prediction of the system response.

The comparison of predicted results with measured results is an important part of the simulation process. Design studies are only of value if the engineer can have confidence that the model is properly representing the actual hardware. The use of finite element models of the various components of the transmission was validated by comparing test and analysis results of the major components of the transmission. Any necessary changes were then made in the modelling assumptions and procedures to ensure that the predicted results correlated with the measured data. Once the components were validated on an individual basis then the system model was assembled from these components.

The mechanism for applying the transmission error was developed to allow for the wide range of interactions between the various planet gear meshes.

There are twelve gear meshes for the planetary gear set. Each of these meshes operate at the same mesh frequency. The transmission error at each mesh can arise from a complex combination of tooth form errors, assembly misalignments, and load deflection terms. Procedures have been developed for determining a set of net transmission errors which take into account the possible worse case phasings. The procedures used in this paper are described in Reference 1.

This validated model was then used to identify design modifications that would reduce the transmission vibration levels, thus reducing the gear noise. These design modifications were implemented in hardware and the modified system was subjectively and objectively evaluated.

CONCERN IDENTIFICATION & TARGET SETTING

A subjective evaluation of several vehicles that use this particular transmission indicated a gear noise concern in 1st gear. A test program to quantify the gear noise and to set targets for an acceptable level of gear noise was performed. The first task was to measure vehicle operating data with the vehicle on a chassis dynamometer.

Acceleration measurements were made at the transmission mount, tailstock housing and the differential case. Sound Pressure Levels (SPL's) were measured inside the vehicle using a binaural head and underneath the vehicle in the vicinity of the transmission and propeller shaft. Torsional vibration data was also taken on the propshaft.

From the measured data, SPL and acceleration targets were developed. Tone on random masking theory was used to determine the acceptable levels of the various gear mesh orders.

Figure 1 shows order track traces of the measured vibration data on the transmission housing. The figure shows the results for both a truck and two door sedan platform. The frequency range of 400 to 800 Hz was the range where the gear whine was most objectionable. Although there are major driveline difference between these two vehicles, the measured case vibration was very similar for frequencies up to 800 Hz. The deviation at the higher frequency range is believed to be due to the exhaust system packaging which had a hanger location off of the transmission mount.

There is a fairly heavily damped resonance in both curves with a 10 dB magnification at about 600 Hz which correlated with the whine noise in the vehicle. Tests performed on the transmission and vehicle pointed toward this resonance being controlled by the transmission rather than the driveline or vehicle. These results suggest that there is a potential for changes internal to the transmission which could provide up to a 10 dB reduction in gear whine.

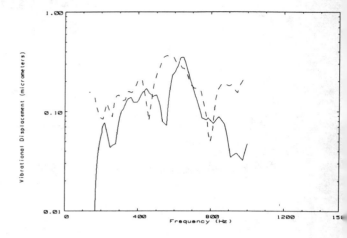

Figure 1: Measured order track plot of transmission case vertical vibration at the gear mesh frequency for the transmission installed in a truck (solid line) and a two door sedan (dashed).

COMPONENT MODELS

Figure 2 shows a schematic of the transmission that was evaluated in this case study, with the corresponding system finite element model geometry shown in Figure 3. The primary focus of the investigation was on simulating the behavior of the transmission in 1st gear.

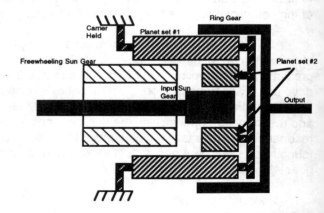

Figure 2: Schematic of the Gear Train layout for first gear operation.

In order to have confidence in the design direction provided by the model it was necessary to ensure that the model predictions could be correlated to experimental data. In this case, models of the various components were developed and validated using SDRC's I-DEAS Master Series software [3]. The frequency range of interest for the system model was up to 800 Hz. Typically the components are validated above the maximum frequency of interest for the system so in this case the goal was to achieve correlation up to at least twice this frequency (1600 Hz) or in the case of the planet carrier to match the frequency of the first mode. The general guidelines for correlation were that the predicted natural frequencies be in agreement to within +/-3%, and that the mass and stiffness lines of the frequency response functions match to within 10% for the frequency range of 100 to 1600 Hz.

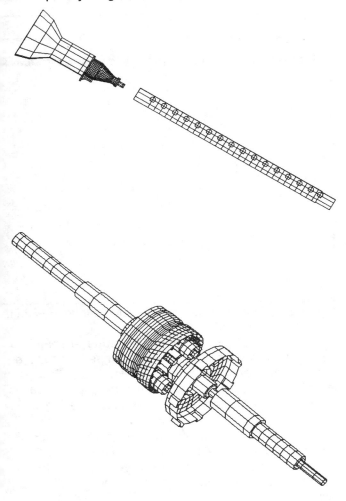

Figure 3: Finite element model geometry plots of the overall transmission case and propeller shaft (upper) and the internal components.

The individual component models which were correlated were the ring gear and output shaft, propeller shaft, transmission tailstock housing, slip yoke, transmission internal center support, and the differential carrier.

Figure 4 illustrates the finite element model of the ring gear and output shaft. The model consisted of shell elements that represented the major portion of the ring gear, beam elements for the output shaft and gears, and spring elements that represent the connection of the back plate to the ring gear shell. Artificial excitation tests were performed on the ring gear and output shaft.

Frequency response functions (FRF) were taken at approximately 20 locations on the ring gear in the free-free condition.

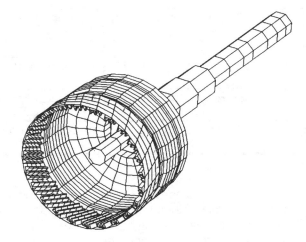

Figure 4: Geometry plot of the ring gear and output shaft finite element model.

Many correlation processes have been developed to quantify the correlation results through the use of Modal Assurance Criteria (MAC) matricies, or modal orthogonality criteria. However, the most meaningful and valid process is the overlay of test and analysis synthesized frequency response functions, and minimizing the error between them. This process was used.

RING GEAR - Figure 5 shows a typical comparison plot of the correlation results for the ring gear. Table 1 summarizes the comparison of natural frequencies of the experimental and analytical modes for the ring gear and output shaft.

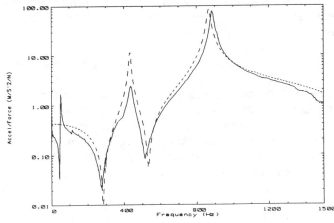

Figure 5: Transfer function correlation for a driving point on the ring gear. Experimental (solid) Analytical (dashed)

Table 1: Ring Gear and Output Shaft Modal Frequency Correlation

Test	Analysis	Description
438 Hz	431,434 Hz	Output Shaft Bending
Not Identified	784 Hz	Fore/Aft mode
884 Hz	863,867 Hz	Ring Gear 1st Lobar
2272 Hz	2215,2225 Hz	Ring Gear 2nd Lobar

Table 2: Percent Modal Strain Energy

Frequency Component	431 Hz	784 Hz	863 Hz	1107 H
Ring Gear Shell	2		93	
Backing Plate	51	57	3	1
Spring	23	39	2	
Output Shaft	24			8

In the development of finite element models of automatic transmissions one of the most difficult aspects is the assumptions regarding component connectivities. The representation of the ring gear, or the output shaft is a fairly straight forward modelling process. However, the ring gear is connected to the backing plate through a spline connection and held in place with a slip ring. This type of connection raises numerous modeling questions. Furthermore, it complicates the testing. This type of loose fit between components will cause rattling and poor quality frequency response functions, thus obscuring the data and aggravating the correlation process.

For components like this, it is often necessary to understand the operating loads and environment, and to simulate them in the static aritificial excitation testing. The operating torque on the ring gear would cause friction in the spline which would help to lock up the joint. Tests have been performed using various methods of providing preload through the joint, while not affecting the dynamics of the subsystem. For this component, the testing was performed using dental cement to eliminate the rattling at this location.

During the correlation of the ring gear one of the important parameters controlling the frequency of several of the modes was the stiffness at the interface of the back plate and the ring gear. This compliance was accounted for by including discrete spring elements between the back plate and the ring gear. These springs were a controlling factor in the fore/aft mode at 784 Hz, as illustrated by the strain energy tabulation in Table 2.

Figure 6 presents the mode shape for the 1st lobar mode at 863 Hz. This type of mode will provide additional compliance at the resonant frequency, thus significantly reducing the dynamic mesh force which plays a very important role in the resulting gear noise. It is important to ensure that the finite element model properly predicts the ring gear lobar modes.

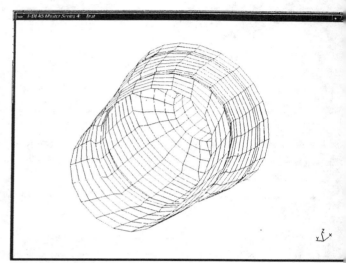

Figure 6: First ring gear lobar bending mode at 863 Hz

PROPELLER SHAFT - The propeller shaft and vehicle driveline control the dynamic impedance at the output of the transmission. It is important that they be included in the system model with sufficient accuracy in order to properly represent this dynamic impedance.

The propeller shaft model is shown in Figure 3. The model consists of beam elements plus lumped masses representing the front and rear yoke. In the initial analysis of the propshaft there were significant errors in the predicted frequencies of the first five modes.

The SDRC CORDS [4] correlation program was used to improve the correlation of the propshaft. This program uses design sensitivity and optimization methods to identify model updates that will minimize the difference in the test and analysis frequencies.

Figures 7 and 8 show a comparison of the vertical and torsional driving point frequency response functions at the front yoke. The comparison between test and analysis was excellent for the vertical driving point. The torsional response was accurately predicted through 900 Hz. The second order torsion mode was still in slight error, however the dynamics were accurately represented for the frequency range of interest.

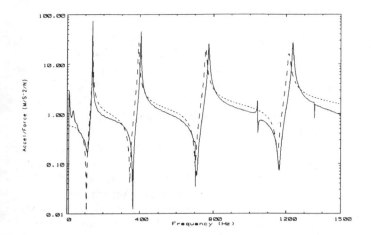

Figure 7: Transfer function correlation for a vertical driving point transfer function on the propeller shaft. Experimental (solid) Analytical (dashed)

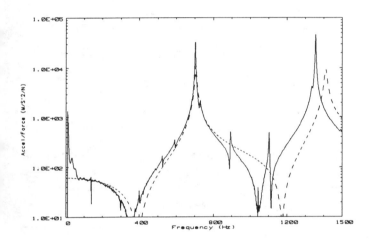

Figure 8: Transfer function correlation for a torsional driving point transfer function on the propeller shaft. Experimental (solid) Analytical (dashed)

The parameter which had the most effect on correlation was the propshaft thickness. A thickness increase of 20% was specified by the CORDS program to match the frequencies of the first five modes. This level of thickness change was actually within the tolerance band on the propeller shaft production parts.

TAILSTOCK HOUSING - The model of the housing was created using shell elements with solid elements added at the mounting flange to the transmission case as shown in Figure 9. Figure 10 presents a comparison of the vertical driving point at the rear of the tailstock housing. For this test, the tailstock housing was restrained at the locations where the housing bolts to the transmission case. To correlate this component, only minor changes were made to the original model to achieve a reasonable comparison.

These changes included adjusting the model weight to match the measured weight.

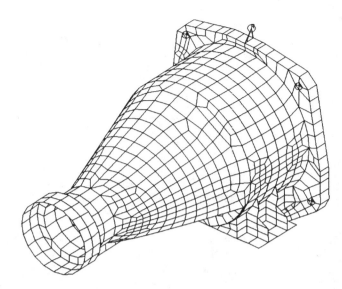

Figure 9: Geometry plot of the tailstock housing finite element model.

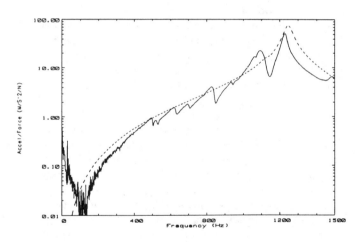

Figure 10: Transfer function correlation for a driving point transfer function vertically at the end of the tailstock housing. Experimental (solid) Analytical (dashed)

The models of the slip yoke and center support structure were also validated using the same techniques that were used for the other components. The slip yoke was important since a torsional damper was part of this component.

CARRIER - The last major component that was validated was the carrier. The first free-free modes of the carrier were in the 4000-6000 Hz frequency range. Since this frequency is well out of the frequency range of interest, a lumped parameter model was used to represent the basic stiffness of the carrier. This model included springs and masses that were adjusted to match the lateral, axial and torsional driving points of the carrier. Figure 11 shows the comparison of a driving point function comparison. The comparison is good in the frequency range of interest.

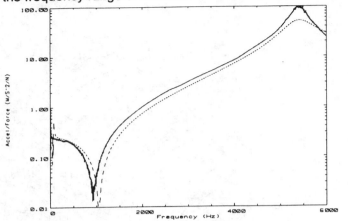

Figure 11: Transfer function correlation for a driving point transfer function on the carrier. Experimental (solid) Analytical (dashed)

Other components included in the model, but not explicitly correlated, were the transmission case, engine, internal shafts, and rear axle. These components were represented by rigid elements and lumped masses except the internal shafts which were modeled as beam elements.

TRANSMISSION SYSTEM MODEL

Once the models of the individual components were validated, the components were assembled into a system model of the transmission as previously shown in Figure 3. The components were assembled using combinations of bearings, splined connections, and gears. The bearings were represented by spring elements and the splined connections were generally modeled with rigid elements.

The gear tooth stiffness was modeled with spring elements for both mating gears. A stiff spring was then connected between the nodes representing the pitch point on the mating gears, and aligned with the line of action. By applying a force across this stiff spring it is possible to excite the model with a specified relative displacement which is equivalent to the transmission error.

The gear tooth stiffness was entered in a local coordinate system that is aligned with the gear mesh force line of action. The line of action is determined from the helix angle and pressure angle. For this particular transmission there were four meshes to consider:

- Short Sun to Short Planet
- Short Planet to Long Planet
- Long Planet to Ring Gear
- Long Planet to Long Sun

Each of the mesh zones contains three mating planet gears.

The validation of the transmission system model involved two separate activities. These are a comparison of results from artificial excitation tests and a comparison of operating response.

ARTIFICIAL EXCITATION VALIDATION

For the artificial excitation test comparison, vibration data was taken at the slip yoke, transmission case and the tailstock housing. Figures 12 and 13 show the comparison of the lateral driving points at the tailstock housing and transmission case up to 1500 Hz.

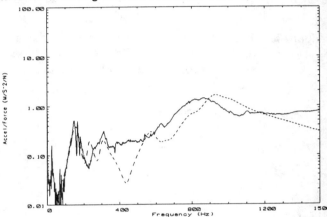

Figure 12: Transfer function correlation for full vehicle driving point transfer function lateral on the tailstock housing. Experimental (solid) Analytical (dashed).

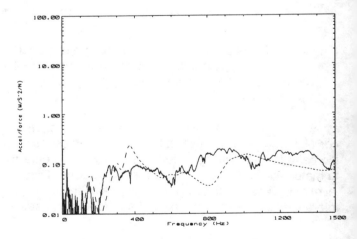

Figure 13: Transfer function correlation for full vehicle driving point transfer function lateral on the transmission housing. Experimental (solid) Analytical (dashed).

System modes were calculated up to 2400 Hz which includes the first 70 modes. In general the lateral and vertical response is the same since the only difference in stiffness in the two directions is the tailstock housing. All the other components have basically the same stiffness and mass in the lateral and vertical direction. In many cases there were two modes at a particular frequency, one involving motion in the lateral direction and the other motion in the vertical direction. In general the comparison between measured and predicted response is excellent in the 400-800 Hz frequency range which is the primary frequency range of interest. There were numerous modifications made to the original model to achieve this level of correlation. Examples of these modifications are modifying bearing stiffnesses and the connection of the components. Of particular use in the correlation process was the strain energy distribution in each of the components.

When modes were identified as contributing to the response, the parameters that controlled the frequency of the mode could be determined from the strain energy distribution. Appropriate changes could then be made to the finite element representation of these components to improve the correlation. Strain energy information is very valuable in trouble shooting and understanding the model results.

OPERATING RESPONSE VALIDATION

The next step in the correlation of the model was to compare operating results. As discussed previously, there are 12 gear meshes that can each produce a unique transmission error. There are a large number of ways these 12 errors can be combined to excite the transmission system. Not only is the amplitude of the error unknown but the phase relationship between each gear is also unknown.

The developed mesh force, and the transmissibility of forces through the transmission are, however, largely controlled by the rigid body motions of the individual gears and carrier. Thus it is possible to take the three individual transmission errors at any given mesh zone location, and perform the appropriate linear combination of these inputs in order to obtain three independent inputs which excite the vertical, lateral, and torsional motions of the carrier. This greatly aids in the physical understanding and interpretation of the dynamics.

The transformation of the individual inputs into net rigid body inputs was accomplished through multi-point constraint (MPC) equations within the finite element model. Using the twelve sources of transmission error, the following set of equations were developed to generate a unit transmission error at the four mesh zones.

$$[T]'\{F\}=\{Te\}$$

Where [T]=Direction Cosine Matrix of
         Gear Mesh Force
     $\{Te\}$= Net Total Transmission Error
     $\{F\}$=Gear Mesh Coefficient

This equation was solved 3 times, for a unit lateral, vertical and torsional transmission error. A degree of freedom was created in the model for each gear mesh transmission error and MPC equations representing the gear mesh coefficient were used to define this degree of freedom. The response at each of the critical locations was calculated for a unit input at these degrees of freedom. With four mesh zones and three transmission errors at each mesh zone this results in 12 response calculations.

The measured vibrations at the gear mesh frequency contain the effects of each of these 12 inputs. It is impossible to separate out the effect of any individual gear. In order to compare the predicted response to the measured operating data it was necessary to perform an appropriate combination of the response from the individual inputs. These 12 responses were combined using the Square Root Sum of the Squares (SRSS) method. Once the response was combined for a unit transmission error, the predicted response was scaled by a constant to match the measured operating vibration levels. A single value of transmission error was chosen for the vertical and lateral transmission errors at all the mesh zones. A separate single value was chosen for the effective torsional transmission error at all mesh zones.

The response from the individual rigid body inputs were inspected in order to better understand the controlling factors in the transmission design. This helped to separate out the effects of system torsional modes from shaft bending modes, and to understand the ramifications of various phasing relationships between the various gears.

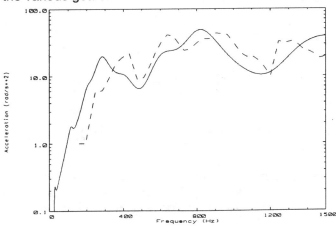

Figure 14: Comparison of experimental (dashed) and predicted analytical (solid) operating gear mesh order track for torsional acceleration at the propeller shaft

Figure 14 shows the comparison of the predicted and measured torsional acceleration at the front of the propshaft. The comparison is reasonably accurate over the frequency range of interest. This comparison is quite satisfying given the simple model for transmission error, and the complexity of the actual transmission error process.

Figure 15 shows the same comparison for the tail stock vertical vibration and Figure 16 for the transmission case vertical vibration. Again, the level of correlation in the 400 to 800 Hertz frequency range is quite acceptable.

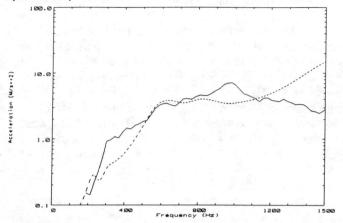

Figure 15: Comparison of experimental (dashed) and predicted analytical (solid) operating gear mesh order track for vertical acceleration at the tailstock housing.

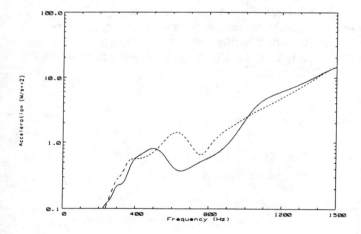

Figure 16: Comparison of experimental (dashed) and predicted analytical (solid) operating gear mesh order track for vertical acceleration at the transmission case.

IDENTIFICATION OF CONTRIBUTING FACTORS

The comparison of the predicted and measured torsional response at the slip yoke for a 1st gear speed sweep was shown in Figure 14. A detailed investigation of the response indicated that the response was due entirely to torsional transmission errors. Figure 17 shows the contribution of the various rigid body transmission error components to the total SRSS level. The torsional components contribute approximately two orders of magnitude more response than the vertical and lateral error components. Furthermore, the system torsional modes dominate the response.

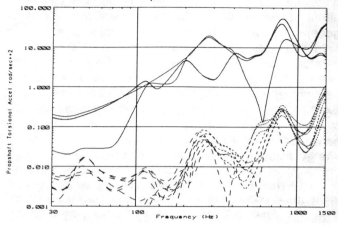

Figure 18: Contributions of the mesh torsional transmission errors (solid) and the mesh translational transmission errors (dashed) to the propeller shaft torsional response.

The response due to lateral or vertical transmission errors was approximately 40 dB lower than that due to torsional transmission errors.

A similar analysis for the case vertical vibration response had shown that the torsional inputs and system modes were contributing very little. This shows that the problems of case vibration and torque fluctuation out the propeller shaft are independent issues, which would require different design modifications to address.

The baseline test data and noise path evaluation had pointed toward the case vibration as being the greater concern.

Figures 15 and 16 illustrate the comparison of the predicted and measured operating response at the transmission case. The comparison of predicted and measured operating response is good. Unfortunately no comparison could be made for the internal components due to the difficulty of locating instrumentation on the internal rotating components. The peaks in the vertical response are due to modes at 576, 656 and 800 Hz. These modes involve the internal shafts, ring gear and the radial bearings. Table 3 summarizes the strain energy distribution for these three modes.

If the response due to these modes is to be reduced then changes should be made to the components shown in the table. This is one of the benefits of evaluating the strain energy distribution of each mode. In general, changes to components that do not contribute significant strain energy to these modes will have no effect on the system response.

Table 3: Percent Strain Energy

Frequency----> Component	576	656	800
Ring Gear	14	43	24
Intermediate Shaft	25	30	16
Clutch Hub	16	7	38
Bearings	14	2	2

DESIGN STUDIES

The operating tests had identified a requirement of a 10 dB reduction to the interior noise levels to meet the 1st gear target levels. The objective of the design studies was to identify changes that would meet the target. The assumption was made that a 10 dB reduction in transmission case vibration response would produce the same reduction in interior noise levels.

Approximately 20 design modifications were evaluated. Changes were made to the FEM of the selected components and a normal mode analysis was performed. Using these results the operating response was predicted and compared to the baseline response. The responses at the tailstock housing, transmission case and slip yoke were monitored and an increase, decrease or no change rating was made for each design study. The design modifications that reduced the response at the transmission case were:

1. Decreased thickness of ring gear shell
2. Increased stiffness of tailstock housing
3. Decreased stiffness of intermediate shaft

These three modifications were tried in combination and the response at the transmission case predicted. Figure 18 presents a comparison of the response for this configuration with the baseline response. The reduction in the response in the frequency range of interest is 12 dB which meets the target of a 10 dB reduction.

Another way to explain the reduction is to examine the effect these changes have on the mesh dynamic forces and the transmissibility of these forces to the case. Gear noise is a result of meshing errors between mating tooth pairs. This meshing error is effectively the same as imposing a relative dynamic displacement between the gear teeth. The inertia of the shafts and gears, the stiffness, and system resonances will determine how much dynamic force is developed as a result of this relative displacement. The modes of the shafts and supporting structure will then determine the transmissibility of this mesh force to a case shaking force. This case shaking force then acts on the mass of the case, causing a vibration acceleration. This process is illustrated in Figure 19. The top plot shows the amount of dynamic force developed at the mesh per unit micrometer transmission error at the long planet to ring gear mesh. Notice the presence of a strong resonance peak at about 680 Hz.

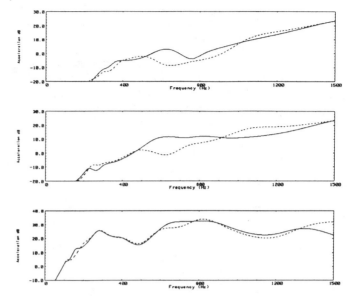

Figure 18: Predicted changes in the transmission case vertical acceleration (upper trace), tail stock vertical acceleration (middle trace), and propeller shaft torsional acceleration (lower trace). Baseline design (solid) Modified deisgn (dashed).

The second plot in Figure 19 shows the transmissibility of this mesh force to the transmission case. The majority of the case vibration was a pitching motion. Thus the transmissibility was expressed in terms of a pitching moment reacted to the case per unit mesh force. This shows three modes between 600-900 Hz, however they are less prominent than the mode in the mesh force.

The lower plot in Figure 19 shows the resulting transmission case acceleration. This is the result of the mesh force per unit transmission error times the force transmissibility, divided by the effective mass of the case. Notice that the peaks and valleys in the predicted case acceleration can be directly related to the peaks and valleys in the upper two plots.

Also shown on Figure 19 are the same three items for the final design modification. This clearly shows that the primary benefit of the design modification was in the reduction of the mesh force within the 500-900 Hz frequency range. This was principally achieved by moving the lobar mode resonance frequency of the ring gear from 920 Hz to 680 Hz. This change in the resonance frequency causes a reduction of the impedance into the ring gear, which accounts for the corresponding dip in the mesh force plot of Figure 19.

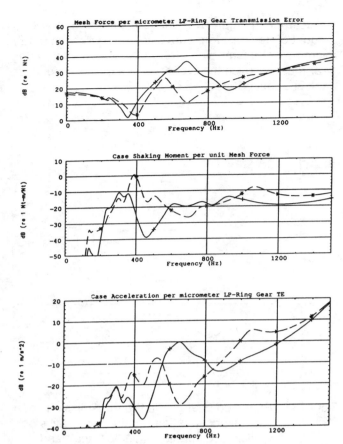

Figure 19: The effect of the design modification on the developed dynamic mesh force (upper trace), the nominal transmission case shaking moment per unit dynamic mesh force (middle trace), and the resulting case acceleration (lower trace). Baseline design (solid) Modified design (dashed)

Table 4: Percent Modal Strain Energy Distribution. Baseline (BL) vs Design Modification (DM)

	DM	BL	DM	BL	DM	BL
Frequency Component	495	576	546	655	741	801
Ring Gear	4	14	74	43	10	24
Input Shaft	18	25	2	7	1	5
Inter. Shaft	52	25	7	29	10	16
C3 Hub	3	16	5	6	56	38

A secondary effect of the design modification was a reduction in the force transmissibility in the range of 600 to 800 Hz. This was achieved through the change to the intermediate shaft.

While the results presented in Figure 19 only illustrate the effect on mesh force and force transmissibility for the long planet to ring gear mesh, a similar effect would be seen in the other three mesh zones.

Another interesting observation results from a comparison of the strain energy distribution for the baseline and the design modification. This comparison is shown in Table 4

For the baseline, the strain energy distribution is more uniformly distributed among the components while for the design modification, the strain energy is more localized in the components. This suggests that the vibrational energy is more contained and less likely to couple with the transmission case. This situation likely arises due to greater impedance mismatch between the transmission components, thus causing less dynamic coupling.

PROTOTYPE VALIDATION

A prototype transmission was built with the recommended design modifications and installed in a vehicle. The chassis roll dynamometer tests were repeated and the operating measurements were compared with the previous baseline results.

The results of the modifications are shown in Figure 20 for the transmission case vibration, and Figure 21 for the rear axle pinion nose vibration. Figure 20 shows that the case vibration has been reduced by approximately 10 dB for frequencies above 400 Hz, as predicted by the model.

The pinion nose vertical vibration is most controlled by the torque fluctuation for a rear wheel drive vehicle. Figure 21 shows that this vibration was essentially unaffected, and even slightly increased, as predicted by the model. These results have shown that the model was able to accurately predict the effect of the composite design modifications.

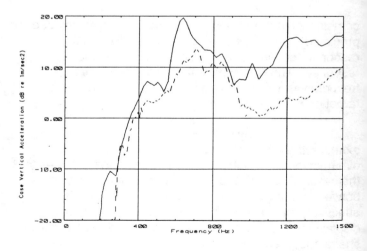

Figure 20: Measured operating gear mesh order track plot for transmission case acceleration for the baseline transmission design (solid) and the final modified transmission (dashed).

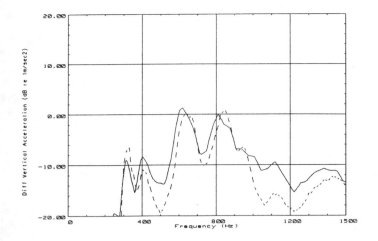

Figure 21: Measured operating gear mesh order track plot for rear axle pinion nose vertical acceleration for the baseline transmission design (solid) and the final modified transmission (dashed).

SUMMARY

Using a combined test and analysis approach, a high degree of confidence can be established in the predicted results of an analytical simulation. Operating test results were used to establish a quantifiable target for the system response. Each of the major components of the transmission were verified and assembled into a system model of the transmission. Operating response levels were predicted and compared to the measured levels. Using this validated model of the transmission, design studies were performed and several design modifications were identified that would reduce the transmission case vibration levels.

In this case study, the key to reducing the transmission case response is to separate the frequencies of the internal shafts bending modes from the ring gear lobar mode. The gear mesh forces are reduced and the response at the transmission case due to the individual transmission errors was reduced. This reduction was achieved by successfully decoupling these components.

REFERENCES

1. M. G. Donley, T.C. Lim, and G.C. Steyer, "Dynamic Analysis of Automotive Gearing Systems" SAE Paper 920762, 1992.

2. M. G. Donley and G. C. Steyer, "Dynamic Analysis of a Planetray Gear System", Proceedings of the 6th International Power Transmission and Gearing Conference, Phoenix, Arizona, pp. 117-128, September 1992.

3. I-Deas Masters Series 4 Users Manual, SDRC, Milford Ohio.

4. SDRC CORDS Version 6 Users Manual, SDRC Operations, San Diego, Ca 1995

971967

Influence of the Valve and Accessory Gear Train on the Crankshaft Three-Dimensional Vibrations in High Speed Engines

Hideo Okamura
Sophia Univ.

Kenichi Yamashita
Mitsubishi Motors Ltd.

Copyright 1997 Society of Automotive Engineers, Inc.

ABSTRACT

In most high-speed engines of the OHC (over head camshaft) type, a number of gears are engaged with the crankshaft gear to drive the valve gear mechanism, the fuel injection pump, and other accessories such as the oil pump or power steering system.

Each of the gears usually has a significant mass and moment of inertia. We investigated the influence of the masses and moments of inertia of the gears on three-dimensional vibrations of the crankshaft system. A four-cylinder in-line diesel engine ($4 - \phi 115 \times 110$, $140ps / N = 3200rpm$) was used for a series of experiments and analyses. The three-dimensional vibrations of the crankshaft system were measured by the hammering tests and the shaker tests. We calculated the vibration behavior by applying an idealized simple modeling for the crankshaft system and the gear train.

From the series of experiments and analyses, we found that the gear train should be engaged to the crankshaft system at the crankshaft rear end (flywheel side), as close as possible to the nodal point of the crankshaft torsional vibrations, so that the influence of the moments of inertia of the gears on the crankshaft vibrations could be minimized.

1 INTRODUCTION

In the most of the vibration analyses for an engine crankshaft system, the influence of the engine gear train has been neglected. However, for recent high speed engines, a large number of gears are arranged as shown in Fig.1 (especially for the OHC (over head camshaft) type engine), to drive the valve mechanism, the fuel-injection system and other accessories from the crankshaft.

Since each of the gears has significant mass and moment of inertia, significant influence can be caused by the moments of inertia of the gears on the vibration behavior of the crankshaft system. We investigated such influence for a four-cylinder in-line diesel engine in which the gear train is driven by the crankshaft (1) at the engine front end and (2) at the engine rear end. For simplicity we will call gear train (1) the FEDGT, and gear train (2) the REDGT, in the following.

Initially, we investigated the torsional vibrations by Holzer's method. Then, we calculated the three-dimensional vibrations by the dynamic stiffness matrix method. The calculated results were compared with the experimental results.

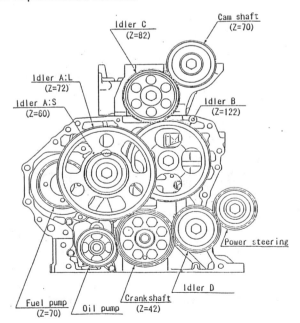

Fig.1 Gear arrangement in a timing gear train

2 ENGINE FOR EXPERIMENTS AND ANALYSES

For our experiments and analyses, we used a four-cylinder in-line diesel engine ($4 - \phi 115 \times 110$, $140ps / N = 3200rpm$). In this engine a gear train of the REDGT type shown in Fig.1 is arranged as shown in Fig.2.

For comparison, the schematic views are shown in Fig.3 for the virtual engine in which the gear train of the FEDGT was arranged in place of the REDGT for the engine in Fig.2. For the two arrangements shown in Fig.2 and Fig.3, we investigated the vibration behavior of the crankshaft systems.

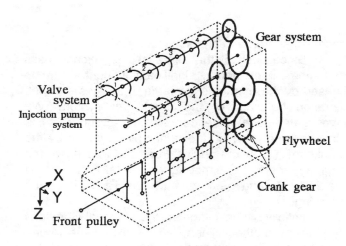

Fig.2 Engine with the REDGT (rear end drive gear train)

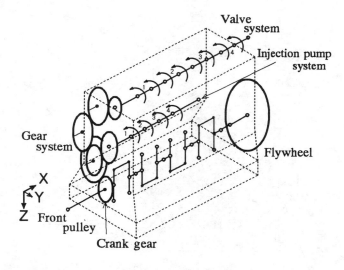

Fig.3 Engine with the FEDGT (front end drive gear train)

3 MODELING AND ANALYSES OF TORSIONAL VIBRATIONS FOR THE CRANKSHAFT SYSTEM AND GEAR TRAIN [1]

3.1 Modeling and analysis for the torsional vibrations

To calculate the torsional vibrations of the crankshaft system (the crankshaft with the pulley, the crankshaft gear and the flywheel attached), and the gear train arranged as shown in Fig.2 and Fig.3, we idealized the torsional models as shown in Figs.4(a), (b).

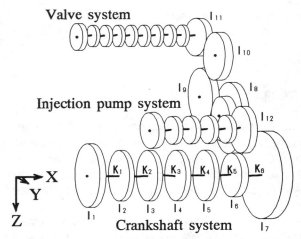

Fig.4(a) Torsional model of engine with the REDGT

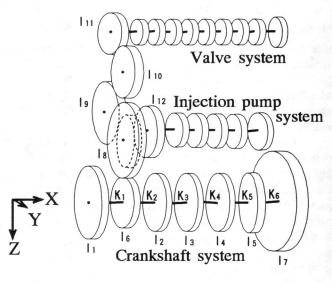

Fig.4(b) Torsional model of engine with the FEDGT

For a preliminary analysis, however, we disconnected the injection pump gear, the power steering drive gear, and the oil-pump gear from the gear train to simplify the analyses and experiments, since both the injection pump system and power the gear train of the FEDGT was engaged, and removed. The calculated results are shown in Table3(b) and Figs.7(a),(b), similarly for the crankshaft systems, to which the gear train of the REDGT type was engaged, and removed.

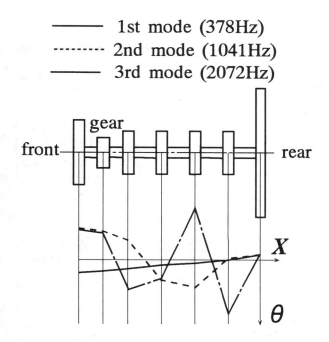

Fig.6(a) Modes of the crankshaft system disengaged from the FEDGT gear train

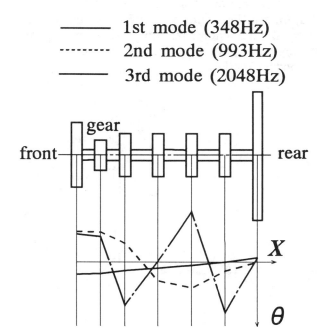

Fig.6(b) Modes of the crankshaft system engaged with the FEDGT gear train

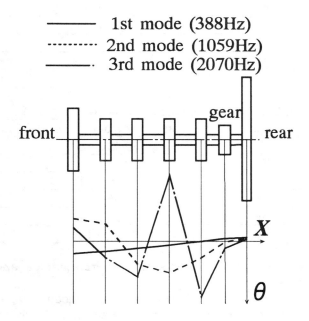

Fig.7(a) Modes of the crankshaft system disengaged from the FEDGT gear train

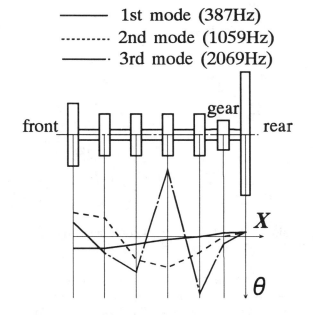

Fig.7(b) Modes of the crankshaft system engaged with the REDGT gear train

steering pump system were so complicated. In the preliminary analysis, we idealized the gears in the gear train, the valve system and the injection pump system in Figs.4(a),(b), by a set of equivalent moments of inertia, as shown in Figs.5(a), (b).

Finally, we further reduced the models shown in Figs.5(a), (b) to the common speed models shown in Figs.6(a),(b) and Figs.7(a),(b). In which all gears are engaged with the crankshaft gear by 1 to 1 gear speeds, preserving equal kinetic and potential energy with the original models.

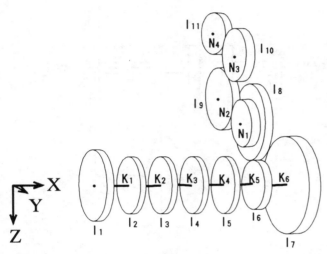

Fig.5(a) Torsional model of the crankshaft system and the REDGT for experimental set-up

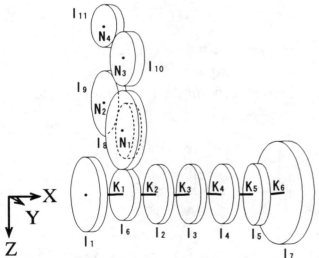

Fig.5(b) Torsional model of the crankshaft system and the FEDGT for experimental set-up

The values of the moments of inertia for each of the crankshaft-throws, the flywheel, the front pulley and the gears in the gear train (see Fig.1) are shown in Tables1(a),(b). Similarly, the values of the torsional stiffness are shown in Tables2(a),(b) for each of the crankshaft throws, and the two crank-journals at the crankshaft ends.

The original values of the moments of inertia for the idle gears A, B, C and the cam gear and the values for their common speed models mentioned above are shown in Table1(b).

Table1(a) Moments of inertia of torsional model for crankshaft system in Figs.4(a),(b) ($kg \cdot m^2$)

Name of inertial mass	Sign	Moment of inertia
Front pulley	I_1	2.16×10^{-2}
1st throw	I_2	2.46×10^{-2}
2nd throw	I_3	2.46×10^{-2}
3rd throw	I_4	2.46×10^{-2}
4th throw	I_5	2.46×10^{-2}
Crank gear	I_6	7.00×10^{-3}
Flywheel	I_7	4.20×10^{-2}

Table1(b) Moments of inertia of gears in Figs.4(a),(b)

Name of gear	Sign	Original ($kg \cdot m^2$)	Common speed models ($kg \cdot m^2$)
Idle gear A	I_8	1.57×10^{-2}	5.34×10^{-3}
Idle gear B	I_9	7.48×10^{-3}	6.62×10^{-4}
Idle gear C	I_{10}	1.76×10^{-3}	3.21×10^{-4}
Cam gear	I_{11}	2.09×10^{-3}	5.23×10^{-4}

Table2(a) Torsional stiffness for model in Fig.5(a)

Sign	Torsional rigidity ($N \cdot m/rad$)
K_1	2.19×10^6
K_2	1.15×10^6
K_3	1.15×10^6
K_4	1.15×10^6
K_5	2.09×10^6
K_6	2.48×10^7

Table2(b) Torsional stiffness for model in Fig.5(b)

Sign	Torsional rigidity ($N \cdot m/rad$)
K_1	2.43×10^7
K_2	2.09×10^6
K_3	1.15×10^6
K_4	1.15×10^6
K_5	1.15×10^6
K_6	2.06×10^6

3.2 Calculation of the natural frequencies and mode shapes

The natural frequencies and mode shapes calculated by Holzer's method are shown in Table3(a) and Figs.6(a), (b) for the crankshaft systems to which

Table3(a) Natural frequencies calculated by Holzer's method: The crankshaft system with FEDGT gear train

Mode	Natural frequency	(Hz)
	With gear train	without gear train
1	348	378
2	993	1041
3	2048	2072

Table3(b) Natural frequencies calculated by Holzer's method: The crankshaft system with REDGT gear train

Mode	Natural frequency	(Hz)
	With gear train	without gear train
1	387	388
2	1059	1059
3	2069	2070

3.3 Experimental results of the torsional vibrations

To investigate the validity of the natural frequencies shown in Tables3(a),(b) and the mode shapes shown in Figs.6(a),(b), and Figs.7(a),(b), we first measured the natural frequencies and mode shapes for the torsional vibrations of the crankshaft system suspended in the free-free conditions. The test set-up is shown in Fig.8(a). The mode shapes of the torsional modes were determined by attaching two accelerometers to the crank-journals, as shown in Fig.8(b). The torsional amplitude θ was calculated by the following equation:

$$\theta = \frac{y(2) - y(1)}{D} \quad (1)$$

Here the $y(1)$ and $y(2)$ are the tangential displacements measured by the two accelerometers, and D is the journal diameter as shown in Fig.8(b).

Table4 Experimental and calculated results of natural frequencies for crankshaft system

The condition of crankshaft	Mode	measured (Hz)	Calculated (Hz)
free-free	1	375	388
	2	1255	1059
	3	1780	2070
in cylinder block	1	390	398
	2	1235	1215
	3	*	*
in cylinder block with REDGT gear train	1	390	397
	2	1235	1215
	3	*	*

In Fig.9, the experimental results of the 1st mode which appear at the natural frequency of 373Hz are shown.

For the crankshaft system assembled in the experimental engine, we measured the vibration behavior by shaker tests. In the shaker test, we applied some preload to the gear train to eliminate the influence of the gear backlash. The excitation point and the measuring points for the accelerometers are shown in Fig.10. The experimental results of crankshaft natural frequencies obtained in the shaker tests are shown in Table4.

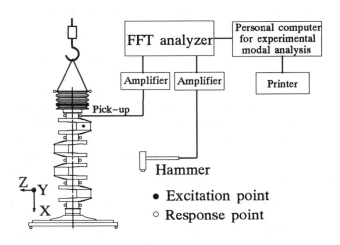

Fig.8(a) Test set-up of hammering tests

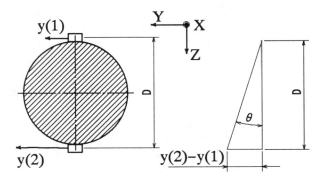

Fig.8(b) Estimation of the torsional amplitude

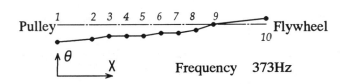

Fig.9 Experimental results of torsional mode of crankshaft mode system (1st mode)

■ excitation point: A
● response points: B and C

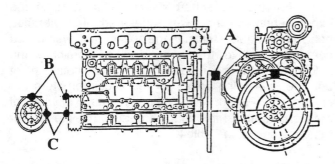

Fig.10 Engine crankshaft system and the gear train for shaker tests

3.4 Discussion about the torsional vibrations

a) Moments of inertia of the crankshaft system and the gear train

One can see from Tables1(a),(b) that the gears in the gear train have significant values of moments of inertia, e.g. the idler gear A has almost the same values of moment of inertia as that of the crankshaft pulley, and the idler gear B has almost one half amount of the moment of inertia as that for the gear A.

However, since the inertia torques for operating gears depend on the gear shaft speeds, and the gear speeds are different for each gear, one must convert the original values of the moments of inertia shown in Table1(a) to those for different gear speeds under common engine operating speed. To do this, by taking the crankshaft rotating speed as the reference speed, we have derived the moments of inertia for common speed models as shown in Table1(b). To do this we multiplied the original values of the moments of inertia in Table1(a) by n^2. Here n is the speed ratio of the gear shafts to the crankshaft. In our analysis, n was calculated from the number of gear teeth for each gear shown in Fig.1.

From Table1(b) one can see that the total value of the moments of inertia for the four gears is almost equal to that for the crankshaft gear, and it is also equal to about 40% of that for the crankshaft pulley.

b) Influence of the gear train on the natural frequencies and mode shapes of the crankshaft system

One can see from Tables3(a), (b) and Figs.(a), (b) and Figs.7(a), (b) that, if the gear trains are not engaged with their corresponding crankshaft system, the natural frequencies for the 1st and 2nd modes were 378Hz and 1041Hz for the FEDGT type crankshaft system and 388Hz and 1059Hz for the REDGT type one. And the mode shapes are very similar for the two crankshaft systems, irrespective of the types of gear train.

If the gear trains were engaged with their corresponding crankshaft systems, the natural frequencies mentioned above would change to 348Hz and 993Hz for the FEDGT type crankshaft system, and to 387Hz and 1059Hz for the REDGT type one. Namely, by the gear train of the FEDGT type, the natural frequencies of the crankshaft system would decrease by 30Hz, namely by 8 % for the 1st mode, and by 48Hz, namely by 4.8 % for the 2nd mode. However, for the gear train of the REDGT type, only negligibly small changes of the natural frequencies were caused by the crankshaft system.

As for the reasons, we suspect that (1) for the gear train of the FEDGT type, since a significant amplitude of torsional vibration can be induced at the crankshaft gear position as shown in Fig.6(b), accordingly significant torsional vibrations and significant inertia torque can be induced in the gear train; (2) for the gear train of the REDGT type, since the crankshaft gear is located so close to the nodal point of the 1st mode of torsional vibration, the gear train would not be excited so significantly by the crankshaft, accordingly, the influence mentioned above may not be induced. Finally, the natural frequencies of the crankshaft system would become higher for the case of the REDGT type than those for the FEDGT type.

In Table4, one can see relatively good agreement between the calculated results and experimental results, for the natural frequencies of the crankshaft system of the experimental engine to which the gear train of the REDGT type is engaged.

We could not perform similar experiments for the crankshaft system to which the gear train of the FEDGT type was engaged. However, we think that we might be able to see similarly good agreement when we install the gear train of the FEDGT type in our experimental engine.

4 MODELING AND ANALYSES FOR THE THREE-DIMENSIONAL VIBRATIONS OF THE CRANKSHAFT SYSTEM AND GEAR TRAIN [2], [3], [4]

In the previous pages, we have investigated the torsional vibrations of the crankshaft system and gear train by simple modeling and analysis by Holzer's method. However, the vibrations of the crankshaft system are three-dimensional. To investigate the influence of the gear train on the three-dimensional vibrations of the crankshaft system and the gear train under engine operating conditions, we idealized the crankshaft system for the firing conditions as shown in the following, and calculated the vibrations of the crankshaft system, the injection pump shaft, and the valve drive camshaft. The dynamic stiffness method was applied for the calculation. The details will be shown in the following.

4.1 Modeling of the crankshaft system and the gear train

a) Modeling of the crankshaft system

In Fig.11 the idealized crankshaft system is shown for the four-cylinder in-line diesel engine, along with the rotating coordinate system x-y-z attached to it, by setting the x-axis along the crankshaft center line.

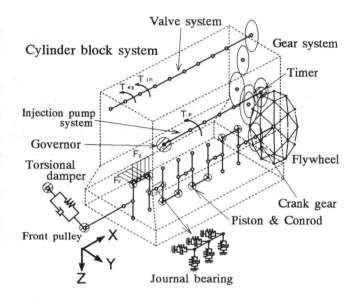

Fig.11 Modeling of the crankshaft system, the gear train, and the accessory driving system

In the modeling,
(1) the crankshaft is idealized by a set of jointed structures consisting of round rods, and rods of rectangular cross-section;
(2) the damper pulley and crankshaft gear are idealized respectively by a set of masses and moments of inertia attached at the centers of gravity;
(3) the oil film between each crank-journal and bearing is idealized by three sets of linear springs and dash pots which are attached to an infinitely stiff bearing body, in the horizontal and vertical directions;
(4) The flywheel is idealized by a set of FEM models consisting of 24 triangular plates, attached at the rear end of the crankshaft. since the flywheel was not as stiff as the front pulley and the crankshaft gear.

(5) The big end mass of the connecting rod was idealized by a set of two equal rotating masses, attached at the two ends of crank-pin.
(6) The reciprocating masses consist of the masses of the piston and the piston-pin, and the mass of the small end of the connecting rod. To take account of the mean kinetic energy that would be induced by the reciprocating masses under running conditions, one-quarter of the total reciprocating masses was attached at the two crank-pin ends.
(7) We assumed that the inertia force due to the reciprocating masses is applied together with the gas force directly to the crank-pin as a set of uniformly distributed forces Ft and Fr, in the tangential and radial directions of the crank-throws as shown as in Fig.11. This is because the rigidity of the connecting rod is much higher than that of the crankshaft system.

b) Modeling of the gear train, the injection-pump drive system, and the valve train

As shown in Fig.11, for the modeling,
(1) the gears in the gear train were idealized by a set of equivalent moments of inertia, taking account of the speed ratio: n, here n=gear shaft speed/crankshaft speed;
(2) the injection-pump drive system was idealized by a set of moments of inertia for the governor and timer, attached to the injection-pump shaft;
(3) the valve train was idealized by a set of moments of inertia attached to the cam shaft;
(4) the injection-pump drive torque T_p, and the camshaft drive torques T_{in}, and T_{ex} were estimated from the experimental values for the T_p, T_{in}, and T_{ex}.
(5) The transmission errors in the gear transmission were neglected.

4.2 Calculation of steady-state vibrations of the crankshaft system under firing conditions

To calculate the steady-state vibrations of the crankshaft system as shown in Fig.11, we assumed that the excitation forces of the No.i cylinder would be exerted directly on the No.i crank-pin as a set of distributed forces F_{ri} and F_{ti} in the radial and tangential directions of the No.i crank-throw. And F_{ri} and F_{ti} would have the same time history as F_{r1} and F_{t1} for the No.1 crank-throw, having a firing phase angle δ_i with respect to the F_{r1} and F_{t1}.

By setting the No.1 crank-throw in the vertical plane (x-z plane) as shown in Fig.11, one can determine the excitation forces: $F_{y1}(\theta)$ and $F_{z1}(\theta)$ in x-y-z coordinates at crank angle $\theta = \omega t$, for any engine operating conditions, and one can see that $F_{y1}(\theta) = F_{t1}(\theta)$ and $F_{z1}(\theta) = F_{r1}(\theta)$. Then the excitation forces F_{yi} and F_{zi} in the x and y directions for the No.i crank-throw can be expressed as follows, by taking into account the geometrical angle β_i between the No.i and No.1 crank-throw in the crankshaft structure:

$$F_{yi}(\theta) = F_{y1}(\theta - \delta_i)\cos\beta_i - F_{zi}(\theta - \delta_i)\sin\beta_i \quad (2)$$

$$F_{zi}(\theta) = F_{y1}(\theta - \delta_i)\sin\beta_i - F_{zi}(\theta - \delta_i)\cos\beta_i \quad (3)$$

By assembling together the excitation forces $F_{yi}(\theta)$ and $F_{zi}(\theta)$ for all crank-throws, one can derive the force vector $\{F(t)\}$ for the total crankshaft system in the time domain. The force vector $\{F(t)\}$ can be expanded into a series of harmonic force components by FFT analysis. Finally, one can derive the following force-displacement equations in matrix form for the nth order harmonic vibrations of the total crankshaft system.

$$[K(n\omega)]\{u_n\} = \{F_n\} \quad (4)$$

where $[K(n\omega)]$ is the dynamic stiffness matrix of the total crankshaft system with angular frequency $n\omega$, $\{u_n\}$ is the displacement vector including phase angle, and $\{F_n\}$ is the force vector for the nth order harmonic force components.

4.3 Calculated results and experimental results

To examine the validity of the modeling of the crankshaft system shown in Fig.11, we calculated the natural frequencies and mode shapes of the crankshaft system.

Next, we calculated the steady vibrations of the crankshaft system, the injection-pump drive system, and the valve drive system for the engine operating conditions. The calculations were performed for the crankshaft system with a torsional damper attached, and for the full-load conditions.

In Tables5(a), (b), the calculated results of the natural frequencies of the crankshaft system are shown for the "axial-bending in the x-z plane modes", and the "torsional-bending in the x-y plane modes". The calculated results are shown for the three cases, when the gear train is disengaged with the crankshaft system, and when the gear trains of the FEDGT type and the REDGT type were engaged respectively with the crankshaft system.

In Tables6(a), (b), the calculated results and experimental results are shown for the natural frequencies of the crankshaft system disengaged with the REDGT gear train shown in Fig.11. In Table6(a), the natural frequencies of the torsional modes are shown in the frames with hatching lines.

Table5(a) Natural frequencies of the crankshaft system (mode: axial and bending in x-z plane)

Mode	without gear train (Hz)	with the REDGT gear train (Hz)	with the FEDGT gear train (Hz)
1	142	142	141
2	273	273	275
3	572	570	571
4	644	644	649
5	1020	1020	1022

Table5(b) Natural frequencies of the crankshaft system (mode: torsional and bending in x-y plane)

Mode	without gear train (Hz)	with the REDGT gear train (Hz)	with the FEDGT gear train (Hz)
1	176	176	175
2	398	397	366
3	698	698	694
4	1215	1215	1128

Table6(a) Natural frequencies of the crankshaft system disengaged with gear train in Fig.11 (mode: axial and bending in x-z plane)

Mode number	Experimental (Hz)	Calculated (Hz)	Relative error (%)
1	135	137	+1.5
2	300	346	+15.3
3	455	455	0
4	775	960	+23.9
5	1040	1140	+9.6
6	1785	1663	−6.8

Table6(b) Natural frequencies of the crankshaft system disengaged with gear train in Fig.11 (mode: torsional and bending in x-y plane)

Mode number	Experimental (Hz)	Calculated (Hz)	Relative error (%)
1	173	193	+11.6
2	373	405	+8.6
3	696	537	−22.8
4	1258	1045	−16.9
5	1753	1597	−8.9

In Fig.12, the calculated results of the crankshaft torsional vibrations at the crankshaft front pulley, which were induced by the 6th order harmonic force are shown. Since the most predominant crankshaft vibrations were induced by the 6th order harmonic forces. The results are shown for the two cases, i.e. when the FEDGT gear train was engaged to the crankshaft, and when the REDGT gear train was engaged to the crankshaft system.

In Fig.13, the calculated results are shown for the torsional vibration at the crankshaft gear induced by the 6th order harmonic forces similarly for the two cases mentioned above for Fig.12.

In Fig.14, the calculated results are shown for the torsional vibrations at the injection-pump drive shaft end (governor side shown in Fig.11), induced by the 6th order harmonic forces, similarly for the two cases mentioned as regards Fig.12.

In Fig.15, the calculated results are shown for the torsional vibrations at the camshaft end opposite to the camshaft gear. These were induced by the 6th order harmonic forces, similarly for the two cases mentioned as regards Fig.12.

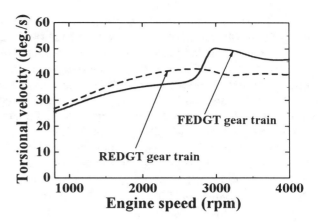

Fig.14 The 6th order amplitudes of torsional velocity at the end of injection pump shaft opposite to gear

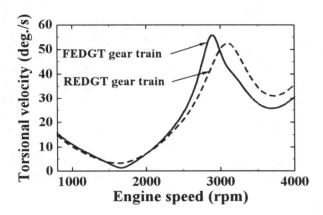

Fig.12 Amplitudes of the torsional velocity at front pulley, due to the 6th order harmonic forces

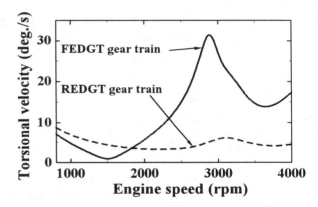

Fig.15 The 6th order amplitudes of torsional velocity at the end of valve drive cam shaft opposite to gear

4.4 Discussion about three-dimensional vibrations of the crankshaft system

From the series of three-dimensional vibration analyses for the crankshaft system, one can see from Tables5(a), (b) that by the gear train of the FEDGT type, significant reduction of the natural frequencies would be induced in the torsional vibrations of crankshaft system. Here the reduction was about 8% for the 1st mode, and 7.2% for the 2nd mode; such values almost coincide with the values obtained by Holzer's method.

From Tables6(a), (b), one can see that the calculated results of the natural frequencies almost agree with the experimental results. However, because of the different modeling methods for the crankshaft system shown in Figs.6(a),(b), and Fig.11, slight differences are seen in the calculated results shown in Tables3(a), (b) and Tables6(a), (b).

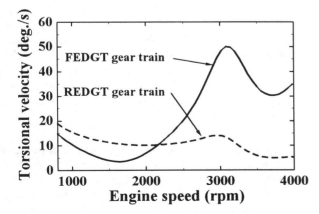

Fig.13 Amplitudes of the torsional velocity at crank gear, due to the 6th order harmonic forces

From Fig.12, one can see that the magnitudes of torsional vibrations of the crankshaft front pulley are almost the same for the two types of the gear train. However, since the resonance peaks appears at the engine speed: N=2850 rpm for the gear train of the FEDGT type, and at N=3070 rpm, for the REDGT type, the natural frequency for the former case: 285Hz is lower than that for the latter one of 307Hz. Here the reduction of 22Hz means reduction of only 6%. The value is also very close to the one obtained from Holzer's method. We suspect that the similar vibration amplitudes might be caused by the predominantly large values of the flywheel of $I_f = 39.4 \times 10^{-2}\ kg \cdot m^2$ compared to that of the front pulley of $I_p = 0.7 \times 10^{-2}\ kg \cdot m^2$.

From Fig.13, one can see the predominant difference in the amplitudes of the torsional vibrations appeared at the crankshaft gears. Since the gear train is driven by the crankshaft gear, one can imagine that significant gear vibrations would be induced when the gear train of the FEDGT type was engaged to it. Accordingly significant gear rattling might be induced in the gear train.

From Fig.14, one can see that, although significant magnitudes of torsional vibrations can be induced in the injection–pump drive shaft, the vibration amplitudes were very similar to the two cases when the two types of gear trains were engaged to the crankshaft system.

From Fig.15, again one can see the predominant difference appeared in the amplitudes of the torsional vibrations at the valve drive camshaft. Since the two resonance peaks appear simultaneously with those for the front pulley torsional vibrations shown in Fig.12, we think that these camshaft resonance vibrations might be caused by the crankshaft torsional vibrations.

CONCLUSION

1) For a four–cylinder in–line diesel engine of OHC type, we studied the influence of the gear train on the vibrations behavior of the crankshaft system. We investigated the influence for the cases when the gear trains of the (engine) front end drive gear train type: the FEDGT type, and of the rear end drive gear train type: the REDGT type were engaged to the camshaft system.
2) We calculated the natural frequencies and mode shapes of the torsional vibrations of the crankshaft system by Holzer's method, and compared the calculated results with the experimental results.
3) We also calculated the three–dimensional vibrations of the crankshaft system, the injection–

pump drive system, and the valve drive camshaft, for the engine operating conditions by the dynamic stiffness matrix methods.

4) From a series of calculation, we found that when the gear train of the FEDGT type is engaged with the crankshaft system, (a) reduction of the natural frequencies of the crankshaft system in the torsional modes, and (b) significant torsional vibrations of the gears in the gear train, can be caused due to the large moments of inertia of the gears in the gear train, and the large torsional amplitude of the crankshaft gear.
5) Therefore, the gear train should be engaged to the crankshaft system at the crankshaft rear end (flywheel side), as close as possible to the nodal point of the crankshaft torsional vibrations.

ACKNOWLEDGMENTS

The authors would like to thank Mr. Kazuyoshi Ikuno of Isuzu Motors Ltd for providing us the test engine, and Mr. Atsushi Takeda and Hiroyuki Shishito, our former students, for their cooperation during the experiments. We also thank Professor F. Scott Howell of Sophia University for his kindness in reading and making corrections during the preparation of this paper.

REFERENCES

[1] W. T. Thomson, "MECHANICAL VIBRATIONS," PRENTICE–HALL, INC., Englewood Cliffs, N. J., 1953.

[2] Okamura, H. and Morita,T., "Influence of Crankshaft–Pulley Dimensions on Crankshaft Vibrations and Engine–Structure Noise and Vibrations," SAE Paper 931303, 1993.

[3] H. Okamura, A. Shinno, T. Yamanaka, A. Suzuku and K. Sogabe, "Simple Modeling and Analysis for the Crankshaft Tree-Dimensional Vibrations, Part 1: Background and Application to Free Vibrations," Transactions of the ASME, Journal of Vibration and Acoustics, VOLUME 117 · NUMBER 1 · JANUARY 1995.

[4] T. Morita and H. Okamura, "Simple Modeling of and Analysis for Crankshaft Three-Dimensional Vibrations, Part 2: Application to an Operating Engine Crankshaft," Transactions of the ASME, Journal of Vibration and Acoustics, VOLUME 117 · NUMBER 1 · JANUARY 1995.

971968

Overview of the Experimental Approach to Statistical Energy Analysis

Benjamin Cimerman
Vibro-Acoustic Sciences Inc.

Tej Bharj
Ford Motor Co.

Gerard Borello
InterAC

Copyright 1997 Society of Automotive Engineers, Inc.

ABSTRACT

Statistical Energy Analysis (SEA) is used to predict wide-bandwidth noise and vibration. That prediction may rely on parameters derived from theory or from test, which essentially means that there are two distinct approaches, analytical SEA and test-based SEA.

The latter is the focus of this paper. Both theory and practice are reviewed, so that the current status of the method can be established. This review also provides some insight on what information can be extracted from the experiment, how the measurements must be conducted and how the results must be interpreted.

Another important aspect of test-based SEA is its interaction with the more widely used analytical SEA method. It is demonstrated that both methods are complementary and that the analytical and test-based parameters can either be compared or mixed in a "hybrid" SEA model. Benefits of the combined use of the methods are discussed.

The discussions are supported by results obtained for automotive applications. This overview paper will provide the reader with an understanding of the method and its applications.

INTRODUCTION

The noise, vibration and harshness (NVH) performance of automobiles is currently a major issue for manufacturers around the world, since global competition is forcing them to reduce cost and weight while improving comfort.

In order to improve the NVH performance of a vehicle without resorting to the costly addition of un-optimized weight, it is essential to first understand the key noise generation mechanisms in the vehicle. This is where SEA-based techniques are attractive since they are fast (implementation and solve time), require a moderate level of detail information on the vehicle design, can be used on a wide frequency band, and offer attractive diagnosis tools such as transmission path analysis and source ranking [1].

GENERAL THEORY OF SEA

Statistical Energy Analysis (SEA) is a noise and vibration prediction method, particularly effective at mid to high frequencies [2].

SEA describes a complex system in terms of a network of connected subsystems, each of which has a highly resonant multi-modal response (or equivalently, a reverberant wavefield). Each subsystem's response is represented by a space-averaged and frequency band-integrated energy level, E_i.

Based on the principle of conservation of energy, a band-limited power balance matrix equation for the connected subsystems can be derived. The equation may be written in two equivalent forms. The symmetric form

$$\omega \begin{bmatrix} \left[\eta_1 + \sum_{i \neq 1} \eta_{1i}\right]n_1 & -\eta_{12}n_1 & -\eta_{1k}n_1 \\ \dots & \dots & \dots \\ -\eta_{k1}n_k & -\eta_{k2}n_k & \left[\eta_k + \sum_{i \neq 1} \eta_{ki}\right]n_k \end{bmatrix} \begin{bmatrix} \dfrac{E_1}{n_1} \\ \dots \\ \dfrac{E_k}{n_k} \end{bmatrix} = \begin{bmatrix} \Pi_1 \\ \dots \\ \Pi_k \end{bmatrix}$$

(1)

or the non-symmetric form

$$\omega \begin{bmatrix} \left[\eta_1 + \sum_{i \neq 1} \eta_{1i}\right] & -\eta_{21} & -\eta_{k1} \\ -\eta_{12} & \dots & \dots \\ -\eta_{1k} & -\eta_{2k} & \left[\eta_k + \sum_{i \neq 1} \eta_{ki}\right] \end{bmatrix} \begin{bmatrix} E_1 \\ \dots \\ E_k \end{bmatrix} = \begin{bmatrix} \Pi_1 \\ \dots \\ \Pi_k \end{bmatrix}$$

(2)

where η_i is the subsystem Damping Loss Factor (DLF)

η_{ij} is the Coupling Loss Factor (CLF) between subsystems

n_i is the subsystem modal density (number of modes per Hertz)

Π_i is the Source Input Power.

The Loss Factor matrix of equation (1) is symmetric because of the reciprocity relationship:

$$\eta_{ij} n_i = \eta_{ji} n_j$$

This form is commonly used for analytical SEA modeling, where the modal densities, coupling loss factors and power inputs are calculated using theoretical derivations [3] (damping loss factors are usually measured). Inversion of the matrix of equation (1) yields the modal power E_i/n_i and consequently the energy E_i of each subsystem.

In test-based SEA, the loss factor matrix is derived from test. While not strictly necessary, it is easier to measure the coupling loss factors in both directions so that equation (2) can be used and modal densities need not be measured. The matrix inversion directly yields the energy E_i. Note that although the modal densities are not directly involved in the evaluation of subsystem energies, the traditional SEA assumptions still apply and therefore each subsystem must have a sufficient number of resonant modes in each frequency band of analysis for the resonant behavior to dominate the response.

For both the analytical and test-based approach, the space-averaged and frequency band-integrated vibration level in a structural subsystem is then calculated from:

$$E_i = m_i \langle V_i^2 \rangle_{sp}$$

where m_i is the total mass of the subsystem and $\langle V_i^2 \rangle_{sp}$ the space-averaged square velocity.

For an acoustic subsystem, the space-averaged and frequency band-integrated sound pressure level (SPL) is calculated from:

$$E_i = \frac{V}{\rho_0 c_0^2} \langle p_i \rangle_{sp}^2$$

where ρ_0, c_0 are the mass density and wavespeed in the fluid, respectively,
V is the volume of the acoustic subsystem, and
$\langle p_i \rangle_{sp}^2$ is the space-averaged square sound pressure level.

TEST-BASED SEA EQUATIONS

Considering a system composed of N subsystems, up to N^2 unknown must be estimated (N damping loss factors and N(N-1) coupling loss factors). Acquiring enough test data to generate N^2 equations is achieved by applying input power on one subsystem only and measuring the response of all subsystems for that specific excitation. This generates N equations and therefore this operation must be repeated for all subsystems in order to generate NxN equations. If N=2, the set of equations can be written in a matrix form as [4]

$$\omega \begin{bmatrix} E_{11} & E_{11} & -E_{21} & 0 \\ 0 & E_{11} & -E_{21} & -E_{21} \\ -E_{12} & -E_{12} & E_{22} & 0 \\ 0 & -E_{12} & E_{22} & E_{22} \end{bmatrix} \begin{bmatrix} \eta_1 \\ \eta_{12} \\ \eta_{21} \\ \eta_2 \end{bmatrix} = \begin{bmatrix} \Pi_1 \\ 0 \\ 0 \\ \Pi_2 \end{bmatrix}$$

(3)

where E_{ij} is the space-averaged energy of subsystem i when subsystem j is excited.

Unfortunately, the matrix of equation (3) is rather ill-conditioned and errors in the calculation get larger with an increasing number of subsystems. Lalor [4] proposes to break down the NxN matrix into N sets of (N-1)x(N-1) matrices giving the coupling loss factors and N equations giving the damping loss factors. For the coupling loss factors related to subsystem i, one obtains

$$\begin{bmatrix} \eta_{1i} \\ \cdots \\ \eta_{Ni} \end{bmatrix}_{j \neq i} = \frac{\Pi_i}{\omega E_{ii}} \begin{bmatrix} \left(\frac{E_{11}}{E_{i1}} - \frac{E_{1i}}{E_{ii}}\right) & \cdots & \left(\frac{E_{N1}}{E_{i1}} - \frac{E_{Ni}}{E_{ii}}\right) \\ \cdots & \cdots & \cdots \\ \left(\frac{E_{1N}}{E_{iN}} - \frac{E_{1i}}{E_{ii}}\right) & \cdots & \left(\frac{E_{NN}}{E_{iN}} - \frac{E_{Ni}}{E_{ii}}\right) \end{bmatrix}^{-1} \begin{bmatrix} 1 \\ \cdots \\ 1 \end{bmatrix}$$

(4)

This matrix is usually well conditioned since the diagonal terms are large (the energy term E_{ii} is larger than E_{ji}).

The damping loss factor of subsystem i is then obtained as [4]

$$\eta_i = \frac{\frac{\Pi_i}{\omega} - \left\{ \sum_{j=1}^{N} (E_{ji}\eta_{ji} - E_{ii}\eta_{ij}) \right\}_{(j \neq i)}}{E_{ii}}$$

(5)

or directly as

$$\begin{bmatrix} \eta_1 \\ \cdots \\ \eta_N \end{bmatrix} = \frac{1}{\omega} \begin{bmatrix} E_{11} & \cdots & E_{N1} \\ \cdots & \cdots & \cdots \\ E_{1N} & \cdots & E_{NN} \end{bmatrix}^{-1} \begin{bmatrix} \Pi_1 \\ \cdots \\ \Pi_N \end{bmatrix}$$

(6)

Lalor [5] has shown that even the matrix equation (4) can lead to aberrations such as negative coupling loss factors. Under certain assumptions the problem can be greatly simplified, so that no numerical matrix inversion is required. The coupling loss factor from subsystem i to subsystem j is then calculated as

$$\eta_{ij} \approx \frac{1}{\omega} \left(\frac{E_{ji}}{E_{ii}}\right) \left(\frac{\Pi_j}{E_{jj}}\right)$$

(7)

The following assumptions must be considered:
1) The modal energy E/N of a directly driven subsystem (here N is the number of modes per band) is greater than that of a subsystem connected to

it. Therefore it is not valid for strong coupling $(E_{ij}/N_i \cong E_{jj}/N_j)$.

2) The expression calculates the CLF between two subsystems from measurements on these subsystems only (the underlying assumption is that most of the power flows directly between the two subsystems, not through a third one). This second assumption means that equation (7) can not be used to calculate indirect coupling loss factors (i.e. coupling loss factors between subsystems that are not physically directly connected), unless the indirect path dominates (such is often the case for panel "mass law" transmission at low frequency).

Although these assumptions may not always be completely verified, they allow for less measurements to be acquired and avoid potentially difficult numerical problems in inverting matrices.

If equation (7) is used to estimate the coupling loss factors, then the damping loss factors can still be obtained from equation (5), with some of the coupling loss factors set to zero. Nevertheless, many practitioners prefer to rely on the traditional decay rate measurement method to derive damping loss factors.

PRACTICAL MEASUREMENT CONSIDERATIONS

Equations (3)-(7) clearly indicate that both power input and subsystem energies must be evaluated in order to estimate loss factors.

POWER INPUT - For a point force excitation of rms amplitude F_i on subsystem i, the corresponding power input is

$$\Pi_i = Re(F_i V_i^*) = F_i^2 Re[Y_i]$$

(8)

where V_i is the rms amplitude of the input velocity and Y_i is the mobility at the input location.

The phase relationship between force transducer and accelerometer is critical and errors can significantly bias the results. Some practitioners claim good power estimates can be obtained by exciting the panel next to the accelerometer, but using phase calibrated and in-line force transducer and accelerometer is more accurate. This can be achieved by using an impedance head whenever appropriate, by placing the accelerometer on the side opposite the excitation or by using a small, light and rigid cap on top of the accelerometer to strike on.

For practical reasons, transient excitation (hammer strike) is preferred. Nevertheless one must make sure that enough power can be injected into the system at all frequencies of interest. High frequency results of good quality may be difficult to obtain for both transient and steady-state excitation methods unless adequate equipment (such as hammer size and tip) is selected.

Since loss factors are independent of the locations where the input power is applied and of the locations where the responses are measured, a space-averaging scheme must be considered. The number of input and measurement locations heavily depends on the structure

tested but three input locations and five measurement locations are quite typical. It is a usual practice to normalize the power input by calculating the average power per unit square force so that no bias due to unequal forcing is introduced.

For acoustic subsystem excitation, it is customary to use a loudspeaker for which the power output per unit square volt has been measured in a reverberant or anechoic environment. The assumption is that the power output is independent of the surrounding structure (interaction between reflected acoustics waves and loudspeaker membrane). An omnidirectional speaker of small size compared to the acoustic cavity dimensions is preferred.

For both structural and acoustic excitation, the power is first calculated on a narrow frequency band, but must then be band-integrated over the frequency band of interest for the SEA analysis (1/3 octave or octave band).

SUBSYSTEM ENERGY - Considering two structural subsystems i and j, the time-averaged total energy of subsystem i for excitation on subsystem j is, in each frequency band of interest

$$E_{ij} = m_i \langle V_{ij}^2 \rangle_{sp}$$

(9)

Equation (9) appears to offer a straightforward relationship between energy and velocity. In practice, deriving the subsystem energy from the measurement of a set of accelerations presents some difficulties.

First, equation (9) actually states that the total energy is equal to twice the kinetic energy. This is correct only if the energy is stored by modes excited at resonance. This is part of the SEA assumptions, but not always verified in practice.

Second, a true average of the velocity is difficult to measure. In practice measurements are made at discrete locations that may not yield an accurate representation of the subsystem space averaged velocity.

Finally, most often only the normal velocity is measured which means that the energy stored by longitudinal/in-plane waves is overlooked.

In order to obtain a correct estimate of the energy of the subsystem, the concept of equivalent mass is introduced. Lalor showed that the equivalent mass of the subsystem can be estimated from an early decay rate measurement [4] as

$$M_i^{eq} = \frac{\Pi_i}{0.23 \gamma_i \langle V_{ii}^2 \rangle_{sp}}$$

(10)

where γ_i is the early decay rate coefficient.

Estimating the slope of the early decay of a measured time signal is not easy. Wu et al. [6] compared calculations of damping loss factors from the energy (equations (5) and (6)) and decay rate methods and made some recommendations on how to evaluate the correct slope. They also discussed the concept of equivalent mass and showed that it may be larger than the mass of the subsystem and even larger than the total mass of the system, something often observed at low frequency. There is agreement that if the subsystem is a homogeneous plate, the equivalent mass should be close to the real mass, especially as the frequency increases. For a subsystem of complex shape, spatial undersampling may lead to an equivalent mass that is very dependent on accelerometer locations. If such is the case it is important to perform all the measurements using the same accelerometer locations.

Some practitioners feel that considering the uncertainties related to the measurement of equivalent mass, it may sometimes be more appropriate to use the real mass in estimating the subsystem energy.

Similarly, an equivalent volume can be derived for acoustic subsystems.

VARIANCE ESTIMATES

In test-based SEA, the measured data is manipulated to obtain damping and coupling loss factors. It is therefore very important to know how an error in the input data (measured power and energy) will affect the validity of the output data (coupling, damping loss factors and ultimately predicted velocities and sound pressure levels).

In [7], De Langhe and Sas propose analytical derivations for the variance of damping and coupling loss factors. For the coupling loss factor between subsystem i and subsystem j they write

$$s^2_{\eta_{ij}} \cong \sum_{l,\, m\, =\, 1}^{N} (\eta^0_{il}\eta^0_{mj})s^2_{E^n_{lm}} \qquad i \neq j$$

(11)

where N is the total number of subsystems, η^0_{pq} the coefficient for row p and column q in the matrix of equation, and $s^2_{E^n_{lm}}$ the variance of the energy normalized to unit power.

For the damping loss factor of subsystem i they write

$$s^2_{\eta_{ii}} \cong \sum_{l,\, m\, =\, 1}^{N} (\eta^0_{mi}\eta^0_{ll})s^2_{E^n_{lm}}$$

(12)

According to equations (11) and (12), the variance will increase with the number of subsystems. This is because this variance calculation is based on the inversion of the complete $N^2 \times N^2$ loss factor matrix. On the other hand, $s^2_{E^n_{lm}}$ will tend to decrease if more measurement locations are considered on each subsystem, which will in turn reduce the variance estimates of the loss factors.

Our understanding is that a probability law is assumed in order to derive algebraic equations for the variances on loss factors.

A numerical alternative is proposed [8]. The approach consists of estimating the variance on measured power and normalized square velocity and then calculating the loss factors as

$$[LF] = \frac{1}{\omega}[E + \Delta E]^{-1}[P + \Delta P]$$

(13)

ΔP and ΔE are randomly varied within the limits of the estimated standard deviations on power and energy, and the loss factors are calculated for each $(\Delta P, \Delta E)$ pair. Mean and variance estimates for the loss factors are finally derived from the set of calculated loss factors. This perturbation method can always be applied, regardless of the method chosen to derive the coupling loss factor (equations (3), (4) and (7)). Also, a numerical method based on first calculating the variance of sub-ensembles can be used to estimate the variance on the normalized square velocity, so that no given probability law is assumed.

Note that both numerical and algebraic approaches do not consider the variance on the measured equivalent mass or volume.

HYBRID SEA

Test-based SEA models are useful but their predictive capabilities are limited. Also, test results can be erroneous and therefore comparison to analytical results can serve as a sanity check. In short, using both test-based SEA and analytical SEA methods can enhance our ability to model the vibro-acoustic behavior of complex systems.

Hybrid SEA modeling can be defined as the integration of test and analytical data in a common SEA model. To start with, introducing measured estimates of damping and input power in an analytical SEA model can significantly improve the accuracy of the predictions. This is quite a common practice today and therefore we may wish to consider a more restrictive definition of hybrid modeling: the mixing of test-derived and analytical coupling loss factors in a common SEA model.

In [9], Bharj et. al. review compatibility issues. For example, test subsystems are usually defined differently to the subsystems in analytical SEA for two reasons:
- test subsystems are often larger in size to limit the number of measurements and avoid equipartition of energy between subsystems (see assumptions following equation (7)),
- measurement of in-plane or torsional response is difficult and therefore test-derived coupling loss factors usually correspond to the coupling of flexural waves.

Therefore many structure-to-structure CLFs are not equivalent between test-based and analytical SEA, unless special care is taken to substructure the system in the same way. Some CLFs that are more nearly interchangeable are structure-to-acoustic (i.e. radiation) CLFs and some acoustic-to-acoustic CLFs. The most benefit can be derived from building an analytical model that corresponds to the experimental one in order to perform a validation exercise, before the analytical model is refined to introduce the level of detail needed for design purposes [9].

RESULTS

In [10], Bharj presents results for the test-based SEA modeling of a passenger vehicle. The system was broken down into 49 subsystems (including 6 acoustic ones) and studied on a [200Hz-5KHz] frequency range. An impact hammer was used to excite structural subsystems at three locations per subsystem. The impact hammer was struck on a rigid cap covering the input accelerometer to ensure a good estimate of power input. For the acoustic subsystems four loudspeaker locations were considered. Damping was derived from decay rate measurements and coupling loss factors from equation (7). The only indirect connections considered were for mass law across dash, floor and package tray. After coupling and damping loss factors were obtained, the operational power inputs were calculated for a number of operating conditions by placing the car on chassis rolls (with engine on) and measuring subsystem energies. The measurement of subsystem energies was limited to the "external input" subsystems, i.e. the ten subsystems receiving external excitation. This approach tends to slightly overestimate the power input but validation proved it effective as shown on Figure 1.

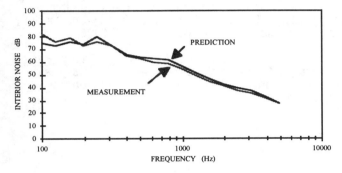

Figure 1 : Interior Noise at 55mph

The validated test-based SEA model was then used as a diagnostic tool to evaluate dominant paths and rank sources. Figure 2 shows that the road input on the right hand-side rear shock tower is the main contributor to interior noise at 250Hz.

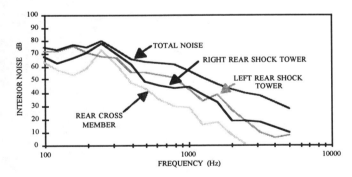

Figure 2: Rear Structure-borne Inputs

The diagnosis led to the design of a sound package that was introduced in a prototype vehicle. The modified coupling loss factors were re-measured and the new prediction was compared with test. Figure 3 shows that good correlation with on-road measurements was obtained.

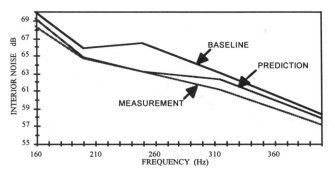

Figure 3: Interior Noise for Optimized Sound Package

This study clearly demonstrated that test-based SEA can be applied to a system as complex as a passenger car. The test-based SEA model was useful in two ways:
- to identify and understand a specific noise problem,
- to help improve an analytical model of the same vehicle and generate a hybrid model.

A smaller-scale project was conducted in Europe on a truck cabin. The objective of the study was to collect enough test data within a few days so that the analytical SEA model of the truck could be validated and improved.

The system was broken down into 18 subsystems and the test data was collected on a [100Hz-2KHz] bandwidth, using transient excitation. In order to acquire the data needed to calculate the input power, the impact hammer was struck next to the input accelerometer. Between 3 and 6 input and measurement locations were used on each subsystem, depending on its complexity. The coupling and damping loss factors were computed independently using equations (4) and (6), respectively. Most indirect coupling loss factors were not measured, which reduced the dimension of the matrix of equation (4) (one row and one column removed for each CLF that is not measured). Results presented in Figure 4 and Figure 5 show that acceptable results were obtained, even though difficult experimental conditions were encountered.

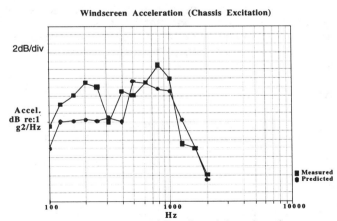

Figure 4: Measured / Predicted Acceleration

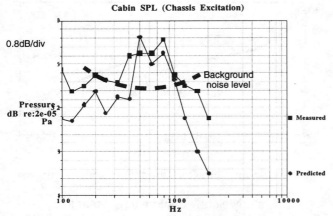

Figure 5: Measured / Predicted SPL

CONCLUSION

Current academic research and application to the automotive industry are helping define the practical field of application of test-based SEA. Results clearly indicate that the method can be turned into a very useful diagnosis tool in the NVH test laboratory, as well as a complement to analytical SEA simulation (playing a role similar to what modal analysis is to finite elements analysis).

Nevertheless some important issues remain to be addressed by more researchers and practitioners. Concerning the equivalent mass, there are different opinions on what it really is (a legitimate quantity or a compensating factor?). Practically it requires some experience to derive it from test because the measurement of early decay rate is error-prone. This also bears consequences on the estimate of the damping loss factor, which is a critical quantity in the power balance approach since it represents the energy dissipation mechanism. Also, the ill-conditioning of the energy matrix tends to limit the application of the method. For example, the measurement of indirect coupling loss factors could be a guide to on-going research on analytical estimates of those (which some believe would significantly improve the accuracy of SEA predictions), but is seldom practised to avoid numerical matrix inversion problems. Finally, precise test guidelines and procedures need to be established, and compatibility/complementarity between test-based and analytical SEA is also of an area of interest that would benefit from additional investigation.

REFERENCES

[1] Dong B., et. al., "Road Noise Modeling Using Statistical Energy Analysis", Proc. SAE N&V Conf., 1995

[2] Lyon R.H., DeJong R. G., "Theory and Application of Statistical Energy Analysis", Butterworth-Heinemann, 1995.

[3] AutoSEA Theory & Q.A. Manual, Vibro-Acoustic Sciences, Doc. AS-100, 1991-1996

[4] Lalor, N., "The Experimental Determination of Vibrational Energy Balance in Complex Structures," Paper No. 108429, Proc. SIRA Conf. on Stress and Vibration, London, 1989.

[5] Lalor, N., ISVR Technical Report No182, 1990

[6] Wu L., Agren A., "Analysis of Initial Decay Rate in Relation to Estimates of Loss FActor and Equivalent Mass in Experimental SEA", Proc. ISMA 21, Leuven, Bel., 1996

[7] De Langhe K., Sas P., "Statistical Analysis of the Power Injection Method", JASA 100(10, 294-303, 1996

[8] Rosen M., Borello G., "Damping and Coupling Loss Factors Estimation in SEA Method: What is Really Measured?", Proc. InterNoise 96, Liverpool, England, 1996

[9] Bharj T., Cimerman B., "Application of Statistical Energy Analysis to a Passenger Vehicle: Combining Analytical and Test-based Prediction in a Hybrid Model", Proc. InterNoise 96, Liverpool, England, 1996

[10] Bharj T., Pham H., "Application of Energy Flow Analysis (EFA) to Reduce Structure-borne Noise Inside a Passenger Vehicle", Proc. InterNoise 96, Liverpool, England, 1996

971969

Methods to Estimate the Confidence Level of the Experimentally Derived Statistical Energy Analysis Model: Application to Vehicles

L. Hermans, K. De Langhe, and L. Demeestere
LMS International NV

Copyright 1997 Society of Automotive Engineers, Inc.

ABSTRACT

The effectiveness of vehicle design modifications based on an experimental SEA model will strongly depend on the confidence one can have in the identified model. This paper deals with the estimation of the variances of the experimentally derived internal and coupling loss factors, allowing to assess the confidence levels of the SEA predictions. The theory concerning the statistical aspects of the Power Injection Method (PIM) is outlined and the derived analytical expressions are validated based on a Monte-Carlo variability analysis. The statistical formulas are then applied to a railway carriage of a high-speed train, illustrating that the calculation of confidence levels is a very useful tool to assess the accuracy of the experimental SEA model.

INTRODUCTION

The Power Injection Method (PIM) is a well-established technique to experimentally derive the SEA loss factor model without the need to disassemble the structure into components. It basically comes down to exciting each subsystem in turn at different excitation points while the responses on all subsystems are measured. By spatially averaging the measured responses and injected powers and using the mass or volume of each subsystem, a matrix containing the subsystem energies normalized to unity input power can be obtained, which allows to derive all loss factors.

The accuracy of the experimental SEA model will depend on the specific number of excitation and response points chosen on each subsystem. There will always be a certain spread associated with the subsystem energies due to the spatial averaging of a finite number of measurement points. This paper addresses how the variance of the normalized energies measured in the power injection process can be derived. The statistical formulas describing the variances of the loss factors in terms of the variances of the normalized

energies are outlined and validated based on a Monte-Carlo variability analysis. The Monte-Carlo approach requires a high computational effort, but additionally allows to have a look at the probability histogram giving a picture of the shape of the distribution of the loss factors. Subsequently, it is described how the predictive capabilities of the SEA model can be analyzed in terms of confidence levels. Finally, the influence of the subsystem mass or volume in the loss factor calculation and the SEA predictions is detailed, showing that the subsystem mass or volume is not critical in the experimental SEA process.

The statistical formulas are validated on some simple subsystems and subsequently applied to a railway carriage of a high-speed train. The examples illustrate that the calculation of confidence levels is a very useful tool to assess the accuracy of the SEA model. The effectiveness of some loss factor modifications is evaluated in terms of the confidence level of the SEA model.

THE BASIC SEA PRINCIPLES

THE POWER BALANCE EQUATIONS - SEA is a high-frequency vibro-acoustic modeling technique which works with the averaged response taken in the time, frequency and spatial domain instead of local response variables. The successful application of SEA relies on a high modal density and high modal overlap to ensure that the average is a useful and reasonably accurate quantity. The technique allows to model the internally dissipated energy and the energy flow between various structural and acoustical subsystems. The following concept is adopted [1] :

- The time averaged net power flow $P_{i \to j}$ from subsystem i to another (adjacent) subsystem j varies at a rate proportional to the difference in modal energy in a given frequency band. This assumption is mostly written in the following form :

$$P_{i \to j} = \omega \left(\eta_{ij} E_i - \eta_{ji} E_j \right) \tag{1}$$

with

$$\eta_{ij} n_i = \eta_{ji} n_j \text{ (SEA reciprocity equation)} \tag{2}$$

where E_i represents the time averaged total energy of the i-th subsystem, ω is the center band frequency, η_{ij} is the coupling loss factor between subsystems i and j and n_i represents the modal density of the i-th subsystem, i.e. the number of modes per unit frequency. For sake of simplicity, no explicit annotation is used in the formulas to denote the time averaging.

- The time averaged power internally dissipated in the i-th subsystem, $P_{dis,i}$, is proportional to the total energy level of this subsystem, giving the following equation:

$$P_{dis,i} = \omega \eta_{ii} E_i \tag{3}$$

where η_{ii} is the internal loss factor of subsystem i.

The global SEA equation can be obtained by balancing the time averaged external power input P_i to subsystem i with the power dissipated in the subsystem and the net power flows to the coupled subsystems j. Hereto, equations (1) and (3) concerning energy flow and respectively energy dissipation are used. This results in the SEA power balance equation :

$$\begin{bmatrix} \sum\limits_{i=1}^{N}\eta_{1i} & -\eta_{21} & \cdots & -\eta_{N1} \\ -\eta_{12} & \sum\limits_{i=1}^{N}\eta_{2i} & \cdots & -\eta_{N2} \\ \cdot & & \cdot & \cdot \\ \cdot & & & \cdot \\ \cdot & & & \cdot \\ -\eta_{1N} & -\eta_{2N} & \cdots & \sum\limits_{i=1}^{N}\eta_{Ni} \end{bmatrix} \begin{bmatrix} E_1 \\ E_2 \\ \cdot \\ \cdot \\ \cdot \\ E_N \end{bmatrix} = \begin{bmatrix} \frac{P_1}{\omega} \\ \frac{P_2}{\omega} \\ \cdot \\ \cdot \\ \cdot \\ \frac{P_N}{\omega} \end{bmatrix} \quad or \quad [L]\{E\} = \frac{1}{\omega}\{P\} \tag{4}$$

where N is the total number of subsystems. The matrix [L] in equation (4), composed of the internal and coupling loss factors, is further referred to as the total loss factor matrix.

THE POWER INJECTION METHOD - The Power Injection Method (PIM) is the most widely used technique to experimentally identify the SEA loss factor model without knowledge of the modal densities [2]. It basically comes down to exciting each subsystem in turn and measuring the injected input power and the response energies of all subsystems. For excitation in each subsystem, these response energies can be normalized by unit input power and the power balance equations can be written down, yielding the following matrix equation :

$$\begin{bmatrix} \sum\limits_{i=1}^{N}\eta_{1i} & -\eta_{21} & \cdots & -\eta_{N1} \\ -\eta_{12} & \sum\limits_{i=1}^{N}\eta_{2i} & \cdots & -\eta_{N2} \\ \vdots & \vdots & \ddots & \vdots \\ -\eta_{1N} & -\eta_{2N} & \cdots & \sum\limits_{i=1}^{N}\eta_{Ni} \end{bmatrix} \begin{bmatrix} E_{11}^n & E_{12}^n & \cdots & E_{1N}^n \\ E_{21}^n & E_{22}^n & \cdots & E_{2N}^n \\ \vdots & \vdots & \ddots & \vdots \\ E_{N1}^n & E_{N2}^n & \cdots & E_{NN}^n \end{bmatrix} =$$

$$\frac{1}{\omega} \begin{bmatrix} 1 & 0 & \cdots & 0 \\ 0 & 1 & \cdots & 0 \\ \vdots & \vdots & \ddots & \vdots \\ 0 & 0 & \cdots & 1 \end{bmatrix} \quad or \quad [L][E^n] = \frac{1}{\omega}[I] \tag{5}$$

where $E_{ij}^n = \dfrac{E_{ij}}{P_j}$ denotes the normalized energy in

subsystem i due to excitation in subsystem j.

THE LOSS FACTOR DERIVATION - It can be clearly seen from equation (5) that the total loss factor matrix [L] can be derived by inverting the normalized energy matrix [E^n]:

$$[L] = \frac{1}{\omega}[E^n]^{-1} \tag{6}$$

Unfortunately, the inversion of the normalized energy matrix is quite sensitive to measurement errors, often resulting in some negative loss factor values which are physically unacceptable. Approximate formulas can be derived on the basis that the energy in non-driven subsystems is significantly lower than in the directly driven subsystem (weak coupling assumption), guaranteeing positive loss factor values [3] :

$$\eta_{ii} \cong \frac{1}{\omega}\frac{1}{\left\langle E_{ii}^n \right\rangle} \text{ and } \eta_{ij} \cong \frac{1}{\omega}\left(\frac{\left\langle E_{ji}^n \right\rangle}{\left\langle E_{ii}^n \right\rangle} \right)\left(\frac{1}{\left\langle E_{jj}^n \right\rangle} \right) \tag{7}$$

An important aspect of equation (7) is that it enables the coupling loss factors between two coupled subsystems to be calculated from measurements made only on those two subsystems, regardless of the other subsystems.

THE MEASUREMENT OF ENERGY - The difficulty that arises in experimental SEA is the measurement of the total resonant subsystem energy E_i. In order to cope with this, the vibrating or acoustical field is sampled at several discrete points. The measured point velocities or pressures are squared and spatially averaged to respectively yield the space averaged square velocity $\left\langle \dot{X}_i^2 \right\rangle$ or pressure $\left\langle p_i^2 \right\rangle$. The brackets <> denote the spatial averaging. The energy of the subsystem is then estimated by multiplying these averaged quantities with the total subsystem mass M_i or the subsystem volume V_i :

$$E_i = M_i \left\langle \dot{X}_i^2 \right\rangle \text{ or } E_i = V_i \left\langle p_i^2 \right\rangle \frac{1}{\rho c^2} \tag{8}$$

where ρ is the medium (mostly air) density and c the sound velocity in the medium.

For sake of simplicity, following variables are introduced for respectively structural and acoustical subsystems :

$$\alpha_i = M_i \text{ or } \alpha_i = \frac{V_i}{\rho c^2} \qquad (9)$$

$$\langle R_i \rangle = \langle \dot{X}_i^2 \rangle \text{ or } \langle R_i \rangle = \langle p_i^2 \rangle \qquad (10)$$

where R_i is further on termed the square velocity/pressure and α_i is called the mass/volume multiplicator. Hence, equation (8) can be compactly rewritten as :

$$E_i = \alpha_i \langle R_i \rangle \qquad (11)$$

Equations (8) or (11) are approximate as they are based on discrete velocity/pressure measurements which are squared and space averaged. The total subsystem mass used in equation (8) can only be used in the calculation of the total energy for uniform structures of constant thickness (plate) or cross-section (beam). It has been suggested to adjust for this discrepancy by applying a correction factor on the mass/volume of the subsystems, referred to as the "equivalent" mass/volume [4].

THE STATISTICAL ASPECTS OF PIM

In the power injection process, each subsystem is excited at 3 or more points chosen at random in order to simulate statistical independence of the modes [2]. The responses at different locations are measured for each subsystem. Typically, about 5 to 10 response points per subsystem are measured. The square velocities/pressures are individually normalized by unit input power, subsequently linearly averaged and multiplied by the mass/volume multiplicator, yielding the normalized energies E^n_{ij} appearing in equations (6) and (7). There will always be a certain spread associated with the normalized space averaged square velocity/pressure $<R^n_{ij}>$ due to the averaging of a finite number of values. Therefore, it is meaningful to calculate the associated confidence levels of $<R^n_{ij}>$ and to comprehend how these confidence levels can be translated into confidence levels of the loss factors.

THE MEAN AND VARIANCE OF THE POPULATION OF THE NORMALIZED RESPONSES -
The population of normalized square velocities/pressures which are dealt with in PIM are infinite because of an infinite number of potential excitation and response points. From the measurements, only a sample of normalized responses is obtained, each corresponding to a specific excitation and response point. If $R^n_{ij,pq}$ denotes the normalized square velocity/pressure measured at response location p of the i-th subsystem due to excitation at location q of the j-th subsystem, the mean value and the variance of the population of the normalized square velocities/pressures of subsystem i for excitation in subsystem j can be estimated as follows [5] :

$$\mu_{R^n_{ij}} \cong \overline{R}^n_{ij} = \frac{1}{N_{resp}N_{inp}} \sum_{q=1}^{N_{inp}} \sum_{p=1}^{N_{resp}} R^n_{ij,pq} \qquad (12)$$

$$\sigma^2_{R^n_{ij}} \cong s^2_{R^n_{ij}} = \frac{1}{N_{resp}N_{inp}-1} \sum_{q=1}^{N_{inp}} \sum_{p=1}^{N_{resp}} (R^n_{ij,pq} - \overline{R}^n_{ij})^2 \qquad (13)$$

where N_{resp} is the number of response locations of response subsystem i and N_{inp} the number of excitation locations of the excited subsystem j. The overbar denotes the estimated mean value. It corresponds to the brackets $< >$ denoting the spatial averaging, i.e. $<R^n_{ij}> = \overline{R}^n_{ij}$. Note the difference in annotation between the (unknown) mean and variance representing the entire population, respectively denoted by the Greek letters μ (population mean) and σ (population variance) and the mean and variance estimated from a sample of quantitative data ($R^n_{ij,pq}$ for $p = 1 \ldots N_{resp}$, $q = 1 \ldots N_{inp}$) respectively designated by the overbar (sample mean) and the Roman letter s^2 (sample variance). The sample standard deviation, s, is the nonnegative square root of the sample variance. It is a measure of spread expressed in the units of the original data. The relative standard deviation which allows comparison of the spread of different data sets on a relative basis is defined as the ratio between the standard deviation and the mean.

THE VARIANCE OF THE NORMALIZED ENERGIES - For experimental SEA, the main result of interest is not as such the estimated mean and variance of the population of the normalized responses, but rather the mean and the variance of the space averaged square velocity/pressure $<R^n_{ij}>$ or the sample mean $\overline{R}^n_{ij}$. In other words, what is e.g. the 95% interval bracketing the unknown mean of the entire (infinite) population of the normalized responses.

The Central Limit Theorem says that, for a large number of samples $N_{resp}N_{inp}$, the sample mean $\overline{R}^n_{ij}$ is approximately normally distributed with the mean $\mu_{R^n_{ij}}$ and a variance given by

$$\sigma^2_{\overline{R}^n_{ij}} = \frac{\sigma^2_{R^n_{ij}}}{N_{resp}N_{inp}} = \frac{1}{N_{resp}N_{inp}} \left(\frac{1}{N_{resp}N_{inp}-1} \sum_{q=1}^{N_{inp}} \sum_{p=1}^{N_{resp}} \left(R^n_{ij,pq} - \overline{R}^n_{ij} \right)^2 \right) \qquad (14)$$

Thus, an interval with center $\overline{R}^n_{ij}$ and endpoints

$$\overline{R}^n_{ij} \pm z \frac{s_{R^n_{ij}}}{\sqrt{N_{resp}N_{inp}}} = \overline{R}^n_{ij} \pm z s_{\overline{R}^n_{ij}} \qquad (15)$$

can be used as an approximate confidence interval bracketing the (unknown) mean $\mu_{R^n_{ij}}$ associated with

excitation in subsystem j and response in subsystem i. Table 1 gives values of z for use in expression (15), for some commonly used confidence levels :

Confidence	z-value
80%	1.28
90%	1.645
95%	1.96
98%	2.33
99%	2.58

Table 1 : z-values for confidence intervals

For PIM, the sample size $N_{reep}N_{inp}$ is often quite small. This implies that using e.g. z= 1.96 generally does not produce the 95% confidence intervals. In the case of a small sample size, it has been shown that the Student t-distribution should in principle be used instead of the normal distribution in order to evaluate the confidence levels. The t-distribution is characterized by a degrees of freedom parameter v. If v goes to infinite, the standard normal distribution is found. In practice, for v larger than about 30, the t distribution and the standard normal distribution are indistinguishable. For practical reasons however and because later on the confidence levels of the SEA parameters are anyway approximated by a first order Taylor expansion, it is found more convenient to determine the confidence levels of all SEA parameters and SEA predictions under the assumption of a normally distributed population.

Assuming then that there is no uncertainty on the mass/volume, the estimated variance of the normalized energies is given by

$$s^2_{E^n_{ij}} = \alpha_i^2 s^2_{\overline{R^n_{ij}}} \qquad (16)$$

Including the uncertainty on the mass/volume gives rise to an additional term $\left(\overline{R^n_{ij}}\right)^2 s^2_{\alpha_i}$ in equation (16), on condition that the mass/volume multiplicator α_i and the square velocity/pressure $\overline{R^n_{ij}}$ are statistically independent variables. The latter is not completely fulfilled in case the equivalent mass/volume is used as in that case, α_i is inversely proportional to $\overline{R^n_{ii}}$. Important to remark is that the uncertainty on the mass/volume multiplicator α_i will affect proportionally all energy values on the i-th row of the normalized energy matrix [E^n] in equation (5). Therefore, including the variance of the mass/volume in the energy variance calculation will cause that the normalized energies can not be considered anymore as statistically independent variables when computing the propagation of the energy variances through the loss factor calculation. Further on in this paper, a section is devoted to the influence of the mass/volume in experimental SEA. It will be shown that, from an experimental point of view, the assumption of no uncertainty on the mass/volume is justified when computing the variances of the loss factors.

THE CALCULATION OF THE VARIANCES OF THE LOSS FACTORS

The objective is now to study the transmission of the variances of the normalized energies through the loss factor calculation. Two methods can be distinguished to calculate the loss factor variances. On the one hand, the application of the statistical propagation of error formulas giving an approximate analytical expression for the loss factor variance as function of the variances of the normalized energies. These expressions can be quickly evaluated during the acquisition process. On the other hand, based on a Monte-Carlo variability analysis which requires much more computational effort and time, but offers the advantage to be more accurate if a sufficiently high number of experiments are evaluated. The statistical Monte-Carlo method also provides the possibility to evaluate the distribution of the loss factors.

ANALYTICAL APPROACH - The variance calculation for the loss factors derived from either full matrix inversion (see equation (6)) or from approximate formulas (see equation (7)) is outlined. In case the approximate formulas are employed to compute the loss factors, it should be stressed that a systematic error is introduced which is not covered in the statistically obtained variance.

Full matrix inversion - The loss factors are estimated through the inverse of the normalized energy matrix. The statistical transmission of variance or propagation of error formula says that the mean of the loss factors can be approximated by evaluating equation (6) for the estimated mean of the normalized energies and that, based on performing a first order Taylor expansion, the variance of the total loss factor l_{ij} is approximately given by :

$$s^2_{l_{ij}} \cong \sum_{k=1}^{N} \sum_{l=1}^{N} \left(\frac{\partial l_{ij}}{\partial E^n_{kl}}\right)^2 s^2_{E^n_{kl}} \qquad (17)$$

It can then be shown that the variance associated with the coupling loss factor between subsystems i and j is given by [6] :

$$s^2_{\eta_{ij}} \cong \frac{1}{\omega^2} \sum_{k=1}^{N} \sum_{l=1}^{N} \left(\overline{l}_{jk}\overline{l}_{li}\right)^2 s^2_{E^n_{kl}} \qquad (18)$$

In case of weak coupling, the major contribution to the variance of the coupling loss factor η_{ij} will occur if k=j and if l=i as then the product of two total loss factors l_{ii} and l_{jj} occurs. This means that it is important that the variance of the normalized energy E^n_{ji} is low to get a high precision for the loss factor η_{ij}. Formula (18) also shows that in case of strong couplings, the variance will be higher in comparison with weak coupling. This implies that, if the variances on the normalized energies are within reasonable limits, the computed confidence intervals of the loss factors allow to assess the effects of rong coupling.

milar to the derivation of expression (18), it can be own that the variance of the internal loss factor of bsystem i is given by [6]:

$$s^2_{\eta_{ii}} \cong \frac{1}{\omega^2} \sum_{k=1}^N \sum_{l=1}^N (\bar{l}_{li}.\bar{\eta}_{kk})^2 . s^2_{E^n_{kl}} \qquad (19)$$

Approximate formulas - The propagation of error rmula can also be applied to equation (7), giving a ood approximation for the loss factors in case of weakly oupled subsystems. By performing the first order Taylor xpansion and computing the derivatives $\frac{\eta_{ij}}{E^n_{ji}}, \frac{\partial \eta_{ij}}{\partial E^n_{ii}}$ and $\frac{\partial \eta_{ij}}{\partial E^n_{jj}}$, the following expressions can be

erived for respectively the internal loss factor and the oupling loss factor :

$$s^2_{\eta_{ij}} \cong \frac{1}{\omega^2} \bar{\eta}^2_{ij} \left(\frac{s^2_{E^n_{ji}}}{\left(\bar{E}^n_{ji}\right)^2} + \frac{s^2_{E^n_{ii}}}{\left(\bar{E}^n_{ii}\right)^2} + \frac{s^2_{E^n_{jj}}}{\left(\bar{E}^n_{jj}\right)^2} \right) \qquad (20)$$

$$s^2_{\eta_{ii}} \cong \frac{1}{\omega^2} \bar{\eta}^4_{ii} s^2_{E^n_{ii}} \qquad (21)$$

n should realize that the formulas (20) and (21) escribe how variability of error is propagated or ansmitted through a mathematical function. In this ase, the mathematical formula is an approximate ethod to compute the loss factors. The formulas will ot take into account the bias error made due to using e approximate loss factor calculation instead of the full atrix inversion. This error is however of second order nd typically, the approximation will result in a slight verestimation of the loss factors. The quality of the pproximate loss factor calculation can be assessed by nthesizing the energies and comparing these with the easured energies obtained from the PIM tests [7].

The approximate analytical expressions ansforming the variances of the normalized energies to variances of the loss factors are extremely helpful to ecide whether or not PIM measurements can be rminated. By analyzing the different contributions to e loss factor variance, it can be seen which critical bsystems require a higher discrete spatial sampling of e response field. The confidence levels also contribute gaining more insight about the sub-division into bsystems, the key-stage in the whole SEA-process. s SEA deals with the reverberant response field, one ould expect that the spread of the normalized point sponses within the same subsystem is quite low. The tter implies that the variance of the population of the ormalized square velocities/pressures given by quation (13), is low. A fairly low number of excitation nd response points is then sufficient to accurately stimate the true average. If high variances are bserved for the normalized energies for a reasonable umber of excitation and response points, this might

indicate that the subsystem division does not comply with the SEA basic assumptions. From equation (14), it can be seen that increasing the number of excitation and response points will narrow down the 95% confidence interval, but precautions have to be taken with this. In case of large spread of the responses within the same subsystem, it might be much more meaningful to re-partition the subsystem into smaller subsystems, having a response population with a lower variance.

THE MONTE-CARLO APPROACH - The Monte-Carlo method computes the loss factor mean and the loss factor variance based on conducting experiments or simulations. In each experiment, the normalized energies are perturbed conform to a normal or Gaussian distribution having the sample mean $\bar{E}^n_{ij}$ and sample variance $s^2_{E^n_{ij}}$. It is assumed that each term of the normalized energy matrix can be independently perturbed. The loss factor calculation is then performed for a vast amount of possible combinations of disturbances of the normalized energies. In case M evaluations are carried out, for each loss factor, M values will be produced, $\eta^1_{ij}, \eta^2_{ij}, \eta^3_{ij}, ... \eta^M_{ij}$. The mean of the loss factor is then estimated by

$$\bar{\eta}_{ij\,MC} = \frac{1}{M} \sum_{k=1}^M \eta^k_{ij} \qquad (22)$$

and the variance is given by

$$s^2_{\eta_{ij}\,MC} = \frac{1}{M-1} \sum_{k=1}^M (\eta^k_{ij} - \eta_{ij\,M-C})^2 \qquad (23)$$

Contrary to the analytical expressions which yield approximations for the loss factor mean and variance, if the number of simulations M is sufficiently high, an accurate result for the loss factor mean and variance is obtained. Additionally, the data set, $\eta^1_{ij}, \eta^2_{ij}, \eta^3_{ij}, ... \eta^M_{ij}$, allows to compute the probability histogram for each loss factor which can be visualized to check whether still a normally distributed population is found and by consequence, whether the normal distribution is adequate to describe the confidence intervals given by Table 1. The disadvantage of the Monte-Carlo approach however lies within the significant amount of computation time in case the number of subsystems included in the SEA-model becomes large. In case of N-subsystems, N^2 variables can be independently perturbed in each experiment and it can be easily understood that the number of experiments, M, should be much larger than N^2 in order to get reliable estimates.

THE VARIANCE OF THE SEA PREDICTIONS

Assuming that the operational power inputs into different subsystems are given, the mean energy level of each subsystem can be estimated by applying the SEA power balance equation :

$$\{\overline{E}_{oper}\} = \frac{1}{\omega}[\overline{L}]^{-1}\{P_{oper}\} \qquad (24)$$

where $\{\overline{E}_{oper}\}$ represents the mean levels of the predicted response energies and $\{P_{op}\}$ is the known operational input power vector. As the loss factors are experimentally identified by inverting the normalized energy matrix as expressed by equation (6), it is more convenient to rewrite equation (24) in terms of normalized energies which were measured in the power injection process :

$$\{\overline{E}_{oper}\} = [\overline{E}^n]\{P_{oper}\} \quad or \quad \overline{E}_{i,oper} = \sum_{k=1}^{N} \overline{E}_k^n P_k \qquad (25)$$

Consequently, the variance of the predicted energy level of subsystem i is given by :

$$s_{E_{oper,i}}^2 = \sum_{k=1}^{N} s_{E_{ik}^n}^2 P_k^2 \qquad (26)$$

As equation (25) is a linear equation, the probability distribution of the predicted operational energy will be normal if the normalized response energies are normally distributed.

Equation (26) assumes that the input powers are exactly known and that the variance of them is zero. In most cases, the input powers cannot be directly measured and need to be indirectly determined from operational energy measurements [8]. Due to the variances of the normalized PIM-energies and the variances of the operational measured responses, there will be an uncertainty on the indirectly calculated input powers which could be included in the variance calculation of the SEA predictions. However, as the normalized energies appear both in the indirect formulation to estimate the input powers and in equation (25), this leads to complex interdependencies of the involved variables, resulting in less straightforward analytical expressions describing the variance of the predicted energies in terms of the variances of the normalized PIM-energies and the measured operational energies. Therefore, for practical reasons, equation (26) will be assumed to be adequate with respect to the variance calculation of the SEA-predictions.

THE INFLUENCE OF THE MASS/VOLUME

The derivations leading to the equations expressing the variances of the loss factors or the variances of the predicted operational energies in terms of the variances of the elements of the normalized PIM-energy matrix do not take into account the variances of the mass/volume multiplicators α_i. However, it is interesting to have a closer look at the role of the mass/volume in the experimental SEA calculations.

Substituting equation (11) into equations (1) and (3) gives the following :

$$P_{i \to j} = \omega\left(\eta_{ij}\alpha_i\langle R_i\rangle - \eta_{ji}\alpha_j\langle R_j\rangle\right) \qquad (27)$$

$$P_{dis,i} = \omega\eta_{ii}\alpha_i\langle R_i\rangle \qquad (28)$$

Based on these two equations and based on the definition of a new mass-or volume-scaled loss factor $\eta_{ij}' = \eta_{ij}\alpha_i$, the power balance equations can be rewritten with the space averaged square velocity/pressure as main variable instead of the subsystem energy level :

$$\begin{bmatrix} \sum_{i=1}^{N}\eta_{1i}' & -\eta_{21}' & \cdots & -\eta_{N1}' \\ -\eta_{12}' & \sum_{i=1}^{N}\eta_{2i}' & \cdots & -\eta_{N2}' \\ \cdot & \cdot & & \cdot \\ \cdot & \cdot & & \cdot \\ \cdot & \cdot & & \cdot \\ -\eta_{1N}' & -\eta_{2N}' & \cdots & \sum_{i=1}^{N}\eta_{Ni}' \end{bmatrix} \begin{bmatrix} \langle R_1\rangle \\ \langle R_2\rangle \\ \cdot \\ \cdot \\ \cdot \\ \langle R_N\rangle \end{bmatrix} = \begin{bmatrix} \frac{P_1}{\omega} \\ \frac{P_2}{\omega} \\ \cdot \\ \cdot \\ \cdot \\ \frac{P_N}{\omega} \end{bmatrix} \qquad (29)$$

From PIM, it can then be easily understood that the mass-or volume-scaled loss factor is only dependent on the normalized responses, i.e. :

$$\eta_{ij}' = fu(\langle R_{11}^n\rangle, ..., \langle R_{ij}^n\rangle, ..., \langle R_{NN}^n\rangle) \qquad (30)$$

By consequence, the true or absolute loss factor is only a function of the normalized responses and the inverse of the corresponding mass/volume multiplicator :

$$\eta_{ij} = \frac{1}{\alpha_i} fu(\langle R_{11}^n\rangle, ..., \langle R_{ij}^n\rangle, ..., \langle R_{NN}^n\rangle) \qquad (31)$$

So, the mass/volume multiplicator α_i can be interpreted in experimental SEA as a scaling effect for the loss factors η_{ij} (j=1..N).

With respect to the influence of the mass/volume on the SEA predictions, equation (29) clearly shows that the predicted space averaged velocity/pressure is independent of the mass/volume. This is also true for the prediction of the power flow between two subsystems and the dissipated energy, given by equations (27) and (28). It should be remarked that this conclusion is based on the assumption that the same mass/volume is used in the SEA predictions as in the PIM-tests. In case the equivalent mass/volume is used to calculate the subsystem energies and other response points are selected in the operational measurements than in the PIM-tests, the equivalent mass/volume used in the PIM-tests and in the operational energy measurements might not be the same anymore.

In conclusion, if space averaged velocity/pressure variables are preferred which are more 'engineer-friendly' than energy variables, no justification can be found for including the variance of the mass/volume in the experimental SEA predictions. The same is true for the loss factor calculation on condition at the experimenter is not interested in the absolute values of the loss factors. The latter is not true anymore when the main interest of the user lies in the correlation of the experimental SEA model with an analytical loss factor model.

TEST RESULTS AND DISCUSSIONS

EVALUATION OF THE ACCURACY OF THE ANALYTICAL VARIANCE EXPRESSIONS BASED ON MONTE-CARLO VARIABILITY ANALYSIS - In order to assess the accuracy of the analytical expressions which are approximate due to the first order Taylor expansion, the derived formulas and the statistical Monte-Carlo approach were applied to two coupled plates, having a thickness of 2mm and dimensions of about 500mm by 600mm. On both plates, 4 excitation points and 6 response points were selected for the power injection method. Two configurations were tested: undamped plates and fairly damped plates. The full matrix inversion (equation (6)) was used to estimate the loss factors. Figure 1 shows the relative standard deviation for the coupling loss factor (equation (18)) between 2 damped plates.

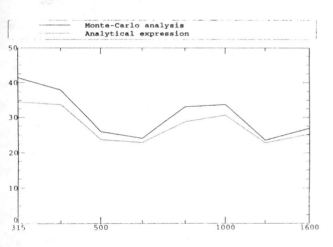

Figure 1: Relative standard deviation of the coupling loss factor between two damped plates. Comparison is made between the Monte-Carlo approach (equation (23)) and the analytical expression (equation (18)).

With respect to the Monte-Carlo calculation, 100.000 evaluations of perturbed 2x2 normalized energy matrices were performed. The analytical expression (equation (18)) seems to slightly underestimate the variance given by the Monte-Carlo approach (equation (23)), but from an engineering point of view, it can be clearly noticed that the analytical expressions yield a very useful result. The histogram of the internal loss factor of the plates have also been evaluated for the undamped case and the fairly damped case, illustrated in figure 2. Clearly, a significant difference between both internal loss factors is revealed. It can also be seen that the loss factor's distributions are not completely symmetric anymore. However, practically speaking, the assumption of a normal distribution with respect to the loss factor confidence calculation is adequate.

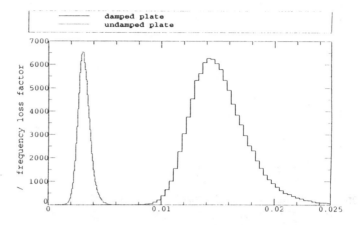

Figure 2: Histogram of an internal loss factor of 2 coupled plates. Damped and undamped case are compared for the one-third octave band 1000Hz.

Similar conclusions were drawn for a larger SEA model consisting of 20 subsystems. Figure 3 gives a typical shape of the histogram for a coupling loss factor. Figures 4 and 5 respectively depict the comparison of the mean and the relative standard deviation which are both analytically computed (equations (6) and (18)) and calculated according to the Monte-Carlo approach (equations (22) and (23)). A reasonable agreement is shown.

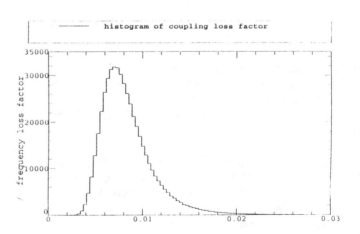

Figure 3 : Histogram of a coupling loss factor of a 20-subsystem model for the one-third octave band 1000Hz.

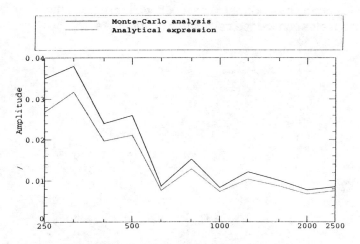

Figure 4 : Comparison between the Monte-Carlo approach (equation (22)) and the analytical expression (equation (6)) with respect to the mean of a coupling loss factor of a 20-subsystem model.

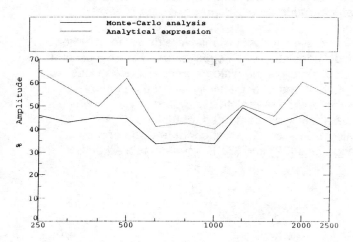

Figure 5 : Comparison between the Monte-Carlo approach (equation (23)) and the analytical expression (equation (18)) with respect to the relative standard deviation of a coupling loss factor of a 20-subsystem model.

THE VARIANCE OF THE NORMALIZED PIM-ENERGY OF THE BONNET AND THE INTERIOR CAVITY OF A MIDSIZE CAR - The relative standard deviations of the normalized PIM-energies have been calculated for the interior cavity and the bonnet of a midsize car. With respect to the bonnet, FRFs between acceleration and force were measured. The investigation of the measured data points out that the variance of the input powers injected at different excitations points is low, but the variance of the population of the velocity responses is quite high as shown in figure 6.

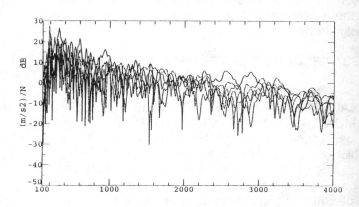

Figure 6 : Variation of FRFs between different response and excitation points of the bonnet.

Figure 7 shows the relative standard deviation of the normalized energy (equation (16)) for a selection of 3 excitation points and 7 responses, which is about 35%. This implies that the population of the normalized velocities has a relative standard deviation of about 160%.

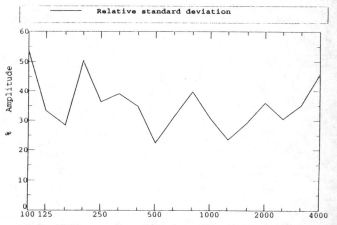

Figure 7 : Relative standard deviation of the normalized energy of the bonnet (equation (16)).

The scatter of the normalized responses is completely related to the choice and the physical characteristics of the subsystem. Increasing the number of excitation and response points will increase the precision of the spatially averaged value in the sense that e.g. the 95% confidence interval bracketing the (unknown) true average becomes smaller.

The same type of analysis is performed for the interior cavity. In this case, 8 microphone pressures were measured for a calibrated sound power injected at 3 excitation points. Figure 8 illustrates the relative standard deviation for the normalized PIM-energy of the interior cavity, which is about 25% on average.

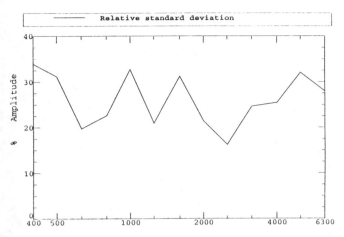

Figure 8 : Relative standard deviation of the normalized energy of the interior cavity (equation (16)).

APPLICATION OF THE STATISTICAL FORMULAS TO A RAILWAY CARRIAGE OF A HIGH-SPEED TRAIN - The experimental SEA model of a railway carriage of a high-speed train has been derived [9]. Justified by the repetitive nature of the structure, only two adjacent vertical sections of the railway carriage were studied, shown in figure 9.

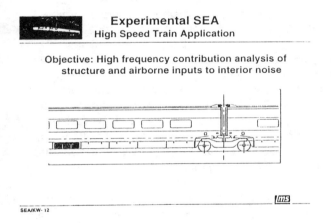

Figure 9 : Side view of the railway carriage of the high-speed train.

Taking two adjacent sections instead of one allowed to investigate the energy transfer in longitudinal direction. The two sections were divided in 18 structural components which were the roof, the roof edge, the upper part of the side wall, the lower part of the side wall, the wooden floor, the supporting steel floor, the supporting beam of the floor, the closing plate of the underfloor cavity, the exterior and interior windows and the inner mask around the windows. Including the passenger compartment and the underfloor cavity containing the auxiliary equipment as 2 acoustical subsystems yielded a SEA model consisting of 20 subsystems.

In the power injection process, each subsystem was excited in 3 different input locations while the responses at 5 to 8 locations chosen differently from the input locations to exclude nearby field effects were measured for all subsystems. The relative standard deviations of the normalized energies (equation (16)) are typically in the range of 25 to 35%. Figure 10 depicts the normalized energy in respectively the upper part and the lower part of the side wall due to excitation in the upper part of the side wall. The 80% confidence levels are computed on the basis of a normal or Gaussian distribution.

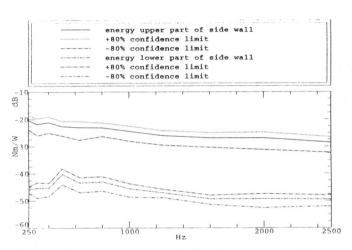

Figure 10 : Normalized energy and associated 80% confidence intervals (equation (16)) for the response in the upper part and lower part of the side wall due to excitation in the upper part of the side wall.

The internal and coupling loss factors were derived based on the approximate equation (7), assuming that the energy matrix normalized by unit input power is diagonal dominant. The obtained loss factors were validated by superposing and comparing the measured and the synthesized response energies, confirming that the approximate methods gave good results. The confidence levels of the loss factors were calculated according to equations (20) and (21). Figures 11 and 12 respectively show the internal loss factor for the interior cavity and the coupling loss factor between the upper part and lower part of the side wall and their corresponding 80% confidence intervals.

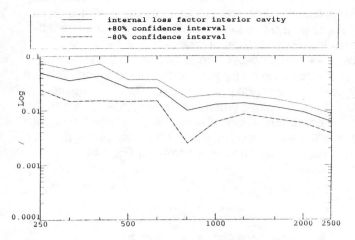

Figure 11 : Internal loss factor of the interior cavity (equation (7)) and associated 80% confidence interval (equation (21)).

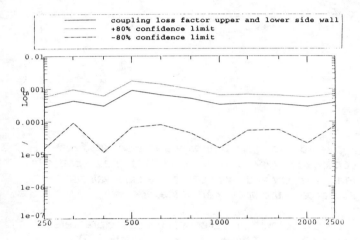

Figure 12 : Coupling loss factor between upper part of side wall and lower part of side wall (equation (7)) and the associated 80% confidence interval (equation (20)).

The derived loss factor model was then used to determine the way power flows through the structure for known power inputs. Figure 13 depicts the predicted energy level (equation (24)) and the associated 80% confidence levels (equation (26)) for the interior cavity when input power is applied at the supporting beams of the floor. Subsequently, two effective modifications in terms of adding damping treatment were determined from sensitivity studies. The internal loss factors of the supporting steel floor and the lower part of the side wall were modified. Figure 13 also shows that the predicted energy level for the modified structure is significantly lower than the original level.

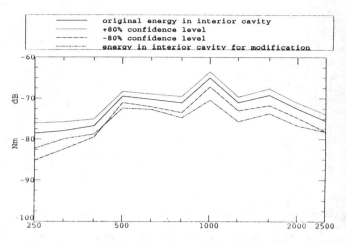

Figure 13 : Predicted energy level (equation (24)) and associated 80% confidence interval (equation (26)) for the interior cavity due to operational input in the beam. The predicted energy level for modification in two internal loss factors is superposed.

SUMMARY AND CONCLUSIONS

This paper has presented the statistical aspects concerning the Power Injection Method, a well-known technique to experimentally derive the loss factors. It has been shown that analytical expressions can be deduced describing the loss factor variance in terms of the variances of the normalized energies. These formulas are approximate as they are based on a first order Taylor expansion. However, a Monte-Carlo variability analysis points out that they are of practical usefulness. The analytical expressions allow a quick assessment of the confidence levels of the variances during the PIM-tests, aiding the user to decide whether the desired accuracy has been achieved and whether or not the PIM-measurements can be terminated.

The role of the subsystem mass/volume in experimental SEA has also been addressed. The study has shown that the experimentally derived loss factors are inversely proportional to the corresponding subsystem mass/volume. Furthermore, the predictive capabilities of the SEA model such as the prediction of the spatially averaged subsystem responses expressed in velocity or pressure units instead of energies and the estimation of power flows and dissipated energies are not effected by the values of the subsystem mass/volume. This implies that, if the experimenter is not interested in the absolute values of the loss factors, the determination of the subsystem mass/volume is not at all critical in the whole experimental SEA process. There is no need then to take into account the uncertainty of the mass/volume in the variance analysis.

The statistical theory has been applied to a railway carriage of a high-speed train, illustrating how the confidence levels of the loss factors and the SEA-predictions can be obtained and how the effectiveness of loss factor modifications can be evaluated in terms of confidence levels.

ACKNOWLEDGMENTS

The presented work has been carried out within the framework of the IWT-project no AUT/940090, supported by the Flemish institute for the promotion of scientific and technological research in Industry. The authors also wish to express their gratitude to F. Aliberti, a thesis student of the University of Naples who did a part of the analyses.

REFERENCES

[1] R.H. Lyon, "Statistical Energy Analysis of Dynamical Systems: Theory and Practice", MIT Press, Cambridge, 1975.

[2] D.A. Bies and S. Hamid, "In situ determination of loss and coupling loss factors by the power injection method, Journal of Sound and Vibration", 70(2), pp. 187-204, 1980.

[3] N. Lalor, "Practical considerations for the measurements of internal and coupling loss factors on complex structures", ISVR Technical Report No. 182, June 1990.

[4] N. Lalor, "The experimental determination of vibrational energy balance in complex structures", Proc. SIRA Conference on Stress and Vibration - Recent Developments in Industrial Measurements and Analysis, London, 1989.

[5] Stephen B. Vardeman, "Statistics for Engineering Problem Solving", IEEE Press - PWS Publishing Company, New York - Boston, 1994.

[6] K. Delanghe, "High Frequency Vibrations : Contributions to Experimental and Computational SEA Parameter Identification Techniques", Ph.D. dissertation, Department PMA, K.U.Leuven, 1996.

[7] L. Hermans, K. Wyckaert, "Experimental Statistical Energy Analysis : Internal and Coupling Loss Factor Matrix Validation", Proc. Inter-noise, Inter-Noise 96, Liverpool, July-August 1996.

[8] U. Fingberg, T. Bharj, B. Cimerman, "High-Frequency NVH Optimization Using Energy Techniques, ISMA21", Proc. of ISMA21, Leuven, 1996.

[9] K. De Meester, L. Hermans, K. Wyckaert, N. Cuny, "Experimental SEA on a Highspeed Train Carriage", Proc. of ISMA21, Leuven, 1996.

971970

Statistical Energy Analysis of Airborne and Structure-Borne Automobile Interior Noise

Alan V. Parrett and John K. Hicks
General Motors Corp.

Thomas E. Burton
Vibro-Acoustic Sciences, Inc.

Luc Hermans
LMS International

Copyright 1997 Society of Automotive Engineers, Inc.

ABSTRACT

This paper describes the application of Statistical Energy Analysis (SEA) and Experimental SEA (ESEA) to calculating the transmission of air-borne and structure-borne noise in a mid-sized sedan. SEA can be applied rapidly in the early stages of vehicle design where the degree of geometric detail is relatively low. It is well suited to the analysis of multiple paths of vibrational energy flow from multiple sources into the passenger compartment at mid to high frequencies. However, the application of SEA is made difficult by the geometry of the vehicle's subsystems and joints. Experience with current unibody vehicles leads to distinct modeling strategies for the various frequency ranges in which airborne or structure-borne noise predominates. The theory and application of ESEA to structure-borne noise is discussed. ESEA yields loss factors and input powers which are combined with an analytical SEA model to yield a single hybrid model. Results from model validation and correlation with measured data are presented.

INTRODUCTION

A brief overview of the theory is presented, and subsequently the paper is divided into 2 parts. Part 1 is focused on ESEA and its application to structure-borne noise prediction. In Part 2, the focus shifts to analytical SEA, and the synthesis of the two, including extension to higher frequencies incorporating airborne noise analysis. Following customary practice, this paper often refers to analytical SEA as simply "SEA".

OVERVIEW OF SEA - The basic equations in SEA describing the net energy flow in a given frequency band between 2 subsystems i and j, and the internal energy dissipation are [1]

$$\left\langle \overline{P}_{i \to j} \right\rangle = \omega \left(\eta_{ij} \left\langle \overline{E}_i \right\rangle - \eta_{ji} \left\langle \overline{E}_j \right\rangle \right) \qquad (1)$$

with

$$\eta_{ij} n_i = \eta_{ji} n_j \qquad (2)$$

and

$$\left\langle \overline{P}_{ii} \right\rangle = \omega \eta_{ii} \left\langle \overline{E}_i \right\rangle \qquad (3)$$

where ω is the center band frequency, n_i the modal density of the i-th subsystem, η_{ij} represents the coupling loss factor between subsystems i and j, η_{ii} represents the internal loss factor of subsystem i, $< \overline{E_i}>$ the total reverberant energy in subsystem i, $< \overline{P}_{i \to j}>$ the net power flow from subsystem i to subsystem j, and $< \overline{P}_{ii}>$ the power dissipated in subsystem i. The brackets <> and the bar ‾ denote space and time averaging respectively. Equations (1)-(3) are combined to represent the internal power flow in a structure consisting of N interconnected subsystems. With M load cases of externally applied powers $< \overline{P}_{ik}>$ (the injected power in the ith subsystem under load case k) the power balance equations yield in matrix form

$$\omega LE = \Pi \qquad (4)$$

where

$$L = \begin{bmatrix} \sum_{i=1}^{N}\eta_{1i} & -\eta_{21} & \cdots & -\eta_{N1} \\ -\eta_{12} & \sum_{i=1}^{N}\eta_{2i} & \cdots & -\eta_{N2} \\ \vdots & \vdots & \ddots & \vdots \\ -\eta_{1N} & -\eta_{2N} & \cdots & \sum_{i=1}^{N}\eta_{Ni} \end{bmatrix}$$

$$E = \begin{bmatrix} E_{11} & E_{12} & \cdots & E_{1M} \\ E_{21} & E_{22} & \cdots & E_{2M} \\ \vdots & \vdots & \ddots & \vdots \\ E_{N1} & E_{N2} & \cdots & E_{NM} \end{bmatrix}$$

and

$$\Pi = \begin{bmatrix} P_{11} & P_{12} & \cdots & P_{1M} \\ P_{21} & P_{22} & \cdots & P_{2M} \\ \vdots & \vdots & \ddots & \vdots \\ P_{N1} & P_{N2} & \cdots & P_{NM} \end{bmatrix}$$

In this case, E_{ij} is the energy in subsystem i when power is applied in subsystem j.

SEA COMPLEMENTARY TO ESEA. In SEA, the matrix of loss factors L is estimated analytically. Then, given a matrix (vector for an individual load case) of input powers Π, the energies of resonant vibration of the subsystems are obtained from

$$E = \frac{1}{\omega}L^{-1}\Pi$$

The matrix L is square and positive definite, so its inverse always exists.

In contrast, the application of ESEA starts with the matrix of input powers Π and measured energies E. Since power is injected separately into each sub-system, $M=N$, and L is calculated using

$$L = \frac{1}{\omega}\Pi E^{-1}$$

Once the loss factor matrix has been obtained, perturbations to the matrix representing structural changes can be evaluated and responses predicted. Another difference between ESEA and SEA is that the reciprocity equation $\eta_{ij}n_i = \eta_{ji}n_j$ describing the relationship between coupling loss factors and modal densities is not explicitly used to experimentally derive the internal and coupling loss factor parameters. As a consequence, the knowledge of the modal densities, usually difficult to accurately derive from tests, is not required for ESEA.

ESEA promises a wider range of application than SEA, including difficult systems where analytical estimates of loss factors are not feasible. On the other hand, ESEA by itself has limitations in predictive capabilities, especially in terms of new designs. Therefore, both ESEA and SEA are necessary to provide the best possible diagnostic and predictive capability for vehicle modeling.

SUBSYSTEMS - The most critical step in the application of SEA and ESEA to a complex structure is the division into subsystems. A subsystem is defined as an element of an SEA model corresponding to a substantial energy storage location. This implies that the choice of subsystems should be such that the response in each frequency band of interest is dominated by resonant modes, hence there is a requirement of high modal density and modal overlap. The modes contained in a subsystem should be similar in energetic terms and have more or less the same order of damping. The subsystems should also be weakly coupled. In addition to these theoretical criteria, there are several practical criteria which are important in vehicle modeling:

- Only subsystems which play a significant role in the energy balance should be taken into account.

- Each subsystem should be a homogeneous structure.

- In a subsystem, there is a relationship between energy and a useful measure of vibration such as RMS velocity.

- Subsystems are small enough to resolve spatial variation of vibration to the desired scale.

SEA APPLIED TO AUTOMOBILES -The application of SEA to automobile noise analysis is made difficult by the vehicle subsystem geometry and joint complexity. Application of the criteria for partitioning into subsystems usually presents conflicts in terms of size, uniformity, and coupling. The concepts of global modes and equivalent mass, discussed in later sections, attempt to address these practical difficulties. ESEA can provide measurements of equivalent mass and of global coupling loss factors.

Good estimates of damping loss factors are crucial to SEA, yet these are notoriously difficult to estimate by analysis. Perhaps the most valuable contribution to SEA from ESEA is the experimental determination of in situ damping loss factors.

1. ESEA

MEASUREMENT THEORY. The Power Injection Method (PIM) [2] is used to experimentally derive the internal and coupling loss factor model without the need to disassemble the built-up structure into components. It is based on exciting each subsystem in turn and measuring the responses on all subsystems. For each subsystem, a number of excitation and response locations are chosen. The number of points should be high enough in order to get a statistical average which is representative for the subsystem energy, and will depend on the size and complexity of the subsystem.

The PIM implementation discussed here is based on FRF measurements instead of absolute power and velocity measurements. The velocities derived from the measured FRFs between the responses and the input force do not correspond to the actual energy but are normalized by the autopower of the force spectrum. The real part of the driving point mobility measurements gives the injected normalized input power. For acoustical excitation, FRFs can also be measured, using the volume velocity or the drive voltage to the loudspeaker as the input signal. The pressures derived from the FRFs and the radiated sound power of the loudspeaker are then normalized by the autopower of the input signal. The calibration of the radiated sound power can be carried out in advance in a semi-anechoic room, assuming that the sound power is not affected when placing the loudspeaker at different locations inside the cavity.

Calculation of the SEA parameters - The narrow-band data are converted to the frequency bands of interest, typically one-third octave bands, as appropriate. The response data are normalized to unit input power and the subsequently averaged over each subsystem, yielding the averaged mean velocity or pressure squared normalized to unit input power in subsystem i due to excitation in subsystem j, designated respectively by $\left\langle \dfrac{\dot{X}_{ij}^2}{P_{jj}} \right\rangle$ or $\left\langle \dfrac{p_{ij}^2}{P_{jj}} \right\rangle$.

The normalized subsystem energy $E_{ij}^{\,n}$ of response subsystem i due to excitation in subsystem j can then be derived from the following equations :

$$\text{structural subsystem}: \left\langle \overline{E_{ij}^n} \right\rangle = M_i \left\langle \frac{\dot{X}_{ij}^2}{P_{jj}} \right\rangle \tag{5}$$

$$\text{acoustical subsystem}: \left\langle \overline{E_{ij}^n} \right\rangle = V_i \left\langle \frac{p_{ij}^2}{P_{jj}} \right\rangle \frac{1}{\rho c^2} \tag{6}$$

In these expressions, M_i and V_i represent the total mass and volume of the i-th subsystem. These equations are approximate, based on discrete velocity or pressure measurements which are space averaged. In addition, the total subsystem mass can be used in the calculation of total energy only for uniform structures of constant thickness or cross-section. In order to overcome the discrepancy between the energy expressed by equation (3) and the 'true' energy of the subsystem, the usage of the equivalent mass/volume [3,4] which can be derived from decay rate measurements:

$$M_{eq,i} = \frac{1}{\beta_i \left\langle \dfrac{\dot{X}_{ii}^2}{P_{ii}} \right\rangle} \text{ and } V_{eq,i} = \frac{\rho c^2}{\beta_i \left\langle \dfrac{p_{ii}^2}{P_{ii}} \right\rangle} \tag{7,8}$$

where β_i represents the decay rate of subsystem i, expressed in neper/sec. No additional decay rate measurements are required since the impulse response functions calculated by performing the inverse FFT of the measured FRFs can be used to derive the decay rate for each subsystem in each analysis band of interest. The influence of coupled subsystems can cause multiple decay curves in the impulse response hence care is required to ensure that only the initial decay rate is estimated. It is of interest to note is that the mass M_i or volume V_i can be interpreted as a scaling effect for the loss factors η_{ij}.

Once all normalized energies are calculated, the total loss factor matrix $[L]$ is derived by inverting the normalized energy matrix $[E^n]$

$$[\mathrm{L}] = \frac{1}{\omega} \left[E^n \right]^{-1} \tag{9}$$

This energy matrix inversion method is sensitive to inaccuracies in the quantification of the injected input powers and subsystem energies and might yield negative loss factors, which are physically unacceptable. In case of weakly coupled subsystems, approximate methods have been suggested [5] to guarantee positive values:

$$\eta_{ii} \cong \frac{1}{\omega} \frac{1}{\left\langle E_{ii}^n \right\rangle} \text{ and } \eta_{ij} \cong \frac{1}{\omega} \left(\frac{\left\langle E_{ji}^n \right\rangle}{\left\langle E_{ii}^n \right\rangle} \right) \left(\frac{1}{\left\langle E_{jj}^n \right\rangle} \right) \tag{10}$$

An important aspect of equation (10) is that it enables the coupling loss factors between two coupled subsystems to be calculated from measurements made only on those two subsystems, regardless of other connections.

The Loss factor matrix validation - Several validation tools are available to assess the quality of the loss factor model [6]. The accuracy of the approximate loss factor calculation described by equation (10) can be evaluated by synthesizing the normalized energy matrix from the calculated loss factor matrix and evaluating the difference between the synthesized and PIM measured normalized energies. The synthesized energy matrix is given by

$$[E_{synth}^n] = \frac{1}{\omega}[L]^{-1} \qquad (11)$$

The estimation of the operational input powers - Direct measurement of the operational input powers is often very difficult in practice: An indirect technique can be applied, allowing the computation of operational input powers from the loss factor matrix and the measured operational response energies [7]. Power is assumed to be input at a number of the subsystems. Operational measurements (from which energies will be derived) are taken on at least as many subsystems (preferably at the PIM locations). The total loss factor matrix equation can then be sub-structured to yield the powers P necessary to produce the responses E.

Given E_{meas_sel} the measured energies of x subsystems and P_{oper} the x unknown input powers.

$$\begin{bmatrix} [P_{oper}] \\ [0] \end{bmatrix} = \omega \begin{bmatrix} [L_{11}] [L_{12}] \\ [L_{21}] [L_{22}] \end{bmatrix} \begin{bmatrix} [E_{meas\_sel}] \\ [E_{predict}] \end{bmatrix} \qquad (12)$$

Solving the second matrix row equation and substituting the found solution for $E_{predict}$ into the first matrix row equation gives the indirectly estimated input powers :

$$[P_{oper}] = \omega \begin{bmatrix} [L_{11}] - [L_{12}][L_{22}]^{-1}[L_{21}] \end{bmatrix} [E_{meas\_sel}] \qquad (13)$$

An approximate method can be derived to compute the operational input powers based on the assumption that most of the power entering a subsystem is dissipated in that subsystem :

$$P_{oper,i} \cong \omega \, \eta_{ii} \, E_{meas,i} \qquad (14)$$

This method requires that the operational energies are measured in the subsystems where the operating power loads are acting and it always gives positive values.

EXPERIMENTAL PROCEDURE

Operational Measurements. The first stage of measurements dealt with operational measurements on the full vehicle. Data were taken both on road (coarse surface at 35 mph to accentuate tire-pavement noise

and minimize wind/powertrain noise), and on a dynamometer to accentuate powertrain noise. Sound pressure levels and accelerations were measured as appropriate on each sub-system. Both narrow band (0-2000 Hz) and 1/3 octave band (200-10 000 Hz) data were acquired.

Power Calculation (Non ESEA). An alternative to the ESEA procedure for determining input powers is to use an impedance approach. Measured operational velocities (v) are combined with impedances (Z) derived from drive point mobilities to yield input power using

$$P_{in} = \frac{1}{2} \mathrm{Re}(Z).\,v^2$$

(15)

Power Injection Method. After operational measurements were completed, the power injection method was applied. The frequency range used for the ESEA procedure was 200-2000 Hz. The structural loss factors were of primary interest and road/powertrain structure-borne noise dominates in the lower part of this range. Limiting the upper frequency meant that impact measurements could be used to a large extent since an impact hammer can supply enough energy to induce reverberant fields in adjacent subsystems at mid frequencies. To obtain meaningful structural FRF's at higher frequencies on a fully assembled structure (with trim in place) is difficult, even when using shakers.

Typically, each subsystem was excited at about 3 to 5 excitation points and about 5 to 10 response points are measured on the excited and adjacent subsystems. Response and excitation locations were different in order to exclude near field effects and ensure correct estimates of subsystem reverberant energy. A speaker with calibrated sound power/voltage was used for acoustical inputs.

All data were acquired narrow-band for later integration to 1/3 octave band in the analysis software. The software used here was the LMS ESEA module. The use of transfer and drive point accelerances on the structure meant that simultaneous power input and response measurements were unnecessary, thus the number of analysis channels could be reduced. There is a trade-off between the number of accelerometers and the time (and wiring complexity) to move and re-mount them. Measurement strategies are discussed extensively in [3].

The exploitation of symmetry significantly reduced the number of FRF's, nevertheless many thousands of frequency response functions resulted. This was not without cost however, since a half model does not yield correct loss factors for systems spanning he vehicle transversely, and data manipulation to reconstruct a full model was not trivial.

Loss factors could be exported from ESEA and imported into the analytical SEA code either singly or as a whole model. Careful naming of the sub-systems saves much time and effort in the comparison with analytical SEA results. The measurement procedure spanned several weeks, and was carried out on a fully trimmed vehicle.

Airborne Noise Measurements. To complement the structure-borne noise model, airborne noise measurements were made over a frequency range of 100-10 000 Hz in 1/3 octave bands under the operating conditions described above. Sound pressure levels were averaged over major panels and cavities to provide input powers, and responses were measured in the passenger compartment. Additional separate measurements for reverberation time, sound transmission loss, and results correlation were also made.

SEA

SEA MODEL - The initial SEA model of the vehicle was made using AutoSEA software. A coarse finite element body model was imported to establish geometry and section properties. Power was input to subsystems at suspension and engine mount attachments (structure-borne) and exterior panels (airborne). The model contains approximately 70 subsystems. A combination of predicted [10] and measured trim properties for the interior were used. At higher frequencies, the sound transmission loss of certain subsystem is characterized experimentally (for example the doors). This was done to save modeling effort in this project.

HYBRID ACOUSTIC MODELING - The ESEA model is validated and test-based damping loss factors are used to replace those in the SEA model. Next, predictions yielded by the analytical model for each of the experimental load cases are compared with test results. Both overall response levels and energy transfer paths are compared. Where there is significant disagreement, consideration is given on a subsystem by subsystem basis whether to substitute the equivalent mass from ESEA for the analytic mass, and whether to substitute (or better approximate) coupling loss factors from ESEA for the analytical estimates. Global coupling loss factors from ESEA are also considered. The concepts of equivalent mass and of global modes are discussed below. After correlating based on structure-borne inputs, the last step is to combine the airborne noise modeling data into the model for final overall correlation.

The result is a hybrid of the analysis and test correlated by essentially the same process described in [11]. The analytical portions relate masses and coupling loss factors to design data (materials and shapes), and can therefore be generalized to new models. The test-based portions are limited in their predictive capabilities. The goal of this exercise is therefore to produce a hybrid model which can reproduce test data to the desired accuracy and be useful in predicting new design configurations.

SUBSYSTEMS AND WAVEFIELDS – Analytically, the resonant vibration of a substructure consists of several types of wavefields (e.g. flexural, torsional, longitudinal waves). In general, each field has a different relationship between the energy and velocity of vibration, and different coupling loss factors. In SEA, each wavefield constitutes a subsystem.

Experimentally, the particular wave fields in a subsystem may be difficult to quantify individually because of transducer limitations. The physical measurement technique may combine the effects of these fields. This limitation tends to lead to different experimental predictions of L (damping loss factors and coupling loss factors) for different load cases, because each load case can generate wave fields in different proportions within a single substructure.

The errors induced by this limitation have been lessened in the current study by interpreting the vibration of a substructure in terms of only the more important of its wavefields, i.e. the one that generates more velocity (hence energy) of vibration, which in turn is usually one that includes a higher number of resonant modes of vibration. In this study, the flexural field in assumed to be the most important for plates and shells. For beams, the flexural field around the weaker of the two principal axes is taken to be most important. Accelerometers supplying vibration data for ESEA are positioned on a structure to favor these fields. In practical terms, certain wave fields may be neglected with little resultant error if the relevant subsystem has a minor effect on the overall power flow of interest.

EQUIVALENT MASS - The earlier discussion on equivalent mass (equation (5)) related to aspects of its measurement. It is typical of strongly coupled subsystems, especially at lower frequencies, to exhibit an equivalent mass that differs significantly from the true mass. The physical interpretation of the difference between the two is that the measured velocities are not a good measure of the RMS resonant velocity of the subsystem. This discrepancy can be due to measurement location (minimized as described in the PIM discussion in the experimental procedure section), and global-mode bias. Some of the measured velocity can be contributed by global modes—modes of vibration extending beyond the subsystem. The equivalent mass tends to be larger than the true mass. The test-based measure of equivalent mass compensates partially for this bias, but the presence of global modes also affects coupling loss factors, as explained below.

GLOBAL MODES - Analytical estimates of coupling loss factors are dependent upon an assumption that subsystems can vibrate somewhat independently from each other, to the extent that the phases of vibration are uncorrelated between subsystems. This assumption fails in automobiles under a combination of two circumstances. First, a subsystem is so stiff that the wavelength of vibration approaches or even exceeds the size of the subsystem. Second, the joint between two subsystems is so stiff that the two subsystems tend to vibrate as a unit. These two circumstances combine to yield patterns of vibration that, when forced into an SEA model, appear to tunnel energy between subsystems not directly connected.

Conventional analytical SEA provided no basis on which to estimate coupling loss factors between a pair of subsystems with no direct physical connection [1]. Therefore, ESEA is relied upon to estimate global coupling loss factors.

RESULTS AND DISCUSSION

OVERALL MODEL RESULTS. Figures 1 and 2 show interior SPL comparisons between measured data and the hybrid model. Overall, agreement between the model and measurements is good. The powertrain noise tests on dynamometer shown in Figure 1 were performed with the vehicle sealed (HVAC ducts sealed off, door cuts taped and sealed with mastic). The SPL maximum/minimum range (per 1/3 octave band) for 6 microphone locations is also shown. As expected the variance in SPL is greater at lower frequencies. Figure 2 shows the results for coarse road noise. Testing on road was performed without any extra sealing. Comparison of the overall shapes of the SPL's at high frequencies between the road and dynamometer show leakage. The third curve in Figure 2 has leaks simulated the model to account for the effect. Figure 3 shows the breakdown of the overall predicted level to structure-borne and airborne components. As expected, the structure-borne noise is dominant at the low frequencies. Investigations with the powertrain noise model show that non resonant airborne noise transmission dominates at high frequencies.

ESEA VALIDATION Loss factors discussed in this analysis were calculated using the approximate method described in equation (10). The normalized response energies were used as the primary validation for the ESEA results. These were available directly from the processing software since they are generated in the loss factor calculation procedure. Figures 4-6 show the comparison of synthesized and measured energies across vehicle subsystems. The results represent a transfer function from unit power injected in one subsystem to response energy in another subsystem.

The same approach was used for SEA model validation. In general, the ESEA model results agree well for panel cavity and panel-panel measurements, in which modal densities are relatively high in both subsystems. Figure shows the transfer from the roof to interior, Figure 5 shows the transfer from the rear floor to the front floor. Errors at high frequencies could be due to high damping and difficulty inducing reverberant fields in both panels however they will not affect the lower frequency structure-borne noise predictions. Figure 6 shows the transfer from the rear strut tower to the package shelf. For most coupled subsystems, errors of the order of 2 dB were considered acceptable. In cases where agreement was poor, the raw normalized energies could be used for SEA validation. Areas of the vehicle which were difficult to model analytically presented challenge for ESEA.

The energy transfer across remote (not directly connected) subsystems was also measured. Figure 7 shows the transfer from the rear rail to the rear glass. The ESEA prediction for these uncoupled systems is in error by up to 12 dB at low frequencies, whereas when direct or global connection is included in the ESEA model, the discrepancy decreases to within 3 dB. Similar effects at lower frequencies were found in transfers between the front to rear of the vehicle. The discrepancy observed in ESEA without global coupling can be attributed to global modes or tunneling, non-measured wave propagation, and growth in the cascading of adjacent subsystem loss factor errors.

As noted in [13], it was found that grouping strongly coupled subsystems into larger subsystems improved ESEA loss factor predictions e.g. the roof rail and roof panel. This imposes limitations on a hybrid model in terms of predictive capability and model detail.

POWER INPUT CALCULATIONS. Input power was calculated based on equations (14) - the ESEA approach, and equation (15), the impedance method. Results are shown for the Front Strut in Figure 8. Agreement in this case was good at low frequencies but at higher frequencies, there are discrepancies. To resolve discrepancies, the ESEA model responses are compared with measurements. The use of velocities to calculate energies hence input powers (ESEA approach) means that there is inherent sensitivity to equivalent mass estimates, so care must be taken to obtain correct equivalent masses.

AIRBORNE NOISE MODELING. Airborne noise sources in the SEA model are characterized by external panel Sound Pressure Levels. Figure 9 shows the levels measured on the exterior of the front of dash both for powertrain noise and road noise. The road noise plot agrees with [12].

Panel trim sound transmission loss can be calculated in the SEA model. Figure 10 shows the predicted (ideal trim model) vs measured transmission loss for the front floor treatment. Experience with STL testing has validated the trim model previously, and [10] discusses model STL predictions at length. The in-car performance of the trim is influenced by incomplete coverage in the damping layer and decoupler, leaks, holes, non-uniformities etc. As shown by the third line in Figure 10, the overall in-car transmission loss is approximated better by mass law rather than a double wall construction.

CONCLUSION

This paper has demonstrated the combination of analytical SEA with ESEA to form a hybrid model for vehicle interior noise analysis in the frequency range of approximately 200-10 000 Hz.

With prior knowledge of typical vehicle structure-borne noise problems, the analysis was performed by separating the airborne and structure-borne regimes to facilitate experimental measurements. Separate analytical SEA and ESEA models were made, with equivalent subsystems. The power injection method was used to generate the ESEA model. Damping loss factors and input powers generated were incorporated in the SEA model and airborne noise subsystems added. In the course of constructing a hybrid model, the ESEA coupling and global loss factors provided direction to analytical SEA for modeling complex areas of the structure. Overall results agree well with measurements.

Structure-borne road and powertrain noise is dominant in the low frequencies (in general up to approximately 300-500 Hz) and airborne noise is dominant at higher frequencies. For many vehicle subsystems, airborne noise transmission is controlled by non-resonant response such as mass law and leakage effects.

The overall goal of this work was to develop a hybrid model which can reproduce test data and be useful in predicting new design configurations. At the present time, further work is ongoing to validate predictive capabilities. The aim of future work is to reduce the degree of hybridization to enhance these predictive capabilities.

REFERENCES

[1] R.H. Lyon and Richard G. DeJong, **"Theory and Application of Statistical Energy Analysis"**, Butterworth-Heinemann, 1995.
[2] D.A. Bies and S. Hamid, **"In situ determination of loss and coupling loss factors by the power injection method,** Journal of Sound and Vibration", 70(2), pp. 187-204, 1980.
[3] K. Delanghe, **"High Frequency Vibrations : Contributions to Experimental and Computational SEA Parameter Identification Techniques"**, Ph.D. dissertation, Department PMA, K.U.Leuven, 1996.
[4] N. Lalor, **"The experimental determination of vibrational energy balance in complex structures"**, Proc. SIRA Conference on Stress and Vibration - Recent Developments in Industrial Measurements and Analysis, London, 1989.
[5] N. Lalor, **"Practical considerations for the measurements of internal and coupling loss factors on complex structures"**, ISVR Technical Report No. 182, June 1990.
[6] L. Hermans, K. Wyckaert, **"Experimental Statistical Energy Analysis : Internal and Coupling Loss Factor Matrix Validation"**, Proc. Inter-noise Inter-Noise 96, Liverpool, July-August 1996.
[7] U. Fingberg, T. Bharj, B. Cimerman, **"High-Frequency NVH Optimization Using Energy Techniques, ISMA21"**, Proc. of ISMA21, Leuven, 1996.
[8] D. De Vis, W. Hendricx, P.J.G. van der Linden, **"Development and Integration of an Advanced Unified Approach to Structure Borne Noise Analysis"**, Second International Conference on Vehicle Comfort, ATA, 1992.
[9] N. Lalor and G Stimpson, **"FEM+SEA+Optimisation = Low noise"**, Proc. ATA Conference on Vehicle Comfort, Bologna, 1992.
[10] B. Cimmerman, P Bremner, Y.Qian, J Van Buskirk,**"Incorporating Layered Acoustic Trim Materials in Body Structural-Acoustic Models"** ,SAE
[11] H. Chen, M. O'Keefe, P. Bremner **"A Comparison of Test-Based and Analytic SEA Models for Vibro-Acoustics of a Light Truck"**, SAE 951329
[12] B. Dong, M. Green, M. Voutyras, P Bremner, P. Kasper,**"Road Noise Modelling Using Statistical Energy Analysis Method"** ,SAE 951327
[13] T. Bharj, B. Cimmerman, **"Application of Statistical Energy Analysis to a Passenger Vehicle: Combining Analytical and Test-based Prediction in a Hybrid Model"**, Proc Internoise 96, 1303-1306

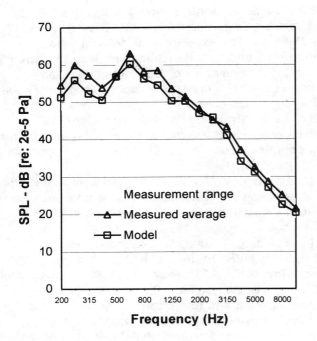

Figure 1. Powertrain Noise: Predicted vs measured interior SPL (front).

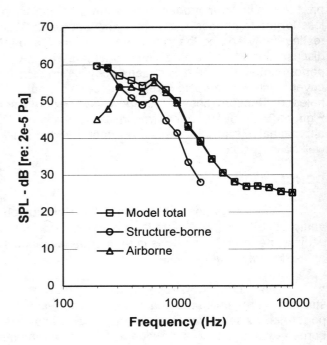

Figure 3. Coarse Road Noise: Predicted interior SPL. Total vs structure-borne and airborne noise components.

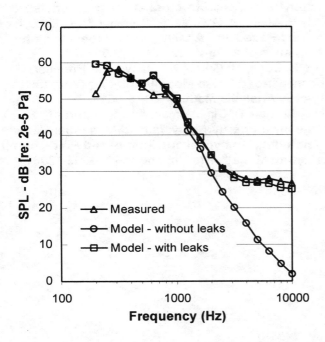

Figure 2. Coarse Road Noise: Predicted vs measured interior SPL including modeled leak effects.

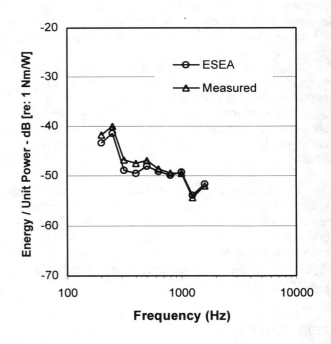

Figure 4. Normalized Response Energy - roof to interior subsystems: ESEA prediction vs measured.

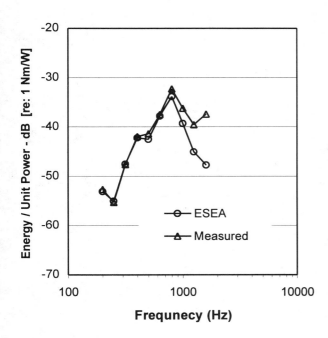

Figure 5. Normalized Response Energy - rear floor to front floor subsystems: ESEA prediction vs measured.

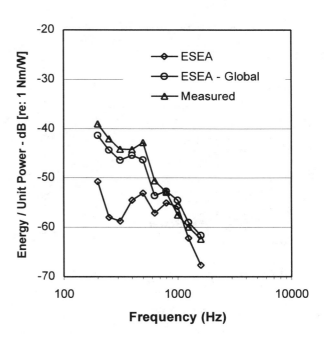

Figure 7. Normalized Response Energy - rear rail to rear glass subsystems: ESEA prediction (with and without global coupling included) vs measured.

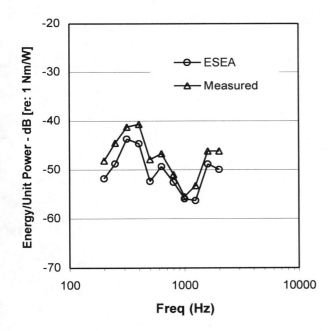

Figure 6. Normalized Response Energy - rear strut tower to package shelf subsystems: ESEA prediction vs measured.

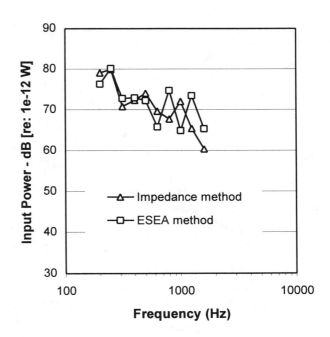

Figure 8. Structure-borne Input Power on Front Strut Tower - Powertrain Noise.

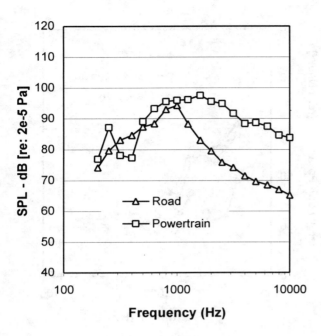

Figure 9. Airborne Exterior Front of Dash SPL - Road and Powertrain Noise

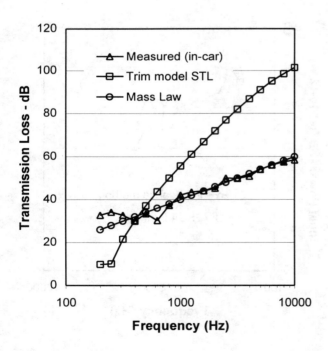

Figure 10. Floor Panel Sound Transmission Loss: Predicted vs Measured.

971972

Statistical Energy Analysis for Road Noise Simulation

Mark J. Moeller and Jian Pan
Ford Motor Co.

Copyright 1997 Society of Automotive Engineers, Inc.

ABSTRACT

Statistical Energy Analysis (SEA) is being actively pursued in the automotive industry as a tool for vehicle high frequency noise and vibration analysis. A D-class passenger car SEA model has been developed for this purpose. This paper describes the development of load cases for the SEA model to simulate road noise on rumble road.

Chassis roll test with rough shells was performed to simulate rumble road noise. Sound radiation from tire patch and vibration transmission through spindles were measured to construct the SEA load cases. Correlation between SEA model predictions and measured data was examined. Test and SEA result comparisons have shown that simulation of airborne road noise requires only a trimmed body SEA model, while simulation of structure-borne road noise may require SEA modeling of chassis components.

INTRODUCTION

As quality and fuel economy improve, operator and occupant comfort and ergonomics are important factors in public buying decisions for passenger vehicles today. Customer expectation for quiet vehicles has increased and vehicles have gotten quieter in response. Noise plays a role in the perceived quality of the vehicle. As a result, the sound quality of the vehicle is as important as the overall level. These technically challenging requirements come at the same time as reduced cycle times are being mandated throughout the automotive industry. Meeting these challenges requires developing new tools for automobile acoustic design.

One of these new tools is Statistical Energy Analysis (SEA). SEA is an analytical technique that is useful for estimating the response of complex structures at high frequency. SEA theory is described in Lyon and DeJong [1].

DeJong [2] applied SEA to cars and considered road, wind and powertrain noise. Interest in building SEA models of vehicles has continued to grow. Recently, effort to build test based SEA models of cars has developed, for example, Shaw [3], Lalor [4] and Bharj [5]. Much of the initial application of SEA to light vehicle modeling has focused on transfer function characterization Steele [6], Chen [7].

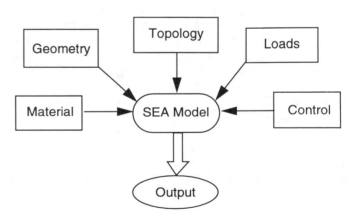

Figure 1. SEA Model Organization

SEA models contain the same general information that any analytical model contains. Figure 1 is an illustration of SEA model information. The basic model is made up of five different sets relating to material, geometry, topology, load and analysis control. The material set specifies what the physical system is made from including information on density, wave speeds/Young's modulus, poisson's ratio, and damping. In SEA, only the local geometry is specified to determine subsystem specifications. The topology contains information on how the subsystems are interconnected. The control information selects the solution sequence, frequency range, output request and other solution parameter information. The material, geometry and topology files are sufficient to describe the transfer functions of the model. Having reliable transfer functions is necessary but not sufficient for simulation of road noise events. The load case information is critical to proper simulation.

Previous investigators have reported road noise simulations. The load cases used have been predominately gathered by making measurements on an operating vehicle and extrapolating to the vehicle model. DeJong [2] had a comparatively coarse SEA model of 36 subsystems imposing wheel well sound pressure levels and spindle vibration as the road noise source. The presence of contamination in the source level measurements was believed to be the main reason for the discrepancy between predicted and measured levels. Dong et. al [8]. had a more detailed model with sound pressure in the wheel wells, shocktower vibration levels, and frame power levels as excitation in the model. Measurements were taken in a two-wheel chassis roll test cell. The Dong model did not explicitly represent the sound package. Enforced vibration and wheel well sound pressures have been used by Bharj [5] and Lalor [4]. Both road tests and chassis roll tests have been used to develop road noise load cases.

The current investigation uses a trimmed body vehicle SEA model and modeling technology described in the literature. The trimmed body model contains about 450 subsystems connected at 1500 point, line and area connections. It consists of acoustic subsystems outside and inside the vehicle, body structural subsystems and sound package subsystems. The baselining of the model is described in Moeller et al [9]. Here the correlation of transfer functions using shaker and speaker is examined. The model was shown to correctly predict transfer functions. The correlation of the model design sensitivities to hardware designed experiments is described in Pan et al [10]. The application of quality technology to SEA models and model development is described in Thomas et al [11] using this model as a case study.

In section 1 of this paper, the correlation of road noise with chassis roll noise is examined. Experiments were conducted in the all-wheel-drive (AWD) chassis roll facility at Ford's Advanced Engineering Center. This facility has a casting of a coarse aggregate road surface maintained at Ford's Dearborn Proving Ground. The assumption of independence of the tire loads is examined in section 2. Section 3 examines airborne noise transmission into a vehicle. Tests were performed in the AWD chassis roll facility. Microphones placed around the tires are used to create the airborne load case for the SEA model. The correlation with exterior sound, interior sound and shell vibration is examined. Structure-borne road noise simulation is discussed in section 4.

ROAD NOISE AND CHASSIS ROLL DATA CORRELATION

Road noise measurements are frequently conducted in a chassis dynamometer laboratory to reduce data contamination from other noise sources and for ease of instrumentation. In an effort to examine the correlation of road noise with chassis roll data, experiments were performed in the AWD chassis dynamometer facility with rough shell as well as on rumble road at Ford's Dearborn Proving Ground. A total of 43 vibration measurements and 14 sound pressure measurements were taken at common locations on a passenger car. Experimental results obtained from chassis roll tests generally exhibits good agreement with those obtained from road tests. Figure 2 shows very good correlation between the road test and the chassis roll test for left front vertical spindle vibration. The left front wheel well microphone has good correlation up to about 3 kHz above which the wheel well sound pressure is consistent with that at the engine compartment for road test (Figure 3). Thus, engine noise is predominant at the left front wheel well above 3 kHz for road test. The driver's head sound pressure has good correlation up to 2 KHz, wind noise starts to affect the driver's head sound above 2 kHz (Figure 4). The good agreement between chassis roll data and road data justifies the use of rough surface chassis roll for studying road noise on rumble road. Road noise experiments for source characterization and SEA modeling described in the next sections were conducted on the AWD chassis roll with rough surface.

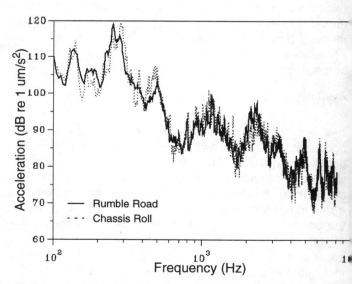

Figure 2. Left Front Spindle Vertical Vibration at 50 mph

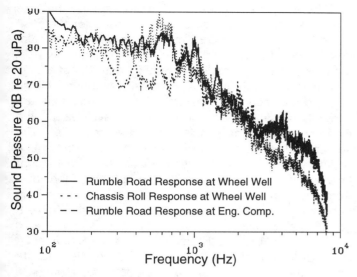

Figure 3. Left Front Wheel Well Pressure at 50 mph

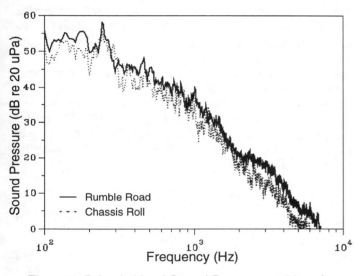

Figure 4. Driver's Head Sound Pressure at 50 mph

ROAD NOISE SOURCE CHARACTERIZATION

Simulation of road noise requires both the characterization of noise sources and transfer functions. Road noise sources include both the structureborne and airborne noise sources. Assuming that (1). structureborne and airborne sources are uncorrelated and (2). road inputs to the four tires are independent, then road noise inside passenger compartment is the power sum of the products of structureborne and airborne noise sources and the appropriate transfer functions. The structureborne sources (vibration) are observed in three directions on each of the four spindles. The airborne sources (sound pressure) are observed around each of the four tires. This results in eight uncorrelated broadband noise sources for road noise.

Experiments were performed to verify the second assumption by testing a car on the AWD chassis roll with four wheels rotating at 50 mph and one wheel at a time rotating at 50 mph. Figure 5 shows that above 100 Hz, the power sum of the driver's head responses for one wheel at a time equals to the measured response for four wheels running at the same time. Pressure as well as vibration measurements at various locations exhibit the same result (Figure 6). Experimental results indicate that phase is not important for road noise above 100 Hz, independent excitations of structureborne and airborne sources at each tire can be applied in the SEA model for road noise simulation.

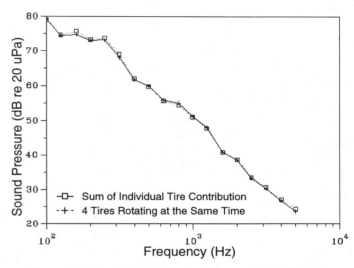

Figure 5. Driver's Head Sound Pressure at 50 mph

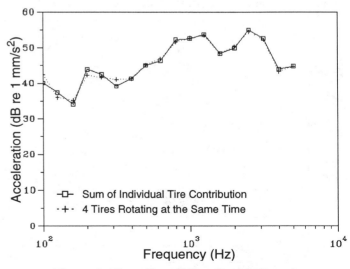

Figure 6. Floor Panel Vibration at 50 mph

SEA SIMULATION OF AIRBORNE ROAD NOISE

Road noise measurements for SEA simulation were conducted on the AWD chassis roll. A total of 88 response measurements, including 57 sound pressure measurements and 31 vibration measurements, were taken at the speed of 50 mph. There were 18 microphone measurements inside the passenger compartment to describe interior sound pressure distribution, 3 acoustic measurements around each tire to characterize airborne noise sources. Figure 7 shows that the acoustic responses inside the passenger compartment could be different by 5 to 10 dB. There were also sound pressure measurements under the vehicle, above the vehicle and on left/right side of the vehicle. Acoustic responses inside wheel well, outside left front door and above the windshield are shown in Figure 8.

The measured acoustic responses around the tires were used to develop airborne excitations for the SEA model. Trimmed body SEA model was used for transfer function simulation. Validation of the trimmed body SEA model was reported in Moeller et al [9] and Pan et al [10]. It is expected that the SEA predicted responses will be close to the measured responses if the excitations are correctly modeled. Acoustic source subsystems for the tires were created to simulate airborne road noise. The source subsystems were excited with measured sound pressure level around the tires and connected to the near field acoustic subsystems with appropriate dimensions. Acoustic responses outside the vehicle depend only on airborne sources and they should correlate well with measurements. Figure 9 shows the comparison between measured and predicted acoustic responses half a meter away from the left front wheel. Good correlations are observed for microphone locations above the windshield and outside the left front door (Figure 10 and Figure 11). Figure 12 shows the vibration comparison at the left front door glass. The SEA predicted vibration underestimates the measured vibration below 600 Hz because the structural excitations are absent from the SEA model. Figure 13 and 14 show the SEA model underestimates the responses at driver's head and right rear passenger head locations below 400 - 500 Hz. This indicates that structureborne noise probably dominates interior acoustic responses and structural vibration below 400 - 600 Hz for road noise described in this data set. The airborne road noise transmission can be sufficiently described with the trimmed body SEA model.

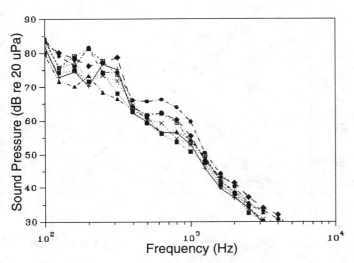

Figure 7. Passenger Comp. Sound Pressure at 50 mph

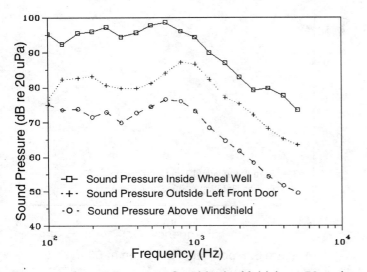

Figure 8. Sound Pressure Outside the Vehicle at 50 mph

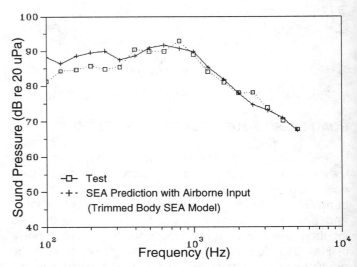

Figure 9. Sound Pressure Outside the Left Front Wheel

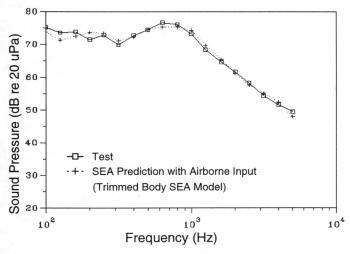

Figure 10. Sound Pressure Above the Windshield

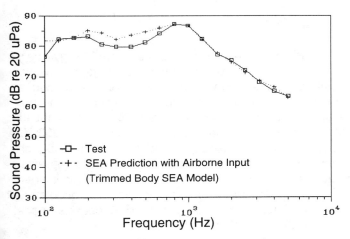

Figure 11. Sound Pressure Outside the Left Front Door

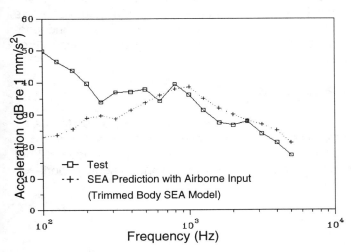

Figure 12. Vibration at the Left Front Door Glass

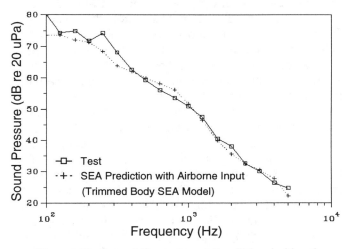

Figure 13. Sound Pressure at the Driver's Head

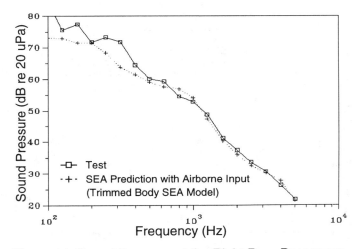

Figure 14. Sound Pressure at the Right Rear Passenger Head

SEA SIMULATION OF STRUCTURE-BORNE ROAD NOISE

The structureborne road noise generated by tires rolling on a road surface is transmitted from tires and wheels via chassis components to body structure through body-chassis attachment points. The radiation of body panels such as the floor panel, the windshield and door glasses contributes to the noise level inside the passenger compartment. Airborne road noise simulation has shown that structural radiation dominates passenger compartment acoustic responses below 400 - 600 Hz. Our first attempt to simulate structural excitation was to apply measured vibration level to all body-chassis attachment points. One accelerometer was placed in the vicinity of each attachment point. Figure 15 and 16 show the response comparison at the driver's head and the left front door glass. The SEA model under-predicts both acoustic and vibration responses by more than 10 dB. One possible explanation is that non-resonant coupling between chassis components and body in-plane subsystems transmitted most of the vibrational energy through attachment points [12] and [13]. The in-plane vibration at attachment points was not recorded by the accelerometers. A preliminary SEA model for chassis components was created and connected to the body model to help understand vibration transmission to body structure. The chassis model mostly consists of beam elements. Elastomer bushings were modeled as spring elements with estimated stiffnesses. Measured spindle vibration was enforced in the model. Figure 17 shows the resulting comparison at driver's head for the four wheel rotation at 50 mph load case. It can be seen that the SEA prediction has significantly improved.

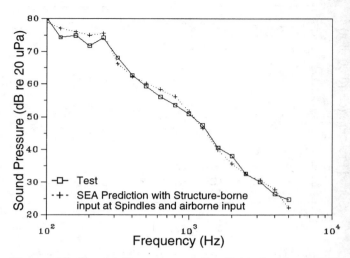

Figure 17. Sound Pressure at the Driver's Head, SEA Model Includes Chassis Components.

CONCLUSIONS

Road noise simulation using Statistical Energy Analysis has been presented. Airborne road noise was modeled using a trimmed body SEA model with acoustic source excitation around each tire. Measured and predicted results show good agreement for interior acoustic responses and structural vibration above 400 - 600 Hz. Measured vibration responses at body-chassis attachment points were shown to be inadequate to represent structure-borne sources. The addition of a preliminary SEA model for chassis components show improved structure-borne noise simulation. The SEA chassis model needs further validation.

Experimental results indicate that a chassis roll with a rough surface is a good platform for studying road noise on rumble road. Road inputs to the four tires are independent above 100 Hz, and contributions from individual tires are additive.

ACKNOWLEDGMENT

The authors would like to thank Dr. James Moore for his constructive comments and assistance with some of the measurements. We thank Mr. Dave Flanigan of Ford Motor Company for his overall management support.

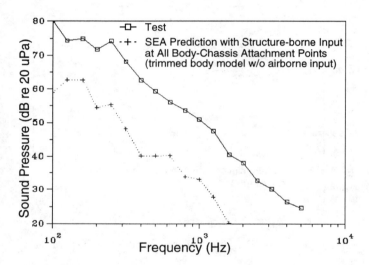

Figure 15. Sound Pressure at the Driver's Head

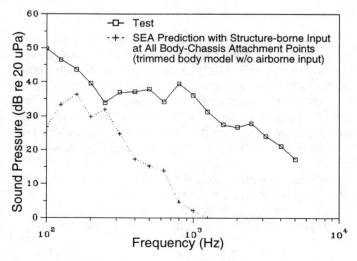

Figure 16. Vibration at the Left Front Door Glass

REFERENCES

1. Lyon, R.H. and DeJong, R.G., "Theory and Application of Statistical Energy Analysis," 2nd Edition, Butterworth-Heinemann, Boston, 1995.

2. DeJong, R.G., "A Study of Vehicle Interior Noise Using Statistical Energy Analysis," SAE paper 850960, 1985.

3. Shaw, S., "Analysis and Prediction of the Acoustical Characteristics of Automobile Bodies Using Statistical energy Analysis," Unikeller Conference, Switz., 1987.

4. Lalor, N. and Bharj, T., "The Application of SEA to the Reduction of Passenger Car Interior Noise,' 27th ISATA, Belgium, 1994.

5. Bharj, T. and Foam, H., "Application of Energy Flow Analysis (EFA) to Reduce Structure-Borne Noise Inside a Passenger Vehicle," Internoise 96, Liverpool, UK, 1996.

6. Steel, J.A., "The Prediction of Structural Vibration Transmission Through a Motor Vehicle Using Statistical Energy Analysis," Journal of Sound and Vibration, Vol 193(3), pp691-703, 1996.

7. Chen, H.Y., O'Keefe, M., Bremner, P., "A Comparison of Test-Based and Analytic SEA Models for Vibro-Acoustics of a Light Truck," SAE paper 951329, 1995.

8. Dong, B. et al, "Road Noise Modeling Using Statistical Energy Analysis Method," SAE paper 951327, 1995.

9. Moeller, Mark J., Pan, Jian and DeJong, Richard, "A Novel Approach to Statistical Energy Analysis Model Validation," SAE paper 951328, 1995.

10. Pan, J., Thomas, R.S. and Moeller, M.J., "Verifying Vehicle SEA Model Predictions for Airborne Noise Transmission Using Designed Experiments," InterNoise 96, Liverpool, UK, 1996.

11. Thomas, R.S., Pan, J., Moeller, M.J. and Nolan, T.N., "Implementing and Improving Statistical Energy Analysis Models Using Quality Technology," Noise Control Engineering Journal, January-February, 1997.

12. Powell, R. E. and Manning, J. E., "The Importance of Non-resonant and In-plane Vibration Transmission in Statistical Energy Analysis," presented at the 59th Shock & Vibration Symposium, Oct. 1988

13. Lyon, R. H., "In-plane Contribution to Structural Noise Transmission," Noise Control Engineering Journal, January-February 1986.

971973

SEA Modeling and Testing for Airborne Transmission Through Vehicle Sound Package

Robert E. Powell
Ford Motor Co.

Jason Zhu
Collins & Aikman Corp.

Jerome E. Manning
Cambridge Collaborative, Inc.

Copyright 1997 Society of Automotive Engineers, Inc.

ABSTRACT

Airborne sound transmission through vehicle panels with penetrations and sound insulation is a major component of high frequency interior noise in cars and trucks. Accurate analytical models of interior noise require high fidelity simulation of these paths in order to perform upfront design of the sound package. This paper describes a modeling approach based on Statistical Energy Analysis (SEA) that provides a general and flexible capability for incorporating sound package parameters within an analytical model of high frequency interior noise. Validation of the model for sound transmission through panels with holes and with typical sound insulation material is achieved through innovative testing methods that reveal dynamics of the decoupler and barrier layers. Refinements of the general approach that consider more deterministic features of the specific decoupler material are also suggested.

INTRODUCTION

Reduced design cycle times are requiring manufacturers to rely more on computer aided engineering (CAE) to verify performance of new designs. Interior noise has become an important performance attribute in modern passenger vehicles as quiet and comfortable interiors are routinely expected by the customer. Accurate high frequency CAE system models are required in order to design vehicles to meet targets for interior noise.

While no existing single technology offers sufficient accuracy to cover vehicle noise prediction over the full frequency range of human hearing, current CAE practice appears to divide the noise, vibration, and harshness (NVH) spectrum into two ranges: low and high frequency. Low frequency models use a deterministic approach, usually based on finite element or boundary element representations of acoustic spaces. High frequency models use a statistical approach, usually based on some form of statistical energy analysis (SEA). The statistical approach works with energy and power, predicting mean-square acoustic responses averaged over spatial volumes and frequency bands. The dividing line between the low and high frequency ranges is quite fuzzy – vehicle excitation and response appear to be mainly deterministic below 80 Hz, and mainly statistical above 250 Hz, with both aspects apparent in the intermediate or middle frequency range.

This paper describes a sound package modeling approach based on SEA that provides a general and flexible capability for incorporating sound package parameters within an analytical model of high frequency vehicle interior noise. The method described has been successfully applied to many vehicles over several years. The current study was initiated to determine what changes could be made to the general approach to improve accuracy and to better differentiate performance of advanced materials. The baseline model is compared to laboratory tests of sound transmission through flat panels, panels with holes, and panels with typical sound insulation material. Detailed acoustic and vibration testing methods were applied that reveal dynamics of the decoupler and barrier layers of the insulator. A refinement of the general approach is described that uses frequency-dependent material properties and more deterministic features of the specific insulator system.

SEA MODELING OF SOUND PACKAGE

BACKGROUND – Statistical energy analysis provides a robust analytical model for vehicle system NVH studies, due to its fundamental reliance on conservation of dynamic energy. The vehicle components are divided into subsystems that represent groups of resonant modes. Each subsystem, whether structural or acoustic, can accept external energy from sources, dissipate energy through internal damping, transmit energy to connected subsystems, and store energy in its resonant modes. The steady state energy distribution of the system is calculated by solution of a set of power balance equations, one equation per subsystem. Average response accelerations and sound pressures are calculated from subsystem energies and material properties. The power exchange between subsystems is also calculated, providing noise path analysis capabilities. The SEA method is fully described in the textbook by Lyon and DeJong [1].

SEA application to automobile interior noise was initially described over ten years ago by DeJong [2]. Recent automotive applications following this analytical approach have been described by Cimerman [3], Dong [4], Moeller [5], and Pan [6]. The acoustical behavior of vehicle panels with sound insulating decouplers and mass barriers can be modeled as a double-wall partition, similar to the design used in building construction. SEA modeling of sound transmission loss, an acoustic performance metric for attenuation of airborne noise, was illustrated for such double-wall topologies over twenty-five years ago by Price and Crocker [7]. Full double-wall modeling with SEA requires consideration of both resonant and non-resonant modal responses of the panels and the decoupler layer. Cimerman et al [3] argued that resonant response of the decoupler and barrier materials in automotive sound package was secondary because of the high damping of typical materials used in this application. Their model used only the base panel (body sheetmetal) as an SEA subsystem, with the insulator affecting panel damping, radiation coupling, and interior acoustic absorption; following the calculation approach for fibrous materials described by Mechel and Ver [8]. Flanking of the sound insulator by leaks was reported to be a significant factor at higher frequencies [3].

CURRENT SEA MODELING – A different approach to modeling trim has been adopted for many SEA models at Ford. A common SEA model topology has been developed to describe a broad range of sound package components, including hard plastic trim and open-cell foam, as well as softer plastics and fibrous materials. In this approach, the decoupler layer is described as an acoustic SEA subsystem and the barrier layer as a bending (panel) subsystem. Using the SEAM software [9], two independent *area junctions* are employed to analytically calculate all of the resonant and non-resonant structural/acoustic coupling factors involved in typical panel transmission and radiation. One of these junctions represents motion of the panel, the other the motion of the barrier. A large database of measured panel transmission loss with different decoupler and barrier arrangements, described recently by Wentzel and Saha [10], has been used to develop a single set of material parameters that best model "typical" automotive sound insulators.

Advantages – There are two primary advantages of this approach: ease of modeling (the SEAM material parameters are published on the Ford intranet in a high frequency modeling guideline) and flexibility in handling penetrations of the panel and sound package. A typical automotive dash panel has more than ten penetrations of the sheetmetal by various mechanical, fluid, and electronic connections to the interior compartment. Although these are treated by grommets and other seals, we have found these invariably to be weaker acoustically than continuous steel, and the larger ones must be modeled explicitly in order to accurately describe the airborne sound transmission. Some of the penetrations are covered by sound insulation, so their junctions can be connected from exterior air to the decoupler acoustic subsystem. Others are mechanically connected to the barrier, requiring additional structural junctions. The ability to perform such detailed modeling, including major penetrations such as the steering column and climate control assembly, has been a key factor in the growing use of CAE models to guide sound package design.

Concerns – The existing trim modeling approach has two concerns that prompt the current study: a limited ability to differentiate between decoupler materials and a lack of explicit handling of the double-wall resonance. By using a common set of material properties to describe all of the sound package decouplers and absorbers, the analyst can assemble a new model rapidly without having to measure performance of candidate materials. However, although most materials used in decoupling layers – slab foam, cast foam, needled fiber, and resinated fiber – have been found to have approximately the same acoustical behavior, one would not want to discourage use of higher-performance materials, such as one recently reported by Watanabe et al [11]. One of the goals of the current study was to demonstrate material testing procedures to permit sound package suppliers to contribute to CAE models during the design process.

Incorporation of double-wall resonance effects – a possible reduction in insulation effectiveness, usually in the 200 Hz to 400 Hz range – was desired if it would improve accuracy of the SEA models. In theory, at double-wall resonance, the mass reactance of the two panels (dash and barrier) is canceled by the stiffness reactance of the decoupler, reducing sound transmission loss (STL) to zero in the absence of damping. In practice with automotive sound insulation systems, even in careful flat panel STL tests, the double-wall resonance region rarely exhibits any dip in STL [10]. The maximum observed effect is usually a reduction in the slope of STL with frequency – the rapid rise in sound package performance may be delayed for a few third octave bands. There is even less observed effect in a vehicle installation, as the decoupler compliance varies with its thickness as the part covers bends and stiffeners in the body structure (of varying surface density), tending to diffuse the frequency range of any resonance.

MEASUREMENTS OF ACOUSTIC AND DYNAMIC RESPONSES OF FLAT SAMPLES

TESTING FOR MODEL VERIFICATION – A study of typical sound package performance compared to that predicted by SEA models was conducted to verify existing modeling guidelines, and to suggest improvements where discrepancies were significant. Flat panel and dash buck measurements were conducted – this paper discusses only the flat panel results. To provide accurate experimental results in support of SEA modeling, several test panels were measured for vibroacoustic response to a reverberant source sound field in the frequency range 100 Hz to 10 kHz. Panels of typical automotive sheet metal thickness – 0.9 mm and 0.7 mm – were tested. Sound transmitted through a 35 mm diameter hole was also measured to assess modeling of body penetrations. To demonstrate accurate modeling of vehicle sound package, we focused on the detailed measurements of a flat sample consisting of three layers: a 0.7 mm steel panel, a 25 mm thick light polyurethane foam decoupler, and a 2.5 mm thick polyvinyl chloride (PVC) barrier with surface density of 5 kg/m^2. Figure 1 illustrates the trimmed panel sample mounted in a concrete fixture in the wall between a source room and a receiver room. The source room, of dimensions 5.5 m by 7 m by 8.8 m, contained a loudspeaker driven by white noise excitation. The receiver room was built to be semi-anechoic, with the same volume as the source room.

Acceleration on the barrier and on the steel panel was measured by small B&K accelerometers (Type 4344). Mass loading by the 2 gram devices was expected to reduce the high frequency vibration on the barrier, but we hoped to see useful results at lower frequencies, especially any evidence of double-wall resonance effects. Sound pressure level (SPL) inside the decoupler was measured by a 6 mm (1/4 inch) B&K microphone (Type 4187) by inserting it through a hole cut in the barrier from the receiver side, as shown in Figure 1. Sound transmission loss was measured by using a B&K intensity probe (Type 4181 dual microphones and Type ZH0354 remote controller). Signals were measured in 1/3 octave bands by a B&K Dual Channel Real-time Frequency Analyzer (Type 2133).

INTENSITY MEASUREMENT FOR STL – The measurement of STL by the sound intensity method [12] involves several experimental considerations or difficulties, particularly in obtaining the extremely high sound transmission loss of good sound insulators at high frequencies. The first consideration is to deal with the spatial variation or uncertainty of SPL inside the source room. The source room in our tests provided a diffuse field down to 80 Hz due to its large size. The measured SPL at twenty-four points inside the source room

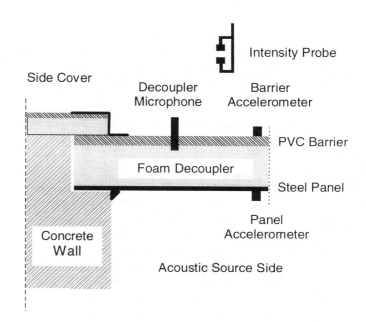

Figure 1. Schematic of test section

showed smaller values of spatial standard deviation than those required by ISO 3741 and SAE J1400 [13].

The second consideration is to control the transmitted acoustic field within the dynamic range of the test equipment. The sound insulation system shown in Figure 1 has an extremely high STL value at high frequencies due to the efficiency of the double-wall with compliant and damped decoupler layer. For example, at 4 kHz, STL is predicted to be 80 dB or higher. With a flat third octave pressure spectrum in the source room, the transmitted intensity level could not be measured within the dynamic range of the equipment simultaneously in all the frequency bands. We performed the measurement twice to obtain transmitted sound intensity at low frequencies and at high frequencies. The final results are the best combination of bands from low frequency and high frequency tests.

The third – and most important – consideration is to control the reactive field, structural flanking paths, and acoustic leakage around sample edges on the receiver side. At high frequencies, the transmitted sound pressure through the trimmed panel may be small in comparison to the background acoustic reactive field as well as the acoustic leakage around the edges. Extensive surrounding treatment was required to make the reactive field and flanking noise very low. As a rule of thumb, the transmitted sound pressure and intensity index (pressure level minus intensity level) must be controlled to be no more than 12 dB. In other words, sound intensity level can be fairly accurately measured down to 12 dB below the sound pressure level. If the index is more than 12 dB, the measured sound intensity can be contaminated by the reactive field.

EDGE TREATMENT – A very careful treatment was constructed around the four edges of the sample adjoining the concrete wall, as shown in Figure 1. In the trimmed panel test, the steel panel was first sealed into the concrete fixture with a thin layer of silicone caulk. A second bead of butyl mastic was applied on the source room side to further reduce acoustic flanking. The slab foam decoupler and barrier were then laid on top of the panel, with the 45 degree angle of the panel fixture allowing gravity to maintain close contact between the layers. A secondary decoupler and barrier system (side cover) was then applied around the edges, with flexible metal foil sealing the gap between the primary and secondary PVC barriers. We verified that sound intensity from the sample edges was lower than the transmitted sound through the sample. The side cover was designed to be capable of blocking and absorbing the inevitable acoustic leakage from the edges, without weighing down the sample or creating unnecessary structural flanking paths.

CALCULATION OF STL – Sound transmission loss is defined as the ratio of incident to transmitted intensity, or sound power, expressed in decibels. Incident sound intensity can be indirectly obtained by measuring the spatially averaged sound pressure level (SPL) inside the source room, and making an assumption that the acoustic field in the source room is diffuse. For a purely reactive or diffusive reverberation room, there is no net flow of acoustic energy at any point. In other words, there is no net sound intensity. However, at a boundary surface, sound intensity normally incident on the sample can be shown to be one quarter that of a freely traveling plane wave with the same rms pressure [1], specifically

$$I_{incident} = \frac{p_{rms}^2}{4 \rho_0 c_0} \quad (1)$$

where p_{rms} denotes the spatially averaged pressure inside the source room, measured in this facility by a rotating microphone boom, and ρ_0 and c_0 are the mass density and acoustic wavespeed of air, respectively.

The transmitted sound intensity (SI) is measured by using a pair of 12 mm (1/2 inch) phase-matched microphones spaced 12 mm apart in an intensity probe. A spatially averaged SI value is obtained by sweeping the probe on a plane parallel to the sample surface with the distance of 100 mm. The standard reference levels for sound intensity level (SIL in dB re 1 pW/m^2) and sound pressure level (SPL in dB re 20 μPa) were chosen historically so that SIL and SPL would match in a freely-traveling plane wave in air at normal conditions. Therefore, STL can be calculated from average source room SPL and transmitted SIL by

$$STL = 10 Log \left[\frac{I_{incident}}{I_{transmitted}} \right] \quad (2)$$
$$= SPL_{source} - SIL_{receiver} - 6 \text{ dB}$$

STL TEST RESULTS – Transmission loss test results for all of the panels tested are shown in Figure There is little difference observed between the sample below the 315 Hz band. The trimmed panel has by far the best performance – highest STL. The values over 90 dB at the highest frequencies may still be limited somewhat by the experimental considerations describ above. At 4 kHz, the trimmed STL is over 40 dB abov that of the bare panel, indicating that the well-sealed tr is effectively eliminating over 99.99% of the sound transmitted through the bare panel at this frequency. The two bare panels follow the theoretical 6 dB/octave *mass law* line over most of the frequency range, indicating that structural damping would have little effe in airborne noise transmission through thin flat panels. The smooth leveling off of STL for the bare panels at t highest frequencies is presumably due to increasing resonant coupling as the acoustic wavelength begins t approach the structural bending wavelength.

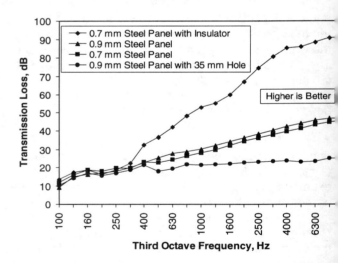

Figure 2. Measured sound transmission loss of 1 m square steel panels

A pseudo-STL for the 35 mm diameter hole wa approximated by applying Equation (2) to SIL measure at 100 mm distance from the panel over a square area of 450 mm edges centered on the hole. The area was reduced over that of the bare panel test to concentrate more of the measurement near the hole. Even so, the rising STL curve observed in Figure 2 may be caused concentrated directivity around the hole that may be spatially undersampled by the intensity probe. In theor the STL through a hole in a thin panel should level off a constant value when the acoustic wavelength is sma compared to the diameter. At high frequencies the transmission coefficient at the hole should approach unity and STL should be proportional to the area ratio

pressed in decibels. At low frequencies, transmission
reduced by the acoustic reactance caused by
trained air mass at the opening, so that there is little
gradation in STL below 400 Hz for this hole. Note that
e hole allows about 15 dB more sound power through
e sample at 4 kHz than the sealed 0.9 mm bare panel,
d a staggering 60 dB (one million times) more power
an the trimmed sample at this frequency. This clearly
ustrates the need for careful treatment of penetrations
reducing airborne noise transmission in vehicles.

ESULTS COMPARED TO BASELINE SEA MODELS

For comparison to the test results, baseline
EAM models of the transmission loss setup and test
amples were constructed following Ford internally
ublished modeling guidelines, including that of panel
amping. An option in modeling panel radiation
uggests including a length parameter when the panel
as line impedance discontinuities. In this case, the 4 m
dge length of the sample was included in the area
nctions as radiating perimeter. STL was calculated
sing Equation (2). Transmitted intensity was calculated
om predicted SPL and damping of the receiving room.

BARE 0.9 MM PANEL – Figure 3 shows the
easured and predicted STL curves of the 0.9 mm bare
teel panel. Excellent agreement is observed over most
f the frequency range, where the STL closely follows
e 6 dB/octave mass-controlled theory. Predicted STL
alls several dB below the test value in a few of the
owest third octave bands and is about 7 dB lower at
0 kHz.

In Figure 4, the third octave acceleration level of
e 0.9 mm panel is plotted, normalized by the average
hird octave SPL in the source room. An 11 by 11 grid
f points was sampled, one at a time to avoid increasing
anel mass, and the average of 121 levels calculated.
his decibel averaging follows the assumption that third
ctave vibration levels are approximately normally
istributed spatially [14]. Again, the CAE model is in
ood agreement with the test data. A third curve on
igure 4 represents the expected panel response if the
ccelerometer mass was zero. This is obtained by
alancing the induced force on the panel with that
equired to accelerate the sensor mass.

$$\frac{\langle a_{free}^2 \rangle}{\langle a_{test}^2 \rangle} \cong \frac{|Z_m + Z_p|^2}{|Z_m|^2} \quad (3)$$

$$Z_m = j\omega m \quad Z_p = 2.3 \rho c_L h^2$$

where the left hand side is the ratio of free to measured
mean-square acceleration, Z_m is the impedance of the
accelerometer mass m, Z_p is the average flexural
impedance of a thin panel of density ρ, compressional
wavespeed c_L, and thickness h. The SEA-predicted

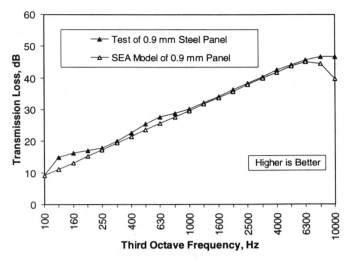

Figure 3. Sound transmission loss comparison of test
data to SEA model for bare 0.9 mm steel panel

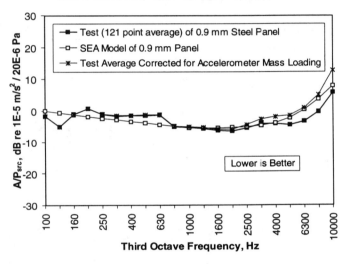

Figure 4. Acceleration/pressure comparison of test data
to SEA model for 0.9 mm bare panel

panel vibration is even closer to the test after this mass-correction is applied.

The deviations in predicted STL observed in
Figure 3 were not explained by the panel vibration data.
This is perhaps not surprising, since STL for such a thin
panel is dominated by the non-resonant mass-law
response, while panel vibration includes resonant modal
contributions that radiate inefficiently. The rise in
acceleration at high frequencies is caused partially by
decreasing damping loss factor (that decreases 5 dB
from 1 kHz to 10 kHz in the Ford guidelines), but mainly
by increasing structural/acoustic coupling, primarily at
the edge discontinuities of the panel, since the flexural
wavelengths are smaller than that of sound in air for this
thin panel up to the coincidence frequency of 14 kHz.

BARE 0.7 MM PANEL – Similar STL and
vibration comparisons for a 0.7 mm thick steel panel are
shown in Figures 5 and 6, respectively. The acoustic
transmission is predicted to be greater than that of the
thicker panel (lower STL), by the ratio of the square of

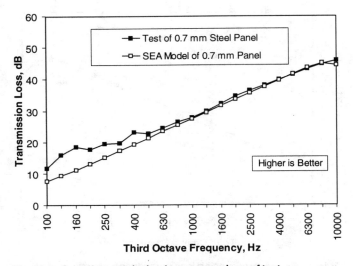

Figure 5. Sound transmission loss comparison of test data to SEA model for bare 0.7 mm steel panel

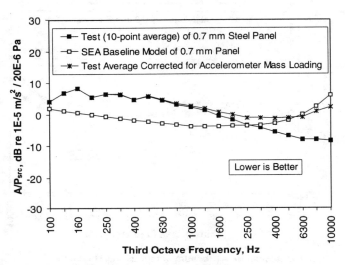

Figure 6. Acceleration/pressure comparison of test data to SEA model for 0.7 mm bare panel

the surface densities – 2 dB – at all but the very highest frequency bands. In this case, the low frequency STL actually measured 1-2 dB higher than that of the thicker panel from 100 Hz to 400 Hz. One possible explanation for higher than mass-law STL at low frequencies is the aperture effect [15] – when the panel dimensions are less than one-half an acoustic wavelength and the panel is in a rigid baffle, the acoustic reactance increases and reduces sound transmission. This effect is identical to the hole effect that was discussed in the STL test results section above. Our 1 m panel width is less than half an acoustic wavelength below 170 Hz. However, since this aperture effect should be identical for all of the test samples, the low frequency crossover in STL between 0.9 mm and 0.7 mm panels must be regarded as a limitation of test accuracy.

The SEAM model can be made to simulate this aperture effect – in fact it is the default area junction behavior – but we have found in vehicle testing that the subdivision of body panels into smaller subsystems works better if the area coupling is forced to be strictly proportional to panel area.

For the vibration of the 0.7 mm panel (Figure 6), prediction agreement with mass-corrected test data is good above 2 kHz. The measured vibration level is an average of 6 dB higher than the SEA prediction below 1250 Hz. One modeling change that would help here is the doubling of radiating perimeter from 4 m to 8 m to account for clamped – rather than supported – edges. However, with only 10 points in the vibration average here compared to 121 points for the 0.9 mm panel in Figure 4, we are inclined to prefer the baseline modeling approach. Interestingly, the 0.7 mm panel tests higher than the SEA model in sound-induced vibration at lower frequencies, but transmits less sound (higher STL) than does the model.

35 MM HOLE – The SEAM model for the 0.9 mm steel panel was supplemented by an area junction directly connecting the source and receiver rooms with the area of the circular hole of 35 mm diameter. As discussed above, the finite-size baffled aperture effect is turned on for holes such as this. STL results – referred to the same 450 mm square area as the test – are compared to test results in Figure 7. Both prediction and test results are similar to the solid panel (Figure 3) below 315 Hz. Agreement is very good through the middle of the frequency range until the test STL continues rising above 2 kHz. As discussed above, we are uncertain as to the cause of this observed behavior, but it may be due to spatial undersampling of the radiated intensity from the hole in the STL test. In any case, the SEA model predicts up to 6 dB less STL (more sound transmission) than the test in the 8 kHz and 10 kHz bands.

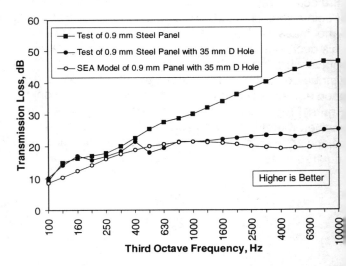

Figure 7. Sound transmission loss comparison of test data to SEA model with 35 mm diameter hole

TRIMMED PANEL – The SEA model for the 0.7 mm panel was then changed to the trimmed panel configuration, as described in the *current SEA modeling*

ection at the beginning of the paper. The physical configuration is shown in Figure 1. In this case, the foam decoupler of 25 mm thickness and the PVC barrier of 5 kg/m² surface density are modeled as SEA subsystems, with one area junction for the panel velocity and another for the barrier velocity. Comparison to test STL results for the trimmed and bare panel is shown in Figure 8. Agreement is excellent, considering that no tuning of the model was performed. From 100 Hz to 160 Hz, the STL follows the mass-law theory of a single panel with the total surface density of the sample. In the 200, 250, and 315 Hz bands, the test shows evidence of double-wall resonance, while the baseline model overpredicts STL by 5-7 dB. There is also a 3 dB overprediction of STL by the model in the 6.3 kHz to 10 kHz bands. Overprediction of STL corresponds to an underprediction of sound transmission, indicating that the baseline model may not be conservative around double-wall resonance frequency of the sound insulator.

Acceleration of the panel and barrier, as well as pressure in the decoupler layer, were studied to identify more clearly the areas of modeling difficulty. The acceleration/pressure transfer function for the trimmed panel is shown in Figure 9. In comparison to the bare panel vibration in Figure 6, both test and model are little changed by the trim above 400 Hz. The test data indicates that below 200 Hz, the trimmed panel vibration decreases relative to that of the bare panel.

SPL in the foam decoupler under the barrier is shown in Figure 10, relative to the SPL in the source room. The test data shows no significant pressure reduction from the source room average from 100 Hz to 315 Hz. This is remarkably similar to the STL performance in Figure 8 – where no significant attenuation from the trim was observed below the 400 Hz band. Measured decoupler SPL shows evidence of double-wall resonance (higher SPL) in the 200, 250, and 315 Hz bands, the same frequencies as in the STL. The SEA model of decoupler SPL appears to trend lower than that observed below 630 Hz. Material tests on the foam, described later in the paper, suggest that the effective specific impedance rises at lower frequencies, while the empirical SEA impedance remains constant.

This modeling error would tend to explain the observed SPL deviation, as higher impedance leads to higher SPL for a given modal energy in the SEA model. Above 6.3 kHz, the measured SPL is again higher than the prediction. This effect is probably caused by acoustic flanking around the sample edges, a path neglected in the SEA model. For STL measured outside the barrier, the side covers helped to attenuate the edge flanking, but there was no equivalent sealing of the foam edges underneath the barrier.

Finally, the PVC barrier vibration is shown in Figure 11. Comparing the test curve with the panel

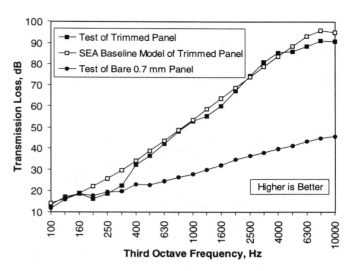

Figure 8. Sound transmission loss comparison of test data to SEA models for trimmed panel

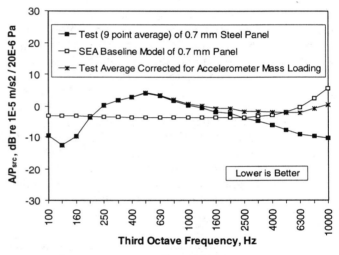

Figure 9. Acceleration/pressure comparison of test data to SEA models for trimmed panel

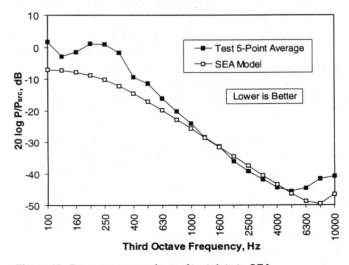

Figure 10. Pressure comparison of test data to SEA model for decoupler layer of trimmed panel

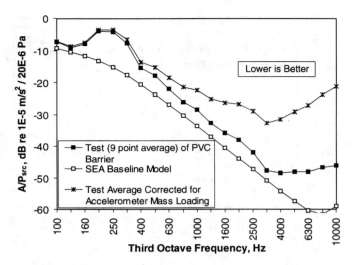

Figure 11. Acceleration/pressure comparison of test data to SEA model for trimmed panel barrier

vibration in Figure 9, there is little vibration attenuation across the decoupler from 100 Hz to 200 Hz –reinforcing the STL behavior that suggests the sample behaves as a single wall panel in this range. In Figure 11 there is about a 6 dB increase in measured barrier vibration above the SEA trend line for the three double-wall resonance bands – 200, 250, and 315 Hz. The leveling off of measured acceleration level above 4 kHz may be caused by the noise floor of the test equipment, as measured acceleration levels in these bands were 47 dB below the peak band levels at 200 Hz and 250 Hz. Correction of barrier acceleration for accelerometer mass is quite approximate in this case, as we do not know the compressional wavespeed parameter very well in Equation (3).

The SEA prediction of barrier vibration tends to run lower than that observed, perhaps indicating that the assumed compressional wavespeed of 300 m/s may be too low, or that the assumed damping loss factor of 0.05 may be too high. Reducing the PVC damping loss factor to 0.02, possibly decreasing to 0.01 at 10 kHz, would improve the vibration correlation with test. This change would not affect STL of the sample at all, as the energy transmitted to the receiver room is almost entirely through non-resonant barrier motion, that is well below the measured resonant vibration, but is coupled much more efficiently to the air.

In summary, the detailed vibroacoustic measurements on the trimmed panel suggest that the baseline SEA model is in fact underpredicting sound transmission in the double-wall frequency range by about 6 dB. Since this effect could be significant for vehicle interior noise from powertrain and road sources, refinements of the model were investigated and are described below. Other minor deviations of the predicted STL from test at the highest frequencies were attributed to edge flanking, especially in light of the measured decoupler SPL (Figure 10).

MEASUREMENT OF ACOUSTIC PROPERTIES OF POROUS MATERIALS

FORMULATION FOR MEASUREMENTS OF ACOUSTIC PROPERTIES – More accurate material properties are required to support refinement of the SEA modeling from the baseline empirical material properties. The decoupler material properties in the baseline model were only approximately correct at higher frequencies. In order to refine the low-frequency model, we needed more accurate properties for the foam decoupler. In this study, a two-microphone transfer function method [16, 17] was used to measure acoustic properties such as characteristic impedance, propagation constant, propagation speed, and effective mass density.

In this approach, only a single wave type is considered. There are additional, slower, waves propagating in the foam frame, but these are generally considered to be of less importance for the very porous decoupler materials used for automotive sound package.

Figure 12 shows the circular standing wave tube with a sample installed at one end opposite the loudspeaker. Two microphones pick up the incident wave from the loudspeaker and the reflected wave from

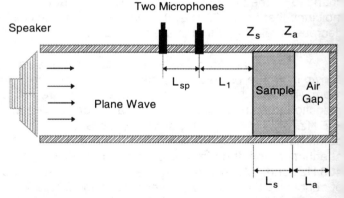

Figure 12. Schematic of decoupler parameter measurement

the sample. The approach to measuring acoustic properties of the sample is to determine the reflection coefficient by decomposing the incident and reflected waves. For plane waves, particle pressure and velocity inside the tube are given by

$$p = \hat{p}e^{-\gamma_0 x+i\omega t} + R\hat{p}e^{\gamma_0 x+i\omega t}$$
$$v = \frac{\hat{p}}{\rho_0 c_0}e^{-\gamma_0 x+i\omega t} - \frac{R\hat{p}}{\rho_0 c_0}e^{\gamma_0 x+i\omega t} \qquad (4)$$

where $\gamma_0 = i\omega/c_0$ and c_0 is the sound speed in air. The first term on the right hand side is the wave form of an incident plane wave and the second term is the wave form of a corresponding reflected wave. R is the complex pressure reflection coefficient defined as the ratio of the incident wave amplitude to the reflected wave amplitude on the sample surface (at x = 0). R can be calculated from the transfer function between the microphones [16],

$$H(\omega) = \frac{p_1}{p_2} = \frac{e^{\gamma_0 L_1} + Re^{-\gamma_0 L_1}}{e^{\gamma_0 (L_1 + L_{sp})} + Re^{-\gamma_0 (L_1 + L_{sp})}}$$

$$R = \frac{1 - H(\omega)e^{\gamma_0 L_{sp}}}{H(\omega)e^{-\gamma_0 L_{sp}} - 1} e^{2\gamma_0 L_1} \tag{5}$$

The normal surface impedance of the sample is [16]

$$Z_s = \frac{p}{v}\bigg|_{x=0} = \rho_0 c_0 \frac{1+R}{1-R}$$

$$= i\rho_0 c_0 \frac{\sin(\omega/c_0 L_1) - H(\omega)\sin[\omega/c_0(L_1 + L_{sp})]}{H(\omega)\cos[\omega/c_0(L_1 + L_{sp})] - \cos(\omega/c_0 L_1)} \tag{6}$$

where L_1 is the distance between the sample surface and the first microphone closer to the sample; and L_{sp} is the spacing between the microphones (Figure 12).

The characteristic impedance Z_c and propagation constant γ of the sample can be expressed in terms of its normal surface impedance Z_s and the acoustic impedance Z_a of the air gap as follows [17, 18]:

$$Z_s = Z_c \frac{Z_a \cosh(\gamma L_s) + Z_c \sinh(\gamma L_s)}{Z_a \sinh(\gamma L_s) + Z_c \cosh(\gamma L_s)} \tag{7}$$

which allows us to solve for γ,

$$\gamma = \frac{1}{2L_s} \ln\left[\frac{(Z_s + Z_c)(Z_a - Z_c)}{(Z_s - Z_c)(Z_a + Z_c)}\right] \tag{8}$$

If the back wall is assumed rigid, Z_a can be analytically found by

$$Z_a = -i\rho_0 c_0 \cot\left(\frac{\omega L_a}{c_0}\right) \tag{9}$$

In Equation (7) or (8), there are two unknowns: Z_c and γ. The additional equation required to solve for both unknowns is obtained by changing the air gap and repeated the test [17]. For the new air gap, the resulting Z'_s is calculated again with Equation (6) and the resulting Z'_a is calculated again with Equation (9). Equation (8) then yields

$$\frac{(Z_s + Z_c)(Z_a - Z_c)}{(Z_s - Z_c)(Z_a + Z_c)} = \frac{(Z'_s + Z_c)(Z'_a - Z_c)}{(Z'_s - Z_c)(Z'_a + Z_c)} \tag{10}$$

Solving for Z_c yields

$$Z_c = \pm\left(\frac{Z_s Z'_s (Z_a - Z'_a) - Z_a Z'_a (Z_s - Z'_s)}{Z_a + Z'_s - Z'_a - Z_s}\right) \tag{11}$$

and inserting Z_c into Equation (8) then provides γ.

Once Z_c and γ are known, the propagation speed c and effective mass density ρ_e of the sample can be readily determined by

$$c = \frac{\omega}{k} = \frac{i\omega}{\gamma} \qquad \rho_e = \frac{Z_c}{c} = \frac{Z_c \gamma}{i\omega} \tag{12}$$

EXPERIMENTAL RESULTS OF FOAM ACOUSTIC PROPERTIES – A B&K large standing wave tube (Type 4206) was used to measure acoustic properties of the polyurethane foam which was applied as the decoupler in the flat sample STL tests. The parameters required in the above calculation are listed in Table 1.

Table 1. Parameters for acoustic property measurement

Sample thickness	L_s	25 mm
Air gaps	L_a	20 mm and 70 mm
Microphone position	L_1	100 mm
Microphone spacing	L_{sp}	100 mm
Sound speed	c_0	343 m/s
Air density	ρ_0	1.2 kg/m^3

We measured the surface normal impedances Z_s and Z'_s by setting air gaps equal to 20 mm and 70 mm, respectively, and processed the transfer functions as per Equation (6). Surface impedances were then inserted into Equations (11) and (8) to calculate the characteristic impedance and propagation constant, respectively; and, finally, the propagation speed and the effective mass density by Equation (12).

Figures 13, 14, 15, and 16 show the measured acoustical properties for the foam decoupler used in the STL tests. The maximum frequency was chosen to be 1600 Hz to avoid lateral cross-modes in the tube. The frequency resolution was 0.5 Hz. Characteristic impedance, normalized by the nominal real value for air $-\rho_0 c_0 -$ is plotted in Figure 13. The real part of this normalized impedance is almost constant at middle and high frequencies, at about 1.3, and the imaginary part tends to zero as the frequency increases. The measured real normalized impedance is higher than the 0.8 value used in the baseline SEA model.

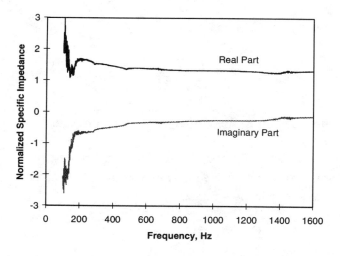

Figure 13. Measurement of normalized specific impedance of sheet foam

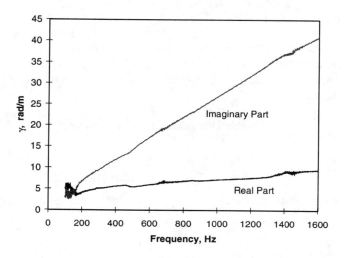

Figure 14. Measurement of complex propagation constant of sheet foam

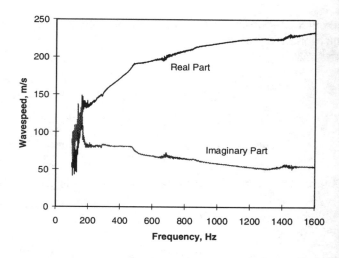

Figure 15. Measurement of complex wavespeed of sheet foam

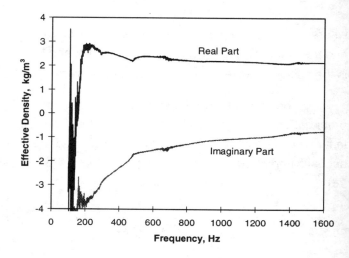

Figure 16. Measurement of complex density of sheet foam

The propagation constant is shown in Figure 14. The real part of γ is also called the attenuation constant that describes the spatial decay rate of the wave. The higher the value of the attenuation constant, the higher the pressure gradient inside the foam. The imaginary part of propagation constant is simply the wavenumber, that would increase linearly with frequency in air.

Figure 15 shows the test-derived propagation speed of an acoustic wave inside the foam. The real part is growing with frequency, but is always smaller than the sound speed of air because of the lateral inertia and viscous effects of the foam frame on acoustic wave propagation.

The effective mass density is shown in Figure 16, where the real part is about 2.2 kg/m^3, about 1.8 times that of air. The imaginary part tends to zero as frequency increases. In view of the real part of effective density going negative below 200 Hz, and of the increased noise in the test data in that region, we decided that the acoustic property data could only be used in the 200 Hz to 1600 Hz frequency range.

IMPROVED SEA MODEL

An improved SEA model for the trimmed panel STL was obtained by refining the baseline model with the effective material parameters for the decoupler foam, and by adding a parallel path that included a forced-wave calculation of double-wall dynamics. This arrangement retains the flexibility of coupling topology provided by the decoupler-as-subsystem approach, while better defining the possible increase in sound transmission in the 200 Hz to 315 Hz third octave bands. The intent is to have the forced-wave junction represent the deterministic lumped-parameter and thickness resonances of the decoupler, while the acoustic subsystem represents the statistical buildup of resonant energy in the decoupler layer.

The foam material properties were imported into the model using a fiber material capability in SEAM [9]. The additional junction makes use of another capability to calculate forced-wave acoustic transmission through arbitrary numbers of infinitely-wide layers of gas, porous materials, flexural plates, and lumped spring and mass elements. The same measured material properties were used for the decoupler in the new junction. Field-incidence averaging over incidence angles was specified, because the source room has a diffuse acoustic field. Normal incidence calculations are much faster, but less accurate, as the effective frequency of double-wall resonance changes with angle of incidence.

Comparison of tested STL to that of the improved SEA model is shown in Figure 17. Contributions from both paths are shown separately, along with the combined result of both. The measured material properties are seen to cause less transmission (more STL) in the resonant subsystem path at lower frequencies, compared to the baseline model results in Figure 8. The predicted STL from 200 Hz to 1000 Hz is dominated by the non-resonant forced-wave path. At the 1250 Hz band, the resonant path crosses over and becomes dominant at higher frequencies. The new model agrees well with the test data for all bands with available material data, except at the 200 Hz frequency, where the model is 4 dB under the tested STL.

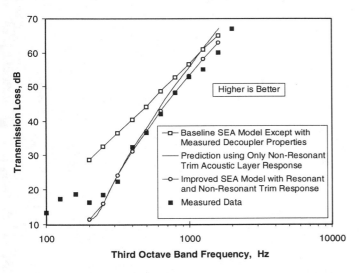

Figure 17. Sound transmission loss comparison of test data to improved SEA models for trimmed panel

CONCLUSIONS

This paper has described an SEA modeling and testing approach for automotive sound insulation components. Detailed testing of flat panel samples with bare panels and with a circular hole served to validate airborne sound transmission behavior of existing modeling guidelines. For a steel panel with typical dash sound insulation treatment, a baseline model using empirical material properties for the decoupler layer was found to provide generally good predictions of sound transmission loss (STL), except for an octave-wide frequency band where transmission was underpredicted.

Detailed testing methods that revealed dynamics of the decoupler and barrier layers indicated that the underprediction was caused by the system resonance where panel and barrier masses resonate on the decoupler compliance. Frequency-dependent material properties of the foam decoupler were measured and used in an improved SEA model that provided excellent agreement with test STL except at the 200 Hz third octave band. Future efforts will be directed at expanding the frequency range of decoupler material measurement, as well as extending the material property testing to other insulation materials. The methodology used in this flat panel study is being investigated to improve SEA simulation of airborne sound transmission and structureborne radiation for actual dash constructions of passenger vehicles and light trucks.

ACKNOWLEDGMENTS

The authors are grateful for important contributions from many sources. The baseline material properties for SEAM models of automotive decouplers were determined by Jian Pan at Ford and Professor Richard DeJong at Calvin College. James Lee maintains the SEAM modeling guidelines on Ford's intranet. Jianmin Guan at Ford conceived of this model improvement project and secured its funding. XianLi Huang and Shaobo Yang initiated the use of measured acoustic propagation data for SEA material properties at Ford. Kent Fung at Collins & Aikman (formerly Perstorp Components) assisted in the measurement effort. Mark Moeller at Ford and Melvin Care at Collins & Aikman provided valuable guidance to the authors throughout the duration of this effort.

REFERENCES

[1] R.H. Lyon and R.G. DeJong, *Theory and Application of Statistical Energy Analysis*, 2nd Edition, Butterworth-Heinemann, Boston, 1994.

[2] R.G. DeJong, "A Study of Vehicle Interior Noise Using Statistical Energy Analysis," SAE Paper 850960, 1985.

[3] B. Cimerman et al., "Incorporating Layered Acoustic Trim Materials in Body Structural-Acoustic Models," SAE Paper 951307, 1995.

[4] B. Dong et al, "Road Noise Modeling Using Statistical Energy Analysis Method," SAE Paper 951327, 1995.

[5] M. Moeller et al, "A Novel Approach to Statistical Energy Analysis Model Validation," SAE Paper 951328, 1995.

[6] J. Pan et al, "Verifying Vehicle SEA Model Predictions for Airborne Noise Transmission Using Designed Experiments," *InterNoise 96 Proceedings*, Liverpool, UK, 1996.

[7] A.J. Price and M.J. Crocker, "Sound Transmission through Double Panels Using Statistical Energy Analysis," *J. Acoust. Soc. Am.*, Vol. 47, pp. 683-693, 1970.

[8] F.P. Mechel and I.L. Ver, "Sound-Absorbing Materials and Sound Absorbers," Chapter 8 of *Noise and Vibration Control Engineering* Edited by L.L. Beranek and I.L. Ver, Wiley, New York, 1992.

[9] *SEAM 3.0 Users Manual*, Cambridge Collaborative, Inc., Cambridge, Massachusetts, 1996.

[10] R.E. Wentzel and P. Saha, "Empirically Predicting the Sound Transmission Loss of Double-Wall Sound Barrier Assemblies," SAE Paper 951268, 1995.

[11] K. Watanabe et al, "Development of a High-Performance Dash Silencer Made of a Novel Shaped Fiber Sound-Absorbing Material," SAE Paper 960192, 1996.

[12] M.J. Crocker, P.K. Raju, and B. Forssen, "Measurement of transmission loss of panels by the direct determination of transmitted acoustic intensity," *Noise Control Engineering Journal*, **17**(1), July-August, 1981, pp. 6-11.

[13] Society of Automotive Engineers Recommended Practice J1400: "Laboratory Measurement of the Airborne Sound Barrier Performance of Automotive Materials and Assemblies."

[14] R.E. Powell, "Response Variability Observed in a Reverberant Acoustic Test of a Model Aerospace Structure," *Proceedings of the Second International Congress on Recent Developments in Air- and Structure-Borne Sound and Vibration*, Auburn University, March 4-6, 1992.

[15] I.L. Ver, "Interaction of Sound Waves with Solid Structures," Chapter 9 of *Noise and Vibration Control Engineering*, Edited by L.L. Beranek and I.L. Ver, Wiley, New York, 1992.

[16] ASTM Standard Test Method E 1050-86, "Impedance and Absorption of Acoustical Materials Using a Tube, Two Microphones, and a Digital Frequency Analysis System."

[17] H. Utsuno, A.F. Seybert, *et al.*, "Transfer function method for measuring characteristic impedance and propagation constant of porous materials," *J. Acoust. Soc. Am.*, **86**(2), pp. 637-43, August, 1989.

[18] C. Zwikker and C.W. Kosten, *Sound Absorbing Materials*, Elsevier,. New York, 1949.

971974

High Frequency NVH Analysis of Full Size Pickups Using "SEAM"

Martin Botz and Bijan Khatib-Shahidi
Ford Motor Co.

Copyright 1997 Society of Automotive Engineers, Inc.

ABSTRACT

The recent surveys of customer satisfaction regarding full size pickup trucks have created new mandates in performance of such vehicles. The customers for this class of vehicles demand new frontiers in attributes such as NVH, ride and handling performance that previously only belonged to the luxury passenger cars. The full size pickup truck in question must retain a tough image and be as durable as the previous generation truck that it replaces. But it also needs to be user friendly in order for one to drive it like an every day passenger car on a daily basis.

The challenge is to design for the NVH performance that matches and surpasses many well behaved and "good" NVH passenger cars without any compromise in durability performance. An NVH 7-8 subjective rating performance is targeted for the design of full size pickup truck during vehicle operation.

In this paper, the authors demonstrate a new method that was used in the design for NVH for the full size pickup truck in the high frequency range. The method is largely based on parametric modeling of the structures in which one can quickly change the size, delete or add structural members, modify or improve the planned sound package to access the NVH performance. A comercially available spreadsheet was used to lay out the structural parameters as well as sound package data. The CAE driven design was verified by prototype testing.

INTRODUCTION

Finite Element Analysis (FEA) is the CAE tool in automotive industry to predict vibrational behavior of cars and trucks at low frequencies but fails to predict noise and vibrations as frequency rises beyond 100 Hz. At high frequencies, the average frequency spacing between adjacent modes decreases making it harder to track single modes, therefore, conventional FEA codes fail to predict NVH and/or correlate to test results. In contrast, Statistical Energy Analysis (SEA) is being utilized as a modeling technique to study the interaction of groups of resonant modes which is described in an average sense. The average energy response of each mode group becomes the system variable and the distribution of the energy among the mode groups is obtained by solving a set of power balance equations.

The analysis tool used herein is SEAM[*] which embodies statistical energy analysis equations and it covers the frequency range at about 250 Hz to about 10 kHz. The basic elements of NVH characteristics such as the occupant ear level average noise in the aforemetioned frequency range and the associated average vibrations were analyzed using a parametric model. The parameters of the SEA model were iterated so that the NVH audible and tactile targets which had been set forth by the NVH team were achievable and met within a short period of time in the design cycle. In what follows, first the basic SEA equations are presented below [1], second the model building and validation process is highlighted and parameter studies are conducted.

BASIC SEA EQUATIONS

A schematic of a single two subsystem SEA model is shown in Figure 1 where $\Pi_{1,in}$ and $\Pi_{2,in}$ are the input powers from external excitation sources, $\Pi_{1,diss}$ and $\Pi_{2,diss}$ are the powers dissipated in the subsystems, and $\Pi_{12} = -\Pi_{21}$ is the power transmitted from subsytem 1 to subsystem 2. The power balance equations are then given by

$$\Pi_{1,in} = \Pi_{1,diss} + \Pi_{12}$$
$$\Pi_{2,in} = \Pi_{2,diss} + \Pi_{21} \quad . \qquad \text{(EQ 1)}$$

The dissipated power at frequency f in subsystem 1, for example, can be written as

$$\Pi_{1,diss} = 2\pi f \eta_1 E_1 \qquad \text{(EQ 2)}$$

with the total dynamical energy E_1 of the subsystem modes and the damping loss factor η_1. For the transmitted power, a similar equation can be evaluated by

$$\Pi_{12} = 2\pi f (\eta_{12} E_1 - \eta_{21} E_2) \qquad \text{(EQ 3)}$$

---

*. SEAM is a registered trademark of Cambridge Collaborative, Inc.

with the coupling loss factors η_{12} and η_{21}. Inserting into EQ 1 finally leads to the system equations in matrix form

$$2\pi f \begin{bmatrix} \eta_1 + \eta_{12} & -\eta_{21} \\ -\eta_{12} & \eta_2 + \eta_{21} \end{bmatrix} \begin{bmatrix} E_1 \\ E_2 \end{bmatrix} = \begin{bmatrix} \Pi_{1,\,in} \\ \Pi_{2,\,in} \end{bmatrix} \qquad \text{(EQ 4)}$$

which, for known input powers and loss factors, can be solved for the unknown system energies E_1 and E_2.

It is convenient to introduce the modal energy

$$\varepsilon_1 = \frac{E_1}{N_1} \qquad \text{(EQ 5)}$$

where N_1 is the number of modes of subsystem 1 in the considered frequency band. Using reciprocity $N_1 \eta_{12} = N_2 \eta_{21}$ the matrix equation EQ 4 can be rewritten with a symmetric system matrix and the new system variables ε_1 and ε_2.

In order to convert subsystem energies to measured response quantities, the relation

$$E = M \langle v^2 \rangle \qquad \text{(EQ 6)}$$

leads to the root-mean-square velocity $\langle v^2 \rangle = v_{rms}^2$ averaged over frequency and over the interior space of a subsystem with total mass M. In case of an acoustic subsystem with impedance ρc and total volume V,

$$E = V \langle p \rangle^2 / \rho c^2 \qquad \text{(EQ 7)}$$

relates subsystem energy to pressure level p.

The first step in defining a SEA model is to divide the system into physical components and to specify subsystems within each component which are groups of modes with similar characteristics. The next step is to evaluate the parameters describing the subsystems, i.e.

- the mode count,

- the damping loss factor,

- the coupling loss factors, and

- the input power from external excitation sources.

The number of resonance modes in a frequency band is represented by N and this is one way of describing the mode count. N depends on the wavenumber k which, for non-dispersive wave types, is linearly dependent on frequency. For regularly shaped systems, expressions for the wavenumber are obtained via theoretical methods and in the case of a rectangular system with lengths L_1 and L_2, for example, this leads to

$$k_{mn} = \sqrt{\left[(m - \delta_1) \frac{\pi}{L_1} \right]^2 + \left[(n - \delta_2) \frac{\pi}{L_2} \right]^2} \qquad \text{(EQ 8)}$$

with the mode numbers m and n and the correction terms δ_1 and δ_2 depending on the boundary conditions. As frequency rises, the correction terms converge to zero which means that at higher frequencies boundary conditions become less and less important.

The damping loss factor can be experimentally determined using decay rate or power balance methods. In a decay rate measurement, the transient response of a subsystem is plotted versus time after an initial excitation is turned off. The decay rate DR, which is defined as the slope of the decay in dB/sec, is related to the damping loss factor by

$$\eta = \frac{DR}{27.3f} \qquad \text{(EQ 9)}$$

The power balance method uses the power balance equation for an isolated subsystem which is solved for the damping loss factor

$$\eta = \frac{\Pi_{in}}{2\pi f E_{tot}} \qquad \text{(EQ 10)}$$

For this measurement, the subsystem is excited by a continuous source and the input power as well as the total energy of the dynamical response is measured.

In order to determine coupling loss factors, the geometry of the connections between adjacent subsystems is important and point, line, or area coupling must be considered. Taking a wave approach the transmission of power across junctions is determined by general solutions of wave equations in semi-infinite systems which meet continuity and force balance conditions at the boundary. The transmission coefficient defined by the ratio of transmitted power to incident power can be calculated which is directly related to the coupling loss factors.

The last parameter to be evaluated is the input power which is an important quantity due to the fact that all responses are proportional to it. In case of point excitation of a structure, the power put into the structure is determined by the time averaged product of the force and the velocity at this point. A similar expression is found for the power put into an acoustic space by a point source where the time averaged product of the volume velocity and the pressure has to be built.

MODEL BUILDING AND VALIDATION PROCESS

A SEA model for a pickup truck was built by using the commercially available code SEAM. In different files, the user has to specify subsystems, junctions, material properties as well as an excitation source and control parameters to request an appropriate analysis and output data. In this model, different element types such as beams, plates, pipes, and acoustic volumes were considered to prescribe the aforementioned full size pickup truck. The excitation source was engine noise. In this case of an acoustic noise source, we prescribed the Sound Pressure Level (SPL) of an acoustic subsystem.

The SEA model for the pickup truck had about 450 subsystems. This includes structural elements, interior acoustic spaces inside cabin and engine compartment, and exterior acoustic spaces. Additional structural and acoustic subsystems were also defined to describe the sound package, i.e. the carpet, underlay, headliner, and plastic trim. At the beginning of the model building process, a trimmed body FEA model of the pickup truck was used to identify geometric dimensions and connections. Based on the FEA model, structural and acoustic display models were built. Figure 2 shows the division of the structure into subsystems where the structure includes front end, cabin, pickup box, and frame. The acoustic display model is shown in Figure 3 which contains engine compartment, underbody and cabin air spaces as well as acoustic subsystems describing the seats. Spaces behind, above, underneath, beneath, and in front of the vehicle are taken out to give a better view.

Display models are used in the early validation process to draw modal energy thermograms and, therefore, to check connectivity of the subsystems [2]. The excited subsystem always has to be the hottest spot with the highest modal energy and energy must flow from hot to cool areas in order to have the connections right. The diagrams should indicate that energy flow paths and directions are reasonable.

In order to further validate the model, a vehicle test was conducted and the test results were compared to SEA model predictions. Damping loss factors were determined using decay rate measurements and EQ 9. Exciting the respective subsystem by a unit power in SEAM and applying EQ 10 led to theoretical values for the damping loss factors which were compared to the experimental results.

Since engine presence, airborne noise, was the major concern, a loudspeaker with a white noise excitation was installed in the engine compartment to simulate the noise source. Microphone locations were chosen to characterize the SPL inside the engine compartment, outside the vehicle, and inside the cabin. The average value of the 4 microphones placed inside the engine compartment was applied to the engine compartment subsystems in the SEA model to give the excitation source. Therefore, the SPL measured at all other locations could directly be compared to the SPL predicted by the model.

Figures 4 and 5 show the comparison of test versus SEAM results in two selected exterior spaces. The interior space SPL validations are shown in Figures 6 and 7. The SEA model predictions match the test results quite well for frequencies above 250 Hz, where the theory is good for. The SEA model was validated for engine induced airborne noise transmission and used for sound package evaluation and design change iterations.

PARAMETER STUDIES

The validated SEA model was used to provide design recommendations which would reduce the driver ear SPL. First of all, the major noise transmission path must be identified by tracing the power flow from the driver ear location. Looking at a power flow chart, it turned out that most of the power is coming from the driver knee location. At very low frequencies, power is transmitted through the windshield and the side window and at around 3000 Hz, bending motion of the side window becomes important. Doubling the thickness of the windshield and the side window did not give the anticipated mass law transmission loss at low frequency. In order to reduce the bending motion of the side glass, glass damping was added which did not lead to any significant reduction in SPL either. Those paths are probably not dominant enough and will become more important when road noise load cases are considered.

Tracing the power flow further back and looking at a driver knee power flow chart led to the conclusion that, at high frequencies, almost all of the power is coming from the instrument panel acoustic subsystem where at lower frequencies, power is also coming through dash and floor trim. This suggested to reduce the openings in the instrument panel and to treat dash and floor by adding a better sound package. The effects of those design changes are shown in Figures 8-11. Going from a very light sound package in the baseline model to better sound packages meant increasing the barrier weight up to 1.5 lb/ft^2 in the luxury version. Figure 11 shows the reduction in SPL by scaling all of the current holes in the instrument panel by 50% of the baseline areas.

CONCLUSIONS

SEA can be used to predict vibrational behavior of full size pickup trucks at frequencies above 250 Hz. The analytical model can help the development engineer to identify major noise transmission paths and to evaluate the effects of design changes on the sound quality of the vehicle. This leads to tremendous cost savings since fewer tests have to be conducted.

Design changes can easily be incorporated using parametric modeling techniques. All quantities depend on a few key parameters which are defined in a spreadsheet. It is set up so that a development engineer with little background in SEA can go in and make changes to structural parameters or sound package data and quickly access the new NVH performance.

ACKNOWLEDGEMENT

The authors would like to thank Mr. Nick Kazan, Chief Chassis Engineer at Ford Motor Company Light Truck Vehicle Center for his encouragement and support of high frequency energy analysis during the years at the NVH department. Sincere appreviations are extended to Mr. Bob Himes, Chief Program Engineer at Ford Motor Company Light Truck Vehicle Center and Mr. Tom Walsh, CAE Manager at Ford Motor Company Light Truck Vehicle Center for the continuation support of high frequency energy analysis applied to full size pickup trucks.

It must also be noted that for the last three years of research, validation, and implemetation in high frequency energy analysis applied to the automotive structures, the authors are indebted to Dr. Mark Moeller, Dr. Bob Powell at Ford Motor Company, Dr. Jim Moore, and Prof. Richard DeJong of Calvin College for their enormous technical support and encouragement.

Finally, the authors would like to thank Dr. David Watts for carefully reviewing the manuscript and providing many useful comments.

REFERENCES

[1] Lyon, R. H. and DeJong, R. G., Theory and Application of Statistical Energy Analysis, Second Edition, Butterworth-Heinemann, ISBN 0-7506-9111-5, Newton, MA, 1995.

[2] Pan, J., Moeller, M. J., and DeJong, R. G., A Novel Approach to Statistical Energy Analysis Validation, 1995 SAE Noise and Vibration Symposium, Traverse City, MI, May 1995.

971975

Statistical Energy Analysis of Noise and Vibration from an Automotive Engine

Cliff Kaminsky
Vibro-Acoustic Sciences, Inc.

Robert Unglenieks
Roush-Anatrol, Inc.

Copyright 1997 Society of Automotive Engineers, Inc.

ABSTRACT

Statistical energy analysis (SEA) is a validated tool for vibro-acoustic modeling of larger systems such as air- and spacecraft, ships, and automobile bodies. In this paper, we report on the application of SEA to model the noise and vibration output of a vee-configuration automotive engine above 1000 Hertz. This approach may complement boundary element methods, which are time consuming to use and can only be applied up to one or two kilohertz. In a feasibility study, experimental data is used to specify power inputs and damping losses, as well as to validate the model. The study shows that SEA can be used to model the vibrational behavior of the engine block at high frequencies that are unattainable with deterministic methods. This result may apply to other structures such as transmissions, pumps, and superchargers.

INTRODUCTION

The powertrain is a primary contributor to high-frequency noise in an automobile. Currently, boundary element analysis is often used to analytically predict radiated noise. This technology, however, is computationally intensive and limited to low frequency analysis. There is a current need for high-frequency analysis techniques to predict powertrain radiated noise.

Statistical energy analysis (SEA), developed since the early 1960s, is a well-validated vibro-acoustic analysis method for modally dense systems, and is now finding more use in stiffer structures such as powertrains. A few SEA studies on automotive powertrain structures exist, but more work, both theoretical and experimental, must be done to validate this method on stiff structures.

The system model, in this case a V-8 automotive engine, is broken into subsystems, discrete sections that are assumed to exhibit uniform resonant response. Several assumptions are implied by this, one of which is

low damping. If the subsystem is highly damped, then local response can vary dramatically over the surface and the average response computed by SEA may not be representative.

Another assumption of SEA is the existence of resonant modes in each analysis frequency band. SEA calculates the energy stored in the subsystem's resonant modes. If there are very few modes in the analysis band, then the SEA result may vary dramatically from the actual system response. On an automotive powertrain structure, therefore, our analysis begins at 1000 Hz to be certain that resonant response exists.

Given the subsystems' geometric properties and the nature of the connections between them, the SEA software will then compute the flow of energy through the system in each frequency band. We can then find the average response of each subsystem. The model can also determine the dominant energy flow paths and the effects of design changes.

SEA MODELING

THEORY - SEA is based on power flow through a resonant structure. The power into a subsystem must equal the sum of the power lost in damping and the power transmitted to neighboring subsystems, as represented in figure 1.

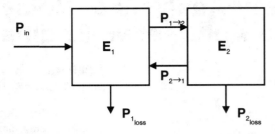

Figure 1. Power balance in a two-subsystem SEA model

This power flow from subsystem 1 to subsystem 2 is summarized in the equation:

$$P_{1\to 2} = \omega \eta_{12} n_1 \left(\frac{E_1}{n_1} - \frac{E_2}{n_2} \right) \quad (1)$$

where $P_{1\to 2}$ is the power flowing from subsystem 1 to subsystem 2, and E_i is the energy level in subsystem i. The symbol η is a unitless power flow coefficient. With a single subscript it represents power lost in damping, and is called a damping loss factor (DLF). With two subscripts, η is a coupling loss factor (CLF) that defines the power flow from one subsystem to another. The letter n denotes the subsystem modal density, the number of resonant modes in the frequency band divided by the bandwidth. As in heat transfer applications, the transfer of acoustic energy is proportional to the difference in energy levels of the two subsystems. In application, the total power balance equation appears in matrix form:

$$\omega \begin{bmatrix} \left(\eta_1 + \sum_{i\neq 1}^{k}\eta_{1i}\right)n_1 & (-\eta_{12}n_1) & \cdots & (-\eta_{1k}n_1) \\ (-\eta_{21}n_2) & \left(\eta_2 + \sum_{i\neq 2}^{k}\eta_{2i}\right)n_2 & \cdots & (-\eta_{2k}n_2) \\ \cdots & \cdots & \cdots & \cdots \\ (-\eta_{k1}n_k) & \cdots & \cdots & \left(\eta_k + \sum_{i\neq k}^{k}\eta_{ik}\right)n_k \end{bmatrix} \begin{bmatrix} \frac{E_1}{n_1} \\ \frac{E_2}{n_2} \\ \cdots \\ \frac{E_k}{n_k} \end{bmatrix} = \begin{bmatrix} P_1 \\ P_2 \\ \cdots \\ P_k \end{bmatrix} \quad (2)$$

Our SEA model was developed and solved with AutoSEA analysis software. The model consists of four major sections: intake manifold, heads, block, and oil pan. Each of these sections is divided into a number of subsystems. A section of the model is shown in figure 2.

SUBSYSTEMS - In general, subsystems were chosen for size and geometric uniformity. For example, the valve cover is a single sheet of stamped steel. It has a uniform thickness and few distinct discontinuities. Therefore, the valve cover is modeled as a single subsystem.

The subsystems available in the SEA software include beams, plates, cylinders, and acoustic cavities. Each subsystem is characterized by its geometric and material properties. The engine model consists primarily of cylindrical shells, plates, and three-dimensional acoustic cavities.

The variables of primary interest in creating the subsystems were modal density and, for those parts that radiate noise, surface area and perimeter. Correctly representing these can be a difficult task when modeling rib-stiffened structures. We made use of AutoSEA's built-in analytical model for rib-stiffened structures in the cylinder head and block. This model simulates the phenomenon of "mode grouping" when the vibrational half wavelength is shorter than the rib spacing.

DAMPING LOSS FACTORS - Because internal subsystem damping is very difficult to predict, the DLFs in our model were measured on a hot, static engine. Data was measured on the cylinder head, valve cover, and skirt. The remaining damping values were estimated

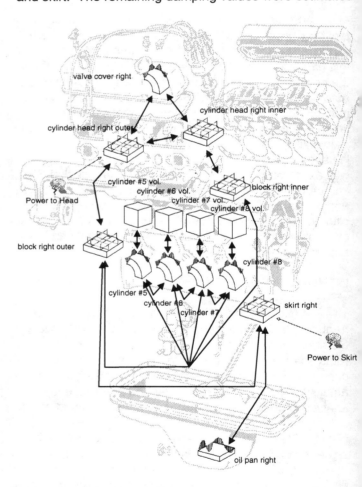

Figure 2. Section of Engine Model

from the measured values and the mentioned references.

COUPLING LOSS FACTORS - All of the CLFs in our model are analytical. This allows us to analyze design changes without retesting. The CLFs are derived from analytical point impedances of ideal infinite systems. This can cause difficulties when modeling small structures that may display finite system characteristics at lower frequencies. This is one reason that our analysis must begin at 1000 Hz. Previous published data has already shown good correlation with *empirical* CLFs on powertrain structures [2]. The effect of gaskets was not considered in this model and may be an area of interest for further study.

POWER INPUTS - The use of an infinite impedance formula to calculate the input impedance can also present a problem when modeling power inputs. For validation of transfer functions, a white noise power input was applied to the input subsystem, and the predicted velocity used to calculate mobility. Using the expression

$$\Pi_{in} = \left(\frac{1}{2}\hat{v}^2\right)\mathrm{Re}\left\{Z_{f\infty}\right\} \qquad (3)$$

and using a unit power input, we can invert the expression to obtain

$$\mathrm{Re}\left\{\frac{1}{Z_{f\infty}}\right\} = \left(\frac{1}{2}\hat{v}^2\right) \qquad (4)$$

where Π_{in} is the injected power, $\hat{v}$ is the peak surface velocity, $Z_{f\infty}$ is the point force impedance, and therefore $1/Z_{f\infty}$ is the point input mobility. In the operating conditions model, the subsystems receiving power input are the skirt, the cylinder heads, and the front cover. For validation of operating vibration levels, the input power was obtained by measuring the point mobility at several locations on each driven subsystem, and the operating velocity at the same locations. We then calculated the average mobility and the average velocity to compute an "average" power input using equation 3. This method assumes that the measured vibration in each driven subsystem is unrelated to the inputs in the other driven subsystems. If this assumption were not true, we would expect to find an overestimation of operating vibration levels.

TESTING

A production V-8 dual overhead camshaft internal combustion engine and front-wheel drive transaxle were tested at Roush Anatrol's Livonia, Michigan facility. Sound pressure data were obtained in a hemi-anechoic test cell and in a reverberant room. Vibration data were collected both in and out of the test cell. The engine was mounted to an isolated steel bedplate using heavy-duty cast iron mounting stands and fabricated steel mounting brackets. Production level vehicle engine mounting components were used throughout the testing.

The transaxle was in neutral throughout the running condition testing. The engine speed, about 750 RPM, and fluid temperatures near 190°F were monitored and maintained throughout the testing. The engine was allowed to idle during the testing and data was collected only after the engine had reached a steady-state operating condition. Measurements made on the static engine were taken immediately after the engine was stopped. Thus, the engine was close to the temperatures observed during operation.

Frequency response function (FRF) measurements were taken at many locations on the engine and transaxle. The FRF measurements were obtained with the engine in the static condition. A modally tuned impact hammer and a 50 pound modal shaker were used to apply the necessary force to the structure. Broadband random noise was used as the input to the shaker during the testing. Accelerometers were placed at strategic locations to measure the vibration spectra of a particular component or group of components. The accelerometers were also used to measure the vibration with the engine running. The static engine data and part of the running engine data were collected and processed using an HP 35670 digital dynamic signal analyzer. The remainder of the running engine data was recorded using a TEAC 21-channel VHS recorder and post-processed using a computer workstation.

Linear weighted sound pressure level (SPL) measurements were taken at the locations specified in the SAE J1074 standard. The microphone locations were 1 meter from the front, rear, left and right sides of the engine and 1 meter above the floor. The top microphone was 1 meter directly above the center of the upper intake manifold. For this testing the transaxle was masked to reduce the contribution of this component to the measured sound pressure. The data was recorded with the TEAC VHS recorder and post-processed using the computer workstation.

RESULTS

The first phase in validation was to compare absolute vibration levels in operating conditions. The results from the model are promising. As shown in figures 3 through 8, with the exception of the oil pan, the predicted velocities were accurate.

The other part of the noise path is the structural-acoustic radiation. We chose to model the reverberant SPL because our SPL measurements were made in a reverberant test facility. A subsystem was created to represent the large room in which the SPL measurements were taken. Because an SEA subsystem represents resonant energy, this measurement is more appropriate for an SEA model than SPL in an anechoic environment. At high frequencies, the two results are similar. Below 3000 Hz, however, the results vary considerably. This data appears in figure 9.

In addition to the reverberant sound pressure level, the SEA model can also predict energy flow paths and individual power contributions. Figure 11 shows the software's prediction of some contributions to the radiated sound power.

CONCLUSIONS

Our work thus far in engine modeling has shown promising results. Some parts of the structure require more attention than others, most notably the oil pan. In this case, the discrepancy between model and test could be caused by incorrect damping values, which in that particular subsystem were estimated values. It is also possible that a power source exists near the oil pan, such as the oil sump system, that was not included in the model.

It has been established that the oil pan can be a primary noise source, and this may explain the disagreement between the predicted and measured radiated SPL below 3000 Hz. To test this hypothesis, the measured velocity was enforced on the oil pan subsystems. We then again compared the reverberant SPL to measured data. The results are shown in figure 10. The remaining difference in SPL may also be due to an incorrect estimation of the acoustic damping in the reverberant room and contributions from engine accessories such as the water pump and alternator, which were not modeled. While measures were taken to reduce accessory and belt noise in the direct field, these components would still be present in the reverberant field. Future testing would involve a more painstaking method either to eliminate tonal accessory contributions in the measured data, or to include accessories in the SEA model. Measured accessory noise could be included as an acoustic power input to the reverberant room and measured vibration as an input to the engine block.

Some of the subsystems display odd behavior above 6300 Hz in the velocity comparisons. This may occur because our input power was measured up to 6400 Hz, and extrapolated to higher frequencies. Overall, the model is performing quite well, especially in the prediction of panel velocities. Future work to improve the model will be to include gasket effects, in-plane energy transfer, and possibly an additional power input at the oil pan. The SEA software also allows the use of measured mobilities to specify the subsystem modal density. This can help to improve correlation at lower frequencies. This technique can be applied to other stiff structures as well, such as compressors, pumps, transmissions, and superchargers.

REFERENCES

1. Lalor, N. and G. Stimpson. SEA Models the Engine Parts FEM Cannot Reach, AJA, 1992

2. Beranek, Leo L. and István Vér. Noise and Vibration Control Engineering, John Wiley and Sons, New York, 1992

3. Lyon, Richard H. and Richard G. DeJong. Theory and Application of Statistical Energy Analysis, Butterworth-Heinemann, Boston, 1995

4. Vibro-Acoustic Sciences, Ltd. AutoSEA Training Course, Version 3.1. 1995

5. Wu, Lie and A. Ågren. Applicability of the Experimental Statistical Energy Analysis to an Engine Structure, Proceedings of the SAE Noise and Vibration Conference, 1995

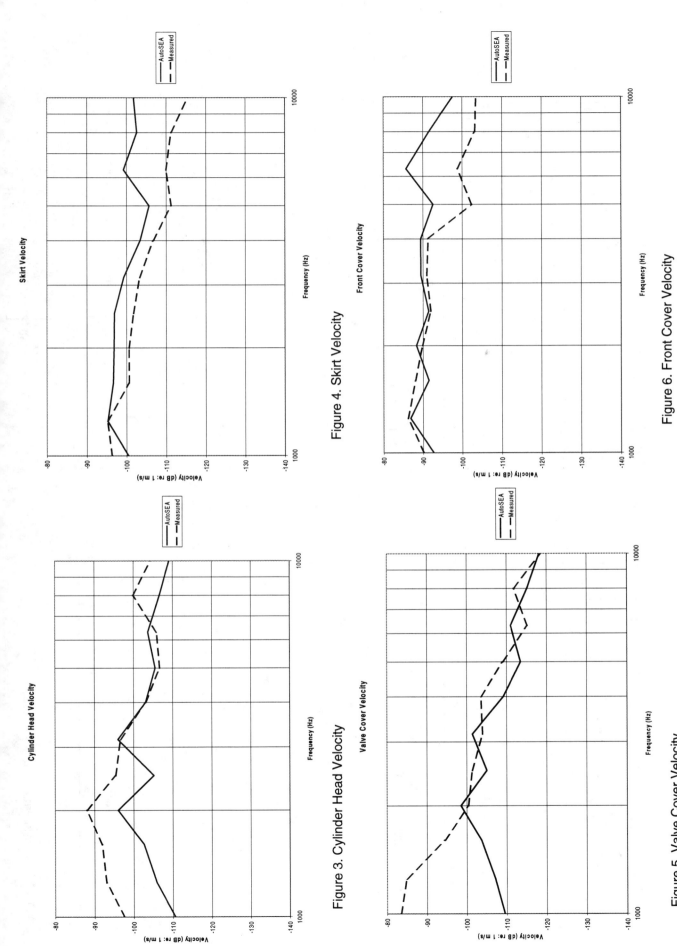

Figure 3. Cylinder Head Velocity

Figure 4. Skirt Velocity

Figure 5. Valve Cover Velocity

Figure 6. Front Cover Velocity

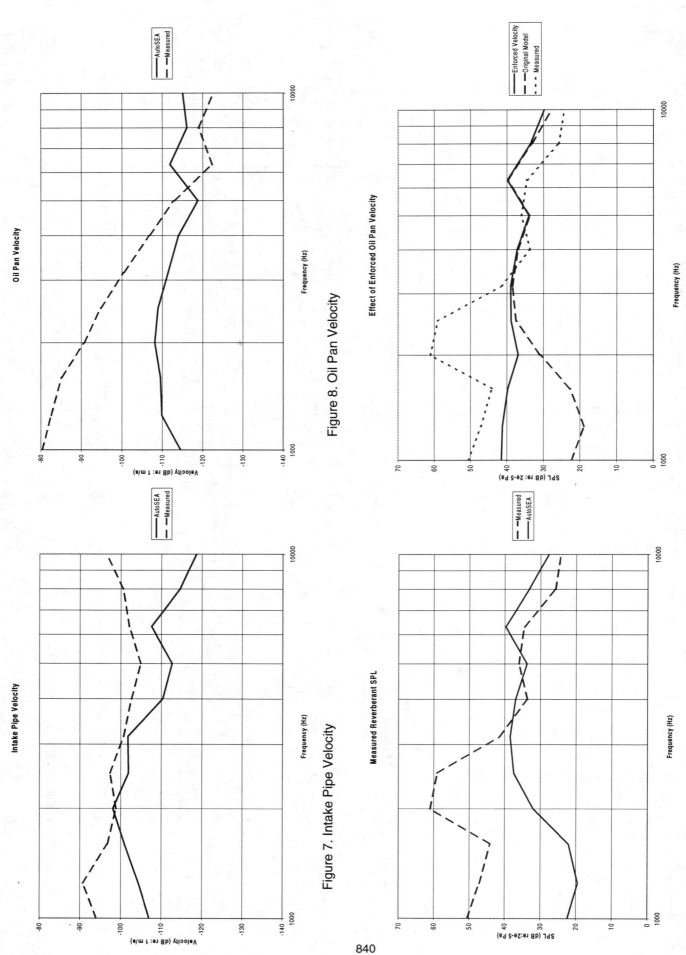

Figure 7. Intake Pipe Velocity

Figure 8. Oil Pan Velocity

Figure 9. Reverberant SPL

Figure 10. Effect of Enforcing Oil Pan Velocity

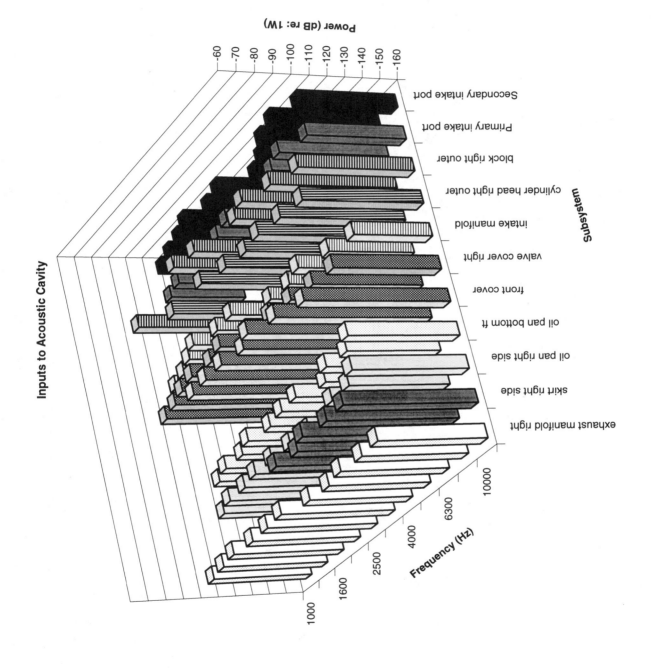

Figure 11. Power Inputs to Reverberant Acoustic Cavity

971976

A Standardized Scale for the Assessment of Car Interior Sound Quality

Rudolf Bisping, Sönke Giehl, and Martin Vogt
S.A.S. Systems

Copyright 1997 Society of Automotive Engineers, Inc.

ABSTRACT

The paper describes the development of a scale for the assessment of car interior sound quality. The scale had to meet three demands:
- Standardization, i.e. the scale values of different sound events should be directly comparable to each other.
- Economy; i.e. the use of the scale should be time efficient.
- Validity, i.e. the scale should be suitable for a precise distinction between sounds of different quality.

A rating scale based on centile standards was constructed enabling the economic and specific measurement of four basic qualities: pleasantness, power, brightness and impulsiveness. In a series of experiments the scale was applied to a representative sample of interior sounds. The results show that the scale differentiates indeed precisely between the individual vehicles. This gives a reliable data base for the selection of reference sounds for optimization purposes.

INTRODUCTION

Optimizing the quality of vehicle interior noise goes far beyond a simple reduction of annoying noise components. Especially in non-stationary driving situations (e.g. acceleration), interior noise should be such that passengers can perceive it as pleasant, attractive or even powerful. This means for purposes of practical application that it is no longer only a question of interior noise reduction in the classical sense (see Klingenberg, 1988), but also a question of the specific and unique characteristics of the sound. The reason for this conclusion - which is seemingly contradictory to noise reduction efforts - is that interior noise does not only have the negative features of an acoustic stress factor, but also a specific signal character for the passengers (see Bisping, 1995). Accordingly, optimization of noise quality also means optimizing the signal characteristics of the noise, with the aim of improving the driver's communication with his vehicle and thereby ensuring a positive influence on his driving behavior by acoustic feedback. The signal characteristics of interior noise should also be acoustically attractive in order to be accepted by the driver and the passengers. This aim leads to the following questions:

- Which features must interior noise have in order to be attractive?

- How can these features be measured in a reliable and valid way?

In the present paper, findings of laboratory investigations are presented that offer at least some answers to these questions, with the aim of showing a economical way of finding a sustainable solution. A representative set of interior car sounds will be experimentally analyzed by comparing an absolute scaling method and a relative one. Both scales consist of several subscales representing various sound quality attributes. For the absolute scale no external comparison stimulus is needed. For the relative one, each sound is compared with a standard stimulus. Both scales have to meet three demands:

1. The scale values should be directly statistically comparable with each other, i.e. the scales are standardized (**standardization**).

2. It should be possible to use the scales in an economical and time-efficient manner (**economy**).

3. The scales should reflect the main features of interior noise quality and should be suitable for precise distinction between different sounds (**validity**).

DEVELOPMENT OF THE SCALE

Recording and playback of experimental sounds

The interior noise of 17 different vehicles were recorded inside the car by means of artificial head technology (HMS 1, HEAD acoustics, Aachen) on R-DAT. Seven 4-cylinder, eight 6-cylinder and two 8-cylinder vehicles were used for the test. The RPM signal was recorded simultaneously with the audio signal for subsequent analysis. The sounds were recorded within 13-15 seconds, at a speed range of 600 to 6000 RPM in the second gear. For relative scaling by means of magnitude estimation, an interior noise was selected as a comparison stimulus (anchor) that had been judged as being neutral in previous tests. Each pair of sounds (anchor stimulus, experimental stimulus) was announced as "pair number one, two, etc.". For absolute scaling by means of categorical partitioning no anchor is needed, therefore the experimental sound was announced as "stimulus number one, two, etc.". The sounds were acoustically presented through free-field equalized headphones.

Subjects

N = 20 men and women took part in the experiments. The average age was 27 years.

Standardization of the absolute scale

The following sound attributes were initially chosen.

Table 1: List of initially chosen sound attributes

> pleasant
> racy
> booming
> bright
> strong
> rough
> loud
> high-pitch
> knocking
> whining
> powerful
> impulsive

During the first phase of scale development it was analyzed whether the twelve attributes could be assigned to certain basic dimensions. For this purpose, twelve individual scales subdivided into categories (Heller, 1985) were made up for all attributes. Fig. 1 shows the example of the "loudness" subscale.

With this type of scaling, the subject first decides on one of the seven categories that best suits the impression of the noise that he/she has at the moment. Then the subject selects the finer classification within the chosen category. Scaling was carried out in the acoustics laboratory of S.A.S. Systems. During one session, a group of six to eight people scaled the experimental sounds simultaneously.

Fig. 2 shows one subject during the experiment.

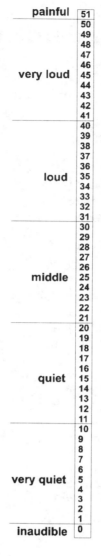

Figure 2: Scaling of experimental sounds in the laboratory

Each attribute was individually assessed for all seventeen vehicles, i.e. all in all, 12(subscales) x 17(vehicles) = 204 assessments were carried out. The factor analysis of the data showed that "pleasant", "booming", "loud" and "whining" form one statistical cluster because of their high loadings and that they belong to the common psychometric factor of "pleasantness". "Racy" and "powerful" are also connected and form the "power" factor. The features "bright" and "high-pitched" are closely connected and express the different perceptions of "brightness". "Impulsive" and "knocking" are also correlated and express the perception of impulse sounds, mainly due to the engine, such as the knocking of diesel engines at low speed. All four factors explain approx. 73% of variance.

Figure 1: Scaling of loudness subdivided into seven categories

Summarizing, it can be said that based on factor analysis, the interior noise quality of the vehicles can be assigned to four basic dimensions:

- **Pleasantness**
- **Power**
- **Brightness**
- **Impulsiveness**

The next step was to choose a reduced list of attributes that describes all relevant aspects of sound quality but is more economical than the original list. A maximum of two attributes was chosen for each factor, only those with loadings higher than 0.70. The following attributes were selected:

Table 2: Reduced list of sound attributes

> pleasant
> loud
> racy
> powerful
> bright
> impulsive

In order to compare them statistically, the six subscales were normalized. A direct comparison is not possible for the raw values (RV) of the scales, since they have different mean values and standard deviations and differ in their statistical distribution. The standard values (SV) resulting from normalizing have the advantage that their statistical distribution follows the form of a normal distribution. This is a prerequisite for carrying out parametric inference-statistical comparisons, e.g. analyses of variance. In addition, the individual scales can be directly compared on the basis of identical means, standard deviations and distribution forms.

Standardization was carried out as follows: For a given subscale, each RV of a subject was transformed to a percentage value in the range between 0 and 100%, describing its relative position within the sample of all RV. Thus, based on the normal distribution function, statistical z-values can be achieved, since a given z-value corresponds to an individual percentage rating (McCall, 1939). Then the resulting z-values were transformed to scale values from 0 - 10 by means of the equation $SV = 5 + 2z$ (the so-called centile scale, see Lienert & Raatz, 1994). In order to remain in the scale range between 0 and 10 all values bigger than the threshold value of 2.5 or smaller than -2.5 were set equal to the corresponding threshold value. Having a set of normally distributed data, this happens very rarely. In the following, the scale will be called CSQ scale. Starting from the four extracted factors of pleasantness, power, brightness and impulsiveness, the scale has six sub-scales. There are two sub-scales for pleasantness and power each, and one for brightness and impulsiveness.

EXPERIMENTAL COMPARISON OF ABSOLUTE AND RELATIVE CSQ-SCALES

In a second experiment it was tested whether measurements with a relative scale by means of magnitude estimation (see Borg & Staufenbiel, 1993; Stevens, 1975) would lead to similar results, or if there were significant deviations from data gained with absolute scaling. The selected features were scaled according to table 2. The experiment took place five days after the first one, with the same sample of subjects.

Other than with the absolute scaling the subject is asked to rate the apparent intensity of various experimental stimuli relative to a fixed standard stimulus (anchor). The anchor is always presented first, to which the number 100 is assigned. Then the experimental stimulus is presented. The subject is asked to assign a number to the experimental stimulus which expresses the ratio between the magnitude of sensation due to the experimental stimulus and the sensation produced by the anchor. For example, if the experimental stimulus appears to be twice as pleasant as the anchor the number 200 would be the correct assignment. If the experimental stimulus appears half as pleasant as the anchor the number 50 would be appropriate. Each attribute was individually scaled for all seventeen vehicles, i.e. all in all 6(subscales) x 17(vehicles) = 102 judgments were given. The raw values of the magnitude estimations were converted into standard values in the same way as described above.

Results

In table 3, the results of the analyses of variance for all 6 subscales are compared. The data show that the vehicles differ highly significant in all subscales[1]. In all cases, the p-values cannot be mathematically distinguished from zero.

Table 3: F and p values of the subscales

absolute CSQ-scale		
attribute	F (16, 304)	p
pleasant	38,52	< 0,001
loud	59,54	< 0,001
racy	13,06	< 0,001
powerful	13,46	< 0,001
bright	6,01	< 0,001
impulsive	22,26	< 0,001

relative CSQ-scale		
attribute	F (16, 304)	p
pleasant	18,23	< 0,001
loud	58,55	< 0,001
racy	11,24	< 0,001
powerful	13,84	< 0,001
bright	11,92	< 0,001
impulsive	27,40	< 0,001

---

[1] The F values describe the ratio of systematic variance - due to mean value differences between the vehicles - and the error variance. The corresponding degrees of freedom for vehicle variance and error variance are indicated in brackets. The higher the F value, the smaller the probability p that the difference between the vehicles is accidental.

The highest F values on both scales (absolute, relative) occur for the two "pleasantness" features ("pleasant", "loud") and for "impulsiveness". This means that the distinction between the vehicles was particularly easy for the subjects with regard to these subscales. Differentiation as to the two "power" subscales ("racy", "powerful") and "brightness" was obviously more difficult, as indicated by the smaller F values. The fact that they are still highly significant means that the vehicles can be clearly distinguished with regard to "brightness" and "power".

Fig. 3 shows the position of the 17 vehicles in the four-quadrant scheme of "pleasantness" and "power" on the basis of the absolute scale. For positioning of the vehicles in the diagram, for each individual car the mean of the two "pleasantness" subscales and the two "power" subscales was taken and one single scale value was calculated. Due to the standardization of the scales a value of 5 is the mean value of a scale; i.e. values greater than 5 are above the average and values lower than 5 are below the average. The figure shows that only 6-cylinder and 8-cylinder vehicles are in the first quadrant ("pleasant"/ "powerful") of the scheme. In the second quadrant ("pleasant"/ "not powerful") vehicles of all three groups are represented. The third quadrant ("unpleasant"/ "not powerful") has only 4-cylinder cars. In the fourth quadrant ("unpleasant"/ "powerful") two 4-cylinder and one 6-cylinder motor sports car can be found.

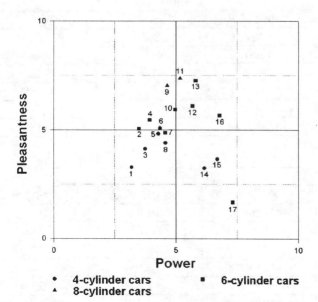

Figure 3: Four-quadrant scheme of "pleasantness" and "power" based on absolute judgments

Fig. 4 shows the result of relative scaling. The mean of the "pleasantness" and "power" subscales was also taken and one single value was formed. Fig. 5 shows a close correspondence between the absolute and the relative scale.

Fig. 6 shows the four-quadrant scheme for "brightness" and "impulsiveness" for the absolute scale. Fig. 7 shows the corresponding diagram for the relative scale.

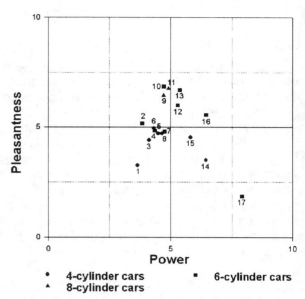

Figure 4: Four-quadrant scheme of "pleasantness" and "power" based on relative judgments

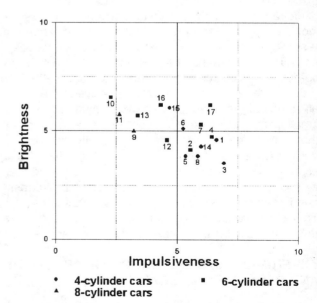

Figure 5: Four-quadrant scheme of "brightness" and "impulsiveness" based on absolute judgments

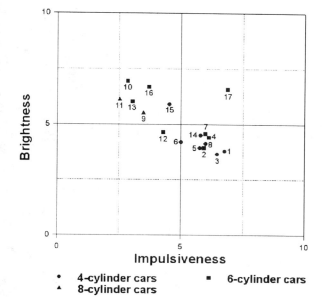

Figure 6: Four-quadrant scheme of "brightness" and "impulsiveness" based on relative judgments

Figure 7 shows the correlation between the absolute and the relative CSQ-scale values for all four subscales.

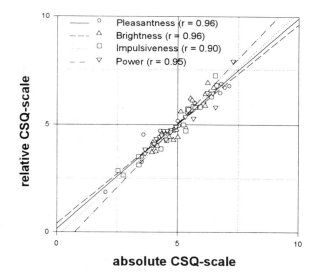

Figure 7: Correlation between the absolute and the relative scale values

For "pleasantness" the correlation between the two scale types is r = 0.96. For "power" the correlation is r = 0.95. Correlation of the two other scales is high, too. For "brightness", the correlation is r = 0.96 and for "impulsiveness" r = 0.90.

CONCLUSION

The analyses of variance revealed that the vehicles differ significantly with regard to the measured attributes. This proves that the two methods (absolute CSQ scale, relative CSQ scale) distinguish extremely well between the individual vehicles. The four-quadrant scheme of "pleasantness" and "power" (Fig. 3 and 4) underline that 6-cylinder and 8-cylinder vehicles differ greatly in their position from 4-cylinder vehicles. The latter are mostly described as being unpleasant, whereas 6-cylinder and 8-cylinder cars (with the exception of the sports car) are considered as comparatively pleasant. The vehicles in the first quadrant are most interesting, since they combine both desirable basic qualities ("pleasantness" and "power"). They are therefore of special interest as reference vehicles for optimization of interior noise. The four-quadrant scheme of "brightness" and "impulsiveness" (Fig. 5 and 6) also show major differences between the individual vehicles and the three vehicle categories. There is high correlation between the two types of scales (absolute, relative) with regard to all four basic features ("pleasantness", "power", "brightness" and "impulsiveness"). It can therefore be concluded that both methods reliably describe the same basic features of sound quality. Due to their standardization, both methods provide statistically comparable values. It is, for example, possible to compare the scaling of "pleasantness" directly with the scaling of "power". This is a great advantage as compared to the usual methods that are only based on raw values.

REFERENCES

Bisping, R. (1995). Emotional Effect of Car Interior Sounds: Pleasantness and Power and Their Relation to Acoustic Key Features. In: Society of Automotive Engineers, Inc. (Ed.), Proceedings of the 1995 Noise and Vibration Conference, Vol. 2 (pp. 1203 - 1209). Warrendale: SAE.

Heller, O. (1985). Hörfeldaudiometrie mit dem Verfahren der Kategorieunterteilung (KU). Psychologische Beiträge, 27, 478 - 493.

Klingenberg, H. (1988). Automobil-Meßtechnik. Band A: Akustik. Berlin etc.: Springer.

Lienert, G., A., & Raatz, U. (1994). Testaufbau und Testanalyse. Weinheim: Beltz Psychologie Verlags Union.

McCall, W., A. (1939). Measurement. New York.

Stevens, S. S. (1975). Psychophysics. New York: Wiley.

971977

Engine Sound Quality in Sub-Compact Economy Vehicles: A Comparative Case Study

Brian Chapnik and Brian Howe
HGC Engineering Ltd.

Copyright 1997 Society of Automotive Engineers, Inc.

ABSTRACT

A comparative study was undertaken to investigate engine noise in sub-compact vehicles under typical highway operating conditions.

Baseline acoustical testing was performed. Typical sound quality metrics indicated that passenger cabin noise levels and sound quality were similar among all vehicles tested. However, poorer subjective sound quality in a vehicle equipped with a 3-speed automatic transmission was related to a higher degree of perceived periodicity than in other models.

Further study of the 3-speed vehicle indicated that installing a 4-speed transmission would reduce noise more effectively than any substantial changes to the vehicle structure or engine mounting system. Damping or barrier treatments applied to the firewall or toe pan did not provide significant benefit.

INTRODUCTION

At highway speeds, engine noise in automobiles is usually of secondary concern to road noise. This is because the engine is normally operated at reduced speeds in conjunction with a high transmission gear. This may not be the case in sub-compact economy vehicles, in which lower gearing ratios are often found to compensate for lower engine capacities. This vehicle class is also characterized by light weight construction materials, which can transmit more noise between the engine compartment and the passenger cabin.

In this study, various models characteristic of the sub-compact class are tested, and the results compared. Typical sound quality metrics (loudness, sharpness, roughness) are used as the basis for comparison, as well as sound pressure spectra and signatures.

In addition, a noise control engineering study was undertaken of one vehicle, to evaluate various strategies for reducing noise transmission between the engine and the passenger cabin noise. The measurements and experimentation performed as part of this study are described herein.

COMPARATIVE DESCRIPTION OF VEHICLES

Four vehicles in all were tested, each a four-door sedan of the 1995 model year:

Car 1, having a 4-cylinder, 1.3L, 8 valve engine and a 3-speed automatic transmission

Car 2, having the same engine as Car 1 and a 5-speed manual transmission

Car 3, having a 4-cylinder, 1.5L, 12 valve engine and a 4-speed automatic transmission

Car 4, having a 4-cylinder, 1.5L, 16 valve engine and a 4-speed automatic transmission

A detailed inspection of each vehicle was undertaken to identify significant differences in the context of this investigation. The results of these inspections are summarized in Table 1.

There are some significant differences between Car 3 and the other vehicles (aside from the obvious differences in engines and transmissions). Firstly, the neutral torque axis engine mounting system in this vehicle provides a second stage structure borne noise isolation between the engine and the firewall. Although lower frequencies are unlikely to be affected, higher frequencies can be significantly reduced across the bolted connection between the supporting strut and the firewall. The overall effectiveness of this scheme of course depends on levels of structure borne noise flanking through the side mounts and air borne noise levels. Car 3 also has a firewall consisting of two sheet steel layers instead of a single layer as in the other vehicles, which would result in increased mass and stiffness, reducing both air borne and structure borne noise. Finally, the interior treatments in Car 3 are better suited to the control of noise than in the other vehicles, including the sealing of holes behind plastic trim panels and a lower dashmat barrier layer of foam instead of shoddy, the former tending to form better into cracks.

The differences between Cars 1, 2 and 4 are more subtle (again, aside from the obvious differences between engines and transmissions). Car 1 has two aft mounts supporting the powertrain from the firewall, while the other vehicles have only one. The fact that Car 2 is equipped with the same engine as Car 1 indicates that the second mount may be required to balance off-centre loading from the automatic transmission. The only other significant difference between the vehicles appears to be the dashmat, which is heavier and more extensive in Cars 1 and 2, and thus likely provides better noise reduction.

COMPARATIVE MEASUREMENTS

Sound measurements were performed in all vehicles under the following conditions:

- Inside the passenger cabin, with the transmission in Park (or Neutral in the case of Car 2), at engine speeds of 2000, 2500, 3000, 3500, and 4000 rpm.

- Inside the passenger cabin while the car was driven on a newly surfaced freeway at speeds of 80, 90, 100 and 110 kph, with the transmission in both Drive (D - 3rd gear) and Overdrive (OD - 4th gear), where applicable. In Car 2, the 4th and 5th gears were compared to Drive and Overdrive, respectively. All vehicles were driven at constant speed on neutral grades.

- Inside the passenger cabin while the car was driven on a newly surfaced freeway at speeds of 80, 90, 100 and 110 kph, with the transmission in Neutral and the engine turned off, to obtain road noise contributions at these speeds.

- Outside the engine compartment, at a distance of approximately 0.6 m from the centreline of the engine head, with the hood up and the transmission in Park (or Neutral in the case of Car 2), at an engine speed of 3000 rpm.

The measurements were performed using a single 1/2" condenser microphone located near the driver's right ear. The signal from the microphone was recorded on Digital Audio Tape (DAT), and post-processed on a Personal Computer.

The results of the benchmark sound testing are summarized in Table 2. The table is divided into four sections, each representing a different aspect or measure for all tests. The first section represents the measured engine speed for each test, in rpm; these speeds may be considered accurate within ±5%. The second section represents the overall measured Zwicker loudness for each test (all inside the passenger cabin except for the measurement outside the engine compartment as described above), expressed in dBphon. The third section represents the roughness, expressed in asper, which is essentially a measure of the amplitude modulation of the sound signal, weighted to correspond to the human perception of roughness. As a rule of thumb, roughness values below 0.1 asper are considered negligible, and above 0.5 asper are considered significant. The fourth section represents the sharpness, expressed in acum, which indicates the relative amount of perceived high-frequency content contained in the signal. Sharpness values above 1 acum may be considered significant.

The algorithms used to obtain the required sound quality metrics from the raw data can be found in the literature. These algorithms were implemented in a series of digital signal processing modules, which were applied to the stored raw signals on a Personal Computer. Algorithms for Zwicker loudness may be found in [1] and [2]. The calculation of Sharpness is also described in [2]. The calculation of roughness is as described in [3] and [4], although the correlation between adjacent critical bands as described in [4] has been neglected. Publications [5] and [6] describe the calculation and interpretation of these and other related metrics, as well as other factors contributing to the subjective perception of noise by humans.

As indicated in Table 2, Car 1 is operated on the highway at higher engine speeds than the other vehicles, even when the other vehicles are operated in Drive instead of Overdrive. This appears to have only negligible impact on the loudness inside the passenger cabin during on-road testing at most highway speeds (except at 110 kph, where Car 1 is 2-3 dB louder than in the other vehicles operated in Overdrive or 5th gear). The reason for this is apparent given that the overall loudness levels generated inside the passenger cabin by the engine during stationary testing only begin to approach road noise levels as the engine speed exceeds 3500 rpm. That is, under most highway operating conditions, irrespective of the gear and speed of travel, road noise is the dominant factor for this vehicle class, and will mask most of the engine noise in all vehicles tested. In fact, at most highway speeds, only the lowest engine orders make any significant contribution to the spectral loudness, especially if driven in Overdrive. Figure 1 illustrates loudness spectra (expressed in sones vs. critical band rate) at 110 kph for the different operating conditions tested in each vehicle.

However, even though most of the total sound energy may be due to road noise, there may still be elements of the engine noise which are audible and thus contribute to the overall subjective perception of noise in the vehicle. Analysis of discrete sound signatures illustrates that, in all vehicles tested, the second engine order typically dominates the engine noise contribution at highway speeds. The higher the frequency of this order (i.e. the higher the engine speed), the higher its perceived periodicity, which reduces its overall pleasantness [6]. As the engine in Car 1 is operated at substantially higher speeds during highway travel, due to its lack of an overdrive gear, the second engine order occurs at a higher frequency, which reduces its pleasantness. In Car 1 there are also some higher orders (between 8th and 10th) which contribute to perceived periodicity of the engine and corresponding reduced pleasantness. As illustrated in Figure 1, the second engine order contributes to a high loudness value near a critical band rate of 2, and higher engine orders contribute to the spectrum near a critical band rate of 4. In the other vehicles, the engine's contribution to the spectrum is not as pronounced, or occurs at a lower frequency (e.g. Car 4 in Overdrive).

The roughness and sharpness metrics do not appear to provide any valuable information for on-road tests. The tire/pavement interaction dominates the overall sound spectrum, and since amplitude modulation associated with such noise is essentially random, the measured roughness is practically constant at all speeds. Moreover, as broadband road noise predominates over primarily low frequency powertrain noise as speed is increased (assuming that the transmission gears up accordingly), the measured sharpness generally increases with speed.

The stationary tests indicate that no vehicle can clearly be considered loudest at all engine speeds. Car 3 appears to be slightly quieter over most of the range, but produces large low order tonalities at engine speeds above 3500 rpm (which is typical of highway operation in Drive).

The calculated roughness values for the stationary tests indicate that amplitude modulation does not play a significant role. The sharpness values, although fairly low in absolute terms given that engine noise is dominated by low frequencies, indicate that there is less high frequency noise in Car 3 than in the other vehicles at equivalent loudness levels. Stationary tests outside the engine compartment indicate that all engines produce almost identical airborne noise levels and spectra.

Some limited vibration testing was performed across the engine mounts of the vehicles equipped with automatic transmissions, in Park and at 3000 rpm.

Calculated mount efficiencies for most engine mounts indicated minimum values between 400 and 700 Hz, particularly the dual aft mounts in Car 1. In Car 3, the longitudinal strut from which the neutral-torque-axis mounts are supported appears to reduce the transmission of structural vibrations to the chassis at all but the lowest frequencies; however, near the fundamental engine order (i.e. 3000 rpm), the strut amplifies vibration.

NOISE CONTROL INVESTIGATION IN CAR 1

Results of the benchmark tests indicated that Car 1, the model with the 3-speed transmission, has a subjectively noisier engine than the other models when operated at highway speeds, despite the fact that the primary sound quality indicators are masked by road noise. Second engine order noise appears to be the dominant contributing factor, although higher orders are also audible. At the high engine speeds at which Car 1 is operated on the highway, these orders contribute to a higher degree of perceived periodicity, and poorer subjective sound quality, than in competing vehicles with 4-speed transmissions.

For this model, additional engineering tests were performed to identify the most important substructures associated with the radiation of engine noise into the passenger cabin, as only by treating primary transmission paths and/or radiating surfaces can perceptible noise reduction be achieved. In this context, the potential to reduce engine noise through treatments applied to the firewall and/or toe pan was evaluated.

VIBRATION TESTS - Vibration measurements on various surfaces were performed with the vehicle in Park, at various engine speeds. Figure 2 illustrates average vibration levels on the various surfaces during stationary testing at 4000 rpm.

These tests indicate that at low frequencies, levels on all surfaces are similar, except at the second engine order at which there appears to be some amplification on the windshield and instrument panel. At mid-band frequencies near 500 Hz, where some higher orders are identifiable in the on-road sound signature, levels on the firewall and toe pan far exceed those on supported structures. Of course, the dashmat and carpet serve to somewhat reduce radiated noise at these frequencies, although experience suggests that they are not massive enough or decoupled enough from the radiating steel to be very effective at these frequencies. Vibration levels at high frequencies (above 1000 Hz) are still highest on the firewall, but the dashmat and carpet are likely very effective in this range, and high frequency noise radiated by the windshield and windows may dominate.

These data indicate that treatments applied to the firewall and toe pan will likely only affect mid-band and, to some degree, high frequency noise. Low frequency engine orders, which dominate the loudness spectrum, are equally present as vibrational energy on the supported structures such as the windshield and instrument panel. These structures are much better sound radiators than the firewall, being undamped and uncovered, and are in closer proximity to the ears of the driver and passenger.

INTENSITY TESTS - Sound intensity measurements along a number of vectors were performed at three representative locations in the front section of the passenger compartment (driver side / centre / passenger side). Near field scans were also performed over various radiating surfaces, including the windshield, instrument panel, side windows, and "knee plane", an imaginary surface between the instrument panel and front seats used to estimate the effective sound radiation from the firewall / floor pan area.

The results of these measurements clearly indicate the following trends:

- Low frequency sound energy (250 Hz and lower) is radiated in approximately equal amounts to the occupants of the passenger compartment from all of the nearest surfaces, including the windshield, instrument panel, side windows and roof. There is no strong directionality to this sound energy, and absolute values of sound intensity vary dramatically with location. This suggests that structural modal responses dominate in the low frequency range.

- All surfaces except the side windows continue to contribute in the mid-frequency range (315 to 800 Hz), but energy radiated up through the knee plane is equally significant.

- At high frequencies (1000 Hz and up), practically all of the energy is radiated from the instrument panel and through the knee plane, in approximately equivalent quantities.

These findings are consistent with the vibration tests, and indicate that treatments applied to the firewall and floor pan will only affect the mid- and high- frequency noise components. Moreover, the intensity data illustrates that the degree of benefit that these treatments may offer is marginal, given that other surfaces radiate a significant amount of noise in all frequency ranges.

DAMPING TESTS - An indication of the existing damping on the firewall and toe pan was obtained by tapping the structures with a small steel-tipped hammer and measuring the resulting vibration decay, digitally time-filtered into octave bands.

These results indicated average structural damping factors of between 0.02 and 0.04 in all octave bands, with a mean value of approximately 0.03. These values indicate that substantial damping already exists on the firewall and toe pan, due likely to joints, penetrations, and supported structures (i.e. the instrument panel), as well as to the existing applied damping treatments and dashmat. Undamped steel would exhibit damping factors on the order of 0.001, while highly damped steel panels with well designed viscous damping treatments might exhibit damping factors on the order of 0.05.

Thus it is unlikely that additional damping treatments applied to these areas would provide much additional benefit.

PROTOTYPE NOISE CONTROL MEASURES

To validate the noise control predictions, some prototype noise control measures were installed in Car 1. Additional damping material was applied to the firewall, toe and floor pan, expandable polymeric foam was used to fill the void areas of the frame at the bottom of the A-pillars and along the side rails, and some heavy barrier material was applied between the existing dashmat and the carpet to cover all accessible areas of the firewall, toe and floor pan up to the front seats. The results of these measures are described below.

DAMPING - Constrained layer damping material, consisting of a 1 mm thick damping sheet faced with an 0.13 mm thick aluminum layer, was applied to the firewall, toe pan and floor pan in areas where melt mats were not applied. Approximately 0.3 m^2 of this material was used.

Sound testing indicated virtually no difference in sound levels between this case and the untreated case, over the entire frequency range of interest.

SEALANTS - Cavities in the front chassis area were investigated, as they seemed to be undamped and resonant. On each side of the car, there is one cavity below the A-pillar, running from the firewall back to the front door, and a second forming the rail which runs directly below the front door. These cavities were completely filled with polymeric expandable foam.

Sound testing indicated very little difference in sound levels between this case and the untreated case. There did appear to be a slight reduction of some orders above 2000 Hz, but these contribute only negligibly to the overall noise spectrum, and are completely masked by road noise at highway speeds.

ADDITIONAL BARRIER - Finally, some heavy barrier material was added to accessible areas of the firewall, toe and floor pans. A heavy mat was used, consisting of a loaded vinyl layer weighing approximately 75 Pa (1.6 psf) with a 6 mm foam decoupling layer and mylar surface. The barrier was adhered to the existing dashmat or directly to the frame. Just over 1 m^2 of the material was used, weighing a total of 85 N (20 lbs.).

Sound testing indicated no appreciable noise reductions at low frequencies. However, there was a reduction in mid-frequency peaks near 500 Hz, and some reduction of high frequency energy above 1000 Hz. Figure 3 illustrates the discrete sound levels at one location in the cabin (driver's right ear) before and after treatment, with the car in Park and the engine operating at 4000 rpm. The average overall loudness in the cabin has been reduced by just over 1 dB (note this applies to the stationary condition only; loudness at highway speeds, which is dominated by road noise contributions, is not affected).

A 1 dB reduction in loudness is not perceptible to the average human hearing. However, the reduction in level of the higher harmonics reduces the overall perception of the composite engine noise. The high cost associated with adding a barrier or improving the dashmat to incorporate its features (i.e. heavier barrier layer, broader extent) does not likely warrant this subtle improvement.

CONCLUSIONS

Benchmark testing of four sub-compact 4-cylinder vehicles indicated the following areas of significance:

- All vehicles tested generate more noise and vibration at higher engine speeds.

- At typical highway speeds, the 3-speed model operates at higher engine speeds than the 4-speed or 5-speed models, even when the 4-speed models are operated in 3rd gear or the 5-speed model in 4th gear. This leads to higher perceived periodicity and poorer subjective sound quality in the 3-speed model, despite the fact that primary sound quality metrics do not indicate that significant differences exist between the models (see Table 2). Thus the vehicles with extended gearing ratio transmissions are subjectively quieter at typical highway speeds, especially when operated in the overdrive gear with which they are equipped.

- Stationary testing indicates that all engines generate comparable levels of noise when operated at similar engine speeds. None of the vehicles is clearly quieter inside the passenger cabin at all engine speeds, although Car 3 is marginally quieter and exhibits less high frequency engine noise at most typical engine speeds, due probably to its neutral torque axis engine mounting system, which provides a second stage of engine vibration isolation.

- The engine of Car 1 is supported by two rear engine mounts from the firewall and floor pan. Transmissibility of engine vibration through these mounts is highest among the vehicles tested.

Engineering studies of Car 1 indicated that additional damping treatments would have negligible benefit. This was borne out by experiment. Additional sealants used to fill hollow cavities in the frame were also found to have negligible benefit.

Barrier materials applied to the firewall, toe and floor pan were predicted to have only a marginal effect on the overall engine noise in the passenger cabin. Prototype testing of an additional heavy barrier between the existing dashmat and carpet illustrated that a modest improvement in noise could be obtained, reducing some mid- and high frequency orders in the engine noise signature and reducing the overall loudness level by approximately 1 dB. This reduction in loudness would not be perceptible to most individuals, but the change in sound character due to the reduction in higher frequency harmonics may make engine noise less noticeable in the passenger cabin. Adding a secondary barrier, or incorporating its features into the existing dashmat by making it heavier and more extensive, may not be cost effective to provide these marginal benefits.

Much of the engine noise impacting the passenger cabin in Car 1 appears to be structure-borne through the engine mounts, and re-radiated by all surfaces according to the modal response of the entire structure. An improved mounting system may provide perceptible reductions in engine noise transmitted to the cabin. By relocating and re-configuring engine support structures, lower cabin noise may be achieved without any significant cost implications.

As the 3-speed model typically generates more engine noise than the 4-speed or 5-speed models when operated on the highway, some reduction in engine noise transmitted to the passenger cabin may be warranted. However, as road noise dominates the spectrum, as in all tested sub-compact vehicles, small reductions may provide a sufficient improvement. In any case, standard sound quality metrics do not appear to yield helpful results, and jury testing may be warranted.

REFERENCES

[1]   Zwicker, E., Fastl, H., Dallmayr, C., "BASIC-Program for calculating the loudness of sounds from their 1/3-oct band spectra according to ISO 532B", Acustica, Vol.55 (1984), 63.

[2]   Zwicker, E., Fastl, H., Psychoacoustics: Facts and Models, Springer-Verlag, Berlin, 1990.

[3]   Fastl, H., "Roughness and Temporal Masking Patterns of Sinusoidally Amplitude Modulated Broadband Noise", in Psychophysics and Physiology of Hearing, Academic Press, London (1977), 403.

[4]   Aures, von W., "A Procedure for Calculating Auditory Roughness", Acustica, Vol. 58 (1985), 268.

[5]   Genuit, K., Gierlich, H.W., "Investigation of the Correlation between Objective Noise Measurement and Subjective Classification", SAE Paper 891154.

[6]   Schiffbanker, H., Brandl, F.K., Thien, G.E., "Development and Application of an Evaluation Technique to Assess the Subjective Character of Engine Noise", SAE Paper 911081.

ABOUT THE AUTHORS

Brian Chapnik and Brian Howe are principals of HGC Engineering, an engineering consultancy specializing in Noise, Vibration and Acoustics. Their clients include automakers and automotive parts suppliers, as well as large industrial facilities and commercial developments. For automotive clients, HGC Engineering has performed testing and analysis in relation to product improvements and improved quality assurance, design of acoustical test facilities, design of measures to reduce plant environment noise impact, assessment and control of plant interior noise levels, machine vibration analysis and design of foundations for sensitive equipment.

For further information, contact us at:

HGC Engineering
2000 Argentia Road, Plaza One, Suite 203
Mississauga, ON
Canada   L5N 1P7

Telephone: (905) 826-4044
Facsimile: (905) 826-4940
e-mail: info@hgcengineering.com
        or visit http://www.hgcengineering.com

TABLE 1: Inspection Results

Car	Engine/ Trans.	Engine Mounts	Firewall	Dashmat	Damping	Other
Car 1	4-cyl., 1.3L, 8 valve, 3-speed automatic	2 side mounts, one circular rubber supporting engine, one rubber donut supporting transmission. 2 aft mounts, one circular rubber supporting engine, bolted to toe pan, one dual-orbit circular rubber supporting transmission, bolted to floor pan.	1 layer thick, welded to toe pan near bottom of dash.	2 layer composite, 2 mm thick medium weight vinyl barrier bonded to 12 mm thick shoddy. Cast molded to firewall shape, extending into toe pan, bolted at various locations.	Extensible damping sheets, baked on in toe and floor pan areas.	Carpeted throughout, with 12 mm shoddy below everywhere except at dashmat. Rubber hood gasket seal. Stiff head liner and door panels, not absorptive.
Car 2	4-cyl., 1.3L, 8 valve, 5-speed manual	2 side mounts, one circular rubber supporting engine, one rubber donut supporting transmission. One aft mount, circular rubber, supporting engine and transmission, bolted to toe pan.	1 layer thick, welded to toe pan near bottom of dash.	2 layer composite, 2 mm thick medium weight vinyl barrier bonded to 12 mm thick shoddy. Cast molded to firewall shape, extending into toe pan, bolted at various locations.	Extensible damping sheets, baked on in toe and floor pan areas.	Carpeted throughout, with 12 mm shoddy below everywhere except at dashmat. Rubber hood gasket seal. Stiff head liner and door panels, not absorptive.
Car 3	4-cyl., 1.5L, 12 valve, 4-speed automatic	Neutral torque axis design. 2 side mounts, circular rubber, one supporting engine, one supporting transmission. Fore and aft mounts, circular rubber, supporting engine, connected to lengthwise support strut which is bolted to frame through rubber grommets.	2 layers thick, welded together, each layer having different relief except at penetrations.	3 layer composite, 2.5 mm thick medium weight vinyl barrier bonded to 6 mm thick shoddy and 6 mm thick foam. Cast molded to firewall shape, extending into toe pan, bolted at various locations.	Extensible damping sheets, baked on in toe and floor pan areas.	Carpeted throughout, with 12 mm shoddy below everywhere except at dashmat. Ungasketed holes behind plastic trim covered with closed-cell neoprene foam with PSA. Rubber hood gasket seal. Stiff head liner and door panels, not absorptive.
Car 4	4-cyl., 1.5L, 16 valve, 4-speed automatic	2 side mounts, circular rubber, one supporting engine, one supporting transmission. One aft mount, circular rubber, supporting transmission, bolted to toe pan.	1 layer thick, fairly thin gauge, welded to toe pan at level of mid-dash.	2 layer composite, 1.5 mm thick light vinyl barrier bonded to 12 mm thick shoddy. Not cast molded, extending to top of toe pan, bolted at various locations.	Extensible damping sheets, baked on in toe and floor pan areas.	Carpeted throughout, 6 mm of shoddy below in toe pan area where dashmat stops, otherwise no underlay apparent. Rubber hood gasket seal. Stiff head liner, door panels part cloth.

TABLE 2: Benchmark Sound Test Results

Test Name	RPM (+/- 5%)				Loudness (dB phon)				Roughness (asper)				Sharpness (acum)			
	Car 1	Car 2	Car 3	Car 4	Car 1	Car 2	Car 3	Car 4	Car 1	Car 2	Car 3	Car 4	Car 1	Car 2	Car 3	Car 4
Park 2000	2000	2000	2000	2000	74	73	71	76	0.21	0.23	0.19	0.15	0.79	0.82	0.62	0.64
Park 2500	2500	2500	2500	2500	82	76	75	78	0.14	0.21	0.21	0.16	0.56	0.78	0.63	0.67
Park 3000	3000	3000	3000	3000	78	77	79	77	0.21	0.23	0.18	0.22	0.81	0.90	0.68	0.85
Park 3500	3500	3500	3500	3500	81	79	79	81	0.21	0.24	0.22	0.19	0.83	0.93	0.80	0.84
Park 4000	4000	4000	4000	4000	86	81	86	83	0.19	0.24	0.13	0.18	0.75	0.96	0.61	0.81
Eng Compartment	3000	3000	3000	3000	104	104	104	105	0.24	0.25	0.28	0.26	1.98	2.00	1.83	1.95
80 kph: coast	NA	NA	NA	NA	84	85	85	83	0.25	0.26	0.26	0.27	0.74	0.74	0.76	0.81
80 kph: OD/5th	NA	2210	1850	1860	NA	86	86	85	NA	0.25	0.26	0.26	NA	0.71	0.74	0.74
80 kph: D/4th	2940	2700	2710	2820	86	85	86	84	0.25	0.24	0.24	0.26	0.75	0.73	0.72	0.76
90 kph: coast	NA	NA	NA	NA	85	85	86	86	0.26	0.26	0.26	0.24	0.76	0.74	0.79	0.81
90 kph: OD/5th	NA	2475	2080	2100	NA	87	87	85	NA	0.25	0.26	0.26	NA	0.75	0.74	0.76
90 kph: D/4th	3225	3000	3015	3190	87	87	88	88	0.24	0.25	0.26	0.26	0.79	0.77	0.76	0.8
100 kph: coast	NA	NA	NA	NA	86	86	87	85	0.26	0.27	0.27	0.27	0.77	0.78	0.81	0.85
100 kph: OD/5th	NA	2740	2310	2325	NA	88	87	88	NA	0.26	0.26	0.25	NA	0.77	0.77	0.75
100 kph: D/4th	3630	3300	3390	3525	88	89	88	87	0.26	0.24	0.27	0.25	0.84	0.77	0.81	0.77
110 kph: coast	NA	NA	NA	NA	88	88	87	89	0.26	0.27	0.28	0.26	0.8	0.79	0.87	0.9
110 kph: OD/5th	NA	3040	2535	2590	NA	89	88	89	NA	0.27	0.26	0.25	NA	0.83	0.80	0.81
110 kph: D/4th	4035	3675	3690	3890	91	90	88	91	0.26	0.26	0.26	0.27	0.78	0.80	0.82	0.85

Authors' Note: These results are based on a limited sample population and a limited number of experiments, and were used only to indicate trends. No statistical analysis was performed to evaluate confidence limits.

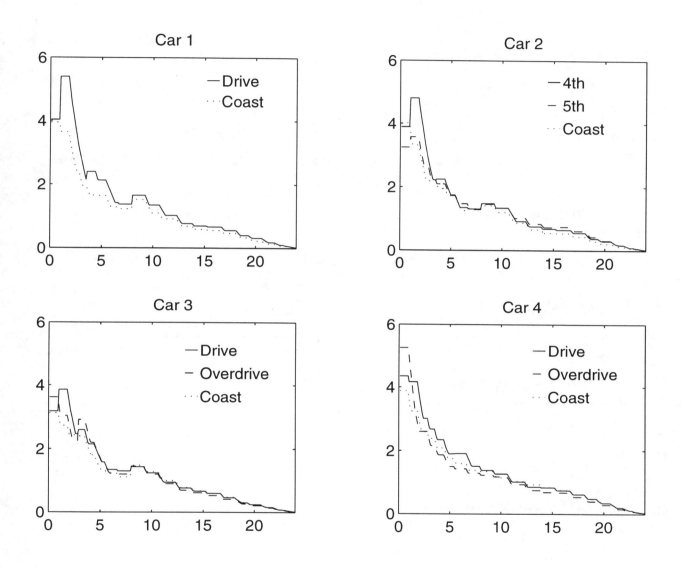

**FIGURE 1: Comparison of Loudness Spectra, 110 kph
(Sones vs. Critical Band Rate)**

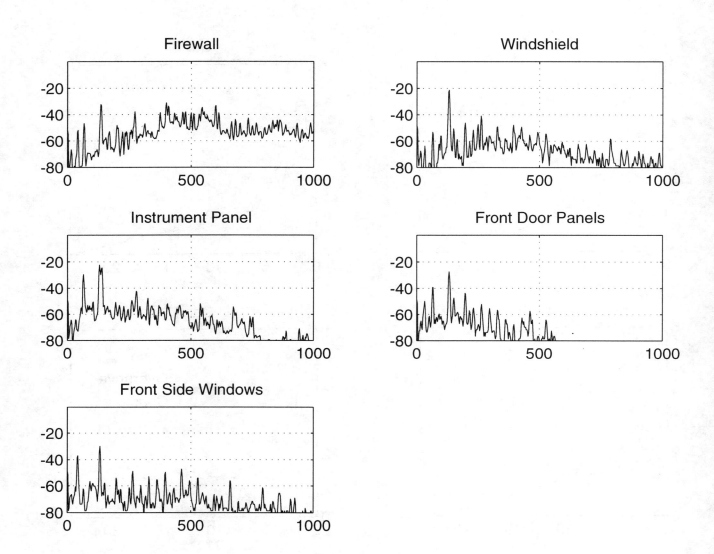

**FIGURE 2: Surface Vibration in Car 1, 4000 rpm in Park
(Vibration level in dB re g vs. Frequency in Hz)**

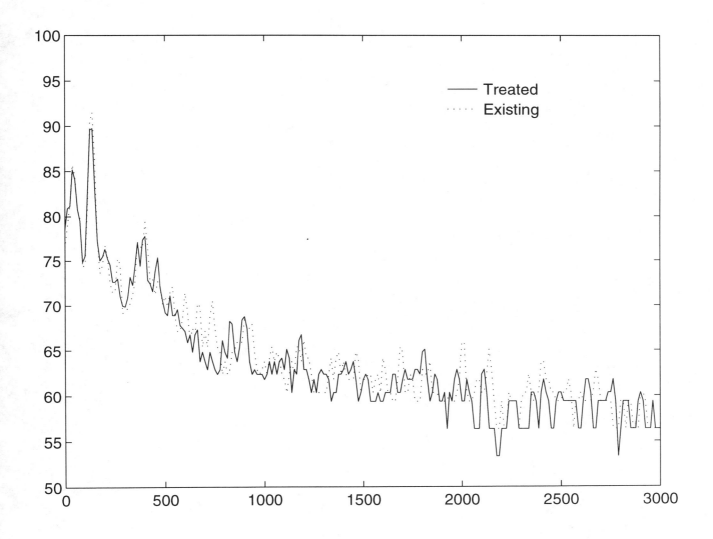

**FIGURE 3: Noise in Car 1, 4000 rpm in Park
(Sound Pressure Level vs. Frequency in Hz)**

971978

Binaural "Hybrid" Model for Simulation of Engine and Wind Noise in the Interior of Vehicles

K. Genuit and N. Xiang
HEAD acoustics GmbH

Copyright 1997 Society of Automotive Engineers, Inc.

ABSTRACT

In the development process of products there is at present only one approved method for the "prediction" of sound quality: It is the subjective evaluation by hearing or by binaural playback of Artificial Head recordings. This method is possible in the prototype stage at the earliest. Due to the fact that the development processes in industry become shorter and the tasks - especially in the fields of acoustics and vibrations - become increasingly complex, there exist strong requests for time- and cost saving prediction of the expected sound quality. For this purpose a method can be used that is based on binaural recordings and the determination of vibro-acoustical characteristics of components on test facilities. The combination with a (binaural) transfer path database allows a simulation that describes the effects of modifications of the particular characteristics or transfer paths on the resulting sound situation for the listener. By using a binaural playback system, the simulation can be evaluated subjectively and - consequently - the resulting sound quality can be predicted. The development of the described procedure for the simulation of engine and wind noise in the interior of vehicles is presented in this paper.

INTRODUCTION

The quality of noise in the interior of vehicles is becoming increasingly significant. This applies not only to top-of-the-range, sophisticated vehicles, but also to vehicles in less expensive price classes. The trend to even shorter development phases for new car models continues, accompanied by an extension of the model range. Previously it was only possible to assess sound quality after producing a prototype. This means, prior to series production only relatively little time remained for optimization measures. This background gave rise to the idea of developing new options. The aim was to allow reliable predictions about the probable sound quality, not only in terms of numbers, but also in terms of acoustic representation.

Within the scope of the presented research project, designated AQUSTA (Improvement of the Structural Acoustic Quality of Transportation Vehicles Using Simulation Techniques of Binaural Analysis), several approaches have been developed, tried and tested, towards achieving the aurally-accurate, binaural simulation of noise created in the interior of a vehicle by wind and engine. The presented model is based on measurements made with a vehicle or engine on a dedicated test rig. It includes the relevant transmission paths of the airborne and structure-borne sound components to the ears of a person sitting on driver's seat and takes into account triaxially vibrational excitation up to 2 kHz from the engine, engine stiffness and the reaction from the vehicle body. Airborne sound components are included at a limit frequency of 8.5 kHz. An extension to the complete audio frequency range is also possible. The application of modified transmission elements, as realized in other calculation programs, or also through measurements, allows for both objective and subjective predictions about expected vehicle interior sound quality based on headphone monitoring.

The model was evaluated for a low-cost front-wheel drive vehicle. Good results for the wind noise component have already been demonstrated. For the engine noise components, the simulation model results are very much in the range of scatter resulting from comparative measurements at several vehicles of the same model /1/. Further improvement of the simulation can be expected by including noise components from drive train, wheel suspension and exhaust system.

AERODYNAMIC NOISE SIMULATION

Aerodynamic noise becomes a dominant noise component above a certain speed, the limit of which depends on the type of vehicles and environmental conditions. In the laboratory, the aerodynamic noise of a vehicle can be investigated in a wind tunnel under definable conditions. Under the experimentally verified hypothesis that different panels are incoherent sources for aerodynamic external noise, a masking procedure /2/ has been applied on the test vehicle in order to estimate transfer paths of different panels. Using this masking procedure, binaural recordings at the driver or co-driver position in the interior of the vehicle can be used for estimating noise transfer paths of different panels. Fig. 1 illustrates the masking procedure used in a wind tunnel for binaural recordings.

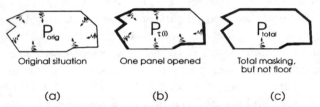

Fig. 1: Masking procedure for separating transfer paths of aerodynamic noise with removal of panel(s).

In the complex aerodynamic noise situation with transmission through all panels as shown in Fig. 1(a), the binaural sound signal P_{orig} is recorded. The masking procedure is then applied to each individual panel from the interior of the vehicle. The recording of $P_{T(i)}$ (Fig. 1(b)) is performed one after another for each panel under test. The rest signal of total masking P_{total} (Fig. 1(c)) has also to be recorded since the floor part of the vehicle can not be easily masked.

These binaural recordings corresponding to the given panels are used to generate an estimation of the magnitude of panel frequency response functions. The magnitude of the individual transfer function (expressed here in the frequency domain) is estimated by:

$$H_{C(i)} = \sqrt{\frac{|P_{T(i)}|^2 - |P_{total}|^2}{|P_{orig}|^2}}. \quad (1)$$

In similar fashion, the magnitude of the unmasked floor transfer function (here for the m-th panel) can be estimated by

$$H_{C(m)} = \frac{|P_{total}|}{|P_{orig}|}. \quad (2)$$

Once the magnitude of these transfer functions is estimated, the corresponding minimum-phase impulse responses $h_{C(i)}(t)$ of FIR filters can be straightforwardly achieved by the HILBERT transformation.

From experimental studies of wind tunnel recordings it turns out that lower frequency components, below a certain frequency, do not show significant differences between individual panels. For this reason, a separation between these two frequency ranges is performed for the aerodynamic noise simulation as follows:

$$P_{simu}(t) = P_{orig}(t) * h_{low}(t) + \sum_{i=1}^{m} h_{C(i)}(t) * \left[P_{orig}(t) * h_{high}(t) \right], \quad (3)$$

where $h_{low}(t)$, $h_{high}(t)$ stand for IIR impulse response of low pass and high pass filter, respectively.

Eq. (3) implies a useful simulation strategy when the difference between one panel currently used and another panel to be exchanged is characterized in terms of the transparency index. In effect, the difference in the sense of the transparency index can be used to construct a minimum-phase impulse response $h_{\tau(i)}(t)$, so that the simulation of exchanging n-th panel is straightforwardly performed using:

$$\begin{aligned}P_n(t) = & P_{orig}(t) * h_{low}(t) + \\ & h_{\tau(n)}(t) * h_{C(n)}(t) * \left[P_{orig}(t) * h_{high}(t) \right] + \\ & \sum_{i=1}^{n-1} h_{C(i)}(t) * \left[P_{orig}(t) * h_{high}(t) \right] + \\ & \sum_{i=n+1}^{m} h_{C(i)}(t) * \left[P_{orig}(t) * h_{high}(t) \right]. \end{aligned} \quad (4)$$

The difference of panel transparency indices used for exchanging the front-door glass is depicted in Fig. 2. Here, the transparency index can be defined as ratio of the transmitted energy to the incident energy of a panel or structure. The general concept of the simulation procedure is illustrated in Fig. 3. A physical modification of one or several panels can then be simulated in terms of FIR filtering by knowing the transparency indices or transmission loss of the panels.

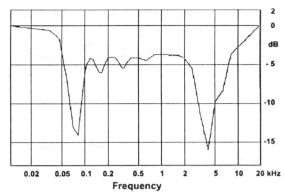

Fig. 2: Difference of the transparency indices between the currently used glass and to be exchanged glass.

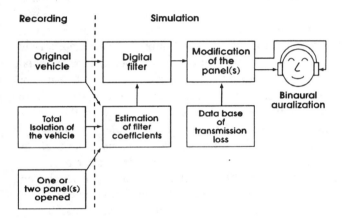

Fig. 3: Hybrid model for aerodynamic noise study in terms of binaural recording and simulation.

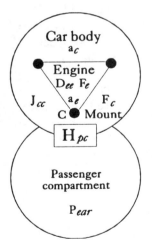

Fig. 4: Schematic description of a vehicle with a three-points suspension system.

ENGINE NOISE SIMULATION

Structure-borne noise generated by the engine, amongst others, excites passenger-compartment panels that radiate sound in the interior of a vehicle. Significant transfer paths of this structure-borne sound involve that of the engine suspension system, of vehicle structure and the airborne paths in the interior. Fig. 4 shows a schematic description of a vehicle including the suspension system of the engine. For the purpose of structure-borne sound simulation, the transfer properties of engine mounts, of the car body particularly between the mounting points of the engine mounts have to be determined /3/. The basic relations between force **F** and acceleration **a** on both sides of the engine suspension are well-established /4/:

$$\begin{pmatrix} \mathbf{F}_c \\ \mathbf{F}_e \end{pmatrix} = \begin{pmatrix} \mathbf{C}_{cc} & \mathbf{C}_{ce} \\ \mathbf{C}_{ec} & \mathbf{C}_{ee} \end{pmatrix} \begin{pmatrix} \mathbf{a}_c \\ \mathbf{a}_e \end{pmatrix} \quad (5)$$

and

$$\mathbf{a}_c = \mathbf{J}_{cc} \mathbf{F}_c \quad (6)$$

where indices c, e represent the car body and engine side respectively. Matrices **C** are defined as engine suspension transfer functions. The car body transfer functions between the mounting points of the engine suspension are represented by the matrix $\mathbf{J}_{cc}$. All elements of force vectors **F**'s, acceleration vectors **a**'s, matrices **C**'s and the matrix $\mathbf{J}_{cc}$ are of triaxial characteristics which should be carefully defined.

The artificial head measurement system can be used to determine the resulting sound inside the car. In addition to this, the transfer functions $\mathbf{H}_{pc}$ between the selected mounting points of the engine mounts on the car body and the ear canal entrance of a 'driver' or 'co-driver' in the interior of the vehicle have to be determined using the artificial head for the purpose of simulation:

$$\mathbf{P}_{ear} = \mathbf{H}_{pc} \mathbf{F}_c \quad (7)$$

where vector $\mathbf{P}_{ear}$ denotes the two sound pressure signals at both ear canal entrances. It is obvious /4/ that

$$\mathbf{P}_{ear} = \mathbf{H}_{pc} \left[\mathbf{I} - \mathbf{C}_{cc} \mathbf{J}_{cc} \right]^{-1} \mathbf{C}_{ce} \mathbf{a}_e \quad (8)$$

which represents the basic relations between acceleration signals on the engine side and the sound pressure signals at the ear canal entrances of a passenger /5/.

In addition to measuring sample of transfer functions, the structure-borne input signals associated with $\mathbf{a}_e$ and individual input airborne sound signals have to be synchronized recorded using an appropriate multi-channel system. In similar fashion, the structure-borne sound arising from other parts of the vehicle can also be simulated. For the simulation of airborne sound, airborne transfer functions between the ear canal entrance and sound sources must also be determined. A first approach is given using three microphones: one at the intake side, one at the exhaust side and the third one at the gearbox side. For the complex superposition of these three binaural signals it is important to consider the coherence between these microphone signals. The sound signals corresponding to the engine orders have a nearby 100% coherence whereas all the other signals are uncorrelated. Another way to analyze the airborne sound components is to subtract all simulated structure-borne sound components from the artificial head recording of total sound events in the passenger compartment.

In the simulation procedure, some representative modifications of certain transmission paths in terms of **signal processing** can be conveniently performed as schematically shown in Fig. 5. For example: If a modified engine mount should be used, the acoustical behavior can be evaluated subjectively by binaural listening using the new calculated dynamic stiffness of the modified engine mount for the simulation. Such a procedure would mean a very rapid and appropriate method of improving sound quality in the interior of vehicles.

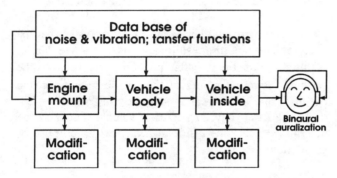

Fig. 5: Hybrid model for engine noise study. The data base of noise and vibration is collected by multi-channel recordings and that of transfer functions can be both measured and analytically modeled.

EXPERIMENT

The test vehicle was a front-wheel driven car of the lower middle range. Its engine suspension is a three-points system. For the aerodynamic study, the binaural recordings were conducted in a wind tunnel using the masking procedure. The separating frequency of $h_{low}(t)$, $h_{high}(t)$ in eqs. (3-4) for this vehicle is approximately 250 Hz. In order to demonstrate the agreement of binaural simulation results with binaural recordings in the original situation, a comparison between P_{simu} and P_{orig} in eq. (3) in terms of a 3nd-octave frequency analysis is illustrated in Fig. 6 for one experimental case. Extensive results from psychoacoustic A-B comparisons /1/ also confirmed satisfactory agreements between responses to $P_{orig}(t)$ and $P_{simu}(t)$ in eq. (3) and between recordings and simulations of individual panels in eq. (4).

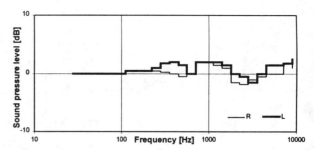

Fig. 6: Difference in level between the simulation results and the original binaural recording of aerodynamic noise.

For the engine noise study, basic recordings were made during acceleration in third gear in a semi-anechoic chamber of our laboratory. A multi-channel system equipped with synchronized 32 channels was used in the recording. Inside the engine compartment, three microphones were mounted at appropriate positions for picking up airborne sound. Two artificial heads were put on the driver and co-driver seat. For the three engine mounts, 9 acceleration signals were measured with 3 triaxial accelerometers on the engine side and on the car body side respectively. Additionally, three airborne sound signals and four ear canal entrance signals were recorded under a precise synchronization. The A/D converter had a resolution of 14 bits, the frequency range for structure-borne sound was limited up to 2.5 kHz for each channel.

All transfer functions associated with matrix elements of **C**'s, **J**$_{cc}$ and **H**$_{pc}$ were determined. A frequency range between 20 Hz and 2 kHz was covered with 1 Hz frequency resolution. To measure the matrix **C** elements (equation 5), a specially designed test-rig has been used as shown in Fig. 7.

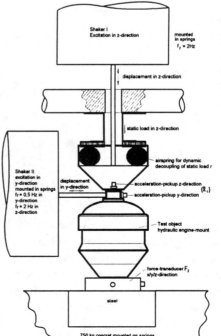

Fig. 7: Test rig for three-directional measurements of dynamic stiffness with static preload.

Both, dynamic stiffness and the influence of the static load on vibrational characteristics of mounts are presented in Fig. 8.

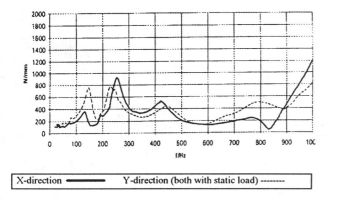

Fig. 8a: Measurement of dynamic stiffness at engine mount

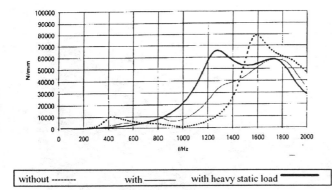

Fig. 8b: Influence of static load on engine mounts' dynamic stiffness (z-direction)

Static load of engine mounts was added to be as close to that found in reality. Great care had to be given to reduce the nonlinearity as much as possible. Measurements of J_{cc} matrix elements were carried out in the semi-anechoic chamber. They were determined between acceleration sensor and force sensor signals mounted at the engine mount points in the car body. The engine was dismounted during the measurements. Care had to be given for simulating the static load corresponding to the engine. In similar fashion, the matrix elements H_{pc} were determined between sound pressure signals picked up by the artificial head microphone and the excitation force signals. Note that at the present stage the measurement procedure of the different structure-borne transfer functions are all limited in the frequency range between 20 Hz and 2 kHz. The binaural recordings and airborne transfer functions measurements were performed within the frequency range up to 8.5 kHz. The simulation was carried out using the basic modeling relations mentioned above. Among them, a pseudoinverse of complex-valued matrices /6/ and a linear convolution procedure /7/ had to be applied. Binaural samples resulting from the simulation were not only used for a number of objective analysis, but also for binaural auralization, allowing the possibility of gaining an impression of how the vehicle will sound in the reality.

Fig. 9 shows the result of the simulation of the interior sound in comparison to the original sound with respect to the 2nd and 4th order.

Fig. 10 shows the total spectrum in dependency on speed of the simulated sound in comparison to the original sound. To estimate the quality of the simulation it is necessary to consider that this model does not consider the transfer path of tire suspension and exhaust system on the one hand and on the other hand the original sound is also influenced by some noise produced from the chassis dynamometer.

SUMMARY

The hybrid modeling of aerodynamic and engine noise is particularly important in automotive engineering because the effect of physically changing one parameter in a vehicle can be predicted at minimal time-cost without reassembling a model of the complete system. The model described here, i.e. the simultaneous measurement of airborne and structure-borne sound signals using a multi-channel recording and simulation has been developed to a user-friendly tool which will be commercially put at the disposal of the acoustic engineer in the near future. The additional possibility of subsequent digital signal processing also allows both objective and subjective predictions about sound quality in the interior of vehicles which brings the objective of higher acoustic comfort a further step nearer.

ACKNOWLEDGMENTS

The present work was funded by the European Union under BRITE EURAM. The authors would like to express their acknowledgment to the European Commission for supporting this BRITE AQUSTA project, particularly to Mr. Andrieu and Mr. Kruppa of the EEC Coordinators. They are also very grateful to Dr. Bohineust of PSA Peugeot Citroen, Dr. Rehfeld of Saint Gobain Vitrage for the excellent cooperation.

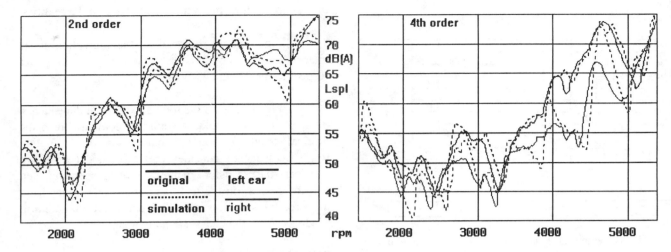

Fig. 9: 2nd (left) and 4th (right) order of left and right ear on the driver seat, comparison of original and simulated sound.

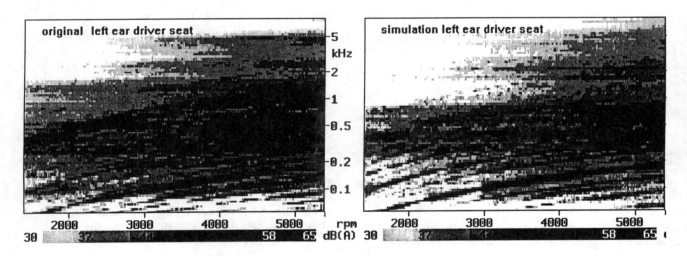

Fig. 10: 12th octave spectrum during acceleration, left ear driver seat.

REFERENCES

/1/ Synthesis Report AQUSTA: Improvement of the structural Acoustic Quality of transportation vehicles Using Simulation Techniques of binaural Analysis, December 1992

/2/ Rehfeld, M.; Bardot, A.; Bohineust, X. and Chanudet, P.: Influence des vitrages sur les composantes aerodynamiques du bruit a l'interieur de l'habitacle, presented on SIA Journées D'études: Aérodyna-mique-Aéroacoustique-Aérothermique Automobile et Ferroviaire, 5-6th Nov. 1996, Courbevoie, France.

/3/ Hanus, K.-H.: Verbesserung des Innengeräusches von Kraftfahrzeugen mit Hilfe von Elastomerbauteilen, Haus der Technik, December 1996, Essen

/4/ Genuit, K. and Xiang, N.: Combining the Artificial Head Measurement with Multi-channel Measurements for Analysing Noise and Vibration, Proc. euro-noise'95, p. 991-996, 1995.

/5/ Mas, P., Wyckert, K., Sas, P.: Indirect Force Identification based upon Impedance Matrix Inversion, A Study on statistical and deterministic Accuracy, ISMA 19 - Tools for Noise and Vibration Analysis, Leuven 1994

/6/ Golub and W. Kahan: Calculating the singular values and pseudo-inverse of a matrix, J. SIAM Numer. Anal. Vol 2, 1965, pp. 205-224.

/7/ Oppenheim and R.W. Schafer, Discrete time signal processing, Printice hall, 1989.

971979

A New Tool for the Vibration Engineer

R. C. Meier Jr., N. C. Otto, W. J. Pielemeier, and V. Jeyabalan
Ford Motor Co.

Copyright 1997 Society of Automotive Engineers, Inc.

ABSTRACT

Significant progress could not have been made in the Sound Quality area without the invention and development of engineering tools. For the sound engineer, the binaural recording head is a primary example of one of those tools. The use of the binaural recording head was crucial to the development of the sound characterization process and has become an essential tool in the Sound Quality areas in Ford Motor Company. A similar tool, The Ford Vehicle Vibration Simulator, has been developed for the vibration engineer. The vehicle vibration simulator (VVS) is unique, consisting of vibration of the vehicle seat (6 degrees of freedom), steering wheel (4 DOF), vehicle floorpan section (1 DOF), and the brake or accelerator pedal (1 DOF). Many vibration test systems have been developed to study human response to vibration, especially for military and space applications. To our knowledge, this is the first multi-axis, fully integrated vibration test system to be used for automotive applications. Initially, the vibration simulator has been used to study vehicle ride and truck idle quality. Vibration time histories measured on different vehicles for a rough road surface and various engine idle conditions were used for playback on the vehicle simulator. Subjective impressions from human evaluators of vehicle ride and idle quality were correlated to objective measures derived from each vibration time history. The purpose of this paper is to provide a brief description of the simulator operation, present the detailed results of these studies, and discuss potential future applications for the VVS system.

BACKGROUND

Ford has been a pioneer in the sound quality field. Several papers have been published [1,2] describing the sound quality process and methods developed by Ford Research. This process begins with the gathering of in-vehicle sound data through the use of a binaural recording head. The recordings are then used for "playback" to obtain subjective ratings as well as analyzed for objective measures. A statistical analysis is performed to determine the correlation between the subjective and objective measures. This method has provided the most significant solution to the complex problem of translating customer "wants and needs" into actual engineering specifications which can then be applied in the vehicle development process.

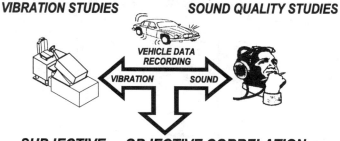

Figure 1. Vibration / Sound Quality Process

In June of 1993, funding was obtained and a project was started to develop a vehicle vibration simulator so that the sound quality process could be applied to vehicle vibration. Figure 1 shows the vibration and sound quality process. In order to apply the same process to vehicle vibration, an analogy to the recording head had to be developed.

THE FORD VEHICLE VIBRATION SIMULATOR (VVS)

Figure 2 shows the major components of the Ford Vehicle Vibration Simulator (VVS). Hydraulic shaker systems were designed and developed to provide motion to the vehicles major points of human contact, namely, the vehicle seat, steering column, and a section of the floorpan underneath the drivers feet. A relocatable actuator was incorporated in the simulator design to be used for other possible points of human contact with the vehicle interior such as, the shift lever, and the brake or accelerator pedals.

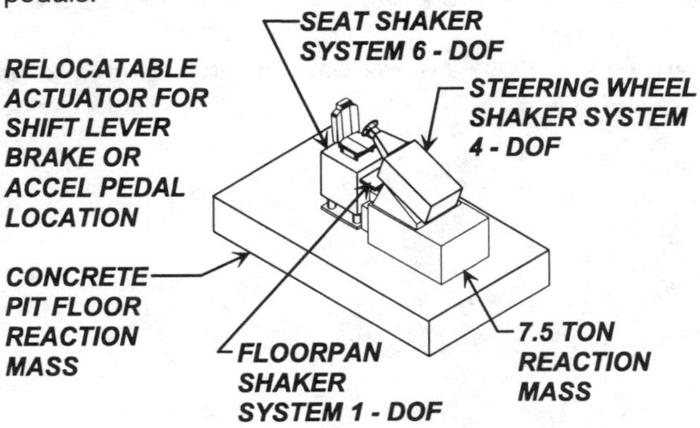

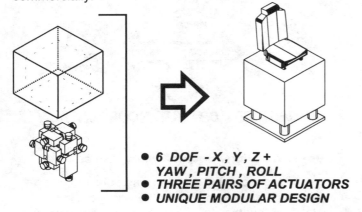

Figure 2. Major Components of VVS

SIMULATOR DESIGN - A common industrial hydraulic actuator provides a single degree of freedom motion (vertical only) for the floorpan section. However, both the seat and steering wheel hydraulic shaker systems are unique and were developed by TEAM Corporation specifically for this application. In fact, the seat shaker is actually prototype number one of a hydraulic shaker system known as the "CUBE" ™ that TEAM Corporation is now marketing commercially.

- 6 DOF - X, Y, Z + YAW, PITCH, ROLL
- THREE PAIRS OF ACTUATORS
- UNIQUE MODULAR DESIGN

Figure 3. Seat Shaker System

Figure 3 shows the unique hydraulic actuator design that was implemented for the seat shaker system. Six degrees of motion was achieved by using 3 pairs of the basic hydraulic actuators assembled inside of a hollow "cube" or block as shown in Figure 3. Each actuator pair provides both linear and rotational motion along or about a single axis as described in Figure 4. When the actuator pairs are mounted on three orthogonal axes (X,Y, Z) and their motions combined, they produce linear translation along the X, Y, and Z axis as well as the primary rotational motions of Yaw, Pitch, and Roll. The steering wheel shaker system is a modular variation of the seat shaker system. Four of the basic actuators were combined with a stationary hydraulic "pivot point' inside of a rectangular shaped block to provide a 4 DOF system for the vehicle steering wheel. Two vertical actuators combine to provide vertical motion as well as rotation about the center of the steering wheel similar to the pair described in Figure 4.

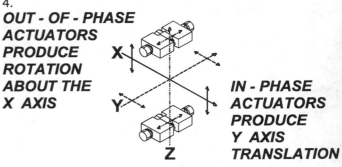

Figure 4. Actuator Motion

Two additional actuators provide the lateral and longitudinal translation motion for the steering wheel shaker system. A small compact moveable hydraulic actuator provides single axis linear motion to other possible points of contact such as the brake pedal, accelerator pedal, or gear shift lever. The entire simulator system is mounted in a three foot deep pit. The floor of this pit is actually four foot of reinforced concrete which provides a reaction inertia mass to absorb vibrations from the floorpan and seat shaker systems. The sides and bottom of this concrete mass are separated from the ground by sheets of vibration isolation material. The steering wheel shaker is mounted on a separate 7.5 ton reaction mass which is isolated from the main inertia mass (concrete pit floor) by a system of air cushions and heavy duty shock absorbers. This "isolation" provides sufficient system decoupling so that each shaker system can be operated simultaneously or independently depending on the testing requirements (hand, feet, or whole body vibration studies).

SIMULATOR CONTROL SYSTEM - A schematic diagram of the fully integrated computer systems used to control the vehicle vibration simulator and acquire vehicle road vibration data from test vehicles is shown in Figure 5. A ruggedized portable data acquisition system along with a portable PC computer is used to obtain vibration (acceleration measurements) time histories from test vehicles operated over various types of road surfaces. A PC/ISA bus type computer contains all of the signal processing cards which are connected to the simulator system actuators (12 output channels) and control accelerometers (12 input channels/32 total). This computer is connected by Ethernet to the main (Digital) host computer. This host computer contains all of the control, signal generation, and processing software. To insure the safety of the human occupant, an iterative approach is used for the actuator drive control scheme. In fact, many limits, condition checks, and features were designed into the system to help ensure the necessary safe environment required for human testing. A complete explanation of the simulator control scheme and details of the system safety features were published [3] in an earlier paper.

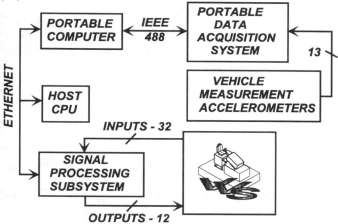

Figure 5. Schematic Diagram of Computer Systems

VIBRATION METRIC DEVELOPMENT

Prior to the development of the Vehicle vibration simulator most vibration metric development has been performed on the test track. In one such study, six (6) mid-size class vehicles were compared for ride quality on a rough road test surface (40 MPH) at the Ford Dearborn Proving Grounds (DPG). During the test track study, time history accelerometer measurements were recorded at the driver's front outboard seat track bolt (vertical and lateral) and the steering wheel 12 O'clk position (vertical and lateral) to provide data to develop objective measures. Subjective impressions were also obtained for each vehicle from human evaluators driven on the test track. Significant correlation of the subjective impressions and objective measures could not be obtained with the test track data. In the Sound Quality area, some of the reasons for the lack of correlation between subjective impressions and objective measurements has been shown to be caused by (1) lack of data repeatability with test track studies, (2) subject bias due to brand name, interior design, etc. (3) inability to perform back-to-back comparisons due to unavoidable logistics of test track studies, and (4) the subjective evaluation methods used for obtaining subjective impressions. To demonstrate the capability of the Vehicle Vibration Simulator (VVS) to overcome some of these problems, the test track study was performed on the VVS using subjective evaluation methods [2] developed in the Sound Quality area.

SIMULATOR DATA PREPARATION - Time history acceleration recordings from the six vehicles used in the test track studies were prepared for use on the simulator. The data was filtered (3 to 200 Hz) for the frequencies of interest and windowed to provide smooth beginning and ending transitions. A typical time history of seat track vertical acceleration for one of the vehicles used in this study is show in Figure 6. All six of the prepared vehicle vibration profiles were then combined randomly into one large "block" test with a 2 second pause between each vehicle to allow time for the subjective evaluations.

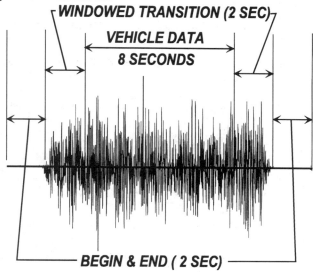

Figure 6. Typical Time History for Vehicle Data

One test was created to acclimate the subjects, and two additional tests were created so that two subjective categories could be evaluated for this experiment.

SUBJECTIVE EVALUATION METHOD - The Semantic Differential (SD) method developed in the Sound Quality area [2] was used in this study. The subjective evaluation consisted of two parts. In the first, subjects were asked to evaluate roughness for each vehicle by placing a mark on an unnumbered line labeled "slightly rough" on one end and "extremely rough" on the other. They were asked to mark the evaluation sheet during the 2 second pause between each vehicle's vibration profile "playback". In the second part of the experiment, annoyance was evaluated, again on an unnumbered line running from "slightly annoying" to "extremely annoying". As mentioned previously, the vibration profiles were presented to the subjects in random order. A total of twenty four mid-size class vehicle owners (company personnel) participated in this study. After the evaluation, a grid of numbers, running from 1 to 10, was superimposed upon the evaluation line. Thus, the subjective results were converted to a number. The results are shown in Table I, where the average rating for each vehicle is given for both parts of the evaluation. The higher the value, the rougher or more annoying the vibration. Looking at the roughness results, we can see that Vehicle C and F seem to stand out as the roughest while the other four vehicles are not significantly different. The annoyance results are somewhat confusing with only Vehicle B being significantly different than the rest (less annoying).

OBJECTIVE MEASURES AND CORRELATION

- The accelerometer measurements were examined for possible development of objective measures and statistical correlation was determined between these measures and the results of the subjective evaluations.

Table 1. Subjective Results for Rough Road Metric Development

	ROUGHNESS	ANNOYING
VEHICLE A	5.375	5.333
VEHICLE B	5.292	3.958
VEHICLE C	7.417	6.042
VEHICLE D	5.542	5.708
VEHICLE E	5.542	4.875
VEHICLE F	6.792	5.208

Objective Measures - Three possible schemes were examined to develop the objective measures for this study. In the first scheme, we simply computed the RMS values of acceleration and velocity for the vertical and lateral axes for both the steering wheel and the seat. These values seemed to correlate well with the subjective data. Vibration dose values (VDV) [4], calculated for each of the four vibration axes did not seem to correlate very well. We then decided to combine the four vibration axes by the following method. Calculating a combined magnitude for the vertical and lateral acceleration vectors, we have:

Let

$A_v(t)$ be the magnitude of the vertical acceleration as a function of time

$A_l(t)$ be the magnitude of the lateral acceleration as a function of time

where $t = 1,....,n$

Calculating the magnitude of the resultant vector for the vertical and lateral accelerations, we have a combined acceleration of:

$$A_c(t) = \sqrt{\{[A_v(t)]^2 + [A_l(t)]^2\}}$$

Calculating the RMS value of the combined acceleration, we have:

$$A_{RMS} = \sqrt{\sum_1^n \left\{ \frac{[A_c(t)]^2}{n} \right\}}$$

Calculating RMS accelerations for the seat (ST) and steering wheel (SW) for each of the six vehicles, we use the general form of the linear regression equation for subjective/objective correlation.

$$Subj. = a \cdot A_{ST}(x) + b \cdot A_{SW}(x) + c$$

Statistical Correlation - Using a two variable linear regression (combined steering wheel and combined seat RMS values as described above), we obtained a correlation coefficient of 0.93. The correlation results are shown in Figure 7.

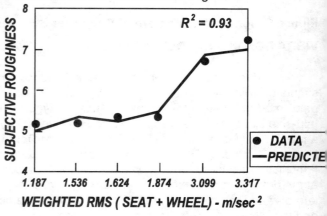

Figure 7. Correlation Results

TRUCK IDLE QUALITY STUDY

Another study, which demonstrated the simulator's ability to accurately reproduce low signal levels, involved an investigation of truck idle quality. In preparation for this study, accelerometer measurements were made on six (6) light trucks during a fully warmed-up idle condition. The measurements were conducted with the transmission in both neutral and drive. The accelerometers were placed at the seat track (six degrees of freedom), the steering wheel (four degrees of freedom - wheel rotation, vertical, lateral, and longitudinal direction in plan of wheel), and vehicle floorpan (vertical only). This placement of transducers provided the maximum allowable set of data for the current simulator configuration. The vehicle vibration data were then transferred to the simulator system's computer and preprocessed by filtering, windowing, and scaling to translate the vehicle measurement coordinate system to that of the simulator's control accelerometer locations. The RMS acceleration values of the preprocessed vehicle data (drive idle condition) for the major axes of the seat, steering wheel, and floorpan systems of the VVS are summarized in Table II. As indicated by this data, the seat average RMS levels ranged from 3 to 18 mG, the steering wheel ranged from 1.5 to 5 mG, and the floorpan ranged from 7 to 14 mG. Due to these relatively low signal levels and narrow range of values for the steering wheel and floorpan, it was decided to drop them from the study and concentrate on the seat only.

Table 2. Idle Quality Vehicle Data

AXIS	AVERAGE RMS ACCELERATION (mG)					
	VEH-1	VEH-2	VEH-3	VEH-4	VEH-5	VEH-6
SEAT						
VERTICAL	7.095	4.300	7.415	7.155	8.760	4.640
LATERAL	1.850	7.735	11.500	12.000	15.500	10.380
LONGITUDINAL	9.710	3.920	9.375	8.285	5.795	5.040
STEERING WHEEL						
VERTICAL	3.800	3.010	3.330	1.720	6.170	3.630
LATERAL	3.870	4.190	4.400	5.720	5.660	4.120
LONGITUDINAL	5.160	4.910	5.960	6.040	9.400	5.850
FLOORPAN						
VERTICAL	11.0	6.420	14.0	9.770	7.630	7.210

In actual practice, since wheel and floorpan vibrations are extremely low at idle conditions, the seat is usually the only thing that is rated for idle quality. As mentioned previously, the simulator was designed so that the systems can run independently. Time history vibration profiles were combined as in the previous study for all six vehicles in random order for the seat data only.

SIGNAL TO NOISE RATIO PROBLEM - The greatest challenge in preparing the idle quality data for subjective testing was handling the low signal levels associated with the idle vibration data. The major problem was dealing with the background noise inherent in the seat shaker system. Initial measurements indicated that the average background noise on all channels of the seat shaker system (hydraulic pump on with no drive signal to actuators) was on average about 4 to 8 mG. Subsequent testing, revealed that the background noise levels could be significantly reduced through a combination of lowering the hydraulic oil reservoir temperature and main hydraulic pump pressure. The relationship of background noise (RMS acceleration) to system pump pressure for the seat shaker system at a constant (95° F) oil reservoir temperature is shown in Figure 8.

SUBJECTIVE EVALUATION METHOD - In this study, the Semantic Differential (SD) method was also used for the subjective evaluations using a bipolar scale. However, instead of asking the subjects to mark on an unnumbered line, the line was divided into discrete choices. An example is shown below for the two attributes that were used in this study for both the neutral and drive idle conditions.

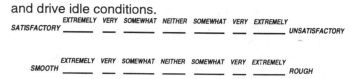

The subjective evaluations for this study consisted of two major parts. In the first, the subjects were asked to evaluate the neutral and drive idle conditions for all six vehicles against the attributes of satisfactory/unsatisfactory and place a mark on the scale shown above.

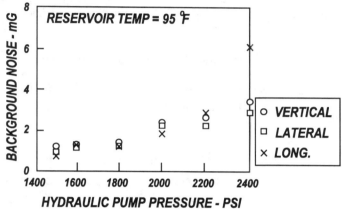

Figure 8. Effect of Hydraulic Pump Pressure

In the second part of the experiment, the idle conditions for all six vehicles were repeated and they were asked to evaluate against the attributes of smooth/rough. The vibration profiles were presented to the subjects in random order. For this experiment, a combined total of 32 subjects (company personnel who owned light truck type vehicles). Nine of the subjects were expert truck idle evaluators from one of the truck development groups at Ford Motor Company. An example of converting the subjective results to a numeric value is shown below.

The subjective results of the idle study are shown in Table III. The average ratings are shown for both parts of the evaluation. For both parts of the study, the ratings of the non-experts was not significantly different than the experts. The combined results for all subjects indicate that:

(1) For the neutral idle condition of the first part of the test (satisfactory/unsatisfactory), Vehicle 2 (-1.806) was rated the most satisfactory, and Vehicle 1 (2.845) was rated the most unsatisfactory. Vehicle order from satisfactory to unsatisfactory for the neutral idle condition was 2, 3, 5, 6, 4, and 1.

(2) For the drive idle condition of the first part of the test, Vehicle 2 (-2.065) also was rated the most satisfactory as in the neutral condition. However, Vehicle 4 (2.645) was rated the most unsatisfactory followed by Vehicle 1 (1.387) for the drive idle condition. Vehicle order for satisfactory to unsatisfactory for the drive idle condition was 2, 5, 3, 6, 1, and 4.

(3) For the neutral idle condition of the second part of the test (smooth/rough), Vehicle 5 (-1.452) was rated the most smooth followed closely by Vehicle2 (-1.323), and Vehicle 1 was rated the most rough. Vehicle order for smooth to rough for the neutral idle condition was 5, 2, 3, 6, 4, and 1.

(4) For the drive idle condition of the second part of the test, Vehicle 2 was rated the most smooth, and Vehicle 4 was rated the most rough. Vehicle order for smooth to rough for the drive idle condition was 2, 5, 3, 6, 1, and 4.

Table 3. Subjective Results for Idle Quality Study

	SATISFACTORY (-3) / UNSATISFACTORY (+3)					
	IDLE NEUTRAL			IDLE DRIVE		
	EXPERT	NON	ALL	EXPERT	NON	ALL
VEHICLE 1	2.778	2.591	2.645	1.667	1.273	1.387
VEHICLE 2	-1.556	-1.909	-1.806	-1.667	-2.227	-2.065
VEHICLE 3	-1.000	-1.091	-1.065	-0.333	-0.045	-0.129
VEHICLE 4	2.000	2.000	2.000	2.667	2.636	2.645
VEHICLE 5	-1.000	-0.818	-0.871	-1.111	-1.864	-1.645
VEHICLE 6	1.000	0.818	0.871	0.778	-0.318	0.000
	SMOOTH (-3) / ROUGH (+3)					
	IDLE NEUTRAL			IDLE DRIVE		
	EXPERT	NON	ALL	EXPERT	NON	ALL
VEHICLE 1	2.667	2.456	2.516	2.444	2.000	2.129
VEHICLE 2	-0.778	-1.545	-1.323	0.111	-1.273	-0.871
VEHICLE 3	0.000	-0.455	-0.323	0.667	0.409	0.484
VEHICLE 4	1.889	1.909	1.903	2.556	2.773	2.710
VEHICLE 5	-0.889	-1.682	-1.452	-0.333	-0.864	-0.710
VEHICLE 6	1.111	0.955	1.000	1.000	0.682	0.774

The preliminary results indicate fairly good consistency for the subjective ratings across all vehicles. Currently, the acceleration data is being examined for possible development of the objective measures for this study.

CONCLUSION

The initial studies discussed in this paper clearly demonstrate that the Ford Vehicle Vibration Simulator (VVS) has the potential to become an extremely valuable tool for the vibration engineer. It has been shown to provide the following benefits:

(1) Provide a more consistent and repeatable environment for human subjective testing of vehicle vibrations.

(2) Allow back-to-back comparison of vibration profiles which are virtually impossible with test track studies.

(3) Allow blind evaluations of different vehicles (eliminates brand bias).

We are confident that the full potential of the Vehicle Vibration Simulator will eventually be demonstrated through the application of more studies. Some of the studies we are currently planning to complete are:

- Simulation of transmission shift.

- Basic research on vibration perception

- Seat and steering wheel component evaluations and comfort studies.

- Vehicle ride and harshness evaluations for metric development.

We feel the ultimate potential of the Vehicle Vibration Simulator will be demonstrated through the "playback" of CAE vibration data generated in the early design stages of the vehicle development process. This would allow the evaluation of the ride of a new vehicle design without the enormous cost associated with building an engineering prototype. Other potential future uses of this new tool are only limited by the imagination of the engineer or scientist.

ACKNOWLEDGMENTS

The authors wish to acknowledge the efforts of extremely talented people at TEAM Corporation, especially Bill Woyski, John Davis, and Bob Tauscher who played key roles in the design and development of the Ford Vehicle Vibration Simulator (VVS). Without their combined talents, this project would not have been possible. We would also like to thank our Ford "internal customers", especially Mike Haffey, Sung-Ping Cheng, Bill Osborne, and Chuck Gray, for their help and assistance in the rough road and truck Idle metric development programs.

REFERENCES

1. N. C. Otto, G. H. Wakefield, and K. C. Wei, "A Subjective Evaluation and Analysis of Automotive Starter Sounds", Inter-Noise, (1992).

2. N. C. Otto, and B. J. Feng, "Automotive Sound Quality in the 1990's", Third International Congress on Air and Structural Borne Sound and Vibration, (1994).

3. R. C. Meier Jr., N. C. Otto, W. J. Pielemeier, and Bill Woyski, "The Ford Vehicle Vibration Simulator", Noise Con (1996).

4. M. J. Griffin, "Handbook of Human Vibration", Page 76, Academic Press Limited, London, (1990).

971980

Commercial Van Diesel Idle Sound Quality

Anthony J. Champagne and Nae-Ming Shiau
Ford Motor Co.

Copyright 1997 Society of Automotive Engineers, Inc.

ABSTRACT

The customer's perception of diesel sounds is receiving more attention since diesel engines are being used more frequently in recent years. This paper summarizes the results of a study investigating the sound quality of diesel idle sounds in eight vans and light trucks. Subjective evaluations were conducted both in the US and the UK so that a comparison could be made. Paired comparison of annoyance and semantic differential subjective evaluation techniques were used. Correlation analysis was applied to the subjective evaluation results to determine annoying characteristics. Subjective results indicated that most annoyance rankings were similar for both the US and UK participants, with some specific differences. Correlation of objective measures to annoyance indicated a high correlation to ISO 532B loudness, dBA and kurtosis in the 1.4 kHz to 4 kHz range (aimed at quantifying the impulsiveness perception).

INTRODUCTION

THE PURPOSE of this paper is to present the results of a study to obtain the most important characteristics influencing perception of commercial van interior diesel idle sounds. Paired comparison of annoyance and semantic differential subjective evaluation techniques were used. The subjective evaluations were conducted in both the US and UK so that comparisons could be made.

The semantic differential results were correlated to the paired comparison of annoyance results to determine the characteristics influencing the annoyance judgments. Objective measures were correlated to the semantic differential averages and paired comparison of annoyance merit values to find the measure best characterizing the annoyance.

SOUND RECORDINGS

Diesel idle sounds from six commercial vans and two light trucks were used in the evaluation. The sound recordings were obtained with a binaural recording head in the front passenger's seat.

PAIRED COMPARISON RESULTS

The paired comparison of annoyance sound evaluation technique was used to determine the most annoying sounds [1]. All possible pairs of the sounds were presented, and evaluators were asked to judge which of the pair they thought was more annoying. Results are compiled using a Bradley-Terry model [2] and presented as *merit values*. More annoying sounds have higher merit values. The data is then checked for validity by calculating the *repeatability* and *consistency* of the evaluators [2].

The evaluation participants in the US study (64 total) were both technical and non-technical Ford employees. Evaluators in the UK study (104 total) were owners and operators of commercial vans.

Table 1 below shows the number of subjects, average repeatability, and consistency of the paired comparison of annoyance evaluations for both the US and UK studies. The table also lists R^2, which is the Bradley-Terry model fit used to generate merit values. The repeatability, consistency, and R^2 are all reasonably high, indicating that the evaluation results are statistically valid and that people perceived definite annoyance differences among the sounds.

	US	UK
# Subjects	64	104
R^2	0.933	0.936
Repeatability	84.2%	80.2%
Consistency	0.890	0.843

Table 1. Paired comparison statistics.

Figure 1 shows the paired comparison of annoyance merit values from both the US and UK evaluations. In general, the same trend existed in both markets. However, the order of annoyance between A and B was different in the UK study than the US study. Similarly, F was ranked significantly more annoying than G in the US study, while this was not the case in the UK study.

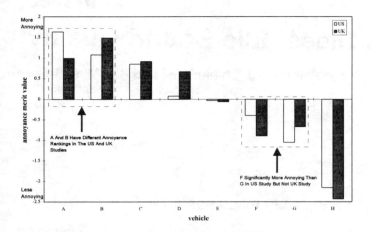

Figure 1. US and UK annoyance merit values.

SEMANTIC DIFFERENTIAL EVALUATIONS

Adjective pair descriptors were used in the semantic differential evaluation. The subject was asked to mark on a horizontal line the degree to which the adjective pair characterizes the sound. For example, the loud/quiet adjective pair was evaluated by having a horizontal line on a sheet of paper, with *loud* at the left end and *quiet* at the right end. A sound was played for three seconds, and the participant put a mark on the line indicating their judgment of that particular sound. The marks were later given a rating from 1 to 11. The average ratings for each sound were calculated separately for each adjective pair.

RESULTS - Figure 2 shows average results for both the US and UK loud/quiet ratings, with sounds A-H as labeled. The order of the sounds is from most annoying on the left to least annoying on the right, based on the respective US and UK paired comparison merit values in Figure 1. Figures 3 and 4 show semantic differential averages for the smooth/rough and impulsive/not impulsive adjective pairs.

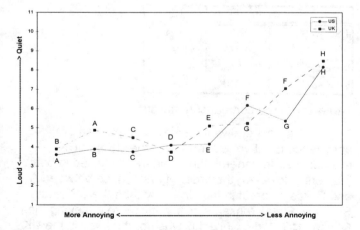

Figure 2. US & UK loud/quiet averages for sounds A-H.

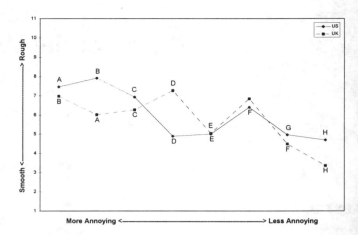

Figure 3. US & UK smooth/rough averages for sounds A-H.

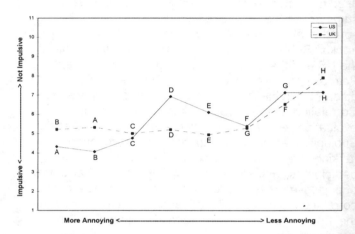

Figure 4. US & UK impulsive/not impulsive averages for sounds A-H.

Examining Figure 2, a trend clearly exists between subjective loudness and order of annoyance. A noticeable trend is also evident with order of annoyance and smooth/rough and impulsive/not impulsive.

The UK semantic differential averages were correlated to the corresponding US averages to investigate differences in responses. Figure 5 shows the correlation coefficients, r. Generally, the two markets evaluated the sounds similarly. However, the US and UK participants evaluated the sounds differently for the smooth/rough and impulsive/not impulsive adjective pairs, giving the low correlation indicated. This may be an indication that the two markets had different opinions or that they had a different understanding of what the adjectives mean.

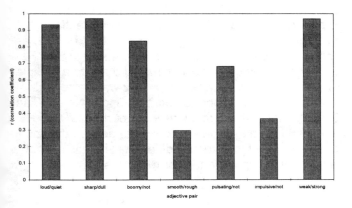

Figure 5. Correlation of US and UK semantic differential averages.

Note from Figure 1 that the order of annoyance in the two groups with A, B and F, G was different in the US and UK studies. The semantic differential results may provide information as to why the differences in the two markets occurred. US participants rated A as being rougher compared to the UK ratings for A. UK participants rated G rougher than the US rated G.

CORRELATION OF SEMANTIC DIFFERENTIAL RESULTS TO ANNOYANCE MERIT VALUES

US STUDY - The average values of the US semantic differential results were used in linear regression equations to predict the US annoyance merit values. Such analysis gives information related to the perceived contributors to annoyance. Table 2 shows the results for the US data. For example, the loud/quiet average results for the eight sounds were used in a regression equation to predict the eight US annoyance merit values. The resulting regression equation had a coefficient of determination $R^2 = 0.81$ and p-value[1] = .002. In fact, the loud/quiet adjective pair had the highest correlation to the annoyance merit values.

In addition to using a single adjective pair in a regression equation, two adjective pair descriptors were used to predict the annoyance merit values, and the results are also shown in Table 2. Loud/quiet averages combined with impulsive/not impulsive averages to predict the US annoyance merit values yielded $R^2 = 0.97$. Similar correlation was obtained when the loud/quiet averages were combined with the smooth/rough averages ($R^2 = .96$). Table 3 below explains why similar results were obtained using the smooth/rough and impulsive/not impulsive adjective pair averages. Smooth/rough averages were highly correlated to impulsive/not impulsive averages, indicating that participants may have evaluated both adjective pairs with the same criteria. Boomy/not boomy was highly correlated to pulsating/not pulsating, again indicating similar judgment criteria. The results point to the fact that loudness and impulsiveness or roughness are contributors to diesel idle annoyance. Boominess and pulsating-ness may also be factors, but the variation of this dimension among the sounds was not large enough to analyze. Table 2 also shows a three variable regression with loud/quiet, smooth/rough, and pulsating/not pulsating.

(Dep. Var.)	Descriptor (Indep. Var.)	R^2	p
annoyance merit value	loud/quiet	0.81	0.002
	sharp/dull	0.2	0.265
	boomy/not boomy	0.49	0.054
	smooth/rough	0.74	0.017
	pulsating/not pulsating	0.56	0.034
	impulsive/not impulsive	0.71	0.008
	weak/strong	0.72	0.008
	loud/quiet and impulsive/not impulsive	**0.97**	**0.001 0.004**
	loud/quiet and smooth/rough	0.96	0.001 0.006
	loud/quiet and pulsating/not pulsating	0.93	0.003 0.027
	loud/quiet and boomy/not boomy	0.89	0.008 0.12
	loud/quiet and smooth/rough and pulsating/not pulsating	0.98	0.002 0.043 0.166

Table 2. Correlation between US annoyance merit values and semantic differential averages.

Descriptors (Dep. Var.)	Descriptors (Indep. Var.)	R^2	p
impulsive/not impulsive	smooth/rough	0.95	<.0005
boomy/not boomy	pulsating/not pulsating	0.94	<.0005

Table 3. Correlation between selected US adjective pair averages.

UK STUDY - As with the US semantic differential data, the average values of the UK semantic differential results were used in linear regression equations to predict the UK annoyance merit values. Table 4 shows the results.

The loud/quiet adjective pair averages correlated the highest of any single variable to the UK annoyance merit values, with $R^2=0.85$. Loudness combined with pulsating-ness increased the R^2 to 0.93. Similar correlation was achieved in the previous US study, where higher correlation was obtained using the two variables loud/quiet and impulsive/not impulsive in the linear regression.

Table 5 below shows the correlation between selected adjective word descriptors used in the UK study. Boomy/not boomy averages were highly correlated to the pulsating/not pulsating averages, indicating that participants may have evaluated both adjective pairs with the same criteria.

---

[1] The p-value is the probability of getting a loudness regression coefficient at least as extreme as what was obtained, under the hypothesis that the regression coefficient is zero.

(Dep. Var.)	Descriptor (Indep. Var.)	R^2	p
annoyance merit value	loud/quiet	0.85	0.001
	sharp/dull	0.48	0.059
	boomy/not boomy	0.58	0.027
	smooth/rough	0.63	0.019
	pulsating/not pulsating	0.82	0.002
	impulsive/not impulsive	0.71	0.009
	weak/strong	0.82	0.002
	loud/quiet and **impulsive/not impulsive**	**0.85**	0.078 **0.852**
	loud/quiet and smooth/rough	0.88	0.025 0.374
	loud/quiet and pulsating/not pulsating	0.93	0.033 0.059
	loud/quiet and boomy/not boomy	0.91	0.009 0.157
	loud/quiet and smooth/rough and pulsating/not pulsating	0.93	0.27 0.876 0.139

Table 4. Correlation between UK annoyance merit values and semantic differential results.

Descriptors (Dep. Var.)	Descriptors (Indep. Var.)	R^2	p
impulsive/not impulsive	smooth/rough	0.66	0.014
boomy/not boomy	pulsating/not pulsating	0.88	0.001
loud/quiet	impulsive/not impulsive	0.86	0.001

Table 5. Correlation between selected UK adjective pair averages.

CORRELATION OF SEMANTIC DIFFERENTIAL RESULTS TO OBJECTIVE MEASURES

US STUDY - To find measures that reflect participant responses, objective measures were investigated and correlated to US loud/quiet, impulsive/not impulsive, and smooth/rough semantic differential averages. N_{50}, the loudness exceeded 50% of the time in a 2.5 s interval, was calculated on the Head Acoustics Binaural Analysis System (BAS) v4.1 (with the analysis set to Filter ISO) and correlated to the loud/quiet adjective averages. Table 6 below shows the correlation results with $R^2 = 0.95$, indicating that ISO532B loudness is highly correlated to the loud/quiet perception results. dB_A also correlates with loud/quiet, yielding $R^2 = 0.92$.

Impulsiveness was shown to be an important factor influencing diesel sound quality in [3], with kurtosis used as a measure in [4, 5, 6]. Kurtosis, a_4, is defined in [7] to be the fourth moment about the mean, m_4, divided by the square of the variance (2nd moment about the mean), m_2. A Matlab program was written to calculate kurtosis according to the following equation:

$$a_4 = \frac{m_4}{[m_2]^2} = \frac{\frac{1}{N}\sum_{i=1}^{N}(x_i - \bar{x})^4}{\left[\frac{1}{N}\sum_{i=1}^{N}(x_i - \bar{x})^2\right]^2} \ .$$

Kurtosis was used as a measure of impulsiveness since it reflects the existence of large signal excursions that occur for a small percentage of the time.

To investigate impulsiveness in various frequency bands of a signal, [4, 5] suggest applying filters in the frequency range of interest before calculating kurtosis. Several filter bands were investigated, based upon [4, 5]. The measure that provided the highest correlation to impulsive/not impulsive and smooth/rough semantic differential averages was kurtosis in the band 1.4 kHz to 4 kHz, as shown in Table 6. [4] recommends 1.4 kHz to 5.6 kHz, but because the data was decimated down to a sample rate of 10 kHz for the analysis, 4 kHz was more appropriate for the upper cutoff frequency. Note in Table 6 that the kurtosis correlations were performed with H removed. H had a high kurtosis, but because the sound was very quiet the semantic differential results did not indicate high impulsiveness or roughness.

Descriptor (Dep. Var.)	Objective Measurement (Indep. Var.)	R^2	p
loud/quiet	loudness (N50)	0.95	<.0005
loud/quiet	dB_A	0.92	<.0005
impulsive/not impulsive	kurtosis (1.4 kHz to 4 kHz) (H not included in regression)	0.86	0.003
smooth/rough	kurtosis (1.4 kHz to 4 kHz) (H not included in regression)	0.96	<.0005

Table 6. Correlation between objective measures and US adjective pair averages.

UK STUDY - The same objective measures shown in Table 6 were correlated to the UK semantic differential averages. Table 7 below shows the correlation results, where N_{50} is highly correlated to the UK loud/quiet perception results with $R^2 = 0.90$. dB_A correlates to a somewhat lesser degree with the loud/quiet results, yielding $R^2 = 0.78$.

As with the US results, kurtosis was correlated to the UK impulsive/not impulsive and smooth/rough adjective pair averages. Both regressions yielded low R^2 values as shown in Table 7, indicating that kurtosis does not follow the impulsiveness or roughness ratings. The reason that kurtosis did not correlate well to the UK averages compared to the US data may be attributed to the two markets giving different impulsive/not impulsive and smooth/rough ratings, as illustrated in Figure 5.

Descriptor (Dep. Var.)	Objective Measurement (Indep. Var.)	R^2	p
loud/quiet	loudness (N50)	0.90	<.0005
loud/quiet	dB_A	0.78	0.003
impulsive/not impulsive	kurtosis (1.4 kHz to 4 kHz) (H not included in regression)	0.035	0.668
smooth/rough	kurtosis (1.4 kHz to 4 kHz) (H not included in regression)	0.00	0.984

Table 7. Correlation between objective measures and UK adjective pair averages.

CORRELATION OF OBJECTIVE MEASURES TO ANNOYANCE MERIT VALUES

US ANNOYANCE MERIT VALUES - The objective measures previously correlated to the US semantic differential pairs were correlated to the US paired comparison of annoyance merit values. Table 8 shows results using two-variable regressions, where the independent variables were the objective measures, used to predict the US annoyance merit values (dependent variable).

Dep. Var.)	Objective Measurement (Indep. Var.)	R^2	p
nnoyance merit value	loudness (N50)	0.92	0.001
	kurtosis (1.4 kHz to 4 kHz)		0.011
	dB$_A$	0.97	<.0005
	kurtosis (1.4 kHz to 4 kHz)		0.003

Table 8. Correlation between objective measures and nnoyance merit values.

From Table 8, loudness (N50) and kurtosis rovide good correlation to the annoyance merit values R^2=.92). Similarly good correlation is achieved with dB$_A$ nd kurtosis (R^2=.97). Roughness from the BAS v4.1 nd loudness together (not shown) correlate slightly less vith the annoyance merit values (R^2=.89). dB$_A$ and oughness have comparable correlation (R^2=.85).

UK ANNOYANCE MERIT VALUES- The bjective measures shown in Table 8 were correlated to ie UK paired comparison of annoyance merit values. able 9 shows results using two-variable regressions.

Dep. Var.)	Objective Measurement (Indep. Var.)	R^2	p
nnoyance merit value	loudness (N50)	0.98	<.0005
	kurtosis (1.4 kHz to 4 kHz)		0.001
	dB$_A$	0.91	0.001
	kurtosis (1.4 kHz to 4 kHz)		0.061

Table 9. Correlation between objective measures and JK annoyance merit values.

From Table 9, loudness (N50) and kurtosis rovide high correlation to the annoyance merit values. he reason that kurtosis correlates to the UK annoyance ierit values may seem unclear since, from Table 7, urtosis does not correlate to the UK impulsiveness verage adjective ratings. However, the UK npulsiveness average ratings, when combined in a near regression with the UK loud/quiet average djective ratings as shown in Table 4, did not provide dditional correlation to the UK annoyance merit values s indicated by the high p-value. Thus while the UK and JS average impulsiveness ratings were not highly orrelated, the overall annoyance judgments were made ased upon the same phenomena, estimated by kurtosis nd ISO532B loudness. From Table 9, high correlation also achieved with dB$_A$ and kurtosis.

CONCLUSIONS

Overall annoyance rankings were similar for both ie US and UK participants. The results from the US tudy indicated that the loud/quiet and impulsive/not npulsive adjective descriptors correlated best with US innoyance judgments. The UK loud/quiet and ulsating/not pulsating adjective descriptors together orrelated best with UK annoyance results. UK espondents rated impulsiveness and smoothness lifferently than the US, indicating either different opinions ir different adjective meanings in the two markets. To educe the amount of different interpretations of the idjectives in future evaluations, experiments should

include training sessions using examples of sounds to illustrate the meaning of each adjective.

Loudness (N50) and dB$_A$ measures both yielded high correlation to the US loud/quiet adjective averages. Kurtosis in the range 1.4 kHz to 4 kHz correlated to the US smooth/rough and impulsive/not impulsive adjective pair averages. Loudness (N50), as well as dB$_A$ to a lesser extent, correlated to the UK loud/quiet adjective pair averages. Kurtosis combined with either loudness (N50) or dB$_A$ provided high correlation to both the US and UK annoyance merit values.

ACKNOWLEDGMENTS

The suggestions provided by John Feng at Ford's Research Laboratory are gratefully acknowledged. The UK sound playback equipment and assistance provided by Jason Blewitt, Geoff Shepherd and Mike Hamilton of Ford's Dunton Facility is greatly appreciated. Thanks also to Roy Reynoldson for assisting with the arrangements of the UK study. Finally, the inclusion of the UK study in a clinic by Ford's Market Research Office is appreciated.

REFERENCES

[1] Otto N., Wakefield G., "A Subjective Evaluation Analysis of Automotive Starter Sounds," *Noise Control Engineering Journal*, **41**, pp. 377-382, 1993
[2] David H.A., *The Method of Paired Comparisons*, 1988
[3] Russell M.F., Worley S.A., and Young C.D., *Towards an Objective Estimate of the Subjective Reaction to Diesel Engine Noise*, SAE Paper 870958
[4] Hughes M., *Subjective Evaluation of Diesel Idle Noise Using Simulation Techniques*, Institute of Sound and Vibration Research, University of Southampton, Ford Motor Company Limited Consultation Report No. 816/94, June 1994
[5] Russell M.F., Worley S.A., and Young C.D., *An Analyser to Estimate Subjective Reaction to Diesel Engine Noise*, IMechE Paper C30/88
[6] Schiffbanker H., Brandl F.K., Thien G.E., *Development and Application of an Evaluation Technique to Assess the Subjective Character of Engine Noise*, AVL LIST Ges.m.b.H., SAE Paper 911081
[7] Spiegel M.R., *Schaum's Outline Series, Theory and Problems of Statistics, 2nd Edition*, McGraw-Hill, 1990

971981

Pitch Matching for Impulsive Sounds

Richard J. Fridrich
General Motors Corp.

Copyright 1997 Society of Automotive Engineers, Inc.

ABSTRACT

Sound Quality work often seeks to bridge the gap between human perception and objective noise measurements. Sometimes a Sound Quality requirement is a concept, such as Pitch, that can have many meanings depending on the type of sound under consideration. The idea of impulsive sounds having indefinite pitch may be unfamiliar to the engineering world, but the field of music has used the idea for decades if not centuries. Indefinite pitch may be of value in Sound Quality work, especially when dealing with door closing sounds. No instrumentation yet exists to measure objectively the indefinite pitch of impulsive sounds. However, a pitch matching technique using sine wave bursts is described which allows the indefinite pitch of impulsive sounds to be measured by ear. Some limitations of the method are discussed as well as the need for basic psychoacoustical research on the topic of indefinite pitch.

INTRODUCTION

This paper is an attempt to capture at least a portion of a "live" presentation of examples of impulsive sounds. The author regrets that the reader cannot experience these sounds for himself, so that he can judge whether pitch matching does occur. The motivation for this work originated with the need to evaluate Door Slam Sound Quality. Because of the complexity of a door slam sound with its multiple impacts [1], the approach taken for this work was to try to reduce the complexity to simpler terms that might be more easily understood. The focus of this paper is to demonstrate a potentially useful technique, not to provide a rigorous psychoacoustic proof. It will also not attempt to discuss what is a "good" or "bad" pitch. Pitch will be considered as a neutral characteristic to be measured for later interpretation.

BACKGROUND

Traditional noise and vibration control requirements are often given in terms of level (dB) and frequency (Hz). Customer requirements based on human perception are usually poorly defined, often relying on adjectives and adverbs such as "solid" or "tinny" to describe what is desired or what is unacceptable. The nouns used to identify impulsive sounds, which range from "click and snap" to "thump," while evocative, are also poorly defined in engineering terms [2]. Sound Quality work often seeks to bridge the gap between human perception and the realm of objective measurement. The objective of traditional N&V work is usually to reduce the level of the sound. In contrast, the objective of Sound Quality work is to make it sound "good." To achieve that objective requires a deeper understanding of human perception and the words used to describe sound. Sometimes a Sound Quality requirement is a word or a concept that can have many meanings depending on the type of sound under consideration. To better understand Sound Quality requirements it will be helpful to revisit the musical roots of acoustics where many of the terms we use originated.

The origins of acoustics date back to the time of ancient Greece where Pythagoras sought to understand the mathematical and physical relationships present in music and musical instruments. The musical roots of acoustics (and by extension noise and vibration control) provide us with words that have nearly universal recognition in our present world as they once did in ancient Greece. Some of the words used in music include loudness, pitch, duration, rhythm, tempo, melody and harmony. Some of these words are more universally recognized than others. Loudness and pitch are terms that seem to have definite meaning for everyone. However, as with many words, pitch can have multiple meanings depending on the type of sound being considered. Some of these meanings are well known within the musical world but not in the engineering world, especially when considering impulsive sounds. If musical terms are used to describe door slam sounds, a simple Sound Quality model can be constructed as shown in Figure 1. However, the model is not as simple as we might like because all the terms are interconnected. It might be argued that terms such as pitch and duration are independent, but an acoustics demonstration can show that pitch and duration are connected [3]. That demonstration will be discussed later.

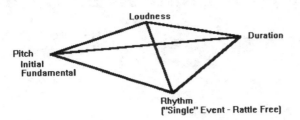

Fig. 1 Door Slam Sound Quality Model

PITCH AND IMPULSES

Many books on acoustics introduce pitch as a subjective characteristic which is primarily determined by its frequency but also is a function of its sound level and waveform [2,4,5,6]. The range of the subjective impression of pitch is often described as extending from "low" to "high." When the discussion turns to pitch as a function of waveform, the usually point of reference is a pure tone or sine wave as shown in figure 2. If Fourier analysis were to be used on the waveform, it would be assumed that the sine wave starts at minus infinity and continues on forever [7]. The results of analysis of such a repeating waveform would be a line spectrum. Because of the physical impossibility of infinity, approximations are made to shorten the time requirement and make Fourier analysis practical. Using practical Fourier analysis of this waveform, the assumption is still made that the waveform repeats as it appears in the window. The resultant spectrum would show some spectral spreading from a simple line spectrum, but it still would be interpreted as representing a pure tone.

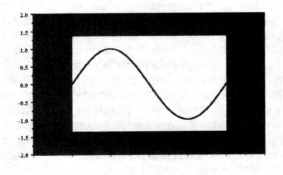

Fig. 2 Pitch = Frequency of a Sine Wave (?)

But what if the waveform did not repeat? What if the waveform of the sound looked like a sine wave of one period in length, as shown in Figure 3? What would the waveform sound like? Because Figure 3 is similar to Figure 2, you might expect that it would still sound something like a pure tone. However, if you listen to one of the auditory demonstrations produced for the Acoustical Society of America [3], a sine wave of one period in length will not have any of the familiar characteristics of a pure tone. It has pitch but musicians would say it has indefinite pitch [8]. Clearly, the duration of a sine wave influences its pitch.

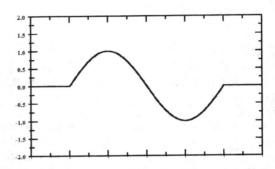

Fig. 3 For Impulsive Sounds Pitch = Frequency & Duration

DEMONSTRATION OF PITCH THRESHOLD

The title of the demonstration on the Acoustical Society of America's Auditory Demonstration CD (Demonstration 13 -Track 29) is "Pitch Salience and Tone Duration." The demonstration consists of a series of sine wave bursts presented at three different frequencies, 300 Hz, 1000 Hz, and 3000 Hz. Each series starts with one period of a sine wave and then increases the duration in a binary fashion (e.g. 1, 2, 4, 8, 16...) up to 128 periods. As the duration increases there is a change from a "click" to a "tone."

This change from a "click" to a "tone" can also be interpreted as the change from indefinite pitch to definite pitch (Note: The term, "indefinite pitch," while sometimes used in musical acoustics, is not generally used in the psychoacoustics). The threshold between indefinite pitch and definite pitch is frequency dependent. The literature (i.e. English language books and journals) of the last 30 years does not seem to contain the details on this frequency dependent. Olson, in his 1967 book, presents a graph showing the threshold of the perception of definite pitch as a function of frequency [9], however, he references the work of others published in the years 1929 and 1936 in European journals (Burck, Kotowski, and Lichte; von Bekesy) and in 1944 in the Journal of Experimental Psychology (Turnbull). From Olson's graph, the threshold for definite pitch at 60 Hz is near 3 periods, at 800 Hz it is 10 periods and at 4 kHz it is near 50 periods. The author's own observations, using less than ideal equipment, place the threshold at about two periods at frequencies below 60 Hz and near 30 periods at frequencies around 4 kHz. Because of difficulties in obtaining the original published reports, the variations in the threshold from subject to subject are not known. Olson also notes that the time required to perceive a definite pitch does not vary greatly from low to high frequencies. He states that the average time is about 13 milliseconds. From his graph it would appear that the upper bounds of the range is about 60 milliseconds.

In order to make good use of this information on the pitch threshold, it is also necessary to have detailed measurements of the just noticeable frequency difference for indefinite pitch. Some of these details already may have been published but are not easily accessible in an automotive engineering library (e.g. psychology journals, non-English language publications, etc.). In lieu of detailed psychoacoustic information, a short demonstration from the world of music can provide some guidance on the just noticeable frequency difference. The smallest pitch interval used in the well tempered musical scale (called a half tone or semitone) represents an increase in frequency of about six percent (i.e. by definition the ratio is two to the one-twelfth power) [9]. To demonstrate that the interval of a semitone is audible for impulses, a series of sine bursts of two periods in duration was generated using the frequencies of the well tempered scale. When this series of sounds have been played for individuals and groups in informal settings, most people have agreed that the series does have the characteristics of a musical scale. Because the series of sine bursts does sound like a musical scale and a difference can be heard between each step of the scale, it can be concluded that the just noticeable difference in frequency is not greater than a six percent increase.

PITCH MATCHING TECHNIQUE

Armed with some knowledge of pitch threshold and the just noticeable frequency difference, useful engineering measurements of pitch appear to be possible by adapting the pitch matching technique to impulsive sounds. In psychoacoustics this pitch matching technique is an example of the method of adjustment [10]. For continuous sounds, the pitch matching technique uses human hearing to compare the test sound to either a simultaneously presented or sequentially presented sine wave sound or pure tone. For musical purposes the sine wave is set to a standard frequency and the musical instrument is adjusted until its pitch matches the pure tone. The instrument is then said to be "in-tune." For engineering purposes, the frequency of the sine wave is adjusted to match the pitch of the test sound. When the match is achieved, the pitch of the test sound is then defined by the frequency of the sine wave.

Since the match for impulsive sounds must be in both frequency and duration, the simple sine generator, used for pitch matching, must be replaced by a sine burst function generator which allows the user control of the frequency and number of periods generated for the reference signal. Because of concerns about masking and problems with synchronizing impulses, the test impulse and the sine burst are not presented simultaneously but instead one after the other with a brief pause in between.

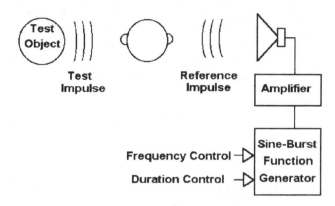

Figure 4 Diagram of Impulsive Pitch Matching
(Test and reference impulses listened to sequentially)

Figure 4 shows a diagram of the equipment needed for the impulsive pitch matching technique. The sine burst function generator used in this project was a Hewlett Packard 8111A. A good quality power amplifier and loudspeaker are essential. (The details of the sound system as well as the short comings of loudspeakers will be discussed later.) A quiet listening environment with a short reverberation time helps improve the user's concentration on the pitch matching task and reduces the influence of the listening environment.

IMPULSIVE SOUND EXAMPLES

The discussion that follows refers to waveform traces for examples of the impulsive sound pitch matching technique. All the sounds that were evaluated by the author, including the reference sine bursts from the loudspeaker, were digitally recorded in an anechoic room using a B&K free field microphone at distances of either 1.0 or 0.5 meters from the source. A Tektronix 2500 Testlab digital data acquisition system was used to acquire the waveform traces from the recordings. Neither the recording system nor the data acquisition system are needed to perform the pitch matching technique but they were found to be useful in assessing the complexity of impulsive sounds. The pitch matching technique seems work best for simple impulsive sounds.

"SIMPLE" IMPULSIVE SOUND - Before discussing an automotive example sound, a "simple" impulsive sound will be discussed to demonstrate the pitch matching technique. The "simple" impulsive sound was produced by the depressing and releasing a Heinz ketchup bottle cap. Its waveform is shown in Figure 5. It was chosen because it is loud and easy to obtain. Another possible "simple" source could have been an old-fashioned "cricket" noise-maker if one had been available. The "simple" sound being evaluated was generated when the cap, having been depressed, was allowed to rapidly return to its undeformed shape. (The act of depressing the cap also creates an impulsive sound which can be evaluated separately, but in this case it is not as loud as when the cap is released.) The waveform of the "simple" sound also looks relatively simple. It shows several periods of a waveform which generally retains its shape as it decays.

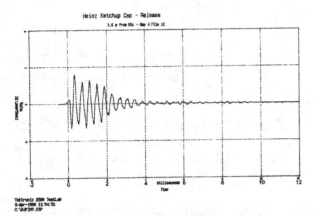

**Fig. 5 "SIMPLE" IMPULSE EXAMPLE
2 msec / Div.**

Using the method for pitch matching for impulsive sounds, a sine burst of 2.56 kHz and about 10 periods was matched to the sound from the Heinz ketchup bottle cap. Its waveform is shown in Figure 6. An interesting observation is that if the waveform traces of Figures 5 and 6 are overlaid, one can see that the spacings at the zero crossing match reasonably well. Note that the waveform in Figure 6 (as well as the other sine burst waveforms to be presented later) shows distortion from the ideal sine wave. This distortion is due to the sound system which will be discussed later.

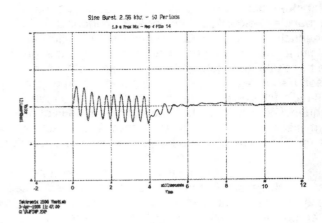

**Fig. 6 SINE BURST MATCH
TO "SIMPLE" IMPULSE EXAMPLE
2 msec / Div.**

AUTOMOTIVE SOUND EXAMPLE - The automotive impulsive sound example is a door latch. The door latch was held by hand using a simple spring clamp. To produce the sound the latch was manually closed using a wood dowel in place of the usual striker. Figure 7 shows the waveform of the door latch sound. As can be seen in the figure it is a complex series of impulses with at least two major impulsive events that could be evaluated, the secondary latch and the primary latch. Because these events can be separated in time they can be evaluated separately. For sake of demonstration, only the secondary latch sound (the first event in the figure) will be discussed here.

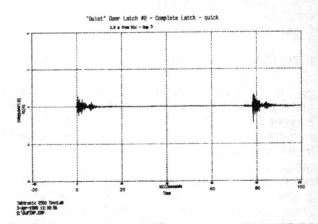

**Fig. 7 DOOR LATCH SOUND WAVEFORM
SECONDARY & PRIMARY LATCH
20 msec / Div.**

Even though the secondary latch sound can be isolated, as can be seen in Figure 8, it is far from a simple sound. It is known that a door latch has many moving parts and it is likely that sounds will be generated by these parts. How these various sounds combine is not known so it is wise to assume that the process is complex.

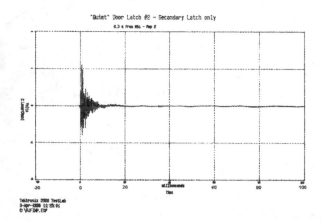

Fig. 8 DOOR LATCH SOUND WAVEFORM SECONDARY LATCH ONLY
20 msec / Div.

Figure 9 shows the waveform of the secondary latch on an expanded time scale. As can be seen, the waveform is far more complex than that of the "simple" ketchup cap example sound.

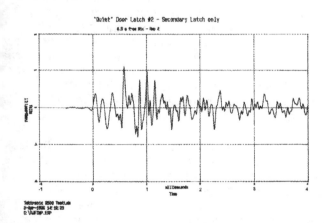

Fig. 9 SECONDARY LATCH WAVEFORM
1 msec / Div.

Applying the impulsive pitch matching technique produced a match using a sine burst with a frequency of 2.66 kHz and a duration of three periods. The waveform of the reference sine burst is shown in Figure 10. Even without overlaying the traces, it can be seen that there is very little similarity between the waveforms of the secondary latch sound and the sine burst match. Recall, however, that the intent of this technique is to find a match for pitch only. Other Sound Quality characteristics, such as timbre, are not part of the focus of this work. In spite of the lack of similarity, if Figures 9 and 10 are overlaid, one can still find at least a few places where the spacings at the zero crossing match reasonably well. There may also be some significance in the fact that the places where the zero crossing match occur early in the event. Little else can be said except to make the observation.

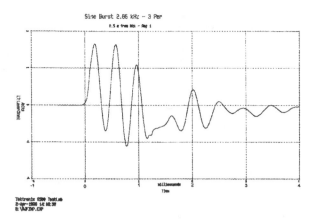

Fig. 10 SINE BURST MATCH TO SECONDARY LATCH SOUND
1 msec / Div.

EQUIPMENT PERFORMANCE ISSUES.

To specify the pitch of a real world impulsive test sound requires specifying the frequency and number of periods of the sine burst whose pitch has been matched to the test sound using human hearing. While the technique has been outlined, it is not complete because the sound system, which is essential, has not been specified. Exactly how the sound system is to be specified has not been determined. In the sine burst waveforms that were previously presented some degree of distortion was clearly visible. Since the loudspeaker is the transducer that produces the sound, it is obvious that the performance of the loudspeaker is an important issue. Most specifications for loudspeaker performance are based on test methods using continuous signals. However since pitch matching for impulsive sounds uses sine bursts as the reference signal, testing the loudspeaker with sine bursts seems more appropriate. Figures 11, 12, and 13 show some example waveform traces from the Acoustical Society's Auditory Demonstration CD. The loudspeaker was an Acoustic Research AR-2xa. It was powered by a McIntosh MC2150 amplifier. (Limited availability of equipment dictated the selection.) Only the first sine burst of each series is presented, that is one period at 300 Hz, 1 kHz, and 3 kHz.

Notice that each waveform is distorted and is far from the ideal, "sine wave of one period in length." Given the nature of a sine wave, where the maximum velocity occurs at the zero crossing, it is unlikely that a distortion-free, ideal sine burst can ever be achieved by a mechanical system such as a loudspeaker. Consequently, the specification for the loudspeaker to be used for impulsive sound pitch matching will need to describe what is "good enough" for the technique to work. Work on such a specification remains to be done.

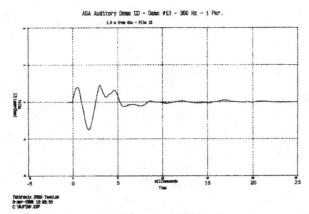

Fig. 11 LOUDSPEAKER RESPONSE WAVEFORM ASA DEMO - 300 Hz - 1 Per. 5 msec / Div.

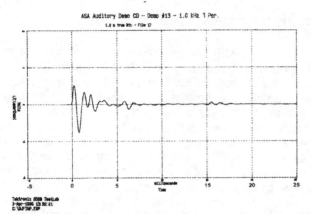

Fig. 12 LOUDSPEAKER RESPONSE WAVEFORM ASA DEMO - 1 kHz - 1 Per. 5 msec / Div.

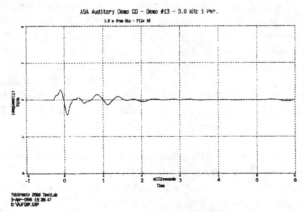

Fig. 13 LOUDSPEAKER RESPONSE WAVEFORM ASA DEMO - 3 kHz - 1 Per. 1 msec / Div.

ADVANTAGES AND DISADVANTAGES

The advantages in using the impulsive sound pitch matching technique are that it produces definite values that might be useful as requirements in the development of a product. The technique produces a reproducible reference sound which provides some stability for the subjectiveness of human hearing. The subjectiveness of human hearing is the major disadvantage of the technique. Individual ability is a factor in using the technique just as it is in performing music. The repeatability between different users is unknown, as is the day-to-day variability of a single individual. Obviously, further research needs to be done.

ROADBLOCKS TO DEVELOPMENT OF AN OBJECTIVE METHOD

The pitch matching technique that has been described is a subjective evaluation. It has the advantage of using a reproducible reference sound, but there are many engineers who prefer completely objective methods. The future development of an objective method for the pitch of impulsive sounds faces a time-frequency resolution problem and an interpretation problem. The time-frequency resolution problem must deal with the uncertainty principle which applies to frequency analyzers based on linear principles. A previous attempt by the author to develop an objective method was found to lack the necessary resolution to handle pitch evaluation of the auditory demonstration sounds discussed in this paper [11]. Human hearing is a non-linear process which somehow outperforms any known analyzer. Hartmann reports that the human "pitch perception process beats the uncertainty principle by a factor of 5." [12] A breakthrough in analyzer capability appears to be needed.

The interpretation problem stems from the lack of information on the topic of impulsive sound pitch. The results of any objective method that might be developed must be compared to psychoacoustical measurements. It would be helpful to have a psychoacoustical model dealing with indefinite pitch. Before such a model can be developed, more detailed psychoacoustical measurements will be needed to supplement the existing data.

CONCLUSION

In conclusion, for the foreseeable future, Sound Quality work dealing with the pitch of impulsive sounds will still require decisions based on subjective evaluation due to the lack of an objective measurement technique. The technique of pitch matching for impulsive sounds has been demonstrated but still needs more investigation and development. It is hoped that the curiosity of the interested reader has been stimulated to try the technique and help advance its development.

REFERENCES

[1] Malen, D.E. & Scott, R.A., Improving Automobile Door-Closing Sound for Customer Preference, Noise Control Engineering Journal, Vol. 41, No. 1, pp. 265-267.

[2] Benade, A.H., Fundamentals of Musical Acoustics, Oxford University Press, 1976, Ch. 2.

[3] "Pitch Salience and Tone Duration" demonstration, Acoustical Society of America Auditory Demonstration CD, 1989.

[4] Kinsler & Frey, Fundamentals of Acoustics, John Wiley & Sons, 1962, pg 399

[5] Beranek, L.L., Acoustics, McGraw Hill, 1954, pg 397

[6] Backus, J., The Acoustical Foundations of Music, W.W. Norton & Company, 1969, Ch.7.

[7] Bracewell, R., The Fourier Transform and Its Applications, McGraw-Hill, 1965, pp. 6, 9, 11.

[8] Apel, W., & Daniel, R.T., The Harvard Brief Dictionary of Music, Washington Square Press, 1961, pg. 221 ("Percussion Instruments").

[9] Olson, H.F., Music, Physics and Engineering, Dover Publications, 1967, pp. 248-251 & 46-49.

[10] Zwicker, E. & Fastl, H., Psychoacoustics, Facts and Models, Springer-Verlag, pg 8

[11] Fridrich, R.J., Investigating Impulsive Sounds - Beyond "Zwicker Loudness," SAE Paper 931329, 1993.

[12] Hartmann, W.M., Pitch, Periodicity, and Auditory Organization, Journal of the Acoustical Society of America, Vol. 100, No. 6, 1996,

971982

Linearity of Powertrain Acceleration Sound

Norman Otto and John Feng
Ford Motor Co.

Robert Cheng
University of Michigan

Eric Wisniewski
Edsel Ford High School

Copyright 1997 Society of Automotive Engineers, Inc.

ABSTRACT

The loudness of powertrain noise generally increases with increasing rpm. In the case of 'linear' powertrain acceleration sound, the loudness versus time relationship is well described by a linear function. Two studies were conducted on powertrain linearity. The first used tests of similarity and preference to determine whether subjects could detect changes in linearity. The second used a subjective test of preference to investigate how subjects' preference varied with differing degrees of linearity. In both studies, stimulus sets were created by artificially introducing a controlled degree of non-linearity into a nominally linear powertrain sound. The results of the first study indicate that linearity is a phenomenon that naive subjects can readily detect, and that it has an effect on overall preference. Furthermore, the second study shows that preference is related to the magnitude and position of nonlinearities in the growth of loudness versus time during an acceleration run. Based on these results, an objective model is developed that predicts the effects of powertrain linearity on subjects' preference.

INTRODUCTION

Powertrain sound quality during vehicle acceleration is a well studied phenomena[3]. Recent work has shown the loudness, roughness, and pitch are important perceptual factors in determining sound quality during WOT (wide-open throttle) accelerations [2]. The influence of temporal behavior over the course of an acceleration is generally ignored. In this paper we discuss two such temporal effects, linearity and resonance.

Linearity is thought to be related to the progression of intensity and/or loudness with time over the course of the acceleration. A linear progression is generally thought to be more desirable than a non-linear progression, although the data in this area are sparse.

Causes of powertrain non-linearity may be combustion variability, body("boom") and mount effects, and engine transmission mismatch. A resonance, on the other hand, is a short duration increase in loudness, or dB, during an engine runup. Resonances are thought to negatively impact the sound quality of the acceleration. In the following sections, the effects of non-linearity and resonances on powertrain sound quality are investigated.

LINEARITY

SUBJECTIVE EVALUATION - To conduct subjective testing, sound samples with varying degrees of linearity are needed. Finding appropriate vehicles would have been very difficult. Instead, we took a vehicle runup in first gear (LS400) which was fairly linear and, using an in house software tool, digitally added increasing amounts of non-linearity. Figure 1 shows some of these synthesized runups. A total of seven sounds were produced (baseline plus six with varying amounts of non-linearity). The loudness of each was then equalized to the baseline, N_{10}, loudness [3].

A paired comparison of preference was then conducted on the synthesized sounds with 24 employees as subjects. In this method, subjects are exposed to pairs of sounds and choose which sound in the pair they prefer. This process is repeated (twice) for all possible pairs. Two measures of subject performance, repeatability and consistency, both had subject averages of 70%, a decent, if not great, result. Preference scores[1], defined as simply the total number of times a sound was preferred (summed over all comparisons), are shown below in Table 1 for each sound. Scores provide a quick look at the subjective results. The higher the sound's score, the more preferred the sound.

Sound	Score
Base	201
2	187
3	184
4	171
5	113
6	84
7	68

TABLE 1. Linearity Preference Scores

In Table 1, linearity decreases with increasing sound number. Sounds grouped together had scores which were not significantly different at the 0.01 level. The results shows that baseline sound and the first three levels of decreasing linearity were found to have equivalent preference scores. It is only at the higher levels of non-linearity (sounds 5,6,7) that any degradation of preference occurs. This suggests a threshold effect, up to a certain level degrading linearity has no effect on preference. Only after this level is exceeded, is non-linearity an issue.

While preference scores provide a quick look at the subjective results, they are not sufficient for correlating subjective impressions to the objective factors underlying those impressions. For this, a PC model (Bradley-Terry) is used to convert the paired comparison data into single value measures of preference called merit values. Generally, if one sound has a higher merit value than another, that sound will be preferred over sounds with lower merit values. The merit values for this study are shown in Table 2.

Sound	Value
Base	0.77
2	0.60
3	0.55
4	0.43
5	0.40
6	-0.8
7	-1.15

TABLE 2. Linearity Merit Values

OBJECTIVE MEASURE - In searching for an objective measure that would account for the subjective results, a number of factors should be considered. First is the assumption that negative deviations from linearity (below the linear line) would be as damaging as positive deviations (above the line). While positive deviations might sound like a surge, negative deviations might sound like a stumble. If we accept this assumption, a linear measure is ruled out, since + and - deviations would cancel. The square of the deviation from linearity is a good choice. The second consideration is whether a point measure of linearity will suffice. Inspection of Figure 1. show that the deviation varies greatly over time. Thus, an valid estimate based on a single time estimate is unlikely. Instead, an integrated measure seems more appropriate. Finally, should linearity be measured using dB or loudness vs. time or vs. RPM? Loudness is more appropriate than dB (or dB(A)) since deviations are heard primarily as loudness variations. During a first gear runup, time and rpm have a linear relationship so the choice is somewhat arbitrary. Loudness vs RPM plots might be useful in comparing the linearity of vehicles with vastly different runup times. These considerations leave us with the following objective measure for non-linearity.

$$\text{NONLIN} = \{ (L - L_b)^2 \}^{1/2} \, dt$$

where

L = loudness value
L_b = baseline loudness value

In our example, the baseline is the LS400. More generally, a baseline can be generated for any sound by simply drawing a straight line from the start to the end of a runnup on a loudness vs. time (or RPM) plot.

Applying the this metric to our preference study show that for NONLIN equal to or less than 1, the effects of non-linearity on preference is negligible. For NONLIN greater than 1, significant degradation in preference is seen. Using regression, the following subjective-objective correlation is found. The correlation is seen to be excellent.

$$\text{Merit Value} = 1.86 - 0.095 \, \text{NONLIN} \quad R^2 = 0.99$$

CONCLUSIONS - This study of linearity has shown that up to a point decreasing linearity had little o no effect on subjective preference. After this point, however, degrading linearity further reduces subjective preference significantly. A measure of non-linearity is shown to be an integrated root square difference between the actual and baseline runup. This objective metric is shown to predict the subjective data very well. Further study is needed to test the robustness of this metric.This is an example of a Main Heading section. This section will include sub-sections.

RESONANCES

A resonance is generally a short duration increases in loudness in accelerations or steady state engine sounds. While these are generally thought to decrease sound quality, a systematic study of the effec of resonances has not been done. Some of the parameters that might influence the sound quality impac of a resonance are it's duration, amplitude, shape, and position in the runup. The effect of amplitude and position in the acceleration have been investigated.

SUBJECTIVE EVALUATION - Using the LS400 acceleration as a baseline, a resonance was added, as in Figure 2. The magnitudes were either 4 or 8 dB above the baseline. The resonance was placed at one of four position in the first gear runup; start, 1/3 of the way, 2/3 of the way and at the end of the acceleration. Each resonance was 1 second long with 400 ms rise and decay times and a 200 ms peak. Nine sounds (2 amplitudes x 4 positions + baseline) were generated in this way for subjective testing. To investigate the effect of these resonances on sound quality, a paired comparison of preference was conducted using 24 Ford employees as subjects.

The resulting merit values for this study are plotted vs. resonance position on Figure 3. The straight line is the merit value for the baseline, the top curve for the 4 dB resonances and the bottom curve for the 8dB resonances. Remember, the higher the merit values the more preferred the sound.

CONCLUSIONS - The effect of the resonance amplitude on preference is clear. The 8 db resonances were less preferred that their 4dB counterparts. It is interesting to examine difference between the 4 dB resonance and the baseline (0 dB resonance). At the first position, the preference merit values for these two are virtually identical. This means that engineers can neglect resonances of 4 dB of less, if they at the start of the runup. From Figure 3, we also see that the position of the resonance in the acceleration also effects sound quality. Preference decreases as the resonance is moved from the start to the end of the runup. This observation may explain why some resonances are annoying to customers while others go unnoticed. Further study is needed to determine the effect of resonance duration and shape on sound quality.

CONCLUSION

In this paper, we have investigated the linearity (or the lack of linearity) of powertrain accelerations. Subjective evaluations show that while a small amount of non-linearity has no sound quality penalties, larger amount severely degrade perceptions. An objective measure of non-linearity was developed and shown to fit the subject results very well. Resonances were also studied as a function of both amplitude and position in the runup. Sound quality decreased with increasing resonance amplitude and resonance position. These results should help engineers to improve the sound quality of future powertrains.

REFERENCES

[1] David, H. A., <u>The Method of Paired Comparisons</u>, Oxford Press, NY 1988.

[2] Otto, N. and Feng, B., "Identification & Characterization of Automotive Powertrain Sound Quality Attributes," 132th Meeting of the Acous. Soc. Am., Honolulu, HI, Dec. 1996.

[3] Yagihashi, W. and Imai, H, "A Study on Sound Quality Evaluation of Car Interior Noise During Acceleration Using a Sound Simulator," JSAE Fall Convention, 1990.

[4] Zwicker, E. and Fastl, H, <u>Psychoacoustics: Facts and Models</u>, Springer-Verlag, Heidelberg, 1990.

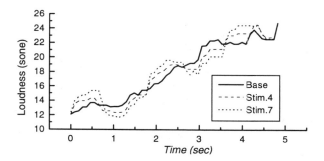

Figure 1: Base and Modified Linearity Sounds

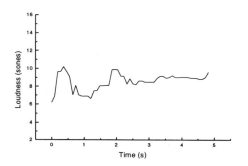

Figure 2a: Original Resonance Sound

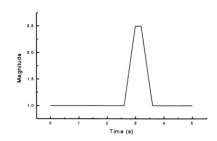

Figure 2b: Non-linear Multiply Factor

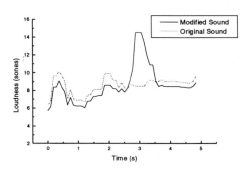

Figure 2c: Original and Modified Resonance Sounds

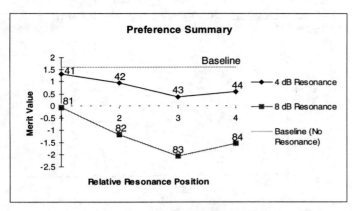

Figure 3: Merit values vs resonance position

971983

The Effect of Powertrain Sound on Perceived Vehicle Performance

Michael A. Blommer, Scott A. Amman, and Norman C. Otto
Ford Motor Co.

Copyright 1997 Society of Automotive Engineers, Inc.

ABSTRACT

One factor believed to influence a customer's perception of vehicle performance is the powertrain sound. However, its influence relative to other factors, such as vehicle acceleration or transmission shift characteristics, remains largely unknown. Past studies of performance perception have either neglected the effect of powertrain sound, studied its effect independent of other factors (e.g., listening to powertrain sounds over headphones), or have confounded its effect with other factors. In this paper, we describe an in-vehicle system for methodically studying the influence of powertrain sound on a customer's perception of performance. The system is beneficial in that it allows the experimenter to electronically modify the sound associated with the powertrain while not affecting other possible performance factors. Results from a corresponding experiment are also presented, in which customers rated their perceived performance of the powertrain sounds. These results suggest that the synthesized powertrain sounds affect performance impressions, and can be used in a planned in-vehicle experiment.

INTRODUCTION

Customer's impressions of vehicle characteristics such as performance, quality, and durability, are affected by sounds associated with the vehicle. Manufacturers of automobiles have recognized this fact as well as the high expectations customers have recently placed on a vehicle's acoustic environment. A goal of sound quality engineering is to determine the characteristics of vehicle sounds that can be engineered to meet these expectations.

One sound believed to effect the perception of vehicle performance is that of the powertrain. Previous research from listening room environments suggests that the loudness and roughness of the powertrain sound are the two main characteristics influencing this perception [1]-[3]. The roughness is caused by physical phenomena such as cylinder-to-cylinder combustion variability [4], unequal length exhaust manifolds and downpipes [3], and intake runner configuration [5]. These phenomena produce amplitude modulation of the sound waveform,

which produce sidebands at half engine orders in the powertrain sound. The amount of roughness is dictated by the levels of these sidebands.

In this paper, we describe an in-vehicle system to evaluate the influence of powertrain sound on a customer's perception of performance. The system allows the loudness and roughness of the powertrain sound to be modified without changing other vehicle characteristics (e.g., transmission shifts or acceleration). Thus, the effect of powertrain sound on vehicle performance can be studied without confounding effects due to changes in these other vehicle characteristics.

We also present results from a powertrain sound evaluation conducted in a listening room environment. The sound recordings were generated from the in-vehicle system. An analysis of the results shows the effect of loudness and roughness on the perception of vehicle performance independent of other perceptual influences. These experimental results can then be compared to those from an in-vehicle experiment where other perceptual factors such as vehicle acceleration, transmission shift, and steering feel are present. This in-vehicle experiment is outlined in this paper, with results to be published in a subsequent paper.

VEHICLE INSTRUMENTATION

HARDWARE - The main in-vehicle hardware consists of an accelerometer and a digital equalizer (DEQ). The accelerometer (Bruel & Kjaer 4135) is mounted on the engine in order to derive an RPM-dependent signal containing half and whole engine orders. The accelerometer signal is passed through the DEQ (Yamaha DEQ7) and subsequently played over the vehicle's audio system. Additional sound-deadening material applied to the vehicle's engine compartment is used to reduce the unmodified powertrain sound in the passenger cabin.

MODIFICATION OF POWERTRAIN SOUND - Empirical listening indicated that the roughness and loudness of the powertrain sound could be modified by changing the 2/3-octave band levels of the DEQ. Precise control and tracking of the engine order components was not required.

A realistic baseline powertrain sound was obtained by equalizing the spectrum of the accelerometer signal to that of the baseline vehicle. Specifically, the vehicle was run through 1st gear under a wide-open throttle (WOT) condition and its unmodified powertrain sound was recorded. A similar recording was made with the accelerometer signal bypassing the DEQ and played over the audio system. Average levels of non-overlapping 1/3-octave bands were calculated from each recording over the same portion of the engine run-up, and their difference was used as the equalization curve. There were 27 1/3-octave band levels between 40 Hz and 16 kHz in the equalization curve.

Since the DEQ has overlapping 2/3-octave band filters, a least-squares method was used to determine the filter gains resulting in a spectrum that approximated the equalization curve. If $\mathbf{d} = [d_1, d_2, ..., d_{27}]^T$ corresponds to the 27 values in the equalization curve, and $\mathbf{g} = [g_1, g_2, ..., g_{14}]^T$ corresponds to the gains of the 14 2/3-octave bands, then the two are related by

$$A\mathbf{g} = \mathbf{d} \quad (1)$$

where the (i,j)-th element of the matrix A contains the gain contribution of the j-th 2/3-octave band filter at the center frequency of the i-th 1/3-octave band. The least-squares approximation for the filter gains is

$$\hat{\mathbf{g}} = (A^T A)^{-1} A^T \mathbf{d} \quad (2)$$

Roughness modification - As described in [6], the perception of powertrain roughness is strongly influenced by the powertrain's spectral content over 0.1-1 kHz. Neighboring engine orders in this frequency region which are above the auditory system's masked threshold can cause varying degrees of roughness. Therefore, the roughness of the synthesized powertrain sound was modified by varying the gains of the 2/3-octave filters covering 0.1-1 kHz. Increased roughness was obtained by boosting the filter gains and creating an upward slope of the DEQ filter spectrum over this spectral region. The upward slope compensates for spectral masking in the auditory system by raising neighboring engine order components above their masked threshold. Similarly, decreased roughness was obtained by attenuating the filter gains and creating a downward slope of the DEQ filter spectrum over 0.1-1 kHz. The downward slope assists in lowering neighboring engine order components below their masked threshold. As an example, the dashed line in Fig. 1 shows spectral *differences* between DEQ filters designed to produce powertrain sounds with high and low roughness. In all, three filters were designed to produce different levels of roughness.

Loudness modification - The loudness of the powertrain sound was varied by changing the overall gain of the DEQ. This was implemented by adding $\mathbf{x} = (A^T A)^{-1} A^T \mathbf{c}$ (cf. (2)) to the 2/3-octave band filter gains, where $\mathbf{c} = [c, c, ..., c]^T$ is a constant vector indicating an overall gain increase of c dB. Three loudness levels were designed for each roughness filter, resulting in a total of nine DEQ filters. The solid line Fig. 1 shows an example of spectral *differences* between DEQ filters designed to produce powertrain sounds with high and low loudness.

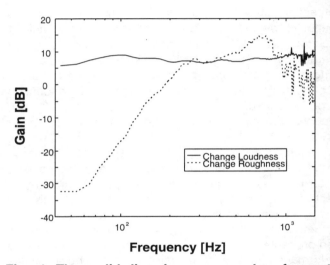

Fig. 1 The solid line is an example of spectral differences between DEQ filters designed to produce powertrain sounds with high and low loudness. Similarly, the dashed line is an example of spectral differences between DEQ filters designed to produce powertrain sounds with high and low roughness.

In theory, the roughness and loudness of the DEQ output are controlled independently for the set of filters discussed in this and the previous sub-section. However, the loudness and roughness of the overall powertrain sound heard in the vehicle will covary due to the presence of the vehicle's unmodified powertrain sound. This effect is discussed further in the following section.

EXPERIMENTAL METHODS

The subjective in-lab listening evaluation was conducted in order to address three questions:

1. Can subjects discriminate among the various roughness and loudness equalizer settings?

2. If the subjects can discriminate among the various levels, do the roughness and loudness levels effect their performance impression?

3. Does the effect of loudness and roughness on performance impression change when going from an in-lab evaluation to an in-vehicle evaluation?

The first two questions will be addressed in this paper through both paired-comparison and semantic differential evaluations. A subsequent customer drive evaluation is planned to address the third question.

SUBJECTS - Thirty-two subjects took part in both the paired comparison and semantic differential evaluations. The subject pool was made up of both technical and non-technical employees.

STIMULI - Sounds were recorded in the vehicle for the nine DEQ filters. A binaural recording head was used to faithfully reproduce the acoustic environment of the vehicle interior. The driving maneuver used for recording and subjective playback was a heavy throttle acceleration from 2000 to 4800 RPM. All accelerations occurred over a 4 second interval.

In the following, roughness and loudness levels are identified by low (L), medium (M), and high (H) levels, and the powertrain sounds are labeled by their roughness/loudness combination. For example, the powertrain sound labeled LH corresponds to low roughness and high loudness. Roughness and loudness statistics corresponding to the nine powertrain sounds are listed in Table I. The roughness statistic is based on the metric described in [6], and the ISO 532B loudness metric [7] was used to calculate the N10 loudness statistic.

Table I. Roughness and loudness statistics for the nine powertrain sounds.

Powertrain Sound Label (Roughness, Loudness)	Roughness [asper*100]	N10 [sone]
LL	59.3	20.3
LM	62.9	22.0
LH	101.2	26.6
ML	61.5	22.5
MM	73.1	25.0
MH	110.1	28.8
HL	101.3	23.8
HM	94.2	26.4
HH	113.6	31.7

The metrics listed in Table I show how the powertrain loudness and roughness covary due to the presence of the vehicle's unmodified powertrain sound. However, the loudness does increase from lower to higher levels for a given roughness. Similarly, the roughness increases from lower to higher levels for a given loudness.

PAIRED COMPARISON EVALUATION - The paired comparison evaluation was conducted in order to determine which combinations of roughness and loudness levels contribute to the impression of performance. All possible paired combinations of the nine sounds were presented randomly to the evaluators. Full replication was performed to check evaluator repeatability. For each pair, subjects were asked to "select the sound out of the pair that gives you the best impression of performance." Results from this evaluation will indicate which roughness and loudness settings have a significant influence on performance impression. This addresses the question of discrimination among the sounds in terms of their performance impression.

SEMANTIC DIFFERENTIAL EVALUATION - The semantic differential evaluation [8] was conducted in order to determine if there were perceived differences in the sounds based on four sets of bipolar adjectives: *quiet/loud*, *smooth/rough*, *weak/powerful* and *conservative/sporty*. The subjects were asked to evaluate all nine sounds based on these adjectives using the seven-point rating scale shown below:

extremely very somewhat neither somewhat very extremely

quiet---------|---------|---------|---------|---------|---------|---------loud

Results from this evaluation will indicate how subjects rate the sounds in terms of the individual adjective pairs. Moreover, results from the *quiet/loud* and *smooth/rough* pairs addresses the question of discrimination among the various loudness and roughness settings.

RESULTS AND DISCUSSION

PAIRED COMPARISON EVALUATION - When considering all 32 subjects, the repeatability for the group was quite low (65%), indicating the difficulty in discriminating between some of the sounds. Further analysis was performed on a reduced set of 19 evaluators having repeatability scores greater than 60%. David [9] outlines a nonparametric test of significance which determines if two objects (sounds) were judged to be significantly different. When testing to a 95% significance level, Fig. 2 shows the groups of sounds that are significantly different from each other. Sounds contained within the same bracketed group were not significantly different in terms of the performance impression. The groups are arranged from highest to lowest performance impression, with the highest performance impression to the left.

high perf. ⟨ HH LH MH HM MM LM ML LL HL ⟩ low perf.

Fig. 2 Groupings of sounds with similar performance impressions. Higher performance impressions are to the left.

It is clear from these observed groupings that the loudness level had a distinct effect on the performance impression. The louder sounds were generally judged to give an impression of higher performance. In support of the results shown in Fig. 2, results from a Bradley-Terry fit [9] to the raw paired comparison data are shown in Fig. 3.

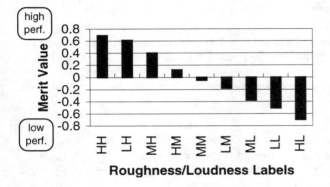

Fig. 3 Merit values from a Bradley-Terry fit to the raw paired comparison data.

SEMANTIC DIFFERENTIAL EVALUATION - After converting the seven possible ratings to numerical scores from -3 to +3, the semantic differential (SD) evaluations were analyzed using analysis of variance (ANOVA) techniques. Loudness, roughness and subjects were used as the possible factors in ANOVA. When subjects are not included as a factor in ANOVA, the subject's absolute ratings are analyzed. Similarly, inclusion of the subjects as a factor in ANOVA corresponds to an analysis of subject's relative ratings. (i.e., ANOVA subtracts out each listener's average response, and the analysis indicates how their ratings *changed* across the different powertrain sounds.)

The reason for including subjects as a factor is due to the possible use of different "internal" scales by different listeners. It is well-known that listeners are better at making comparison judgments than they are at making absolute judgments [10]. For example, a sound that is "extremely loud" to one listener may be "somewhat loud" to other listeners. Yet, when this sound is compared to another sound, both listeners are likely to agree as to which sound is louder.

Quiet/Loud - ANOVA indicated that loudness and subject factors were significant significantly ($p = 0.05$). Fig. 4 shows the mean response and 95% confidence interval for each powertrain sound. The powertrain sounds are characterized by their N10 loudness statistic. A +3 rating corresponds to "extremely loud" and a -3 rating corresponds to "extremely quiet". The circular groupings indicate mean responses that are not significantly ($p = 0.05$) different from each other. Although some sounds were similar in their loudness, the trend in Fig. 4 shows that subjects were able to correctly identify the louder sounds.

When each subject's responses are adjusted to be 0-mean, the mean response and 95% confidence interval for each powertrain sound is shown in Fig. 5. The mean responses are merely offset by a constant from those in Fig. 4, but the 95% confidence intervals are smaller than those in Fig. 4. Also, more of the mean responses are significantly different from each other. This suggests that different subjects were using different "internal" scales when rating the sounds. Similar differences between absolute and relative ratings were observed in the results for the other SD pairs. In the following, only the relative ratings are presented.

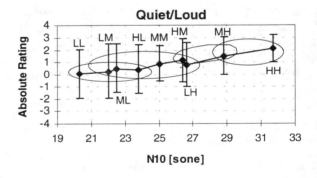

Fig. 4 Mean response and 95% confidence interval for each powertrain sound. The powertrain sounds are characterized as a function of their N10 loudness statistic. A +3 rating corresponds to "extremely loud" and a -3 rating corresponds to "extremely quiet". The circular groupings indicate mean responses that are not significantly ($p = 0.05$) different from each other.

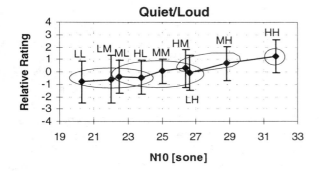

Fig. 5 Same as Fig. 4, except ratings are relative.

Rough/Smooth - ANOVA indicated that roughness and subject factors were significant. Fig. 6 shows the mean response and 95% confidence interval for each powertrain sound after each subject's responses were adjusted to be 0-mean. The powertrain sounds are characterized by their roughness values. As indicated in this figure, subjects were able to classify the sounds along the smooth/rough dimension. Combined with the results from the *quiet/loud* SD evaluation, these results suggest that subjects were able to classify the sounds along the dimensions for which the sounds were characterized (i.e., loudness and roughness). However, the relatively small changes in the mean responses of Fig. 5 and Fig. 6 also indicate that subjects thought many of the sounds were similar in loudness and roughness.[1]

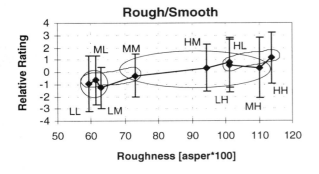

Fig. 6 Mean response and 95% confidence interval for each powertrain sound after each subject's responses were adjusted to be 0-mean. The powertrain sounds are characterized as a function of their roughness. The circular groupings indicate mean responses that are not significantly different from each other.

Weak/Powerful, Conservative/Sporty - ANOVA indicated that loudness and subject factors were significant for both the *weak/powerful* and *conservative/sporty* SD pairs. Fig. 7 shows results for the *weak/powerful* pair, and Fig. 8 shows results for the *conservative/sporty* pair. The mean responses shown in Fig. 7 indicate that subjects thought many of the sounds were similar along the weak/powerful dimension. However, the mean responses shown in Fig. 8 indicate that subjects thought the sounds spanned a larger range of the conservative/sporty dimension. The mean responses of this SD pair are also the most correlated ($R^2 = 0.88$) to the merit values obtained the paired-comparison evaluation. This suggests that the impression of performance is highly correlated with the conservative/sporty dimension.

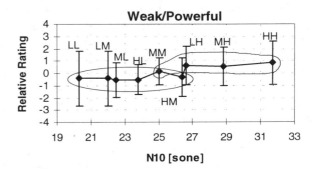

Fig. 7 Mean response and 95% confidence interval for each powertrain sound after each subject's responses were adjusted to be 0-mean. The powertrain sounds are characterized as a function of their N10 loudness statistic. The circular groupings indicate mean responses that are not significantly different from each other.

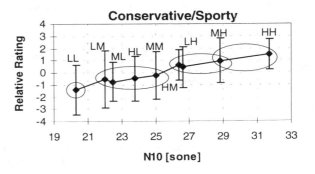

Fig. 8 Same as Fig. 7, except relative ratings are for the conservative/sporty SD pair.

FUTURE WORK

A customer drive is planned to investigate the effect of powertrain sound on perceived vehicle performance. The in-vehicle system will be used to modify the loudness and roughness of the powertrain sound, and subjects will rate their perceived performance of the vehicle during five different maneuvers. Results from this evaluation can be compared to those presented in this paper. From this comparison, the interaction of

---

[1] Some subjects verbally indicated that many of the engine run-ups sounded similar.

powertrain sound with other factors (e.g., acceleration, transmission shift) can be ascertained.

CONCLUSION

An in-vehicle system for the purpose of evaluating the influence of powertrain sound on a customer's performance impression has been described. The system is beneficial in that it allows for investigating effects of loudness and roughness on performance impression in the absence of other confounding effects. Powertrain recordings with various loudness and roughness characteristics were made with the in-vehicle system, and a listening evaluation was conducted. The paired-comparison evaluation showed that loudness had a distinct effect on the performance impression, with louder sounds giving higher performance impressions. The semantic differential evaluation showed that although subjects could discriminate the loudness and roughness of the sounds, only loudness affected their ratings for *weak vs. powerful* and *conservative vs. sporty*. The louder sounds were perceived as more powerful and more sporty. In comparing the paired-comparison and semantic differential ratings, the *conservative/sporty* ratings were most correlated to the merit values from the paired-comparison evaluation. These results provide a useful base from which to compare in-vehicle evaluations of performance impressions.

ACKNOWLEDGMENTS

The authors would like to acknowledge people associated with this paper. B. John Feng of Ford Motor Company for his assistance with this project. Nae-Ming Shiau and David Johnson of Ford Motor Company for their initial development of the in-vehicle system to synthesize powertrain sounds. Also, the Performance Feel QFD team at Ford Motor Company.

REFERENCES

[1] N. C. Otto, B. J. Feng, "Automotive sound quality in the 1990's," *3rd Intl. Conf. on Structure-born Sound and Vibration*, Montreal, 1994.

[2] M. Maunder, "Practical application of sound engineering," *Sound and Vibration*, May 1996.

[3] P. Garcia, K. Genuit, H. Strassner, B. Fuhrmann, "New approach regarding the objective evaluation of an exhaust system," *Inter-noise*, Newport Beach, CA, 1995.

[4] J. Hartog. *Mechanical Vibrations*. Dover Publications Inc. New York, NY, 1985.

[5] T. Suzuki, F. Kayaba, "The analysis and mechanism of engine intake rumbling noise," *SAE Paper 901755*.

[6] B. J. Feng and G. H. Wakefield, "A model for roughness of powertrain noise," *Proc. of Noise-Con 96*, Oct., 1996.

[7] M. A. Blommer, G. H. Wakefield, S. Jones, N. C. Otto, and B. J. Feng, "Calculating the loudness of impulsive sounds," *Proc. of SAE Noise and Vibration*, May, 1995.

[8] N. K. Malhotra. *Marketing Research-An Applied Orientation*. Prentice Hall. Englewood Cliffs, New Jersey, 1993.

[9] H. A. David. *The Method of Paired Comparisons*. Oxford University Press. New York, NY, 1988.

[10] M. Meilgaard, G. V. Civille, and B. T. Caar. *Sensory Evaluation Techniques, 2nd Ed*. CRC Press. Boca Raton, FL, 1991.

971984

The Investigation of a Towed Trailer Test for Passenger Tire Coast-By Noise Measurement

James K. Thompson
Automated Analysis Corp.

Thomas A. Williams
Hankook Tire

Copyright 1997 Society of Automotive Engineers, Inc.

ABSTRACT

It is difficult to quantify the portion of coasting vehicle noise that is due to tire-pavement interaction alone. There are often contributions from aerodynamic noise of the vehicle, transmission whine, noise from suspension components, and other miscellaneous sources. The towed-trailer method used in the revised SAE J57 standard has been shown to be an effective means of isolating tire-pavement noise for truck tires. This paper reports the results of a test program conducted by SAE Tire Noise Standards Committee to evaluate the feasibility of towed-trailer coast-by testing of passenger and light truck tires.

The results of tests conducted in April 1996 at the Ohio Transportation Research Center are described and they indicate that accurate measurements are possible for towed-trailer testing of passenger tires. It is shown that a key aspect of performing such a test is reducing the noise of the tow vehicle and that sufficient reduction is possible even for extremely quiet test tires.

During the time in which the SAE committee was preparing to conduct this test, the ISO/ TC 31/ WG 3 working group developed a draft trailer test method for evaluating tire noise. This test procedure was also evaluated during the April 1996 testing sequence. The results show the proposed test method to be highly accurate and a viable means for tire noise evaluation.

INTRODUCTION

In investigating the appropriateness of developing a recommended practice for the measurement of passenger and light truck tire noise, the SAE Tire Noise Standards Committee found there were no standard test procedures for passenger tire coast-by testing. The committee considered a number of possible measurement procedures. These included a vehicle coast-by test, a standard tire in a comparative test, and a towed trailer test.

It was known that the ISO/TC 31/ WG 3 [1,2] working group was developing a towed trailer coast-by test for passenger, light truck, and medium truck tires. No measurements had been done to assess the validity of this test procedure. It was hoped that this procedure could also be evaluated for usage as an SAE standard practice.

After considering all these alternatives, the committee concluded, based on its experience with revising SAE J57 Sound Levels of Highway Truck Tires [2,3], the best method would be a towed-trailer test specification. The nature of this test procedure would be similar to that developed for J57 in that a single axle trailer would be used and multiple passes with the tow vehicle alone and the tow vehicle and trailer combination would be necessary to compute a tire sound level. However, there was much concern as to whether such a procedure was feasible for passenger tires. Past experience in developing the revised J57 procedure and the tests to assure its validity [2] had shown that it is crucial that the towing vehicle noise be kept as low as possible. The J57 procedure specifies a tow vehicle alone coast-by level 3 dBA below that of the tow vehicle and trailer combination. The validation tests for J57 showed that this is not easily achieved with the quietest medium truck tires.

Therefore, an investigation was undertaken to determine whether it would be possible to develop a tow vehicle that was sufficiently quiet to test passenger tires and that would have the capability to tow the loads necessary for the evaluation of the largest light truck tires. To determine if such a test procedure was feasible, a test was to be conducted exploring the worst case scenario: a tow vehicle equipped to pull the largest loads expected for light truck tires and the quietest passenger tires. In addition, the ISO/TC 31/WG 3 draft procedure would be evaluated as a part of the test. This investigation is reported here.

TEST DEVELOPMENT

The first step in developing this test was to define the critical parameters. After going through possible test scenarios loosely based on SAE J57, but with loads more appropriate for passenger and light truck tires, it was clear that the critical tests parameters were the allowable tow vehicle coast-by noise level, the minimum expected passenger tire coast-by noise level, maximum towing weight, and acceleration and braking capabilities.

MINIMUM PASSENGER TIRE SOUND LEVEL - The first critical parameter was to determine what were the minimum passenger tire sound levels which would be encountered. Based on input from 5 of the largest tire companies in the world, it was determined that the minimum tire sound level at 80 Km/h for present and anticipated tire designs in the foreseeable future would be approximately 70 dBA when measured at the standard 7.5 m. distance. This level would represent a small summer design passenger tire with relatively low load carrying capacity. The Goodyear Tire and Rubber Company provided a P145/80R13 tire with extremely low noise levels. Goodyear personnel anticipated sound levels of approximately 71 dBA at 80 Km/h for this tire. A benchmark of 70 dBA at 80 Km/h was used for the quietest tire.

ALLOWABLE TOW VEHICLE SOUND LEVEL - Based on the anticipated tire sound levels it was clear that the tow vehicle sound level at 80 Km/h would have to be less than 70 dBA. If it were 70dBA then the combined tow vehicle and trailer coast-by level would be a maximum of 73 dBA. This would just meet the SAE J57 criterion the tow vehicle alone being 3 dBA quieter than the tow vehicle and trailer combination. Therefore, the goal was to achieve a sound level of approximately 67 dBA at 80 Km/h.

MAXIMUM TOWING WEIGHT - One of the critical issues for developing a standard practice covering passenger and light truck tires was the large span in load range for these tires. At the low end the passengers tires have a maximum load of approximately 454 Kilograms (1000 pounds). At the upper end of the range the highest load was approximated using the maximum load for a P275/60R17 tire. Using 70% of the rated load as described in SAE J57, the maximum load would be 742 Kg. (1636 Lb.) per tire. This number was rounded to give a maximum towing capability requirement of 1542 Kg. (3400 Lb.) load.

TOW VEHICLE ACCELERATION AND BRAKING CAPABILITIES - It was not possible to specify limits for braking distance or acceleration distance to reach 80 Km/h by the tow vehicle with the maximum load. Since there are wide variations in the facilities available for such tests, there are different requirements for different test tracks. Instead, it was determined that the tow vehicle used would have to be rated for towing such a load in normal operation and that nothing could be done in quieting the tow vehicle to compromise this capability.

VEHICLE SELECTON AND TREATMENT - There were several possible vehicles that met the towing requirements. Based on vehicle availability a 1995 Ford Crown Victoria was selected for use in this test.

To achieve the required reduction in the vehicle coast-by level, sound barriers with absorptive interior surfaces were placed around each wheel well, as shown in Figure 1. In addition the sides of the vehicle between the front and rear wheels were enclosed with the same sort of treatment. The rear of the vehicle below the bumper was enclosed in a similar fashion. The bottom edge of each of these enclosures was made up of a flexible material that minimized the open area. The enclosures around the wheels were large enough to prevent interference with steering. An aluminum skin was used on the exterior of these enclosures for protection from the weather, to provide the necessary strength, to provide the required sound transmission loss, and to minimize turbulence and aerodynamic noise generation. The sound enclosures and their application are explained in further detail in a separate paper by Harris and Williams [4].

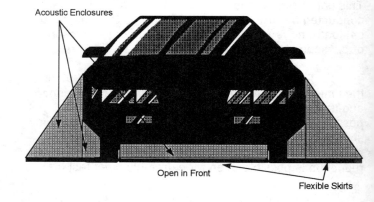

Figure 1 - Tow Vehicle with Sound Enclosures

TRAILER - The last required item to perform coast-by testing was a trailer for the test tires. A trailer was designed and its construction overseen by Brad Dibble of the Cooper Tire & Rubber Co. Using a twist-beam axle for suspension this trailer was simple in design providing a variable length drawbar and platform for adding ballast as shown in Figure 2.

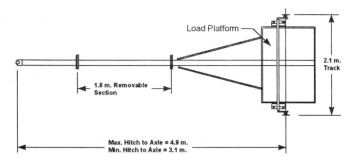

Figure 2 - Tire Test Trailer Layout

COAST-BY TESTING

Once all the components were assembled a series of tests were conducted at the Transportation Research Center in Marysville, Ohio. All tests were conducted on the ISO 10844 compliant surface since this is required by both the SAE J57 and ISO/TC 31/WG 3 procedures.

The first series of test runs were conducted in accordance with the practices of SAE J57. Although this procedure is written for truck tire testing, the basic practices apply equally well to this passenger tire test.

As specified in the SAE procedure the test section was configured according to Figure 3. To avoid interference and safety concerns with tests going on in an adjacent portion of the test track, measurements were made for the vehicle traveling in only one direction. Only one microphone was used on the passenger side of the vehicle during coast-by.

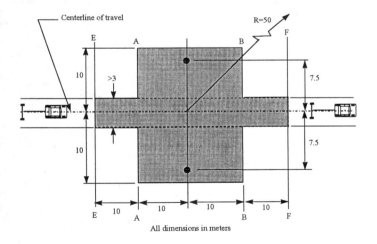

Figure 3 - Standard SAE/ISO Test Section

The tests conducted are outlined in Table 1. One difficulty encountered from the beginning and through all tests reported here was high wind speeds. The average wind speed was well in excess of the 5 m/s specified by the J57 procedure and the ISO draft. Unfortunately, limited track availability prevented waiting until the winds diminished. However, the measurements made are still valid for comparative purposes. The impact of the high wind speeds on results will be discussed further in a later section.

Table 1 - Collected Data for Tests at TRC

Event	Speed, Km/h	L_t	L_{tp}	L_{tire}	Notes
1	60	55.6			TVO treated
2	80	60.7			TVO treated
5	60	55.6	61.4	60.1	ISO Config.
6	80	60.7	65.5	63.8	ISO Config.
13	60	61.5			TVO untreated
14	80	64.1			TVO untreated
15	60	55.7			TVO, ISO Configuration
16	80	61.9			TVO, ISO Configuration

TVO = Tow vehicle only

ISO Config. = ISO configuration, timing system shifted by d_t

L_t = average time history sound level of the tow vehicle without trailer

L_{tp} (t) = average time history sound level of the test pass (tow vehicle- trailer combination)

ISO/TC31/WG 3 TESTS - The test procedure specified in the ISO document [1] is somewhat unusual in that it requires changes in the timing of noise measurements to compensate for the length of the tow vehicle and draw bar. The concept is to insure that measurements of the tow vehicle alone and the tow vehicle and trailer combination are aligned correctly in time. The subtraction of these time histories permits an accurate calculation of the tire sound level without additional corrections.

The time history measurement begins with the definition of lines A'A' and B'B' as shown in Figure 4. These lines are defined as using the length d_t, the distance from the center of the test tire to the trigger point on the tow vehicle, see Figure 4. The trigger point is the point on the vehicle which will cause a mark on the time record when it passes lines A'A' and B'B'. Sound recording begins and ends at these times, respectively. The same recording procedure is used for passes of the tow vehicle and trailer combination and for the tow vehicle alone. This causes sound level measurements to be made as if the trailer with test tires was traveling through the test zone even when the tow vehicle alone is tested.

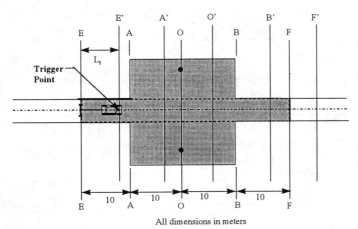

Figure 4 - ISO/TC 31/WG 3 Test Layout

The subject tyres' sound levels reported are the differences between the averaged sound pressure level of the tow vehicle-trailer combination and the averaged sound pressure level of the tow vehicle alone. To compute this difference, the averaged sound level time history for the tow vehicle must be subtracted from that for the tow vehicle and trailer combination. The averaged sound levels are calculated from the data obtained for the five runs in which the maximum sound levels are within ±0,5 dB of the mean of the maximum levels. An example of time history sound levels are shown in Figure 5.

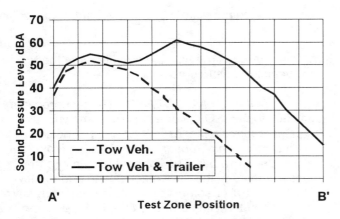

Figure 5 - Example of ISO/TC 31/WG 3 Time History

After the alignment of the time histories with respect to O'O', a key parameter in the analysis is the difference between the maximum average time history level for the combination (trailer and tow vehicle) and the tow vehicle alone average time history level at the same point. This difference is highlighted in Figure 5.

If this difference is equal to or greater than 10 dB, the levels measured for the tow vehicle and trailer combination are the correct values for the tyre under test. If this difference is less than 10 dB, then the tyre sound level is calculated by the logarithmic subtraction of the combined and the tow vehicle levels as shown below. This logarithmic subtraction is performed using the average time history levels noted above and shown in Figure 5.

$$L_{tyre} = 10 \log_{10}\left[10^{\left(L_{tp}/10\right)} - 10^{\left(L_{t}/10\right)}\right]$$

The reported tire sound level is the result from the above equation.

DISCUSSION OF RESULTS

The first series of tests was made with the tow vehicle alone to assess how well the treatment was working. The results of these tests as shown in Table 1 were encouraging. At 60 Km/h the measured coast-by level was 55.6 dBA. At 80 Km/h the level was well below the target of 67 dBA at 60.7 dBA. These levels are the average of 5 runs according to the SAE J57 procedure [3]. Although runs were made with the treated tow vehicle alone and the timing points shifted according to the ISO procedure these cannot be compared with Events 1 and 2. For the ISO configuration the sound level measurements are made with the tow vehicle at different positions in the test section. Looking at Figure 6 it can also be seen that the tow vehicle will be at a slightly higher speed near the microphone position which will tend to give slightly higher sound levels. This should be a minimal increase in sound level since the speed range from maximum to minimum is small for this coast down. It is interesting to note that, comparing events 1 and 2 to events 15 and 16, the measured levels for the ISO configuration are slightly higher.

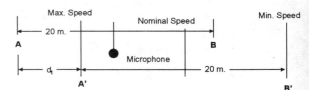

Figure 6 - Illustration of Timing and Tow Vehicle Position Differences for SAE and ISO Procedures

Measurements were also made of the tow vehicle alone without the sound enclosures attached to the vehicle. These results are also shown in Table 1. At 60 Km/h the measured coast-by level was 61.5 dBA and at 80 Km/h the level was 64.1 dBA. Comparing this to the levels for the treated tow vehicle, a reduction of 5.9 dBA was achieved at 60 Km/h. At the 80 Km/h speed the reduction in sound level is 3.4 dBA. The effectiveness of the enclosures appears to diminish with higher speeds. It is possible that this is an indication that aerodynamic noise is more dominant at higher vehicle speeds. Based on these measurements the tow vehicle could be used for tires with sound levels as low as approximately 64 dBA at 80 Km/h. A more detailed discussion of the effectiveness of the vehicle treatment can be found in a separate paper by Harris and Williams [4].

SAE J57 TEST PROCEDURE - Using the measured tow vehicle alone levels and the data from events 5 and 6 it is possible to calculate the tire sound levels as shown in Table 1 [2,3]. The tire was found to have a sound level of 60.1 dBA at 60 Km/h and 63.8 dBA at 80 Km/h. These values are lower than those expected by the manufacturer. At 80 Km/h the combination of the tow vehicle and the trailer produces only 65.5 dBA. This would indicate that the tire levels were overestimated by manufacturer. Fortunately, the vehicle treatment worked well enough to maintain the necessary 3 dB difference between L_{tp} and L_t.

Comparing the tow vehicle alone sound level to the combined tow vehicle and trailer levels it is clear that a valid test was obtained. The differences in these levels were 5.8 dBA at 60 Km/h and 4.8 dBA at 80 Km/h. In both case the level differences are greater than the required minimum of 3 dBA.

ISO PROCEDURE - Using the data collected in Events 15 and 16 for the tow vehicle in combination with the data from Events from 5 and 6 for the tow vehicle and trailer combination it is possible to compute tire sound levels from time history subtraction. This procedure is described in detail in the ISO draft [1]. For the levels in this case, the procedure consisted of finding the point in time representing the highest level for the tow vehicle trailer combination. The tow vehicle alone level at this same point in time is subtracted logarithmically from the combination level to give the tire level. The calculated level at 60 Km/h was 60.7 dBA. The calculated level at 80 Km/h was 63.3 dBA. These compare quite well with the results of the SAE method computation. The level difference between the ISO and SAE procedures is 0.6 dB or less.

The average time histories for the tow vehicle alone, the tow vehicle and trailer, and the computed tire sound level are shown in Figures 7 and 8 for 60 and 80 Km/h, respectively. For these and subsequent figures showing ISO data, the 20 m test zone begins at 0.48 seconds and ends at 1.72 seconds for the 60 Km/h runs and at 1.42 seconds for the 80 Km/h runs. Figures 7 and 8 represent the average of 5 passes all within ± 0.5 dBA of their mean. In both these cases the maximum level for the tow vehicle is just inside the test zone and the tow vehicle noise has declined greatly by the time the maximum tow vehicle and trailer combination level is reached. The tow vehicle levels at this maximum point are down 1.5-2.0 dBA from their peaks. As noted previously the differences in the maximum levels for the tow vehicle alone and the combination are 5.8 and 4.8 dBA. However, at the critical time of maximum level for the combination, the differences are 7.6 and 5.0 dBA. Thus an accurate calculation of the tire coast-by level would be anticipated assuming all other aspects of the test were conducted properly.

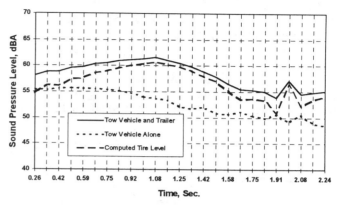

Figure 7 - Measured Time History Data at 60 Km/h.

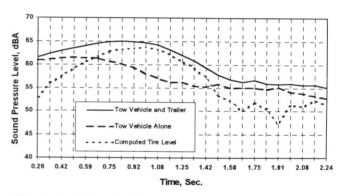

Figure 8 - Measured Time History Data at 80 Km/h.

Variation among test runs - To examine the multiple runs that make up one of these averages the runs encompassed in the average for the tow vehicle and trailer combination for 80 Km/h are shown in Figure 9. There is a great deal of variation in the sound levels of runs at the entry to the test zone. However near the region where the maximum level is achieved there is much greater consistency between runs.

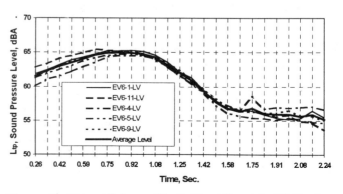

Figure 9 - Time History of Individual Runs In 80 Km/h Average.

Figure 10 provides the same type of data for the tests of the tow vehicle alone at 60 Km/h. It is clear that there is much greater variation for these runs at the lower speed. Again the variation seems large as the test tire enters the test zone and gets smaller near the point of the maximum level. In this case there appears to be one run that is consistently higher than the others. Also, at the end of the test zone large variations appear. These are most likely due to wind gusts since the measured level has dropped to roughly the ambient background sound level. There is clearly more variance in the lower speed sound levels which is most likely due to the lower levels being closer to the magnitude of the background level fluctuations due to wind gusts.

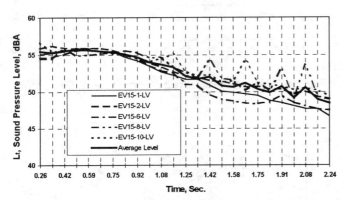

Figure 10 - Time History of Individual Runs In 60 Km/h Average.

CONCLUSION

Based on this study there are several conclusions which can be drawn concerning tire coast-by noise testing.

It is possible to sufficiently quiet a tow vehicle to the point that coast-by testing can be conducted for the smallest and quietest passenger tires to the highest load range light truck tires.

The SAE J57 recommended practice is a viable method for conducting such coast-by tests. With minimal adaptation, this truck test procedure can be used to quantify passenger tire sound levels.

The draft ISO/TC 31/WG 3 towed trailer test procedure is also a viable means for quantifying tire coast-by noise.

Unless the difference between the maximum tow vehicle alone and the maximum combination tow vehicle and trailer sound level is greater that 10 dB, these two methods, SAE and ISO, should not give the same results. This is true since in one case the maximum levels are being subtracted and in the other the level at a corresponding point in time is being subtracted from the combination maximum. When there is greater than 10 dB difference between the levels as noted above there is no need to perform the subtraction, and thus, the results will be the same.

The tests reported here should be repeated under better environmental conditions to better define the best method for tire sound level evaluation. A more extensive array of tests should be included to provide definitive comparisons between the methods with differing ranges between the tow vehicle alone and combination sound levels.

The requirements of the SAE and ISO procedures for runs within ± 0.5 dB of the mean to be averaged to determine the tire level is clearly necessary. As was shown in the plots of multiple runs, the averaging process eliminated possible errors due to sound level variations.

ACKNOWLEDGMENTS

The authors wish to acknowledge the support and participation by the members of the SAE Tire Noise Standards Committee who worked so diligently to make these tests possible and prove that sufficient tow vehicle quieting could be achieved in a practical fashion. Key members of this committee who contributed in a major way were: John R. Harris of Continental-General Tire, R. Brad Dibble and Craig Selhorst of Cooper Tire & Rubber, Alan R. Hartke of Goodyear Tire & Rubber, R. Ellis Johnson of Michelin America, and Steven Butcher of the Rubber Manufacturers Association.

In addition Rich Adrian of EAR Inc. provided acoustical foams and other materials for the sound enclosures on the vehicle. He also provided advice as to the best approaches to reach the level of attenuation required. Hankook Tire and Roush Anatrol provided financing for the many additional pieces that were required to assemble the vehicle sound enclosures.

The Rubber Manufacturers Association funded the manufacturing of the trailer used in these tests.

The Goodyear Tire and Rubber Company transported much of the material and equipment to and from the Transportation Research Center.

The support and cooperation of the Transportation Research Center is greatly appreciated. Milt Dunlop and Doug Bobb provided vital assistance and support during the committee's testing. TRC provided use of their facilities and ISO 10844 test surface for three days at no charge to further this investigation.

John Harris provided an extraordinary amount of effort in developing and fabricating the enclosures for the vehicle. His company, Continental-General Tire, supported this effort and provided the Ford Crown Victoria used for these tests.

REFERENCES

1. ISO Draft Test Procedure, WD 13325 as revised 9/16/96, see ISO document ISO/TC 31/WG 3 N35.

2. Thompson, J K, "Confirmation Tests for SAE J57 Test procedure Revision," 951353, Society of Automotive Engineers, Warrendale, Pennsylvania, 1995.

3. SAE J57 "Sound Levels of Highway Truck Tires," Society of Automotive Engineers, Warrendale, Pennsylvania.

4. Harris, JR and Williams, TA, "Tire Noise Reduction Treatment for a Passenger Car Used as a Tow Vehicle for Passby Noise Testing," 97NV171, Society of Automotive Engineers, Warrendale, Pennsylvania, 1997.

971985

Prediction of Vehicle Radiated Noise

Luigi Pilo
FIAT Auto

Francesco Gamba and Bernard J. Challen
Cornaglia Research Center

Copyright 1997 Society of Automotive Engineers, Inc.

ABSTRACT

Engineers need the most capable design support tools to reconcile the demands of the market, legislation, economic design and production. Simulation tools are a very effective support system.

The optimization of the complete vehicle system for external noise radiation is very complex due to the interaction of the engine performance with the vehicle positioning. Simple predictive methods and assumptions are often not usefully accurate. The work described here covers an approach to combining rapid predictions of engine and vehicle performance with those of radiated noise from the various significant subsystems, to produce a useful, rapid, prediction tool for complete vehicle noise. A mono-dimensional gas flow code is used to provide a sufficiently accurate prediction of the engine power for vehicle performance prediction and also the noise output of intake and exhaust systems.

Various approximate prediction methods are used to assemble an overall vehicle radiated noise prediction. This takes account of vehicle performance, size, component location and sub-sources such as the exhaust shell noise. The derivations of the approximations and the assumptions in the code are given. The results of various examples illustrate the overall effectiveness of this combined approach to vehicle noise modeling. Continuing development is underway which uses an improved flow code (GTPower) which effectively covers both the engine pipe gas flows and the performance modeling, including the acoustic radiation from the intake and exhaust.

INTRODUCTION

It is becoming increasingly important to reduce the noise generated by vehicles, their engines, transmissions and necessary subsystems such as intake and exhaust. Both environmental and commercial pressures mean that improvements must be made while at the same time leaving the engine performance unaffected.

Market demands require higher levels of comfort and there are increasing requirements of pollution legislation. In 1973 the pass-by noise limit was 82dBA; it was reduced to 80dBA in 1980, to 77dBA in 1987 and now is 74dBA. It is probable that the level will soon be reduced to 71dBA.

At these lower levels the noise from the intake and exhaust systems makes a more significant contribution to the overall noise level of the vehicle. Engineers must design systems using sensible compromises to achieve better performance within the limits set for the total radiated noise.

There are great pressures to reduce development time and thus costs of new models. To support this lead-time reduction, computer aided design and simulation is used to reduce prototype testing in test-beds and cars, and to increase confidence in the final design.

Noise and performance development work has employed considerable testing, taking time and extending overall program time and costs. Predictive calculation and design tools can assist powertrain system design and be a vital requirement in maximizing system efficiency.

Systems must be fast and accurate simulations of the real case and readily usable by engineers in design and development programs.

Combining the car acceleration, speed and position data with the radiated noise levels from the intake and exhaust systems allows us to develop an analytical method to predict the complete vehicle pass-by noise level, based on real values. There have been various developments in this field over the years [1, 2, 3, 4] and this present work extends the capabilities of such models.

The present model can also take into account the aerodynamic behavior of the exhaust system flow noise, which becomes more important at higher frequencies. A new prediction method has been developed where the sound pressure level is proportional to the Mach number to the power of 4.5.

Combining the acoustic radiation calculation from the intake and exhaust systems with the kinetics of the vehicle, we can predict the peak vehicle noise level and its position during the test.

From this information we can calculate the vehicle acceleration under test and so find the engine speed in the relevant gear which is most significant for the overall optimization of external noise radiation.

PASS-BY TEST

European law requires manufacturers to produce cars which comply with the ISO 362 pass-by test at a level of 74dbA - this is at a distance from the centerline of 7.5m. The pass-by test specified in ISO 362, for normal performance cars, is carried out in 2nd and 3rd gear for manual transmission versions.

The car approaches the test area (fig.1) at a constant speed of 50 km/h. At the start of the test area, point A, the car accelerates at wide-open throttle (WOT) conditions, up to the end of the test area at point B. The microphones, located at positions C and C1, are at 7.5 m from the centerline, in the middle of the test area and at 1.2 m from the ground. They measure the instantaneous noise level. The test is carried out in both 2nd and 3rd gears and the maximum noise levels noted for each.

To determine the homologation value, the maximum values are arithmetically averaged. The European limit today is 74dBA.

VEHICLE PERFORMANCE

In order to determine the acoustic behavior in the pass-by test it is necessary to know the dynamic behavior of the car (specifically the longitudinal acceleration). Part of the present prediction code was developed to simulate the acceleration of the car along the pass-by test starting from a known speed.

We have movement of the accelerating car when the engine power available at the wheels (Pm) is greater than the sum of the retarding power due to the rolling and aerodynamic forces.

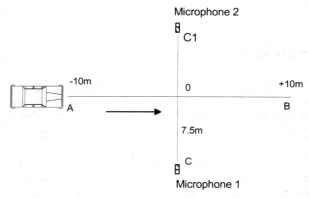

Fig. 1   ISO pass-by test

The kinetic energy of the car is the sum of the translating and rotating energy :

$$E_c = \frac{1}{2} m v^2 + \frac{1}{2} J_M \frac{Z_P^2 Z_M^2}{R_0^2} v^2 + \frac{1}{2} J_T \frac{Z_P^2}{R_0^2} v^2 + \frac{1}{2} \sum_K \frac{J^{(K)}_R}{R_0^2} \quad (1)$$

where :
- m = car mass [kg]
- J_M = Engine inertial moment [kg *m2]
- J_T = Transmission moment of inertia [kg *m2]
- J_R = Wheel moment of inertia [kg *m2]
- Z_P = total gear ratio
- Z_M = transmission gear ratio
- R_0 = rolling radius [m]
- v = speed [m/s]

We define translation mass as :

$$M_T = m + \frac{1}{2} J_M \frac{Z_P^2 Z_M 2}{R_0^2} + \frac{1}{2} J_T \frac{Z_P}{R_0^2} + \frac{1}{2} \sum_K \frac{J^{(K)} R}{R_0^2} \quad (2)$$

and equation (1) becomes :

$$E_c = \frac{1}{2} m_T v^2 \, . \qquad (3)$$

Deriving the equation (3) versus time we have :

$$\frac{dE_c}{dt} = m_T v \frac{dv}{dt} = P_{ex} \qquad (4)$$

where :

$$P_{ex} = P_d - (P_{aer} + P_{rol}) \qquad (5)$$

The vehicle accelerates forward if the first part of (5) is $P_{ex} > 0$, while if $P_{ex}=0$ we have constant velocity motion.

In equation (5) we define Available Power as:

$$P_d = \eta \, P_m \qquad [kW] \qquad (6)$$

the engine power related to the wheels and η is the transmission efficiency.

We define Rolling Power as :

$$P_{rol} = A v + B v^3 \qquad [kW] \qquad (7)$$

that is, the resistance power associated with the rolling mode. The coefficients A and B are taken from the SAE standard for rolling resistance.

We define the aerodynamic power as :

$$P_c = \frac{1}{2} \rho \, C_x S v^3 \qquad (8)$$

where :

ρ = air density $[kg/m^3]$
C_x = the aerodynamic coefficient
S = the vehicle frontal area $[m^2]$

The Available Power $P_{ex} = P_{ex}(v)$ is a function of the vehicle speed, so from (9),

$$P_{ex} = P_{ex}(v) = m_T v \frac{dv}{dt} \, . \qquad (9)$$

We integrate, choosing speed as the independent variable and then we have :

Time [s]

$$T = T_0 + m_T \int_{v_{in}}^{v_{out}} \frac{v dv}{P_{ex}(v)} \qquad (10)$$

Distance [m]

$$S = S_0 + m_T \int_{v_{in}}^{v_{out}} \frac{v^2 dv}{P_{ex}(v)} \qquad (11)$$

Acceleration $[m/s^2]$

$$\frac{dv}{dt} = \frac{P_{ex}(v)}{m_T v} \qquad (12)$$

Engine speed [RPM]

$$RPM = \frac{60 Z_P Z_M}{2\pi R_0} v \qquad (13)$$

From equations (10), (11), (12) and (13) we can connect the motion of the car during the pass-by test with the speed of the engine (RPM) and so we can correlate the noise sources with the vehicle position.

NOISE SOURCES

Aerodynamic noise can be neglected without serious error during the pass-by test since the vehicle speed is low. Even with full acceleration, speed increases typically from 50 to some 60km/h.

For this reason phenomena where the noise increases very rapidly with speed are very important. The radiated noise from the vehicle during the pass-by test depends principally on these sources :

1. engine
2. tires
3. intake system
4. exhaust noise

For each of these sources we have developed an approximation approach, using various techniques. The geometric position in the vehicle, in all three axes, is normalized to the vehicle dimensions for distance effects to be calculated.

ENGINE NOISE - The characterization of the engine source is complicated because it is not possible to consider it a point source. It is therefore difficult to have accurate information on the sound power level since generally only measurements of engine sound pressure level are available and this information is not completely sufficient; it is always preferable to measure the engine sound power. To avoid this limitation we use measurements (or an empirical relationship) and make an arithmetic average of all data at the measured points.

In this prediction code we use the formulae developed by Challen and others [2, 9] for different types of engines:

- Naturally aspirated DI Diesel
- Turbocharged DI Diesel
- High speed IDI Diesel
- Gasoline

These use a simple approximation based on the combustion system, the cylinder bore and the speed. The specific formula used here for gasoline engines depends on speed according to $I \propto RPM5$ where I is the sound intensity.

We use a plane source model to describe the engine noise propagation as illustrated in figure 2 (see also [6, 7, 8]). The engine Sound Pressure Level at R2 is given by:

$$(P_{rms})^2 = \frac{WZ_0}{\pi hb} arct(\frac{h}{2d}) arct(\frac{b}{2d}) \quad (14)$$

where h is the block height, b the block length, W the radiated sound power and Z_0 the specific acoustic impedance.

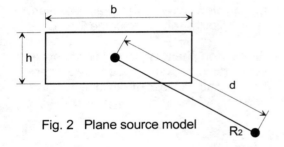

Fig. 2 Plane source model

If $\xi = h/2d$ and $\zeta = b/2d$ we have:

$$(P_{rms})^2 = \frac{WZ_0}{\pi hb} arct(\xi) arct(\varsigma) \quad (15)$$

We can divide the field of application for the analytical expression of noise sources into three ranges:

near field where h >> d and b >> d then

$$\xi >> \text{ and } \zeta >> 1, \text{ and}$$

$$arct\ \xi \sim \pi/2 \quad \text{and} \quad arct\ \zeta \sim \pi/2$$

So (15) becomes:

$$(P_{rms})^2 = \frac{WZ_0}{\pi hb} \frac{\pi}{2} \frac{\pi}{2} = \frac{WZ_0 \pi}{4hb} \quad (16)$$

From (16) we see that the sound pressure level near to the source does not depend on measurement range.

Intermediate field where $\xi << 1$ and $\zeta >> 1$ and (15) becomes

$$arct\ \xi \sim \pi/2 \text{ and } arct\ \zeta \sim \zeta$$

the SPL becomes

$$(P_{rms})^2 = \frac{WZ_0}{\pi hb} \frac{\pi}{2} \frac{h}{2d} = \frac{WZ_0}{4bd} \quad (17)$$

which is linear with distance (cylindrical wave).

Far field where $\xi << 1$ and $\zeta << 1$ then

$$arct\ \xi \sim \xi \text{ and } arct\ \zeta \sim \zeta$$

so that (15) becomes:

$$(P_{rms})^2 = \frac{WZ_0}{\pi hb} \frac{h}{2d} \frac{b}{2d} = \frac{WZ_0}{4\pi d^2} \quad (18)$$

which is the normal attenuation of a spherical wave.

The link between the measurement point 1 and at pass-by point of measure microphone 2 fig. (3) is given by the following model:

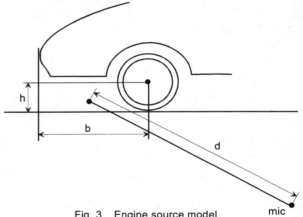

Fig. 3 Engine source model

For a spherical wave the pressure relationship is:

$$P(R) = \frac{E_0}{R} \quad (19)$$

where E_0 is the wave amplitude and the SPL is:

$$SPL(R) = 10 \log \frac{P(R)}{P_0} \quad (20)$$

where P_0 is the reference pressure (2×10^{-5} Pa). The relationship between two points 1 and 2 becomes:

$$SPL_{(R2)} = SPL_{(R1)} + 10\log(\frac{R_1}{R_2})^2 + \chi \quad (21)$$

where we can define an attenuation figure:

$$att = 10\log(\frac{R1}{R2})^2 + \chi$$

- the geometric attenuation, where:
$R_2 = 1 + \delta$ and

$$\delta = \frac{h}{\pi} + \frac{b}{\pi} \quad (22)$$

and χ is a -3 dB / doubling distance attenuation due to δ (end effect).

We also take into account the diffraction effect *diff* due to the sum of reflected and direct pressure waves with different path lengths, using the method of an imaginary point source, as illustrated in figure 4. In this model we have not taken into account the ground resistivity.

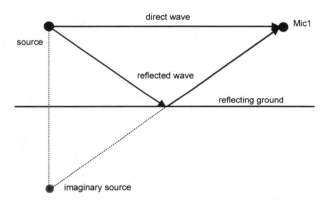

Fig. 4 Point source model

We also account for the source directivity *dir* of the engine using the data shown in figure 5.

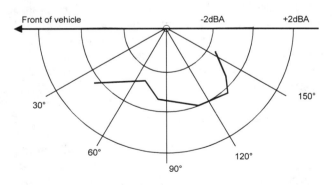

Fig. 5 Engine source directivity

SOURCE POSITION - The contribution from engine noise in the pass-by test is then given by the following formula, which uses the various factors accounted for above:

$$SPL_{M2} = SPL_{M1} + att + diff + dir \quad (23)$$

ROLLING NOISE - The noise generated from the contact between the tires and the ground is described by the empirical formula [2]:

$$SPL_{M1} = n \log v + q \quad (24)$$

where n is 32 and q is 11.2 *log (m_T) - 20.

It remains to correlate the noise at the real distance between the moving car and microphones at 7.5m.

The Rolling noise during the pass-by test is :

$$SPL_{M2} = SPL_{M1} + att \quad (25)$$

where *att* is the geometrical attenuation:

$$att = -20\log\sqrt{1 + (\frac{x}{R_1})^2} \quad (26)$$

Using the *apparent translation mass* (depending on the gear used, which changes Zm) instead of the real weight, since during the pass-by test the dynamics of the car are generated by acceleration.

INTAKE NOISE - The noise from the intake orifice is another significant vehicle source. The measurement location for the intake system is shown in figure 6.

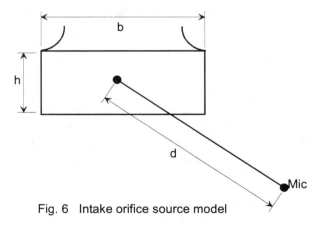

Fig. 6 Intake orifice source model

For the intake source we use a model of a finite plane source as shown in figure 6. The dimensions of the intake orifice are h and b, with d the distance between source and the microphone at 7.5m.

The equation used to calculate the SPL at distance during pass-by test is :

$$SPL_{M2} = SPL_{M1} + att + diff + dir \quad (27)$$

where *diff* is derived in the same way as for the engine. The directivity of the source in the frequency range between 30 to 500 Hz is shown in figure 7.

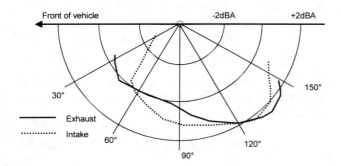

Fig. 7   Intake and exhaust source directivity

EXHAUST NOISE - The exhaust noise model at the tail pipe end is shown in figure 8.

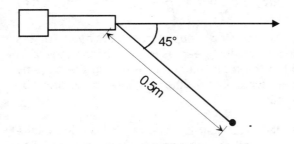

Fig. 8   Exhaust tailpipe model

At this distance (0.5m) we consider the exhaust outlet to be a point source and the equation used for the exhaust SPL is :

$$SPL_{M2} = SPL_{M1} + att + diff + dir \quad (28)$$

The directivity of the exhaust orifice is important for frequencies between 0.5 and 10 kHz (jet and flow noise components). A typical distribution is illustrated in figure 7. This typical data can be used or may be substituted by other available information.

EXHAUST SHELL NOISE - There are many sources of the noise radiation from the external shell of the exhaust system including direct vibration excitation and acoustic excitation: among these the most important is generally the fluctuations of the gas pressure inside the exhaust, with high rates of change, generating impulsive excitations.

The amplitude of the internal pressure waves decreases towards the end of exhaust system, as can be seen in figure 9, because of the attenuation of the silencers. For this reason we use the excitation level of the catalytic converter and of the first silencer.

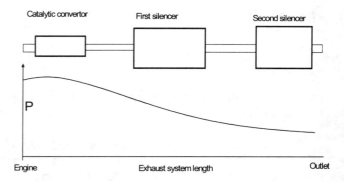

Fig. 9   Exhaust pressure distribution

The external noise propagation model used in the code is a finite line source model.

The shell noise will be :

$$SPL(R_1) = SPL(R_2) + att + diff \quad (29)$$

where *att* is given by:

$$att = 10 \log \frac{R2}{R1} + 10 \log arct(\frac{L}{2R1}) \quad (30)$$

The second term takes in account the actual line dimension.

To account for acoustic diffraction effects, the imaginary source model technique employed is always employed.

INTAKE SHELL NOISE - The same considerations, in principle, as for the exhaust system are used for the intake shell noise. Clearly the excitation levels are different but a similar balance between the various sources of excitation exists.

DISCUSSION AND RESULTS

We have validated the code with some experimental tests using pass-by results of a car with a 1.6 liter, 4 cylinder, 16 valve, gasoline engine. In order to provide the analysis of each separate noise source we silenced all car noise sources except the one for which we wanted to measure the noise level.

The techniques for achieving this type of analysis are all well known, requiring care in implementation and accurate testing, but can be used effectively.

In order to acquire accurate data, to ensure that the acquisition and analysis are carried out according to the specific requirements of the program, significant time is required. This present predictive technique allows us to focus more rapidly on the specific areas which require attention and to concentrate engineering effort on these.

The predictive code describes directly the acoustic propagation of noise generated by six different noise sources in the car.

These sources can be simulated either from any available experimental data or from calculated data alone, combined with experienced judgment.

ENGINE - We use to predict the engine noise source the empirical formula specified above. In the validation example the engine has a bore stroke ratio of 1.28.

This formula is related to an average of measurements made with a number of microphones at 1 meter of distance of the source and the curve is shown in figure 10.

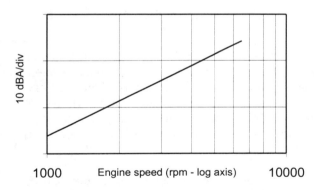

Fig. 10 Engine noise source (1m)

The comparison between calculated and test results - for this source alone - during the pass-by test are given in figure 11.

The comparison shows a good level of agreement: it must be borne in mind also that it is always very difficult, during the practical source testing, to isolate completely the engine source from other sources.

The prediction formula is a semi-empirical expression derived from a best fit of many engines. The tested powertrain was not included in the data.

TIRES - As mentioned earlier, the test vehicle aerodynamic noise is low because of the relatively low speed of the car.

The radiated tire noise depends by the car speed; under the present European car noise legislation, which imposes a 74 dBA limit, the noise contribution from the tires is now significant, especially in third gear.

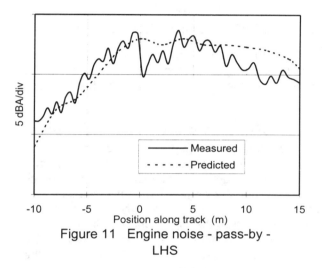

Figure 11 Engine noise - pass-by - LHS

During the pass-by test in both gears (second and third) the speed of the car is similar, so the maximum noise level from the tires is similar.

A reduction of the legal limit to 71 dB(A) will be introduced in the near future, possibly in the year 2000. In this case the contribution of the tires to the overall noise will be very important.

The comparison between calculated and test results for the tire source during pass-by testing is shown in fig. 12.

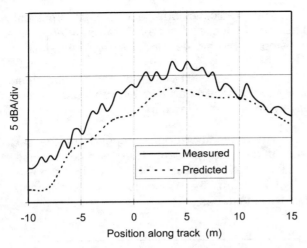

Figure 12 Tire noise - pass-by - LHS

The comparison again shows a good agreement, taking in account the difficulties in evaluating the contribution of the tires as a source, both in test and through calculation.

EXHAUST OUTLET - The noise radiated from the exhaust tail pipe outlet depends on a number of different physical phenomena: the two most important are the acoustic volume source noise generated by engine exhaust pulsations and aero-acoustic noise generated by the turbulence mechanisms inside the pipes and mufflers (both jet and flow noise). The first contribution is relevant in a frequency range between 30-800 Hz (from the second to eighth engine order); the second for frequencies above 1kHz, which are not directly connected with engine orders. The tail pipe orientation is very important in this source since this mechanism generates an intrinsic geometrical acoustic directivity. Tail pipe orifice noise source is illustrated in figure 13, measured at a range of 0.5m and at a 45° angle.

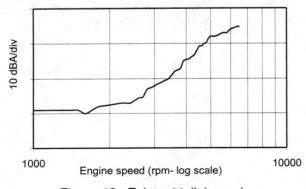

Figure 13 Exhaust tailpipe noise

Analyzing the measurements we find two different slopes in looking at the engine speed against noise:

- the first is up to 3000 RPM (breathing noise range) where the acoustic intensity I varies as the second power of RPM (specially in third gear);
- the second, above 3000 rpm, increases at the sixth power of RPM (a more rapid increase than engine noise, which is proportional to the fifth power).

The comparison between calculated and test results for this source during pass-by testing in figure 14. The comparison shows a satisfactory agreement.

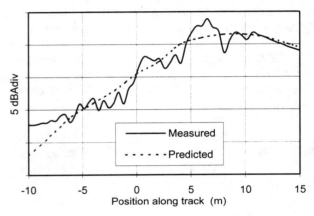

Fig. 14 Exhaust orifice - pass-by - LHS

EXHAUST SHELL NOISE - The noise radiated from the walls of the exhaust system depends principally by the vibrations of the pipe and muffler walls. Sources of exhaust surface noise are distributed throughout the exhaust system, so that the noise level varies from one part to another along the exhaust line.

The results of the simulation of this source are seen in figure 15.

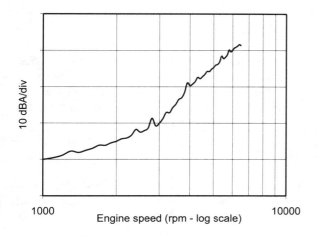

Fig.15 Predicted exhaust shell noise

Again in this case we find that the acoustic intensity has a strong dependence on speed, the rate of which also changes as the speed increases. Up to 3000 rpm, it increases at the third power and above this speed, at the 6.5^{th} power.

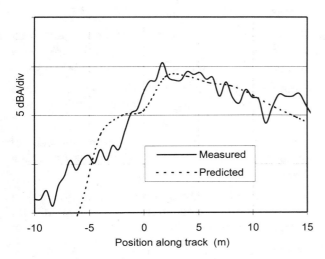

Fig. 16 Exhaust shell - pass-by - LHS

The comparison between calculated and test results for this source during pass-by test can be seen in figure 16. It is reasonably encouraging.

INTAKE ORIFICE - The noise radiated by the intake orifice depends on the pressure pulsations of the engine exciting the acoustical resonances of the intake system.

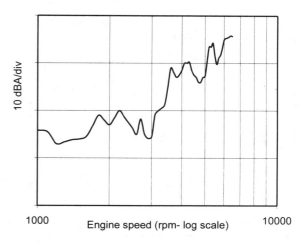

Fig. 17 Exhaust orifice noise

The intake orifice source noise contribution is seen in figure 17. Again in this case we find two different slopes in the curve against engine speed: up to 3000rpm it is independent of speed but above this it increases as the 5.5th power of speed.

The comparison of predicted and test results for this source during the pass-by test can be seen in figure 18: again there is good agreement.

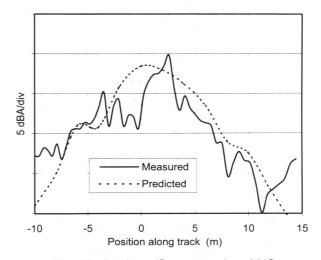

Fig. 18 Intake orifice - pass-by - LHS

The overall noise levels from the drive-by tests are shown in figures 19 and 20, comparing tests with our predictions, for both left and right sides.

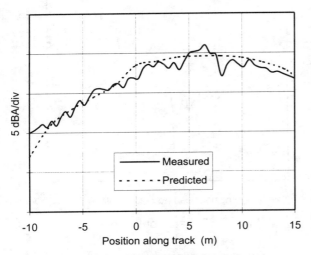

Fig. 19 Total vehicle pass-by - LHS

We believe the comparison shows a sufficiently good agreement to make the technique useful for design and development purposes.

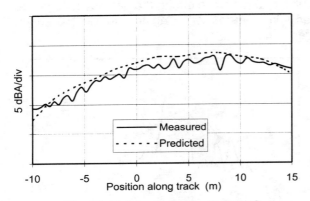

Fig. 20 Total vehicle pass-by - RHS

CONCLUSIONS

Based on a series of predictions for the various vehicle sources, we have developed an approach to the overall vehicle noise prediction which accurately models the vehicle performance and so the instantaneous engine speed and position in the drive-by test zone. Using this information and some approximations based on simple acoustics, existing information and engineering experience, we have developed a series of predictive tools which allow us to assess the overall vehicle behavior and to examine the source contributions to the total.

Although numerous approximations and assumptions have been made, the technique is usefully accurate and provides a reliable predictive tool for vehicle development.

ACKNOWLEDGMENTS

The authors would like to thank their respective colleagues at FIAT Auto and Cornaglia Central Research for their assistance in various areas of this work and to their companies for permission to publish this paper.

REFERENCES

1. Simulation of acceleration Pass-by noise considering the Acoustic radiation characteristics of a vehicle body. - K. Fujita, T. Abe, Y. Hori JSAE Review Vol. 7, 3 Oct.1996

2. On factors of noise emitted by a small vehicle and noise level simulation of pass-by test - Masuko and Takeshi SAE 770011

3. Computer simulation of vehicle fuel economy and performance - Y Hori, Fukeda and Kobyashi SAE 860364

4. A Survey of Passenger Car Noise Levels - D. Morrison and B.J. Challen - SAE 790442

5. Passenger Car Noise Control Measures and their Effects on Fuel Economy, Weight and Cost. - D. Morrison, B.J. Challen, T. Trella - SAE 800439

6. Note on two common problems of sound propagation - Journal of Sound & Vibration, 1969 10(3), 472-479

7. Noise reduction by distance from sources of various shapes - Z. Maekawa, Applied Acoustics 1970

8. A method of predicting engine noise in vehicle pass-by noise Hashimoto, Nogami, Najkata, Andou JSAE Review April 1985

9. A Review of Recent Progress in Diesel Engine Noise Reduction - B.J. Challen and D.M. Croker - SAE 820517

ADDITIONAL SOURCES

- Test cell simulation of drive-by noise test SAE 870967
- Noise emission of road vehicles - evaluation of simple source models Favre - Journal of Sound & Vibration, 1983 91(4), 571-582
- Relation between engine performance and noise - D. Anderton and J. Dixon - Inst. of Sound and Vibration Research, Automotive Group, Uni. of Southampton.
- Drive-by noise predictable by computer simulation - Bierman - FISITA 1996

971986

Time Dependent Correlation Analysis of Truck Pass-by-Noise Signals

Herman Van der Auweraer, Luc Hermans, and Dirk Otte
LMS International NV

Manfred Klopotek
Scania Trucks

Copyright 1997 Society of Automotive Engineers, Inc.

ABSTRACT

The data measured during an ISO 362 pass-by-noise test are strongly non-stationary due to the fast acceleration of the vehicle and its moving position with respect to the ISO microphone position. Nevertheless, one would like to obtain an understanding of the relative contribution of the various noise generating components during the test.

Since the classical signal analysis procedures based on the FFT calculation and auto/crosspower averaging for coherence/correlation analysis are no longer applicable, as they implicitly assume signal (and process) stationarity, an approach based on Autoregressive Vector (ARV) modelling of a set of measurement signals was developed and applied.

An ARV model is calculated directly from a set of time data of limited duration. The auto- and crosspower functions are directly calculated from the ARV model, avoiding the classical averaging procedure and allowing a repetitive calculation over multiple data segments, even for a short duration phenomenon as the pass-by test.

From these spectra, a time varying principal component and ordinary as well as virtual coherence calculation can be performed, attempting at describing a causal relationship between reference measurements on the truck (tyre, engine, exhaust, ...) and the pass-by microphones. One of the main features of the ARV-approach is that this description takes the form of a time/frequency plot, allowing to assess which component contributes the most at which moment.

The method has been validated extensively by a series of truck pass-by-noise tests, which are discussed in detail.

1. INTRODUCTION

Operating response analysis of mechanical structures refers to the examination of the system's vibro-acoustic response, observed under operating conditions.

Critical analysis of the power spectrum of selected transducers as a function of operating conditions or as a function of design measures may provide valuable clues to the main causes of the vibration or the radiated noise. More advanced approaches include the analysis of the correlation between source-related reference transducers and receiver-related target transducers. In environments where multiple uncorrelated phenomena interact in such a way that no pure reference to the individual phenomena can be isolated, multivariate analysis techniques such as principal component analysis can be applied [1]. However, all these techniques have in common that they require the estimation of auto- and/or crosspower spectra of the references, and between the references and the targeted responses. This estimation process assumes stationarity of the data and is based on the use of the (segment-averaged) Discrete Fourier Transform (DFT).

For rapidly varying processes, attempts have been made to use a repetition of short-segment DFTs, which can be plotted as a function of time. Whereas the main effect to spectral analysis is the reduced segment duration- and hence coarse frequency resolution-, this procedure is also not fit to estimate crosspower spectra between multiple signals. For, a high number of averages, during stationary operation, is required to obtain statistically reliable estimates.

To properly address this problem, a methodology to estimate auto- and crosspower spectra from short duration data segments of a multivariate data set is required. An approach based on autoregressive vector (ARV) time series modelling, which is a simplified version of ARMAV (autoregressive moving average vector) modelling, was developed hereto.

2. MULTIVARIATE TIME SERIES MODELLING

2.1. ARV FORMULATION - For a multivariate time series, described by an m-dimensional vector $\{x\}$, an ARMAV model can be developed [2, 3]:

$$\{x_t\}+[\phi_1]\{x_{t-1}\}+...+[\phi_p]\{x_{t-p}\}= \\ \{a_t\}+[\theta_1]\{a_{t-1}\}+...+[\theta_{p-1}]\{a_{t-p+1}\} \tag{1}$$

in which p is the model order; $[\phi_i]$ is an $m \times m$ matrix, containing the autoregressive parameters of the model; $[\theta_i]$ is an $m \times m$ matrix, containing the moving average parameter matrix of the model; $\{a_t\}$ is the model residual vector, an m-dimensional white noise vector function of time, satisfying the following equation

$$E\left[\{a_i\}\{a_j\}^T\right]=[\sigma_a^2] \quad (i=j); \quad =0 \quad (i \neq j) \tag{2}$$

with $[\sigma_a^2]$ the covariance matrix of $\{a_t\}$.

The parameter estimation of the ARMAV model is a non-linear least squares procedure and requires some skill as well as a large computation effort. In practice, an ARV model is often preferred, because of the linear procedure of the involved parameter estimation. Theoretically, an ARMAV model is equivalent to an ARV model with infinite order. However, if high order parameters are very small, and they can be neglected, a truncated ARV model can be used to approximate the ARMAV model.

$$\{x_t\}+[\phi'_1]\{x_{t-1}\}+...+[\phi'_q]\{x_{t-q}\}=\{a_t\} \tag{3}$$

with q the order of the truncated ARV model. Let

$$[\phi'(B)]=([I]+[\phi'_1]B+...[\phi'_q]B^q) \tag{4}$$

in which B is the backward operator:

$$\{x_t\}B=\{x_{t-1}\} \quad and \quad \{a_t\}B=\{a_{t-1}\} \tag{5}$$

then the ARV model can be expressed by a matrix notation:

$$[\phi'(B)]\{x_t\}=\{a_t\} \tag{6}$$

The ARV method in fact models the operating data as the output of a system excited by white noise inputs. In accordance to their system-modelling nature, AR-models are in general not well suited to identify single line (sinusoidal) spectral components.

From the ARV model above, the cross-spectral matrix of the time series $\{x_t\}$ can be derived [3]:

$$\left[S_{xx}^{ARV}(\omega)\right]=\frac{\Delta}{2\pi}\left[\phi'\left(e^{-j\omega\Delta}\right)\right]^{-1}\left[\sigma_a^2\right]\left[\phi'\left(e^{-j\omega\Delta}\right)\right]^{-H} \tag{7}$$

where ^{H} denotes Hermitian transpose and Δ is the sampling interval of the time series. The evaluation of such spectra can be done for any frequency below the Nyquist frequency, not limited by any signal observation restrictions.

2.2. ARV MODELLING ASPECTS - Spectral analysis by means of ARV modelling may yield spectra that are not biased by typical FFT-related errors such as leakage. No averages are required, giving possibility to deal with transient signals.

However, ARV modelling is less straightforward than the FFT and requires a lot of skill from the user, while no failure-proof guidelines are available. The following critical issues are to be addressed during the modelling procedure :

-        estimation of the model order
-        estimation of the model parameters
-        validity check of the model

The model order can be estimated by means of two functions, the residual variance and the final prediction error (FPE).

The residual variance is defined by the norm of the variance matrix , $[\sigma_a^2]$, of the model residuals. This residual variance, calculated for increasing orders will decrease significantly once a proper model order has been reached.

The FPE is defined as [4]:

$$FPE=\frac{1+n/N}{1-n/N}\cdot\varepsilon_a \tag{8}$$

with   n : total number of estimated parameters
N : number of observations of the time series
ε_a : residual variance

The FPE should reach a minimum for the proper model order (Akaike's FPE criterion).

Once a suitable order has been chosen, the model parameters can be estimated. Using linear least squares principles, the model parameters are estimated:

$$\begin{Bmatrix} \{x_{p+1}\}^T \\ \{x_{p+2}\}^T \\ \vdots \\ \{x_N\}^T \end{Bmatrix} = \begin{bmatrix} \{x_p\}^T & \{x_{p-1}\}^T & \cdots & \{x_1\}^T \\ \{x_{p+1}\}^T & \{x_p\}^T & \cdots & \{x_2\}^T \\ \vdots & \vdots & \vdots & \vdots \\ \{x_{N-1}\}^T & \{x_{N-2}\}^T & \cdots & \{x_{N-p}\}^T \end{bmatrix} \begin{bmatrix} -\{\phi'_1\}^T \\ -\{\phi'_2\}^T \\ \vdots \\ -\{\phi'_p\}^T \end{bmatrix}$$
$$+ \begin{Bmatrix} \{a_{p+1}\}^T \\ \{a_{p+2}\}^T \\ \vdots \\ \{a_N\}^T \end{Bmatrix} \qquad (9)$$

where p is the order of the model and N the number of observations of the time series. The above equation can be rewritten in a compact matrix form

$$\{x\} = [X]\{\phi\} + \{a\} \qquad (10)$$

The singular value decomposition of [X] leads to the pseudo-inverse and the model parameters are obtained in a least squares sense:

$$\{\phi\} = [X]^+\{x\} \qquad (11)$$

The model residues are then obtained through:

$$\{a_t\} = \{x_t\} + [\phi'_1]\{x_{t-1}\} + \ldots + [\phi'_p]\{x_{t-p}\} \qquad (12)$$

Finally, a validation of the estimated model needs to be carried out. A useful validity check of the ARV model is to evaluate the white noise character of the model residual time series. Two methods are currently in use.

A first approach is based on the autocorrelation functions $R_{ii}(\tau)$ of the model residuals a_t^i:

$$R_{ii}(\tau) = \frac{1}{N} \sum_{t=1}^{N-\tau} a_t^i a_{t+\tau}^i \qquad (13)$$

The autocorrelation coefficient $r_i(\tau)$ is given by

$$r_i(\tau) = \frac{R_{ii}(\tau)}{R_{ii}(0)} \qquad (14)$$

Except for zero lag, the autocorrelation coefficients should approach zero. If $r_i(\tau)$ remains within the bounds of the 95% confidence interval, the concerned model residual is said to have passed the whiteness test [5].

Another whiteness test is given by the so-called Q-function [4]. If a_t^i is indeed a white noise time series then

$$Q_i(M) = N \sum_{\tau=1}^{M} (r_i(\tau)) \qquad (15)$$

should be asymptotically $\chi^2(M)$ distributed. So it should be checked for each model residual a_t^i whether

$$Q_i(M) < \chi_\alpha^2(M) \qquad (16)$$

with α the level of the $\chi^2(M)$ distribution. Let α=0.05 and M=30 then the validation condition is $Q_i < 44$.

3. MULTIVARIATE COHERENCE ANALYSIS

Once the investigated dataset is condensed into an auto- and cross-spectral format, the classical coherence and principal component calculations can be performed:

Let [S_{xx}] denote the auto/crosspower matrix between all reference signals {x} and [S_{xy}] denote the crosspower matrix between the references {x} and the targets {y}.

The principal component autopower spectra $\lceil S_{x'x'} \rfloor$ are then derived from an Eigenvalue analysis of [S_{xx}]:

$$[S_{xx}] = [U] . \lceil S_{x'x'} \rfloor . [U]^H \qquad (17)$$

Using the same transformation matrix [U], the "virtual" crosspower spectra between physical signals and the principal components can be calculated:

$$[S_{xx'}] = [S_{xx}] [U] \qquad (18)$$

$$[S_{yx'}] = [S_{yx}] [U] \qquad (19)$$

Using this notion of virtual crosspowers, the corresponding virtual coherences are calculated [6]:

$$\gamma_{x_i x'_j}^2 = \frac{\left| S_{x_i x'_j} \right|^2}{S_{x_i x_i} . S_{x'_j x'_j}} \qquad (20)$$

$$\gamma_{y_i x'_j}^2 = \frac{\left| S_{y_i x'_j} \right|^2}{S_{y_i y_i} . S_{x'_j x'_j}} \qquad (21)$$

4. TEST PROCEDURE

Above described methodology was applied to analyse the noise generating mechanism of a heavy truck during a pass-by-noise test.

The truck was instrumented with 22 reference microphones near the suspected noise generating components such as exhaust, gear box and tyres. Especially for the tyres, multiple reference locations were defined (in front, behind or lateral to the tyre, at different heights,...). By telemetry, the noise from the exterior (ISO) microphones at left and right side of the test track (at 7.5m) was acquired simultaneously to this of the references. The objective of the test was to investigate the nature of the sound field around the truck, the intercorrelation between references and between references and ISO-microphones. Three different truck configurations and two gear settings were tested. The configurations consisted of:

- standard tyres and standard gear box
- standard tyres and encapsulated gear box
- deep profile tyres and standard gear box.

A major complication to the data analysis was caused by the fact that pass-by-noise tests are typically of very short duration (2..4 seconds). During the test, the truck accelerates with full throttle from a given starting speed over a distance of 20m. As engine speed, wheel speed and position vary considerably during the test, the assumption of stationarity over different subsegments of the test data is clearly not fulfilled and a classical segment-averaged FFT-based estimation of the power spectral densities cannot be applied.

Hence, two alternative analysis procedures were applied. In a first approach, the whole test sequence is considered as one single time segment of which a large size (high resolution) FFT spectrum can be calculated. By repeating the pass-by test 10 times, the data of each of the pass-by runs can be considered as one averaging segment of a segment-averaged auto- and crosspower calculation. The results are single auto/crosspower spectra for the whole test. When discussing the results, this approach will be referred to as "long-FFT". Although not stationary on a short duration scale, the repetition of the near-identical tests can be considered as a stationary process itself. The drawback is of course that only averaged results over the whole run, and no time-localised information, are available. In the second approach, the above discussed ARV approach is repeatedly applied on small segments of data, all belonging to a single pass-by run. As a result, a frequency/time or frequency/distance representation can be generated, illustrating the non-stationary noise elements during one run. Moreover, they allow to show ordinary coherence, principal component and virtual coherences as a "waterfall" plot with time (or distance or RPM) as a parameter.

5. TEST RESULTS

Figure 1 shows the results of the autopower calculation of the ISO-microphone using the long-FFT approach for the three configurations (standard, deep profile tyre and encapsulated gear box). The effect of the encapsulation at high frequencies can be observed. At low frequencies, an encapsulation resonance even increases the noise. The tyre effect is less discernible.

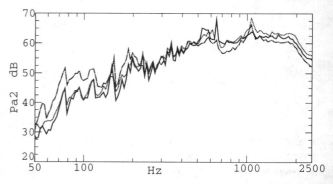

Figure 1 : ISO-microphone long-FFT spectrum for the 3 configurations : deep profile tyre (solid line), standard (dotted line), encapsulated gear box (dashed line)

Figure 2 shows the corresponding ARV results (order 80) as a frequency-time plot, again for the three configurations (standard, deep profile tyre and encapsulated gear box). The effect of the tyres is now more clearly visible as a time-localised phenomenon at the start (slip).

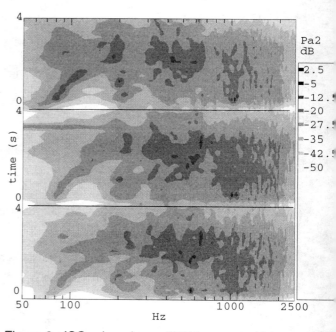

Figure 2 : ISO-microphone ARV frequency/time spectra for the 3 configurations : deep profile tyre (bottom), standard (middle), encapsulated gear box (top)

The corresponding results for a microphone near the gear box and at the rear of the right tyre are shown in Figures 3 to 6. For performing the multivariate correlation analysis, six references were selected for the principal component calculation. These references were selected in such a way as to represent as well as possible the whole noise field (exhaust, gear box, left and right tyre).

Figure 7 shows the principal components for the long-FFT analysis of the deep profile tyre.

Figure 8 shows the ARV-based first principal component for the three configurations (order 80). The peaks occurring in the beginning of the pass-by test around 1000 and 2000 Hz, physically corresponding to the slip of the tyres, are clearly visible.

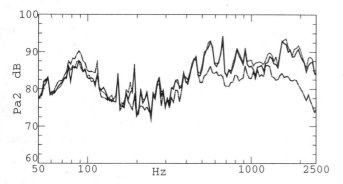

Figure 3 : Gear box long-FFT spectrum for the 3 configurations : deep profile tyre (solid line), standard (dotted line), encapsulated gear box (dashed line)

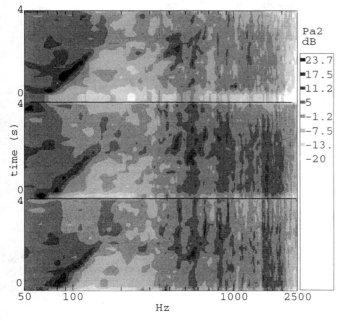

Figure 4 : Gear box ARV frequency/time spectra for the 3 configurations : deep profile tyre (bottom), standard (middle), encapsulated gear box (top)

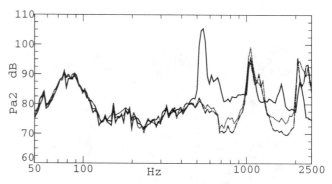

Figure 5 : Tyre long-FFT spectrum for the 3 configurations : deep profile tyre (solid line), standard (dotted line), encapsulated gear box (dashed line)

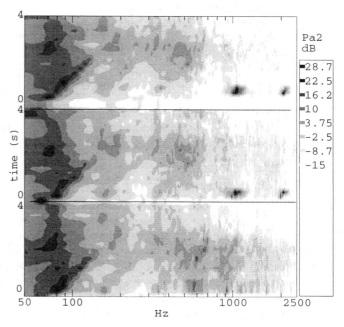

Figure 6 : Tyre ARV frequency/time spectra for the 3 configurations: deep profile tyre (bottom), standard (middle), encapsulated gear box (top)

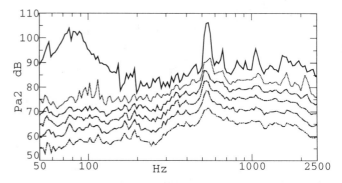

Figure 7 : Principal component spectra for the long-FFT analysis of the deep profile tyre

Finally, the virtual coherences between the physical signals and the principal components were calculated. The virtual coherences between the references and the principal components reveal which references dominantly contribute to it.

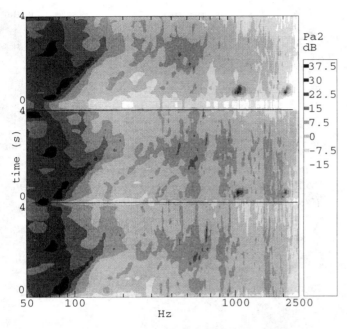

Figure 8 : First principal component frequency/time spectra (ARV) for the 3 configurations : deep profile tyre (bottom), standard (middle), encapsulated gear box (top)

The long-FFT results are shown in Figures 9a and 9b for the first principal component for the case of deep profile type. The dominance of exhaust at low-frequencies, gear box at high frequencies and tyres at very specific frequencies can be observed. The high contribution of the tyre microphones at low frequencies is probably due to the large influence of the exhaust, even at the (not so distant) tyre location. Experience allows to conclude that the highest virtual coherence in that region (even if the others are close in value) is the one related to the physical cause.

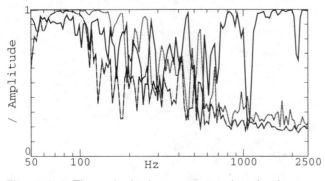

Figure 9a : First principal component virtual coherence spectra (long-FFT) : microphone at front of gear box (solid line), microphone at exhaust (dotted line), microphone at back of left tyre (dashed line)

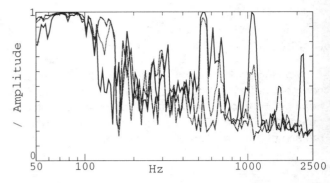

Figure 9b : First principal component virtual coherence spectra (long-FFT) : microphone at back of right tyre (solid line), microphone beside right tyre (dotted line), microphone beside left tyre (dashed line)

In Figure 10, the ARV-results are shown for three references (gear box, tyre and exhaust), clearly indicating the frequency -and time- localisations of each dominant contribution.

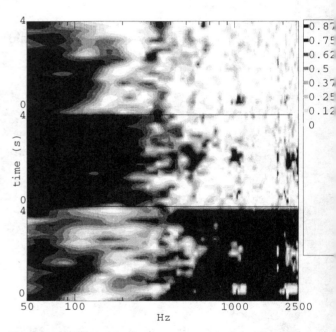

Figure 10 : First principal component virtual coherence frequency/time spectra (ARV) : microphone at front of gear box tyre (bottom), microphone at exhaust (middle), microphone at back of right tyre (top)

A similar analysis however with respect to the ISO-microphones gave unsatisfactory results. Unlike the clear observability of the slip of the tyres occurring in the beginning of the pass-by noise test in the time/frequency spectra, the time/frequency coherence analysis between the ISO-microphone and the reference microphones located on the truck close to the different contributing sources did not really give satisfactory results. A generally low coherence, decreasing with data segment length was observed. The difficulties with respect to analyzing the far field ISO-microphone data in relation to references on the truck did however not seem to be due to the ARV-modeling technique itself, as similar conclusions were found for the long-FFT results. At first, it was thought that the time delay could be held responsible for the low coherence. However, performing a time delay and Doppler correction on the ISO-microphones did not give better results. Therefore, the authors believe that the main reason for the unsatisfactory results lies with the especially phase-related non-stationarity of the acoustic propagation path itself during the test. Similar problems have been observed with the measurement of (also phase related) acoustic frequency response functions over large distances.

In order to gain better insights in the capabilities of ARV for real-world coherence analysis, an additional analysis was performed, limited to the near field data. The test was performed using the deep profile tyres and the 4th gear was selected. In total, 8 microphones located on the truck were taken into account in the ARV-model. 7 microphones served as references while the eighth microphone located under the back of the gear box was the targeted response. The 7 references included 4 microphone positions close to the tyres, 1 microphone near to the exhaust and 2 microphones respectively above and under the front of the gear box. Figure 11 shows the ARV-based multiple coherence between the target and the 7 reference microphones. High multiple coherence values imply a high degree of causality between the reference signals and the target signal. The plot clearly illustrates the high coherence for the low frequencies and the order or speed related phenomena. In total, four order components can be observed in the multiple coherence plot, of which two are closely spaced. At about 1800 Hz, a small zone of high coherence which is not order related is also visible.

of gear box, under front of gear box, exhaust, back of right tyre, back of left tyre). In figure 12, the virtual coherence for the target (under back of gear box) is also depicted. These figures clearly reveal that the exhaust is contributing the most to the target microphone in the low frequency range. For the first order component of the truck, the exhaust remains the dominant source, although all other references are very much correlated with it. Earlier experience with virtual coherence allows to conclude that, even if several virtual coherences are high in a certain frequency range, it is the one with the highest value which corresponds to the physical component responsible for the noise. The second order component is clearly dominated by the gear box, especially by the noise radiated above the gear box. To a less extent, all other references are correlated with this order. Dominant sources for the third order component are the left and right tyres and to a much less extent the exhaust. The two gear box references show a drop-out, illustrating that the gear box is not dominant for this order. The fourth order component which is closely spaced to the third order component is much less related to the tyres. Dominant contribution is now again coming from the gear box. The virtual coherences for the tyres reveal a fifth order component. This order can however not be observed in the multiple coherence for the target microphone positioned at the gear box (see figure 11). Finally, figures 12 and 13 show that the gear box references are dominant in the zone of high multiple coherence at about 1800 Hz.

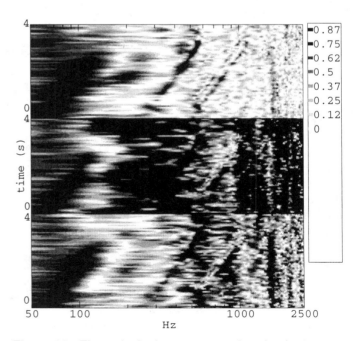

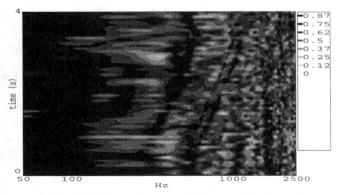

Figure 11 : Multiple coherence between the target microphone (under back of gear box) and 7 reference microphones on the truck (ARV)

Figure 12 : First principal component virtual coherence frequency/time spectra (ARV) : microphone under front of gear box (bottom), microphone above front of gear box (middle), microphone under back of gear box (top)

In order to better understand which references dominantly contribute to the target microphone in the different zones of high multiple coherence, the virtual coherences between the references and the first principal component were calculated. Figures 12 and 13 show its virtual coherence for 5 references (above front

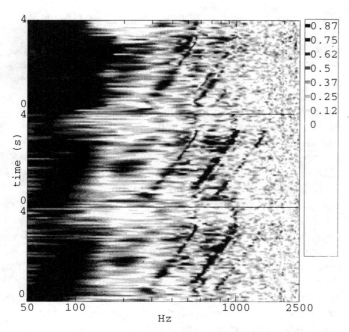

Figure 13 : First principal component virtual coherence frequency/time spectra (ARV) : microphone at back of left tyre (bottom), microphone at back of right tyre (middle), microphone at exhaust (top)

CONCLUSION

An ARV-based spectral analysis methodology has been applied to derive principal component and virtual coherence functions of a set of non stationary signals, resulting from a truck pass-by-noise test. The time localisation of the spectral contribution as well as the zones of high virtual coherence between the principal components and the reference signals significantly help interpreting the results and developing the physics of the generating process. For analyzing the causal contribution of reference signals to a target response, satisfactory results were only obtained with near field target responses. The non-stationarity of the acoustic propagation field related to the -outdoor- propagation of the noise on the moving truck to a distant (7.5 m) ISO microphone did however not permit to derive quantitative conclusions as to the partial contributions of each of the noise components to the measured exterior noise.

ACKNOWLEDGMENTS

The presented research was performed in the framework of Brite/Euram project 5415, "PIANO". The financial support of the EC-DGXII is gratefully acknowledged.

REFERENCES

[1] Otte D., "Development and Evaluation of Singular Value Analysis Methodologies for Studying Multivariate Noise and Vibration Problems", Ph.D. Dissertation, Kath. Univ. Leuven, Dept. of Mech. Eng., 1994.

[2] Marple Jr. S.L., "Digital Spectral Analysis with Applications", Prentice-Hall, 1987.

[3] Pandit S.M., "Modal and Spectrum Analysis: Data Dependent Systems in State Space", John Wiley and Sons, 1991.

[4] Ljung L., "System Identification - Theory for the User", Prentice-Hall, 1987.

[5] Box G.E.P., Jenkins G.M., "Time Series Analysis, Forecasting and Control", Holden-Day, 1970.

[6] Price S.M., Bernhard R.J., "Virtual Coherence: A Digital Signal Processing Technique for Incoherent Source Identification", Proc. 4th Int. Modal Analysis Conference, Los Angeles - CA (USA), Feb. 3-6, 1986, pp. 1256-1262, ed. Union College, Schenectady - NY (USA)

971987

A Doppler Correction Procedure for Exterior Pass-By Noise Testing

Renaat Vancauter
LMS International NV

Copyright 1997 Society of Automotive Engineers, Inc.

ABSTRACT

During a pass-by noise test , the effect of the relative speed of the vehicle is translated into the Doppler phenomenon, causing positive (approaching) or negative (receding) frequency shifts (typically 2 to 5 %) in the signal received by both microphones, which are positioned at both sides of the test track at a certain distance from the center line. These experimental realities can lead to considerable errors for order tracking (used to determine engine/powertrain contributions) and coherence analysis. A Doppler correction process has been developed to deal with these effects. The correction is based on a resampling and interpolation of the target microphone time records, using recorded information on the instantaneous vehicle position measured by means of a radar. With the classical analysis tools, the Doppler correction process can improve result interpretation. It is shown that the combination of advanced techniques like adaptive resampling, and the Doppler correction process has a positive influence in proper signal interpretation.

INTRODUCTION

As part of an international research project "PIANO", a de-Dopplerization process has been developed, based on a resampling and interpolation of the time record busing recorded information on the instantaneous vehicle position. The benefits are shown later on. In most experimental noise and vibration studies, a problem is identified in the frequency domain by applying a Fast Fourier Transform (FFT) to the measured signals. For most applications the FFT is perfectly acceptable,. as long as the limitations inherent in this technique are well understood. However, one of the fundamental premises of the FFT is that the signal is stationary, in other words, does not vary in spectral content in time. In the case of pass-by noise, where the signal spectra are varying rapidly, this technique yields poor results due to spectral smearing and spectral resolution limitations.

This article explains the theory of several alternative techniques, together with practical examples to show how they should be applied to solve real-life problems.

DOPPLER CORRECTION

When analyzing the Pass-by Noise data of a vehicle, it is often necessary to perform an order analysis by cutting the required orders out of the microphone data waterfalls, obtained by repeated Fast Fourier Transforms (FFT), in order to evaluate the engine/powertrain contributions to the vehicle signature. However, these analysis methods are very much disturbed by the Doppler effect, as well as by the continuously varying time delays. Both those effects also largely influence the capability of accurately tracking orders with Computed Order Tracking (section 3.2) and Kalman (section 4). The particular effects of speed and displacement of the vehicle under test lead to two types of errors :

- The effects of the relative speed is translated into the Doppler phenomenon and causes positive (approaching) and negative (receding) frequency shifts in the microphone signals. If the instantaneous vehicle position is recorded, this phenomenon can be corrected by appropriately dilating (= increasing the time step) and compressing (= reducing the time step) the observed data.
- The continuously varying distance between the vehicle and the ISO microphones causes noticeable (and continuously varying) time delays due to the finite speed of sound.

The de-Dopplerization software corrects both the Doppler and the time delay effects. The principle of the signal correction is illustrated in Figure 1.

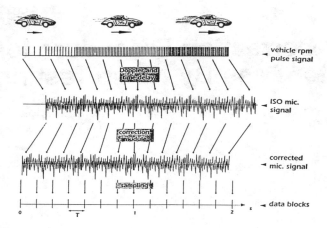

Figure 1 : Principle of the signal correction.

A signal that is emitted by moving source at time τ is received by the fixed observer at time t. If the instantaneous distance between the source and the receiver is denoted by $|d(\tau)|$, the relationship between receiver time t and emitter time τ is expressed as:

$$t = \tau + \frac{|d(\tau)|}{c_o} \quad (1)$$

with c_o = speed of sound

Note that it is the instantaneous distance at emitter time τ that has to be considered, rather than at receiving time t. Further, the equation accounts for both the Doppler phenomenon (signal compression and expansion) and the time delay effect. The issue of the signal correction of the receiver's time record is to express the receiver signal as a function of emitter time τ. Practically, this requires a resampling of the receiver time history, using the equation (1). A time history containing information on the instantaneous position of the source is needed to calculate $d(\tau)$.

The correction proceeds then as follows. From a starting emitter time τ_o, a corresponding receiver time t_o is estimated. The receiver data are in digital format, consisting of equally spaced samples, so that receiver signal value at time t_o is to be estimated by means of linear interpolation between neighboring sampler $t_i < t_o < t_i+1$. This t_o value is then placed at the τ_o position of the correction signal. Incrementing τ by the designated sample rate Δt ($\Delta t \cong \Delta \tau$) and putting the received signal values from corresponding time t at position τ, generates an equally spaced receiver time history, corrected for Doppler shift and time delay.

EXAMPLE - A pass-by noise test was performed according to the ISO 362 standard. The instantaneous vehicle position was recorded using radar. The engine speed was measured by an optical device which generates two pulses per revolution of the crank axle. These pulses were transmitted to the analysis station by telemetry mounted on the vehicle. The vehicle under test was powered by a four cylinder combustion engine and was taken through the speed range of 3900 rpm to 5000 rpm in two seconds. To demonstrate the effect of the Doppler phenomenon and time delay, an order map of the left ISO microphone is shown in Figure 2. The upper plot contains the original, uncorrected data.

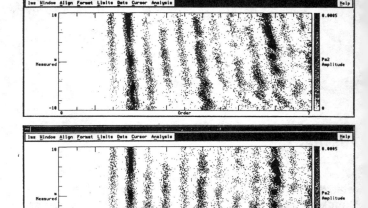

Figure 2 : Order map of the left ISO microphone.

The second, fourth and sixth order evolutions clearly sweep around their 'real' value. For the negative positions (vehicle is approaching the ISO microphone) the order shift is positive, while for positive positions (vehicle is receding) the order shift is negative.

The theoretical frequency/order shift in terms of percentage can easily be calculated and is given by the formula :

$$\left|\frac{\Delta f}{f}\right| = \left|\frac{\Delta O}{O}\right| = \frac{v}{c} \quad (2)$$

with   v = speed of the vehicle
         c = speed of sound

For a velocity of 50 km/h, this shift is approximately 5%. This means that the sixth order should have a theoretical shift of 0.3 orders, which corresponds quite well to the experimental data. The lower plot of Figure 2 shows the order map of the corrected microphone data. The sweeps of the orders around their real values are entirely removed. The presence of the Doppler phenomenon and the time delay in the microphone data can be the cause of a lot of confusion during the analysis and can even lead to misinterpretation of the results. A method to estimate which orders are dominant is to calculate the signal's summed order. Figure 3 shows the summed order traces for the uncorrected signal (full line) and the same data after de-Dopplerization (dotted line).

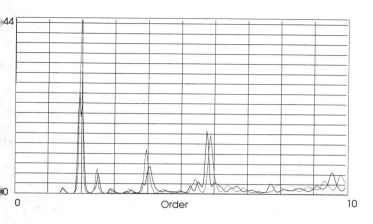

Figure 3 : Summed order traces for the uncorrected signal (full line) and the same data after de-Dopplerization (dotted line).

Note the double peak at the second order in the uncorrected trace, due to the sweep of the order around their 'real value' which translates to a single peak after correction. A second important effect resulting from the Doppler effect is that two nearby orders are swamped and therefore it is impossible to separate them, as one can see in Figure 3 for the orders 5.4 and 5.9. After Doppler correction both orders are clearly separated.

A second important effect resulting from the Doppler effect is that two nearby orders are swamped and therefore it is impossible to separate them, as one can see in Figure 3 for the orders 5.4 and 5.9. After Doppler correction both orders are clearly separated.

Also note the shift from a 'virtual' 5.7th order to an almost 'real' 5.95th order, which can lead to a misinterpretation of results.

The frequency/order shift is also a source of errors when calculating order and frequency sections. In order to compensate for the shift one has to use band integration, which has the big disadvantage that closely spaced orders can not be separated.

Figure 4 shows the evolution of the second order. An order band integration of 0.2 has been used. There exists quite a large difference between the uncorrected order trace (full line) and the corrected (dotted line). This difference can be diminished by using a higher order integration width as shown in Figure 5 (order band width of 1.0).

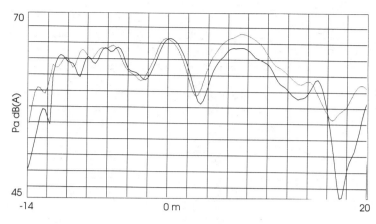

Figure 4 : Evolution of the second order Uncorrected order trace (full line) and the corrected (dotted line).

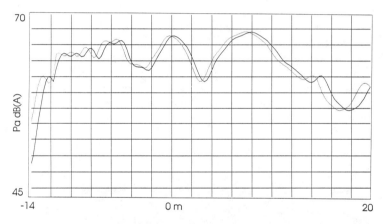

Figure 5 : Diminished difference by using a high order integration.

The Doppler effect also causes considerable errors in the order tracking process, as will be explained in more details in next section.

CLASSICAL SPECTRAL ANALYSIS VERSUS COMPUTED ORDER TRACKING. -

CLASSICAL SPECTRAL ANALYSIS - If the speed variation is relatively slow, e.g. 1000 rpm to 4000 rpm over 60 seconds, the classical method of repeated Discrete Fourier Transforms (DFT's) followed by order sections ("cuts") across the waterfall map, will probably give acceptable order data. For faster run-ups, e.g. over 2 seconds instead of 60 seconds, post processing of data obtained with repeated DFT analysis of samples taken at a fixed sample often yields poor results. This is due to the noticeable variation of the momentary signal spectrum throughout one analysis block, due to a change of the engine RPM, which results in significant spectral smearing. Reducing the analysis blocksize, in order to reduce the change of the engine RPM and consequently diminish the spectral smearing, results in a lower spectral resolution, which is in most cases highly unwanted. Using a large analysis blocksize, required for sufficient spectral resolution, reduces the effective temporal resolution due to a large overlap of consecutive analysis blocks, which entirely removes the dynamics of the order section evolutions obtained from those waterfall maps.

For a sampling frequency of 25 kHz and a rate of change of 550 RPM/sec for the engine RPM (from 3900 rpm to 5000 rpm in 2 sec):

Blocksize	T(sec)	T(%)	ΔF (Hz)	ΔRPM	Overlap (%)
2K	0.082	4 %	12.2	45	56.6
4K	0.164	8 %	6.1	90	79.3
8K	0.328	16 %	3.0	180	91.6
16K	0.656	32 %	1.5	360	96.2

For the calculation of the overlap, an analysis block has been taken each 20 RPM.

From the above table one can easily deduce that for a spectral resolution of 3 Hz an analysis blocksize of 8K is necessary, which causes important spectral smearing due a change of 180 rpm in the engine rpm. This effect is illustrated in Figure 6. The upper plot shows the whole waterfall map. The low plot contains a zoom around the sixth order.

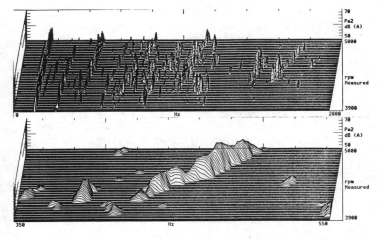

Figure 6 : upper plot the whole waterfall map lower plot contains a zoom around the sixth order

Theoretically a spectral smearing of 18 Hz of the sixth order is expected (Δf = order X ΔRPM = 6 X 180/60 (Hz) = 18 Hz). From the lower plot of Figure 6 one can see that this agrees well with the experimental data. As a consequence of the spectral smearing, one has to use large band integration when doing order analysis in order to compensate for this smearing effect. Figure 7 shows the influence of the order integration width. An analysis blocksize of 8K has been used for the waterfall map calculation. It is necessary to use a rather large order bandwidth such that the entire order sweep lies within this band. At least an order bandwidth of 0.5 is required.

However, it will be difficult to identify closely spaced orders, because any nearby order would be swamped.

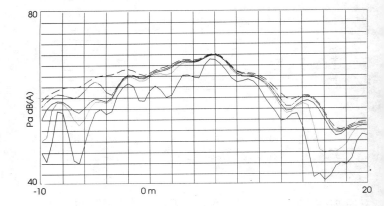

Figure 7 : influence of the order integration width

A second harmful effect of this large blocksize of 8 K is the reduction of the effective temporal resolution due an overlap of more than 90 %. This particular effect is illustrated in Figure 8, which shows the evolution of the sixth order for several blocksizes. Note the unwanted smoothing effect with an increase of the analysis blocksize, entirely removing the dynamics of the order trace. By using a blocksize of 16 K, each analysis block takes approximately 1/3 of the complete time history.

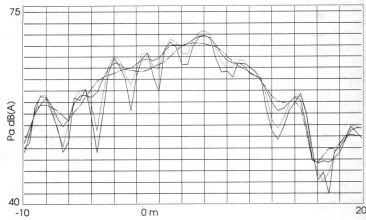

Figure 8 : Evolution of the sixth order for several blocksizes.

Summarizing, following limitations arise with the DFT technique ;

• a large analysis blocksize is required for acceptable spectral resolution

• a large analysis blocksize reduces effective temporal resolution, due to the large overlap and therefore reduces the dynamics of order and frequency cuts.

• a large analysis blocksize, compared to the complete time history, causes important spectral smearing due to a change in the engine speed.

COMPUTED ORDER TRACKING - On-line testing methods have been proposed where the data acquisition sample rate is adapted synchronously with a reference signal describing the underlying signal periodicity. Special analogue or digital electronics (phase or frequency locked loops) are needed to realize this. Since the time axis of the original signal is rescaled to the equidistant "period" (or shaft angle) domain, rotational harmonics calculated by the DFT will always appear at the same "spectral" component, which now takes the signal order as its independent variable.

Although very popular and successful for describing the dynamic behavior during engine run-up tests, for very fast changes in the periodicity these methods suffer from important stability and transitory problems. In addition, on-line calculation performance limitations, the lack of flexible overlapped processing and the dependency on the acquisition parameters set for the test restricts their applicability.

Adaptive Data Resampling is a solution to reconcile the benefits of this "order" domain analysis with the possibilities of on-line digital signal processing This technique is implemented by adaptive digital resampling of the original data. This approach is referred to as "Computed Order Tracking".

In most practical test set-ups, the rotation speed is derived by the measurement of a representative reference (or "tacho") signal - a signal which normally consists of a train of predictable events occurring at a fixed number of times per revolution. Because the tacho event rate is invariably much lower than the (fixed) measurement channel sample rate, the instantaneous speed of each data point is not precisely known. In order to give sharp results, the Computed Order Tracking process requires knowledge of the instantaneous rotation speed for every measurement channel sample, so some form of speed interpolation needs to be made. Assuming that the speed of a mechanical system varies as a continuous function of time, the RPM/time evolution is mathematically modeled by fitting, a quadratic polynomial curve through the measured speed samples from several tachometer events. This model is then used to calculate the intermediate "sample-by-sample" speed estimates, before proceeding to the next observation window. One major benefit of the quadratic polynomial model for the RPM time evolution is that accelerating shafts will be accurately tracked. This is often a practical limitation with classical analogue tracking filter techniques, where only linear speed changes tend to be assumed. Using this momentary periodicity information, the time instance values of a new equidistant shaft-angle axis can be derived (e.g. 32 or 64 samples per revolution).

The next step is to resample the original noise and vibration signals in this new temporal axis. The major difference with the classical "upsampling" problem is that here the up-sample factor is not an integer but is a function of time. A practical solution to this problem is to upsample the original data by a large integer factor and then to perform a linear interpolation around the new sample time instances, obtained by updating the sample frequency and thus the delta time, synchronously with the engine speed. The sample frequency follows the formula :

$$F_s = 2 \times Maxorder \times RPM \left[Hz \right] \qquad (1)$$

where: F_s sample frequency
         MaxOrder   maximum order of interest
         RPM [Hz]   RPM evolution obtained through the spline interpolation.

The spectral characteristics of this linear interpolation step require the upsampling factor to be equal to or greater than 15. This procedure is schematically illustrated in Figure 9. Starting from the adaptively resampled data, a straightforward DFT calculation can be used to extract the order values as a function of time or RPM. The DFT length of the analysis block corresponds to :

$$DFT\ length\ =\ 2 \times MaxOrder \times OrderRes \qquad (2)$$

where : MaxOrder :   max order of interest
         Order Res :   order resolution

The order resolution determines the number of revolutions taken into account in the analysis block, while the maximum order corresponds to the number of equi-angle points per revolution of the shaft. One has to use a larger analysis blocksize to obtain a finer order resolution. Remark the correspondence with the Classical Spectral Technique where a larger time blocksize is required to have a better frequency resolution.

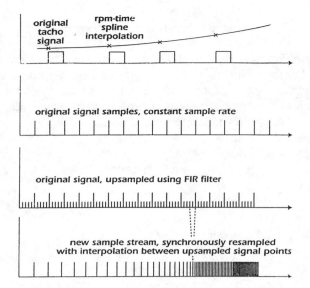

Figure 9 : Scheme for upsampling factor procedure.

Figure 10 shows the contrast between the order map obtained in Classical Spectral Analysis (upper plot) and the same data by Computed Order Tracking (lower plot). The data used for the calculations was not Doppler corrected. A analysis blocksize of 8 K was used for the DFT technique, resulting in a spectral resolution of 3 Hz. For a rpm change of 3900 rpm to 5000 rpm, this corresponds to a change in order resolution from 0.046 to 0.036. The Computed Order Tracking used following processing parameters : max order = 32 and order resolution 1/32, resulting in a DFT size of 2K. Note that an order resolution of 1/32 means that 32 resolutions are taken into account for the DFT calculation. Higher orders are difficult to identify with the DFT technique. The Computed Order Tracking results show more clearly the higher orders, although not as good as expected.

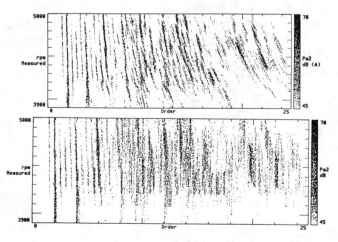

Figure 10 : contrast between the order map obtained in Classical Spectral Analysis (upper plot) and the same data by Computer Order Tracking (lower plot)

The reason is that the data is not yet Doppler corrected, whereby the orders sweep around their 'real' values. These orders are not anymore recognized as super-harmonics of the fundamental frequency and are therefore in some way discriminated by the Computed Order Tracking technique.

After Doppler correction, the same maps are once more estimated. The results are presented in Figure 11. Observing the lower plot, it becomes clear that especially the higher orders are very well estimated, in contrast with the Classical Spectral Analysis (DFT) (upper plot). One can conclude that the Doppler correction is an important pre-processing technique, before the Computed Order Tracking is applied in order to be able to correctly estimate the order evolutions.

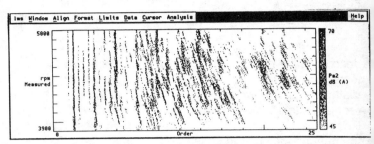

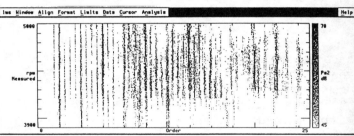

Figure 11 : Same results as in Figure 10 after Doppler correction.

The DFT technique, as well as the Computer Order Tracking is capable of accurately tracking the lower orders, which is obviously the result of less spectral smearing at lower orders (the amount of spectral smearing is proportional to the order number). At orders higher than 15, the Classical DFT technique completely breaks down due to the spectral smearing. The main advantage of the COT is the fact that this technique does not suffer from leakage and ΔRPM smearing due to the inherent synchronous resampling process. Indeed, the specified order has a periodic evolution within the analysis block.

Figure 12 shows the evolution of the second order calculated with both techniques. The Classical Spectral Analysis technique used an analysis blocksize of 2K and an order integration of 0.5, while the Computer Order Tracking DFT size was chosen to be 512 (max order : 64, order resolution : 0.25) and an order integration of 0.5.

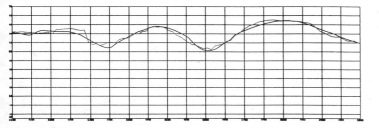

Figure 12 : evolution of the second order calculated with both techniques.

There is a good global agreement between both techniques, although the COT order evolution shows a more dynamic behavior, due to the smaller DFT size and consequently a much lower overlap.

However, if one tries to track less dominant orders using a small bandwith integration, to be able to separate nearby orders, the limitations of the classical DFT technique again appear. Those orders can be e.g. fractional orders causing modulation effects in the perceived sound, which are sensed as stationary but very unpleasant rough tones. From the order map of Figure 13, one can extract that the order 10.95 is present in the data. Cutting this order using both techniques results in the order evolution shown in Figure 14, showing the incapability of the classical DFT technique due to the spectral smearing. Both techniques used comparable order resolutions for the order map calculations (DFT : blocksize = 8K, resulting in order resolution changing from 0.045 till 0.037 for the rpm range 4000-5000 rpm, COT : max order = 64 and order resolution = 1/16).

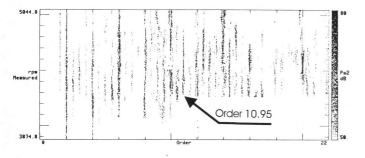

Figure 13 : Presence of order 10.95.

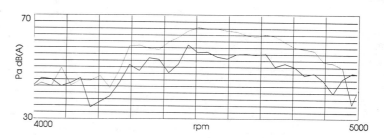

Figure 14 : Order cut with both techniques.

Summarizing, the major advantages of the COT technique are :
- the technique is leakage free
- no smearing due to a change in the engine speed

CONCLUSION

Traditional Discrete Fourier Transform (DFT) analysis techniques do not often allow the correct analysis of non-stationary sound and vibration signals. In general the DFT should be used as a first step in identifying the specific nature of the problem. When further analyzing the problem Computer Order tracking can be applied-track to achieve clearer results. These advanced processing techniques will enhance the results when the original data has been Doppler corrected.

REFERENCES

S.L. Marple Jr. "Digital Spectral Analysis with Applications", Prentice Hall Inc., Englewood Cliffs-NJ, 1987.

H. Van der Auweraer, P. Van de Ponseele, "Digital Signature Techniques for Noise and Vibration Analysis of structures with rotating shafts", Proc. 20 th ISATA, pp. 997-1018, Firenze (I), June 1989.

R. Potter, M. Gribler, "Computed Order Tracking Obsoltes Older Methods", SAE paper 891131, Proc. 1989 SAE Noise and Vibration conference, pp. 63- 67, May 16-19, 1989.

971988

Two-Microphone Measurements of the Acoustical Properties of SAE and ISO Passby Surfaces in the Presence of Wind and Temperature Gradients

Troy J. Hartwig and J. Stuart Bolton
Purdue Univ.

Copyright 1997 Society of Automotive Engineers, Inc.

ABSTRACT

It has been noted that there are consistent differences between sideline sound levels measured on the two track types used for standardized motor vehicle passby testing: i.e., ISO and SAE surfaces. When the two-microphone transfer function method was first used in conjunction with a two parameter ground model to characterize the acoustical properties of these asphalt surfaces it was found that there were significant acoustical differences between the ISO and SAE surfaces. However, it was also noted that environmental conditions, e.g., wind and temperature gradients, affected the estimates of surface properties obtained by using that method. In the present work, a ray tracing algorithm has been used to model the effects of environmental refraction on short range propagation over asphalt, and a physically-based single parameter ground model has been used to characterize the asphalt surfaces. It has been found that when the ray tracing procedure is used in conjunction with the two-microphone method, surface acoustical properties can be estimated under a wide variety of environmental conditions. The ray tracing and ground impedance models introduced here can also be used to predict the effects of varying surface properties and environmental conditions on passby measurements.

INTRODUCTION

Regulatory agencies in the United States and abroad set maximum acceptable levels for noise radiated by accelerating motor vehicles under specified operating conditions. During those tests, the peak A-weighted sound levels are measured at a sideline microphone while the vehicle accelerates through the test section of a "passby site": i.e., a large planar surface, usually asphalt, built specifically for the purpose of conducting

acoustical tests. In the past, the vehicle noise radiated during acceleration tests of this type has been dominated by powertrain-related noise. However, efforts in recent years to reduce the radiated levels of powertrain and exhaust noise have revealed tire/road interaction noise to be a significant contributor to overall vehicle passby levels [1]. Thus, the ability to consistently measure and predict the propagation of tire/road interaction noise accurately in standardized tests has become a subject of concern.

It has been reported that there are consistent differences between the tire/road interaction noise levels measured at sites surfaced according to the prevailing US and international standards: i.e., Society of Automotive Engineers (SAE) surfaces [2], and International Organization of Standards (ISO) surfaces [3]. In particular, the sideline level of passenger vehicle tire/road interaction noise at ISO sites was found to be approximately 1.5 dB lower on average than the levels measured at SAE sites under "cruise" and "coast" conditions (when tire/road interaction noise is assumed to dominate sideline sound levels) [4, 5]. Preliminary work has shown that approximately 1 dB of that difference could be attributed to differences in the tire/road interaction noise generation process on the two surface types, and that perhaps 0.5 dB could be attributed to propagation differences [5]. The latter estimate was based, in part, on the use of the so-called two microphone transfer function method in conjunction with a two parameter ground impedance model to estimate the acoustical properties of ISO and SAE asphalt surfaces [5, 6].

The overall objective of the present work was to improve the quality of the propagation difference estimates by improving the methodology used to estimate asphalt surface acoustical properties. That method has been improved in two ways. First, the effects of near-surface wind and temperature gradients have been accounted for by using a ray tracing technique to model refractive effects. Secondly, the optimization procedure used to estimate the surface acoustical properties was made more robust by the use of a simplified, physically-based, model for the ground impedance.

Note finally that the theoretical and numerical tools that were used to make the estimates of surface properties reported here can also be used in a predictive capacity: i.e., to predict the level (and variability) of the tire/road interaction noise contribution to sideline sound levels under realistic environmental conditions.

SHORT RANGE SOUND PROPAGATION AND THE TWO MICROPHONE TRANSFER FUNCTION METHOD

INTRODUCTION - During a passby test, motor vehicle noise must propagate some distance over a plane asphalt surface before reaching the sideline microphone, and the sound level measured under those conditions is partially determined by the acoustical characteristics of the ground surface. Techniques for predicting that effect are described in this section.

Sound energy incident on a plane surface is partially reflected and partially transmitted (i.e., absorbed). At near-grazing angles of incidence, the measured attenuation of propagating sound can be significantly in excess of that resulting from the inverse square law and atmospheric absorption [7]. Piercy et al. have suggested that a locally reacting ground model is usually adequate for predicting sound propagation over plane outdoor surfaces [7].

Van Wyk et al. [6, 8], developed a two parameter, local reaction, version of Attenborough's four parameter ground model [9], and the former was shown to successfully reproduce grazing incidence sound propagation over asphalt passby surfaces. In work directed towards the development of sound absorbing road surfaces, Berengier has developed a model based on four physical parameters (layer thickness, porosity, specific airflow resistance and shape factor) that has also been shown capable of characterizing asphalt surfaces [10]. Surface models such as these can be used in conjunction with an appropriate sound propagation theory to predict the detailed behavior of sound propagation over asphalt surfaces.

PROPAGATION THEORY - Initially it is assumed that the atmosphere above the reflecting surface is quiescent and at a constant temperature, and that the source is an omni-directional point source. It is also assumed that the reflecting surface can be modeled as a uniform and infinitely deep equivalent fluid medium having a complex characteristic impedance and complex wave number.

For a given configuration, as depicted in Fig. 1, the complex sound pressure at a particular receiver location at a single frequency, ω, can be expressed as [7],

$$P_t = A\left[\frac{e^{ik_1 r_d}}{r_d} + R_p\left(\frac{e^{ik_1 r_r}}{r_r}\right) + \frac{(1-R_p)F(w)e^{ik_1 r_r}}{r_r}\right], \quad (1)$$

where A is an amplitude factor, r_d is the length of the direct path joining the source and the receiver, r_r is the specularly reflected path length, k_1 is the complex wave number in air, R_p is the plane wave reflection coefficient, i is the square root of -1, and a time dependence of $e^{-i\omega t}$ has been suppressed. Here $F(w)$ is a function of w, the numerical distance, given by Piercy et al. [7] as,

$$F(w) = 1 + i\sqrt{\pi}\sqrt{w}e^{-w}\text{erfc}(-i\sqrt{w}). \quad (2)$$

In Eq. (2), erfc denotes the complimentary error function.

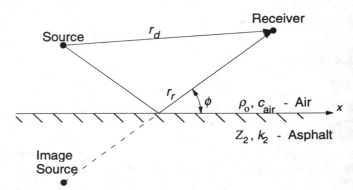

Figure 1. Schematic representation of sound propagation over a reflecting surface.

The first term in Eq. (1) represents the direct contribution to the total received sound pressure, the second term represents the specularly reflected contribution, and the third term accounts for non-specular contributions resulting from ground and surface waves.

When it is assumed, as here, that $|k_2/k_1|^2 \gg 1$, (where k_2 is the wave number in the reflecting medium, i.e., the asphalt) the reflecting medium becomes locally reacting, in which case w can be expressed as

$$w = i\frac{2k_1 r_r}{(1-R_p)^2}\left(\frac{Z_1}{Z_2}\right)^2, \quad (3)$$

where Z_1 and Z_2 are the characteristic impedance of the air and the surface impedance of the asphalt, respectively. When the ground layer depth is assumed to be infinite, Z_2 becomes the characteristic impedance of the asphalt surface, and R_p can be expressed as

$$R_p = \frac{Z_2 \sin\phi - Z_1}{Z_2 \sin\phi + Z_1}. \quad (4)$$

where ϕ is the grazing angle of the specularly reflected ray.

The total sound pressure at the receiver, as expressed by Eqs. (1) to (4), depends on the configuration of the source and receiver arrangement (r_d, r_r, and ϕ) and the characteristic properties of the propagating and reflecting media (k_1, Z_1, and Z_2). In the sub-sections immediately following, it will be shown that the theory outlined here can be combined with two-microphone transfer function measurements to estimate the parameters that define the impedance of an asphalt surface.

PARAMETRIC SURFACE IMPEDANCE MODEL - Initially, the two parameter model adapted by Van Wyk [6] from Attenborough's theory [9] was used to calculate the characteristic impedance, Z_2, of the ground surface. According to this model, the characteristic impedance of a porous medium (normalized by the characteristic impedance of the fluid medium) can be written as,

$$\frac{Z_2}{\rho_o c} = \gamma_1 \sqrt{\left[1 + \frac{2(\gamma-1)}{\sqrt{N_{pr}}\sqrt{\frac{\omega}{\omega_{cr}}}\sqrt{i}} T\left(\sqrt{N_{pr}}\sqrt{\frac{\omega}{\omega_{cr}}}\sqrt{i}\right)\right]^{-1} \left[1 - \frac{2}{\sqrt{\frac{\omega}{\omega_{cr}}}\sqrt{i}} T\left(\sqrt{\frac{\omega}{\omega_{cr}}}\sqrt{i}\right)\right]^{-1}}, \quad (5)$$

where γ is the ratio of specific heats (here assumed to be 1.4), N_{pr} is the Prandtl number (assumed to be 0.713), and the two parameters of the model are

$$\gamma_1 = \sqrt{q^2}/\Omega, \quad (6)$$

and

$$\omega_{cr} = \Omega \phi_e / 8\rho_o q^2, \quad (7)$$

when expressed in terms of physical surface parameters. In Eqs. (6) and (7), Ω is the surface porosity, q^2 is the tortuosity of the porous medium, ϕ_e is its effective flow resistivity, and ρ_o is the ambient air density. In addition, the function $T(x)$ is defined as:

$$T(x) = J_1(x)/J_0(x), \quad (8)$$

where J_0 and J_1 are Bessel functions of the first kind of zero and first order, respectively.

The two parameter ground model represented by Eq. (5) may be used in conjunction with Eqs. (1) to (4) to predict sound propagation over planar, locally reacting surfaces.

THE TWO MICROPHONE TRANSFER FUNCTION METHOD FOR THE DETERMINATION OF SURFACE PROPERTIES - To use the theory presented in the preceding sub-sections to predict the variability in sideline sound levels resulting from differences in surface type, the values of γ_1 and ω_{cr} for ISO and SAE surfaces must be known. Those values can be estimated by using an adaptation of the two microphone transfer function procedure originally developed by Glaretas [11].

In the free-field, two microphone transfer function method, an omni-directional source and two microphones are arranged over a surface as shown in Fig. 2. The source is placed at height h_s above the surface and the two microphones are located in the same vertical plane at a horizontal distance x: microphones 1 and 2 are placed at heights of h_{r1} and h_{r2}, respectively. The source is then made to emit broadband random noise, and the transfer function between the two microphone outputs, $H_{21,meas}$, is measured. Note that here the signal from microphone 1 is considered to be the input, and the signal from microphone 2 is considered to be the output.

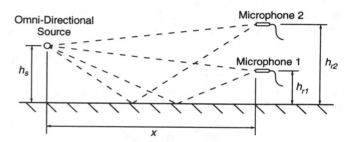

Figure 2. Geometry and notation for two microphone method.

Given the source and receiver configuration, it is possible to derive a theoretical expression for that transfer function,

$$H_{21,theo} = \frac{P_{t_2}}{P_{t_1}}, \quad (9)$$

based on the propagation theory and the two parameter ground impedance model outlined above. The theoretical transfer function is then a function of the geometrical parameters that define the source and microphone arrangement, the physical parameters of the air at the time of measurement, and the ground impedance, which has been expressed in terms of two parameters. When the configuration of the source and microphones and the physical properties of the air are known, only the ground parameters are unknown. A non-linear optimization procedure can then be used to adjust the ground parameter values to obtain the best agreement between the theoretical and measured transfer functions. The parameter values found in this way are said to characterize the asphalt surface. Those parameters may then be used to calculate the normal absorption coefficient of the surface, for example.

In order to perform the numerical optimization, an objective function must be defined which quantifies the "closeness" of the fit between the measured and theoretical transfer functions. That function, $E_1(\gamma_1, \omega_{cr})$, was defined here to be,

$$E_1(\gamma_1, \omega_{cr}) = \sum_j \left| H_{21,theo}(\omega_j, \gamma_1, \omega_{cr}) - H_{21,meas}(\omega_j) \right|^2, \quad (10)$$

where the theoretical transfer function is evaluated at the discrete frequencies, ω_j, at which estimates of the measured transfer function are available. Minimization of E_1 by adjustment of the ground parameters, γ_1 and ω_{cr}, results in identification of the optimal values of the latter.

In previous work, some difficulty was encountered while attempting to obtain a good fit between theoretical and the measured data [5]. The measurements in question were performed when the ground temperature was approximately 10 °C higher than the air temperature at a height of 2 m, and it was thought that this gradient may have been responsible for the fitting difficulties [5]. That it, it was hypothesized that refraction resulting from the near-ground temperature gradients through which the sound must travel from the source to the microphones may have caused both the path lengths between source and receivers, and the grazing angles at which those paths intersect the ground, to differ from the values they would have in a homogeneous atmosphere.

In a first attempt to account for that effect, the values of A, the amplitude factor, and x, the horizontal distance between the source and receivers, were allowed to vary along with the two ground parameters during the optimization. The relatively small modifications of A and x from their nominal values that resulted from that procedure caused there to be a significant improvement in the fit between the measured and theoretical transfer functions [5]. The latter results provided the motivation for developing a ray tracing technique to account more directly for the effect of wind and temperature gradients on the two microphone transfer function method.

RAY TRACING IN THE PRESENCE OF GRADIENTS

INTRODUCTION - It is well known that gradients in atmospheric temperature and wind can cause sound rays to refract as they propagate [7, 12]. When sound propagates upwind through a wind boundary layer or through a temperature lapse (i.e., a region where the air temperature decreases with elevation) sound rays are bent upwards: i.e., towards the region of lower sound speed. When sound propagates downwind through a wind boundary layer or through a temperature inversion (i.e., a region where the air temperature increases with elevation), sound rays are bent downward. Over long distances, it is well known that refraction can significantly affect measured sound pressure levels [7]. In this work, it is of interest to understand how the effects of wind and

temperature gradients manifest themselves over very short distances.

GENERAL APPROACH - It has been assumed here that the primary effects of environmental gradients are four fold: i.e., (i) to alter the length of the direct and reflected paths by which sound propagates from the source to the receiver; (ii) to alter the grazing angle of specular reflection; (iii) to alter the sound pressure along the path as a result of wavefront divergence or convergence; and (iv) to change the speed of sound (and hence the total phase change) along the ray paths. The propagation prediction based on Eqs. (1) to (4) has been modified here in an attempt to account for these effects.

The nature of the individual effects listed above are accommodated as follows:

(i) First, the change in the path lengths affects the amount of spherical spreading between the source and receivers through the $1/r$ expressions in Eq. (1).

(ii) The change in the grazing angle manifests itself as a change in the plane wave reflection coefficient through Eq. (4), where the refracted value for ϕ is substituted for the value determined from a straight line geometry.

(iii) When propagation transverse to the plane defined by the source and two microphones is neglected, the effect of wavefront convergence or divergence can be calculated by considering two rays that subtend an angle θ at the source. Under homogeneous conditions, those rays will be separated at the receiver by a (nearly) perpendicular distance ds, which, under a small angle approximation, can be expressed as,

$$ds = \theta r \quad (11)$$

where r is the length of the rays. If convergence or divergence occurs, the perpendicular separation between adjacent rays will differ from ds. Since sound power is conserved within any ray tube, the amplitude of the refracted wave front will be related to the unrefracted amplitude by the square root of the ratio of the perpendicular separations at the receiver [13]: i.e.,

$$A_{ref} = \sqrt{\frac{\theta r}{ds_{ref}}}, \quad (12)$$

where ds_{ref} is the perpendicular separation of the refracted rays at the receiver location.

(iv) In a homogenous atmosphere, the phase of the signal at the receiver with respect to that at the source is simply the product of the wave number and the path length. In an inhomogenous atmosphere, the phase speed along any path becomes a function of position along that path. Hence, when the atmosphere is refracting, the wave number must be integrated along the path to find the phase of the received signal. Note that the local phase speed at a point on the path is a function of the local temperature and the component of the local wind velocity directed along the path. The line integral that must be evaluated along the path of the ray to determine the phase of the signal arriving at the receiver is,

$$(k_1 r)_t = \int_C \frac{\omega}{c + \bar{v} \cdot \bar{n}} ds \quad (13)$$

where $(k_1 r)_t$ is the total phase change between source and receiver, ω is the frequency, c is the phase speed, $\bar{v}$ is the wind velocity vector, $\bar{n}$ is a unit vector tangent to the path of the ray, ds is the differential arc length along the ray, and C denotes integration along either the direct or the reflected ray path.

Note that the method used here to account for refraction is an approximate one since it is based on the assumption that the effects of refraction can be accounted for purely by adjusting in Eq. (1) the ray path lengths, the level of the pressure along that path, the phase change along the paths, and the grazing angles. While this assumption is reasonable when considering the direct and specularly reflected components, (i.e., the first two terms of Eq. (1)), it is not obvious that this approach correctly accounts for the effect of sound speed gradients on the ground and surface waves (i.e., the last term of Eq. (1)).

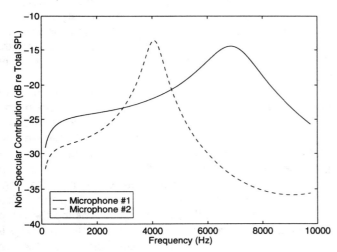

Figure 3. Non-specular contribution to sound field.

Table 1. Surface and geometrical parameters used to calculate results shown in Fig. 3.

γ_1	ω_{cr}	h_s (m)	h_{r1} (m)	h_{r2} (m)	x (m)
13	3×10^4	0.3	0.15	0.3	5

However, in the present instance, it is felt that the relatively simple approach adopted here is justified since the direct and specularly reflected components together dominate the received sound field. For example, for the configurations used in the present measurements, it has been estimated that the sound pressure components due to the ground and surface waves typically lie 15 to 30 dB below the total sound pressure level. For example, Fig. 3 is a plot of the magnitude of the non-specular contribution, in decibels referenced to the total sound pressure, for a typical configuration and surface condition under homogeneous atmospheric conditions. The parameters used in this calculation are given in Table 1.

In situations where the non-specular contribution is not sufficiently small to allow the present approach to be used, it would be necessary to employ a more exact means of accounting for refractive effects: e.g., the Fast Field Program [14] or a parabolic equation approach [15].

RAY CURVATURE FORMULATION - Chessel [16] developed an expression for the path of a sound ray as it propagates through gradients in sound speed and wind. If the direction of propagation and the wind velocity are co-planar, as is assumed here, and there is no vertical component to the wind velocity, Chessel's system of differential equations can be reduced to a closed form expression for the curvature of the sound ray. Under these restrictions, the local curvature of a ray can be expressed as,

$$K = \frac{(dc/dz)[\sin\theta - (v/c)\cos 2\theta] + (dv/dz)[1 + (v/c)\sin^3\theta]}{c[1 + (2v/c)\sin\theta + (v/c)^2]^{3/2}}, \quad (14)$$

where v is the horizontal wind velocity parallel to the direction of propagation, c is the local speed of sound, θ is the elevation angle of the propagating ray, and dv/dz and dc/dz are the gradients of the wind velocity and sound speed with respect to elevation, respectively.

In general, for the ray tracing procedure to be accurate, the wavelength of the propagating sound must be small with respect to the radius of curvature of the ray paths [13]. This restriction imposes a lower frequency limit on the applicability of the equation for curvature. In the calculations reported below, the radius of curvature is typically greater than (and usually much greater than) 10 m, so that the present approach might be expected to work well at frequencies above approximately 300 Hz.

Note finally that since the sound speed in air can be expressed as a function of the air temperature, the local curvature of the ray can be expressed in terms of the local wind velocity, temperature, and their gradients.

RAY TRACING METHODOLOGY - Here, it has been assumed that the wind speed and air temperature vary only with elevation, that the wind direction is parallel to the plane defined by the source and the two microphones, and that the wind direction does not change over the region of interest. Thus, the effects of any propagation transverse to the direct path between the source and the receiver are not accounted for. These assumptions seem reasonable given the very short propagation distances under consideration here: e.g., several meters. In addition, all measurements reported here were conducted when wind speeds were less than 1.5 m/s. When considering sound propagation over asphalt, and especially asphalt heated by the sun, temperature gradients are expected to have a larger effect than wind gradients.

When the wind speed and temperature profiles are known, a ray path can be traced by using the expression for curvature developed by Chessel. To begin, the expression for the curvature is inverted to give an expression for the local ray path radius of curvature. The ray path can then be traced by stepping through the atmosphere in small increments. At each step, the radius of curvature is calculated based on the environmental conditions at that point, and that radius is used to project the ray to the next point: this process is illustrated schematically in Fig. 4.

In Fig. 4, the ray originates at point A with a propagation angle θ_A. Based on the environmental conditions at point A, the radius of curvature is calculated to be ρ. A step size of length Δx is chosen which subtends an angle ϕ, defined by ρ. A ray traveling along the arc AB would terminate at point B with a propagation angle $\theta_B = \theta_A - \phi$; the location of point B can be computed by using the law of cosines. This procedure is then repeated to move from point B to point C, etc. A specified level of accuracy can be achieved by changing the step size, Δx.

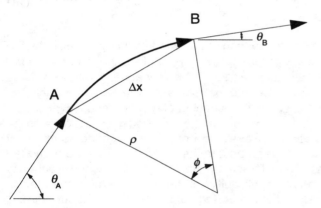

Figure 4. Ray tracing diagram.

The total path length between the source and receiver can then be computed simply by summing the individual step lengths over the path of the ray. Additionally, Eq. (13) can be evaluated numerically as the ray path is traced. At each step in the ray path, the phase speed is evaluated at that location based on the temperature profile and the component of the wind velocity tangent to the ray path based on the wind velocity profile and the direction of the ray. The resultant phase speed is used to compute the wave number, which is then multiplied by the step length, and the result is summed to give the total phase change along the path.

The grazing angle is computed numerically by recording the angle at which the ray intersects the ground surface. The reflected ray is assumed to leave the surface at the same angle (i.e., reflection is assumed to be specular).

WIND AND TEMPERATURE PROFILES - A knowledge of the wind and temperature profiles above the ground surface are required to compute the radius of curvature of the sound ray (to compute dc/dz and dv/dz, for example). To facilitate calculation of the gradients, the wind and temperature profiles were modeled here by using logarithmic expressions whose constants are determined by measurement. One commonly used form for the profile is [17],

$$s(z) = a_1(\ln z)^2 + a_2 \ln z + a_3, \qquad (15)$$

where s can be either temperature or wind speed, z is the height above the surface, and a_1, a_2, and, a_3 are constants that can be adjusted to fit the experimental data. Since this curve is not well behaved as z approaches zero, the profile is extrapolated to the ground as a straight line below some critical height, z_o, taken here to be 0.1 meters.

EFFECTS OF REFRACTION ON THE TWO MICROPHONE TRANSFER FUNCTION

INTRODUCTION - In the absence of wind and temperature gradients, sound rays travel in straight lines from the source to the receivers, in which case the path lengths and grazing angles that are required to calculate the two microphone transfer function are known explicitly. In the presence of gradients, the rays are curved, and the rate of that curvature can be space-dependent, with the result that it is difficult or impossible to derive an analytical expression for the path lengths and grazing angles, and a numerical approach must be followed.

ITERATIVE DETERMINATION OF PATH - The path followed by a ray depends on both the position from which the ray is launched and the initial direction of the ray. When predicting the two microphone transfer function, the starting location of the ray is always known (i.e., the source location), but the initial direction of a ray that will arrive at a prescribed receiver location (i.e., one of the two microphones) is not known *a priori*. This situation presents a difficulty when determining the effect of refraction on the two microphone transfer function. In the absence of closed form expressions for the ray paths, an iterative process is required to determine the ray launch angle that will cause a ray to pass through the receiver. In the present work, an initial guess was made for the starting direction, then the ray tracing procedure was used to calculate the distance between the ray path (at the point of closest approach) and the receiver location. The initial guess for the ray direction was successively refined until the ray passed within one step size of the receiver location (typically less than 1 mm). The length and grazing angle of this ray were used in subsequent calculations.

EFFECT OF PATH LENGTH AND PHASE CHANGE - Refraction causes both the direct and reflected path lengths to increase. Under the effect of reasonable wind and temperature profiles, the changes are small in an absolute sense: i.e., on the order of 1 percent. Note that the effect is typically greater for the reflected paths than for the direct paths, since the strength of the environmental gradients typically increase in the boundary layer near the ground. The change in path length has two effects on the measured pressure. First, the amount of spherical spreading over the ray path is changed, and secondly, the phase difference between the direct and reflected signals arriving at the receiver is altered. The latter effect is amplified by the change in sound speed along the path, and this effect has proven to be far more significant than the change in the relative amount of spherical spreading. Therefore, it is not the change in the absolute length of the paths that is significant, but rather it is the difference between the phase change along the direct and reflected paths that will most strongly affects the inter-microphone transfer function.

EFFECT OF DIVERGENCE OR CONVERGENCE OF WAVEFRONTS - Refraction can also have the effect of causing a wavefront to focus or diverge. This effect will be different for the direct and reflected rays since those rays typically pass through different gradients. Under a temperature lapse condition, as encountered here, this effect results in a focusing of the direct sound and a divergence of the reflected sound. This effect will be most significant in frequency ranges where the direct and reflected rays nearly cancel. Amplitude corrections of up to 20 percent for individual rays (with respect to spherical spreading) were found to occur in the present measurements.

EFFECT OF GRAZING ANGLE CHANGE - In addition to phase and focusing affects, the refraction also changes the angle at which the reflected rays intersect the surface, i.e., the grazing angle. For any finite impedance surface the reflection coefficient is a function of the grazing angle. For very hard surfaces, the reflection coefficient changes very rapidly as the incidence angle approaches grazing. In the two microphone setup as used here, the source and receivers are usually arranged such that the reflected rays are near grazing in order to increase the sensitivity of the measurements to the ground impedance.

Figure 5 is a plot of the magnitude of the reflection coefficient, as defined in Eq. (4), at three frequencies, for a surface whose impedance is defined in terms of the two parameter ground impedance model, Eq. (5), with $\gamma_1=14$ and $\omega_{cr}=3 \times 10^4$. It can be seen from Fig. 5 that the reflection coefficient changes very rapidly for grazing angles below 5 degrees. Also in Fig. 5, "+" and "o" markers are used to indicate the grazing angles that result from the three configurations used in the two microphone measurements reported here (i.e., horizontal separations of 1.8 m, 3 m, and 5 m, and average source and receiver heights as given in Table 1). A small marker is used to indicate the grazing angle calculated for unrefracted propagation, and a large marker is used to indicate the value of the grazing angle after refraction through a logarithmic profile where the ground is approximately 3 °C warmer than the ambient air temperature (note that this is not a particularly strong gradient). From this figure it can be seen that even though the environmental gradients cause only slight changes (in absolute terms) in the grazing angle, these small grazing angle changes can translate to significant changes of the reflection coefficient, especially at longer ranges and higher frequencies.

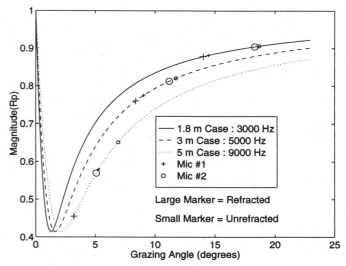

Figure 5. Changes in the magnitude of the reflection coefficient with changing grazing angle.

SAMPLE ANALYSIS - To demonstrate first that refraction can have a significant effect on the two microphone transfer function, theoretical transfer functions calculated with and without environmental gradients were calculated and are plotted in Fig. 6. The values of the environmental parameters used in the calculations are summarized in Table 2, while the configuration and the ground parameters in all cases were those given in Table 1. Note that even the most severe environmental conditions used in this simulation are easily achieved on a hot sunny day in summer.

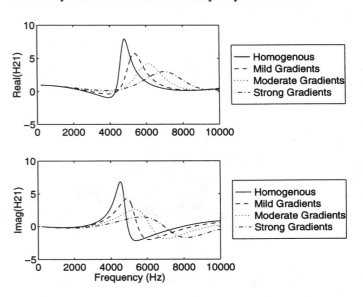

Figure 6. Effect of refraction on H_{21}

SINGLE PARAMETER GROUND IMPEDANCE MODEL

Van Wyk [8] observed while using the two parameter ground model that it was often difficult to determine a unique combination of parameters to describe a surface. It was typically found that a range of γ_1 and ω_{cr} combinations would produce nearly identical values for the ground impedance. Figure 7 is a plot of Eq.(10) when simulated propagation data was used to calculate $H_{21,meas}$ so that the objective function could be plotted for a range of γ_1 and ω_{cr} values. In this case, the "correct" values for γ_1 and ω_{cr} were 18 and 4×10^4, respectively. The configuration used in these calculations is tabulated in Table 1. It can be seen from Fig. 7 that a range of values of γ_1 and ω_{cr} along the "valley" nearly minimize the objective function. This observation implies that in this region of parameter values, the parameters γ_1 and ω_{cr} are not independent, and thus the ground acoustical properties are dependent on only a single parameter.

Table 2. Parameters for demonstration of refractive effects on two microphone transfer function.

Transfer Function	No Gradients	Mild Gradients	Moderate Gradients	Strong Gradients
Ground Temp. (°C)	32	35	39	43
Air Temp. (°C)	32	32	32	32
Logarithmic Parameter a_1	32	32	32	32
Logarithmic Parameter a_2	0	-0.14	-0.64	-1.54
Logarithmic Parameter a_3	0	0.24	0.45	0.59

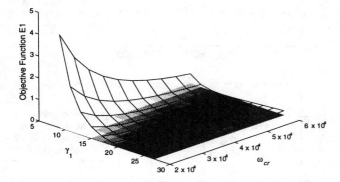

Figure 7. Effect of ground parameters on objective function for H_{21}.

A single parameter ground model can be identified as follows. By returning to the model for the ground impedance, Eq. (5), a simplification can be effected by assuming that $\omega/\omega_{cr} \ll 1$. This approximation is valid at low frequencies and/or when the effective flow resistivity is high. Under this assumption, the Bessel functions can be approximated by the first terms in a Taylor series expansion. When the higher order terms are discarded, Eq. (8) can be expressed in terms of a ratio of polynomials,

$$T(x) = \frac{x(8-x^2)}{4(4-x^2)}. \quad (16)$$

The latter expression can be substituted into Eq. (5) which can be further simplified by setting equal to zero the terms involving second and higher powers of (ω/ω_{cr}). The characteristic impedance of the reflecting surface, $Z_2/\rho_0 c$, can then be expressed as

$$\frac{Z_2}{\rho_0 c} = \Gamma \left(\frac{8i}{\gamma\omega} \right)^{1/2}. \quad (17)$$

A new ground parameter, Γ, is here defined in terms of the two previous parameters as

$$\Gamma = \gamma_1 \sqrt{\omega_{cr}} = \sqrt{\frac{\phi_e}{8\rho_0 \Omega}}. \quad (18)$$

Thus under the conditions listed here, the ground model reduces to a single parameter model. The single parameter model is, of course, an approximation of the two parameter model, so an error is introduced by its usage. Note that the role of γ_1 remains unchanged in the single parameter model, but the two models will differ as a function of ω_{cr} and ω. Note also, that this final form is the same as a low frequency approximation previously derived by Attenborough [9].

A function quantifying the difference between the one and two parameter models can be defined as

$$E_2(\omega_{cr}, \omega) = \frac{|Z_{1,\text{single-param}} - Z_{1,\text{two-param}}|^2}{|Z_{1,\text{two-param}}|^2}, \quad (19)$$

in which the subscript indicates whether the impedance has been calculated using the one or two parameter model.

Figure 8 is a plot of E_2 over a range of values of ω_{cr} and ω. From Fig. 8 it can be seen that the models are nearly equivalent below 8000 Hz when ω_{cr} is greater than 5×10^4. As long as the optimization is performed in this region, the single parameter ground approximation can be considered valid. The use of the single parameter model speeds the optimization procedure since the number of unknowns is reduced by one. Unfortunately, however, information about the individual values of γ_1 and ω_{cr} is lost.

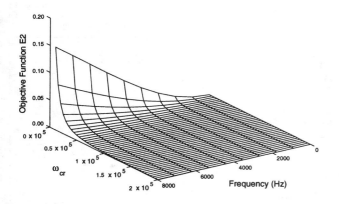

Figure 8. Difference between two parameter and single parameter ground models.

EXPERIMENTAL APPARATUS

In order to predict the effects of refraction on the two microphone transfer function, meteorological data must be collected at the same time as the acoustical data. Two independent data acquisition systems were used to accomplish this.

The air temperature was recorded by using an array of eight thermocouples ranging in elevation from 0.005 m to 1.8 m above the ground, spaced logarithmically. Seven of the thermocouples were shielded from ground and solar radiation with metal plates that allowed the free flow of air around the transducers. Note, however, that the thermocouple nearest the ground was shielded only from solar radiation. The wind speed and direction were measured at two heights, 0.4 m and 1.3 m, above the surface using a pair of rotating cup anemometers. All of these signals were collected and stored by a portable PC-based data acquisition system.

A PC-controlled Fourier analyzer (Tektronix Model 2630) was used to generate the random noise source signal and to collect the acoustical data. The acoustical source consisted of a driver (ElectroVoice Model 1829) enclosed in a wooden box with an outlet at the end of a tube (inner diameter of 0.005 m) providing a compact source at a distance 0.9 m from the nearest reflecting surface on the box. Two low noise, ½ inch microphones (Brüel and Kjær Model 4190) were mounted on a stand at the end of 1 meter long extensions, again to place them a significant distance from the nearest reflecting surface. Figure 9 is a schematic representation of the experimental apparatus.

Three nominal configurations were used. In all three arrangements, the source was placed approximately 0.3 m above the asphalt surface, while microphones 1 and 2, were, as indicated in Fig. 2, placed approximately 0.15 m and 0.30 m above the surface, respectively. The three configurations differed only in the nominal horizontal separation between the source and the microphones: distances of 1.8 m, 3 m, and 5 m were used.

The acoustical signals were sampled at 25600 Hz for 80 ms per record giving a measurement bandwidth of 10 kHz and a frequency resolution of 12.5 Hz. Each data record was weighted by a Hann window, and sequential records were overlapped by 50 percent when performing the cross spectral averaging. A total of 30 records were averaged per measurement.

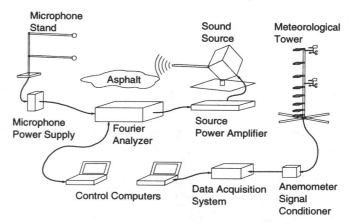

Figure 9. Experimental apparatus schematic.

MEASURED RESULTS

Two-microphone data were taken at two sites in the southwestern United States at a number of ISO and SAE surfaces. Data were taken under a variety of meteorological conditions.

PREPROCESSING DATA - The two microphone transfer function measurement is very sensitive to interfering reflections. The experimental apparatus was designed to reduce the early reflections as much as possible. It was intended that the first reflections from the source enclosure or the microphone stand arrive at the microphones at least 5 ms after the direct signal. These late reflections, although much lower in amplitude than the direct signal, still cause significant distortion in the measured transfer function, especially in frequency regions in which the direct and specularly reflected components nearly cancel at a particular microphone location.

In order to obtain the cleanest possible data for the optimization process, a post-collection time domain windowing technique was used to reduce the effect of late reflections. A window, centered around t = 0 seconds, was applied to the measured impulse response (i.e., the inverse Fourier transform of the two microphone transfer function). The window consisted of a sinusoidal ramp from zero to unity, followed by a uniform section of unity value occupying 80 percent of the total window length, followed by a sinusoidal ramp down from unity to zero. The total window length was varied from 5 ms to 10 ms depending on the nature of the data.

Figure 10 is a plot of a typical impulse response with a 10 ms window superimposed. Figure 11 is a plot of the real and imaginary components of the transfer function corresponding to the impulse response of Fig. 10 before and after the time domain windowing was applied. This windowing technique was applied to a number of simulated impulse responses to verify that a window of this length would not significantly distort the transfer function in the frequency range of interest.

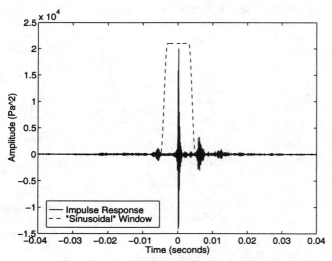

Figure 10. Impulse response and overlaid "sinusoidal" taper time window.

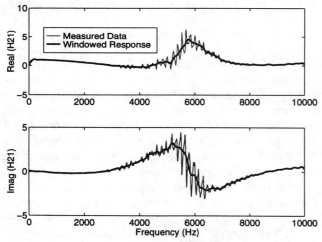

Figure 11. Effect of time domain windowing on measured transfer function.

WIND AND TEMPERATURE PROFILING - The wind and temperature were measured at discrete heights at the same time that the acoustical data was recorded. The ray tracing approach requires a continuous representation of the wind and temperature, so a standard profile shape (Eq. (15)) was fitted to the measured data. Figure 12 is a plot of a typical set of meteorological data and the fit provided by the logarithmic profile. The individual dots in Fig. 12 indicate the scatter of the data that occurred during a typical propagation measurement (lasting approximately 30 seconds).

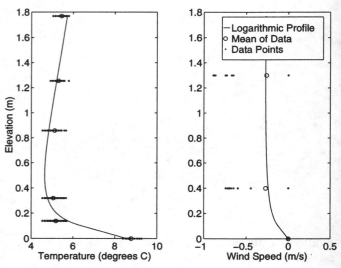

Figure 12. Meteorological data with fitted profile.

Analysis of the temperature data revealed that the air temperature measured by the thermocouple nearest the ground may have been significantly increased by radiation from the asphalt: this error was accounted for in the optimization process, described next.

OPTIMIZATION - Equipped with a knowledge of the environmental conditions, the ray tracing methodology can be applied to predict the effect of refraction on the measured two microphone transfer function. Once this effect has been accounted for, the ground parameter can be systematically adjusted to achieve an optimal fit between the measured transfer function and the transfer function predicted by using the sound propagation equations in conjunction with the ground model.

As previously mentioned, there was some concern about the accuracy of the air temperature measurement closest to the ground. To account for this uncertainty, the ground temperature was treated as an adjustable parameter during optimization and was varied along with the ground parameter to obtain the best possible fit with the measured data.

The inclusion of the ground temperature as an unknown did result in significant additional computation, however, since at each step in the optimization, it was necessary to generate a new temperature profile to accommodate the adjusted ground temperature. New ray paths were traced through the new profile, and finally the two microphone transfer function was calculated in terms of the ground parameter. The inclusion of the ground temperature in the optimization process did, however, result in very good agreement between the measured data and the predicted transfer functions. The required downward adjustment in the air temperature close to the ground was generally than 2 degrees Celsius.

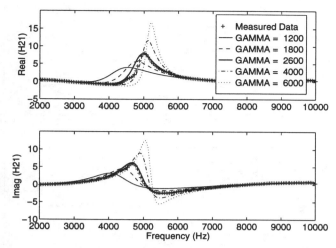

Figure 13. Effect of a change in Γ on the calculated transfer function.

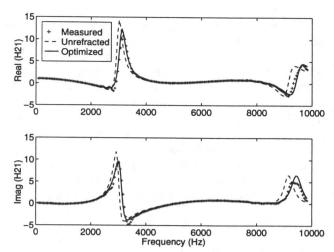

Figure 14. Sample optimization results for ISO surface #1, 1.8 meter configuration.

If the optimization is to be robust, it is also necessary that the ground parameter have a significant impact on the theoretical transfer function: i.e., a variation of the ground parameter over a reasonable range should result in changes of the transfer function that are large compared to the uncertainties in the measured data. Figure 13 is a plot of measured transfer function data for ISO surface #1 in the 3 m configuration overlaid with plots of predicted transfer functions based on a range of values of the ground parameter Γ. The predicted transfer functions have been corrected for refraction based on the measured environmental conditions. This plot demonstrates that changing the ground parameter has a significant effect on the two microphone transfer function.

Figures 14 to 16 show the results of optimizations for a set of measurements on ISO surface #1. The real and imaginary components of the transfer functions for the 1.8 m, 3 m, and 5 m, configurations are plotted in Fig. 14, 15 and 16, respectively. The small crosses represent the measured data (after the time domain windowing technique has been applied). The solid line represents the optimal fit achieved by adjusting both the ground parameter, Γ, and the air temperature at the ground while using ray tracing to model the effects of refraction. The dashed line represents the predicted transfer function based on the optimal value for Γ but without including the effects of refraction. It can be seen from these plots that the effect of refraction becomes progressively more significant as both the frequency and propagation range increase, and that the refracted prediction comes much closer to the measured data than does the unrefracted prediction.

The optimized values of Γ for the SAE and ISO surfaces at three horizontal separations are shown in Fig. 17. Each line represents a different surface, and the abscissa represents the horizontal separation between the source and the microphones. From this plot it can be seen that the optimized value for Γ appears to be a function of propagation range, and not just surface type. This behavior suggests either that some aspect of the

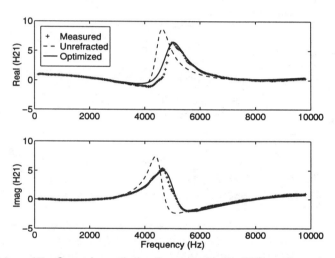

Figure 15. Sample optimization results for ISO surface #1, 3 meter configuration.

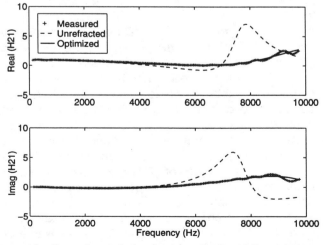

Figure 16. Sample optimization results for ISO surface #1, 5 meter configuration.

propagation process is not being modeled accurately, or that an effect that has not been accounted for, e.g., turbulence or surface roughness, is systematically influencing the result. Based on preliminary work, it appears that the apparent range-dependence of Γ may be due to the relatively poor resolution of the temperature profile near the ground in the present measurements. In future measurements, all the thermocouples will be distributed between the ground and the highest level likely to be reached by the direct ray, i.e., approximately 0.3 m, so that the near-ground temperature profile can be identified accurately.

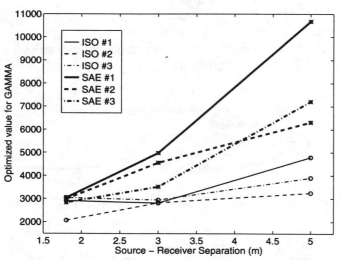

Figure 17. Optimized values for Γ for ISO and SAE surfaces for each of the measurement configurations.

Although the values of Γ determined from the individual optimizations were found to be a function of measurement configuration, there nonetheless appears to be a consistent difference between the results for the two surface types. Note in particular that the values of Γ for the SAE surfaces were in all cases larger than those for the ISO surfaces, and that the Γ values estimated for the SAE surfaces show a stronger dependence on range than do those for the ISO surfaces. Note that the result denoted SAE #1 was measured under the most extreme gradient conditions of any of the surfaces.

The range dependence of Γ notwithstanding, it is of interest to obtain a single value of Γ to characterize a surface. To obtain a single value for Γ in the present case, the data from all three measurements on a single surface were used in the optimization process simultaneously: i.e., theoretical transfer functions based on the same value of Γ were fitted to the three measurements made on a particular surface, and the objective function to be minimized was the sum of the squared differences between the predicted and measured data at all three ranges. In this way a single value for Γ could be determined that would provide the best possible fit to all three sets of measurements. Figures 18 to 20 are plots of the results from such an optimization for SAE surface #3 for the 1.8 m, 3 m, and 5 m configurations, respectively. A reasonable fit was

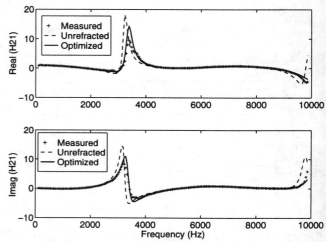

Figure 18. Multi-measurement Optimization Result for 1.8 meter Configuration, SAE Surface #3.

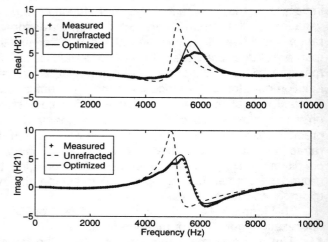

Figure 19. Multi-measurement Optimization Result for 3 meter Configuration, SAE Surface #3.

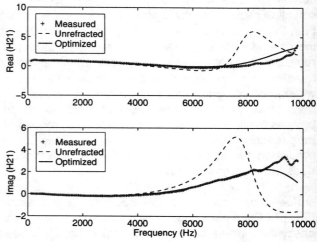

Figure 20. Multi-measurement Optimization Result for 5 meter Configuration, SAE Surface #3.

achieved for all three measurements using the same value for the ground parameter (except above 8 kHz in the 5 m configuration), but the resultant multi-measurement fit was slightly worse than the fit obtained by optimizing at each configuration dependently (i.e., the residual error in the multi-measurement fit was larger than the sum of the individual residual errors when the results at each range were optimized independently).

The Γ values resulting from the multi-measurement optimization technique for all six surfaces are given in Table 3. Again, these results reveal a distinction between the two surface types: i.e., the Γ values for all the SAE surfaces are larger those for all of the ISO surfaces. Thus it is possible to conclude, at least tentatively, that the value of Γ is consistently larger for SAE than for ISO surfaces.

An inspection of Eq. 18 shows that an increase in Γ is associated with an increase in the surface flow resistivity, and/or a decrease in the surface porosity. Thus it might be expected that the application of a sealant to the asphalt, as is normally done to SAE surfaces, would cause Γ to increase by increasing the flow resistivity of the surface as a result of sealing the surface pores. The current findings are thus consistent with the physical nature of the SAE and ISO surfaces.

Table 3. Parameter value results for multi-measurement optimization.

Surface	Optimized Parameter Value, Γ
ISO #1	3600
ISO #2	3400
ISO #3	2800
SAE #1	4000
SAE #2	4100
SAE #3	5300

Given the values for Γ, it is possible to calculate the normal incidence absorption coefficient for each surface. The normal incidence absorption coefficient, α_n, is defined in terms of the reflection coefficient as,

$$\alpha_n = 1 - \left| R_p(0) \right|^2 \quad (0.20)$$

where $R_p(0)$ is the normal incidence reflection coefficient, which is itself a function of Γ and frequency. Figure 21 is a plot of minimum and maximum normal incidence absorption coefficients calculated from the Γ's resulting from the multi-measurement optimizations for the SAE and ISO surfaces. This plot again demonstrates the distinction between the two surface types: i.e., the SAE surfaces are consistently less absorptive than the ISO surfaces. In addition, "+" and "o" symbols are used to indicate impedance tube absorption measurements made for an ISO and an SAE site in Southern Michigan [18]. The current results are consistent with these previous findings.

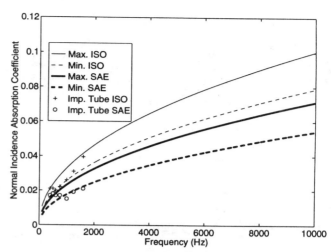

Figure 21. Minimum and maximum normal incidence absorption coefficients for ISO and SAE surfaces.

EFFECTS OF REFRACTION - The data presented here show that environmental refraction can have a significant impact on short range propagation over asphalt. That is, significant differences can be observed in the measured two microphone transfer function when a site is measured under different meteorological conditions. To illustrate this point clearly, data were collected at two sites at different times (and hence under different temperature and wind gradient conditions) in order to assess whether the estimation of Γ would be stable against the effects of differing meteorological conditions. Figure 22 is a plot the optimization results for the ISO surface #1 measured with a 3 meter separation between the source and microphones. The predictions in this plot are based on a Γ value determined from a multi-measurement optimization, taking into account the different meteorological environment for both measurements. The resultant best fit of the analytical expressions, both calculated with a Γ value of 2800, is overlaid with the data. The differences in the two best fit transfer functions arises entirely from the differences in the effects of refraction. In addition, a fifth trace indicates the transfer function that would be predicted if the effects of refraction were not accounted for. This unrefracted result does not fit either of the measurements as well as the results that have been corrected for refraction.

Figure 23 shows similar results for the ISO surface number #3 in the 5 meter configuration. Again, the two sample measurements were taken at the same location, in the same source - receiver configuration, but under different meteorological conditions. The theoretical predictions were made using a single value of Γ (3700). Again, it can be seen that it is necessary to account for refraction to achieve reasonable fits between predictions and measurements.

We believe that these results demonstrate that it is necessary to allow for the effects of refraction when using the two-microphone method to characterize relatively high impedance outdoor surfaces.

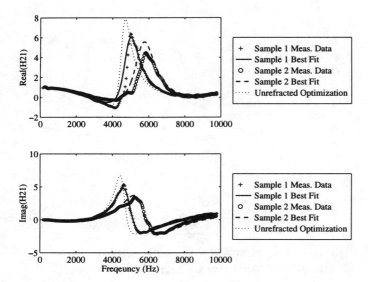

Figure 22. Optimization of ISO surface #1, 3 meter configuration, under different meteorological conditions.

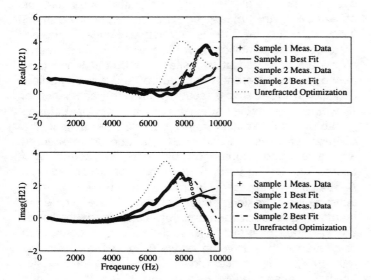

Figure 23. Optimization of ISO surface #3, 5 meter configuration, under different meteorological conditions.

CONCLUSION

It has been suggested that some of the variability in motor vehicle passby measurements originates in propagation differences over different types of passby site surfaces. This study supports previous findings that the two primary standardized surface types have different acoustical properties, and that there may be significant site-to-site variations between surfaces of the same type. Additionally, the results of this study have shown that environmental refraction of sound due to gradients in the wind and temperature above asphalt surfaces can have a significant impact on short range acoustical measurements. A ray tracing method has been used to successfully model these refractive effects. The inclusion of refractive effects in the two microphone method for estimating surface acoustic properties has resulted in improved accuracy when environmental gradients exist. At present, the methodology described here is able to distinguish between different surface types, but there appears to be some dependence on the measurement configuration. Several factors may account for the remaining uncertainty, including turbulence, surface roughness, and oversimplification of the temperature profiles and/or the ground surface models. The effects of these various factors will be the subject of future work.

ACKNOWLEDGMENTS

This work was supported by the American Automobile Manufacturers Association (Grant monitors: Ronald J. Wasko and Barbara Wendling), and by a fellowship from General Motors. The authors are especially grateful to Richard Schumacher (General Motors) and Brian Rees (Chrysler) for their logistical support during the measurements reported here.

REFERENCES

[1] J.S. Bolton, H.R. Hall, R.F. Schumacher and J. Stott, "Correlation of Tire Intensity Levels and Passby Sound Pressure Levels," Paper 951355, *Proceedings of the SAE Noise and Vibration Conference*, pp. 981-989 (1995).

[2] SAE J1470 MAR92 Measurement of Noise Emitted by Accelerating Highway Vehicles (1992).

[3] ISO 10844 Draft 1992 Acoustics - Specification of Test Tracks for the Purpose of Measuring Noise Emitted by Road Vehicle (1992).

[4] R.F. Schumacher, K.G. Phaneuf and W.J. Haley, "SAE and ISO Noise Test Site Variability," Paper 951361, *Proceedings of the SAE Noise and Vibration Conference*, pp. 1037-1044 (1995).

[5] D.M.B. Yim, "Influence of Surface Variables on Motor Vehicle Passby Noise Measurements," *MSME Thesis*, Purdue University (1995).

[6] K.D. Van Wyk, "Models for the Characterization of the Acoustical Properties of Asphalt Surfaces," *MSME Thesis*, Purdue University (1991).

[7] J.E. Piercy, T.F.W. Embleton and L.C. Sutherland, "Review of Noise Propagation in the Atmosphere," *Journal of the Acoustical Society of America*, vol. 61, pp. 1403-1418 (1977).

[8] K.D. Van Wyk, J.S. Bolton and P.J. Sherman, "Multi-parameter Models for the Characterization of Asphalt's Acoustical Properties," *Proceedings of Internoise 91*, pp. 459-462 (1991).

[9] K. Attenborough, "Acoustical Impedance Models for Outdoor Ground Surfaces," *Journal of Sound and Vibration*, vol. 99, pp. 521-544 (1985).

[10] M. Berengier, J.F. Hamet and P. Bar, "Acoustical Properties of Porous Asphalts: Theoretical and Environmental Aspects," Transportation Research Record 1265 (1990).

[11] C. Glaretas, "A New Method for Measuring the Acoustic Impedance of the Ground," *PhD Thesis*, Pennsylvania State University (1981).

[12] B. Hallberg, C. Larsson and S. Israelsson, "Outdoor Sound Level Variations Due to Fluctuating Meteorological Parameters," *Applied Acoustics* vol. 26, pp. 235-240 (1989).

[13] A. D. Pierce, Chaper 8 "Ray Acoustics," in *Acoustics: An Introduction to Its Physical Principles and Applications*, pp. 371-423 (1991).

[14] S. J. Franke, G. W. Swenson, Jr., "A Brief Tutorial on the Fast Field Program (FFP) as Applied to Sound Propagation in the Air," *Applied Acoustics* vol. 27, pp.203-215 (1989).

[15] G. A. Daigle, M. R. Stinson, D. I. Havelock, "Use of the PE Method for Predicting Noise Outdoor," *Proceeding of Internoise 96*, pp. 561-566 (1996).

[16] C. I. Chessell, "Three Dimensional Acoustic-Ray Tracing in an Inhomogeneous Anisotropic Atmosphere Using Hamilton's Equations," *Journal of the Acoustical Society of America,* vol. 53(1), pp. 83-87 (1973).

[17] B. Hallberg, C. Larson, "Some Aspects on Sound Propagation Outdoors," *Acustica*, vol. 66, pp. 107-112 (1988).

[18] R. F. Schumacher, Personal Communication.

971989

Methods of Passby Noise Prediction in a Semi-Anechoic Chamber

S. -H. Park and Y. -H. Kim
KAIST

B. -S. Ko
DaeWoo Motors Co. Ltd.

Copyright 1997 Society of Automotive Engineers, Inc.

ABSTRACTS

This paper addresses the indoor passby noise measurement techniques satisfying SAE J1470 Recommended Practice. The restrictions of the suggested methods are also investigated. We suggested the passby noise measurement method in a large semi-anechoic chamber. We also have tried two prediction methods using array microphone system to predict the passby noise in a small-size semi-anechoic chamber. The first method is based on the line array microphone method. This method is found to be only applicable in the nearfield, because spherical Hankel function is used as a basis function. This is just simple consequence of fact that the basis functions are not separable in the farfield. The second one is the well-known nearfield acoustic holography(NAH). Not only the detailed procedures of applying NAH to the passby noise prediction but also the drawback of this method are discussed.

INTRODUCTION

Passby noise has been one of major concerns to car manufactures, not only because it is related to the environment noise problems but also it is controlled by national regulations. Several measuring standards, for example, ISO 362 standard and SAE J1470 Recommended Practice[1] specify the method for measuring passby noise. This includes operating conditions of vehicles and the acoustical environment; an extensive open space of a radius 50m around the center of the track, which is regarded to be free of large reflecting objects, is required. This paper addresses the possibility to substitute these measurement requirements to other easier form; indoor measurement.

The overall configuration of outdoor measurement of passby noise is as follows. A-weighted overall sound pressure level(SPL) at 7.5m

away from the center of the track is normally used as a measure of passby noise. The track is 20m long. A vehicle under test runs along a track with full throttling condition, therefore not only speed of run but also RPM(rotational engine speed) of the vehicle will vary.

If one wishes to measure passby noise in an semi-anechoic chamber, then one must establish equivalent condition with these outdoor measurements. Note that the rear tire noise and air intake noise cannot be considered indoor situation, because a dynamometer is usually used to run the vehicle. This equivalent condition can be readily implemented in semi-anechoic chamber whose lateral dimension is sufficient to allow 7.5m from the center of the side view of a vehicle. This means one needs at least 15m plus lateral dimension of a vehicle under test. This gives us idea of the size of the chamber, which is quite large and requires non-trivial investment. If one has a large chamber, in other words, unlimited-size semi-anechoic chamber, one can measure passby noise easily. However, if one has the chamber whose lateral dimension is smaller than 15m, one must devise appropriate prediction method.

We have tried two prediction methods of passby noise. One is based on the measurements using line array microphone. The free space sound field is decomposed into its eigenfunctions in the spherical coordinates and rearranged according to the order of spherical Hankel function. However, this method is found to be only applicable in the nearfield because the basis functions are separable only in the nearfield.

The other would be utilizing nearfield acoustic holography.[2-4] Major drawback of this method is that one has to measure acoustic pressure on hologram with regard to corresponding RPM of running vehicle. This RPM must be continuous in time, therefore one must sample this continuous RPM, and measure acoustic pressure corresponding to sampled RPM's. The passby noise prediction at one specified RPM is carried out to ensure the applicability of this method. The predicted passby noise level shows good agreement with the actual one.

PASSBY NOISE MEASUREMENTS IN AN UNLIMITED-SIZE SEMI-ANECHOIC CHAMBER.

Before one investigates the indoor passby noise measurement method, it is necessary to review passby noise measurements specified in SAE J1470 in detail. According to it, while a test vehicle is running along the 20m straight track, A-weighted SPL shall be measured at 7.5m away from the center of the track. It shall be measured using sound level meter installed at 1.2m above the ground as shown in Figure 1. During each passages of the vehicle when it is operated specified conditions, the maximum A-weighted SPL is measured. The passby noise level is regarded to be legitimate when SPL differences between two consecutive measurements do not exceed 2dB. In particular, vehicles having a mass rating of not more than 4540kg and fitted with a transmission having more than four forward gears, shall be tested in both second and third gears, then the average value shall be calculated as a passby noise measure.[1]

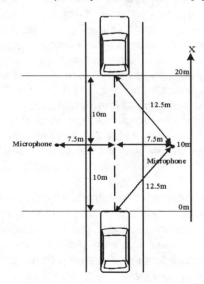

Figure 1. Microphone position for passby noise measurements and measurement distances. [SAE J1470]

When one has a large-size semi-anechoic chamber whose lateral dimensions is large enough to allow 15m plus the lateral dimension of the test vehicle, one can measure passby noise in the semi-anechoic chamber. In the first place, it is required to measure the velocity and RPM profiles along the 20m track. Then the dynamometer runs the vehicle exactly following the RPM profile. Figure 2 depicts the method. The microphone must be moving with vehicle's speed. As previously mentioned the drawback of this method is that it requires non-trivial investment.

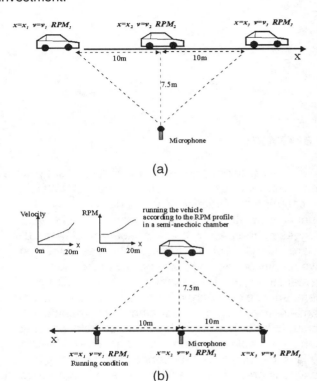

Figure 2. (a) Simplified measurement configuration of outdoor passby noise measurement (b) its equivalent situation of passby noise measurements in an unlimited-size semi-anechoic chamber. If one knows the RPM and velocity profiles, one can measure passby noise.

PASSBY NOISE PREDICTION METHODS IN A SMALL-SIZE SEMI-ANECHOIC CHAMBER

When one only has the smaller semi-anechoic chamber than required, the proposed method has to be appropriately modified.

THE SIMPLEST PREDICTION METHOD AND ITS FEASIBILITY

Let's start with very simple case first. When the noise sources of the vehicle can be represented by a monopole, one can predict whole sound field by measuring pressure at only one point. If the noise sources are monopole-like one and the radiation pattern of the noise sources depends on distances but not directions, the passby noise prediction method can be easily constructed. One measures sound field on the measurement line as shown in Figure 3 by a microphone array, then the sound pressure along the prediction line can be simply predicted using the 1/r propagator and the measured sound pressure. However, it is known that the main noise sources of the vehicle consist of engine, exhaust orifice, air intake system, cooling fans and tires. Therefore, noise source model must be the sum of distributed point sources. Therefore, the assumption is too far to express the practical situation.

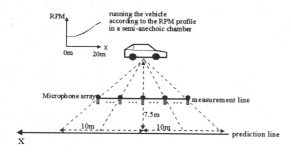

Figure 3. The simplest passby noise prediction method. However it cannot be applied in the real situation because noise sources of vehicle must be the sum of distributed point sources.

SOUND PRESSURE PREDICTION METHOD USING LINE ARRAY MICROPHONE

If one assumes free space sound field, one can decompose the sound field into its eigenfunctions in the spherical coordinates. If the azimuthal and the polar angles are fixed, the free space sound field can be expressed as

$$p(\vec{r},\omega) = \sum_{m=0}^{\infty}\sum_{n=0}^{m} \hat{P}_{mn} Y_{mn}(\theta,\phi) h_m^{(1)}(kr)$$
$$= \sum_{m=0}^{\infty} A_m h_m^{(1)}(kr) \quad (1)$$

where, $Y_{mn}(\theta,\phi) = P_m^n(\cos\theta)e^{in\phi}$ represents spherical harmonics, $h_m^{(1)}(kr)$ represents spherical Hankel function of the first kind of order m and A_m represents complex coefficient of spherical Hankel function when θ,ϕ is fixed.

From this expansion, sound pressure can be predicted when all complex coefficients A_m(m=0,1,2,...) are known. In practice, all of them cannot be known, but only some of them ($A_0, A_1,...,A_M$) can be obtained using M+1 line array microphone placed in the fixed θ, ϕ directions, then the sound pressure can be predicted using the complex coefficients A_m(m=0,1,2,...,M) as shown in Figure 4. However, this method must be applied in the nearfield where spherical Hankel functions have the ability to distinguish each other: spherical Hankel functions in far field are identical. If the complex coefficients were obtained from the farfield measurements, then it will have serious and round-off errors due to the characteristics of the spherical Hankel function in the farfield.

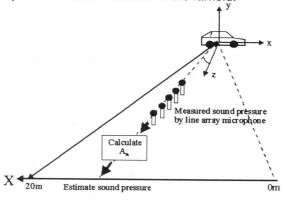

Figure 4. The measurement configuration of sound pressure prediction method using line array microphone.

PREDICTION OF SOUND PRESSURE USING NEARFIELD ACOUSTIC HOLOGRAPHY[2-4]

This method stems from Kirchhoff-Helmholtz integral equation,

$$P(\vec{x}) = -\frac{1}{4\pi}\int_S \left(G(\vec{x}|\vec{x}_S)\nabla P(\vec{x}_S) - P(\vec{x}_S)\nabla G(\vec{x}|\vec{x}_S)\right)\cdot \vec{n}_S dS \quad (2)$$

where, P is sound pressure, $\vec{x}$, $\vec{x}_S$ and $\vec{n}_S$ represent position vector of the prediction point, position vector of the boundary surface and normal vector over the boundary surface shown in Figure 5 respectively. $G(\vec{x}|\vec{x}_S)$ represents Green's function. It satisfies the following inhomogeneous Helmholtz equation.

$$(\nabla^2 + k^2)G(\vec{x}|\vec{x}_S) = -4\pi\delta(\vec{x}-\vec{x}_S) \quad (3)$$

Kirchhoff-Helmholtz integral equation states that sound pressure can be predicted at an arbitrary position in the volume of interests when a Green's function satisfying Dirichlet boundary condition and boundary values of source plane S_1 are known and Sommerfeld radiation condition is satisfied at infinity. Then, we can predict the sound pressure at any point in the volume of interests,

$$P(x,y,z) = \int\int_{-\infty}^{\infty} \hat{P}(k_x,k_y,z_H)\frac{\hat{G}'(k_x,k_y,z-z_S)}{\hat{G}'(k_x,k_y,z_H-z_S)}e^{i(k_x x+k_y y)}dk_x dk_y \quad (4)$$

where,

$$\hat{G}'(k_x,k_y,z) = \begin{cases} e^{iz\sqrt{k^2-k_x^2-k_y^2}} & k_x^2+k_y^2 \leq k^2 \\ e^{-z\sqrt{k_x^2+k_y^2-k^2}} & k_x^2+k_y^2 > k^2 \end{cases}$$

$$\hat{P}(k_x,k_y,z_H) = \int_{-\infty}^{\infty}\int_{-\infty}^{\infty} P(x,y,z_H)e^{-i(k_x x+k_y y)}dxdy$$

$P(x,y,z_H)$: measured sound pressure field in hologram plane

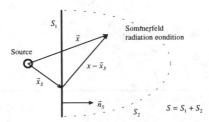

Figure 5. Region of interests and boundary surfaces for prediction of sound pressure by Kirchhoff-Helmholtz integral equation

Although one should measure the infinitely large hologram, one usually obtains the finite-size hologram in practice. Thus finite aperture size and sampled measurements will limit spatial resolution, and produces spatial aliasing and wraparound error. The sampling space must be less than 1/2 of the shortest wavelength to prevent the spatial aliasing. The wraparound error will not be ignorable in forward prediction, because imaginary sources resulting from FFT have influence on the predicted sound pressure. One can apply zero-padding method to the measured hologram to reduce the wraparound error.

Strictly speaking, the passby noise prediction requires the measurement of holograms at each specified RPM. The predicted SPL's from these measured holograms give us the discrete SPL profile along the 20m track. The maximum value of this result is regarded as passby noise level. This means that a large number of measurements are required at this sampled RPM's. This would be laborious. This is the major drawback of this method. Once we find out the RPM at which maximum A-weighted SPL occurs, we can easily apply NAH to predict the passby noise. However, this RPM can be found only from prior outdoor passby noise measurements. We investigated the validity of applying NAH to the passby noise prediction by means of experiment. As an illustration, we performed the passby noise prediction at the RPM where maximum SPL occurs instead of predicting every SPL's along the 20m track discretely. This RPM is selected from the outdoor passby noise results. The predicted result is compared with the actual passby noise level.

We measured the outdoor passby noise. It is required because it gives us not only the actual passby noise results that can be compared with predicted ones but also RPM and velocity profiles of the vehicle along the 20m track, which are needed in the prediction by NAH. The vehicle under test was a small-size passenger car with 5 forward gears. The result shown in Figure 6(a) illustrates that the maximum RHS(right hand side) SPL measured on the right side of the vehicle in the second gear is 76.9dBA at 10.7m from the start position along the 20m track. Similarly, the result in the third gear is 71.47dBA at 10.7m in Figure 6(b). Note that passby noise at the right side of vehicle is required in Korea. The corresponding RPM's at this position(10.7m) are 4500RPM in the second gear and 3000RPM in the third gear.(See second graph in Figures 6(a) and 6(b))

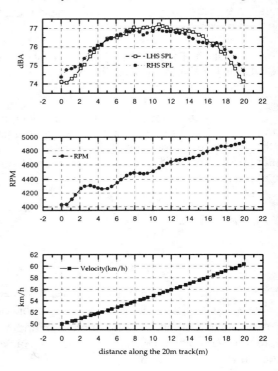

Figure 6(a). Outdoor passby noise measurement results in the second gear.

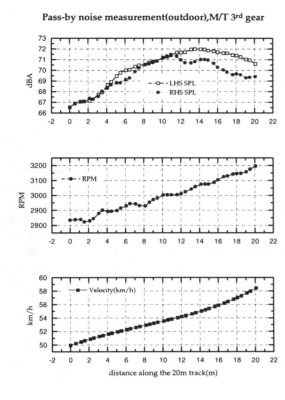

Figure 6(b). Outdoor passby noise measurement results in the third gear.

Experimental set-up and procedures

We measured hologram to predict passby noise of the identical test vehicle in the semi-anechoic chamber by conventional scanning technique at 3000RPM and 4500RPM respectively. Hologram is measured on the right side of vehicle only.

For data acquisition, we used HP3566A signal analyzer which can sample 32 channels simultaneously. We used 30 PCB acousticel 130A microphones for array. B&K type 4134 microphone used for reference and monitoring channel. B&K type 2232 sound level meter and sound level calibrator are also used. We used two line arrays. They consist of 15 PCB microphones in each array to save the measuring effort. The two arrays are 0.1m apart and microphone spacing is also 0.1m. We obtained 0.1m spatially sampled hologram to avoid the spatial aliasing below 1kHz. The horizontal aperture size of the hologram is 5.2m(53 spatial samples). While the original vertical aperture size is 1.45m(15 spatial samples), it should be doubled by virtue of mirror image. It is the reason why nearfield acoustic holography is usually applied in the free field. The distance between the vehicle and hologram plane is 0.565m in the second gear experiment and 0.575m in the third gear experiment respectively. The reference microphone is placed near the vehicle engine, because it is the main noise source of the vehicle in indoor experiment.(See figure 7)

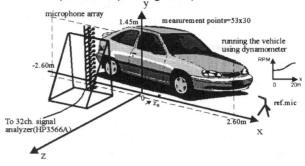

Figure 7. Experimental set-up for passby noise prediction by NAH in a semi-anechoic chamber.

Results of passby noise prediction at the selected RPM and its comparison with the actual results from outdoor passby noise measurements

After measuring hologram by running the vehicle at the selected RPM, we have measured power spectrum and overall A-weighted SPL at monitoring position, 3.54m away from the vehicle surface(see Figure 8). By doing this we can compare the actual SPL measured at the monitoring position and the predicted SPL using NAH. As it'll be discussed later in this paper, two results shows good agreement. This justifies the use of NAH for the passby noise prediction.

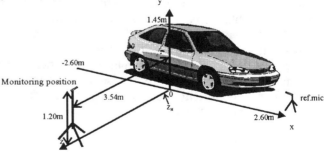

Figure 8. Monitoring position. SPL and power spectrum are measured for comparison purpose

It is noteworthy that NAH predicts sound pressure distribution based on the signals on measurement plane. The signals are coherent signals. However, our goal is to calculate the overall A-weighted SPL of the vehicle. This means that we not only need the pressure data on the coherent frequencies, but also on the incoherent frequencies. The overall A-weighted SPL is calculated from the A-weighted sound pressure in 1/3 octave band.[5]

Therefore, we need assumptions to calculate the pressure data of incoherent frequencies. There are two possible cases which cause incoherence.(See Figures 9(a) and 9(b)) The first one is that these components are not originated from the noise sources of the vehicle. In other words, they are due to background noise. The other one is due to strong directivity of the source. In this case, sound propagates toward only array microphones and does not reach the reference microphone or vice versa.

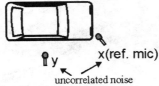

Figure 9(a). The first possible situation of producing incoherence. The incoherent peaks are not originated from the noise sources of the vehicle. They are due to background noise.

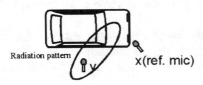

Figure 9(b). The second possible situation of producing incoherence. The incoherent peaks are due to the strong directivity of the vehicle.

First, we predicted SPL's of coherent frequencies at the monitoring position and compared them with the measured ones as shown in Figure 10. Zero-padding is applied to hologram to reduce the wraparound error. The errors are less than 3dB below 700Hz, but more than 3dB above 800Hz. There can be two possible explanations for this phenomenon. The first one is the aliasing, which takes place in the wave number domain. The other possible reason is that all the noise sources are not included in hologram due to the finite aperture size. Dense spatial sampling and employing large aperture on the hologram are required to reduce this error.

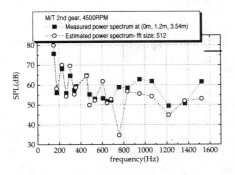

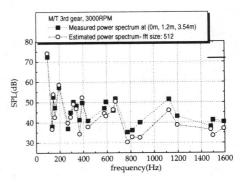

Figure 10. SPL prediction results by NAH at the monitoring position(0m, 1.2m, 3.54m)

With the above results we calculated the overall A-weighted SPL considering two possible situation of incoherent signals. First, we assume the incoherent frequency components are due to only the background noise, then its level can be obtained by averaging the incoherent power spectral densities measured at the monitoring position. The overall SPL can be obtained by adding this value to the coherent power spectral densities which are predicted from hologram. The calculated background noise levels are 47.1dB at 4500RPM and 40.6dB at 3000RPM respectively. Second assumption is that the incoherent peaks are due to the strong directivity of the source. One can also estimate the overall SPL at the monitoring position from the hologram; all the frequency components were assumed to have originated from the vehicle. The background noise level is of course regarded to be zero.

The predicted results and the measured ones are compared in Table 1. The calculated SPL's based on the two different assumptions show good agreement with the actual value except the last one in the table. A careful observation of the results shows that the assumption of the background noise is more realistic than the assumption of the strong directivity of the source.

Table 1. Passby noise and monitoring position prediction results and its comparison with measured results

Gear (RPM) estimation position	The way of treating of incoherent frequencies	estimated SPL(dBA)	measured SPL(dBA)	Prediction error(dBA)
2nd (4500) monitoring position	background noise level 47.1dB	79.62	78.5	+1.12
2nd (4500) monitoring position	processing all frequency components	78.89	78.5	+0.39

2nd (4500) passby noise prediction	background noise level 47.1dB	78.66	76.91	+1.75
2nd (4500) passby noise prediction	processing all frequency components	75.11	76.91	-1.80
3rd (3000) monitoring position	background noise level 40.6dB	71.59	71.30	+0.29
3rd (3000) monitoring position	processing all frequency components	70.94	71.30	-0.36
3rd (3000) passby noise prediction	background noise level 40.6dB	71.35	71.47	-0.12
3rd (3000) passby noise prediction	processing all frequency components	65.15	71.47	-6.32

CONCLUSIONS

In this paper, we have suggested several methods of indoor passby noise measurements and investigated their limitation and applicability. Of these suggested method, NAH proved to be a possible method for predicting the passby noise. However, as previously mentioned, this requires non-trivial investment in the measurements.

ACKNOWLEDGMENTS

This work is supported by DaeWoo Motors Co. Ltd.

REFERENCES

[1] The Vehicle Sound Level Committee-Light Exterior Sound Level Subcommittee, Measurement of noise emitted by accelerating highway vehicles, SAEJ1470, OCT84, 1984

[2] J. D. Maynard, E. G. Williams and Y. Lee, "Nearfield acoustic holography: I.Theory of generalized holography and the development of NAH", J. Acoust. Soc. Am., Vol78(4), 1985, pp.1395-1412

[3] W. A. Veronesi and J. D. Maynard, "Nearfield acoustic holography(NAH) II. Holographic reconstruction algorithms and computer implementation", J. Acoust. Soc. Am., Vol81(5), 1987, pp.1307-1322

[4] H.-S. Kwon, "The forward prediction of radiation sound field using acoustic holography", KAIST M.S. Thesis, 1992

[5] J. D. Irwin and E. R. Graf, *Industrial Noise and Vibration Control,* Prentice-Hall, Englewood Cliffs, 1979, chap.2

971990

An Assessment of the Tire Noise Generation and Sound Propagation Characteristics of an ISO 10844 Road Surface

Paul R. Donavan
General Motors

Copyright 1997 Society of Automotive Engineers, Inc.

ABSTRACT

A road surface complying with the new International Standards Organization (ISO) specification was installed at an Arizona test facility (DPG site) in the winter of 1995/96. As part of the acoustic qualification of this site, comparative tests were conducted between this new surface, a Society of Automotive Engineers (SAE) sealed asphalt surface and an existing ISO surface in Michigan (MPG site). Initial testing with one vehicle and tire combination indicated that the new ISO surface produced ISO 362-1994 passby and coastby levels about 2 dB lower than sealed asphalt. Relative to the Michigan surface, the levels for the new Arizona ISO surface were 3 to 3½ dB lower. These differences were much greater than expected based on previously published studies of these two test surface types. Since the new surface was constructed to the ISO specification and meet the physical requirements for sound absorption coefficient, porosity, and surface texture, further investigation was conducted to determine if sound propagation or tire noise generation differences accounted for the differences. Experimental work to understand this difference included the use of on-board sound intensity measurements to isolate tire noise generation under both acceleration and coast and static sound propagation tests to isolate surface reflective properties. Analytically, a sound reflection model was developed to predict differences in attenuation based on measured surface impedance data. Taken together, the results of this investigation support the conclusion that a majority of the differences observed are due to tire noise generation. However, in comparing the new ISO surface to the SAE, a significant portion was also found to be attributable to sound propagation differences.

INTRODUCTION

In recent years, a number of test tracks in the United States have installed ISO 10844 road surfaces [1] for purposes of light vehicle passby noise testing in compliance with the ISO 362-1994 procedure [2]. To understand the differences between this new surface and that specified by the SAE [3, 4, 5], a SAE Cooperative Research Project was undertaken in 1994-95 [6]. From testing at a number of sites, it was found that on average an ISO surface will produce levels 0.7 dB lower for passby tests and 1.8 dB lower for coastby tests relative to SAE sealed asphalt surfaces. Also, the standard deviation in levels between ISO sites was found to be 1.0 dB for passby and an average of 1.1 dB for coastby tests. Based on this study, it was expected that when a new ISO test surface was constructed at the DPG facility that the differences between the new surface and exisitng SAE and ISO surfaces would fall within these ranges.

As an initial validation of the new surface, a single passenger car was tested under passby and coastby procedures at both the MPG and DPG facilities. In comparing the results for the two ISO surfaces, the new surface at the DPG yielded overall A-weighted sound pressure levels which were 3.0 dB lower for coastby and 3.5 dB lower for passby. Comparing the SAE and new ISO surfaces at the DPG, the coastby levels were 1.9 dB lower on the ISO surface while passby levels were 1.8 dB lower. Relative to the expected differences, only the comparison of the SAE and ISO surfaces for coastby fell within the previously determined range. Of particular concern was the very large differences between the two ISO surfaces. As the MPG surface was included in the SAE research project and fell within the range of other test tracks, it was concluded the levels produced by the new surface were too low.

Review of the construction of the new ISO surface indicated that it did conform to the material specifications of ISO 10844. Further, the final surface did meet the surface specifications for sound absorption coefficient, porosity, and surface texture. However, it was found that to achieve the surface texture requirement, the contractor had to "roll-in" additional aggregate after the asphalt was first laid. Not knowing if this construction practice influenced the final acoustic perfomance of the surface, a more extensive program of passby and coastby testing was performed. These tests were designed to separate tire noise generation differences from sound propagation. For logistical reasons, this study was concentrated on comparison of the SAE and ISO surfaces at the DPG facility. Follow up testing was also conducted at the MPG site.

INVESTIGATION OF ISO AND SAE TEST SURFACE DIFFERENCES

VEHICLE TESTS - A series of tests using a single test vehicle and five different tire sets were conducted at the DPG facility on both the new ISO and SAE surfaces. The vehicle was a rear wheel drive, two door sport coupe equipped with a V6 engine and automatic transmission. The tires included a full depth blank tread tire and production tires spanning all seasons, touring, performance, and speed rated categories (Table 1). All tires were tested at inflation pressures of 30 psi with the all seasons tire being tested at the additional pressures of 20 and 40 psi. For each tire set, the vehicle was tested on both surfaces under the ISO 362-1994 procedure and a coastby procedure where the engine remained on with the transmission in neutral. Sound pressure level (SPL) was measured a distance of 7.5 meters from the centerline of vehicle travel with a microphone height of 1.2 meters. Data were acquired simulataneously on both sides of the vehicle as a function of vehicle position and reduced into overall and 1/3 octave band A-weighted sound pressure levels. For both the passby and coastby tests, data corresponding to the maximum overall level from four consecutive runs were averaged together. Only consecutive runs within 2 dB of each other were used. The averages from both sides of the vehicle were also averaged to yield a single values for each test configuration and procedure. Coastby tests were performed at a nominal vehicle speed of 56 km/hr.

To isolate differences in tire noise generation, sound intensity levels (IL) were measured very close to the tire/pavement interface on board the vehicle. The data were acquired using a sound intensity probe consisting of two 12.5mm diameter microphones and preamplifiers in a "side-by-side" configuration spaced 16mm apart. The data acquistion, processing and reduction were identical to that documented previously [7] and very similar to that used in the SAE research project [8]. In the current study, ILs measured opposite both the leading and trailing edges of the tire contact patch were averaged together. For comparison to the sidelne SPL data, coastby ILs from consecutive runs were averaged together for the nominal speed of 56 km/hr. Passby ILs were averaged for a vehicle speed of 58 km/hr which corresponded to the speed at which peak passby level typically occurred.

VEHICLE TEST RESULTS - The complete test results are summarized in Table 1. Included are overall A-weighted levels for both sideline SPL and on-board IL for each tire configuration on each pavement surface for both passby and coastby test conditions. Also indicated is the difference in level between the two test surfaces for each case and the average differences for each test procedure and overall. Review of these data indicate several trends which are expected from previously published results. First, the both the 7.5m SPLs and the

Table 1: Summary of ISO 362-1994 Passby and Coastby Test Results for DPG Site

Tire	Test	Sound Pressure Levels, dB(A)			Tire Sound Intensity Levels, dB(A)		
		SAE	ISO	Difference	SAE	ISO	Difference
Goodyear Eagle GA	Passby	75.6	73.6	2	95.4	94.6	0.8
	Coast	69.6	67.7	1.9	93.4	92.5	0.9
Goodyear Eagle RS-A	Passby	75.5	73.9	1.6	94.8	94.5	0.3
	Coast	69.9	68.1	1.8	93	92	1
Goodyear Eagle GS-C	Passby	75.6	73.6	2	95.5	94.3	1.2
	Coast	69.5	67.4	2.1	92.6	91.6	1
Goodyear Blank	Passby	74.5	72.5	2	90.3	88.9	1.4
	Coast	64.5	62.4	2.1	85	84	1
Michelin XW4 @ 40 psi	Passby	76.0	73.7	2.3	96.9	95.7	1.2
	Coast	67.7	65.6	2.1	89.2	88.5	0.7
Michelin XW4 @ 30 psi	Passby	75.2	73.5	1.7	95.4	94.4	1
	Coast	67.1	65	2.1	88.9	88.1	0.8
Michelin XW4 @ 20 psi	Passby	74.7	72.8	1.9	94.3	93.4	0.9
	Coast	66.3	64.2	2.1	88.1	87.1	1
Averages	Passby			1.9			1
	Coast			2.1			0.9
	Overall			2.0			0.9

tire ILs are consistently higher for the accelerating operating condition. As indicated by the increased intensity level, the increase in sideline SPLs is due to both an the increase in driveline related sources and an increase in tire noise generation under acceleration. The second observation is that both the SPLs and ILs are consistently greater on the SAE surface. For the SPLs, the differences are consistently about 2 dB. As noted from the initial testing cited above, this difference is greater than the expected 0.7 to 1.8 dB determined in the SAE research project [6]. For the ILs, the values for the SAE surface are always greater, but the differences display a larger variance. Further, the average difference for the ILs is about 1 dB or about 1 dB less than the difference for the SPLs. This suggests that the difference in SPLs between the surfaces can only be partially ascribed to differences in tire noise source strength for both operating conditions.

The overall SPLs and ILs of Table 1 are plotted against each other for the coastby operating condition in Figure 1. For comparsion, data from an earlier, similar study performed on a different non-sealed asphalt surface are also plotted along with the 1 to 1 linear relationship developed for that data [7]. Also included in Figure 1 is an indication of the average range in run-to-run variation comprised of ±1.2 dB for SPL and ± 0.6 dB for IL as determined in that study. With the exception of the blank tread tire data, most of the new data points fall within the range of the previous investigation. Data from the higher sound level, non-all seasons tire compare very well with the older data. For the quieter blank tread and all season tire configurations, the divergence from the 1 to 1 linear relationship established in Reference 7 may be due "background" noise from the vehicle itself. During the "coast" condition, for safety reasons, the engine was not shut off. This left a residule engine speed of about 1100 RPM. Since the purpose of the current study was not to re-establish results of the

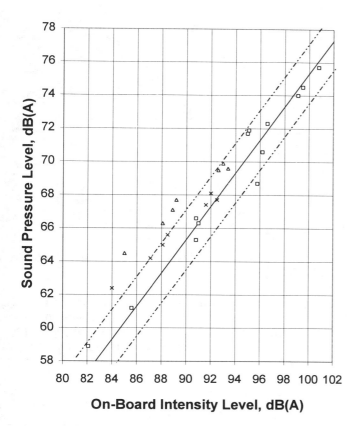

Figure 1: Comparison of Tire Sound Intensity Levels to Sideline Sound Pressure Levels at 7.5m - Coastby Conditions

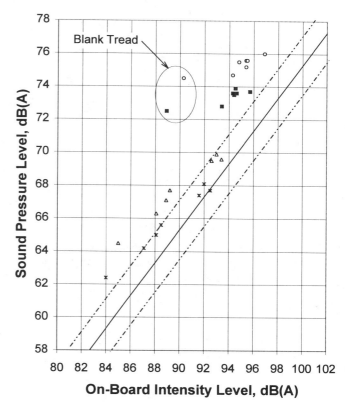

Figure 2: Comparison of Tire Sound Intensity Levels to Sideline Sound Pressure Levels at 7.5m - Coastby and ISO 362-1994 Passby Conditions

previous study, but rather to examine test road surface differences, these data were not repeated with the engine off. For the current investigation, the more significant observation from Figure 1 is the consistent offset between the results on the two surfaces relative to the 1 to 1 line. This offset indicates that for the same tire IL, the resultant SPL would be higher on the SAE surface further suggesting that sound propagation is a contributor to the differences between the surfaces.

Data from the 362-1994 test procedure are presented in Figure 2 along with the coastby data of Figure 1. This presentation of the data yields insight not only to the surfaces, but also to the relative contribution of tire and driveline noises to the passby test. By comparing the passby data points to the 1 to 1 tire noise line, the increase in level under passby conditions due to noise sources other than tire noise is indicated. For the blank tread tires, this increase is about 8 dB indicating that the driveline sources are much more significant than the blank tread tire noise under acceleration. For the treaded tires, the increase over the 1 to 1 line it is about 4 dB. In this case, the noise from the treaded tires under acceleration is almost equal to the driveline sources. Considering the data of Figure 2 in regard to the surfaces, a consistent offset between the ISO and SAE surfaces for the passby data as well as the coastby data is apparent.

One-third octave band spectral comparisons between the five different tires are provided in Figure 3 for the ISO surface and in Figure 4 for the SAE surface for the coastby operating condition. In comparing these figures, the overall spectral shapes for both the SPLs and ILs are similar between the two surfaces. Also, the test tires rank order similarly between the two surfaces for both SPL and IL. For each surface, differences seen between tires for the IIs are also seen in the SPLs. There is, however, some "compression" of the SPLs most likely due the coastby method used. SPL and IL spectra for the touring tire are presented in Figure 5 for both passby and coastby and both surfaces. This plot also indicates a very good correspondence between SPL and IL data. It also shows that the difference between the surfaces for both the on-board and sideline data start at about 800 hertz and continue through the higher frequencies. Further, the differences between the surfaces are almost identical for coast and passby test conditions.

The tendencies illustrated in Figure 5 were found to occur identically for each of the other test tires. A more compact method of comparing the tires is provided in Figure 6 where SPL difference between the surfaces is plotted for each tire along with the average of tires. Because the effect of the surface was similar for both coastby and passby, the spectral difference for each tire

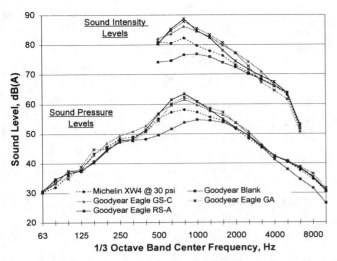

Figure 3: Sound Pressure Levels at 7.5m and On-Board Tire Sound Intensity Levels for Coast Conditions on ISO Surface at DPG Site

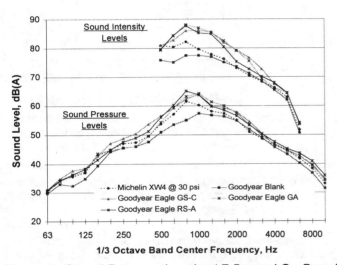

Figure 4: Sound Pressure Levels at 7.5m and On-Board Tire Sound Intensity Levels for Coast Conditions on SAE Surface at DPG Site

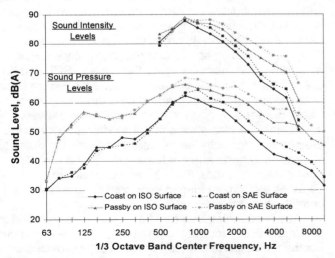

Figure 5: Sound Pressure Levels and On-Board Sound Intensity Levels for Coast & ISO 362-1994 Passby on ISO & SAE Surfaces - Touring Tire

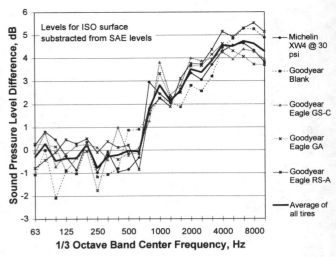

Figure 6: Difference Between Sound Pressure Levels on ISO and SAE Surfaces for Various Tires

in Figure 6 is the average of the two test procedures. As noted for Figure 5, for all tires, consistent difference in the test road surfaces begins at 800 hertz and grows to about 4 or 5 dB for the higher frequencies. A similar plot of the tire ILs is provided in Figure 7. Although the IL

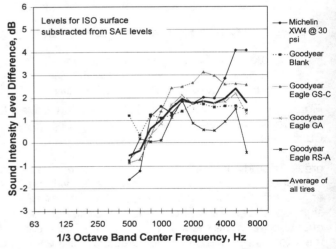

Figure 7: Difference Between On-Board Sound Intensity Levels on ISO and SAE Surfaces for Various Tires

data display more scatter, they also indicate that consistent higher differences in surfaces begin at 800 hertz and continue into the higher frequencies. The SPL and IL averages are plotted together in Figure 8. As expected from the results of Table 1, the differences in sideline SPL are greater than the differences in on-board tire IL. This leads to the conclusion that the increased SPLs measured for the SAE surface can not be accounted for completely by increases tire noise source strength.

SOUND PROPAGATION INVESTIGATION - In order to determine if some of the SPL difference between the ISO and SAE test surfaces was attributable to differences in sound propagation over those surfaces, additional experimental and analytical investigations were conducted. These consisted of measuring the attenuation of sound propagating over the surfaces from

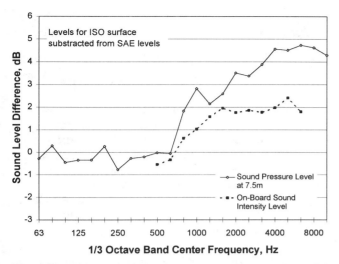

Figure 8: Difference Between Levels Measured on ISO and SAE Surfaces Averaged for All Tires

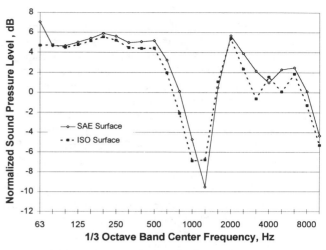

Figure 9: Measured Increase in Sound Pressure Level over Free Field Due to Reflection for ISO and SAE Surfaces at DPG Site - Source Height 0.45m

a stationary, electronic sound source, measuring the normal impedance of core samples from each surface, and calculating the expected differences in excess sound attenuation based on the impedances.

Sound Propagation Measurements - To determine the differences in sound attenuation over the ISO and SAE road surfaces, a Bruel and Kjaer (B&K) Reference Source, Type 4204, was placed at the center of each test section positioned along the centerline of vehicle travel. SPLs near the source and at the normal sideline distance of 7.5m were measured with 12.5mm diameter B&K Type 4165 microphones pointed toward the source. The source was measured at two heights, 8.5cm and 45cm above the road surface representing near ground level sources such as tire/pavement interaction noise and higher powertrain sources. The measurements were made under overcast weather conditions to minimize pavement to air thermal gradients and under virtually calm wind. The data were acquired on the two surfaces within the same hour to obtain nearly identical environmental conditions and assure that ambient air temperature differences were negligible.

The SPL data from the near and 7.5m microphones were reduced to a normalized attenuation. This was formed by first subtracting the far data from the near to get a measured attenuation. The theoretical, spherical spreading attenuation for a point source in a free field (6 dB/doubling of distance) was then subtracted from the measured attenuation. As a result, the normalized attenuations indicate the increase or decrease of the measured SPL over what would be expected in the absence of any reflections from the road surface. For the higher source source location (Figure 9), the normalized attenuation is characterized by increases of almost 6 dB at the lower frequencies up 500 hertz corresponding to the region of source strength doubling[9]. In the region around 1250 hertz, a large "dip" in the attenuation occurs due destructive interference of the direct and reflected sound waves. At the very high frequencies, 8000 and 10,000 hertz, the attenuation is greater than expected suggesting the presence of additional mechanisms of excess attenuation [10]. Comparing the two surfaces, the levels are generally lower for the ISO surface. A portion of this difference appears to be a slight shift of the interference dip to lower frequency for the ISO surface.

With the source near ground level, the normalized attenuation is more complex (Figure 10). As expected

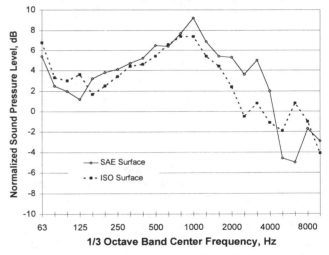

Figure 10: Measured Increase in Sound Pressure Level over Free Field Due to Reflection for ISO and SAE Surfaces at DPG Site - Source Height 8.5cm

with a shorter path length difference, the interference dip has moved up in frequency apparently occuring in the 3150 to 6300 hertz region. In the lower frequencies between 80 and 250 hertz, the normalized attenuation is lower than the 6 dB expected for doubled source strength. At 800 and 1000 hertz, the a normalized attenuations are higher than 6 dB by 1 to 3 dB. This may be due to difficulties in accurately quantifying the sound source. Resting on the ground, the source is more acoustically complex particularly at lower frequencies where the soucre and its image are separated by less than a wavelength. As a result, the near microphone location may not accurately represent the source levels. However, differences in the surfaces are still accurately captured. Comparing the two surfaces, from the 160 to 4000 hertz, the ISO surface consistently produced levels at or below those of the SAE surface. In the higher frequencies 2000 to 8000 hertz, there appears to be a more pronounced lowering of the frequency of the interference dip for the ISO surface. This creates higher levels for the ISO surface at 5000 and 6300 hertz where the interference dip occurs for the SAE surface.

Road Surface Impedance Characterization - For purposes of immediate and long term documentation of the surfaces, several 10 cm diameter core samples were taken from the ISO and SAE surfaces. One method used to characterize these samples was to measure their normal acoustic impedance. This was accomplished using a B&K Two-Microphone Impedance Measurement Tube Type 4206. This instrument provided the magnitude and phase of the reflection coefficients of both surfaces over the frequency range from 400 to 1600 hertz (Table 2). These data indicate some differences

Table 2: Impedance Data of SAE and ISO Test Surfaces at DPG Test Site

Frequency Hz	Reflection Coefficient ISO Surface	Reflection Coefficient SAE Surface	Relative Phase shift (Degrees)
400	0.96	0.99	1.6
500	0.96	0.99	1.7
630	0.95	0.99	1.8
800	0.94	0.99	2.1
1000	0.94	0.99	2.4
1250	0.92	0.99	2.2
1600	0.91	0.99	1.7

between the surfaces. While the SAE surface maintains a very high reflection coefficient of .99 throughout the frequency range, the coefficients for the ISO are lower and decrease with increasing frequency. Also a consistent, small phase shift between the samples occurs averaging 1.9 degrees over the measurement range.

Sound Propagation Calculations - To determine if differences in reflection coefficient magnitude and phase on the order of those of Table 2 could account for the propagation differences of Figures 9 and 10, a image point source model of the reflection from the road surface was developed. This model was framed in a manner such that the reflection coefficient data of Table 2 could be used directly. From the onset, it was not expected that the results of this model would exactly reproduce the measured data as available impedance data was for normal incidence only and impedance matching conditions over the entire half space above the ground plane were not imposed [11]. However, it was expected that the results would provide some indication whether that the impedance differences could reproduce some of the observations made.

For purposes of calculating the total pressure sound at a point above a half space from the direct and reflected paths of a single point source, the following expression was used with the notation of Kinsler and Frey [12]:

$$p_t = p_1 + p_2 = (A_1/r_1)e^{i(\omega t - kr_1)} + (RA_1/r_2)e^{i(\omega t - kr_2 - \phi)}$$

where p_1 = sound pressure from the direct path
p_2 = sound pressure from the reflected path
A_1 = pressure amplitude of the source
R = magnitude of the reflection coefficient
r_1 = length of the direct path
r_2 = length of the reflected path
ωt = product of angular frequency and time
k = acoustic wavenumber at frequency ω
ϕ = phase shift occuring upon reflection

To put this expression in a form for calculating normalized attenuations, the time varying terms were factored out assuming a time invariant source and the real part of the expression was taken. This resultant was squared and integrated over a specifiable frequency band width. This term was then divided by the corresponding expression for the mean square pressure of the direct path, p_1^2. This ratio was then calculated in a spread sheet on a 1/3 octave band basis with source/receiver geometry, magnitude of the reflection coefficient, and phase shift upon reflection as variables. Ten times the log of this ratio was taken for comparison to the experimentally determined normalized attenuations.

Calculated attenuations are compared to the test results for the higher source location in Figure 11. For

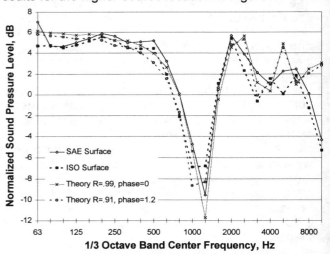

Figure 11: Measured and Calculated Increase in Sound Pressure Level over Free Field Due to Reflection for ISO and SAE Surfaces at DPG Test Site - Source Height 0.45m

the SAE surface, based on the impedance data, the magnitude of the reflection coefficient was taken to be 0.99 while the phase shift was assumed to be zero. Over most of the frequency range, these calculated values compare well to the experimental with the exception of the two highest frequency bands. Also, the occurrence of the interference dip is replicated quite accurately. For the ISO surface, the experimental results could not be replicated without a phase shift. Lowering just the magnitude of the reflection coefficient to 0.91 produces no shift in the interference dip frequency and has only a very small effect on the magnitude of the normalized attenuation. Introducing a frequency independent phase shift of 1.2 degrees along with the 0.91 reflection coefficient magnitude produced the results shown in Figure 11. This phase shift was chosen as it produced the best agreement with the experimental results and was consistent with the measured impedance properties in Table 2. Except for the higher frequencies, use of these values in the calculation replicates the experimental behavior of the ISO surface quite well. They also yield differences between the calculated values for the two surfaces which were noted for the experimental differences.

Calculations were also made representing the propagation tests with the sound source near ground level (Figure 12). To represent the SAE surface, the

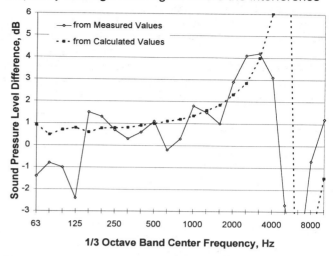

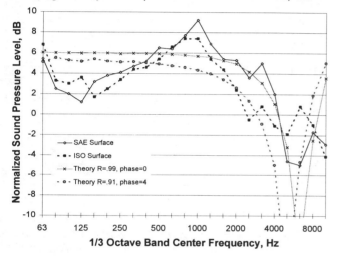

Figure 12: Calculated and Measured Increase in Sound Pressure Level over Free Field Due to Reflection for ISO and SAE Surfaces at the DPG Test Site - Source Height 8.5cm

magnitude of the reflection coefficient was again taken to be 0.99 with no phase shift. Except for those low and very high frequencies discrepancies noted previously in the discussion of the experimental results, the calculated values replicate the test results reasonably well particularly in predicting the occurrence of the interference dip. To approximate the ISO surface, the magnitude of the reflection coefficient was again taken as 0.91. In the absence of any near grazing impedance data, the phase shift for this case was taken as 4 degrees. This choice was based primarily on fitting the data and keeping the phase shift relatively small and consistent with the normal impedance values. In regard to comparing the surfaces, the difference in the calculated normalized attenuations match those noted for the differences in the measured data. This can be more clearly seen in Figure 13 which plots the experimental and analytical differences in attenuation between the ISO and SAE surfaces. In the region from 160 to 3150 hertz, the calculated differences approximate the experimental differences reasonably well on average. Above 3150 hertz, although there is frequency shifting in the region where the interference

Figure 13: Measured vs Calculated Differences Between ISO and SAE Surfaces at the DPG Site for Sound Propagation with Source Height of 8.5 cm

dips occur for the two the surfaces, the calculated differences do follow the same tendency of the experimental values.

APPLICATION OF SOUND PROPAGATION RESULTS TO VEHICLE TESTS - From the investigation of the sound propagation characteristics of the ISO and SAE surfaces, it appears that some of the difference in SPLs for the vehicle passby and coastby tests can be attributed to sound propagation differences. To examine this further, attenuations for the two surfaces were calculated using the values for reflection coefficient magnitude and phase shift that were used for the calculated values in Figure 12. A source height of 2.5cm was used to represent the tire/pavement noise source. Because the interference dip occurs at such a high frequency, the calculated attenuations for this case show the SAE surface producing higher levels over the full frequency range up to 10,000 hertz (Figure 14). To

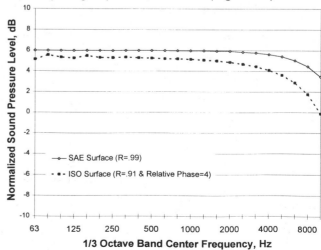

Figure 14: Calculated Increase in Sound Pressure Level over Free Field Due to Reflection from ISO and SAE Surfaces at DPG Site - Source Height of 2.5cm

compare these calculated results to the experimental SPL and IL differences of Figure 8, the difference in normalized attenuation between the surfaces was taken. These values were added to the differences in on-board, tire noise intensity to produce the combined effect of tire noise source strength and propagation differences between the surfaces. By combining these effects, the measured differences in SPL between the two surfaces is accounted for almost completely (Figure 15).

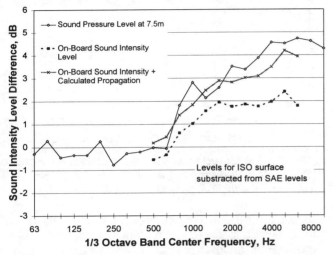

Figure 15: Difference Between Levels Measured on ISO and SAE Surfaces at DPG Test Site Averaged for All Tires with Calculated Propagation Effects Added

INVESTIGATION OF ISO TEST SURFACE DIFFERENCES AT TWO SITES

Some of the methods used to evaluate the differences in passby and coastby SPLs for the SAE and ISO surfaces at the DPG site were also used to evaluate the ISO surfaces at the MPG and DPG sites. Measurements at the MPG were limited to passby and coastby tests for the same vehicle and same production touring tires used in the DPG tests. Comparing the SPL and IL spectra between the two ISO surfaces for the two types of tests (Figure 16) yields observations that are

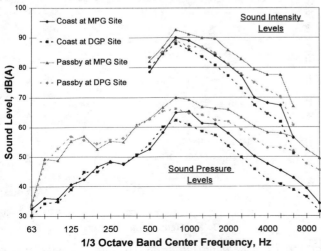

Figure 16: Sound Pressure Levels and On-Board Sound Intensity Levels for Coast & ISO 362-1994 Passby on ISO Surfaces at DPG and MPG Test Sites

very similar to those in comparing the SAE and ISO surfaces at the DPG site (Figure 5). In both cases, the differences between the surfaces is limited to frequencies of 800 hertz and above. Also, the SPL and IL data correspond well to each other. However, in examining the site to site differences in SPL and IL, a different result is obtained than in comparing the SAE and ISO surfaces. For the ISO and SAE surfaces compared in Figure 8, the differences in SPL measured at 7.5m were greater than the differences in tire noise intensity measured near the source. In that case, differences in tire noise generation alone could not account for all of the differences measured at the far microphone. For the two ISO sites, however, the SPL differences between the surfaces at 7.5m are matched by almost equal differences in tire noise IL (Figure 17).

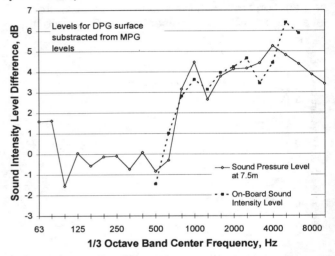

Figure 17: Difference Between Levels Measured on ISO Surfaces at DPG and MPG Test Sites for a Touring Tire

In this case, the SPL differences between the ISO surfaces can be accounted for almost entirely by differences in tire noise generation. Directionally, this is supported by a comparison of the normal impedance data for the two ISO surfaces (Table 3). Although the

Table 3: Impedance Data for ISO Surfaces at the DPG and MPG Test Sites

Frequency Hz	Reflection Coefficient DPG Site	Reflection Coefficient MPG Site	Relative Phase shift (Degrees)
400	0.96	0.98	1.1
500	0.96	0.98	0.9
630	0.95	0.98	1.0
800	0.94	0.98	1.6
1000	0.94	0.97	2.1
1250	0.92	0.97	2.0
1600	0.91	0.96	1.1

values are higher, the magnitude of the reflection coefficients for the MPG site decrease with increasing frequency in a manner similar to those of the DPG ISO surface and not seen for the SAE surface. Also, although there is a relative phase shift between the two

ISO surfaces, it is smaller and less consistent than that between the DPG SAE and ISO surfaces.

Normalized attenuation values were calculated for the two ISO surfaces (Figure 18) using the magnitude of the reflection coefficient data from Table 3. This was

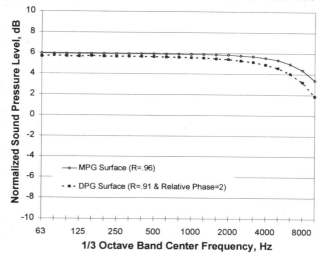

Figure 18: *Calculated Increase in Sound Pressure Level over Free Field Due to Reflection from ISO Surfaces at DPG and MPG Sites - Source Height of 2.5cm*

done assuming near grazing incident and a phase shift half of that used for the SAE surface. Relative to the SAE and ISO attenuations in Figure 14, the differences in attenuation for the two ISO surfaces are small. This indicates that propagation differences should be less of a contributor to differences measured between the ISO surfaces than it was for SAE and ISO surfaces. For the ISO surfaces, calculated differences in propagation are about ½ dB or less between 1000 and 4000 hertz (Figure 18) while differences in tire noise generation are about 3 to 5 dB in this range (Figure 17). By contrast, for the SAE and ISO surfaces, propagation differences are as much as 1½ dB (Figure 14) with tire noise generation differences of 1 to 2 dB in the 1000 to 4000 hertz range (Figure 8). As was done for the SAE to ISO comparison of Figure 15, the differences in propagation for the two ISO surfaces were added to the differences in tire noise IL. These are compared to the differences in SPL at 7.5m in Figure 19. Because of the smaller role of propagation differences for the ISO surfaces, addition of propagation differences did little to improve the prediction of the SPL data. Comparing Figures 19 and 15, it is apparent that changes in tire noise source strength are a more dominate factor in the differences in sideline SPL for the two ISO surfaces than it is for the SAE and ISO surfaces. Conversely, propagation effects appear to be more of a factor in the differences between the SAE and ISO surfaces than between the two ISO surfaces.

CONCLUSION

The primary conclusion of this investigation was that the newly constructed ISO surface could not be reliably used for ISO 362-1994 testing. Further, although

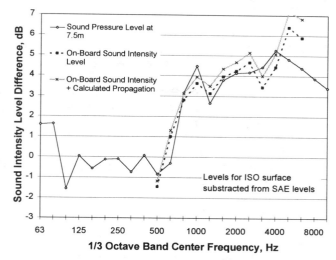

Figure 19: *Difference Between Levels Measured on ISO Surfaces at DPG and MPG Test Sites Averaged for Touring Tire with Calculated Propagation Effects Added*

differences in tire noise generation were the primary reason for the differences measured at 7.5m, there was a definite possibility that some of the difference could be attributed to progapation differences. As result, the decision was made to repave the entire test section, not just the lane of vehicle travel. The reason for the large difference in tire noise generation between the two ISO surfaces was not established. However, it is speculated that by rolling-in aggregate after laying the initial asphalt, pavement surface differences were created relative to other, more conventionally constructed ISO test surfaces. Because this surface met all of the ISO 10844 specifications in its final state but performed significantly different than other ISO surfaces, tighter contols in the pavement specification should be considered.

A second conclusion is that accurate and more complete methods for measuring the acoustic impedance of road surfaces need to be developed. From the analysis performed in this investigation, it is apparent that even small amounts of phase shift upon reflection from a road surface can significantly effect measurements of coastby and passby levels at 7.5m. For this reason, methods to measure the complex impedance of a surface over a wide range of frequency and angle of incidence are needed so that the performance of road surfaces can be better predicted and validated.

ACKNOWLEDGMENTS

Contributions by a number of people were invaluable during the course of this investigation and are very much appreciated. Jeff Stott, Mark Gehringer, and Dick Schumacher were of particular assistance in collecting and processing the tire noise intensity levels, sound propagation data, and passby levels. The efforts of Rich Boles, George Henry, Bill Massie, and Maurice Evans were also an essential part of this work during the data acquisition phases.

REFERENCES

1. ISO 10844 Test Road Specification
2. ISO 362-1994 "Acoustics - Measurement of Noise from Accelerating Road Vehicles - Engineering Method, 1994.
3. SAE Recommended Practice J1470, "Measurement of Noise Emitted by Accelerating Highway Vehicles", March 1992.
4. SAE Recommended Practice J986, "Sound Level for Passenger Cars and Light Trucks",
5. SAE Recommended Practice J366, "Exterior Sound Level for Heavy Trucks and Buses",
6. R. F. Schumacher, K. G. Phaneuf, and W. J. Haley, "SAE and ISO Noise Test Site Variability", SAE Paper 951361, May 1995.
7. P. R. Donavan, "Tire/Pavement Interaction Noise under Vehicle Operating Conditions of Cruise and Acceleration", SAE Paper 931276, May 1993.
8. J. S. Bolton, H. R. Hall, R. F. Schumacher, and J. R. Stott, "Correlation of Tire Intensity Levels and Passby Sound Pressure Levels", SAE Paper 951355, May 1995.
9. P. M. Morse and K. U. Ingard, Theoretical Acoustics, McGraw-Hill Book Company, 1968, pp 367-369.
10. L. L. Beranek, Noise and Vibration Control, McGraw-Hill Book Company, 1971, pp 184-191.
11. R. H. Lyon, Transportation Noise, Grozier Publishing, Cambridge, MA, 1972, pp 29-31.
12. L. E. Kinsler and A. R. Frey, Fundamentals of Acoustics, John Wiley & Sons, Inc, 1962, p 157.

971991

Temperature Dependency of Pass-By Tire Road Noise

Satoshi Konishi, Toshiaki Fujino, Naotaka Tomita, and Toshio Ozaki
Bridgestone Corp.

Copyright 1997 Society of Automotive Engineers, Inc.

ABSTRACT

Coast-by tire road noise is much effected by temperature. As temperature goes up, coast-by tire road noise level becomes lower in the case of passenger car tire. Temperature gradient of coast-by tire road noise is around -0.07 dB(A)/°C using air temperature. On the other hand, the tire is under torque during a pass-by test and some amount of extra noise is radiated from the tire pavement interface. Generally speaking, the level of tire road noise during pass-by mode becomes higher than coast-by mode. How much tire road noise is increased due to the applied torque and how its extra noise is effected by air temperature are reported here.

Pass-by tire road noise consists of two components. The first component is the coast-by noise and the second component is the extra noise by the torque that is obtained by the noise level calculation of pass-by minus coast-by. The second component of the extra noise comes from the slip vibration of tire tread and its noise level becomes higher as temperature goes up. As the first component and the second component have the opposite tendency with the temperature and the two components tend to cancel each other, the total temperature dependency of the pass-by tire road noise is small.

INTRODUCTION

It is well known that tire road noise is influenced by road surface, vehicle type and environmental factors such as temperature, humidity, wind velocity and so on. In particular temperature effects on tire road noise have been given considerable attention. In order to evaluate the noise level more accurately, some temperature correction methods in a coasting condition have been suggested internationally. On the other hand, when the tire is under the torque in an accelerating condition, some amount of extra noise is radiated. This paper investigates how much tire road noise is increased and how its extra noise is effected by air temperature.

NOISE GENERATION MECHANISM

Noise generation mechanism in an accelerating condition is explained in Table 1. Basically, traffic noise consists of vehicle noise and tire road noise in a coasting condition. But when the driving torque is applied to the tire, longitudinal slip between tire and road surface is generated. The slip ratio in an accelerating condition is depend on the engine torque, the gear ratio, the throttle operating condition and so on. Usually it is between 1 to 4 % for passenger car tires.

Table 1: Noise generation mechanism in an accelerating condition

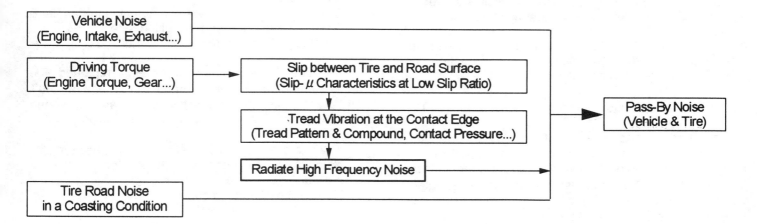

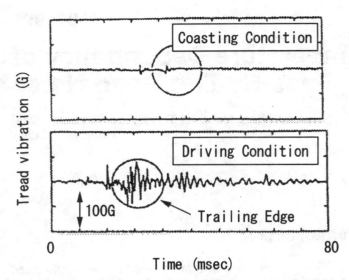

Figure 1: Tread vibration between coasting and driving condition

The driving stiffness which is the slip-μ gradient at such a low slip ratio becomes a very important characteristic for tires. The lower driving stiffness generates much more slip. This slip causes tread vibration at the contact edge, especially when the tread is away from the road surface. Figure 1 shows the comparison of longitudinal tread vibration between coasting and driving condition using micro accelerometer. Tread vibration is extremely increased by the driving torque. And finally it is radiated as high frequency extra noise.

EVALUATION METHOD OF PASS-BY TIRE NOISE

Thus, pass-by tire road noise consists of two components and it is described as the following equation.

$$Lta = Ltc + \Delta Lta \quad (1)$$
  Lta : Tire road noise in an accelerating condition
  Ltc : Tire road noise in a coasting condition
  ΔLta : Extra tire road noise by the driving torque

On the other hand, total pass-by noise is

$$La = Lta + Lv \quad (2)$$
  La : Pass-by noise (Total noise of vehicle and tire)
  Lv : Vehicle noise in an accelerating condition

Our interest is how much tire road noise is increased by the driving torque. From the equation (1) and (2), the extra tire road noise is

$$\Delta Lta = (La - Lv) - Ltc \quad (3)$$

Here the noise level of La and Ltc can be measured directly, so the point is how to reduce or estimate the vehicle noise "Lv". We have two methods to evaluate ΔLta. One is to use special noise insulated vehicle for which the noise level is much smaller than La. Using this special vehicle, ΔLta is given by

$$\Delta Lta \fallingdotseq La - Ltc \quad (4)$$

The other one is called "carpet method". On the carpet, pass-by vehicle noise with smooth tire is given by

$$Lv = Lca - Lcc - \Delta Lca + Labs \quad (5)$$
  Lca : Pass-by noise on the carpet
  Lcc : Coast-by noise on the carpet
  ΔLca : Extra tire road noise by the driving torque on the carpet
  $Labs$: Absorption level of the carpet

The carpet absorbs the pass-by vehicle noise and the absorption level is as much as 1 dB(A) compared with dense asphalt concrete surface. So the absorption level of the carpet $Labs$ should be corrected and it is calculated by the following equation.

$$Labs = Lasp - Lcpt \quad (6)$$
  $Lasp$: Noise level in a racing condition on the asphalt surface
  $Lcpt$: Noise level in a racing condition on the carpet

Most of the cases, Lcc is much smaller than Lca and it is supposed that the extra noise component ΔLca by the driving torque is very small on the carpet. So Lcc and ΔLca are negligible on the carpet.

$$Lv \fallingdotseq Lca + Labs \quad (7)$$

Finally, the extra tire road noise ΔLta is calculated by the following equation.

$$\Delta Lta \fallingdotseq La - (Lca + Labs) - Ltc \quad (8)$$

TEST METHOD

The noise measurements by vehicle accelerating method (ISO 362) and vehicle coasting method (JASO - C606) were carried out during one year on a dense asphalt concrete surface. Tires used for the measurement are as follows;

- Passenger car tire A, B which have different tread compound for drive axle.
  Elastic modulus E' of tire A is much higher than tire B.
- Smooth pattern tire for free rolling axle.
- Smooth pattern tire for carpet method.

Each measurement has been repeated four times through the year in air temperature range of 5-30°C and the relation between noise level and air temperature was compared. Measuring Items are as follows.

o Sound pressure level at each vehicle position
o Vehicle speed at each vehicle position
o Air and road surface temperature
o Wind speed and direction

Figure 2: Photograph of the carpet method

The carpet is 10 mm thick, 450 mm wide and about 30 m long and it is fixed tightly by a strong sticky adhesive tape on the asphalt surface. Figure 2 shows the photograph of the carpet method.

CALCULATION STEPS OF THE CARPET METHOD - For more accurate correction, noise measurement was accomplished using Bruel & Kjaer Pass-by Noise System at each vehicle position and the data was calculated as the next steps.

Step.1 - Pass-by noise measurement "Lca" with smooth tire on the carpet.
The noise measurement is repeated more than five times and averaged at each vehicle position.

Step.2 - Absorption measurement "Labs" of the carpet.
Noise level in a racing condition is measured on the carpet and the asphalt surface, and absorption level Labs is calculated by the equation (3).

Table 2: Test condition of pass-by noise measurement

Vehicle Type	Sports Utility Vehicle (3000cc)
Tire Size	265/70R16 Block & Smooth
Road Surface	Dense Asphalt Concrete and Carpet
Test Speed	50 km/h (Approach Speed)
Gear	2nd , 3rd
Load	GVW 1950kgf
Pressure	200kPa

Step.3 - Pass-by noise measurement "La" with tire A & B on the asphalt surface.

Step.4 - Calculation of pass-by tire road noise "Lta" at each vehicle position.
$$Lta = La - (Lca + Labs) \quad (9)$$

Step.5 - Coast-by noise measurement "Ltc" with tire A & B on the asphalt surface.
Ltc is measured at different vehicle speed and the speed gradient "m" is calculated by first regression analysis.
$$Ltc = m * V_{coast-by} + b \quad (10)$$

Step.6 - Correction of the speed difference between pass-by and coast-by condition at each vehicle position.
$$Ltc' = Ltc + m * (V_{pass-by} - V_{coast-by}) \quad (11)$$

Step.7 - Calculation of the extra noise component "ΔLta" by the driving torque at each vehicle position
$$\Delta Lta = Lta - Ltc' \quad (12)$$

TEST CONDITION - Tire size, vehicle type, road surface and the other test conditions are shown in Table 2. The exhaust noise of the test vehicle is insulated with the extra muffler and the absorption material to make a more accurate measurement.

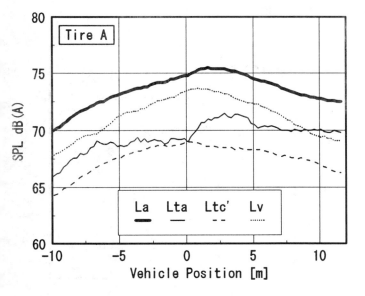
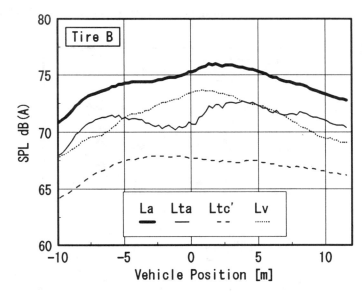

Figure 3: Measured and calculated noise level of each noise component

EXPERIMENTAL RESULT

EXTRA NOISE COMPONENT BY DRIVING TORQUE - Typical measured and calculated data are shown in Figure 3. The peak level of coast-by tire road noise Ltc' is appeared before the microphone position and the peak noise level of tire A is almost 1 dB(A) bigger than tire B. Tire A which has higher elastic modulus E' of tread compound generates much more impact vibration. Extra noise component △Lta which is calculated by subtracting Ltc' from Lta has two peaks. It is considered that the first peak is generated by the impact between block pattern and road surface, and the next one is frictional vibration noise when tire tread is away from the road surface. Figure 4 shows the frequency analysis of the extra noise at each vehicle position. Both peak appear in the high frequency range around 1-2 kHz. The noise increase from coast-by noise is almost 2 dB(A) for tire A, but bigger than 4 dB(A) for tire B. It is explained that higher elastic modulus E' of tire A generates lower slip and frictional vibration between tread and road surface. Then the total pass-by tire road noise Lta of tire A is smaller than tire B.

TEMPERATURE DEPENDENCY OF EACH NOISE COMPONENT - Figure 5 and Table 3 show temperature effect for each noise component and the values of the first regression coefficient (i.e. temperature gradient) averaged in 2nd and 3rd gear. In most cases, temperature gradient of coast-by tire road noise has a negative value except for lug pattern of commercial vehicle tire in high speed range. Tire A and B also have negative gradient and their averaged value is -0.05 dB(A)/°C. On the other hand, extra noise component by driving torque has positive gradient and the values are in

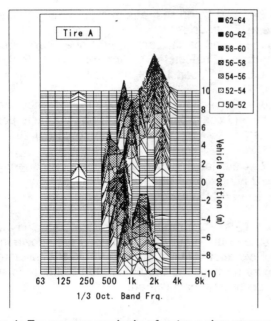 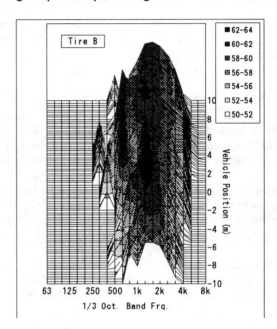

Figure 4: Frequency analysis of extra noise component at each vehicle position

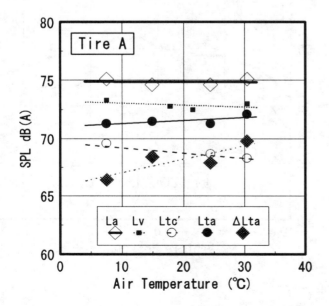 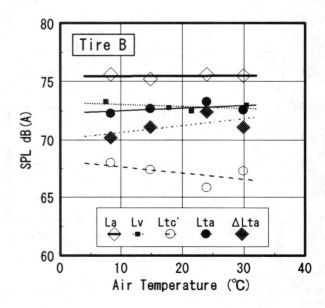

Figure 5: Temperature dependency of each noise component

Table 3: Temperature gradient of each noise component

Tire	Temperature Gradient dB(A)/°C			
	Pass-by noise	Vehicle noise	Pass-by tire noise	Extra tire noise
A	0.001	-0.016	0.026	0.119
B	0.003	-0.016	0.022	0.059
Average	0.002	-0.016	0.024	0.089

tires. As temperature goes up, elastic modulus E' has lower value then temperature gradient of the extra noise has a positive value. These two noise components have the opposite tendency with the temperature and they tend to cancel each other, the total temperature dependency of pass-by tire road noise is small as +0.02 dB(A)/°C. We have also calculated the temperature gradient of vehicle noise and it has a negative gradient as -0.02 dB(A)/°C. So the range from +0.06 to +0.12 dB(A)/°C for both the pass-by noise including vehicle and tire is not much influenced by temperature.

RELATION BETWEEN EACH TEMPERATURE GRADIENT AND TREAD COMPOUND - It is considered that temperature gradient of pass-by tire road noise is described by the following equation.

$$\Delta Tta = (1-\gamma) * \Delta Tc + \gamma * \Delta Ta \qquad (13)$$

ΔTta : Temperature gradient of pass-by tire noise
ΔTc : Temperature gradient of coast-by tire noise
ΔTa : Temperature gradient of extra noise component by driving torque
γ : Contribution coefficient of extra noise component to total pass-by tire road noise

If the contribution coefficient of the extra noise component γ is very small, temperature gradient of pass-by tire road noise ΔTta is almost same value as that of coast-by tire road noise ΔTc. Temperature gradient of each component is related to elastic modulus E' of tread compound. It has been investigated that temperature gradient of coast-by tire road noise ΔTc for air temperature can be described simply by the following equation. [1]

$$\Delta Tc = -6.2 * \Delta E'30 \qquad (14)$$

$\Delta E'30$: Temperature gradient of E'30/E' (Constant value for each tread compound)
E'30 : The value of E' at 30°C

Figure 6 shows the relation between temperature gradient of tread compound $\Delta E'30$ and the extra noise level of tire A, B at different temperature. Temperature effect of the extra noise is well proportional to $\Delta E'30$ for both tires and they have almost same gradient. Then temperature gradient of the extra noise component ΔTa is given by,

$$\Delta Ta = +4.6 * \Delta E'30 \qquad (15)$$

Finally, the total temperature gradient of both components ΔTta is calculated as follows.

$$\Delta Tta = (1-\gamma) * (-6.2 * \Delta E'30) + \gamma * (4.6 * \Delta E'30)$$
$$= (-6.2 + 10.8 * \gamma) * \Delta E'30 \qquad (13)'$$

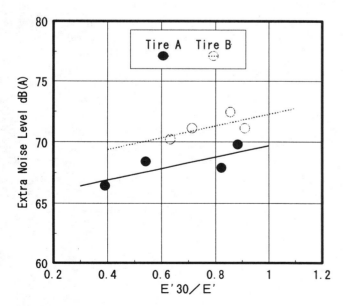

Table 6: Relation between E'30/E' and extra noise component

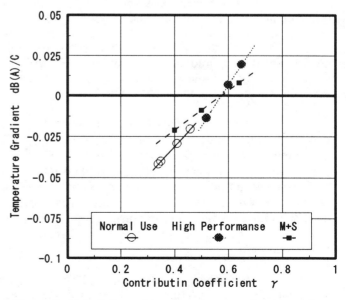

Figure 7: Estimation of temperature gradient for other category of tires

ESTIMATION OF TEMPERATURE GRADIENT FOR OTHER CATEGORY OF TIRES - Our investigation is based on the limited number of tires and a single vehicle. So we try to estimate the variation of temperature gradient ΔTta by measuring γ and ΔE'30 for other category of tires. Each contribution coefficient γ using different type of vehicles is measured by the carpet method and ΔE'30 is averaged for each category of tires. The averaged value of ΔE'30 for high performance tires is higher than other tires. Figure 7 shows the relation between γ and ΔTta calculated by the equation (13)'. Normal use tires are applied less driving torque compared with high performance tires, so the contribution coefficient γ is small and they have more negative values of ΔTta as -0.02 to -0.04 dB(A)/℃. High performance tires have higher values of γ and those temperature gradients are in the range from +0.02 to -0.02 dB(A)/℃. Tires for sports utility vehicles have lower tread pattern stiffness, so they show higher values of γ than normal use tires. The variation of ΔTta is as same as that of high performance tires. The total variation for those category of tires are between +0.02 to -0.04 dB(A)/℃. The averaged value of γ is calculated 0.48, then the temperature gradient of pass-by tire road noise ΔTta is estimated as -0.02 dB(A)/℃.

CONCLUSION

From the experimental approach and the consideration on tread compound characteristics, the following conclusions were obtained

(1) Extra noise component by the driving torque has been estimated accurately using the carpet method. The noise increment level is around 3dB(A) for block type passenger car tires.

(2) It has been investigated that temperature gradient of coast-by tire road noise is described using temperature dependence of elastic modulus E' of tread compound. It is also explained that the extra noise component is proportional to the factor of E'30/E'. The proportional coefficient is -6.2 for coast-by tire road noise and +4.6 for the extra noise component.

(3) Temperature gradient of pass-by tire road noise for block pattern is calculated as +0.02 dB(A)/℃. The variation of temperature gradient for other category of tires are in the range from +0.02 to -0.04 dB(A)/℃ and the averaged temperature gradient is estimated as -0.02 dB(A)/℃. It means there is no necessity of temperature correction for pass-by tire road noise.

REFERENCES

1. S.Konishi, T.Fujino, N.Tomita and M.Sakamoto, "Temperature effect on tire/road noise", Inter-noise 95, pp.147-15

971992

An Integrated Numerical Tool for Engine Noise and Vibration Simulation

B. Loibnegger and G. Ph. Rainer
AVL List GmbH

L. Bernard, D. Micelli, and G. Turino
Fiat Research Centre

Copyright 1997 Society of Automotive Engineers, Inc.

ABSTRACT

The development of low noise engines and vehicles, accompanied by the reduction of costs and development time, can be obtained only if the design engineer is supported by complex calculation tools in a concurrent engineering process. In this respect, the reduction of vibrations (passenger comfort) and of vehicle noise (accelerated pass by noise) are important targets to meet legislative limits. AVL has been developing simulation programs for the dynamic-acoustic optimization of engines and gear trains for many years. To simulate the structure-born and air-born noise behavior of engines under operating conditions, substantial efforts on the mathematical simulation model are necessary.

The simulation tool EXCITE, described in this paper, allows the calculation of the dynamic-acoustic behavior of power units. EXCITE considers different excitations, as well as the nonlinear behavior of the crankshaft (stiffness and mass distribution depending on the crank angle) and of the slider bearings (elastohydrodynamic effects).

The simulation code of EXCITE was further improved, refined, and validated through cooperative effort between major European car manufacturers. The code was validated thoroughly by experimental tests on a reference powertrain, a FIAT 1.8-liter 4-cylinder gasoline engine. This paper describes the results obtained from the experimental tests.

The comparison between experimental and computed results shows that the level of accuracy attained with the use of this EXCITE is excellent.

INTRODUCTION

The development of low noise and low vibration emitting engines, especially for passenger cars, is in strong demand in the automotive industry [1]. On one hand, the legislative limits for the exterior noise of a vehicle (accelerated pass by noise) can be achieved only if the noise emission of the combustion engine is reduced considerably. On the other hand, the vibrations transferred from engine mountings into the passenger compartment must be minimized to fulfill comfort criteria and to make a car successful in the market.

Contrary to the requested noise and vibration reduction, however, are the increasing use of lightweight materials and the close approach to the limits of the admissible stresses and strains. These factors require a shortened processing time for the calculations, and it is also necessary to apply more and more complex calculation methods to obtain reliable results from the calculations. This means acquiring the capability to develop low noise powertrains in the Design Offices more frequently than in the Test Laboratories, to "design-right-first-time," and avoiding the need of costly design modifications in the last phases of the development process.

AVL and Centro Ricerche FIAT (CRF) have been developing and applying mathematical methods based on the Finite Element Mothods (FEM) for dynamic-acoustic optimization of engine structures for many years [2,3,4,5].

Due to the demand to judge investigated design modifications qualitatively and to acquire results which are comparable to measurement data, additional improvements concerning the mathematical formulation of the calculation procedure, the FE-mesh fineness, and the consideration of realistic operating loads are necessary.

To meet these additional improvements and requirements, simulation software packages were developed in cooperation with major European car manufacturers (eight car manufacturers entered into partnership to support this work). AVL implemented this software, on the basis of a proper existing calculation procedure, and performed the calculations on the FIAT reference engine to validate the methodology. CRF generated the engine FE mesh and carried out the experimental work on the same engine. PSA, ROVER, VOLKSWAGEN, and VOLVO performed assessment work on their own powertrains [6].

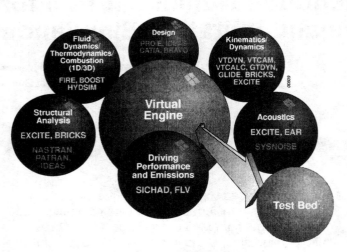

Figure 1. Simulation Tools

The following Simulation Procedure that supports design and development of low noise engines is explained and the validation by measurements is shown. This procedure is an important module in the mosaic of the design and simulation methods, which are necessary for the realization of an "Virtual Engine," Figure 1.

SIMULATION PROCEDURE

Figure 2 shows the procedure for noise and vibration simulation.

The procedure consists of the following software parts:

- CAD Software (geometric modeling)
- FE-Pre- and Post-processing Software (mesh generation and result presentation)
- FE-Kernel System (static and modal analysis)
- EXCITE (rotating crankshaft, elastohydrodynamics, load application, dynamic result calculation)
- an Acoustic Code, for example AKUPOS, SYSNOISE (acoustic result calculation)

These software packages are used to

- create the FE-models
- calculate natural modes and natural frequencies
- calculate forced vibrations
- evaluate the results

The calculation procedure was applied to a reference engine of FIAT. For this engine. CRF performed extensive measurements to validate the calculation procedure.

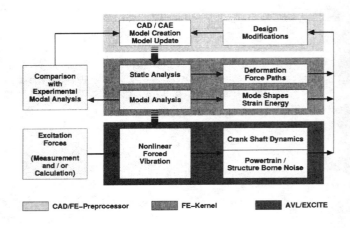

Figure 2. Procedure for noise and vibration simulation

CREATION OF THE FE-MODELS - The FE-grid influences the accuracy of the calculation results by means of the representation exactness of the component geometry, and by the local mesh density and the selection of the element types used. For the representation of component geometries by Finite Elements, various techniques may be applied [7,8]. The most exact idealization of the geometry is achieved by the use of volume elements. Moreover, this kind of modeling offers the advantages that three-dimensional geometries can be taken directly from the CAD-system, and that techniques for an semi-automatic grid-generation may be used.

The FE-model of the reference engine main structural components was generated by CRF using isoparametric linear solid elements and attempting to use sufficiently small size elements in order to not produce large errors due to the meshing.

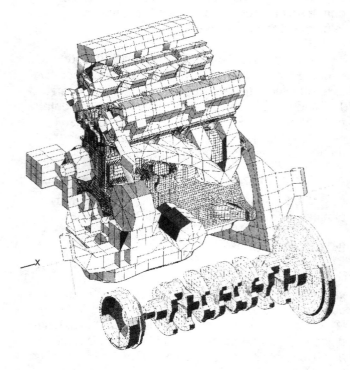

Figure 3. FE-model of the entire power train

Figure 3 shows the FE-model of the entire powertrain of the test engine. The fine meshes for crankcase (64180 nodes and 42040 elements) and cylinder head (55060 nodes and 39440 elements) were set up to acquire reliable results for the comparison with measurements performed on these engine parts. All other engine parts were modeled using a coarser mesh to save degrees of freedom (DOF). The full FE-model consists of about 370000 DOF's.

NATURAL MODE ANALYSIS - The natural mode analysis offers the possibility to find out whether different components of the engine show natural frequencies within a narrow frequency band.

The FE-modal analysis of the engine main structural components was carried out using the super-element reduction technique available with the MSC/NASTRAN code. The comparison with the results coming out from the experimental modal analysis performed on the same components showed extremely close. Tables 1 and 2 show both the frequency and MAC comparison of the cylinder block natural mode.

Mode	Test (Hz)	FEmodel (Hz)	FE/Test (%)	Mode description
1	495	473	-4.4	1st Torsional
2	830	780	-6.0	1st Lateral Bending
3	918	895	-2.5	Bearing cap bending
4	948	913	-3.7	"
5	1003	986	-1.7	"
6	1052	1008	-4.2	"
7	1138	1116	-1.9	"
8	1242	1232	-0.8	Complex mode
9	1556	1473	-5.3	"
10	1599	1532	-4.2	"

Table 1. Frequency comparison between experimental data and FE analysis

The agreement is good not only in terms of frequency values (the highest difference is less than 5%), but also in terms of mode description (both a visual interpretation and a MAC values calculation was performed). Especially good are the bearing cap bending results. These are very important since they affect the dynamic coupling between the crankshaft and the block. Figure 4 shows a typical bearing cap mode - first bending mode - of the reference engine.

0.96	0.00	0.00	0.00	0.00	0.00	0.00	0.00	0.00	0.00
0.01	0.94	0.00	0.00	0.00	0.00	0.00	0.01	0.00	0.00
0.00	0.00	0.90	0.02	0.00	0.00	0.00	0.00	0.00	0.00
0.00	0.01	0.01	0.88	0.00	0.00	0.00	0.00	0.00	0.00
0.00	0.01	0.00	0.00	0.32	0.45	0.11	0.00	0.00	0.00
0.00	0.00	0.00	0.00	0.45	0.46	0.00	0.01	0.00	0.00
0.00	0.00	0.00	0.00	0.04	0.04	0.83	0.00	0.00	0.00
0.00	0.01	0.00	0.00	0.01	0.00	0.02	0.91	0.14	0.01
0.00	0.00	0.00	0.00	0.00	0.00	0.00	0.00	0.85	0.00
0.00	0.00	0.00	0.00	0.00	0.00	0.00	0.00	0.01	0.87

Table 2. MAC matrix for the cylinder block: FE model/TEST

Besides the correct mesh generation for the single components, it is of primary importance to properly reproduce the mechanical connection between the various parts. For the powertrain, the most critical links are those between the cylinder head and the block, the block and the oil sump (especially in case of structural sump), the block and the gearbox.

CRF has been developing and applying, for many years, numerical linking techniques in order to well reproduce the whole powertrain dynamic behavior [1,5].

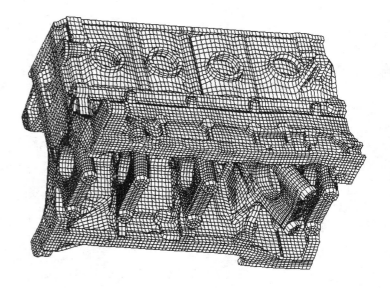

Figure 4. Eigenmode f=895 Hz

EXCITATION FORCES

Forces and moments which are an external input for the simulation of forced vibrations must be determined by precalculations with different software or by measurements. See Figure 5 for an illustration of these.

GAS FORCES - The determination of the gas pressure distribution can be performed by engine cycle calculation or by pressure measurements. These forces are applied at the piston node and at nodes of the combustion chamber of each cylinder. CRF performed the in-cylinder pressure measurements for all cylinders and for two running conditions (3000 and 6000 rpm). Due to the pressure scattering, three cycles of measured in-cylinder pressure were used as input data for the calculation.

FORCES OF THE CRANKTRAIN - To model the dynamic behavior of the engine's cranktrain is a very complex task due to the superimposition of its kinematic movement and deformation vibrations. In addition, the nonlinearity of the cranktrain bearings affects the vibrations. In the running engine, an interaction between the dynamics of piston and conrod on the one hand and vibrations of the rotating crankshaft on the other exists. The forces in the bearings (oilfilm forces) are nonlinear depending on relative movements and velocities. Therefore, the coupling of the structure parts in the conrod and main bearings is performed by nonlinear stiffness and damping forces or EHD forces. For the reference engine, the forces of the cranktrain were calculated based on the gas forces and on the geometric conditions.

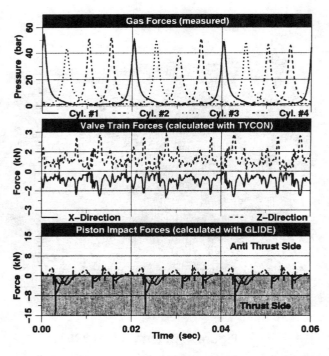

Figure 5. Applied dynamic loads to the power train

PISTON IMPACT - The piston impact excitation derives from the kinematic conditions and the piston secondary motion. To calculate the piston movement AVL uses the software GLIDE. The piston motion consists of oscillation on one hand, tilting and radial motions in the clearance between piston and liner on the other hand. The movement of the piston is simulated in a thrust and antithrust side plane of the liner.

The calculation carried out considers forces due to gas load, inertia, friction, and elasticity of the piston structure, as well as constraints due to the deformed contour of the liner. The piston slap calculation results in piston impact forces, which are transformed to forces at the cylinder liner nodes. Figure 5 shows the impact forces, which act on the thrust and antithrust side over the height of each cylinder liner.

VALVE TRAIN FORCE - The excitation forces in the timing drive are caused by impacts in the valve train and gear train on one hand, and by the rotating mass forces at the camshafts on the other. The impacts due to the valve train dynamics act on the cylinder head at the valve seats and in the bearings of the camshafts (Figure 5). At the present time AVL uses the software VTDYN for the calculation of the dynamic behavior of valve trains. This program is used to predict impacts and to improve the mechanical behavior of valve trains considering a multi-mass system and a static determined cam profile [9,10].

NON-LINEAR FORCED VIBRATION CALCULATION

The basis for the simulation and analysis of the load transfer from the combustion chamber via the crankshaft and its main bearings to the structure of the block surface is the software EXCITE [2,11,12,13]. The calculation program EXCITE enables the calculation of the coupled vibrations of the crankshaft in torsional, bending, and longitudinal directions caused by gas and mass forces.

For the vibrations of the crankshaft, the gyroscopic effects of the flywheel and of the counterweights are considered, as the crankshaft is simulated rotating having an actual deformed shape in each time step of the calculation.

The vibration transfer from the crankshaft to the bearing housing considers effects of the bearing oil film based on elastohydrodynamic theory, taking into account bearing housing flexibility and crankpin skewness. Thus interaction effects of the vibration transfer between crankshaft and bearing structure can be simulated and analyzed in the radial and the axial directions.

The engine block, including cylinder head, adjacent parts and auxiliaries, is considered by means of its dynamic properties modeled by Finite Elements.

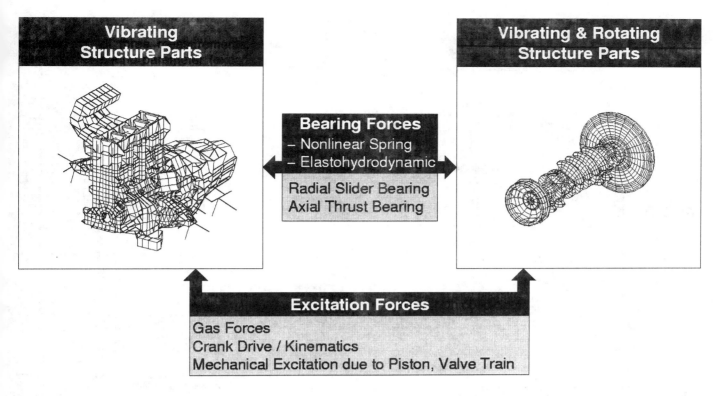

Figure 6. Schematic coupling of vibrating and moving structure parts

Several different damping effects are included in the model. The damping properties of the main bearings act locally in the calculated elastohydrodynamic oil film forces; thus exciting forces acting upon the crankshaft (rotating mass forces and reaction forces of the engine structure vibration) are indirectly affected by these forces. The damping effects are also simulated locally for the torsional damper and the engine mounts. Other damping properties such as effects of the material itself, contact areas and gaskets are currently represented by a modal damping applied on the whole structure.

In this case, the simulation systems have to be solved by the calculation of time history only, i.e. by time integration methods, the system described above has a major disadvantage: the non-linear structure matrices must be inverted at each time step, which means an enormous effort of CPU-time even on high performance computers.

To eliminate this disadvantage the following approach was found to be useful: the block structure is separated from the cranktrain in the bearings, Figure 6. The forces resulting from the oilfilm of the bearings at each time step operate on both structures. The structure matrices become time independent and have to be inverted only once. The rotation of the cranktrain can be treated as a transformation of coordinates and dynamic components. Non-linear parts of the structural matrices change to a vector of external forces and moments acting on both the block and the crankshaft structure. This approach assumes that the dynamic reaction of one structure works upon the second structure not before the following time step.

This can be controlled by choosing short time steps related to the required accuracy of solution over the frequency range. According to AVL experience, time steps of 0.00002 sec. are adequate for calculations considering frequencies up to 3000 Hz.

RESULTS

Results distortions, velocities and accelerations in the time domain are calculated.

To perform acoustic evaluations a transformation of the results into the frequency domain is necessary.

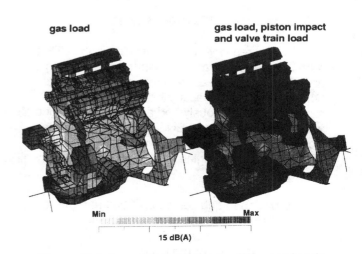

Figure 7. Integral levels due to different loads

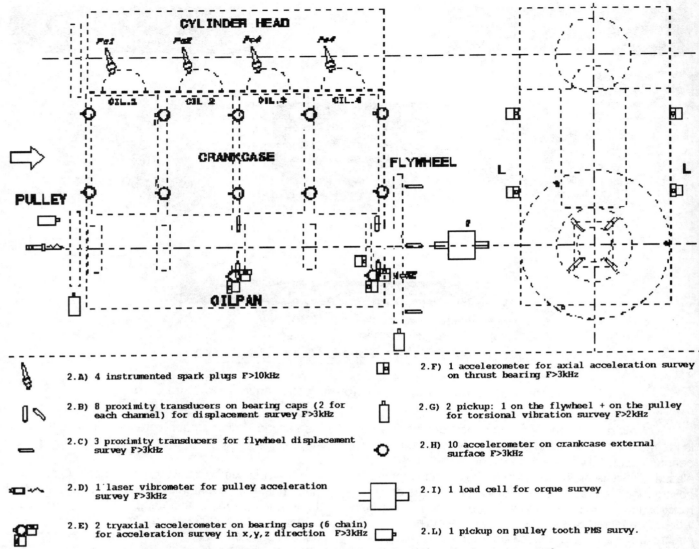

Figure 8. Schematic diagram of the reference engine instrumentation

This is done via FFT (Fast Fourier Transform). Mainly, velocity levels or acceleration levels on the engine surface are used to judge the quality of the design. Figure 7 shows integral levels from 0 to 2500 Hz for different load cases, the left one with only gas load; and the right one with gas load, piston impact, and valve train loads.

EXPERIMENTAL MEASUREMENTS

The real engine was installed in the CRF test room. The engine was fully instrumented (Figure 8) and metrological measurements were carried out before and after the tests. All the signals were recorded simultaneously and analyzed in time domain with a special software (ENGTDA - ENgine Time Domain Analysis) developed internally. At the end, 100 synchronized engine cycles were available for all the transducers.

For the two operating conditions 3000 rpm (full load) and 6000 rpm (max. speed), the following measurements were carried out:
- in-cylinder pressure
- crankshaft axial and torsional velocity
- crankshaft displacement on bearing journal
- acceleration on cylinder walls
- acceleration on the engine block external surface

COMPARISON MEASUREMENT/CALCULATION

To determine the accuracy of the calculation results of the power unit, comparisons with measurement results obtained on the 1.8 liter gasoline engine have been made. The comparison was carried out for two operating conditions - 3000 rpm (max. load) and 6000 rpm (max. speed). Additionally measuring results made on a 11 liter truck diesel engine were compared with calculations.

The following four examples of the evaluated results are chosen to illustrate the close correlation of the data.

JOURNAL DISPLACEMENTS - One criterion to illustrate the accuracy of the dynamic crankshaft behavior is the movement of the crankshaft journal in the main bearings.

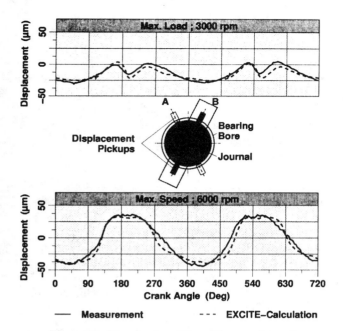

Figure 9. Displacements of journal no. 3

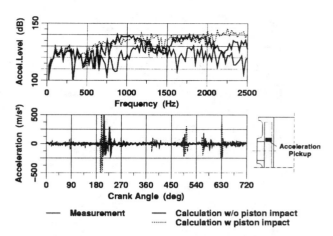

Figure 11. Acceleration at cylinder liner due to piston impact

Figure 9 shows the comparison of journal no. 3 displacements, measured and evaluated from the calculation in direction B. The path of the journal is dominated by the effect of inertia forces and the gas forces of the adjacent cylinders. For low speed operating conditions, the influence of the gas load of cylinder no. 2 and 3 at 180 and 540 degrees crank angle can be observed. For high speed operating conditions, only the inertia forces are relevant for the displacement of the journal. The results correlate closely.

PISTON IMPACT - EXCITE itself considers only the kinematic excitation due to the crank train motion. The influence of the piston impact due to the secondary motion of the piston can be calculated with the program GLIDE and the forces are applied as external forces acting on the cylinder liner.

Figure 10 shows the influence of the piston impact forces on the accelerations at the cylinder liner in comparison to the measurement. Whereas the consideration of the impact forces leads to a good correlation to the measured results, the consideration of the piston kinematic only makes an inapplicable comparison.

SURFACE VELOCITIES - Other evaluation criterion are the velocity levels representing the structure-borne noise at discrete structure points.

Figure 11 shows the comparison of the velocities on a measuring point in the area of the oil sump flange for a 11 liter truck diesel engine. Both in

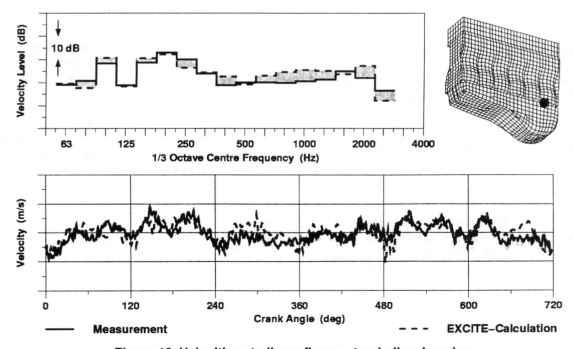

Figure 10. Velocities at oil pan flange - truck diesel engine

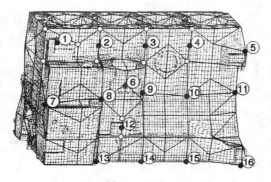

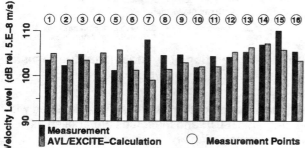

Figure 12. Integral velocity levels at crankcase surface

the time and in the frequency domain, the correspondence is extremely good.

Another comparison about the structure borne noise emission can be made by summing up the velocity levels in the acoustic relevant frequency range to integral levels.

Figure 12 shows the integral velocity levels from 0 to 2500 Hz of the selected measurement points at the engine block surface on the exhaust side. For the majority of the measurement points, the calculated results are very close to the measured results.

CONCLUSIONS

The comparison of calculated and measured results shows that the developed calculation procedure is a reliable tool to support the design process. The establishment of general rules for the FE models' generation, allowing the reproduction of the actual dynamic behavior with a high degree of accuracy, contributed to the achievement of the target of a reliable numerical simulation.

The use of this simulation procedure in the powertrain design can substantially reduce its development time and cost, allowing the engineer to define the best design solutions for noise/vibration reduction as well.

The following points must be considered in order to further improve the simulation procedure for its effective use:

- An intensive use of the software is necessary to build up a database of results and to interpret the results in a reliable way.
- More powerful computer hardware is necessary to use the tool for daily optimization calculations.
- Considerable effort, especially on the measurement side, has to be investigated to get a good correlation between measurement and calculation.

ACKNOWLEDGEMENTS

The work was funded by the CEC within the **Brite Euram Project** BE4126-Subproject „Engine FEM" and by the **Austrian Industrial Research Promotion Fund.** Participants of the European Automotive Industry were PEUGEOT, CRF, VOLVO, ROVER, VOLKSWAGEN, and MERCEDES BENZ.

REFERENCES

[1] Bernard, L.; Piccone, A.; Rinolfi, R.: "New Design Methodologies for Reducing the Development Cycle of Low Noise Powertrains," FISITA '96, Prague

[2] Rainer, G.Ph., B. Loibnegger: "Simulation des Schwingungs- und Geräuschverhaltens von Antriebseinheiten," 8. Internationaler Kongress "Berechnung und Simulation im Fahrzeugbau," 16./17. September 1996, Würzburg

[3] Priebsch, H.H.; Affenzeller, J.; Kuipers, G.: "Structure Borne Noise Prediction Techniques," SAE Technical Paper Series 900019, International Congress and Exposition, Detroit, 1990

[4] Rainer, G.Ph.: "CAD/CAE/CAM-Integrated Applications in Engine Development" Proceedings, Automotive Simulation 91, Schliersee, October 91, Springer Verlag

[5] Cattani E., Micelli D., Sereni L., Cocordano S.: "Numerical Simulation and Experimental Comparison of an Engine Structure Dynamic Modal Behavior," ATA International Conference on New Design Frontiers for More Efficient, Reliable and Ecological Vehicles, Florence 1994

[6] Bernard L. et al.: "Final Reports of the Brite/Euram project BE-4126: Powertrain Noise & Vibration Prediction," 1995

[7] Rainer, G.Ph.; Gschweitl, E.; Pramberger, H.: "Dynamic Analysis of Engine Components using Large FE-Models," IAT'95 Conference, Radenci/Slovenia, 6-7 April 1995

[8] Ott, W.; Kaiser, H.J.; Mayer, J.: "Finite Element Analysis of the Dynamic Behavior of an Engine

Block and Comparison with Experimental Modal Test Results" Proceedings, MSC World Users Conference, L.A., 26-30 March 1990

[9] Hellinger, W.; Priebsch, H.H.; Landfahrer, K.; Mayerhofer, U.: "Valve Train Dynamics and its Contribution to Engine Performance," Paper C389/129 to be published at FISITA, London, 1992

[10] Priebsch, H.H.; Hellinger, W.; Loibnegger, B.; Rainer, G.Ph.: "Application of Computer Simulation for the Prediction of Vibration and Noise in Engines," ATA 3rd. International Conference, Firenze, 1992

[11] Gran, St.: "Mehrkörperdynamik elastischer Körper - Kurbeltrieb" (Multibody Dynamics of Elastic Bodies - Crank Train), Dissertation, Technical University-Graz, May 1994

[12] Priebsch, H.H.; Affenzeller, J.: "Prediction Techniques for Vibration and Stress of Nonlinear Supported Rotating Crankshafts, Paper D59, CIMAC, Florence, 1991

[13] B. Loibnegger, G. Ph. Rainer, "Design Supporting Analysis for Powertrain Noise and Vibration," Internoise 1996, Liverpool

[14] Micelli D., Turino G.,Grasso C., Zappala' A.: "A New predictive Tool to Design the Silent Powertrain," ATA International Conference on The Virtual Automobile and the Role of Experimentation, Florence 1997

971993

Valvetrain Unbalance and Its Effects on Powertrain NVH

Joseph L. Stout
Ford Motor Co.

Copyright 1997 Society of Automotive Engineers, Inc.

ABSTRACT

As vehicle NVH continues to improve, automotive engineers have continually worked at identifying and ameliorating smaller and smaller sources of engine vibration. In this search, what has become an extremely important factor is often overlooked, namely valvetrain related engine unbalance effects.

This paper will present a simple analytical approach to the calculation of valvetrain unbalance effects. Theses unbalances will be compared in magnitude to the well understood inherent cranktrain unbalances. The paper goes on to explain the significant factors that control the directions and magnitudes of these forces and moments. Experimental validation of the analytical procedures will also be presented.

INTRODUCTION

Valvetrain unbalance in this paper is the resultant set of forces acting on the engine caused by the rigid-body kinematic motion of the valvetrain system. This definition assumes that all components remain in contact, i.e. no impact loads due to valve seating or separation. The effect of dynamic impact forces on engine NVH has been studied in detail [1,2,3,4,5] and has been shown to be very important in the high frequency engine radiated noise domain.

The significance of rigid-body kinematic valvetrain forces has been discussed as a source previously [6] and their role in the low-frequency structural-borne noise domain has been identified. Experimental validation identifying the valvetrain as a source of a significant amount of powertrain vibration has also been documented [7].

This paper presents a simple analytical model based on the motion of a single valve as the fundamental building block and shows how this model can be extended to calculate the entire engine valvetrain unbalance.

SINGLE VALVE MODEL

The equation used to solve for the reaction force due to kinematic valve motion of interest in this paper is simply:

$$F = m\ddot{x} \qquad (1)$$

The simplest type of valvetrain to apply this equation to is a direct acting overhead cam type as shown in Figure 1.

Direct Acting Overhead Cam Valvetrain

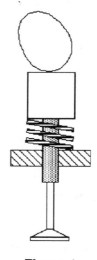

Figure 1

What makes this type simple is the fact that all of the motion of interest is in a single direction. The solution to equation (1) can be determined by differentiating the valve lift curve twice to determine the acceleration and by calculating the effective mass associated with that motion using the Raleigh energy method [8]. The application of the energy method becomes more complex for other styles of valvetrain designs (push rods, finger follower, etc.), but it is still a relatively easy task if the mechanism is well understood.

In the calculation of valvetrain unbalance we are not interested in knowing the force acting on the valve mechanism, but rather the force which is reacted on the engine structure. This reaction force will be in a direction opposite of the acceleration of the valve. A plot of a typical valve acceleration, velocity and displacement (lift) is shown in figure 2.

With this knowledge (effective mass, acceleration, and equation 1) we have all of the information that is needed to calculate valvetrain unbalance. The work from here on out is mainly the 'bookkeeping' involved in summing up each of the valve forces for the engine over a complete 720° crankangle cycle during which each valve will act in a 4-stroke engine.

Valve Displacement, Velocity and Acceleration

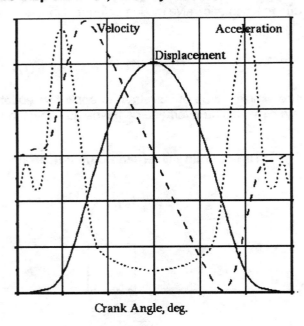

Figure 2

TOTAL ENGINE MODEL

Continuing with the concept of a building block approach requires the development of a common coordinate system that all valvetrain forces can be expressed in. The engine fixed coordinate system shown in figure 3 will be used in the development of this model. Any quantities that are defined in a local valve coordinate system can be mapped to the engine coordinate system using the angles α and β as defined in figure 4.

The reaction force time history for a single valve will have the shape given by the valve acceleration curve in figure 2. The magnitude of the force is the effective mass of the valve system multiplied by the acceleration component mapped into the engine coordinate system.

Engine Fixed Coordinate Frame

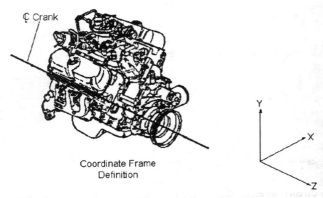

Figure 3

The next step is to include the phasing between the valves of a single cylinder, and then when a single cylinder valvetrain force model is calculated, this model can be used to sum up the forces and moments for the entire engine valvetrain. This is most easily done by taking advantage of the symmetry that exists in the engine configuration being analyzed.

Valve Angle Relative to Engine Fixed System

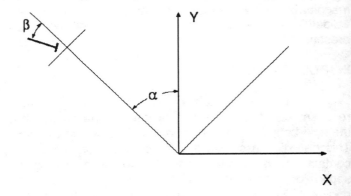

Figure 4

For a V engine you might have mirror symmetry about the y-z plane. In this case the y forces from a cylinder in one bank are identical except for the firing phase from the y forces in the other bank. The x forces however are mirrored and so on opposite banks the forces will be different by the firing order phasing plus an additional 180°. Once the valvetrain unbalance forces are determined the moments can be calculated about any fixed reference point by taking the cross product of the vector from the reference point to the valve force line of action with the vector of the valve reaction force.

RESULTS FOR A 90° V6 ENGINE

The analytical procedure described was applied to a 90° V6 engine with a pushrod type valvetrain. The results of the analytical study are summarized in table 1. The engine analyzed used a front to back / bank to bank firing order with the firing events equally spaced. The moments were summed at the crankshaft centerline at a point mid way between the front and rear cylinders.

The data in table 1 is displayed for each of the first 6 harmonic orders. The harmonic orders are relative to the crankshaft rotation. The existence of the half orders is due to the fact that a complete set of valvetrain events occurs every 2 engine revolutions so the fundamental valvetrain order is 0.5.

The zero value items in the table can be explained and predicted by equation 2.

$$A_n = \sum_i B_n \cos(n(\theta + i(4\pi / N))) \qquad (2)$$

Valvetrain Forces (N) and Moments (N-mm)

calculated at 4000 engine rpm

order	Fx	Fy	Mx	My	Mz
0.5	0	0	5,640	864	0
1.0	0	0	5,940	3,140	0
1.5	175	0	12.000	33,100	163,000
2.0	0	0	11,800	5.570	0
2.5	0	0	3,750	240	0
3.0	0	270	51,100	1,860	0

Table 1

The variable descriptions of equation 2 as it relates to the results in table 1 are:

A_n = the resultant magnitude
B_n = the magnitude of a single cylinder quantity that is equal in magnitude for all cylinders
n = the harmonic order
i = the numerical place in the firing order for a cylinder (e.g. in the firing order 1 3 4 2 cylinder 3 has the numerical place i=2)
N = the number of cylinders

The significant portion of equation 2 is the quantity $n \bullet i(4\pi/N)$ where the 4π represents the fundamental period. This term is significant because when it is summed over all cylinders A_n will equal 0 for all orders except when $(4\pi n/N) = 2\pi j$, where j takes on integer values. This equation can be simplified to n = jN/2.

This explains why for this V6 example the y-direction forces (Fy) equal 0 except for the 3rd order (or any orders that are multiples of 3). As explained earlier, the forces in the y direction are the only quantities identical for each cylinder, so the equation as stated only holds for that quantity.

A very similar phenomena is taking place for the x-direction forces (Fx) and the moments about the z-axis (Mz). These quantities are opposite in sign for each bank of the engine due to the mirror symmetry. The equation which represents these values is

$$A_n = \sum_{i=odd} B_n \cos(n(\theta + i(4\pi / N))) + \sum_{i=even} -B_n \cos(n(\theta + i(4\pi / N))) \qquad (3)$$

Since the value i increments by 2 in each equation, the orders which perfectly add for each bank individually are determined by $(8\pi n/N) = 2\pi j$, or n=jN/4. In this 6 cylinder engine that corresponds to the orders which are multiples of 1.5. The two banks will vary from each other by a phase angle of $n4\pi/N$, or $n2\pi/3$ in this 6 cylinder example. Because the magnitudes in each bank are opposite, the banks will add when their phase difference is an odd multiple of π. That means that for 1.5 order the banks add and for 3.0 order the banks subtract. All of this information is verified by the values in table 1.

The Mx and My values do not easily simplify in this model. The reason that they do not is that their magnitudes for each cylinder are each a function of the distance along the z-axis from the reference point to the cylinder in question. This makes the quantity B_n in equation 2 vary and so the cancellation of effects can not be predicted by phase angle relationships alone. By the same reasoning, there also will not be cases of total addition of effects either, so it can be expected that these quantities (Mx, My) will in general assume lower values than those (Mz) for which cases of perfect addition exist.

COMPARISON TO CRANKTRAIN UNBALANCE

The unbalanced forces and moments due to the motion of the cranktrain are well understood [9]. To create a point of reference for the significance of the valvetrain forces a simple comparison of the magnitudes can be made.

In this 90° V6 engine there exist elliptical 1st and 2nd order rotating couples (Mx and My) due to the reciprocating mass of the piston and connecting rod. In this engine the first order couple is partially eliminated through crankshaft counterweights. The portion of the 1st order that is unbalanced and the 2nd order magnitudes are given in table 2.

By comparing the values in table 1 and table 2 it is easily seen that the 1.5 order valvetrain unbalance moment about the z axis is nearly as large as the cranktrain unbalance.

Cranktrain Moments (N-mm)

calculated @ 4000 engine rpm

order	Mx	My
1.0	61,600	560,000
2.0	243,000	421,000

Table 2

EXPERIMENTAL VALIDATION

Experimental data has been collected which validates the analytical findings for this engine. In this experiment accelerometer measurements were taken on this engine during a neutral engine run-up and as it was being motored with the valvetrain deactivated by removing the pushrods and rocker arms. The neutral engine and motoring tests were chosen in order to pick up the large 1.5 order excitation predicted from the valvetrain without adding any additional half orders of excitation that might be caused by the combustion loads.

The accelerometer data for the two tests is shown in figure 5. This data clearly shows that the presence of the operating valvetrain is the source of the 1.5 order component of engine vibration.

Additional validation was performed by using the analytical data to predict the engine acceleration. To perform this analysis a finite element model of the engine and its mounts was created. Since only the relatively low frequency 1.5 order response was of interest, the model used was a rigid, lumped mass model consisting of only 6 total degrees of freedom. The correlation between the predicted results of that model and the experimental data is shown in figure 6.

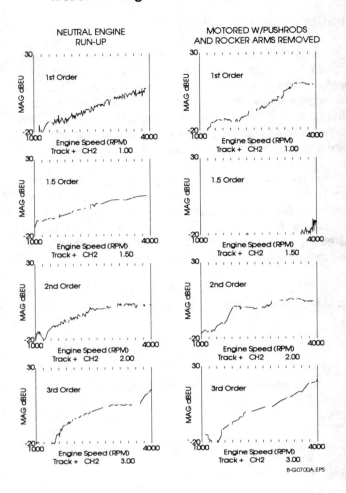

Figure 5

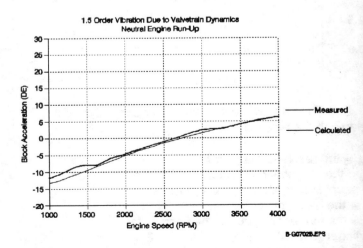

Figure 6

The level of correlation achieved both validates the valvetrain unbalance analytical prediction and shows that powertrain vibration analysis methods that use only a simplistic equation (1) to represent the powertrain motion can be extremely useful. The completeness of the excitation model which includes valvetrain unbalance and not the detail of the powertrain representation is the key factor in this correlation.

VALVETRAIN UNBALANCE DESIGN CHANGES

Following the description of the valvetrain unbalance calculation laid out in this paper there are several opportunities for the engine design engineer to reduce its unwanted effects. By examining equation 1 it can be seen that reductions in either the effective mass or the valve acceleration will reduce the magnitude of the unbalance. By this papers description of the role of phase relationships between valvetrain events it is clear that it also plays an important role.

MASS REDUCTION - The reduction of the masses in the valvetrain system seems to be a relatively straight forward approach to reducing the effects of valvetrain unbalance, and for the valvetrain type shown in figure 1 it is the only option available. Other more complicated valvetrain mechanisms, which usually carry along with them the burden of greater mass, also provide additional opportunities for effective mass reduction. The key element here is 'effective mass'. The application of this concept can be illustrated by the valvetrain system shown in figure 7.

In this valvetrain configuration the mass of the pushrod and tappet move in a direction different than that of the valve and spring. By considering the mass of each, the vector directions involved and the rocker arm ratio a configuration can be deployed which minimizes the effective mass.

ACCELERATION REDUCTION - The reduction of the valve acceleration is a complicated task. Since it is usually a certain order of valvetrain unbalance that is objectionable, like the 1.5 order in our example, reducing the overall acceleration may not reduce the specific order of interest. For this reason we will describe the valvetrain acceleration in terms of its specific orders.

The valve displacement as shown in figure 2 can be described as a summation of harmonic orders of displacement as in equation 4.

$$x = \sum_n C_n \cos(n\theta + \phi_n) \quad (4)$$

Pushrod Valvetrain

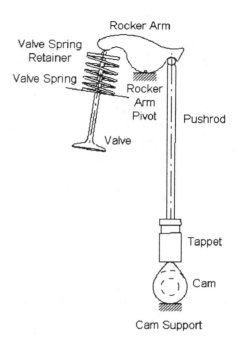

Figure 7

The acceleration can be described by differentiating each of the harmonic orders as shown in equation 5.

$$\ddot{x} = \sum_n -n^2\dot{\theta}^2 C_n \cos(n\theta + \phi_n) \quad (5)$$

It should be noted that the magnitude of the harmonic acceleration terms are the displacement terms multiplied by the order squared. For this reason the largest harmonic magnitudes will be different for the displacement and the acceleration.

This difference in harmonic magnitudes is shown in table 3. The magnitudes in this table are normalized by the root summed squared value of the first 100 orders.

The effect of multiplying the lift magnitudes by the order squared shifts the largest values for the displacement from the smallest orders, 0.5, 1.0 etc., to the slightly higher orders, 1.5, and 2.0 for acceleration.

Harmonic Content Comparison

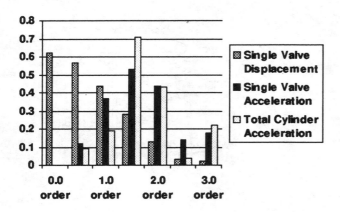

Table 3

Another important quantity compared in table 3 is the total acceleration for the cylinder (exhaust and intake valve). It can be seen that the 1.5 order which has the largest value in the single valve acceleration takes on an even greater portion of the total cylinder acceleration. The explanation of this phenomena comes from an understanding of how two harmonics shifted by a phase angle combine.

The two nth harmonic components to be added can be described as follows:

$$A_n \cos(n\theta + \phi_n) + A_n \cos(n(\theta - \gamma) + \phi_n) \quad (6)$$

These harmonics represent the nth order components due to the intake and exhaust valves and the angle γ represents the phase shift in crankangle degrees between the two events. Equal magnitudes of lift or accelerations, A_n, are assumed here.

To add these, both cosine terms need to be rewritten using the double angle formula into sine and cosine terms of the angles (nθ), (ϕ), and (ϕ-nγ). Then after collecting the sine and cosine terms, the magnitude of the combined nth order term can be expressed as:

$$A_n\{[\cos\phi_n + \cos(\phi_n - n\gamma)]^2 + [\sin\phi_n + \sin(\phi_n - n\gamma)]^2\}^{1/2} \quad (7)$$

This expression takes on a minimum value when (nγ) = j•180° and a maximum value when (nγ) = j•360° where j is equal to the non-zero integers. This means that 1.5 and 3rd order terms will take a maximum value for a phase shift of 240°, the 4th order terms will cancel at 225°, and that 2.5 order terms will cancel at 216°, all values close to the 225° shift used in this example. These effects can be seen in table 3. By properly choosing the phase angle used in the design the additive effects of certain undesirable orders may be minimized.

PHASING EFFECTS - The phasing effects discussed in this section relate to cylinder-to-cylinder phasing, not the intake to exhaust phasing described in the previous section. The way to alter this phasing is to change the firing order. A change in firing order (assuming that all firing orders will be even firing) will not change the values for those components that are identical for all cylinders, like Fy in our V6 example. What a firing order change can do is break-up the symmetry that causes other force or moment components to add perfectly - if a firing order that breaks that symmetry is possible.

In the V6 case one possible firing order without changing the crankshaft configuration (for a given crankshaft configuration there are always as many possible firing orders as there are cylinders in a 4-stroke engine) allows for all cylinders to fire on one bank and then the other. This firing order breaks up the mirror image symmetry for the Fx and Mz unbalance terms. Making this change only, the valvetrain unbalances become as shown in table 4.

Revised V6 Firing Order

Valvetrain Forces (N) and Moments (N-mm)

calculated at 4000 engine rpm

order	Fx	Fy	Mx	My	Mz
0.5	19	0	7,990	3,820	17,000
1.0	0	0	5,940	3,140	0
1.5	58	0	20,700	4,820	54,300
2.0	0	0	11,800	5.570	0
2.5	5	0	4,000	1,050	14,600
3.0	0	270	51,100	1,860	0

Table 4

By comparing this data to the data in table 1 it can be seen that the Fy values are unchanged as expected and the Mz terms for which the symmetry is broken are significantly altered. The 1.5 order Mz term is reduced from 163,000 to 54,300. This demonstrates the significant changes that can be achieved by altering the cylinder-to-cylinder phasing of the valvetrain unbalance components.

CONCLUSION

Valvetrain unbalance can be a significant contributor to engine unbalance. Techniques are available to calculate the valvetrain related unbalance values and to predict their response on engine motion. A complete understanding of the mechanisms leading to the unbalance points to numerous techniques available for reducing the unwanted effects of this unbalance.

ACKNOWLEDGMENTS

I would like to acknowledge Ron Blaha of Ford Motor Co. who performed the experiments mentioned in this paper and initiated many of the investigations that have lead to our understanding of valvetrain unbalance today.

REFERENCES

1. Phlips, P.J., Schamel, A.R., and Meyer, J., "An Efficient Model for Valvetrain and Spring Dynamics," SAE 890619

2. Kaiser, H.J., Phlips, P., Schamel, A., and Adams, W., "Engine Noise Reduction by Camshaft Modifications in Multi-Valve Engines," 2. Aachener Colloquium Automobile and Engine Technology, Oct. 1989

3. Kaiser, H.J., Deges, R., Schwarz, D., and Meyer, J., "Investigations on Valve Train Noise in Multi-Valve Engines," SAE 911062

4. Phlips, P., and Schamel, A., "The Dynamics of Valvetrains with Hydraulic Lash Adjusters and the Interaction with the Gas Exchange Process," SAE 910071

5. Ernst, R. and Meyer, J., "An Optimization Algorithm for Improving Cam Lobes towards Reduced Valve Train Vibrations in Automotive Engines, " ISATA Conference, 1992

6. Uchida, T. and Katano, H., "Honda New In-Line Five Cylinder Engine - Noise and Vibration Reduction," SAE 900389

7. Eichhorn, U., and Schonfeld, H., "The Valve-Train of Internal Combustion Engines as a Source of Vibrations - Experimental Results and a Method of Calculation," SAE 905172 , 23rd FISITA Congress (1990), Torino, Italy

8. Thomson, W. T., *Theory of Vibration with Applications,* Prentice-Hall, Inc.,1981, pp 21-24.

9. Taylor, C.F., *The Internal-Combustion Engine In Theory and Practice, Volume 2: Combustion, Fuels, Materials, Design,* MIT Press, 1968, pp. 240-263

971994

The Effect of Cranktrain Design on Powertrain NVH

J. Querengaesser and J. Meyer
Ford-Werke AG

E. Schaefer and J. Wolschendorf
FEV Motorentechnik

Copyright 1997 Society of Automotive Engineers, Inc.

ABSTRACT

In the last few years the requirement to optimize powertrain noise and vibration has increased significantly. This was caused by the demand to fulfill the vehicle's exterior noise legislative limits in Europe, and by increased customer awareness for high ride comfort. Much effort concentrated on the engine and the powertrain as prime sources of noise and vibration in a vehicle.

The cranktrain with its moving components is a significant source of noise and vibration excitation within the engine. This paper describes results of investigations to evaluate various design alternatives in respect to NVH. The influences of crankshaft material, of balancing rate and of secondary shaking forces are discussed, with the aim to evaluate these various design options.

INTRODUCTION

Over the last few years the customer accepted noise level decreased while the comfort expectations increased. In order to fulfill these requirements efforts are continuing to improve the powertrain NVH characteristics and the vehicle noise transfer paths. Regarding powertrain NVH improvements, a literature review of the past years shows that a lot of effort was focussed on the bottom end design of powertrains /1,2,3,4,5,6/. These papers deal with bedplate, ladderframe, bearing beam etc., as bottom end stiffening devices for the engine block and the powertrain. Regarding cranktrain design, e.g.

crankshaft material and balancing rate, only very few information is available /7,8,9/. The authors of /7/ describe a comparison of crankshaft materials, and state that for a real comparison the design of the forged steel crankshaft and cast iron crankshaft should be identical. An additional parameter in cranktrain design is the effect of secondary shaking forces. On an in-line 4-cylinder engine, a well-established technology is the Lancaster drive with balancing shafts to eliminate the effect of 2nd order mass forces as they act on the powertrain. But still, the mass forces of each cylinder are acting on the individual bearings. To reduce this noise excitation, the conrod counter mass can be increased /10/. Only some information is available about the effect of reduced secondary shaking forces by mass reduction on NVH.

The aim of this paper is to provide additional information regarding the effect of cranktrain design on Noise, Vibration and Harshness. The monitored data includes radiated noise of the engine and the acceleration of the powertrain at the isolators. The measured parameters should give indications for the resulting vehicle interior noise.

TESTBED SET UP

The noise radiated from the surfaces of the engine is normally measured by installing the engine (or powertrain) in an anechoic test cell. However, when evaluating the vibration excitation of the vehicle body through the engine rubber mounts, only the installation of the full powertrain (i.e. engine with transmission) gives valuable results. The simulation of a vehicle installation is normally not critical in case of a rear-wheel drive, where only one driveshaft exits the transmission. The driveshaft is in that case connected directly to the power absorbing unit, e.g. an eddy-current brake. In case of a front-wheel driven vehicle, either both driveshafts or only one (with locked differential) can be used. In order to simulate the vehicle installation, investigations show that the power take-off with both driveshafts seem to be necessary.

Figure 1 shows the comparison in vibration level at the powertrain mount for both versions, i.e. one vs. two driveshafts. For both versions, the identical in-line 4-cylinder powertrain was used. The vibration level of the 2nd engine order shows no significant difference; also, 4th and 4.5^{th} engine orders are almost identical for both versions. However, the vibration levels for all other engine orders are distinctly different, leading to generally higher vibration levels if only one driveshaft is used. Therefore, it was concluded that an installation with a power take-off via two driveshafts was required for these investigations in order to properly simulate the vehicle installation.

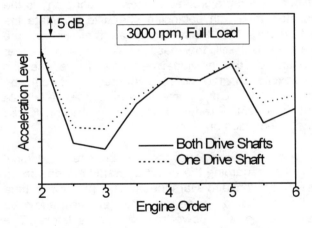

Figure 1: Vibration levels at the powertrain mounts comparing the effect of power take-off with one sideshaft versus two sideshafts

Under these boundary conditions, a powertrain testbed set-up was designed which simulates powertrain installation in a front-wheel driven vehicle even under the constraints of an anechoic test cell. Still, only one power absorption unit is required (see Figure 2). Via a common (bottom-mounted) transmission shaft, both driveshafts are connected, and drive a single shaft which is connected to an externally installed power absorption unit. By using encapsulated V-belts for the power transfer between the driveshafts, no noticeable noise disturbance has been identified with this arrangement.

With this powertrain installation in an anechoic test cell, it is possible to measure radiated powertrain noise and powertrain mount vibrations simultaneously, without the need to use two different test cells for such investigations.

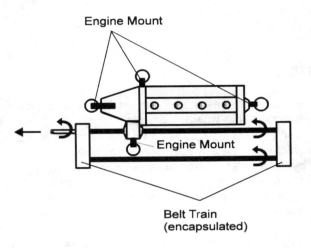

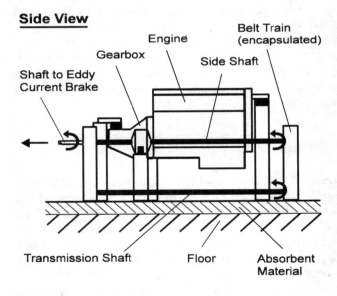

Figure 2: Schematic layout of the front-wheel drive powertrain NVH test rig

The noise radiated from the powertrain was measured by means of sound pressure around the engine. The vibration excitation of the vehicle body was evaluated by means of acceleration measurements on each of the mount locations. Aim was to use these measurements as a basis for further analysis of potential pass-by noise and vehicle interior noise improvements.

The powertrain used for these investigations is an in-line 4-cylinder gasoline engine (4-valve version) with manual transmission.

MEASURED PARAMETERS

To determine the emitted noise of the engine the sound pressure was measured. Aim was to evaluate the exterior pass-by noise and the vehicle's air-borne noise transferred through the firewall into the passenger compartment. Figure 3 shows schematically the location of the microphones in the anechoic test cell.

For the evaluation of the engine radiated noise, the transmission was encapsulated (mineral wool and lead) to suppress as much as possible any disturbing noise by the transmission. The microphone positions are based on pure engine measurement box definitions.

For structure-borne noise evaluation, the overall vibration level was defined as the sum of the acceleration in all three directions in the frequency range of 200Hz to 800Hz third-octaves, without the second engine order. The reason for the extraction of the second engine order is that the second engine order would dominate the overall level and could mask other effects.

The passenger compartment noise is a result of the powertrain radiated noise and the structure-borne noise of the powertrain combined with the transfer path of the vehicle. The structure-borne noise analysis focussed primarily on the frequency range 200 to 800 Hz, as this frequency range determines the sound quality characteristics. The objective was to evaluate and quantify the effect of design modifications on the vehicle interior noise, particularly on the psycho acoustic parameter of 'rumble'. Typically, 'rumble' is a noise phenomena that appears during acceleration in the speed range of 3600 to 5000rpm engine speed /5/. It has been reported that 'rumble' can be caused by crankshaft bending /5,11/ and flywheel tilting /12/, and is mainly caused by the 4.5^{th}, 5th and 5.5^{th} engine orders /12/. Earlier publications indicate that the engine orders identified in the radiated noise are not necessarily identical with the exciting crankshaft orders /14,15/. An engine order shift of ±1 can occur.

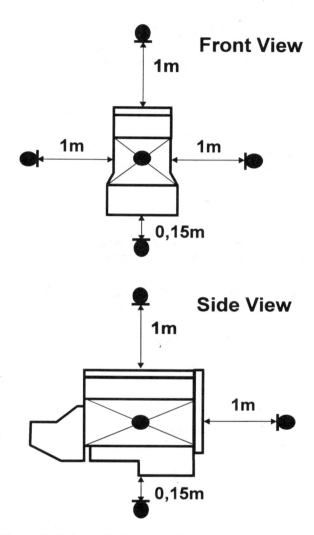

Figure 3: Schematic location of the microphones used for the investigations

INFLUENCE OF CRANKSHAFT MATERIAL

In most passenger car gasoline engines a cast iron crankshaft is used, primarily because of cost reasons. These engines do not require the high bending stiffness of the crankshaft which is necessary with highly loaded diesel engines.

Regarding NVH, a forged steel crankshaft could offer some advantages even on gasoline engines. In order to quantify the influence on a modern gasoline engine, two crankshafts of similar geometric dimensions were manufactured, one a cast iron and the other a forged steel crankshaft. By keeping the dimensions similar for both versions, the comparison focussed on the effect of the material rather than on the effects due to design differences. However, due to the higher weight of the steel crankshaft the counterweights were reduced to give identical balancing rates

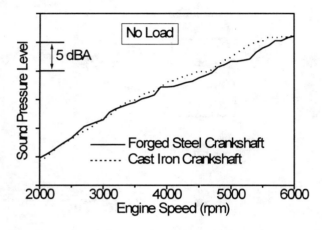

Figure 4: Influence of Crankshaft Material on Engine Radiated Noise (averaged over five microphones) under No Load Condition

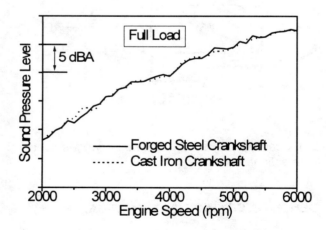

Figure 5: Influence of Crankshaft Material on Engine Radiated Noise (averaged over five microphones) under Full Load Condition

Figure 4 shows the overall sound pressure level versus engine speed under no load condition. Noticeable is that the steel crankshaft has advantages compared to the cast iron crankshaft in the medium and higher speed range by up to 2 dBA.

Figure 5 documents the overall sound pressure level versus engine speed under full load condition. The steel crankshaft has almost no benefits under this condition.

To understand this behavior the frequency characteristics were evaluated. As an example, the 'characteristic third-octave spectra' are shown in Figure 6 for the engine noise under no load condition. (The 'characteristic noise spectrum' is determined by the average of the normalized frequency spectra for all engine speeds multiplied with the average noise level). The advantages of the forged steel crankshaft can be identified over almost the whole frequency range.

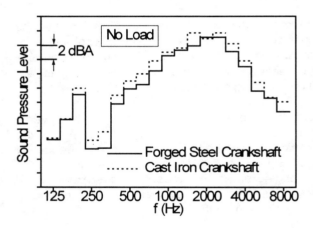

Figure 6: Influence of Crankshaft Material on 'Characteristic Third-Octave Spectra' under No Load Condition.

The 'characteristic third-octave spectra' for the engine under full load condition are shown in Figure 7. In contrast to the no load condition, no significant differences between the two crankshafts can be identified. The frequency content of the forged steel crankshaft is similar to the cast iron crankshaft over nearly the whole frequency range. Only at very low frequencies (below 320 Hz third-octave) the forged steel crankshaft exhibits some advantages. One reason why no advantage under full load condition can be observed is probably the fact that the high combustion noise level on this engine masks some of the improvement of the forged steel crankshaft.

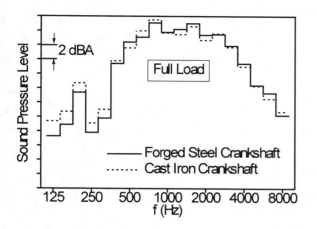

Figure 7: Influence of Crankshaft Material on 'Characteristic Third-Octave Spectra' under Full Load Condition.

In addition to the third-octave analysis, an order analysis was performed. The order analysis concentrated on the no load condition as this condition showed the largest noise differences. The order analysis of the 6th engine order is shown in Figure 8. In most of the speed ranges, the cast iron crankshaft causes a higher noise level than the forged steel crankshaft. The largest differences occur at about 4000 rpm (first torsional mode of the crankshaft) and 4700 rpm engine speed.

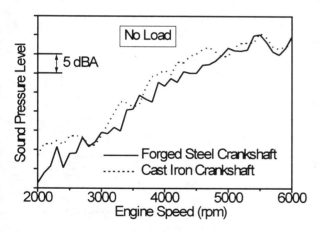

Figure 8: Influence of Crankshaft Material on 6th Engine Order Noise Level.

Figure 9 shows the 7th engine order noise level versus engine speed. Detectable is that the 7th engine order of the cast iron crankshaft is higher than of the forged steel crankshaft in the speed range of 5000 to 6000rpm. In the corresponding frequency range higher order bending vibration modes were identified for the two crankshafts.

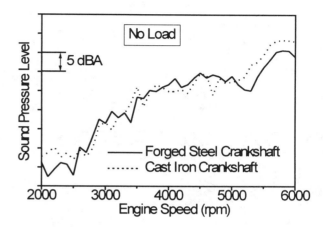

Figure 9: Influence of Crankshaft Material on 7th Engine Order Noise Level.

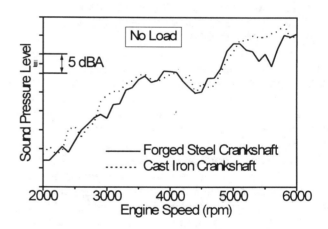

Figure 10: Influence of Crankshaft Material on 8th Engine Order Noise Level.

For the noise level of the 8th engine order (see Figure 10), the forged steel crankshaft shows lower levels than the cast iron crankshaft in the speed range of about 3000 rpm, and between 5100 to 5900 rpm. At 3000 rpm, the improvement is related to the 400 Hz band, which is the region of the first crankshaft torsional mode. The higher speeds correspond to the frequency band of the 2nd torsional vibration mode.

To complete the NVH evaluation, the vibration at the powertrain mounts was measured with the two crankshafts. Figure 11 documents the overall vibration level versus engine speed under full load condition.

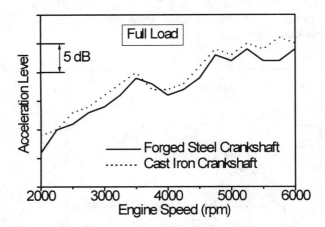

Figure 11: Influence of Crankshaft Material on Overall Vibration Level at a Powertrain Mount.

The forged steel crankshaft leads to slightly lower (about 2 dB) structure-borne noise excitation than the cast iron crankshaft, over the whole engine speed range, even under full load. Similar results were also reported earlier /8,13,14,16/, and were explained with the reduction of crankshaft vibrations.

In addition to the objective analysis, the data was also evaluated regarding influence on noise quality. This was based on psychoacoustic parameters as well as subjective assessments. Aim was to correlate these parameters with the human perception. The comparison showed that the curtosis and the articulation index correlated with the subjective impression. An evaluation of only one psychoacoustic parameter does not reflect the human perception. The subjective assessment showed that none of the crankshaft versions was regarded as optimum under all operating conditions. The advantages of either of the crankshafts were depending on the engine speed and load condition. This could indicate that the differences between the forged steel crankshaft and the cast iron crankshaft are relatively small on this engine. But overall, the forged steel crankshaft showed slightly more benefits regarding NVH than the cast iron crankshaft.

INFLUENCE OF CRANKSHAFT BALANCING RATE

The crankshaft balancing rate is regarded as a very important design feature of a crankshaft. Previous investigations indicate that a crankshaft with higher balancing rate improve the engine radiated noise behaviour /8,9/. Based on the definition of the balancing rate:

$$\text{Bal. Rate (\%)} = M_{rot,\ counterweight} / M_{rot,\ cranktrain},$$

a 100 % value represents a complete balancing of the rotating mass (assuming 8 counterweights on a 4-cylinder engine). General perception is that a balancing rate of at least 100 % is required for good NVH behaviour. However, this leads in many cases to unacceptable weight increase, or is even not feasible due to restrictions by the engine block. Therefore, most modern engines nowadays have a balancing rate of less than 100 %.

In order to quantify the influence of the balancing rate on the engine NVH behaviour, two cast iron crankshafts were manufactured with different balancing rates. Design modifications were restricted to the size of the counterweights, achieving balancing rates of 100 % and 82%, respectively. When testing the crankshafts in the fired engine it was ensured that other noise relevant parameters, as e.g. bearing clearances, remained identical.

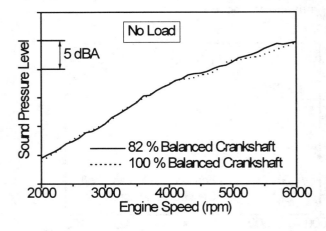

Figure 12: Influence of Crankshaft Balancing Rate on Radiated Engine Noise (averaged over five microphones) under No Load Condition

Figures 12 and 13 show the engine radiated noise, averaged over five microphones, under no load and full load conditions. In general, the radiated noise shows only very small differences.

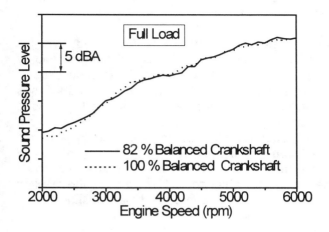

Figure 13 : Influence of Crankshaft Balancing Rate on Radiated Engine Noise (averaged over five microphones) under Full Load Condition

However, when examining the noise radiation to the bottom side of the engine (Figures 14 and 15), advantages for the crankshaft with higher balancing rate exist (up to 2.5 dBA). The improvement with the 100% balancing rate is more substantial for the engine full load condition than for the engine no load condition.

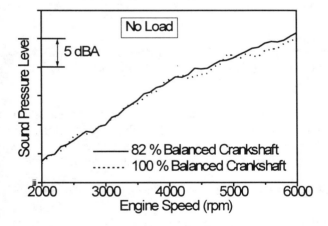

Figure 14: Influence of Crankshaft Balancing Rate on Radiated Engine Noise to the Bottom Side under No Load Condition

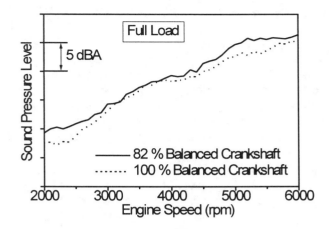

Figure 15: Influence of Crankshaft Balancing Rate on Radiated Engine Noise to the Bottom Side under Full Load Condition

Figure 16 shows a comparison of the powertrain mount vibrations for the two crankshaft balancing rates. No clear benefit for either of the two crankshafts can be identified. The differences are small and there is no clear trend. One reason why the structure-borne noise excitation shows no significant differences is that the total engine forces are similar for both versions. There is no change in exterior resulting force (only interior). Considering these results, the effect on the vehicle interior noise would be minor. This correlates with the findings in /9/.

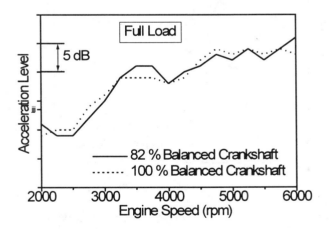

Figure 16: Influence of Crankshaft Balancing Rate on Overall Vibration Level at a Powertrain Mount under Full Load Condition

INFLUENCE OF SECONDARY SHAKING FORCES

For an in-line 4-cylinder engine the 2nd order mass forces are the main excitation forces from the cranktrain. They can partly be compensated by the crankshaft counterweights, or by balancing shafts rotating with twice engine speed /18/. Rather than using secondary measures, a reduction of these forces by a reduction of the moving masses is more effective.

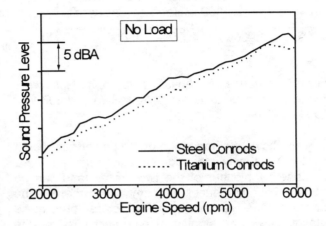

Figure 17: Influence of Conrod Weight on Radiated Engine Noise (averaged over five microphones) under No Load Condition

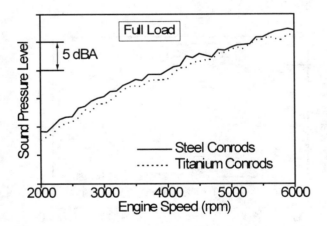

Figure 18: Influence of Conrod Weight on Radiated Engine Noise (averaged over five microphones) under Full Load Condition

Within these investigations, the conrod weight was reduced by 30 % using titanium instead of sintered steel. The piston, piston pin and piston rings remained identical. The modification influenced both, the rotating as well as oscillating mass. But as the crankshaft remained identical for both versions, the reduced conrod mass lead to a higher balancing rate.

Sound pressure measurements (averaged over all engine sides) reveal that the reduction in secondary shaking forces lead to significant noise reductions in the entire engine speed range by 1-1.5 dBA, see Figures 17 and 18. The improvement under no load condition seems to be higher than under full load. Further frequency analysis showed that primarily the higher frequency content (above 2 kHz) is influenced by this modification. Therefore, the improved NVH behaviour of the lighter conrods is probably the result of reduced impacts in the conrod and main bearings.

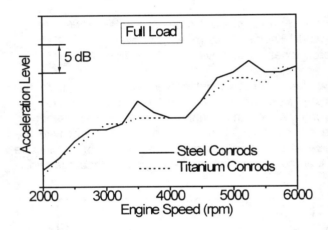

Figure 19: Influence of Conrod Weight on Overall Vibration Level at a Powertrain Mount under Full Load Condition

Figure 19 shows the comparison of the overall vibration level at the powertrain mount over the engine speed for the two different conrods. The lighter conrod has some slight advantages in the overall vibration level.

Also, the 2nd order vibration at the engine mounts can be significantly improved, see Figure 20. As expected, the vibration reduces by about 20-30 % with the lower weight conrod. Although this reduction is still far from what can be achieved with an additional mass balancing drive, it does not have the disadvantages of complexity.

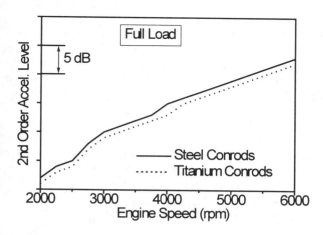

Figure 20: Influence of Conrod Weight on 2nd Engine Order Powertrain Mount Vibration under Full Load Condition

CONCLUSION

Various cranktrain design modifications were investigated in respect to their influence on engine/powertrain NVH. The effect of crankshaft material, crankshaft balancing rate and secondary shaking forces were evaluated on an in-line 4-cylinder gasoline engine.

A comparison of the crankshaft material (cast iron vs. forged steel) revealed differences in engine noise only under no load condition. Noise benefits of up to 2 dBA could be identified with the forged steel crankshaft compared to a geometrically similar cast iron component. Under full load, both versions exhibited the same noise behaviour. The vibration excitation on the powertrain mounts showed also a improvements over the whole engine speed range for the forged steel crankshaft. But these could be also identified under full load condition. Most of the improvements with the forged crankshaft could be related to the crankshaft bending and torsional vibration frequencies, where the cast iron version has disadvantages due to the lower stiffness.

The investigations into the influence of the balancing rate on NVH revealed surprisingly low differences. In the overall engine noise the effect is very small, both under no load and full load. However, the noise radiation to the bottom side improved with the higher balancing rate, particularly under full load (up to 2.5 dBA). The vibration level did not show a clear benefit.

Based on these results, it seems that the necessity of a high crankshaft balancing rate depends on the individual situation. The noise advantage to the bottom side need to be compared with the higher mass or / and increased crankshaft web dimensions.

A reduction in secondary shaking forces (e.g. through reduced conrod or piston weights) should be a primary design goal, also in respect to NVH. For the case of a 30% lighter conrod, the engine noise reduced by 1-1.5 dBA. At the same time, the vibration excitation at the powertrain mounts was decreased by 20-30 %.

Noise quality evaluations using subjective assessments and psychoacoustic parameters during the crankshaft comparison revealed correlations for the curtosis and the articulation index. But no single psychoacoustic parameter can reflect the complex subjective human perception. No crankshaft version showed advantages for all noise examples. But overall, the forged steel crankshaft showed slightly more benefits in noise quality than the cast iron crankshaft.

REFERENCES

/1/ Vorwerk C., Busch G., Kaiser H.-J., Wilhelm M.: Influence of bottom end design on noise and vibration behavior of 4-cylinder in-line gasoline engines, SAE 931315

/2/ Busch G., Maurell R., Meyer J., Vorwerk C.: Investigations on influence of engine block design features on noise and vibration, SAE 911071

/3/ Willenbockel O., Indra F.: Der neue Opel 3,0 Liter 24-Ventil-Motor; 2. Aachener Kolloquium Fahrzeug- und Motorentechnik, 1989

/4/ Basshuysen R., Stock D., Bauder R.: Entwicklung eines DI-Turbo-Diesel-Motors; 2. Aachener Kolloquium Fahrzeug- und Motorentechnik, 1989

/5/ Teramoto T., Deguchi H., Shintani H.: Improvement in Vehicle Interior Sound Quality by Newly Power Plant Members IMechE C420/027, 1990

/6/ Ochiai K., Hickling R., Kamal M.: Dynamic Behaviour of Engine Structure Vibrations, Int. Symposium 'Engine Noise - Excitation, Vibration and Radiation'

/7/ Albright M.F., Staffeld D.F.: Noise and Vibration Refinement of the FORD 3.8L Powertrain, SAE 911073

/8/ Eichhorn U., Gaßmann S., Schönefeld U.: Einfluß des Massenausgleichs der Kurbelwelle eines Ottomotors (Influence of Crankshaft Balancing Rate for a Gasoline Engine), MTZ 50 (1989) 10

/9/ Schwärzel W., Bartsch G.: Die neuen 4-Zylinder-OHC-Motoren von Opel mit 1,6 Litern Hubraum (The new 1.6L Opel 4-cylinder OHC engines), MTZ 42 (1981) 9

/10/ Krüger H.: Massenausgleich durch Pleuelgegenmassen (Mass Balancing by Conrod Counterweights), MTZ 53 (1992) 2

/11/ Nakada T., Tonosaki H., Yamashita H.: Excitation Mechanism of Half Order Engine Vibrations JSAE Vol. 49 No.6, 1995 (59-64)

/12/ Aoki H., Ishihama M., Kinoshita A.: Effect of Power Plant Vibration on Sound Quality in the Passenger Compartment During Acceleration, SAE 870955

/13/ Fujita T., Onogawa K., Kawai H., Ymazaki T.: Development of new approach to noise reduction through investigation of crankshaft behaviour, IMechE C389/237, 1992

/14/ Schmillen K.: Geräuschanalyse und Rechentechniken im Motorenbau II (Noise Analysis and Simulation Techniques II), Vorlesungsumdruck RWTH Aachen 1993

/15/ Katano H., Iwatono A., Saito T.: Dynamic Behavior Analysis and Rumbling Noise of SI Engine Crankshaft, JSAE 46(6) 1992

/16/ Nagayama I., Araki Y., Kakuta K., Usuba Y.: Engine Noise Reduction by Structural Study of Cylinder Block, SAE 800441

/17/ N.N.: BAS Handbook HEAD acoustic 12/95Rev3

/18/ Huegen S., Warren G., Menne R., Wolschendorf J., Schwaderlapp M., Schoenherr C.: A New 2.3L DOHC Engine with Balance Shaft Housing - Steps of Refinement and Optimization, SAE 970921

971995

Experiments and Analysis of Crankshaft Three-Dimensional Vibrations and Bending Stresses in a V-Type Ten-Cylinder Engine: Influence of Crankshaft Gyroscopic Motions

Jouji Kimura and Kazuhiro Shiono
Isuzu Motors Ltd.

Hideo Okamura and Kiyoshi Sogabe
Sophia Univ.

Copyright 1997 Society of Automotive Engineers, Inc.

ABSTRACT

Torsional dampers have been attached to engine crankshafts only for the control of the crankshaft torsional vibrations. However, a torsional damper is a mass-spring system of three-dimensions, so the torsional damper could exert some influence on the three-dimensional vibrations of the crankshaft system. Since the inertia ring of the torsional damper has moments of inertia and it rotates with the crankshaft, gyroscopic vibrations of the inertia ring can also be generated.

For a V-type ten-cylinder diesel engine (V-10, $\phi 119 \times 150$), the three-dimensional vibrations of the crankshaft system were calculated by the dynamic stiffness matrix method, taking account of the influence of the gyroscopic vibrations of the inertia ring of the torsional damper. The dynamic bending stresses were measured at the fillets of both the No.1 crank journal and the No.1 crank pin in the No.1 crank throw plane.

From the comparison of the experimental and calculated results, we investigated the influence of the three-dimensional mass-spring system of the torsional damper and the influence of gyroscopic vibrations of the torsional damper inertia ring on the bending stresses of the crankshaft.

INTRODUCTION

To reduce the weight of diesel engines used for trucks and buses, the weight of the crankshaft must be reduced. Recently, the turbocharged diesel engines with intercooler are widely used, therefore the BMEP

(brake mean effective pressure) in these engines is increasing year by year. On the other hand, since (1) some pulleys with relatively large size must be attached to the crankshaft front end to drive many engine accessories and (2) sometimes a clutch damper must be attached to the rear end, significant weights must be attached at the two ends of the crankshaft. Accordingly, significant excitation forces can be induced by the complicated vibrations of the crankshaft system.

The authors found that unexpectedly serious bending stresses can be caused in the crankshaft by two kinds of bending vibrations : one is due to the coupling with the torsional vibrations [1] [2], the other is caused by the gyroscopic vibrations by the front pulley and the rear flywheel. [3]~[5]

In this paper, a V-type 10-cylinder diesel engine was used for the experiments and analyses of the three-dimensional vibrations of the crankshaft system. We calculated the vibrations by the dynamic stiffness matrix method, taking account of the influence of the gyroscopic vibrations.

The dynamic bending stresses were also measured at the fillets of both the No.1 crank journal and the No.1 crank pin in the No.1 crank throw plane under the operating conditions.

From the comparison of the experimental and calculated results, we investigated (1) the generation mechanism of the bending stresses of the crankshaft caused by (a) the mass-spring system of the torsional damper and (b) the gyroscopic vibrations of the inertia ring of the torsional damper, respectively, and (2) the vibration behavior of the crankshaft front end due to the gyroscopic vibrations of the forward whirl and the reverse whirl.

COMPUTATION OF THREE-DIMENSIONAL VIBRATIONS FOR CRANKSHAFT SYSTEM

MODELING OF CRANKSHAFT SYSTEM [6]~[8]–We used a V-type 10-cylinder natural aspirated diesel engine for this study for the analyses and experiments. The engine specifications are shown in Table 1. Figure 1 shows the coordinate systems and the idealized models for the crankshaft system of the test engine. For a series of calculations, the locations of the joints at the No.1 crank throw are shown in Figure 2.

The outline of the idealizations for the crankshaft system is as follows.

(1) Crankshaft body: For each crank throw, the crank journals and crank pins were idealized as rods with circular cross-sections, and the crank arms were idealized as beams with a rectangular cross-section. The dynamic stiffness matrix for each beam element was derived from Euler's equations for a beam. The dynamic stiffness matrix for the total crankshaft system was constructed by stepwise superposition of the dynamic stiffness matrices of each member element.

(2) Front pulley, balancing weight, flywheel and clutch: They were idealized as rigid bodies. A set of masses and moments of inertia for each element was attached at the corresponding joint in the crankshaft system.

(3) Piston and connecting-rod: They were resolved into equivalent reciprocating masses and rotating masses. Then the equivalent rotating masses [9] were connected at the two joints of each crank pin.

(4) Main bearing: Each of the main bearings was idealized by an isotropic oil film supported by infinitely rigid bulk-heads and bearing caps. Each crank journal was supported by three sets of oil film that were idealized by a set of linear springs and dash-pots. The spring constants and damping coefficients were determined from Sommerfeld numbers under operating conditions.[10]

(5) Torsional damper: A viscous rubber damper was attached to the crankshaft. Figure 3 shows the rough sketch of the torsional damper. The inertia ring of the torsional damper was idealized by masses in three directions (X, Y, Z), and moments of inertia about the three axes (θ_x, θ_y, θ_z). The silicone oil and rubber were idealized by a set of linear springs and dash-pots in the six directions (X, Y, Z, θ_x, θ_y, θ_z). For the torsional damper, the masses and the moments of inertia of the inertia ring, and the three-dimensional stiffness are shown in Table 2 and Table 3, respectively.

Table 1 Engine specifications

Engine form	V-90°, 10-cylinder
Cycle	4 cycle
Cylinder bore × Stroke	$\phi 119 \times 150\ mm$
Displacement	$16.683\ \ell$
Firing order	1-8-7-6-5-4-3-10-9-2

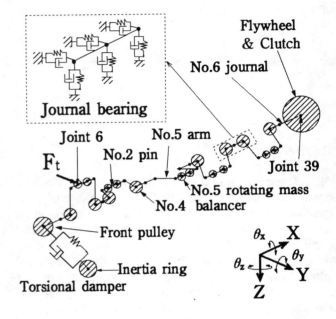

Fig.1 Idealized models of the crankshaft system of V-type 10-cylinder engine

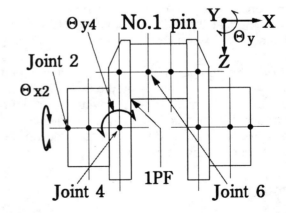

Fig.2 Locations of Joints in the No.1 crank throw

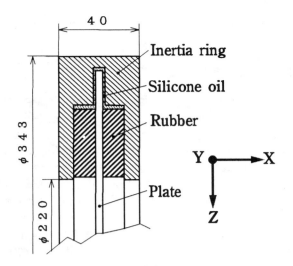

Fig.3 Rough sketch of the torsional damper attached to the engine

Table 2 Mass and moments of inertia of the inertia ring of the torsional damper

Mass : m (kg)	9.70
Moment of inertia about X axis : J_x $(kg \cdot m^2)$	0.22
Moments of inertia about Y, Z axes : J_y, J_z $(kg \cdot m^2)$	0.11

Table 3 Stiffness of the torsional damper

Stiffness in the X direction : K_x (N/m)	1.99×10^8
Stiffness in the Y, Z directions : K_y, K_z (N/m)	9.02×10^6
Stiffness about the X axis : $K_{\theta x}$ (Nm/rad)	1.93×10^5
Stiffness about the Y, Z axes : $K_{\theta y}, K_{\theta z}$ (Nm/rad)	1.79×10^4

EXCITATION FORCE AND DISPLACEMENTS AT EACH JOINT–The displacements in the six directions at each joint can be derived by solving Equation (1) [2] ; here the displacements in the six directions at the joint i can be expressed in the form as shown in Equation (2):

$$\{F(\omega)\} = [K(\omega)]\{A(\omega)\} \quad (1)$$

$$\{A(\omega)\}_i = \{U, V, W, \Theta x, \Theta y, \Theta z\}_i^T \quad (2)$$

where

$[K(\omega)]$: the dynamic stiffness matrix of total crankshaft system at frequency ω

$\{A(\omega)\}$: the vector for the amplitude of the displacements at frequency ω

$\{F(\omega)\}$: the vector for the amplitude of the forces at frequency ω

U, V, W : the amplitude of the displacements in the X, Y, Z axes directions

$\Theta x, \Theta y, \Theta z$: the amplitude of the angular displacements about the X, Y, Z axes

In the following analysis, the displacements in the six directions at each joint were computed for the case that when a sinusoidal unit force Ft of angular frequency ω was applied only in the Y axis direction only at joint 6 of the No.1 crank pin (see Figures 1 and 2). So $\{A(\omega)\}_i$ can be as the compliance vector, when the above unit force was exerted at joint 6, only in the Y direction with an angular frequency ω.

Since Θ_{y4} (the angular displacements about the Y axis at joint 4) could become the representative characteristic values for the stresses at the front fillet of the No.1 crank pin (1PF), we calculated the Θ_{y4} values. [2] Since Θ_{x2} (the torsional vibrations about the X axis at joint 2) were the values of the ordinary torsional vibrations, we also calculated the Θ_{x2} values. (see Figure 2)

GYROSCOPIC MOTIONS–In Figure 4, the three-dimensional motions of a rigid body are shown for the motions in the six directions (X, Y, Z, θx, θy, θz). The equations of motions can be expressed as shown in Equations (3) and (4). Here Equations (4) are Euler's equations of motion, and the second terms on the right side of Equations (4) show the gyroscopic effects.

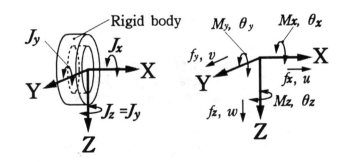

Fig.4 Three-dimensional motions of rigid body

$$\left. \begin{array}{l} f_x = m \cdot \ddot{u} \\ f_y = m \cdot \ddot{v} \\ f_z = m \cdot \ddot{w} \end{array} \right\} \quad (3)$$

$$Mx = Jx \cdot \ddot{\theta}x - \left(Jy - Jz\right)\dot{\theta}y \cdot \dot{\theta}z$$
$$My = Jy \cdot \ddot{\theta}y - \left(Jz - Jx\right)\dot{\theta}z \cdot \dot{\theta}x \qquad (4)$$
$$Mz = Jz \cdot \ddot{\theta}z - \left(Jx - Jy\right)\dot{\theta}x \cdot \dot{\theta}y$$

where:

fx, fy, fz : the applied forces in the X, Y, Z axes directions

Mx, My, Mz : the applied moments about the X, Y, Z axes

Jx, Jy, Jz : the moments of inertia about the X, Y, Z axes

$\theta x, \theta y, \theta z$: the angular displacements about the X, Y, Z axes

u, v, w : the displacements in the X, Y, Z axes directions

m : the mass

If the excitation forces were harmonic forces of angular frequency ω, the motions were steady state vibrations and $\dot{\theta}x$ (the rotating speed of the crankshaft) in the second terms in Equation (4) was a constant value: Ω ($\dot{\theta}x = \Omega$). Then Equations (3) and (4) can be expressed as follows:

$$Fx = -m\omega^2 U$$
$$Fy = -m\omega^2 V$$
$$Fz = -m\omega^2 W$$
$$Mx = -Jx\omega^2 \Theta x \qquad (5)$$
$$My = -Jy\omega^2 \Theta y - j\left(Jz - Jx\right)\omega\Omega\Theta z$$
$$Mz = -Jz\omega^2 \Theta z - j\left(Jx - Jy\right)\omega\Omega\Theta y$$

where:

Fx, Fy, Fz : the amplitude of the applied forces in the X, Y, Z axes directions

Mx, My, Mz : the amplitude of the applied moments about the X, Y, Z axes

$\Theta x, \Theta y, \Theta z$: the amplitude of the angular displacements about the X, Y, Z axes

U, V, W : the amplitude of the displacements in the X, Y, Z axes directions

ω : the angular frequency of the excitation force

Ω : the angular velocity of the crankshaft

$j = \sqrt{-1}$

$Jy = Jz$

We applied Equations (5) to the inertia ring of the torsional damper, and we calculated the displacements at each joint by the dynamic stiffness matrix method.

CALCULATED RESULTS

CALCULATED RESULTS FOR THE CASES WITHOUT THE GYROSCOPIC EFFECTS–To investigate the gyroscopic effects due to the inertia ring of the torsional damper, first, we calculated $\Theta x2$ (the torsional vibrations at joint 2) and $\Theta y4$ (the angular displacements at joint 4) when the gyroscopic effects (the second terms on the right side of Equations (4)) were neglected. Figures 5 and 6 show the values of $\Theta x2$ and $\Theta y4$ in the cases when the gyroscopic effects were neglected at the engine speed of 1200 rpm.

Since the resonant frequency $f_{\Theta x2 - 1}$ of the $\Theta x2$ which appears in Figure 5 is the same as the resonant frequency $f_{\Theta y4 - 2}$ of the $\Theta y4$ which appears in Figure 6, one can see that the bending vibration is induced along with the torsional vibration [2] : namely, the two vibrations are induced in a coupled way.

In Figure 6, one can see the two other resonant frequencies: $f_{\Theta y4 - 1}$ and $f_{\Theta y4 - 3}$. To investigate the generation mechanism for the two resonant frequencies $f_{\Theta y4 - 1}$ and $f_{\Theta y4 - 3}$, we calculated the $\Theta y4$ for the different values of the stiffness of the torsional damper about the Y axis, and those in the Y direction: $K\theta y$ and Ky.

The values of stiffness shown in Table 3 are denoted as original stiffness . The original stiffness in the Y axis direction is symbolized by Kyo and the original stiffness about the Y axis is symbolized by $K\theta yo$. In Figure 7, the resonant frequencies of $f_{\Theta y4 - 1}$ and $f_{\Theta y4 - 3}$ are shown for the ratio $K\theta y / K\theta yo$. Here we assumed that $Ky = const.$, and $Ky = Kyo$, for the different values of $K\theta y$ at an engine speed of 1200 rpm. In Figure 8, the resonant frequencies of $f_{\Theta y4 - 1}$ and $f_{\Theta y4 - 3}$ are shown in a similar way for the constant values of $K\theta y$.

From Figures 7 and 8, one can see that the resonant frequency $f_{\Theta y4 - 1}$ depends on the $K\theta y$ rather than the Ky, whereas the resonant frequency $f_{\Theta y4 - 3}$ depends on the Ky rather than the $K\theta y$.

Thus, one can see that the resonant frequency $f_{\Theta y4 - 1}$ in Figure 6 is caused by the mass–spring system about the Y axis of the torsional damper, and the resonant frequency $f_{\Theta y4 - 3}$ in Figure 6 is caused by the mass–spring system in the Y direction of the torsional damper, respectively.

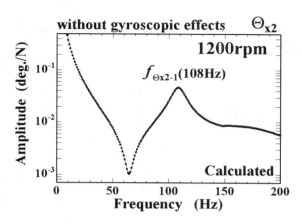

Fig.5 Torsional vibrations about the X axis at Joint 2 Θ_{x2} without gyroscopic effects

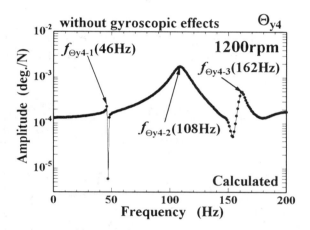

Fig.6 Angular displacements about the Y axis at Joint 4 Θ_{y4} without gyroscopic effects

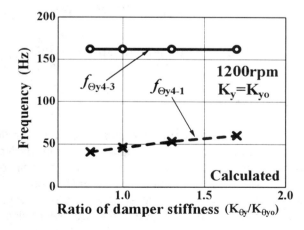

Fig.7 Influence of the stiffness of the torsional damper $K_{\theta y}$ on the $f_{\Theta y4-1}$ and $f_{\Theta y4-3}$

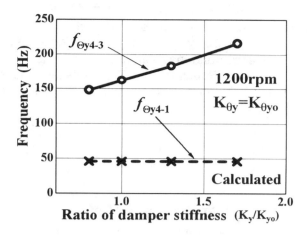

Fig.8 Influence of the stiffness of the torsional damper K_y on the $f_{\Theta y4-1}$ and $f_{\Theta y4-3}$

CALCULATED RESULTS FOR THE CASES WITH THE GYROSCOPIC EFFECTS—Figure 9 shows the calculated values of Θ_{y4} at the engine speed of 1200 rpm, for which the gyroscopic effects were taken into account. Figure 10 shows the zooming plot in the frequency range between the 43Hz and 50Hz in Figure 9, with every 0.1Hz step. From Figures 6, 9 and 10, one can see that the resonant frequency $f_{\Theta y4-1}$ in Figure 6 is separated into the two resonant frequencies $f_{\Theta y4-11}$ and $f_{\Theta y4-12}$.

To investigate the influence of the engine speed on the resonant frequencies $f_{\Theta y4-11}$ and $f_{\Theta y4-12}$ in Figure 10, we calculated the Θ_{y4} for different values of engine speeds. We applied the Nyquist plot for the compliance of the Θ_{y4} to evaluate the resonant frequencies $f_{\Theta y4-11}$ and $f_{\Theta y4-12}$. An example of the Nyquist plot of the compliance Θ_{y4} values is shown in Figure 11 for the engine speed of 1200 rpm. In Figure 11, one resonant frequency $f_{\Theta y4-11}$ is given by the intersection of the circle C1 and the line from the origin 0 to the center of the circle P1, and the other resonant frequency $f_{\Theta y4-12}$ is given by the intersection of the circle C2 and the line from the origin 0 to the center of the circle P2.

Figure 12 shows the resonant frequencies $f_{\Theta y4-11}$ and $f_{\Theta y4-12}$ for the different values of the engine speeds. From Figure 12, one can see that the resonant frequencies depend on the engine speeds. In Figure 12, the resonant frequencies $f_{\Theta y4-12}$ increase with the engine speeds and one calls this the forward whirl phenomenon; the other resonant frequencies

$f_{\Theta y4-11}$ decrease with the engine speeds and the phenomenon is called the reverse whirl.

The relationships among the engine speed Ne, the frequency f and the harmonic order number n can be shown by Equation (6). Their relationships are plotted with the dotted lines for the 1st and 1.5th order in Figure 12.

$$f = \frac{Ne \cdot n}{60} \quad (6)$$

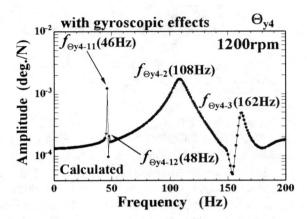

Fig.9 Angular displacements about the Y axis at Joint 4 Θ_{y4} with gyroscopic effects

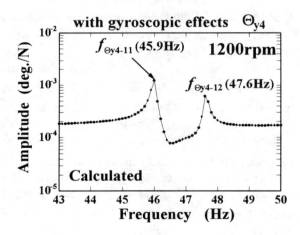

Fig.10 Angular displacements about the Y axis at Joint 4 Θ_{y4} with gyroscopic effects (zooming plot in the frequency range between 43Hz and 50Hz in Fig. 9, with every 0.1Hz step)

From Figure 12, we can predict that the resonance of the reverse whirl would occur for the 1.5th order at the engine speed of 1820 rpm, the resonance of the forward whirl would occur for the 1.5th order at the engine speed of 1920 rpm and the resonance of the reverse whirl would occur for the 1st order at the engine speed of 2700 rpm.

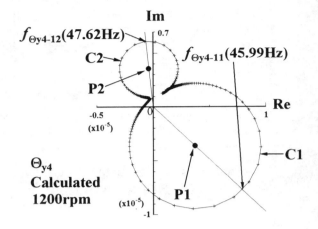

Fig.11 Nyquist plot of the angular displacements about the Y axis at Joint 4 Θ_{y4} at 1200 rpm

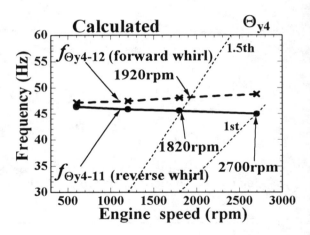

Fig.12 Influence of the engine speeds on the resonant frequencies $f_{\Theta y4-11}$ and $f_{\Theta y4-12}$

EXPERIMENTS

In the experiments, we measured the bending stresses caused by the bending vibrations in the No.1 crank throw plane (the X-Z plane) at the rear fillet of No.1 crank journal and the front fillet of No.1 crank pin, as shown in Figure 13. The stresses were measured by the strain gauges attached at the two fillet positions, denoted by the '1JR' and '1PF' in Figure 13.

During the experiments, the engine was operated under full load conditions for the engine speed from 1000rpm to 2700rpm, and the measurements mentioned above were performed under constant engine speed for every 20rpm step.

The measured strains at the 1JR (the rear fillet of the No.1 crank journal) and at the 1PF (the front fillet of the No.1 crank pin) are shown in Figures 14 and 15, respectively. In Figures 14 and 15, the harmonic order components of the most predominant strains in the measured ones are shown.

From Figures 14 and 15, we can see that the bending strains at the 1JR and 1PF are caused by the same excitation source in the crankshaft system. The predominant peaks of the 3.5th and 2.5th order strains in Figures 14 and 15 are caused by the bending vibrations coupled with the torsional vibrations.[2]

We suspected that the peaks of the 1st and 1.5th order strains in Figures 14 and 15 might be caused by the gyroscopic effects due to the damper inertia ring.

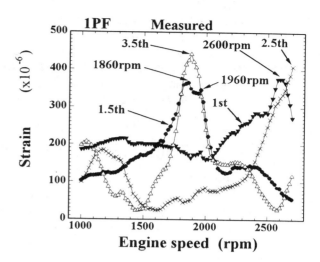

Fig.15 Measured strains at 1PF (the front fillet of the No.1 crank pin)

DISCUSSION

The above suspicion can be verified by comparing the calculated resonant frequency for the Θ_{y4} shown in Figure 12 with the resonant peaks of the bending strains shown in Figures 14 and 15.

One can see from Figure 12 that the calculated resonant engine speed for the 1st order of the reverse whirl is 2700 rpm, and the experimental results are 2620 rpm and 2600 rpm in Figures 14 and 15. Therefore the relative errors of +3% and +4% can be seen.

The calculated resonant engine speed for the 1.5th order of the reverse whirl is 1820 rpm in Figure 12. The experimental values are 1860 rpm and 1860 rpm in Figures 14 and 15. Therefore the relative errors of −2% and −2% are seen.

The calculated resonant engine speed for the 1.5th order of the forward whirl is 1920 rpm in Figure 12, and the experimental values are 1960 rpm and 1960 rpm in Figures 14 and 15. Therefore the relative errors of −2% and −2% are seen.

From the results mentioned above and in Section 3, we can see that the bending vibrations induced by the mass-spring system about the Y axis of the torsional damper are separated into the bending vibrations of the forward whirl and those of the reverse whirl by gyroscopic vibrations of the inertia ring of the torsional damper. Therefore one can conclude that the resonant peaks of the strains for the 1st and 1.5th order in Figures 14 and 15 are generated by the bending vibrations of the forward whirl and the reverse whirl.

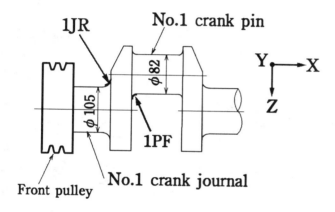

Fig.13 Locations of measuring positions for bending stresses at No.1 crank throw

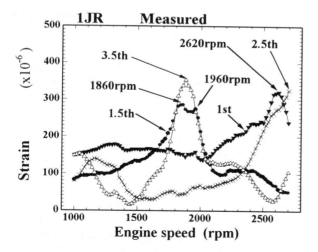

Fig.14 Measured strains at 1JR (the rear fillet of the No.1 crank journal)

LOCUS OF THE FRONT END CENTER OF CRANKSHAFT-We tried to calculate the locus of the front end center of the crankshaft at engine speeds of 1920 rpm (the resonant engine speed for the 1.5th order in the forward whirl mode) and of 1820 rpm (the resonant engine speed for the 1.5th order in the reverse whirl mode). The locus of the front end center of the crankshaft can indicate the displacements of joint 2 (see Figure 2) in the Y-Z plane.

Figure 16 shows the idealized models of the crankshaft system of the engine under operating conditions. For the calculation of the locus of the front end center of the crankshaft, the excitation forces under the full load operating conditions were applied to the Y and Z axis directions at each crank-pin.[2]

The locus in Figures 17 and 18 correspond to the ones at the front view of the engine under the operating conditions and the coordinate system in Figures 17 and 18 corresponds to it of Figure 16. The crankshaft rotations clockwise at the front view of the engine.

From Figures 17 and 18, one can see that, in the forward whirl mode, the center of the crankshaft front end shows an elliptical form in the direction of the crankshaft rotation, whereas it shows one in the opposite direction in the reverse whirl mode.

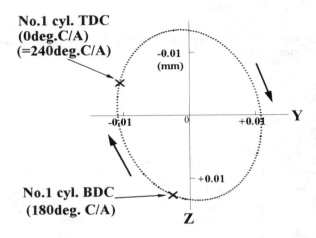

Fig.17 Locus of the front end of the crankshaft in the forward whirl at 1920 rpm

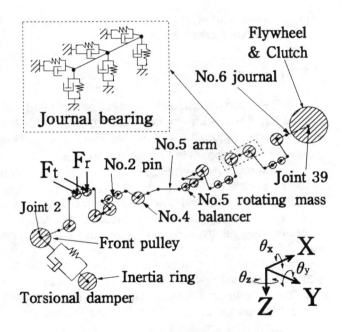

Fig.16 Idealized models of the crankshaft system of the engine under operating conditions

Figure 17 and Figure 18 show the locus of the front end center of the crankshaft at engine speeds of 1920 rpm (the resonant engine speed for the 1.5th order in the forward whirl mode) and of 1820 rpm (the resonant engine speed for the 1.5th order in the reverse whirl mode), respectively. In Figures 17 and 18, the displacements at the joint 2 in the Y-Z plane are shown for 240° crank-angle (one cycle for the 1.5th order vibration) from TDC (top dead center) of No.1 cylinder to 240° of the crank-angle with every 2° crank-angle step. In Figures 17 and 18, the locus are shown for the two cases when the crankshaft resonance vibrations were induced by the forward whirl mode and by the reverse whirl mode, respectively.

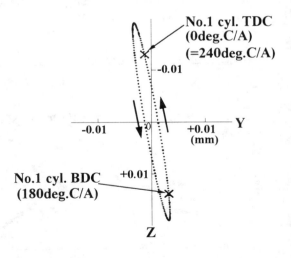

Fig.18 Locus of the front end of the crankshaft in the reverse whirl at 1820 rpm

SUMMARY

For a V–type ten–cylinder natural aspirated diesel engine (V–10, $\phi 119 \times 150$), the three-dimensional vibrations of the crankshaft system were calculated by the dynamic stiffness matrix method, taking account of the influence of the gyroscopic vibrations. The dynamic bending stresses were measured at the fillets of both the No.1 crank journal and the No.1 crank pin in the No.1 crank throw plane. From the comparison of the experimental and calculated results, we found that:

(1) In the experimental results, we found predominant harmonic components of the bending stresses which were induced by the bending vibrations of the crankshaft system. The bending vibrations of the crankshaft system induced by the mass–spring system of the torsional damper about the axis orthogonal to No.1 crank throw plane are separated into the bending vibrations of the forward whirl and the reverse whirl by gyroscopic vibrations of inertia ring of the torsional damper, and significant bending stresses are generated by the bending vibrations of the forward whirl and the reverse whirl.

(2) For the crankshaft system with the gyroscopic vibrations, the resonant frequencies of the predominant bending stresses of the crankshaft depend on the engine speed.

(3) Both the forward whirl and the reverse whirl are induced in the crankshaft system, due to the gyroscopic vibrations of the inertia ring of the torsional damper. In the forward whirl mode, the center of the crankshaft front end shows an elliptical form in the direction of the crankshaft rotation, whereas it shows one in the opposite direction in the reverse whirl mode.

ACKNOWLEDGMENTS

The authors would like to thank Professor F. Scott Howell of Sophia University for his kindness in reading and making corrections during the preparation of this paper.

REFERENCES

[1] Shimoyamada, K., Kodama, T., Honda, Y., and Wakabayashi, K., " A Numerical Computation for Vibration Displacements and Stresses of a Crankshaft with a Shear Rubber Torsional Damper," SAE Paper 930197, 1993.

[2] Kimura, J., Okamura, H., and Sogabe, K., "Experiments and Computation of Crankshaft Three–Dimensional Vibrations and Bending Stresses in a Vee–Type Ten–Cylinder Engine," SAE Paper 951291, 1995.

[3] Martin, I. T. and Low, B., " Prediction of Crankshaft and Flywheel Dynamics," I. Mech. E. Autotech 89, C382/046, November 1989.

[4] Ross, K., Brown and Charles, R., Mischke, "Whirling of a Four–Cylinder Engine Crankshaft," SAE Paper 700121, 1970.

[5] Dennis, Hodgetts, "The Dynamic Response of Crankshafts and Camshafts," FISITA, 865025, vol.1, p1.177–1.184, 1986.

[6] Okamura, H., Shinno, A., Yamanaka, T., Suzuku, A. and Sogabe, K., "A Dynamic Stiffness Matrix Approach to the Analysis of Tree-Dimensional Vibrations of Automobile Engine Crankshafts : Part1 – Background and Application to Free Vibrations," Proceedings of Vehicle Noise, ASME, NCA–Vol.9, 1990.

[7] Morita,T. and Okamura, H., "A Dynamic Stiffness Matrix Approach to the Analysis of Three-Dimensional Vibrations of Automobile Engine Crankshafts : Part 2 – Application to Firing Conditions," Proceedings of Vehicle Noise, ASME, NCA–Vol.9, 1990. 11

[8] Okamura, H. and Morita, T., " Influence of Crankshaft–pulley Dimensions on Crankshaft Vibrations and Engine–Structure Noise and Vibrations," Proceedings of the 1993 Noise and Vibration Conference (p–264), SAE paper 931303, 1993.

[9] Den hartog, J. P., " Mechanical Vibrations," McGraw–Hill, 1956.

[10] Edited by Japan Society of Mechanical Engineers "Data book on the statical and the dynamic characteristics of journal bearings " Japan Industrial Publication 1984. (in Japanese)

971996

Experiments and Analyses of the Three-Dimensional Vibrations of the Crankshaft and Torsional Damper in a Four-Cylinder In-Line High Speed Engine

Takeo Naganuma
NOK-MEGULASTIK Co., Ltd.

Hideo Okamura and Kiyoshi Sogabe
Sophia Univ.

Copyright 1997 Society of Automotive Engineers, Inc.

ABSTRACT

Crankshaft torsional dampers are increasingly being used for the gasoline engines of compact cars as well as for ordinary high speed diesel and gasoline engines. Recently, so-called bending dampers are sometimes attached to the torsional dampers to reduce the bending and axial vibrations.

To investigate the influence of such crankshaft torsional and bending dampers on the crankshaft vibrations, we first designed three kinds of dampers, each for the reduction of the crankshaft vibration, in the torsional, axial, and radial directions. Next, we developed two kinds of dampers for the simultaneous reduction in the torsional and axial modes, and in the torsional and radial direction modes. We measured the three-dimensional vibrations for both the dampers and the crankshaft, under engine operating conditions.

A four-cylinder in-line diesel engine （4 - ϕ 115 × 110） was used for the experiments. In the series of experiments, we found that, in some cases, only by attaching one simple torsional damper, significant reduction of the axial and bending vibrations could be obtained together with the reduction of torsional vibrations.

The effectiveness of the axial dampers and the bending dampers (i.e. radial dampers) was investigated carefully. The coupling behavior of the torsional, axial, and bending vibrations was also investigated for the dampers and the crankshaft.

1 INTRODUCTION

A large number of theoretical analyses have been performed on the torsional vibrations of crankshaft and the torsional damper. Such theoretical analyses have been compared with experimental results. Recently, three-dimensional vibration analyses of the crankshaft are being investigated by many researchers, and computer software for FEM analysis such as MSC/NASTRAN is widely applied. However FEM analysis takes time, and it is expensive. To simplify the modeling procedure and analysis, the authors have proposed an approach which applied the dynamic-stiffness matrix method[1,2]. Moreover, the coupling vibration behavior of the torsional vibrations and the axial vibrations has been investigated[3].

As for the crankshaft bending dampers, recently, Kinoshita, Sakamoto and Okamura have investigated the vibration behavior of the dual function torsional-bending damper pulleys applied to a gasoline engine[4]. However, the correlation between the crankshaft three-dimensional vibrations and the three-dimensional vibrations of the crankshaft dampers of torsional and bending types was not investigated.

In this research, the influence of the crankshaft pulley damper (rubber) on the three-dimensional vibrations of the crankshaft was investigated for firing conditions. We attached a series of different dampers to the crankshaft of a four-cylinder in-line diesel engine, and performed the following vibration measurements :
(1) torsional vibration at the crankshaft front end
(2) axial vibration at the crankshaft front end
(3) radial vibration at the crankshaft front end

2 EXPERIMENTS

To investigate the differences in the three-dimensional vibrations of the crankshaft under firing condition due to the crankshaft pulley dampers, we performed a series of firing tests for the cases when (1) a solid pulley is attached to the crankshaft (2) each of the dampers is attached to the crankshaft. We measured the torsional, axial and radial vibrations of the crankshaft front end for the pulley dampers, and investigated how the torsional vibration, the axial vibration, and the radial vibration would influence each other.

2.1 ENGINE FOR EXPERIMENTS

We used a four-cylinder in-line diesel engine for this research.

The engine specifications are shown in Table 1.

Table 1 Engine specifications

Engine form	4-cylinder in-line diesel engine
Cycle	4 cycle
Cylinder bore	φ115
× Stroke	×110 (mm)
Displacement	4.570 (l)
Firing order	1-3-4-2
Rated output	140 / 3,200 (PS/rpm)

2.2 THE SOLID PULLEY AND THE DAMPERS FOR EXPERIMENTS

In our experiments, we used one kind of solid pulley, and five kinds of dampers which we will call A, B, C, D, and E dampers in the following.

For the design of these experimental dampers, initially, we determined the natural frequencies of the crankshaft system to which the solid pulley is attached, from a series of firing tests.

The values of the masses, moments of inertia, and the natural frequencies in the torsional, axial, and radial directions are shown in Table 2 for the dampers A, B, and C. The values of the masses and moments of inertia of the solid pulley and the flywheel are shown in Table 3. The cross-sectional views of the solid pulley and the experimental dampers attached to the crankshaft are shown in Fig.1. In the shadowed frames of Table 2, the objective frequencies for the vibrations are shown for the dampers A, B, and C.

Table 2 Masses and moments of inertia, and the natural frequencies of experimental dampers

	Mass M (kg) [ratio]	Moment of inertia I_x (kgm^2) [ratio]	f_t (Hz) (Torsional)	f_a (Hz) (Axial)	f_r (Hz) (Radial)
Damper A (Torsional)	3.37 [1]	0.0197 [1]	280	(450)	----
Damper B (Axial)	0.845 [0.25]	0.00115 [0.06]	(150)	430	(140)
Damper C (Radial)	0.845 [0.25]	0.00115 [0.06]	(1250)	(2200)	450

As may be seen from Fig. 1, the dampers: A, B, and C, were designed respectively for the reduction of the torsional, axial, and radial vibrations in the crankshaft system. Here, damper D consists of the dampers A and B, and it was designed for the reduction of both the torsional and axial vibrations. Similarly, damper E consists of the dampers A and C, and it was designed for the reduction of the torsional and radial vibrations.

Table 3 Masses and moments of inertia of the solid pulley and the flywheel

	M (kg)	Moment of inertia I_x (kg·m^2)
Solid pulley	5.1	0.025
Flywheel	19.2	0.42

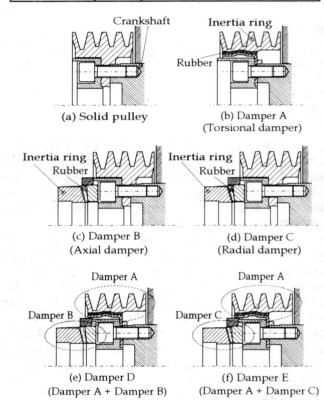

(a) Solid pulley
(b) Damper A (Torsional damper)
(c) Damper B (Axial damper)
(d) Damper C (Radial damper)
(e) Damper D (Damper A + Damper B)
(f) Damper E (Damper A + Damper C)

Fig.1 Cross-sectional views of the solid pulley and the experimental dampers attached to the crankshaft

2.3 THE TEST SET-UP AND THE COORDINATE SYSTEM FOR EXPERIMENTS

The set-up for the experiments is shown in Fig. 2, along with the coordinate systems x-y-z, and X-Y-Z attached to the crankshaft and the cylinder block. The brief specifications of each instrument are shown in Table 4.

The experiments were performed for the engine speeds from 1000rpm to 3600rpm under full load conditions. The signals were measured as follows:

(1) the torsional vibrations at the crankshaft front end, by seismic type torsional pick-up;
(2) the axial vibrations along the X-axis at the crankshaft front end, by a Laser Doppler velocity meter;

(3) the radial vibrations at the crankshaft front end, in the direction shifted 45 degrees from the T.D.C. of No.1 crank-throw as shown in Fig. 2, by a Laser Doppler velocity meter;

(4) the vibration accelerations on the cylinder block surface in the X, Y, Z directions, as shown in Fig. 2, by acceleration pick-ups in the 3-axes attached at the No.4 main bearing position.

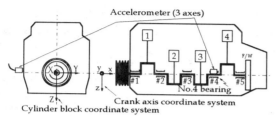

(a) Cylinder block and the coordinate systems, x-y-z and X-Y-Z

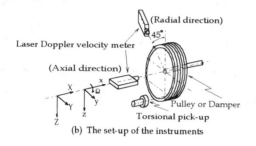

(b) The set-up of the instruments

Fig.2 The set-up for the experiments

Table 4 Specifications of the instruments

Name	Model
FFT	ONO SOKKI CF-8800
Accelerometer	PCB 303A02
Torsional pick-up	Seismic mass pick-up
Laser Doppler velocity meter	ONO SOKKI LV-1100

3 EXPERIMENTAL RESULTS

3.1 CRANKSHAFT SYSTEM WITH SOLID PULLEY

The experimental results are shown in Fig. 3.

(1) The torsional vibrations at the crankshaft front end:

From the resonance peaks of the 6th, 6.5th, 8th, and 10th order components in Fig. 3, one can estimate that the four predominant resonance peaks of the crankshaft torsional vibrations system are induced similarly at 360Hz. The 6th order component is the most remarkable, with the amplitude of 0.36deg. at the engine speed of N=3600rpm (Fig. 3-(a)).

(2) The axial vibrations at the crankshaft front end:

The resonance frequencies of the axial vibrations of the crankshaft system were estimated to be 330Hz, 360Hz, 450Hz, and 500Hz from the resonance peaks of the 2nd, 4th, 6th, 8th, and 10th order components appeared in Fig. 3-(b).

The vibrations of the 6th and the 8th order components were predominant, each with the amplitude of 0.043m/s at the engine speed of N=3300rpm, and of 0.047m/s at of N=3380rpm respectively.

(3) The radial vibrations at the crankshaft front end:

One can see from Fig.3-(c) that the 2nd order component was the largest. We considered that this might be caused by the vibrations of the engine main moving parts under combustion explosion.

The 1st, 5th, 7th, and 9th order components are overlapped at the engine speed of N=3500rpm. The 7th order component predominates with the amplitude of 0.065m/s.

Here, one may notice that, as for the vibrations of the crankshaft front end, the axial vibrations are induced by the predominant components of the even number orders. However, the radial vibrations mostly consist of the dominant components of the odd number orders. This is because the vibrations of the nth order component in the rotating coordinate system attached at the crankshaft would appear as $n \pm 1$th order components in the stationary system attached at the cylinder block [5].

(4) The vibrations on the cylinder block surface in the directions of X, Y, Z-axes:

The vibrations on the cylinder block surface are shown for the components in the directions of the X, Y, and Z axes, in Fig. 3-(d), (e), (f). Here, we noticed the following facts.

• *The vibrations in the X direction:*

In Fig. 3-(d), one can see that, in addition to the predominant components of the odd number, there are several of the predominant components of the even order number. Here these components of even number coincide with those for the torsional and axial vibrations at the crankshaft front end. However, the most remarkable predominant peak was the 7th order component. The peak appeared at the engine speed of N=3600rpm, with the amplitude of 1.5G.

• *The vibrations in the Y direction:*

One can see a few of the 5th, 6th, 7th, 8th, and 9th order components in Fig. 3-(e). However, the most remarkable peak was the one of the 7th order. The peak appeared at the engine speed of N=3550rpm, with the amplitude of 5.5G.

• *The vibrations in the Z direction:*

In Fig. 3-(f), one can see that the 2nd order component was the most predominant. Other components were much less predominant, as were the components in the Y direction. Again, one can see that the predominant peak of the 7th order component appeared at the engine speed of N=3550rpm with the amplitude of 2.5G.

3.2 THE CRANKSHAFT SYSTEM WITH DAMPER A

The experimental results are shown in Fig. 4, and one can see

(1) The torsional vibrations at the crankshaft front end:
In Fig. 4- (a), one can see that, by the damper A, the torsional vibrations of the crankshaft system can be reduced drastically compared to the case with the solid pulley. In Fig. 4- (a), the 6th order component is reduced to about the 1/8 of the value in the case with the solid pulley.

(2) The axial vibrations at the crankshaft front end:
One can see from Fig.4- (b) that, by the damper A, the 8th and 6th order components of the axial vibrations were reduced to about the 2/5 and the 1/4 of the values in the case with the solid pulley, at the engine speed of around N=3600rpm. However, the predominant peak of the 6th order component still appeared at the engine speed of N=3300rpm.

(3) The radial vibrations at the crankshaft front end:
One can see from Fig. 4- (c) that, by the damper A, the 5th, 7th and 9th order components of the radial vibrations, which appeared at the engine speed of N=3500rpm, are reduced to about half of the values for the case with the solid pulley.

(4) The vibrations on the cylinder block surface at the No. 4 main bearing position:
Compared with the case with the solid pulley, the vibrations on the cylinder block surface were reduced to about one- third for the 7th and 10th order components in the X direction, and to about half for the 7th order components in the Y and Z direction (see Fig. 4- (d),(e),(f)).

3.3 THE CRANKSHAFT SYSTEM WITH DAMPER B

The damper B was designed for the reduction of the axial vibrations. The experimental results are shown in Fig. 5.

From Fig. 5- (b) and Fig. 3- (b), one may notice that the values of the axial vibration of the crankshaft front end were almost the same as for the cases with the solid pulley. Namely, by the damper B, only slight reduction of the 8th order components was seen, and no other reduction was seen for any of the other vibrations.

3.4 THE CRANKSHAFT SYSTEM WITH DAMPER C

The damper C was designed for the reduction of the radial vibration. The experimental results are shown in Fig. 6.

From Fig. 6- (c) and Fig. 3- (c), one may notice that the overall values of the front end radial vibration are almost the same for the cases with the solid pulley and with the damper C, at the crankshaft front end; however, significant reduction was seen for the 7th order components. The 7th order cylinder block surface vibrations was also reduced in the X direction.

3.5 THE CRANKSHAFT SYSTEM WITH DAMPER D

The damper D was designed for the reduction of the torsional and axial vibrations. The experimental results are shown in Fig. 7.

One can see from Fig. 7 and Fig. 4 that the effectiveness of the damper D is very similar to that for the damper A, except for a slight difference in the reduction effect for the axial vibrations.

3.6 THE CRANKSHAFT SYSTEM WITH DAMPER E

The damper E was designed for the reduction of the torsional and radial vibrations. The experimental results are shown in Fig. 8.

Again, one can see from Fig. 8 and Fig. 4 the very similar effectiveness for dampers A and E, except the slight difference in the effectiveness for the reduction in the radial vibrations of the crankshaft.

4 DISCUSSION

4.1 VIBRATION BEHAVIOR OF CRANKSHAFT SYSTEM

For the crankshaft system with the solid pulley, one can see from Fig. 3- (a),(b),(c), that

(1) The crankshaft torsional vibrations are induced at certain definite frequencies: e.g., at 360Hz at the engine speed of N=3600rpm by the 6th order harmonic force; at the engine speed of N=2700rpm by the 8th order harmonic force.

(2) The axial vibrations are induced at the frequencies of the even order harmonics of the 2nd, 4th, 6th, and 8th order harmonics at the frequencies of 330Hz, 360Hz, 450Hz, 500Hz, and so on.

(3) However, the radial vibrations were induced at the engine speed of N=3600rpm, with the predominant peaks of the odd order number harmonics of the 1st, 3rd, 5th, and 7th. Since these predominant peaks appeared simultaneously with the resonance peak of the torsional vibrations, we expect that these radial vibrations might be induced together with the torsional vibrations.

(4) One may notice the predominant components of the 2nd order in the torsional and radial vibrations. Since the 2nd order frequencies coincide with the firing

frequencies of our experimental engine, we suspect that these components might be caused by the impulsive gas force in the cylinder and the 2nd order inertia force due to the reciprocating masses of the pistons and the connecting rods.

4.2 VIBRATION BEHAVIOR OF CYLINDER BLOCK

For the crankshaft system with the solid pulley, one can see from Fig. 3- (d),(e),(f), that the three significant peaks of the 7th order vibrations of the cylinder block surface vibrations appeared simultaneously in the X, Y, and Z directions. Since these peaks appeared simultaneously with the crankshaft torsional vibrations, one can see easily that these peaks are caused by the crankshaft torsional vibrations, which appeared at the engine speed of N=3600rpm due to the 6th order harmonic of the frequency of 360Hz.

4.3 EFFECTIVENESS OF TORSIONAL DAMPERS

From Fig. 3- (a),(b),(c), and Fig. 4- (a),(b),(c), one can see that, by the (torsional) damper A, the predominant peaks of the crankshaft torsional and radial vibrations were suppressed drastically, and some of the peaks of the axial vibrations were also suppressed.

And, from Fig. 3- (d),(e),(f) and Fig. 4- (d),(e),(f), one can see that, together with the suppression of the crankshaft system vibrations, the peaks of the vibrations on the cylinder block surface were also suppressed drastically. To clarify the reasons, further investigation will be necessary.

4.4 EFFECTIVENESS OF AXIAL, RADIAL DAMPERS

From Fig. 3- (a),(b), ~ (f) and Fig. 5- (a),(b), ~ (f), one can see that, by the (axial) damper B, only slight effectiveness was seen for the reduction of the crankshaft axial vibrations. Since no other effectiveness was seen for the reduction of the crankshaft vibrations and the cylinder block surface vibrations, one may conclude that the effectiveness of the axial damper might be "very limited".

From Fig. 3- (a),(b), ~ (f) and Fig. 4- (a),(b), ~ (f), one can deduce almost the same conclusion for the (radial) damper C.

To summarize the experimental results, the influence of the experimental dampers on the vibrations of the crankshaft system and the cylinder block was shown in Fig. 9- (a),(b), ~ (f) for the characteristic order numbers.

CONCLUSIONS

(1) By a well- designed torsional damper, the crankshaft axial and radial vibrations were reduced together with the torsional vibration.

(2) In the crankshaft for four- cylinder in- line engine (in which all crank throws are in the same plane), the torsional vibrations can affect both vibration behavior of the axial and radial vibrations in the crankshaft, but the axial and radial vibrations hardly affect the vibration behavior of the torsional vibrations in the crankshaft[3].

(3) The axial damper and the radial damper were effective for the reduction of the axial and radial vibrations of the crankshaft system; however, the effects are slight and no reductions were seen in other vibrations in the crankshaft system and cylinder block.

(4) When the crankshaft system had the resonance frequencies which were very close to each other, one should put the most importance on the design of the torsional vibration dampers.

(5) By the reduction of the crankshaft vibrations, the vibration of the cylinder block can also be reduced simultaneously.

ACKNOWLEDGMENTS

We would like to thank ISUZU MOTORS Ltd. for providing us with the test engine, and NOK-MEGULASTIK Co., Ltd. for arranging the firing tests and the dampers design. We also want to thank Professor Scott Howell of Sophia University for kindness in reading and making corrections during the preparation of this paper.

REFERENCES

[1] Okamura, H., Sogabe, K., and Shinno, A.,
Dynamic Stiffness Matrix Method for the Three Dimensional Analysis of Crankshaft Vibrations, c23/88c/Mech.E., 1988

[2] Morita, T., and Okamura, H.,
Analysis of Crankshaft Tree- Dimensional Vibrations in a Rotating Coordinate System, SAE Paper 951292

[3] Jenzer, J., and Welte, Y.,
Coupling Effect between Torsional and Axial Vibrations in Installations with Two- Stroke Diesel Engines,
New Sulzer Diesel Paper, May,1991

[4] Kinoshita, M., Sakamoto, T., and Okamura., H.,
An Experimental Study of a Torsional/Bending Damper Pulley for an Engine Crankshaft, SAE Paper 891127

[5] Morita, T., and Okamura., H.,
Simple Modeling and Analysis for Crankshaft Tree- Dimensional Vibrations,
Part 2: Application to an Operating Engine Crankshaft, Transactions of the ASME, 80/1.117, January, 1995

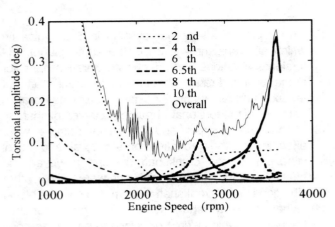

(a) The torsional vibration at the crankshaft front end

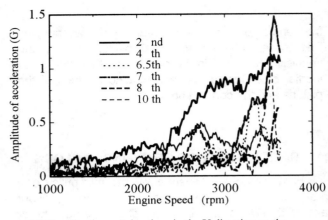

(d) The vibration accelerations in the X direction on the cylinder block surface at the No.4 main bearing position

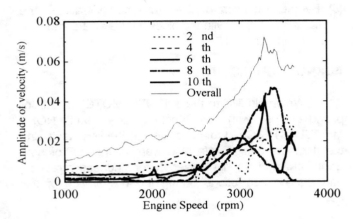

(b) The axial vibration at the crankshaft front end

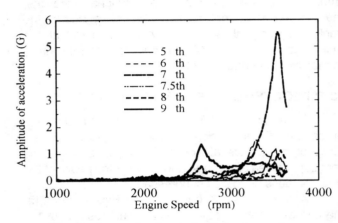

(e) The vibration accelerations in the Y direction on the cylinder block surface at the No.4 main bearing position

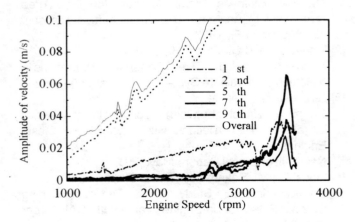

(c) The radial vibration at the crankshaft front end

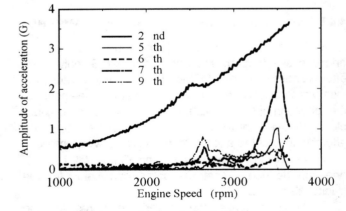

(f) The vibration accelerations in the Z direction on the cylinder block surface at the No.4 main bearing position

Fig. 3 The vibrations of the crankshaft front end and the vibration accelerations on the cylinder block surface at the No.4 main bearing position in the case with solid damper

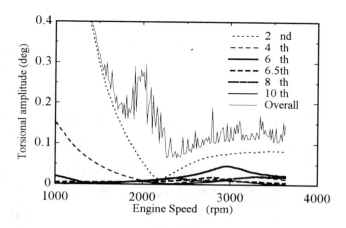

(a) The torsional vibration at the crankshaft front end

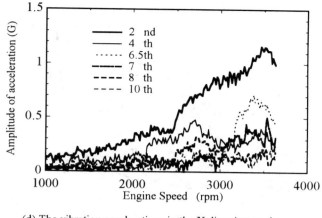

(d) The vibration accelerations in the X direction on the cylinder block surface at the No.4 main bearing position

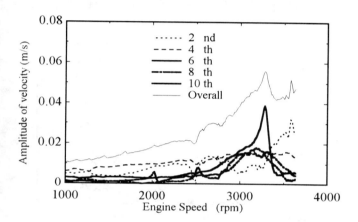

(b) The axial vibration at the crankshaft front end

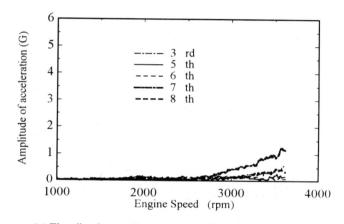

(e) The vibration accelerations in the Y direction on the cylinder block surface at the No.4 main bearing position

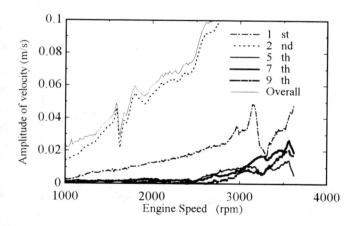

(c) The radial vibration at the crankshaft front end

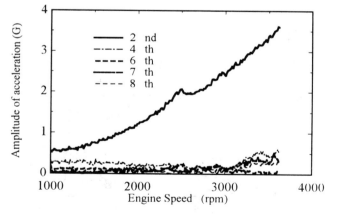

(f) The vibration accelerations in the Z direction on the cylinder block surface at the No.4 main bearing position

Fig. 4   The vibrations of the crankshaft front end and the vibration accelerations on the cylinder block surface at the No.4 main bearing position in the case with damper A

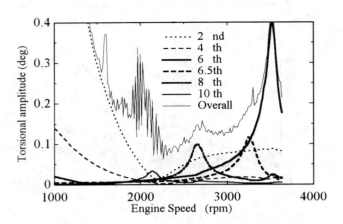

(a) The torsional vibration at the crankshaft front end

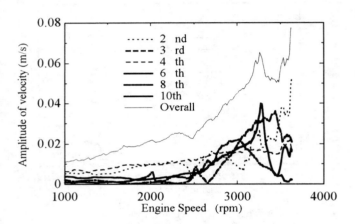

(b) The axial vibration at the crankshaft front end

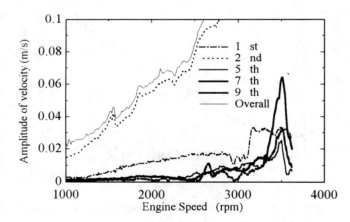

(c) The radial vibration at the crankshaft front end

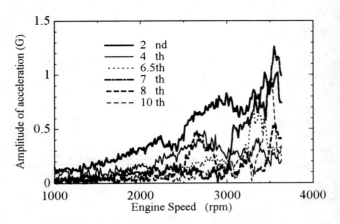

(d) The vibration accelerations in the X direction on the cylinder block surface at the No.4 main bearing position

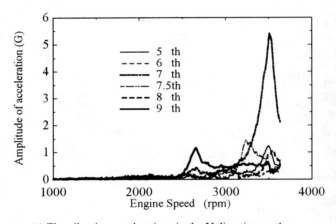

(e) The vibration accelerations in the Y direction on the cylinder block surface at the No.4 main bearing position

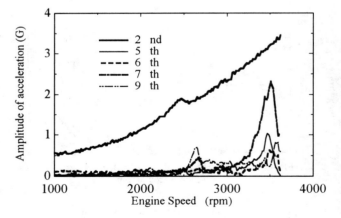

(f) The vibration accelerations in the Z direction on the cylinder block surface at the No.4 main bearing position

Fig. 5 The vibrations of the crankshaft front end and the vibration accelerations on the cylinder block surface at the No.4 main bearing position in the case with damper B

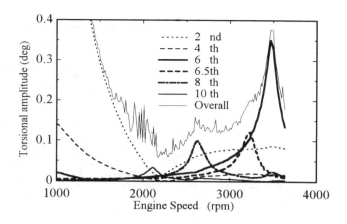

(a) The torsional vibration at the crankshaft front end

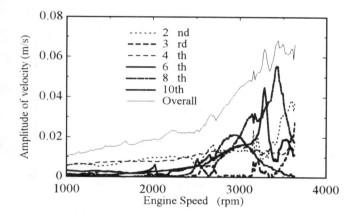

(b) The axial vibration at the crankshaft front end

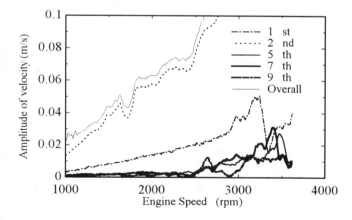

(c) The radial vibration at the crankshaft front end

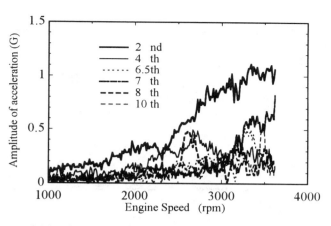

(d) The vibration accelerations in the X direction on the cylinder block surface at the No.4 main bearing position

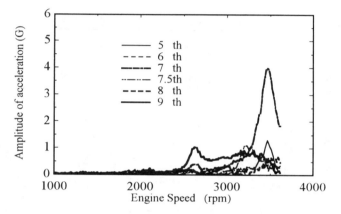

(e) The vibration accelerations in the Y direction on the cylinder block surface at the No.4 main bearing position

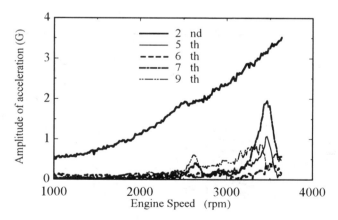

(f) The vibration accelerations in the Z direction on the cylinder block surface at the No.4 main bearing position

Fig. 6 The vibrations of the crankshaft front end and the vibration accelerations on the cylinder block surface at the No.4 main bearing position in the case with damper C

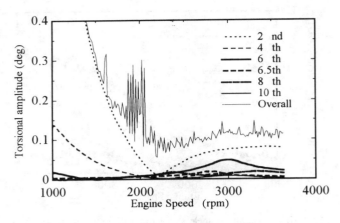

(a) The torsional vibration at the crankshaft front end

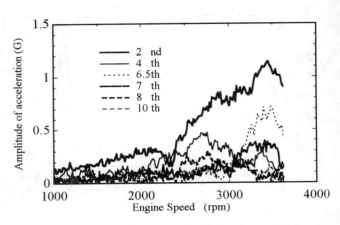

(d) The vibration accelerations in the X direction on the cylinder block surface at the No.4 main bearing position

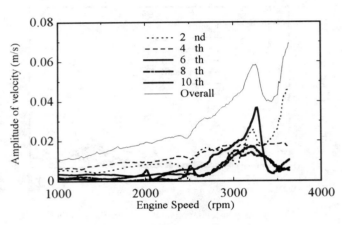

(b) The axial vibration at the crankshaft front end

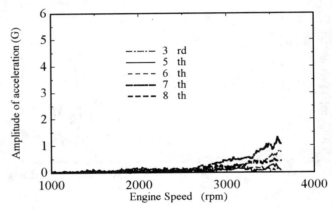

(e) The vibration accelerations in the Y direction on the cylinder block surface at the No.4 main bearing position

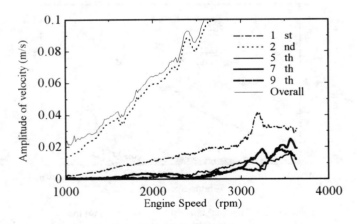

(c) The radial vibration at the crankshaft front end

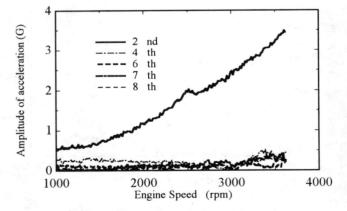

(f) The vibration accelerations in the Z direction on the cylinder block surface at the No.4 main bearing position

Fig. 7 The vibrations of the crankshaft front end and the vibration accelerations on the cylinder block surface at the No.4 main bearing position in the case with damper D

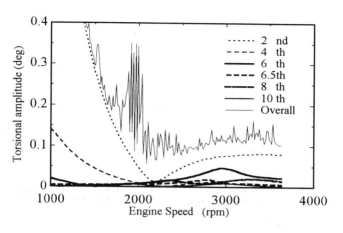

(a) The torsional vibration at the crankshaft front end

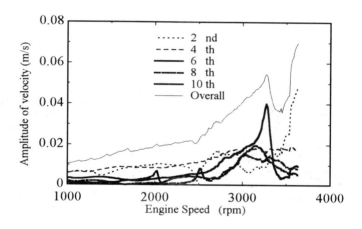

(b) The axial vibration at the crankshaft front end

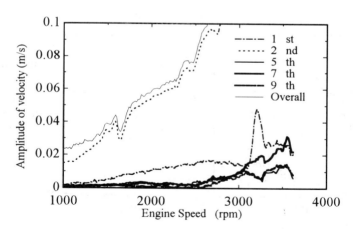

(c) The radial vibration at the crankshaft front end

Fig. 8 The vibrations of the crankshaft front end in the case with damper E

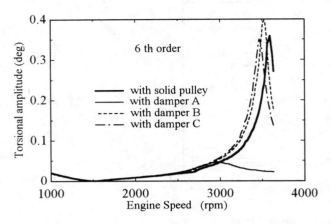

(a) The torsional vibration at the crankshaft front end

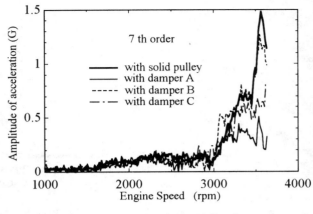

(d) The vibration accelerations in the X direction on the cylinder block surface at the No. 4 main bearing position

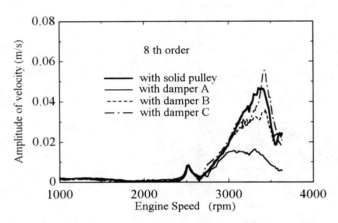

(b) The axial vibration at the crankshaft front end

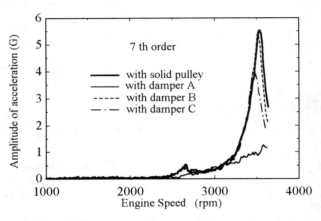

(e) The vibration accelerations in the Y direction on the cylinder block surface at the No. 4 main bearing position

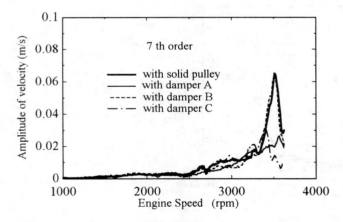

(c) The radial vibration at the crankshaft front end

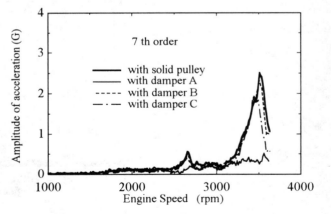

(f) The vibration accelerations in the Z direction on the cylinder block surface at the No. 4 main bearing position

Fig. 9 Comparison of the various vibrations for various kinds of dampers

971998

A New Generation of Condenser Measurement Microphones

Gunnar Rasmussen
G.R.A.S. Sound and Vibration

Ernst Schøntal
PCB Piezotronics

Copyright 1997 Society of Automotive Engineers, Inc.

ABSTRACT

New technologies and new materials have changed the manufacturing process of measurement microphones within the past few years. The basic construction is internationally standardized, ensuring that the acoustic performance of the new generation of microphones is unchanged relative to the old types, but with improved long term stability and electrical performance.

INTRODUCTION

Modern measurements microphones have remained nearly unchanged since their invention almost 40 years ago [1]. The basic construction have now been internationally standardized ensuring a uniform interface to auxiliary equipment like preamplifiers etc. However the advent, within the past few years, of new high technology materials have enabled the manufacturing of a new generation of measurement microphones with improved performance and long term stability. Also as the tolerances for high precision machining tools have been narrowed the production specifications have been tightened ensuring a better reproducibility. The range of microphones, fig.1, includes both ¼" and ½" microphones as well as both externally polarized and prepolarized types. In order to maintain continuity, the basic performance of the new generation of microphones is unchanged relative to the old types. This means that the enormous amount of existing acoustical data, measured with the previously available microphones, can be directly compared and interchanged with data measured with the new generation of microphones.

Figure 1 The G.R.A.S. range of precision measurement microphones

STANDARDS

Some of the basic characteristics of measurement microphones are to a large degree determined simply by their physical dimensions. These dimensions have been standardized in the international standard IEC 1094 'Measurement Microphones, Part 4 : Specifications for working standard microphones' [2], and ensure that for example threads are uniform so that different types of auxiliary equipment, like preamplifiers, can be used on different types of measurement microphones. The standard defines three classes of microphones commonly referred to as 1", ½" and ¼" microphones, fig. 2. As can be seen, dimensions such as the outer diameters of the microphone casings and the threads are precisely defined, while the length of the microphone is not specified. This means that the internal volume in the microphone can be optimized by changing the overall length of the microphone casing, depending on the type of microphone within a class. One of the parts which is not standardized is however the connector from the preamplifiers to the analyzing instrument like frequency analyzer etc.

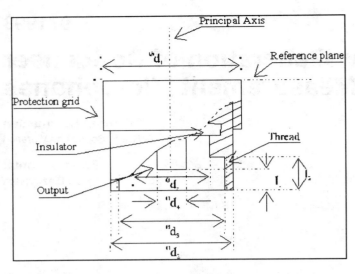

Figure 2 Standardized dimensions for measurement microphones.

Traditionally this connection have been made with some rather bulky 7-pin microphone connectors, originally designed in 1957. They were at that time optimal in providing excellent insulation for the high polarization voltages and screening against disturbing fields, with the materials available. Today, these connectors are in general too big compared to the size of modern measurement equipment and the general trend is to use a 7-pin Lemo connector instead. Although this is not officially standardized it is about to become an industry de-facto standard for measurement microphones. While the IEC 1094 standard determines the dimensions for measurement microphones, the performance of the microphones are normally related to the IEC Standard 651 'Sound level meters' [3], or the corresponding ANSI S1.4 [4]. As these are generally accepted standards for acoustical measurements, they have ensured the creation of a homogeneous database of acoustical measurement results throughout the world. It is important however to notice that these standards do not give specifications for the microphones themselves, but specifies the requirements for the full measurement system, including microphone, preamplifier and analyzer or sound level meter. This means that when for example IEC 651 specifies that the frequency response should be within ±1dB for a Type 1 Sound level meter, in the frequency range from 100Hz to 4kHz, then it is not enough that the microphone in itself satisfies this conditions. The conditions have to be fulfilled for the total measurement system and therefore the requirements for the microphone itself are stronger. Although measurement microphones are often stated to be of Type 1 if they themselves fulfill the requirements, this will allow for no additional tolerances for the rest of the instrumentation. Therefore the requirements for the individual components should be tightened relative to the IEC 651 requirements, so that the overall system will be within the requirements.

It is therefore common practice to specify tighter requirements for type approval of individual components, such as microphones, which will be incorporated into a measurement system. An Outdoor Microphone Unit for permanent noise monitoring, is a typical example where the requirements have been tightened. Such a system would be required to fulfill the IEC 651 requirements for a Type 1 sound level meter. To ensure this, the Outdoor Microphone Unit as a stand alone unit, is specified to be within ±0.7dB in the frequency range from 100Hz to 4kHz. When the unit is then connected to an analyzer or sound level meter, which have been type approved with similar stringent requirements it is ensured that the total measuring system fulfills the IEC 651 requirements.

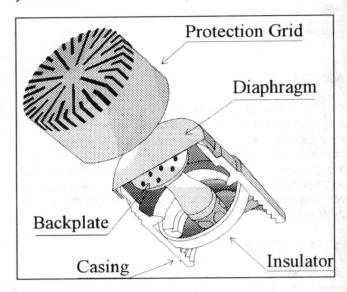

Figure 3 Basic elements of measurement microphone

CONSTRUCTION

The basic construction principle of measurement microphones have remained the same for the last 40 years. The measurement microphone consists basically of 5 elements: the protection grid, microphone casing, diaphragm, backplate and insulator, see Fig. 3. The diaphragm and the backplate form the parallel plates of an air capacitor. This capacitor is polarized with a charge either from an external voltage supply (externally polarized type) or by an electrical charge injected directly into an insulating material on the backplate (prepolarized type). When the sound pressure in a sound field fluctuates, the distance between the diaphragm and the backplate will change, and thereby change the capacitance of the diaphragm/backplate capacitor. As the charge on the capacitor is kept constant, the change in capacitance will generate an output voltage on the output terminal of the microphone. The acoustical performance of a microphone is determined by physical dimensions such as the diaphragm area, the distance between the diaphragm and the backplate, the stiffness and mass of the suspended diaphragm and the internal volume of the microphone casing. These dimensions have over the years been kept constant and thus the acoustical properties of the microphones have been preserved. The environmental properties of a typical measurement microphone, such as temperature stability, long term stability and insensitivity to rough handling, is on the other hand determined mostly by the choice of materials and assembling techniques. Traditionally the microphone diaphragm has been made of a nickel alloy, which was electrogalvanic deposited on a highly polished base material. This resulted in a compromise between a brittle and a ductile diaphragm and also a less corrosion resistant diaphragm, than could be made from stainless steel. Despite this, the nickel diaphragm was preferred, because stainless steel materials were rolled during the production process and therefore had different mechanical properties parallel to and perpendicular to the rolling direction. Also stainless steel diaphragms had problems with varying thickness and stiffness, and occurrence of pin-holes. This has been changed with the advent of new extreme high-quality, isotropic stainless alloy materials. These have maintained the good properties of nickel and at the same time improved the corrosion resistance and mechanical durability of the diaphragm. Furthermore it has made it possible to mount the diaphragm on the microphone casing with a heat-shrinking technique rather than the traditional thread-on method. As the nickel was brittle and very ductile it was necessary to screw the diaphragm on to the casing. The inherent tolerance in the thread made the assembled microphones susceptible to mechanical shocks, so that the acoustical characteristic of the microphone could change.

With the improved assembling technique the microphones have become much more resistant to mechanical shocks. The typical result of a mechanical shock to a conventional measurement microphone, is a slight change in the sensitivity of for example 0.2 dB. This change may be proved simply by checking the sensitivity of the microphone with for example a pistonphone and compare the result with the value stated on the Calibration Chart. This simple test will, however, not reveal anything about possible changes in the frequency response. Even if the sensitivity at the typical calibration frequency of 250Hz has only changed by a small amount, the changes at higher frequencies like 10kHz may be much more dramatic. The effect of small changes at higher frequency is especially critical for a set of sound intensity microphones. For a typical Type 1 sound intensity microphone pair, the phase difference between the two microphones must be less than 0.015° at 250Hz. To achieve this, it is necessary that the frequency responses for the two microphones at higher frequencies are very similar. In particular the diaphragm resonance frequency must be the same for the two microphones and even a small change in the resonance frequency of one of the microphones, due to a mechanical shock, may completely change the phase characteristic at low frequencies. Therefore the improved resistance to mechanical shocks obtained with the new materials and assembling techniques are extra important for intensity microphones. Also the use of stainless steel protection grids, with improved high frequency acoustical transparency, instead of brass protection grids, as has been common in the past, adds extra protection to the microphone diaphragm against physical damage when dropped on a hard floor. The most common fatal damage in the past to measurement microphones dropped on the floor, would be that the protection grid would be deflected inwards and pinch the diaphragm. Traditionally the protection grids have been produced in brass in order to reduce the price, but with new machining techniques the price difference between a brass grid and stainless steel grid is negligible compared to the advantages.

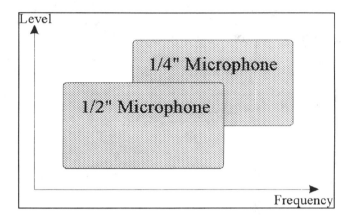

Figure 4 Frequency and range levels for ½" and ¼" microphones

DESIGN PARAMETERS FOR MICROPHONES

The basic characteristics of a measurement microphone is determined by factors such as the size of the microphone, the diaphragm tension, the distance between the diaphragm and the backplate and the acoustical damping in the microphone. These factors will determine the frequency range of the microphone, the sensitivity and the dynamic range. The sensitivity of the microphone is described as the output voltage of the microphone for a given sound pressure excitation, and is in itself of little interest for the operation of the microphone, except for calibration purposes. However, the sensitivity of the microphone (together with the electrical impedance of the cartridge) also determines the lowest sound pressure level, which can be measured with the microphone, fig. 4. For example with a ¼" microphone with a sensitivity of 2.5mV/Pa, the lowest level which can be measured is around 40 dB(re. 20μPa), while a ½" microphone with a sensitivity of 50 mV/Pa can measure levels down to approximately 15 dB(re. 20μPa). The exact lower limits cannot be given, as these depend also on the noise floor and input impedance of the preamplifier used in the measurements. The size of the microphone is the first parameter determining the sensitivity of the microphone. In general the larger the diaphragm diameter, the more sensitive the microphone will be.

There are, however, limits to how sensitive the microphone can be made by simply making it larger. The polarization voltage between the diaphragm and the backplate will attract the diaphragm and deflect this towards the backplate. As the size of the microphone is increased the deflection will increase and eventually the diaphragm will be deflected so much that it will touch the backplate. To avoid this, the distance between the diaphragm and the backplate can be increased or the polarization voltage can be lowered. Both of these actions will, however, decrease the sensitivity, so that the optimum size of a practical measurement microphone for use up to 20 kHz is very close to ½"(12.6mm). For some special applications, involving very low level measurements, 1"(25.2mm) microphones may be required, but for the majority of applications ½" or smaller is the optimum choice.

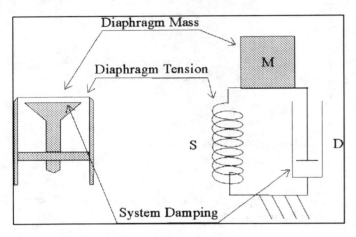

Figure 5 Spring-Mass-Damper analogy of microphone

As the size of the microphone is decreased, the useful frequency range of the microphone is increased. The frequency range, which can be obtained, is determined in part by the size of the microphone. At high frequencies, when the wavelength of the sound waves becomes much smaller than the diameter of the diaphragm, the diaphragm will stop behaving like a rigid piston (the diaphragm 'breaks up' - this is not a destructive phenomenon). Different parts of the diaphragm will start to move with different magnitude and phase, and the frequency response of the microphone will change. To avoid this the upper limiting frequency is placed so that the sensitivity of the microphone drops off before the diaphragm starts to break up. This gives for a typical ½" microphone an upper limiting frequency in the range from 20kHz to 40kHz depending on the diaphragm tension. If the diaphragm is tensioned so that it becomes more stiff, the resonance frequency of the diaphragm will be higher - on the other hand the sensitivity of the microphone will be reduced as the diaphragm deflection by a certain sound pressure level will be smaller.

The frequency response of the microphone is determined by the diaphragm tension, the diaphragm mass and the acoustical damping in the airgap between the diaphragm and the backplate. This system can be represented by the mechanical analogue of a simple mass-spring-damper system as in fig. 5. The mass in the analogy represents the mass of the diaphragm and the spring represents the tension in the diaphragm. Thus if the diaphragm is tensioned to become more stiff, the corresponding spring will become more stiff. The damping element in the analogy is representing the acoustical damping between the diaphragm and the backplate. This can be adjusted by for example drilling holes in the backplate. This will make it easier for the air to move away from the airgap when the diaphragm is deflected, and therefore decrease damping.

The frequency response of the simple mechanical model of the microphone is given in fig. 6 together with the influence of the different parameters. At low frequencies (below the resonance frequency) the response of the microphone is determined by the diaphragm tension, and as described above, the sensitivity will increase if the tension is decreased. The resonance frequency is determined by the diaphragm tension and the diaphragm mass, with an increased tension giving an increased resonance frequency and an increased mass giving a decreased resonance frequency. The response around the resonance frequency is determined by the acoustical damping, where an increase in the damping will decrease the response.

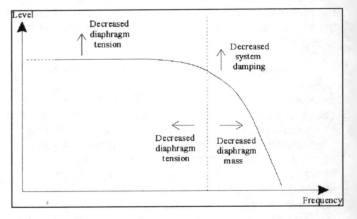

Figure 6 Influence of microphone parameters on frequency response

TYPES OF MICROPHONES

Although the material selection and assembling techniques have changed during the last few years, the basic types of microphones have remained unchanged. The basic types are free-field microphones, pressure microphones and random incidence microphones. They have been constructed with different frequency characteristics, corresponding to different requirements.

The pressure microphone is meant to measure the actual sound pressure as it exists on the diaphragm. A typical application is the measurement of the sound pressure in a closed coupler or as in fig. 7, the measurement of the sound pressure at a boundary. In this case the microphone forms part of the wall and measures the sound pressure on the wall itself. The frequency response of this microphone should be flat in as wide a frequency range as possible, taking into account that the sensitivity will decrease as the frequency range is increased. The acoustical damping in the airgap between the diaphragm and the backplate is adjusted so that the frequency response is flat up to and a little beyond the resonance frequency.

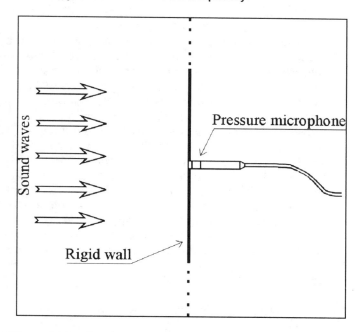

Figure 7 Application of pressure microphone

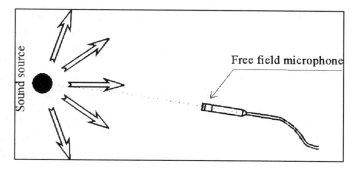

Figure 8 Application of free field microphone

field will result in a pressure increase in front of the diaphragm, see fig. 8a, depending on the wavelength The free-field microphone is designed to essentially measure the sound pressure as it existed before the microphone was introduced into the sound field. At higher frequencies the presence of the microphone itself in the sound field will change the sound pressure. In general the sound pressure around the microphone cartridge will increase due to reflections and diffraction. The free-field microphone is designed so that its frequency characteristics compensates for this pressure increase. The resulting output of the free-field microphone is a signal proportional to the sound pressure as it existed before the microphone was introduced into the sound field. The free-field microphone should always be pointed towards the sound source ('0° incidence'), as in fig. 8. In this situation the presence of the microphone diaphragm in the sound of the sound waves and the microphone diameter. For a typical ½" microphone the maximum pressure increase will occur at 26.9kHz, where the wavelength of the sound ($\lambda = 342 ms^{-1}/26.9 kHz \approx 12.7mm \approx ½"$) coincides with the diameter of the microphone. The microphone is then designed so that the sensitivity of the microphone decreases the same amount as the acoustical pressure increases in front of the diaphragm. This is obtained by increasing the internal acoustical damping in the microphone cartridge, to obtain a frequency response as in fig. 9b.

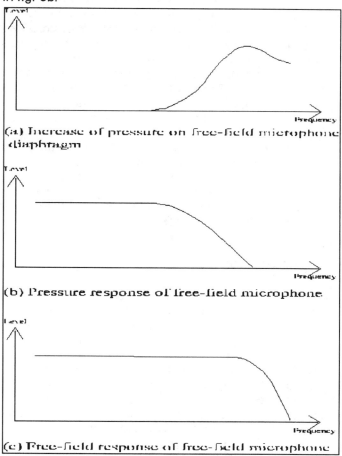

Figure 9 Frequency response of free field microphone

The result is an output from the microphone, fig. 9c, which is proportional to the sound pressure as it existed before the microphone was introduced into the sound field. The curve in fig.10a is also called the free-field correction curve for the microphone, as this is the curve which have to be added to the frequency response of the microphone cartridge to obtain the acoustical characteristic of the microphone in the free field.

The free-field microphone requires in principle to be pointed towards the sound source and that the sound waves are traveling in essentially one direction. In some cases, e.g. when measuring in a reverberation room or other highly reflecting surroundings, the sound waves will not have a well defined propagation direction, but will arrive at the microphone from all directions simultaneously. The sound waves arriving at the microphone from the front will cause a pressure increase as described for the free-field microphone, while the waves arriving from the back of the microphone will cause a pressure decrease to a certain extent due to the shadowing effects of the microphone cartridge. The combined influence of the waves coming from different directions depends therefore on the distribution of sound waves from different directions. For measurement microphones a standard distribution has been defined, based on statistical considerations, resulting in a standardized random incidence microphone. As mentioned earlier, measurement microphones can be either of the externally polarized type or the prepolarized type. The externally polarized types are by far the most stable and accurate microphones and should be preferred for precision measurements. Prepolarized microphones are, however, preferred in some cases, in that they do not require the external polarization voltage source. This is typically the case when the microphone will be used on small hand held devices like sound level meters, where a power supply for polarization voltage would add excessively to cost, weight and battery consumption. Still it should be realized that prepolarized microphones in general are much less stable to environmental changes than externally polarized microphones.

CHARACTERISTICS

The two most important parameters for measurement microphones are the sensitivity and the frequency response. The sensitivity is easily calibrated and checked by using a calibrator, pistonphone or other calibration methods like reciprocity calibration. The calibration methods and resulting accuracy are described in various international standards and is a well established technique. The measurement of the frequency characteristic of the microphone should also be a natural part of the calibration of the microphone and subsequent measurement equipment. This has so far, not been an integral part of the normal calibration procedure, even though it is quite simple to perform with modern measurement equipment. It is often assumed that one needs an elaborate set-up in an anechoic chamber in order to measure the frequency response of a microphone. This set-up is, however, only necessary in order to measure the acoustical parameters like the free-field correction curve for the specific microphone type, as described above. This depends on the geometry of the microphone cartridge and is measured once and for all for the type of microphone, and will be the same for all the individual microphones of this type.

Once the acoustical performance of the microphone type have been established, the performance of the individual microphone can be measured with a simple electrostatic actuator set-up.

This requires an FFT analyzer with white noise generator (or a separate white noise generator), an Actuator Voltage supply and an Electrostatic Actuator. The Electrostatic Actuator is mounted on the microphone instead of the normal microphone protection grid and excites the diaphragm by applying a force in a way similar to sound pressure. When supplied with the white noise input signal through the Actuator Voltage supply, the Electrostatic Actuator will generate a frequency independent signal on the microphone diaphragm. The electrical output from the microphone is then measured with the FFT analyzer to give the Actuator-response curve of the microphone. When the Actuator-response curve have been measured, the free-field response or pressure response can be obtained by adding the free-field or pressure response correction curve for the type of microphone to the Actuator-response curve for the particular microphone. The sensitivity of a microphone depends on a number of factors such as polarization voltage, temperature and barometric pressure. Also the microphone preamplifier and cable between the preamplifier and analyzer will in some cases influence the response. Typical values for these changes may be stated in the documentation for the microphone type, but by actually measuring the Actuator-response for the microphone in-situ, the true influence of all the external factors can be determined.

CONCLUSION

New technologies and new materials have changed the manufacturing process of measurement microphones within the past few years. This have lead to a new generation of microphones with improved performance and long term stability. Still the most important factors in producing high quality measurement microphones are experience and craftsmanship. Microphones have become more stable and rugged, still they are and should be treated as high precision delicate instrument. Just to give an example of the dimensions involved : for a standard ½" microphone, measuring a sound pressure level of 40 dB (corresponding to the level in a quite living room), the diaphragm will move approximately 10^{-11}m. This is too small to appraise, but imagine that the microphone diameter was the same as the diameter of the Earth (12700Km), then the microphone diaphragm would move only 10mm.

REFERENCES :

[1] G. Rasmussen, "A New Condenser Microphone", Brüel & Kjær *Technical Review*, No. 1, 1959.

[2] IEC Standard 1094 'Measurement Microphones, Part 4 : Specifications for working standard microphones',

[3] IEC Standard 651 'Sound level meters'

[4] ANSI S1.4 - 1983 'Specification for Sound level meters'

971999

Accelerometer Calibration

Ernst Schonthal and David M. Lally
PCB Piezotronics, Inc.

Copyright 1997 Society of Automotive Engineers, Inc.

Abstract

With the introduction of ISO 9001, periodic recalibration of accelerometers and other transducers has become a requirement. Many companies have elected to purchase calibration systems that might result in a savings of time and financial resources, but without a thorough understanding of the equipment and its operation, the user may receive incorrect readings and make costly errors. This paper provides a broad overview of the Back-to-Back Calibration technique, a quick and easy method for determining the sensitivity of a test accelerometer over a wide frequency range, as well as an alternative method for making the same determination.

1.0 Introduction

Accelerometer calibration provides, with a definable degree of accuracy, the necessary link between the physical quantity being measured and the signal exiting the accelerometer. In addition, other useful information concerning the operational limits, physical parameters, electrical characteristics, or environmental influences is determined. Without this information, analyzing data becomes a nearly impossible task. Fortunately, sensor manufacturers generally provide a calibration certificate that documents the exact characteristics of each sensor. (The type and amount of data varies, depending on the manufacturer, sensor type, special requirements, or contract regulations.)

Under normal conditions, piezoelectric sensors are extremely stable and their calibrated performance characteristics do not change over time. However, under harsh environments or other unusual conditions causing the sensor to experience dynamic phenomena outside of its specified operating range, the sensor may be temporarily or permanently affected. This change may manifest itself in the following ways: as a shift of the resonance due to a cracked crystal; a temporary loss of low-frequency measuring capability due to a drop in insulation resistance; or total failure of the built-in microelectronic circuit due to a high mechanical shock.

It is therefore recommended that a recalibration cycle be determined for each accelerometer. This schedule is unique and is based on a variety of factors, such as the extent of use, environmental conditions, accuracy requirements, trend information obtained from previous calibration records, contractual regulations, frequency of "cross-checking" against other equipment, manufacturer recommendation, and the risk associated with incorrect readings. Various standards, such as ISO 10012-1 or ANSI/NCSL Z540-1, provide insight and methods for determining recalibration intervals for most measuring equipment. With the above information in mind and under "normal" circumstances, this author conservatively suggests a 12- to 24-month recalibration cycle for piezoelectric accelerometers.

NOTE: *It is good practice to verify the performance of each accelerometer with a handheld shaker or other calibration device before and after each measurement.*

2.0 Sensor Recalibration

Sensor recalibration is typically performed by internal metrology laboratories. (International and national laboratories are also available.) The laboratory should be certified by ISO 9001, should comply with ISO 10012-1 (comparable to MIL-STD-45662A), and should use equipment directly traceable to National Institute of Standards and Technology (NIST), USA. This assures a calibration of relevant specifications.

Many companies choose to purchase the equipment necessary to perform the recalibration procedure themselves. While this may result in both a savings of time and money, it can also be attributed to costly errors and incorrect readings later. In an effort to prevent some of the common errors, this document provides a broad overview of the common Back-to-Back Calibration Technique. It is not, however, intended to replace the manufacturer's operating manual.

3.0 Back-to-Back Calibration Theory

Back-to-Back Calibration is the most common method for determining the sensitivity of accelerometers. This method relies on a simple comparison to a previously calibrated accelerometer, typically referred to as a reference standard. These high-accuracy devices, which are directly traceable to a recognized standards laboratory, are designed for stability, as well as configured to accept a test accelerometer. See Figure 1.

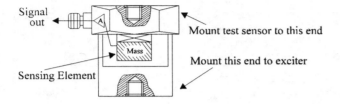

Figure 1. Reference Standard Accelerometer

By mounting a test accelerometer to the reference standard and then connecting this combination to a suitable vibration source, it is possible to vibrate both devices and compare the data. A typical test set-up is shown in Figure 2. (Test set-ups may be automated and vary, depending on the type and number of accelerometers being calibrated.)

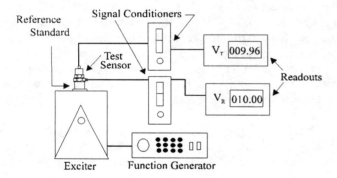

Figure 2. Typical Back-to-Back Calibration System

Because the acceleration is the same on both sensors, the ratio of their outputs (V_T / V_R) must also be the ratio of their sensitivities. With the sensitivity of the reference standard (S_R) known, the exact sensitivity of the test sensor (S_T) is easily calculated by using the following equation:

$$S_T = S_R (V_T / V_R) \qquad \text{EQ (1)}$$

Finally, by varying the frequency of the vibration, the test accelerometer may be calibrated over its entire operating frequency range. The frequency response of a typical unfiltered accelerometer is shown in Figure 3.

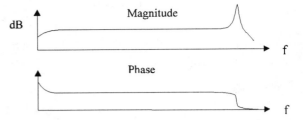

Figure 3. Typical Test Accelerometer Response

4.0 Typical Calibration Procedure

Numerous precautions are taken at a certified laboratory to insure accurate and repeatable results. This section provides an overview of the primary areas of concern.

Most importantly, since the Back-to-Back Calibration method relies on each sensor experiencing an identical acceleration level, proper mounting of the test sensor to the reference standard is critical. Sensors with mounting holes are attached directly to the reference standard with a mounting stud tightened to the recommended mounting torque. A shouldered mounting stud is typically used to prevent the stud from "bottoming out" in the hole. In all cases, both mounting surfaces are precision machined and lapped to provide a smooth, flat interface according to the manufacturer's specification. A thin layer of silicone grease is placed between the mating surfaces to fill any imperfections and increase the mounting stiffness. Sensor cables are "stress relieved" and tied to a stationary location near the shaker armature. This reduces cable motion, which is especially important when testing charge output sensors; stress relief helps prevent extraneous motion or stresses from being imparted into the system. A typical set-up is shown in Figure 4.

Mounting surfaces are smooth and flat with a small portion of grease on the interfaces. Sensors are mounted with a torque wrench to specified requirements.

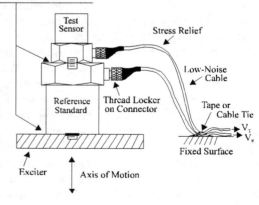

Figure 4. Typical Calibration Set-Up

Adhesively mounted sensors use similar practices. However, in this case, a small portion of "quick-bonding" gel or similar temporary adhesive is used to attach the test sensor to a reference standard designed with a smooth, flat mounting surface.

In addition to mounting, the selection of the proper equipment is critical. Some of the more important considerations include: 1) the reference standard must be specified for and previously calibrated over the frequency and / or amplitude range of interest; 2) the shaker should be selected to provide minimal transverse (lateral) motion and minimal distortion; and 3) the quality of the meters, signal generator and other devices should be selected so as to operate within the limits of permissible error.

5.0 Common Mistakes

For stud-mount sensors, ALWAYS mount the accelerometer directly to the reference standard. ALWAYS use a coupling fluid, such as silicone grease, in the mounting interface to maintain a high mounting stiffness. DO NOT use any intermediate mounting adaptor, as the mounting stiffness may be reduced and compromise high-frequency performance. If necessary, use adaptor studs. See Figure 5.

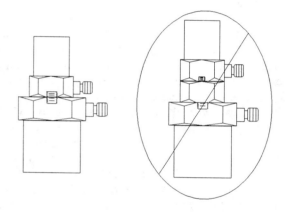

Figure 5. Stud Mounting

For adhesive-mount sensors, ALWAYS use a thin, stiff layer of temporary adhesive such as "quick-bonding" gel or Superglue. DO NOT use excessive amounts of glue or epoxy, as the mounting stiffness may be reduced and compromise high-frequency performance. It may also damage the sensor during removal. See Figure 6.

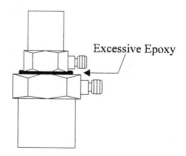

Figure 6. Incorrect Adhesive Mounting

Triaxial accelerometers should ALWAYS be mounted directly to the reference standard. Unless absolutely required, DO NOT use adaptors to reorient the sensor along the axis of motion, as the mounting stiffness may be altered. Also, the vibration at the test sensor sensing element may differ from the vibration at the reference standard due to a type of "cantilever" effect. See Figure 7.

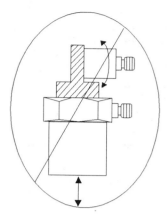

Figure 7. Mounting Triaxial Sensors (Incorrect)

UNDERSTAND Back-to-Back Calibration limitations. DO NOT expect the accuracy of calibration to be any better than ±2%. (In fact, the accuracy may be as high as ±3% or ±4% at frequencies <10 Hz or >3 kHz.) Since large sensors may affect high-frequency accuracy, ALWAYS verify that the test sensor does not mass load the standard. ALWAYS validate your calibration system with another accelerometer prior to each calibration session. Check with the manufacturer for exact specifications.

6.0 Method for Verification

It is recommended that accelerometer calibration be performed by skilled, well-trained people. Is there a way one can verify that the performed calibration is correct? One can normally obtain (calibrate) the reference sensitivity with good accuracy; generally, ±2.0% is considered acceptable. With the reference sensitivity, the electrical output of the accelerometer, as a function of frequency, may be calculated using the following formula:

EQ (2)
$$S(f) = S(f_{ref}) \times \frac{1}{1 - \left(\frac{f}{f_m}\right)^2}$$

where:

$S(f)$ = Sensitivity of unknown accelerometer

$S(f_{ref})$ = Reference sensitivity of unknown accelerometer

EQ (3)
$$\frac{1}{1 - \left(\frac{f}{f_m}\right)^2} = \text{Amplification factor}$$

The relative change in electrical output from an accelerometer is shown in Figure 8.

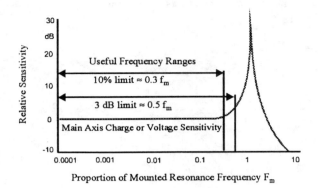

Figure 8. Relative Sensitivity of an Accelerometer vs. Frequency

A frequency response curve of this kind shows the variation in the accelerometer's electrical output when it is excited by a constant vibration level over a wide frequency range. To obtain an approximation of the mounted resonance frequency of the accelerometer, the accelerometer is mounted onto a 180-gram exciter head. The free-hanging natural resonance frequency depends heavily on the ratio of the total seismic mass to the mass of the rest of the accelerometer but primarily to that of the base. As a general rule, the total seismic mass of an accelerometer is approximately the same as the mass of the base, which gives the relationship:

EQ (4)
$$\frac{\text{Mounted Resonance Frequency}}{\text{Free Hanging Resonance Frequency}} \approx \frac{1}{\sqrt{2}}$$

The mounted resonance frequency is normally part of the specifications for a given accelerometer. The voltage output from an accelerometer is usually a straight line up to the point where the mounted resonance starts to influence the response. The charge sensitivity for an accelerometer with ceramic sensing elements has a slope of -2.5%/frequency decade and the calculation becomes a little more involved. The calculation of the charge output vs. frequency is done in the following way:

EQ (5)
$$S(f) = S(f_{ref}) \times \left(1 - 0.025 \times \log\left[\frac{f}{f_{ref}}\right]\right) \times \frac{1}{1 - \left(\frac{f}{f_m}\right)^2}$$

where:

$S(f_{ref})$ = Reference sensitivity at 100 Hz

(f_{ref}) = Reference frequency

(f_m) = Mounted resonance frequency (100 Hz)

f = Frequency of interest

$S(f)$ = Sensitivity at frequency (f)

The charge sensitivity vs. frequency is shown below:

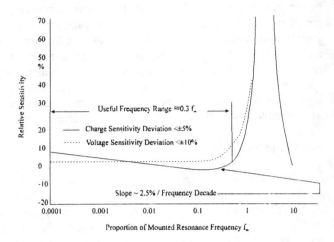

Figure 9. Charge and Voltage Sensitivity vs. Frequency for Accelerometer Using Ceramic Piezoelectric Material

7.0 Conclusion

Although we have faith in calibration laboratories, do-it-yourself calibrations are becoming more popular; therefore, it is wise to do the homework necessary to verify that the performed calibration is correct. There is normally good correlation between the calibrated values and the calculated results. If the calculated values do not correlate with the calibrated values, there is reason to believe that the accelerometer under test has been damaged. It is not rare to see an accelerometer that has been dropped. If it is a small one, it likely survives; however, larger accelerometers quite often get damaged, in that the stiffness of the mechanical system changes and the mounted resonance moves down, causing a reduced frequency response. Damage to the mechanical system in the accelerometer seldom manifests itself as a change in the reference sensitivity.

Without an adequate understanding of dynamics, determining what, when, and how to test a sensor is a difficult task. Each user must weigh the cost, time and risk of self-calibration versus an accredited laboratory.

Reference

Serridge, Mark, BSc, and Licht, Torben R., MSc, *Piezoelectric Accelerometers and Vibration Preamplifiers,* Brüel and Kjær, Naerum, Denmark, 1987.

Biography

Ernst Schonthal joined PCB in March of 1996 as manager of vibration products. Prior to that, Ernst had been employed by Brüel & Kjær for 44 years, during which he was responsible for the training of the worldwide service staff. In the early eighties, he was put in charge of upgrading Brüel & Kjær's worldwide calibration service. This work resulted in well-known calibration systems for microphones and accelerometers.